The Real Number System

Natural numbers	$N = \{1, 2, 3, \ldots\}$
Whole numbers	$W = \{0, 1, 2, 3, \ldots\}$
Integers	$J = \{\ldots, -3, -2, -1, 0, 1, 2, 3, \ldots\}$
Rational numbers	$= \left\{\dfrac{a}{b} \,\middle|\, a \in J, b \in J, b \neq 0\right\}$
Irrational numbers	$= \{\text{nonterminating, nonrepeating decimals}\}$
Real numbers	$= \{\text{rational numbers}\} \cup \{\text{irrational numbers}\}$
Absolute value	$\lvert x \rvert = \begin{cases} x & \text{if } x \geq 0 \\ -x & \text{if } x < 0 \end{cases}$

Properties of Real Numbers

If a, b, and c are real numbers, the following properties are assumed to be true:

	Addition	Multiplication
Commutative	$a + b = b + a$	$a \cdot b = b \cdot a$
Associative	$a + (b + c) = (a + b) + c$	$a(b \cdot c) = (a \cdot b)c$
Identity	$a + 0 = 0 + a = a$	$a \cdot 1 = 1 \cdot a = a$
Inverse	$a + (-a) = (-a) + a = 0$	$a \cdot \dfrac{1}{a} = \dfrac{1}{a} \cdot a = 1,\ a \neq 0$
Distributive		$a(b + c) = ab + ac$

Properties and Definitions

Zero-factor property	If a and b are real numbers and if $a \cdot b = 0$, then either $a = 0$ or $b = 0$.
Definition of subtraction	$a - b = a + (-b)$
Double negative	$-(-a) = a$

Operations with Fractions

$$\frac{a}{b} \cdot \frac{c}{d} = \frac{ac}{bd}, \quad b, d \neq 0 \qquad \frac{a}{b} \div \frac{c}{d} = \frac{a}{b} \cdot \frac{d}{c} = \frac{ad}{bc}, \quad b, c, d \neq 0$$

$$\frac{a}{b} = \frac{ax}{bx}, \quad b, x \neq 0 \qquad \frac{a}{c} + \frac{b}{c} = \frac{a + b}{c}, \quad c \neq 0$$

$$-\frac{a}{b} = \frac{-a}{b} = \frac{a}{-b} = -\frac{-a}{-b}, \quad b \neq 0$$

Operations with Equations

Addition property of equality	If $a = b$, then $a + c = b + c$.
Subtraction property of equality	If $a = b$, then $a - c = b - c$.
Multiplication property of equality	If $a = b$, then $ac = bc$.
Division property of equality	If $a = b$, then $\dfrac{a}{c} = \dfrac{b}{c},\ c \neq 0$.

Operations with Inequalities

If $a < b$, then $a + c < b + c$.

If $a < b$, then $a - c < b - c$

If $a < b$, then $\begin{cases} \dfrac{a}{c} < \dfrac{b}{c} \text{ if } c > 0; \\ \dfrac{a}{c} > \dfrac{b}{c} \text{ if } c < 0. \end{cases}$

If $a < b$, then $\begin{cases} ac < bc \text{ if } c > 0; \\ ac > bc \text{ if } c < 0. \end{cases}$

Absolute Value Equations and Inequalities

If p is a positive number, then

$\lvert x \rvert = p$	means that $x = p$ or $x = -p$
$\lvert x \rvert < p$	means that $-p < x < p$
$\lvert x \rvert > p$	means that $x < -p$ or $x > p$
$\lvert 0 \rvert = 0$	

Properties of Exponents

If a and b are real numbers and m and n are integers, then the following rules hold:

Product rule $\quad b^m \cdot b^n = b^{m+n}$

Power rules $\quad (b^m)^n = b^{mn}; \quad (ab)^m = a^m b^m; \quad \left(\dfrac{a}{b}\right)^m = \dfrac{a^m}{b^m}, \quad b \neq 0$

Quotient rule $\quad \dfrac{b^m}{b^n} = b^{m-n}, \quad b \neq 0$

Negative exponents $\quad b^{-n} = \dfrac{1}{b^n}, \quad b \neq 0$

Zero as an exponent $\quad b^0 = 1, \quad b \neq 0$

Special Products and Factors

$(a + b)^2 = a^2 + 2ab + b^2$

$(a - b)^2 = a^2 - 2ab + b^2$

$(a + b)(a - b) = a^2 - b^2$

$a^3 + b^3 = (a + b)(a^2 - ab + b^2)$

$a^3 - b^3 = (a - b)(a^2 + ab + b^2)$

Properties of Radicals

$\sqrt[n]{ab} = \sqrt[n]{a} \cdot \sqrt[n]{b}$

$\sqrt[n]{\dfrac{a}{b}} = \dfrac{\sqrt[n]{a}}{\sqrt[n]{b}}$

$b^{m/n} = \sqrt[n]{b^m} = (\sqrt[n]{b})^m$

Elementary Algebra

Elementary Algebra

Gilbert M. Peter
Cuesta College

C. Lee Welch
Cuesta College

WEST PUBLISHING COMPANY
Minneapolis/St. Paul   New York   San Francisco   Los Angeles

PRODUCTION CREDITS

Copyediting	*Susan Gerstein*
Text Design	*Geri Davis, Quadrata, Inc.*
Chapter Opening Art	*Nancy Wirsig McClure, Hand to Mouse Arts*
Text Art	*Ed Rose, VGS*
Answers Art	*Ross Rueger*
Proofreading	*Sylvia Dovner, Technical Texts*
Indexer	*Bernice Eisen*
Composition	*Bi-Comp Incorporated*
Problem Checking	*Chuck Heuer*

WEST'S COMMITMENT TO THE ENVIRONMENT

In 1906, West Publishing Company began recycling materials left over from the production of books. This began a tradition of efficient and responsible use of resources. Today, up to 95 percent of our legal books and 70 percent of our college and school texts are printed on recycled, acid-free stock. West also recycles nearly 22 million pounds of scrap paper annually—the equivalent of 181,717 trees. Since the 1960s, West has devised ways to capture and recycle waste inks, solvents, oils, and vapors created in the printing process. We also recycle plastics of all kinds, wood, glass, corrugated cardboard, and batteries, and have eliminated the use of Styrofoam book packaging. We at West are proud of the longevity and the scope of our commitment to the environment.

Production, Prepress, Printing and Binding by West Publishing Company.

British Library Cataloguing-in-Publication Data. A catalogue record for this book is available from the British Library.

COPYRIGHT ©1996   By WEST PUBLISHING COMPANY
610 Opperman Drive
P.O. Box 64526
St. Paul, MN 55164-0526

All rights reserved
Printed in the United States of America
03 02 01 00 99 98 97 96   8 7 6 5 4 3 2 1 0
Library of Congress Cataloging-in-Publication Data

Peter, Gilbert M.
    Elementary algebra / Gilbert M. Peter, C. Lee Welch.
      p. cm.
    Includes index.
    ISBN 0-314-04385-3
    1. Algebra.   I. Welch, C Lee., 1928–   . II. Title.
QA152.2.P463   1996
512.9—dc20   94-10481
     CIP

DEDICATION

We dedicate this book to our wives Marlene and Pauline and to the re-entry students whose desire to change their lives and careers through education has been an inspiration to both of us.

Brief Contents

1	A Review of Arithmetic	2
2	Algebra and the Real Numbers	50
3	Linear Equations and Inequalities in One Variable	106
4	Linear Equations and Inequalities in Two Variables; Functions	174
5	Systems of Linear Equations and Inequalities	230
6	Polynomials and the Laws of Exponents	280
7	Products and Factors	330
8	Rational Expressions and Fractions	380
9	Roots and Radicals	440
10	Quadratic Equations	488

ANSWERS	A-1
INDEX	I-1

Contents

Preface   xv

CHAPTER 1

A Review of Arithmetic   2

- **1.1** Basic Concepts of Arithmetic and Algebra   3
- **1.2** Common Fractions   11
- **1.3** Decimals and Percents   21
- **1.4** Some Concepts from Geometry   33
  - SUMMARY   43
  - COOPERATIVE EXERCISE   46
  - REVIEW EXERCISES   47
  - CHAPTER TEST   49

CHAPTER 2

Algebra and the Real Numbers   50

- **2.1** Sets and Real Numbers   51
- **2.2** Symbols, Variables, and Absolute Value   61
- **2.3** Addition and Subtraction of Real Numbers   68
- **2.4** Multiplication and Division of Real Numbers   77
- **2.5** Properties of Real Numbers   86
- **2.6** More Applications   94
  - SUMMARY   100
  - COOPERATIVE EXERCISE   102
  - REVIEW EXERCISES   103
  - CHAPTER TEST   104

CHAPTER 3

Linear Equations and Inequalities in One Variable   106

- **3.1** Simplifying Algebraic Expressions   107
- **3.2** Solving Linear Equations Using the Addition and Subtraction Properties of Equality   113
- **3.3** Solving Linear Equations Using the Multiplication and Division Properties of Equality   122
- **3.4** Literal Equations and Formulas   131
- **3.5** The Addition and Subtraction Properties of Inequality   137
- **3.6** The Multiplication and Division Properties of Inequality   145
- **3.7** Solving Absolute Value Equations and Inequalities   153
- **3.8** More Applications   160
  - SUMMARY   167
  - COOPERATIVE EXERCISE   168
  - REVIEW EXERCISES   169
  - CHAPTER TEST   170
  - CUMULATIVE REVIEW FOR CHAPTERS 1–3   171

CHAPTER 4

Linear Equations and Inequalities in Two Variables; Functions   174

- **4.1** The Rectangular (Cartesian) Coordinate System   175
- **4.2** Graphing Linear Equations in Two Variables   182
- **4.3** The Slope of a Line; Equations of a Line   189
- **4.4** Graphing Linear Inequalities in Two Variables   204
- **4.5** Relations; Functions; and Function Notation   215
  - SUMMARY   224
  - COOPERATIVE EXERCISE   225
  - REVIEW EXERCISES   226
  - CHAPTER TEST   228

CHAPTER 5

Systems of Linear Equations and Inequalities   230

- **5.1** Solving Systems of Linear Equations by the Graphing Method   231
- **5.2** Solving Systems of Linear Equations by the Elimination Method   240

- 5.3 Solving Systems of Linear Equations by the Substitution Method 249
- 5.4 Solving Systems of Linear Inequalities 257
- 5.5 More Applications 266
  - SUMMARY 273
  - COOPERATIVE EXERCISE 275
  - REVIEW EXERCISES 276
  - CHAPTER TEST 278

CHAPTER 6 Polynomials and the Laws of Exponents 280

- 6.1 Multiplication and Division with Exponents 281
- 6.2 More on Exponents; Scientific Notation 288
- 6.3 Adding and Subtracting Polynomials 295
- 6.4 Multiplying Polynomials 302
- 6.5 Dividing Polynomials 311
- 6.6 More Applications 318
  - SUMMARY 322
  - COOPERATIVE EXERCISE 324
  - REVIEW EXERCISES 324
  - CHAPTER TEST 326
  - CUMULATIVE REVIEW FOR CHAPTERS 4–6 326

CHAPTER 7 Products and Factors 330

- 7.1 Factoring; The Greatest Common Factor 331
- 7.2 Factoring Trinomials of the Form $x^2 + bx + c$ 338
- 7.3 Factoring Trinomials of the Form $ax^2 + bx + c$ 344
- 7.4 Special Factors; Factoring by Grouping 350
- 7.5 A Summary of Factoring 355
- 7.6 Solving Factorable Quadratic Equations 359
- 7.7 More Applications 368
  - SUMMARY 374
  - COOPERATIVE EXERCISE 376
  - REVIEW EXERCISES 376
  - CHAPTER TEST 377

CHAPTER 8

Rational Expressions and Fractions — 380

- 8.1 Simplifying Rational Expressions 381
- 8.2 Multiplying and Dividing Rational Expressions 388
- 8.3 Adding and Subtracting Rational Expressions; The Least Common Denominator 394
- 8.4 Complex Fractions 402
- 8.5 Solving Rational Equations 408
- 8.6 Ratio and Proportion 419
- 8.7 Variation 427
  - SUMMARY 433
  - COOPERATIVE EXERCISE 436
  - REVIEW EXERCISES 436
  - CHAPTER TEST 438

CHAPTER 9

Roots and Radicals — 440

- 9.1 Roots and Radicals 441
- 9.2 Simplified Forms of Radicals 448
- 9.3 Addition and Subtraction of Radicals 455
- 9.4 Multiplication and Division of Radicals 460
- 9.5 Solving Equations Containing Radicals 468
- 9.6 Rational Exponents and Radicals 475
  - SUMMARY 480
  - COOPERATIVE EXERCISE 482
  - REVIEW EXERCISES 482
  - CHAPTER TEST 483
  - CUMULATIVE REVIEW FOR CHAPTERS 7–9 484

CHAPTER 10

Quadratic Equations — 488

- 10.1 Solving Equations by Factoring and the Square Root Method 489
- 10.2 Solving Quadratic Equations by Completing the Square 495
- 10.3 Solving Quadratic Equations by the Quadratic Formula 501
- 10.4 Solving Quadratic Equations with Complex-Number Solutions 510

10.5 Graphing Quadratic Equations; The Parabola 516
10.6 More Applications 527
    SUMMARY 533
    COOPERATIVE EXERCISE 534
    REVIEW EXERCISES 535
    CHAPTER TEST 536
    CUMULATIVE REVIEW FOR CHAPTERS 1–10 537

ANSWERS A-1

INDEX I-1

PREFACE

This text is intended for a one-term course in elementary algebra and is the companion text to *Intermediate Algebra* by the same authors. Our primary goal is to prepare students for success in intermediate algebra or in an associated field of study that requires elementary algebra as a prerequisite. We have included all of the topics that are normally found in an elementary algebra course, but we have made the text flexible enough that some topics can be omitted without loss of continuity. For example, Chapter 1 can be excluded from a briefer course. However, if it is excluded from the instruction phase of the course, we strongly suggest that it be assigned as outside review for those students whose arithmetic skills are found to be lacking. In addition, Sections 3.7 (Solving Linear Equations and Inequalities Involving Absolute Value), 5.4 (Solving Systems of Linear Inequalities), 8.6 (Ratio and Proportion) and 8.7 (variation) can also be eliminated for a briefer course of instruction.

Many years of teaching both elementary and intermediate algebra—as well as more advanced courses—have taught us that students have a difficult time making the transition from skills-oriented courses, in which they learn how to manipulate algebraic expressions, to more concept-oriented courses in which they have to apply their acquired skills to problem solving. Instructors in these courses find that students often are not adequately prepared, even though their past record seems to indicate that they have mastered the concepts. Reviewers have expressed concern that texts at this level can present a false picture of what mastery means. This text has been written to address this problem. It makes reasonable demands of the students while it prepares them for success in intermediate algebra and other courses for which elementary algebra is a prerequisite.

In order for students to master skills and to apply them at a more conceptual level, in a number of different circumstances, they must have experience using them in a number of different circumstances. In this way they will develop both an intuitive understanding of algebra and a reasoning-oriented ability to use mathematics. For example, we believe that graphing and the concept of a function is central to mathematics and have included these topics in Chapter 4. This early coverage gives the students many opportunities during the remainder of the course to use the concepts so that they can better understand mathematical relationships. Graphing gives them the ability to visualize the mathematical relationships between two quantities. This naturally leads to the development of systems of equations, which are introduced in Chapter 5. Throughout we have included references to and examples of the use of graphics calculators and computers but have chosen not to make their use invasive of the textual development. We feel that mathematical understanding should precede the use of these tools, which are more beneficial in more advanced courses.

A major theme of this text is problem solving. The emphasis appears in a variety of ways, but it is most evident in the number of applications and applied problems throughout the book, and in the "More Applications" sections at the end of Chapters 2, 3, 5, 6, 7, and 10. While problem solving is by no means limited to these sections, they do provide many ways in which the mathematical concepts and skills being presented can be used to model and solve real-life applied problems. In addition, the early introduction of graphing and systems of equations provides the students with powerful tools that they can use to visualize and solve some of the applied problems that appear in succeeding chapters.

It has been said that a good mathematics text can be measured by its ability to use examples, rather than a lengthy presentation of theory, to show the how and why of an algebraic concept. We believe that both theory and practice are important and have tried to use both so that they compliment each other. We have made a concerted effort with the use of an abundance of examples to show students how to apply the skills and concepts; this gives them a real opportunity to see mathematics in action. There are over 5500 exercises, supported by nearly 800 in-text worked examples. In addition, there are over 1200 review exercises and 320 items in the chapter tests. Experienced instructors know that exercise sets with a large variety of problems ranging from the routine to the more challenging are essential in a good text. This is where the students have the opportunity to practice the mathematics that they have just learned and to discover where additional work is needed. The exercises are structured to give instructors many options in assigning problems, including group problems and exercises that require the students to do some writing to demonstrate their knowledge.

Key Features

Applied Problems

Applied problems appear early in Chapter 2 and are included in most exercise sets, as well as in the separate sections called "More Applications" at the ends of most chapters. Experience has shown that students often learn how to manipulate algebraic expressions without knowing how to use them as a tool for solving applied problems. The introduction of such problems early in the text helps them to see how mathematics is used while helping them to develop the logical thinking necessary for problem solving. Applied problems present a challenge that must be met to be successful in working with the more difficult material found in subsequent courses. To make the problems "meaningful," and not contrived, we have selected problems from numerous fields that may or may not be in the area of the students' aspirations or experiences. Some of these problems may use units of measurement that are not familiar to the student, but such familiarity is not necessary to solve the problems successfully.

Systems of Equations and Graphing

Linear equations and inequalities are introduced in Chapter 3, expanded upon in Chapter 4, where graphing first occurs, and then reinforced in Chapter 5, when

Key Chapter Pedagogy

CHAPTER OPENERS
Every chapter opens with an applied problem that appears later in the chapter. This gives the student an idea of the kind of problems that they will learn to solve by using the concepts covered in the chapter.

HISTORICAL COMMENTS
We are all indebted to those who preceded us in mathematics, so we have included many historical comments throughout the text. Some are brief glimpses into the lives of early mathematicians and describe the part they played in developing the topic at hand. Others trace the paths that have led to the mathematical symbolism that we use today. Still others present problems from ancient times and from texts that are over 200 years old and that—except for the language used to express them—are still used today to teach important algebraic concepts. In every case, care has been taken to keep the historical comments brief enough that they will be read by the students and interesting enough that some students want to learn more.

SECTION OBJECTIVES
The objectives to be accomplished in each section are clearly stated at its beginning. They are written in a format that tells the student exactly what they are expected to learn by the time they complete the section.

KEY CONCEPTS, DEFINITIONS, AND THEOREMS
Important theorems and other key concepts are boxed to emphasize their importance and to make them easier to find for future reference and review.

CAUTION WARNINGS
Students are warned about common misconceptions with special caution notations in the margin of the pages next to the material where such mistakes may occur. These warnings help the students avoid the frustration that naturally occurs when these errors are made and they are unable to discover what is wrong.

DO YOU REMEMBER?
A short matching quiz concludes the text discussion in many sections. These self tests give students the opportunity to check whether they are ready to begin the exercise set for that section. Answers are supplied at the end of the quiz, providing them with immediate feedback about their level of understanding.

EXERCISE SETS
Thousands of class-tested exercises appear at the section ends and chapter ends. An instructor has a choice of assigning just odd-numbered problems, which have answers in the back of text and in the Student's Solution Manual, or just even-numbered problems, with answers only in the Instructor's Solution Manual.

WRITE IN YOUR OWN WORDS
We believe that students have not fully mastered a mathematical concept until they can discuss and write about what they have learned. Where appropriate, the exercise sets contain a "Write In Your Own Words" section that directs the student

systems of linear equations and inequalities are developed. This organization provides the students with the option of using systems of equations or inequalities to solve applied problems throughout the remainder of the text. The introduction of graphing techniques early in the course gives the students a means of visualizing the mathematics that they are learning, which helps them to solve later problems.

Early Functions

The concept of a function is central to the development of mathematics and as such it is introduced in Chapter 4, earlier than many elementary algebra texts. They are then used in the remainder of the text where appropriate.

Applications of Geometry

Many students who enroll in an elementary algebra course in college have never taken a course in plane geometry. They are, however, aware of the geometry of the world about them, so we have included applications of algebra to geometry together with sufficient explanation, so that they can understand. Common geometric figures together with the formulas for finding their perimeters, areas, and volumes are contained in the Appendix. For students who are reviewing algebra before moving forward and who have completed a course in geometry, the problems will give them a chance to refresh and renew concepts they have already learned.

Calculators

Although we have chosen not to focus on the calculator throughout the book, we are certainly aware of its utility and have included problems where its use may be helpful. Such problems may be identified by words or by a graphics calculator icon beside the particular exercise or group of exercises. Periodically, instructions are given that show how to use a calculator to carry out an operation or a specific order of operations.

Theory versus Practice

We maintain a careful balance between the "how" and the "why" of each mathematical concept. The discussion is generally intuitive but also based on carefully stated mathematical definitions and theorems that are supported by a wealth of examples.

Set Notation

Set notation is used to show the relationship between the components of the number system. It is also used to provide a consistent means of identifying solution sets to equations throughout the text.

Arithmetic and Fractions

Chapter 1 contains a review of the concepts of arithmetic. The review either can be covered in class in part or in whole or it can be used as a reference by students who feel they need additional work in this area.

AMATYC and NCTM Guidelines

This text was written with the AMATYC and NCTM guidelines in mind. There is an abundance of exercises that require critical thinking, writing, and some use of modern technology, as well as cooperative group exercises.

to write about a particular concept. Responses can range from a single sentence to a paragraph or more, depending on the question.

FOR EXTRA THOUGHT

Many of the exercise sets contain problems that are designed to challenge the student by requiring them to think critically. The problems use the concepts of the section and require that the student extend the concept a step or two further to arrive at a solution.

CHAPTER SUMMARIES

A summary of the important concepts in each chapter is included at its end. The items discussed in the summaries are keyed to the chapter sections for quick reference.

COOPERATIVE EXERCISES

Each chapter contains a cooperative exercise after the chapter summary. Whenever possible, we have tried to coordinate the exercise with the concepts learned in the chapter. The content of these exercises reflect situations relevant to everyday life.

REVIEW PROBLEMS

Since algebraic concepts are best mastered and retained with constant practice, many exercise sets contain problems from previous sections and chapters. At the end of each chapter there is a complete set of review problems covering the topics in that chapter. A cumulative review is included at the end of every third chapter for Chapters 3, 6, and 9, and there is a cumulative review of all of the chapters at the end of Chapter 10. Complete solutions to all of the problems in the chapter reviews and the cumulative reviews are included in the Student's Solutions Manual.

Ancillaries

INSTRUCTOR'S SOLUTION MANUAL

An instructor's manual is available free of charge to all qualified adopters of the text. It contains the solutions to all problems not solved in the Student's Solution Manual.

TEST BANK

A test bank manual is available free of charge to qualified adopters of the text. It contains 80 questions for each chapter, half of which are multiple choice and half problems that the student is required to solve. The test bank manual is keyed to West Test, a computer test generator that creates tests based on the instructor's selections from the manual.

GROUP LEARNING RESOURCE MANUAL

A Group Learning Resource Manual for Elementary and Intermediate Algebra by Vi Ann Olson of Rochester Community College contains 20 group-learning exercises for both in-class and homework assignments. Each exercise is accompanied by suggestions for the instructor on how best to use that exercise. A section at the beginning of the manual contains guidelines on how to use group-learning exercises.

The exercises are designed to be photocopied by instructors for distribution to the students.

Student's Solution Manual

A Student's Solutions Manual, prepared by Ross Rueger of the College of the Sequoias, is available for purchase by the student. It contains complete solutions to the odd-numbered exercises in the exercise set as well as *all* of the answers to the in-text chapter tests, the chapter reviews, and the cumulative reviews.

West Math Tutor

The *West Math Tutor* by Mathens is available for IBM and IBM-compatible PCs, as well as for Apple Macintosh computers. This software contains algorithmically based tutorials for most of the topics in the text. The tutorials are interactive in that they give feedback to the students when errors are made and provide hints that help the students when they try the questions again. The student is shown the correct solution after two incorrect attempts.

West Math Videos

The West Math Videos are designed to provide additional instruction for the students in the areas in which they may be experiencing difficulty. Worksheets accompany the videos so that the student can be actively involved in the lesson.

Acknowledgements

We would like to thank the many reviewers of this text whose invaluable comments and suggestions greatly improved the final revision:

Karen E. Barker, Indiana University at South Bend
Dennis Carrie, Golden West College
Mitzi Chaffer, Central Michigan University
Sally Copeland, Johnson County Community College
Elwyn Davis, Pittsburg State University
Gregory J. Davis, University of Wisconsin at Green Bay
Michelle Diel, University of New Mexico, Valencia Campus
Robert Farinelli, Community College of Allegheny County, South Campus
Bonnie Gimbel, University of Louisville
Alvin G. Kaumeyer, Pueblo Community College
Jane Keller, Metropolitan Community College
John Martin, Santa Rosa Junior College
Sheila McNicholas, University of Illinois at Chicago
Cynthia Miller, State Technical Institute at Memphis
ViAnn Olson, Rochester Community College
Deborah J. Ritchie, Moorpark College
Ross M. Rueger, College of the Sequoias
Barbara Sausen, Fresno City College

Ray Stanton, Fresno City College

Sandra Vrem, College of the Redwoods

Tom Walsh, City College of San Francisco

George Witt, Glendale Community College

In particular, we want to thank Marvin Johnes of the Cuesta College mathematics department for his line-by-line review of the manuscript and his insistence that we add additional clarifying comments where his teaching experience has shown that students have difficulty. We would also like to thank Richard Mixter and Keith Dodson of West Publishing, who have been involved with this project since its inception, for their helpful suggestions and encouragement; Susan Gerstein, whose superb copyediting skills greatly improved the readability of the text; and Christine Hurney, our production editor, who smoothly directed the transition from manuscript to textbook.

Preface to the Student

Learning mathematics is hard work and requires dedication on your part. Try to adhere to the 2-for-1 rule, which states that every hour in class will be followed by two hours of *quality study* outside of class.

When you registered for the class of which this text is an integral part, you probably had few planned conflicts. It is worth your effort to try to ensure that none develop. In order for you to be successful, we offer the following six suggestions:

1. Except in cases of emergency, attend every class session. No text is designed to cover every detail that you may question. Your instructor can.
2. Ask questions in class. There is no such thing as a stupid question. What qualifies as stupid is the failure to ask a question about something you don't understand when the opportunity presents itself.
3. Take advantage of your instructor's posted office hours. Failure to understand the content of one section will often hinder your understanding of the sections that follow. College classes tend to be less personal than those you experienced in high school and it is up to you to get to know your instructor well.
4. Begin to do the assigned homework as soon as possible after class. Time is your enemy. If you have limited time and can do only a part of the assignment before your next class, do it. The concepts you have learned will be solidified so that it will be much easier to complete the rest of the assignment when your schedule permits.
5. Rework all problems that you miss on any test until you understand what you did wrong. This will keep you from making the same mistakes again.
6. The greatest threat to success in a mathematics class is procrastination. Many college students hesitate to admit that they don't understand a concept, and they wait and hope that something magic will happen. What *will* happen is that insurmountable problems will develop, causing you to withdraw from the class or, worse yet, to fail the class. Get help as soon as you need it.

Elementary Algebra

CHAPTER 1
A REVIEW OF ARITHMETIC

1.1 Basic Concepts of Arithmetic and Algebra
1.2 Common Fractions
1.3 Decimals and Percents
1.4 Some Concepts from Geometry

Application

A swimming pool is 30 feet long, 20 feet wide, and filled to a depth of 5 feet. How many cubic feet of water are in the pool? How many gallons of water are in the pool if there are approximately $7\frac{1}{2}$ gallons of water in one cubic foot?

See Exercise Set 1.4, Exercise 60.

ADA BYRON (1815–1852)
Many people are aware that Charles Babbage worked from 1820 to 1856 trying to develop a calculating machine, but few are aware that it was a woman, Ada Byron, who developed the instructions, in advance, for operating the machine. Essentially, she was what we today call a computer programmer. Unfortunately, the machine—called an analytical engine—*never ran, because the technology of the day was not good enough to construct the parts to the tolerances required.*

Mathematics, like any language, is a form of communication. You must know the meaning of the words and symbols you are using or you won't be able to communicate effectively. This chapter will acquaint you with many of the ideas and skills that are necessary to be successful throughout the rest of the text. We think that you will enjoy learning about—or in many cases reviewing—the information it contains.

The difference between the language of mathematics and other languages, such as English, French, or Spanish, is that the "words" are numbers and symbols. Before two people can communicate in a language, they must understand the meaning of the words and be able to put them together in a meaningful order. For example, the statements "You are looking good" and "You are good-looking" convey two different meanings, even though they use exactly the same words. Similarly, the expression $10 \div 2 + 3$ can take on entirely different meanings if we don't agree how it is to be interpreted. Here are two interpretations.

1. First divide 10 by 2 to get 5. Then add 3 to 5 to get 8.
2. First add 3 to 2 to get 5. Then divide 10 by 5 to get 2.

As you will discover in this section, the first interpretation is correct.

1.1 Basic Concepts of Arithmetic and Algebra

OBJECTIVES In this section we will learn to
1. express algebraic statements using symbols;
2. simplify expressions using the order of operations;
3. translate verbal statements into symbols; and
4. work with numbers by writing them in prime factored form.

One of the main reasons for studying algebra is its usefulness in solving *applied problems*. However, first we must learn how to translate phrases into symbols. The following list gives many common phrases and their symbolic interpretation using the symbols of operation for addition, subtraction, multiplication, and division, as well as symbols for comparison, such as equal to ($=$), less than ($<$), greater than ($>$), less than or equal to ($\leq$), greater than or equal to ($\geq$), and is not equal to ($\neq$). Recall also that the symbol "$\sqrt{}$" is used to indicate the positive square root of a number.

	EXPRESSION IN WORDS	EXPRESSION IN SYMBOLS
Symbols of operation	3 plus 5 the sum of 3 and 5 3 increased by 5 5 more than 3	$3 + 5$
	5 minus 3 the difference between 5 and 3 3 less than 5 3 subtracted from 5	$5 - 3$
	5 times 3 the product of 5 and 3	$5 \cdot 3,\ 5(3),\ (5)(3),\ 5 \times 3$
	$\frac{1}{2}$ of 7	$\frac{1}{2} \cdot 7,\ \frac{1}{2}(7),\ \frac{7}{2}$
	5 divided by 3 the quotient of 5 and 3	$5 \div 3,\ \frac{5}{3},\ 3\overline{)5}$
Symbols of comparison	The sum of 5 and 3 **is equal to** 8. The sum of 8 and 9 **is greater than** 16. Two times the square root of 36 **is less than** 15. The sum of 7 and 5 **is greater than or equal to** 12.	$5 + 3 = 8$ $8 + 9 > 16$ $2\sqrt{36} < 15$ $7 + 5 \geq 12$

We can also express the operation of multiplication in symbols by using exponents when we're multiplying by the same number repeatedly. For example,

2^3 means $2 \cdot 2 \cdot 2 = 8$ —Three times, or three factors of 2

5^2 means $5 \cdot 5 = 25$ —Two times, or two factors of 5

DEFINITION
Exponent

An **exponent** is a number placed slightly above and to the right of another number, called the **base**.

Base —⟶ b^n ⟵— Exponent

When an exponent is a counting number (1, 2, 3, 4, 5, . . .), it indicates the number of times the base is to be used as a factor.

When $n = 2$ and $b = 3$, for example, 3^2 is read as "3 squared" or "3 to the second power."

Using mathematics successfully depends in part on being able to translate written statements into symbols. Such translation can require one or more symbols.

EXAMPLE 1   (a) *The sum of 3 and 11 is 14* is written as $3 + 11 = 14$.
                 (b) *Twenty-three equals the sum of 15 and 8* is written as $23 = 15 + 8$.
                 (c) *The product of 15 and 4 is 60* is written as $15 \cdot 4 = 60$.

(d) *The sum of 20 and 19 decreased by 17 equals 22* is written as $20 + 19 - 17 = 22$.

(e) *The difference between 6^2 and 3^2 is 27* is written as $6^2 - 3^2 = 27$.

Understanding punctuation is often the key to proper mathematical interpretation of verbal statements, just as it is in the correct interpretation of the meaning of a sentence. For example,

Bob said, "Tom is gone."   (Tom is gone, Bob is here)

has an entirely different meaning than

"Bob," said Tom, "is gone."   (Bob is gone, Tom is here)

even though they use exactly the same words in the same order.

EXAMPLE 2
(a) *Two multiplied by 3, plus 5* is written as $2 \cdot 3 + 5$.
(b) *Two multiplied by 3 plus 5* is written as $2(3 + 5)$.

Commas are not used in mathematical expressions. Instead, we must agree upon a definite order for carrying out the operations when simplifying an expression.

> **Order of Operations**
>
> Simplify a problem from left to right in the following order:
>
> **1.** Carry out the operations within the grouping symbols working from the innermost symbols outward. Grouping symbols are parentheses, ( ), brackets [ ], braces { }, the fraction bar, ——, and the square root symbol, $\sqrt{}$.
> **2.** Evaluate numbers with exponents.
> **3.** Carry out multiplication and division from left to right.
> **4.** Carry out addition and subtraction from left to right.

Now let's look at some examples using the order of operations just outlined.

EXAMPLE 3   Simplify the following expressions by carrying out the indicated operations.

(a) $2 \cdot 3 + 5 = 6 + 5$ Carry out multiplication first: $2 \cdot 3 = 6$.
          $= 11$ Add: $6 + 5 = 11$.

(b) $5 + 4^2 - 3 = 5 + 16 - 3$ Evaluate numbers with exponents first: $4^2 = 16$.
          $= 18$ Add and subtract from left to right.

(c) $(9 - 4)(3 + 6) = (5)(9)$ Work inside grouping symbols first: $9 - 4 = 5$; $3 + 6 = 9$.
          $= 45$ Multiply.

(d) $2 + (9 - 4)^2 = 2 + 5^2$ Work inside grouping symbols first: $(9 - 4)^2 = 5^2$.
$ = 2 + 25$ Evaluate numbers with exponents: $5^2 = 25$.
$ = 27$ Add.

(e) $68 - [3(4 - 2)]^2 = 68 - [3(2)]^2$ Work within innermost grouping symbol first: $3(4 - 2) = 3(2)$.
$ = 68 - [6]^2$ Inside grouping symbol first: $3(2) = 6$.
$ = 68 - 36$ Exponents before subtraction: $6^2 = 36$.
$ = 32$ Subtract: $68 - 36 = 32$.

(f) $24 \div 6 \cdot 2 + 5 = 4 \cdot 2 + 5$ Work from left to right, dividing first: $24 \div 6 = 4$.
$ = 8 + 5$ Multiply: $4 \cdot 2 = 8$.
$ = 13$ Add: $8 + 5 = 13$.

EXAMPLE 4   Simplify: $5^2 - 24 \div 4 \cdot 2$.

SOLUTION   Particular care must be taken to follow the order of operations from left to right, as they occur.

$5^2 - 24 \div 4 \cdot 2 = 25 - 24 \div 4 \cdot 2$ Evaluate numbers with exponents first.
$ = 25 - 6 \cdot 2$ Division precedes subtraction: $24 \div 4 = 6$.
$ = 25 - 12$ Multiplication precedes subtraction: $6 \cdot 2 = 12$.
$ = 13$ Subtract: $25 - 12 = 13$.

EXAMPLE 5   Simplify: $\dfrac{28 - 4 \cdot 6}{7^2 - 9 \cdot 5}$.

SOLUTION   The fraction bar is a grouping symbol. Therefore we must carry out the operations in the numerator and the denominator independently before dividing.

$$\frac{28 - 4 \cdot 6}{7^2 - 9 \cdot 5} = \frac{28 - 24}{49 - 45}$$
$$= \frac{4}{4} = 1$$

In order to use mathematics to help us solve problems, we must be able to translate from either written or verbal statements into mathematical symbols.

EXAMPLE 6   On five successive days, a leading stock market indicator showed gains of $22, $13, $2, $7, and $3. How much did the market gain that week?

SOLUTION   To find out how much the market gained, we have to find the sum of its daily gains:

$$\$22 + \$13 + \$2 + \$7 + \$3 = \$47$$

The gain for the week was $47.

EXAMPLE 7   Jose received 3 units (credits) of A, 6 units of B, and 1 unit of C during his first semester in college. If each unit of A is worth 4 grade points, each unit of B is worth 3 grade points, and each unit of C is worth 2 grade points, how many grade points did he earn that semester? What was his grade point average (GPA)?

SOLUTION   To determine the number of grade points Jose earned, we must find the sum of the grade points he earned with each grade.

Three units of A at 4 points each = $3 \cdot 4$ = 12 grade points
Six units of B at 3 points each = $6 \cdot 3$ = 18 grade points
One unit of C at 2 points each = $1 \cdot 2$ = 2 grade points

Jose's total grade points are

$$3 \cdot 4 + 6 \cdot 3 + 1 \cdot 2 = 12 + 18 + 2 = 32$$

To compute his GPA, we divide the number of grade points he earned by the number of units he completed. Jose completed $3 + 6 + 1 = 10$ units, so

$$\text{GPA} = \frac{32 \text{ grade points}}{10 \text{ units}} = 3.2$$

Before we leave this section, let's return to the idea of factors and exponents. Notice that the number 25 can be written

$$25 = 5 \cdot 5 = 5^2 \quad \leftarrow \text{Two factors of 5}$$

Notice also that the number 8 can be written

$$8 = 2 \cdot 2 \cdot 2 = 2^3 \quad \leftarrow \text{Three factors of 2}$$

Not all numbers can be written this way. The number 15 can't be written using exponents, but it can be written in terms of 3 and 5:

$$15 = 3 \cdot 5$$

The number 15 can be divided by both 3 and 5 without a remainder occurring.

DEFINITION **Divisibility**	When one number can be divided by another without a remainder, it is said to be **divisible** by the second number.

Some very useful tests for divisibility follow.

> **Tests for Divisibility**
>
> **1.** A number is divisible by 2 if it is even.
> **2.** A number is divisible by 3 if the sum of its digits is divisible by 3.
> **3.** A number is divisible by 5 if it ends in a zero or a 5.

EXAMPLE 8 Check each of the following numbers for divisibility by 2, 3, and 5:

(a) 28   (b) 36   (c) 156   (d) 2175.

SOLUTION
(a) 28 is divisible by 2 because it's even.
(b) 36 is divisible by 2 because it's even. It is also divisible by 3 because the sum of its digits, $3 + 6 = 9$, is divisible by 3.
(c) 156 is divisible by both 2 and 3 because it's even and the sum of its digits, $1 + 5 + 6 = 12$, is divisible by 3.
(d) 2175 is divisible by 5 because it ends in 5. It is also divisible by 3, since the sum of its digits is 15, which is divisible by 3.

Some numbers are not divisible by any numbers other than themselves and 1, such as 7 and 23. Such numbers are called *prime numbers*.

> A **prime number** is a counting number greater than 1 that is divisible only by itself and 1. The first fifteen prime numbers are 2, 3, 5, 7, 11, 13, 17, 19, 23, 29, 31, 37, 41, 43, and 47. The only *even* prime number is 2.

When simplifying both arithmetic and algebraic expressions, it is often useful to write them in **prime factored form.** This means that each factor is a prime number. For example, 15 written in prime factored form is $3 \cdot 5$, since both 3 and 5 are prime numbers. In each of the following examples, we used divisibility to find the factors. It is best to begin by using 2 as a divisor followed by 3 and then other successive prime numbers.

EXAMPLE 9 Write 36 in prime factored form.

SOLUTION
$36 = 2 \cdot 18$ 2)36
                          ↓
$36 = 2 \cdot 2 \cdot 9$ 2)18 $36 \div 2 = 18$
                          ↓
$36 = 2 \cdot 2 \cdot 3 \cdot 3$ 3)9 $18 \div 2 = 9$
                                  ↳3 $9 \div 3 = 3$

We follow the arrows to find the prime factored form of 36, which is

$$2 \cdot 2 \cdot 3 \cdot 3 = 2^2 \cdot 3^2$$

EXAMPLE 10 Write 252 in prime factored form.

SOLUTION

$252 = 2 \cdot 126$

$252 = 2 \cdot 2 \cdot 63$

$252 = 2 \cdot 2 \cdot 3 \cdot 21$

$252 = 2 \cdot 2 \cdot 3 \cdot 3 \cdot 7$

$2\overline{)252}$
$2\overline{)126}$ $252 \div 2 = 126$
$3\overline{)63}$ $126 \div 2 = 63$
$3\overline{)21}$ $63 \div 3 = 21$
7 $21 \div 3 = 7$

The prime factored form of 252 is $2 \cdot 2 \cdot 3 \cdot 3 \cdot 7 = 2^2 \cdot 3^2 \cdot 7$.

Do You Remember?

Can you match these?

_____ 1. A word meaning addition   a) difference
_____ 2. A word meaning multiplication   b) sum
_____ 3. A word meaning subtraction   c) quotient
_____ 4. A word meaning division   d) product
_____ 5. An example of a prime number   e) ≠
_____ 6. $2 \cdot 3 + 12 \div 2 \cdot 3$ f) 24
_____ 7. The base in 4^3 g) less than or equal to
_____ 8. ≤   h) 4
_____ 9. Is not equal to   i) 8
     j) 13

Answers: 1. b 2. d 3. a 4. c 5. j 6. f 7. h 8. g 9. e

Exercise Set 1.1

Express each statement using symbols.

1. The sum of 6 and 8 equals 14.
2. The sum of 5 and 11 equals 16.
3. The difference of 16 and 9 is 7.
4. The difference of 37 and 19 is 18.
5. Twenty-six equals the sum of 12 and 14.
6. Seventeen equals the difference of 24 and 7.
7. The product of 6 and 9 is 54.
8. The product of 7 and 11 is 77.
9. Five is the quotient of 30 and 6.

10. Seven is the quotient of 91 and 13.
11. Fifty-six divided by 7 equals 8.
12. Thirty-nine subtract 15 equals 24.
13. Sixty-seven less 51 is more than 11.
14. Eighteen more than 16 is less than 36.
15. Five squared equals 25.
16. The square root of 36 equals 6.
17. One subtracted from 2 squared is 3.
18. The sum of 3 squared and 4 is 13.
19. Seven is less than the sum of 5 and 4.
20. Nineteen is 3 more than 4 squared.
21. The difference of 5 squared and 4 squared is 3 squared.
22. The sum of the square root of 16 and 3 is 7.
23. The product of the square root of 9 and the square root of 25 is 15.
24. Eighteen is more than 4 squared.
25. Sixty-seven is less than 9 squared.
26. Fifty-three is less than the product of 7 and 9.
27. Eighteen is more than the quotient of 72 and 6.
28. The difference of 15 and 6 is not equal to 7.
29. The sum of 5 and 13 is not equal to 19.
30. The product of 9 and 8 is not equal to 73.

Simplify each of the following mathematical expressions.

31. $8 \cdot 2 + 3$
32. $9 \cdot 4 + 4$
33. $6 \cdot 5 - 5$
34. $7 \cdot 3 - 8$
35. $16 \div 4 - 3$
36. $6(3 - 1)$
37. $2(5 + 3)$
38. $3 + 2 \cdot 5^2$
39. $19 - 2 \cdot 3^2$
40. $68 \div 34 - 1$
41. $81 \div 3 + 4$
42. $3 \cdot 5 - 2 \cdot 4$
43. $7 \cdot 6 + 2 \cdot 3$
44. $5^2 - 3^2$
45. $7^2 + 4^2$
46. $(3 - 2)(5 + 4)$
47. $(4 + 7)(11 - 3)$
48. $(3 - 1)[3(5 - 2)]$
49. $(3 + 2)[5(7 - 6)]$
50. $6 + 5^2 \cdot 3$
51. $56 \div 7 \div 4$
52. $69 \div 3 \cdot 2$
53. $27 \div (7 + 2)$
54. $7 - [2(5 - 3)]$
55. $8 - [3(7 - 5)]$
56. $5(4^2 - 3^2) \div 35$
57. $7(5^2 - 4^2) \div 9$

State whether each of the following is divisible by **(a)** *2,* **(b)** *3, and/or* **(c)** *5.*

58. 6
59. 9
60. 10
61. 33
62. 111
63. 210
64. 75
65. 32
66. 30
67. 132

Factor each of the following into prime factored form.

68. 8
69. 12
70. 20
71. 62
72. 84
73. 210
74. 360
75. 225
76. 480
77. 540

Simplify each of the following mathematical expressions.

78. $\dfrac{5 + 3 \cdot 2}{8 + 3}$
79. $\dfrac{16 - 5 \cdot 3}{4^2 - 15}$
80. $\dfrac{16 \div (5 - 1)}{3^2 - (7 - 2)}$
81. $\dfrac{6[5 \div (2 + 3) \cdot 4]}{2[15 \div (7 - 2) \cdot 2]}$
82. $\dfrac{8^2 - 7^2}{5(2 - 1)}$
83. $\dfrac{(8 \cdot 7 + 6^2)2}{2[3^2 + 1]}$

State whether the first number is less than or greater than the second number.

84. 1 and 4
85. 3 and 5
86. 5 and 1
87. 6 and 4

Answer each question.

88. Susan has $36. Mandy has $8 less than Susan. How many dollars does Mandy have?
89. Jose had $67 in his checking account. He deposited $53 and wrote a check for $86. How much is left in his checking account?
90. Sam buys a ream (500 sheets) of paper. If the paper is divided evenly among four people, how many sheets will each person have?
91. Jessica studies 4 hours (hr) per day every day of the week except Sunday. How many hours does she study in one week?
92. Jorge drives 17 miles (mi) round trip to school. If he makes the round trip once on Monday, Wednesday, and Friday and twice on Tuesday and Thursday, how many miles does he cover in one week?
93. Lisa usually spends 7 hr at school on Monday and Wednesday, 5 hr on Tuesday, 3 hr on Thursday and 6 hr on Friday. If she got sick on Wednesday and stayed home that day, how many hours did she spend at school that week?
94. Matt earned 13 units, 15 units, 12 units, and 16 units his first four semesters of college. Seven of these units were not transferrable when he moved. How many units were transferrable?
95. Maria earns 4 grade points for each unit of A, 3 grade points for each unit of B, and 2 grade points for each

unit of C. If she has 15 units of A, 22 units of B, and 17 units of C, how many grade points does she have? What is her GPA?

96. A basketball team has 24 points at halftime, and by the end of the game, they've tripled their halftime score. How many points do they have at the end of the game?

97. Don cuts a very large submarine sandwich into 6 pieces. If each piece is 4 inches (in.) long, how long was the original sandwich?

98. Iva plans to work 54 hr/month. If she works 18 days/month, how many hours does she work each day? Assume she works the same number of hours each day.

99. Les has a car that averages 23 miles per gallon (mpg). If the gas tank holds 13 gal, can he travel 315 mi on one tank of gas? How far can he travel on one tank of gas?

100. Cindy earns $5/hr. She works 5 days each week. Can she earn enough to pay her rent of $87/week if she works no more than 4 hr/day? What is the most she can earn in one week if she works no more than 4 hr/day?

WRITE IN YOUR OWN WORDS

101. $3 \cdot 2 + 5 = 11$ **102.** $35 \div (3 + 2) = 7$

103. $12 = 2 \cdot 2 \cdot 3$

104. How can you tell whether a number is divisible by 2?

105. How can you tell whether a number is divisible by 3?

106. How can you tell whether a number is divisible by 5?

107. $3^2 = 9$ **108.** $5\sqrt{36} = 30$

109. $7 < 15$ **110.** $7 - 5 \cdot 1^3 \neq 2^3$

111. $12 \div 3 \cdot 2 = 8$ **112.** $8 \div (6 + 2) = 1$

113. $6 + 18 \div 3 = 12$

114. $(5 - 3) \div (7 - 3) \geq \frac{1}{4}$

115. What is the answer to a division problem called?

116. What is the answer to a multiplication problem called?

117. What is the answer to an addition problem called?

118. What is the answer to a subtraction problem called?

119. In 4^5, what is the 5 called? What is the 4 called? What does the 5 tell us?

120. What does "simplify" mean?

121. $8 \leq 10$ **122.** $3 \geq 0$ **123.** $5 > 1$

124. Define "prime number."

125. Define "prime factored form."

126. $3 + 4 \neq 8$

1.2 Common Fractions

HISTORICAL COMMENT   In early Greek mathematics, fractions were thought of as the ratio of two numbers. Some later writers preferred to use the symbolism 3 : 4 for a ratio, instead of the more common form $\frac{3}{4}$. The Egyptians considered a fraction to be a part of a number. The word fraction has its roots in the Latin verb *frangere*, which means "to break." Chaucer (1342–1400) used the word *fraccion* and Pacioli (1445–1514) used *fracti* and *fractioni* in their writings. The Greeks had two sets of fractions, one in which the denominators were limited to powers of 60, which were used in astronomy, and everyday fractions, called *minutae vulgares*. The English translation of this name is "vulgar fractions," which eventually took on the American name "common fractions."

OBJECTIVES   In this section we will learn to
1. add, subtract, multiply, and divide fractions;
2. simplify fractions;
3. identify proper and improper fractions; and
4. use fractions to solve applied problems.

A fraction consists of two parts, the **numerator** and the **denominator**:

$$\frac{\text{numerator}}{\text{denominator}}$$

In the fraction $\frac{3}{4}$, the number 3 is the numerator and 4 is the denominator. The denominator tells us how many parts the whole is divided into, while the numerator tells us how many of those parts we have. A **proper fraction,** such as $\frac{15}{22}$, is one in which the numerator is *less than* the denominator. The fraction $\frac{11}{3}$ is said to be **improper,** since the numerator is *greater than or equal to* the denominator. Improper fractions can also be written as **mixed numbers,** which is a counting number and a proper fraction. For example, the improper fraction $\frac{11}{3}$ can be written $3\frac{2}{3}$, where 3 is the quotient obtained when 11 is divided by 3, and 2 is the remainder.

When the numerator and denominator of a fraction contain common factors, the fraction can be simplified by dividing out the common factors. The resulting fraction is **equivalent** to the original fraction. For example,

$$\frac{14}{10} = \frac{\overset{1}{\cancel{2}} \cdot 7}{\cancel{2} \cdot 5} = \frac{7}{5} \quad \text{Divide out the common factor, 2.}$$

Also, we can obtain an equivalent fraction from a given fraction by multiplying the numerator and the denominator of the fraction by the same number. For example,

$$\frac{2}{3} = \frac{2 \cdot 5}{3 \cdot 5} = \frac{10}{15} \quad \text{Multiply both numerator and denominator by 5.}$$

These two examples illustrate the **fundamental principle of fractions.**

The Fundamental Principle of Fractions	If $\dfrac{a \cdot c}{b \cdot c}$ is a fraction, with $b, c \neq 0$, then $$\frac{a \cdot c}{b \cdot c} = \frac{a}{b}$$ If $\dfrac{a}{b}$ is a fraction, with $b \neq 0$, then $$\frac{a}{b} = \frac{a \cdot c}{b \cdot c}, \quad c \neq 0$$

EXAMPLE 1 We can simplify each of the following fractions by first writing the numerator and denominator as the product of prime numbers and then using the fundamental principle of fractions.

(a) $\dfrac{15}{25} = \dfrac{3 \cdot \overset{1}{\cancel{5}}}{5 \cdot \cancel{5}} = \dfrac{3}{5}$ Divide out the common factor, 5.

(b) $\dfrac{162}{72} = \dfrac{\cancel{2}\cdot\cancel{3}\cdot\cancel{3}\cdot 3\cdot 3}{\cancel{2}\cdot 2\cdot 2\cdot \cancel{3}\cdot \cancel{3}}$ Divide out the common factors, $2 \cdot 3 \cdot 3$.

$= \dfrac{3\cdot 3}{2\cdot 2} = \dfrac{9}{4}$

(c) $\dfrac{140}{294} = \dfrac{\cancel{2}\cdot 2\cdot 5\cdot \cancel{7}}{\cancel{2}\cdot 3\cdot 7\cdot \cancel{7}}$ Divide out the common factors, $2 \cdot 7$.

$= \dfrac{10}{21}$

It is not necessary to write the numerator and the denominator in prime factored form in order to simplify a fraction.

EXAMPLE 2 Write each of the following fractions in simplest form: **(a)** $\dfrac{36}{27}$ **(b)** $\dfrac{24}{60}$

SOLUTION **(a)** By inspection, both 36 and 27 are divisible by 9, so a common factor of each is 9.

$$\dfrac{36}{27} = \dfrac{4\cdot \cancel{9}}{3\cdot \cancel{9}} = \dfrac{4}{3}$$ Divide numerator and denominator by 9.

(b) By inspection, both 24 and 60 are divisible by 12, so a common factor of each is 12.

$$\dfrac{24}{60} = \dfrac{2\cdot \cancel{12}}{5\cdot \cancel{12}} = \dfrac{2}{5}$$ Divide numerator and denominator by 12.

We can use the fundamental principle of fractions to write fractions that are equivalent to a given fraction. For example, to write $\tfrac{3}{5}$ as an equivalent fraction with a denominator of 35, we have to multiply the numerator and denominator by 7:

$$\dfrac{3}{5} = \dfrac{3\cdot 7}{5\cdot 7} = \dfrac{21}{35}$$

EXAMPLE 3 Replace ? with a number so that the fraction on the left is equivalent to the one on the right: **(a)** $\dfrac{7}{11} = \dfrac{?}{44}$ **(b)** $\dfrac{11}{27} = \dfrac{?}{81}$

SOLUTION **(a)** Since $44 = 11 \cdot 4$, we multiply the numerator and denominator of $\dfrac{7}{11}$ by 4:

$$\dfrac{7}{11} = \dfrac{7\cdot 4}{11\cdot 4} = \dfrac{28}{44}$$

(b) Since $27 \cdot 3 = 81$, we multiply the numerator and denominator of $\frac{11}{27}$ by 3:

$$\frac{11}{27} = \frac{11 \cdot 3}{27 \cdot 3} = \frac{33}{81}$$

To Multiply Fractions

If $\frac{a}{b}$ and $\frac{c}{d}$ represent fractions, with $b \neq 0$ and $d \neq 0$, then

$$\frac{a}{b} \cdot \frac{c}{d} = \frac{a \cdot c}{b \cdot d}$$

In simplest terms, the product of two fractions is found by writing the product of the numerators over the product of the denominators. Common factors should be divided out *before* the fractions are multiplied.

EXAMPLE 4 Find each of the following products:

(a) $\frac{3}{5} \cdot \frac{2}{7}$ (b) $\left(\frac{5}{13}\right)\left(\frac{39}{54}\right)$ (c) $\left(\frac{8}{3}\right)\left(\frac{2}{5}\right)$ (d) $\left(1\frac{3}{5}\right)\left(2\frac{7}{8}\right)$.

SOLUTION

(a) $\frac{3}{5} \cdot \frac{2}{7} = \frac{3 \cdot 2}{5 \cdot 7} = \frac{6}{35}$ There are no common factors to be divided out.

(b) $\left(\frac{5}{13}\right)\left(\frac{39}{54}\right) = \left(\frac{5}{\cancel{13}}\right)\left(\frac{\cancel{3} \cdot \cancel{13}}{\cancel{3} \cdot 18}\right)$ Divide out the common factors of 3 and 13 before multiplying.

$= \frac{5}{18}$

(c) $\left(\frac{8}{3}\right)\left(\frac{2}{5}\right) = \frac{8 \cdot 2}{3 \cdot 5} = \frac{16}{15}$ There are no common factors to be divided out.

(d) $\left(1\frac{3}{5}\right)\left(2\frac{7}{8}\right) = \left(\frac{8}{5}\right)\left(\frac{23}{8}\right)$ Change each mixed number to an improper fraction.

$= \frac{23}{5}$

In Example 4(d) the answer $\frac{23}{5}$ can be written as $4\frac{3}{5}$, where the whole-number part, 4, is the quotient obtained when 23 is divided by 5. The 3 in $\frac{3}{5}$ is the remainder from that division. In this text, however, we will not generally change an answer to a mixed number when it involves an improper fraction.

To Divide Fractions

If $\dfrac{a}{b}$ and $\dfrac{c}{d}$ represent fractions, with b, c, and $d \neq 0$,

$$\frac{a}{b} \div \frac{c}{d} = \frac{a}{b} \cdot \frac{d}{c}$$

In simplest terms, the quotient of two fractions is found by inverting the divisor and following the rules for multiplication.

EXAMPLE 5   Find the following quotients: **(a)** $\dfrac{2}{3} \div \dfrac{8}{9}$ **(b)** $\dfrac{2}{5} \div \left(\dfrac{3}{4}\right)\left(1\dfrac{1}{8}\right)$

SOLUTION   **(a)** $\dfrac{2}{3} \div \dfrac{8}{9} = \dfrac{\cancel{2}^1}{\cancel{3}} \cdot \dfrac{\cancel{9}^3}{\cancel{8}_4}$ Simplify before multiplying.

$\qquad\qquad\qquad = \dfrac{3}{4}$

(b) We first have to change $1\dfrac{1}{8}$ to $\dfrac{9}{8}$ and then follow the order of operations.

$$\frac{2}{5} \div \left(\frac{3}{4}\right)\left(1\frac{1}{8}\right) = \frac{2}{5} \div \left(\frac{3}{4}\right) \cdot \left(\frac{9}{8}\right)$$

$$= \frac{2}{5} \cdot \frac{4}{3} \cdot \frac{9}{8} \qquad \text{Work from left to right.}$$

$$= \frac{8}{15} \cdot \frac{9}{8}$$

$$= \frac{\cancel{8}^1}{\cancel{3} \cdot 5} \cdot \frac{\cancel{3} \cdot 3}{\cancel{8}} \qquad \text{Divide out the common factors, 3 and 8.}$$

$$= \frac{3}{5}$$

To Add and Subtract Fractions

If $\dfrac{a}{c}$ and $\dfrac{b}{c}$ represent fractions, with $c \neq 0$, then

$$\frac{a}{c} + \frac{b}{c} = \frac{a+b}{c} \qquad \text{and} \qquad \frac{a}{c} - \frac{b}{c} = \frac{a-b}{c}$$

To add or subtract two or more fractions with a common denominator, we add or subtract the numerators and write the result over the common denominator.

EXAMPLE 6 Carry out the following operations: (a) $\dfrac{9}{14} - \dfrac{3}{14}$ (b) $\dfrac{2}{11} + \dfrac{7}{11} - \dfrac{4}{11}$

SOLUTION (a) $\dfrac{9}{14} - \dfrac{3}{14} = \dfrac{9-3}{14} = \dfrac{6}{14} = \dfrac{3}{7}$

(b) $\dfrac{2}{11} + \dfrac{7}{11} - \dfrac{4}{11} = \dfrac{2+7-4}{11} = \dfrac{5}{11}$

When the denominators of two fractions are not the same and we want to add or subtract them, we can use the fundamental principle of fractions to write each of them in terms of a common denominator. For example, to add $\frac{2}{7}$ to $\frac{2}{3}$, we must find a common denominator. Such a common denominator, called the **least common denominator (LCD),** is the smallest counting number that is divisible by both 3 and 7. The LCD for 3 and 7 is 21.

$$\dfrac{2}{7} + \dfrac{2}{3} = \dfrac{2}{7} \cdot \dfrac{3}{3} + \dfrac{2}{3} \cdot \dfrac{7}{7}$$

$$= \dfrac{6}{21} + \dfrac{14}{21}$$

$$= \dfrac{20}{21}$$

EXAMPLE 7 Simplify: (a) $\dfrac{3}{5} + \dfrac{2}{3}$ (b) $\dfrac{1}{7} + \dfrac{3}{14} - \dfrac{1}{4}$

SOLUTION (a) $\dfrac{3}{5} + \dfrac{2}{3} = \dfrac{3 \cdot 3}{5 \cdot 3} + \dfrac{2 \cdot 5}{3 \cdot 5}$ The LCD is 15; write each fraction in terms of 15.

$\qquad = \dfrac{9}{15} + \dfrac{10}{15}$ The denominators are the same.

$\qquad = \dfrac{19}{15}$ Add the numerators.

(b) $\dfrac{1}{7} + \dfrac{3}{14} - \dfrac{1}{4} = \dfrac{1 \cdot 4}{7 \cdot 4} + \dfrac{3 \cdot 2}{14 \cdot 2} - \dfrac{1 \cdot 7}{4 \cdot 7}$ The LCD is 28, write each fraction in terms of 28.

$\qquad = \dfrac{4}{28} + \dfrac{6}{28} - \dfrac{7}{28}$ Multiply.

$\qquad = \dfrac{4 + 6 - 7}{28}$ Write as a single fraction.

$\qquad = \dfrac{3}{28}$ Evaluate the numerator.

Now let's put our computation with fractions together with our previous work on the order of operations.

EXAMPLE 8 Simplify: $\left(\dfrac{1}{3}+\dfrac{2}{5}\right)^2+\dfrac{3}{7}\div\dfrac{15}{14}$.

SOLUTION

$\left(\dfrac{1}{3}+\dfrac{2}{5}\right)^2+\dfrac{3}{7}\div\dfrac{15}{14}=\left(\dfrac{11}{15}\right)^2+\dfrac{3}{7}\div\dfrac{15}{14}$ Grouping symbols first.

$=\dfrac{121}{225}+\dfrac{3}{7}\div\dfrac{15}{14}$ Operations with exponents next.

$=\dfrac{121}{225}+\dfrac{3}{7}\cdot\dfrac{14}{15}$ Rewrite division as multiplication.

$=\dfrac{121}{225}+\dfrac{2}{5}$ Simplify and multiply.

$=\dfrac{121}{225}+\dfrac{90}{225}$ $\dfrac{2}{5}=\dfrac{2}{5}\cdot\dfrac{45}{45}=\dfrac{90}{225}$

$=\dfrac{211}{225}$ Add the numerators.

EXAMPLE 9 Divide the difference between $\dfrac{1}{2}$ and $\dfrac{1}{3}$ by the product of $\dfrac{1}{2}$ and $\dfrac{1}{4}$.

SOLUTION In symbols we can write this as

$\dfrac{\dfrac{1}{2}-\dfrac{1}{3}}{\left(\dfrac{1}{2}\right)\left(\dfrac{1}{4}\right)}=\dfrac{\dfrac{3}{6}-\dfrac{2}{6}}{\dfrac{1}{8}}$ The fraction bar is a grouping symbol.

$=\dfrac{\dfrac{1}{6}}{\dfrac{1}{8}}$ Simplify the numerator.

$=\dfrac{1}{6}\cdot\dfrac{8}{1}=\dfrac{4}{3}$

In Section 1.1 we examined the symbols of comparison. We now examine how they apply to the comparison of fractions.

When two fractions have a common denominator, the one with the greater numerator is the greater fraction. Thus $\frac{5}{8}>\frac{3}{8}$ and so $\frac{5}{8}$ appears to the right of $\frac{3}{8}$ on a number line. To locate the two fractions on a number line, we divide the distance between 0 and 1 into eight equal parts, each $\frac{1}{8}$ unit long; see Figure 1.1.

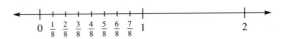

FIGURE 1.1

EXAMPLE 10 Which fraction is greater, $\frac{2}{5}$ or $\frac{3}{7}$?

SOLUTION The least common denominator is 35:

$$\frac{2}{5} = \frac{14}{35} \quad \text{and} \quad \frac{3}{7} = \frac{15}{35}$$

Since $\frac{15}{35} > \frac{14}{35}$, we see that $\frac{3}{7} > \frac{2}{5}$.

EXAMPLE 11 How many $1\frac{1}{2}$ gallon sprayers can be completely filled from a 20-gallon tank of insecticide?

SOLUTION To determine the number of sprayers, we divide 20 by $1\frac{1}{2}$:

$$20 \div 1\frac{1}{2} = 20 \div \frac{3}{2} = 20 \cdot \frac{2}{3}$$

$$= \frac{40}{3} = 13\frac{1}{3}$$

Thirteen sprayers can be *completely* filled.

Do You Remember?

Can you match these

____ 1. The condition for an improper fraction
____ 2. An example of a mixed number
____ 3. The fundamental principle of fractions
____ 4. The quotient of $\frac{a}{b} \div \frac{c}{d}$
____ 5. The condition for a proper fraction
____ 6. The product of $\frac{a}{c} \cdot \frac{c}{d}$
____ 7. $\frac{1}{3} \div \frac{1}{2} \cdot \frac{2}{3}$

a) $\frac{a \cdot c}{b \cdot c} = \frac{a}{b}, \quad b, c \neq 0$

b) $\frac{4}{9}$

c) denominator > numerator

d) $\frac{a \cdot d}{b \cdot c}$

e) $\frac{a}{d}$

f) $7\frac{2}{5}$

g) 1

h) numerator $\geq$ denominator

i) $\frac{1}{9}$

Answers: 1. h, 2. f, 3. a, 4. d, 5. c, 6. e, 7. b

EXERCISE SET 1.2

Use the fundamental principle of fractions to simplify each of the following fractions.

1. $\dfrac{16}{25}$
2. $\dfrac{28}{30}$
3. $\dfrac{48}{60}$
4. $\dfrac{56}{64}$
5. $\dfrac{30}{105}$
6. $\dfrac{168}{24}$
7. $\dfrac{90}{15}$
8. $\dfrac{272}{68}$

Use the fundamental principle of fractions to rewrite each of the following fractions with the given denominator.

9. $\dfrac{1}{2} = \dfrac{?}{4}$
10. $\dfrac{3}{8} = \dfrac{?}{16}$
11. $\dfrac{5}{7} = \dfrac{?}{21}$
12. $\dfrac{3}{5} = \dfrac{?}{30}$
13. $\dfrac{9}{11} = \dfrac{?}{44}$
14. $\dfrac{2}{7} = \dfrac{?}{98}$
15. $\dfrac{3}{4} = \dfrac{?}{64}$
16. $\dfrac{2}{9} = \dfrac{?}{126}$

Multiply or divide as indicated, and simplify.

17. $\dfrac{1}{2} \cdot \dfrac{2}{3}$
18. $\dfrac{3}{4} \cdot \dfrac{8}{9}$
19. $\left(\dfrac{3}{7}\right)\left(\dfrac{5}{4}\right)\left(\dfrac{20}{30}\right)$
20. $\left(\dfrac{3}{2}\right)\left(\dfrac{7}{4}\right)\left(\dfrac{8}{5}\right)$
21. $\left(1\dfrac{1}{2}\right) \div \left(1\dfrac{1}{4}\right)$
22. $\left(1\dfrac{5}{8}\right) \div \left(1\dfrac{3}{4}\right)$
23. $\left(2\dfrac{1}{8}\right)\left(3\dfrac{1}{2}\right) \div \left(\dfrac{17}{7}\right)$
24. $\left(5\dfrac{1}{4}\right) \div \left(3\dfrac{1}{2}\right)\left(2\dfrac{1}{8}\right)$

Add or subtract as indicated, and simplify.

25. $\dfrac{1}{4} + \dfrac{3}{4}$
26. $\dfrac{1}{2} + \dfrac{3}{2}$
27. $1\dfrac{3}{10} + \dfrac{7}{10}$
28. $1\dfrac{5}{11} + \dfrac{6}{11}$
29. $\dfrac{2}{3} + \dfrac{3}{4}$
30. $\dfrac{5}{8} + \dfrac{1}{2}$
31. $\dfrac{11}{12} - \dfrac{1}{2}$
32. $\dfrac{15}{16} - \dfrac{1}{4}$
33. $1\dfrac{7}{8} + 5\dfrac{1}{6}$
34. $1\dfrac{1}{6} + 3\dfrac{1}{9}$
35. $2 + 1\dfrac{1}{10} - 1\dfrac{2}{15}$
36. $3 - 1\dfrac{1}{8} + 2\dfrac{5}{6}$

Perform the indicated operations and simplify.

37. $5\left(\dfrac{2}{3}\right)^2 \div \dfrac{10}{9}$
38. $\left(\dfrac{1}{2}\right) \div \left(\dfrac{1}{3}\right)^2$
39. $\left(1\dfrac{1}{2}\right)^2 \cdot \left(1\dfrac{1}{3}\right)^2$
40. $\left(2\dfrac{1}{3}\right)^2 \cdot \left(3\dfrac{1}{2}\right)^2$
41. $\left(\dfrac{1}{4} + \dfrac{4}{3}\right) \cdot \dfrac{3}{4}$
42. $\left(\dfrac{1}{2} + \dfrac{2}{3}\right) \cdot \dfrac{5}{9}$
43. $\left(\dfrac{3}{4} - \dfrac{1}{5}\right)^2$
44. $\left(\dfrac{1}{3} + \dfrac{1}{4}\right)^2 \div \dfrac{1}{2}$

45. Subtract $\dfrac{5}{6}$ from the product of $1\dfrac{1}{2}$ and $\dfrac{3}{4}$.
46. Add $4\dfrac{1}{2}$ to the quotient of $5\dfrac{1}{2}$ and $1\dfrac{1}{4}$.
47. Add $1\dfrac{1}{7}$ to the difference of 7 and $5\dfrac{1}{8}$.
48. Subtract $3\dfrac{1}{3}$ from the sum of 3 and $2\dfrac{3}{5}$.
49. Add the square of $\dfrac{2}{3}$ to $7\dfrac{3}{4}$.
50. Multiply the difference of $5\dfrac{3}{4}$ and 4 by the square of $1\dfrac{1}{3}$.
51. Divide the sum of $1\dfrac{1}{2}$ and $2\dfrac{1}{3}$ by the sum of $2\dfrac{1}{2}$ and $1\dfrac{1}{4}$.
52. Divide the difference of 5 and $\dfrac{3}{4}$ by the difference of $3\dfrac{7}{8}$ and $2\dfrac{2}{3}$.
53. Find the square of the sum of $\dfrac{1}{2}$ squared and 3.
54. Find the square of the difference of 3 squared and $1\dfrac{2}{3}$ squared.
55. Find the difference of $7\dfrac{1}{3}$ and $4\dfrac{1}{2}$ multiplied by the sum of $2\dfrac{1}{5}$ and 5.
56. Find the product of 8 and $2\dfrac{1}{3}$ divided by the product of $6\dfrac{1}{2}$ and 5.
57. Subtract $4\dfrac{3}{4}$ from the sum of $5\dfrac{1}{8}$ and $1\dfrac{3}{4}$.
58. Subtract 6 from the difference of $15\dfrac{1}{2}$ and $7\dfrac{1}{3}$.

Place a "<" or ">" between each pair of numbers by first finding a common denominator and then comparing the numerators.

59. $\dfrac{1}{2}$ and $\dfrac{1}{3}$
60. $\dfrac{2}{3}$ and $\dfrac{5}{6}$
61. $\dfrac{3}{4}$ and $\dfrac{5}{7}$
62. $\dfrac{5}{8}$ and $\dfrac{11}{17}$
63. $1\dfrac{1}{2}$ and $1\dfrac{1}{3}$
64. $2\dfrac{1}{5}$ and $2\dfrac{2}{11}$
65. $3\dfrac{1}{2}$ and $3\dfrac{4}{7}$
66. $5\dfrac{3}{7}$ and $5\dfrac{7}{15}$

Plot each of the following fractions on a number line. Label each point.

67. $\dfrac{1}{2}$ **68.** $\dfrac{3}{4}$ **69.** $1\dfrac{1}{4}$

70. $2\dfrac{5}{8}$ **71.** $\dfrac{7}{8}$ **72.** $2\dfrac{1}{4}$

73. $2\dfrac{1}{8}$ **74.** $\dfrac{1}{8}$

Answer each of the following questions. Leave all answers in either proper or improper fraction form.

75. A blueberry pie is cut into 7 equal pieces and you eat 2 of them. What fraction of the pie is left?

76. Rich cuts a large pizza into 9 equal pieces and eats 2 pieces. What fraction of the pizza is left?

77. It takes $1\dfrac{2}{3}$ yards (yd) of top soil to cover one-half of the lawn. How much topsoil is needed to cover the whole lawn?

78. A cookie recipe calls for $1\dfrac{1}{3}$ cups (c) of sugar. How much sugar is necessary to double the recipe?

79. A rectangle is $\dfrac{3}{4}$ yd in length on two sides and $\dfrac{5}{7}$ yd in width on the other two sides. What is the distance around the rectangle?

80. A square has a length of $\dfrac{25}{8}$ meters (m) on each of its four sides. What is the distance around the square?

81. Mel worked 40 hr last week: $8\dfrac{3}{4}$ hr Monday, $8\dfrac{1}{4}$ hr Tuesday, $7\dfrac{2}{3}$ hr Wednesday, and $7\dfrac{1}{2}$ hr Thursday. How many hours did he work on Friday?

82. A dump truck is loaded with $10\dfrac{7}{8}$ cubic yards (yd³) of sand. The driver delivers $1\dfrac{1}{2}$ yd³ at his first stop, $2\dfrac{1}{4}$ yd³ at the second stop, and $4\dfrac{7}{12}$ yd³ at the third stop. How many cubic yards are left in the truck?

83. A taxi driver charges 95 cents for the first $\dfrac{1}{4}$ mi and 70 cents for each additional $\dfrac{1}{4}$ mi. How much is the fare for a 2-mi trip?

84. A consultant charges $18 for the first $\dfrac{1}{3}$ of an hour and $12 for each additional $\dfrac{1}{3}$ of an hour. How much does she charge for 3 hr of work?

85. A board $12\dfrac{1}{2}$ feet (ft) long is cut into 5 equal pieces. What is the length of each piece?

86. If 1 in. on a map represents 50 mi on land, how many miles on land are represented by $4\dfrac{1}{10}$ in. on the map?

87. In the town of Duran, 600 home fires were reported in one year. If 225 of these fires were caused by smoking in bed, what fraction of the home fires were caused by smoking in bed?

88. How many $5\dfrac{1}{3}$-ounce (oz) glasses can be filled from a pitcher with 48 oz of water? Is there any water left? If so, how many ounces?

89. If $\dfrac{1}{4}$ of an inch on an architectural drawing equals 2 ft on a building, how many feet on a building is represented by $5\dfrac{3}{4}$ in. on an architectural drawing?

90. A student needs a book that costs $28. If he has $22, what fraction of the cost of the book does he have?

91. Sylvia has $2\dfrac{1}{3}$ c of sugar on hand. She plans to make a recipe that calls for $2\dfrac{3}{4}$ c of sugar. What fraction of the necessary amount of sugar does she have?

92. A chemist has $1\dfrac{1}{6}$ liters (ℓ) of a certain solution. If she adds $\dfrac{3}{8}$ ℓ of water, how much liquid does she have?

93. Water flows through a shower head at the rate of $10\dfrac{1}{2}$ gallons per minute (gal/min). How much water flows through the shower head in $9\dfrac{1}{2}$ min?

94. In a certain algebra class, $\dfrac{1}{9}$ of the students get a A on the first test, $\dfrac{2}{7}$ get a B, and $\dfrac{2}{5}$ get a C. What fraction of the students get *at least* a C on the first test?

95. Leola is making a blouse. She wants to put 7 buttons on the front of the blouse, which is $15\dfrac{1}{2}$ in. long. How far apart should she place the buttons so that they'll be equally spaced?

For Extra Thought

Simplify each of the following.

96. $\dfrac{\dfrac{1}{2}+\dfrac{1}{3}}{\dfrac{1}{2}-\dfrac{1}{3}}$

97. $\dfrac{3+\dfrac{7}{8}}{\dfrac{4}{3}+\dfrac{5}{7}}$

98. $\dfrac{\dfrac{1}{2}+\dfrac{1}{4}\left(3-\dfrac{1}{3}\right)}{\dfrac{3}{5}+\dfrac{1}{2}\left(2+\dfrac{1}{4}\right)}$

99. $\dfrac{1\dfrac{1}{3}+1\dfrac{1}{4}\left(6-5\dfrac{1}{3}\right)}{2\dfrac{3}{4}+1\dfrac{1}{5}\left(2-\dfrac{3}{5}\right)}$

Write In Your Own Words

100. What are equivalent fractions?

101. How do you "simplify" a fraction?

102. Explain the fundamental principle of fractions.

103. What is a mixed number?

104. How do you multiply two mixed numbers?

105. How do you divide two fractions?

106. How do you add two fractions with different denominators?

1.3 Decimals and Percents

HISTORICAL COMMENT It is surprising that the development of the decimal system was so long in coming and even more surprising that the symbolism for it took so long to be developed. Simon Stevin (1546–1620) wrote an essay in 1585 in which he said that the decimal system had been thoroughly tried by practical men and found so superior to earlier methods of computation that they discarded their own shortcuts to adopt this new system. Stevin continued to make the point throughout his essay that this new system allowed business computations to be performed without resorting to the use of fractions.

OBJECTIVES In this section we will learn to
1. convert common fractions to decimal fractions;
2. add, subtract, multiply, and divide with decimals;
3. convert from fractions to percents; and
4. solve applied problems involving decimals and percents.

In our system of numeration, each digit in a number has a value according to the place it occupies in the number. When we write a number in **expanded notation**, we display the place value of each of its digits. For example,

$$5734 = 5(1000) + 7(100) + 3(10) + 4(1)$$

The 5 is in the thousands place, 7 is in the hundreds place, 3 is in the tens place, and 4 is in the ones place.

Notice that as we move from left to right, the place value of each digit is one-tenth as large as the one to its left. If we continue to the right using the same approach, the place values will represent *fractional* parts of numbers with denominators that are powers of 10. To indicate this transition from whole numbers to fractions, we insert a **decimal point** (a period) in the number.

$$265.427 = 2(100) + 6(10) + 5(1) + 4\left(\frac{1}{10}\right) + 2\left(\frac{1}{100}\right) + 7\left(\frac{1}{1000}\right)$$

Decimal point

Recall that to add two fractions, they must have the same denominators. We add two decimal numbers by placing the numbers in column form, taking care to ensure that the decimal points in each number are aligned. For example, consider the sum of 1.23 and 3.56:

1.23
3.56
4.79

The logic of aligning the decimal point is evident when 1.23 and 3.56 are written in expanded notation so that the value of each digit is shown.

$$1.23 = 1(1) + \frac{2}{10} + \frac{3}{100} \quad \text{and} \quad 3.56 = 3(1) + \frac{5}{10} + \frac{6}{100}$$

$$1.23 + 3.56 = 1(1) + \frac{2}{10} + \frac{3}{100} + 3(1) + \frac{5}{10} + \frac{6}{100}$$
$$= (1+3)(1) + \left(\frac{2}{10} + \frac{5}{10}\right) + \left(\frac{3}{100} + \frac{6}{100}\right)$$
$$= 4(1) + \frac{7}{10} + \frac{9}{100}$$
$$= 4.79$$

> **To Add and Subtract Decimals**
>
> 1. Place all numbers in column form, aligning the decimal points.
> 2. Write zero(s) to the right of the right-most nonzero digit on any numbers that are shorter, so that all numbers have the same quantity of decimal places.
> 3. Add or subtract in the usual manner while maintaining the position of the decimal point.

When adding or subtracting decimals, we can write zeros after the right-most nonzero digit of a number without affecting its value. For example,

$$0.4 = \frac{4}{10} = \frac{40}{100} = \frac{400}{1000} \quad \text{so that} \quad 0.4 = 0.40 = 0.400$$

EXAMPLE 1 Carry out each of the indicated operations.

(a) Add 0.8, 3.25, 4.12, and 11.22.
(b) Subtract 6.242 from 15.04.

SOLUTION (a) We think of 0.8 as $\frac{8}{10} = \frac{80}{100} = 0.80$, so that all four numbers will have the same number of places after the decimal point. Then

```
 0.80
 3.25
 4.12
11.22
-----
19.39
```

(b) 15.040   15.04 = 15.040
    −6.242

     8.798

Multiplication of decimal numbers is accomplished in the same manner as multiplication of whole numbers. The only additional task is locating the decimal point in the product. We can see how to do this if we first write the decimals as common fractions. In the next example, we must count the number of digits after the decimal point in each part.

EXAMPLE 2   Find each of the following products by first writing the decimals as common fractions: **(a)** $(0.5)(0.3)$ **(b)** $(1.52)(1.2)$

SOLUTION   **(a)** $(0.5)(0.3) = \dfrac{5}{10} \cdot \dfrac{3}{10} = \dfrac{15}{100} = 0.15$

$(0.5)(0.3) = 0.15$
$1 + 1 = 2$ places

(b) $(1.52)(1.2) = \left(1\dfrac{52}{100}\right)\left(1\dfrac{2}{10}\right) = \left(\dfrac{152}{100}\right)\left(\dfrac{12}{10}\right) = \dfrac{1824}{1000} = 1\dfrac{824}{1000} = 1.824$

$(1.52)(1.2) = 1.824$
$2 + 1 = 3$ places

In practice, we multiply 1.52 and 1.2 by using a calculator or by using a vertical arrangement, as follows:

```
  1.52
  1.2
  ─────
  304
  152
  ─────
  1.824
```

There are $2 + 1 = 3$ places after the decimal point. The product is 1.824.

To Multiply Decimals

1. Multiply the numbers as if they were whole numbers.
2. Locate the decimal point in the product so that the number of places to the right of the decimal point is equal to the sum of the places to the right of the decimal points in the factors.

Dividing decimal numbers uses the same processes we use to divide whole numbers except we need to determine where to locate the decimal point in the quotient. The easiest way to do this is to think of the division problem in fractional form. For example, when we write the division of 24.453 by 6.27 as a fraction, we get

$$\dfrac{24.453}{6.27}$$

If we multiply both numerator and denominator by 100, we move the decimal point in the denominator two places to the right. Now we're dividing by a whole number:

$$\dfrac{24.453}{6.27} = \dfrac{(24.453)(100)}{(6.27)(100)} = \dfrac{2445.3}{627}$$

When we divide by a whole number, we place the decimal point in the quotient directly above the decimal point in the dividend:

```
                3.9      ←Quotient
Divisor→  627)2445.3    ←Dividend
               1881
                5643
                5643
                   0    ←Remainder
```

> **To Divide Decimals**
>
> 1. Count the number of places to the right of the decimal point in the divisor.
> 2. Move the decimal point to the right in both the divisor and the dividend the number of places determined in Step 1.
> 3. Locate the decimal point in the quotient directly above the decimal point in the dividend.
> 4. Carry out the division in the same manner as when dividing whole numbers.

EXAMPLE 3   Divide 41.4 by 1.725.

SOLUTION   There are three places after the decimal point in the divisor, so we move the decimal point three places to the right in both the divisor and the dividend and divide:

```
         24
1725)41400        Move the decimal point three places to the right.
     3450
     6900
     6900
```

We can change common fractions into decimal fractions by dividing the denominator into the numerator. Some decimal equivalents for common fractions obtained in this manner are given in the following table. Notice that some of them terminate while others fall into a repeating pattern. A bar over the digit(s) indicates that the pattern repeats forever.

Terminating	Repeating
$\frac{1}{8} = 0.125$	$\frac{1}{3} = 0.\overline{3}$
$\frac{1}{4} = 0.25$	$\frac{1}{7} = 0.\overline{142857}$
$\frac{1}{2} = 0.5$	$\frac{2}{11} = 0.\overline{18}$
$\frac{3}{5} = 0.6$	$\frac{1}{12} = 0.8\overline{3}$
$\frac{5}{16} = 0.3125$	$\frac{1}{13} = 0.\overline{076923}$

The reader is encouraged to use a calculator or long division to verify the results of these conversions. We will say more about the conversion of fractions to decimals when we discuss the set of real numbers in Section 2.1.

We must be careful to interpret verbal statements correctly when translating them into mathematical symbols. In the next three examples, there is only one correct order in which we can write the terms and carry out the operations.

EXAMPLE 4   Find the difference between 3.25 and $1\frac{3}{4}$.

SOLUTION   First we convert $1\frac{3}{4}$ to a decimal fraction,

$$1\frac{3}{4} = 1.75$$

then we subtract it from 3.25:

$$\begin{array}{r} 3.25 \\ -1.75 \\ \hline 1.50 \end{array} \quad \text{or} \quad 1.5$$

EXAMPLE 5   Find the difference between the sum of 6.02 and 2.35 and the difference of 12.563 and 7.56.

SOLUTION   The sum of 6.02 and 2.35 is $6.02 + 2.35 = 8.37$. The difference between 12.563 and 7.56 is $12.563 - 7.56 = 5.003$.

$$8.37 - 5.003 = 3.367$$

EXAMPLE 6   A standard method of computing medical dosages for children 2–12 years old is

$$\frac{\text{age}}{\text{age} + 12} \cdot (\text{the adult dose})$$

If the average adult dosage of a drug is 0.250 grams, what is the dosage for an eight-year-old child?

SOLUTION
$$\frac{\text{age}}{\text{age} + 12} \cdot (\text{the adult dose}) = \frac{8}{8 + 12}(0.250)$$
$$= \frac{8}{20}(0.250)$$
$$= (0.4)(0.25) \quad \frac{8}{20} = 0.4 \text{ and } 0.250 = 0.25$$
$$= 0.1 \text{ gram}$$

> **Calculator Note**
>
> Calculators are designed to locate the correct placement of the decimal point automatically when arithmetic operations are carried out. As long as the numbers are entered correctly and the correct order of operations is followed, the display will show the decimal point in the correct position.

Square roots involving decimals can be particularly troublesome if a calculator is not used. Special care must be taken to locate the decimal point properly. Recall that

$$\sqrt{36} = 6 \quad \text{because} \quad 6 \cdot 6 = 36$$

Similarly,

$$\sqrt{0.09} = 0.3 \quad \text{because} \quad (0.3)(0.3) = 0.09$$

Notice that *the square root of a number between 0 and 1 is greater than the number itself.*

EXAMPLE 7 Find each of the following square roots without a calculator:

(a) $\sqrt{0.64}$ (b) $\sqrt{1.69}$

SOLUTION (a) $\sqrt{0.64} = 0.8$ because $(0.8)(0.8) = 0.64$.
(b) $\sqrt{1.69} = 1.3$ because $(1.3)(1.3) = 1.69$.

As stated in the calculator note, calculators automatically keep track of the decimal point for us.

EXAMPLE 8 Simplify $\sqrt{12.56} - \sqrt{9.81}$ by using a calculator.

SOLUTION To use a scientific calculator, we use the following keystrokes:

12.56 √ − 9.81 √ =

The display reads 0.411917076.

One of the most common places that we encounter decimals in everyday life is in computing percents. It's difficult to go through a day of reading a newspaper, listening to the radio, or watching television without encountering the concept of percent. The following are typical examples of how percents may occur.

1. Newspaper ads: "Attend our giant year-end sale. Everything in the store is marked down at least 30 percent. Some items are discounted . . . ."
2. TV news: "The unemployment rate in California is $2\frac{1}{2}$ percent above the national average. Analysts attribute it . . . ."
3. Radio news: "Economists report that the cost of living increased four-tenths of one percent in October. If inflation continues at this rate . . . ."

| DEFINITION
Percent | One **percent** of a number is one one-hundredth of that number. The symbol for percent is %.

$$1\% = \frac{1}{100} = 0.01$$ |

The concept of percent can be extended to any number if we think of the percent symbol as meaning the number times $\frac{1}{100}$. Examples 9 and 10 illustrate this concept.

EXAMPLE 9   Change each of the following percents to decimals:

(a) 24%   (b) 127%   (c) 5%   (d) 0.4%

SOLUTION   (a) $24\% = 24 \cdot \frac{1}{100} = \frac{24}{100} = 0.24$

(b) $127\% = 127 \cdot \frac{1}{100} = \frac{127}{100} = 1.27$

(c) $5\% = 5 \cdot \frac{1}{100} = \frac{5}{100} = 0.05$

(d) $0.4\% = (0.4)\frac{1}{100} = \frac{0.4}{100} = 0.004$

The process for changing percents to decimals can be further simplified by observing that dividing a number by 100 moves the decimal point two places to the left.

> **To Change a Percent to a Decimal**
> To change a percent to a decimal, move the decimal point two places to the left and omit the percent symbol.

EXAMPLE 10   Change each percent to a decimal:

(a) 29%   (b) 156%   (c) 1200%   (d) 0.01%

SOLUTION   (a) $29\% = 29.0\% = 0.29$
                           ↑   ↑
                    Two places to left

(b) $156\% = 1.56$
(c) $1200\% = 12$
(d) $0.01\% = 0.0001$

EXAMPLE 11   Change $7\frac{1}{4}\%$ to a decimal.

SOLUTION   First we change $7\frac{1}{4}$ to a decimal:

$$7\frac{1}{4}\% = 7.25\% = 0.0725$$

Changing from a decimal to a percent is just as easy as changing from a percent to a decimal. All we need to do is reverse the procedure we just used.

> **To Change a Decimal to a Percent**
> To change a decimal to a percent, move the decimal point two places to the right and attach a percent symbol.

EXAMPLE 12   Change $7\frac{1}{5}$ to a percent.

SOLUTION   First we change $7\frac{1}{5}$ to a decimal and then we move the decimal point two places to the right:

$$7\frac{1}{5} = 7.2 = 720\%$$

EXAMPLE 13   Change each decimal to a percent.

(a) $1.23 = 123\%$
(b) $0.34 = 34\%$
(c) $0.001 = 0.1\%$
(d) $5 = 500\%$

One of the primary uses of percent is to determine parts of numbers. For example, if 50% of a class of 42 students are females, then 21 are females. This is the same as saying that one-half of the class consists of females.

$$50\% = \frac{50}{100} = \frac{1}{2}$$

Thus

$$50\% \text{ of } 42 = \frac{1}{2} \text{ of } 42 = \frac{1}{2} \cdot 42 = 21$$

Also remember that 50% can be expressed as 0.5, so that

$$50\% \text{ of } 42 = 0.5 \times 42 = 21$$

Notice that the word "of" is interpreted as multiplication.

The Percent Equation

The relationship between the percent of a number, called the **amount,** A, and the original number, called the **base,** B, is given by the formula

Amount = Percent × Base

or

$$A = P \cdot B$$

EXAMPLE 14 Use the percent equation to determine each of the following percents:

(a) 25% of 36   **(b)** 17% of 41   **(c)** 200% of 12   **(d)** $5\frac{3}{4}\%$ of 19

SOLUTION **(a)** We want 25% of 36, so here $P = 25\%$, $B = 36$, and

$A = P \cdot B = 25\%$ of 36

$= 0.25 \times 36 = 9$ Change 25% to 0.25 and multiply.

(b) $P = 17\%$ and $B = 41$, so

$A = 17\%$ of $41 = 0.17 \times 41 = 6.97$ Change 17% to 0.17 and multiply.

(c) $A = 200\%$ of $12 = 2 \times 12 = 24$ Change 200% to 2 and multiply.

(d) $A = 5\frac{3}{4}\%$ of $19 = 5.75\%$ of 19   Change $5\frac{3}{4}$ to a decimal.

$= 0.0575 \times 19$ Change 5.75% to a decimal.

$= 1.0925$

If any two of the three parts of the percent equation are known, we can find the third part by using the various forms of the percent equation:

$$A = P \cdot B \qquad B = \frac{A}{P} \qquad \text{and} \qquad P = \frac{A}{B}$$

EXAMPLE 15 What percent of 80 is 20?

SOLUTION The base B is 80 and the amount A is 20, so

$P = \dfrac{20}{80} = 0.25$ Change to a decimal by dividing the numerator by the denominator.

Now we change the decimal to a percent:

$$P = 0.25 = 25\%$$

EXAMPLE 16 Thirty-two percent of what number is 20?

SOLUTION The amount A is 20 and the percent P is 32. We are to find the base B, so we use

$$B = \frac{A}{P}$$
$$= \frac{20}{0.32} \quad 32\% = 0.32$$
$$= 62.5$$

The number is 62.5.

EXAMPLE 17 The combined state and city sales tax on a new car at Auto City is $7\frac{1}{4}\%$. How much sales tax will be charged on a car that sells for $26,500?

SOLUTION The base B is $26,500 and the percent P is $7\frac{1}{4}\% = 0.0725$.

$$A = P \cdot B$$
$$= 7\frac{1}{4}\% \cdot 26{,}500 = 0.0725 \cdot 26{,}500 = \$1921.25$$

EXAMPLE 18 An item in a furniture store is selling for $297 off the listed price. If everything in the store is marked down 40%, what is the regular price?

SOLUTION The regular price is the base B, the amount A is $297, and the mark-down percentage, 40%, is the percent P.

$$B = \frac{A}{P}$$
$$= \frac{297}{0.4} \quad 40\% = 0.4$$
$$= \$742.50$$

The regular price of the item is $742.50.

EXAMPLE 19 An item costs $75, which includes a sales tax of 6%. What is the cost of the item before tax?

SOLUTION The cost of the item plus 6% of the cost of the item (the tax) totals $75. Thus $75 is 106% of the cost. The amount A is 75 and the percent P is 1.06.

$$B = \frac{A}{P}$$

$$= \frac{75}{1.06} = 70.7547$$

The cost of the item before tax is $70.75 to the nearest cent.

Do You Remember?

Can you match these?

____ 1. The base in 20% of 50 = 10
____ 2. The number of decimal places in 2.321 × 7.51
____ 3. To convert a decimal to a percent, you move the decimal point to the right ____ places before attaching the percent symbol
____ 4. Fifteen percent written as a decimal
____ 5. Two written as a percent
____ 6. One-half percent written as a decimal
____ 7. 0.4 written as a percent

a) 4
b) 0.15
c) 200%
d) 5
e) 20%
f) 40%
g) 0.005
h) 50
i) 100
j) 2

Answers: 1. h 2. d 3. j 4. b 5. c 6. g 7. f

Exercise Set 1.3

Rewrite each fraction in (a) decimal form and (b) percent form.

1. $\frac{1}{4}$
2. $\frac{1}{2}$
3. $\frac{3}{5}$
4. $\frac{3}{8}$
5. $2\frac{1}{4}$
6. $3\frac{3}{20}$
7. $4\frac{7}{50}$
8. 1
9. 2
10. $5\frac{7}{35}$
11. $\frac{1250}{2000}$
12. $\frac{1120}{1792}$

Rewrite each of the following percents in decimal form.

13. 30%
14. 60%
15. 100%
16. 110%
17. 225%
18. 300%
19. 118%
20. 95%
21. 0.8%
22. 0.5%
23. $5\frac{1}{2}$%
24. $7\frac{1}{8}$%

Perform the indicated operation(s) for Exercises 25–42.

25. Add 3.25 to 1.71.
26. Add 6.94 to 0.04.

27. Subtract 2.19 from 5.21.
28. Subtract 1.87 from 3.001.
29. Multiply 21.2 by 1.2.
30. Multiply 0.02 by 3.5.
31. Divide 0.01 by 0.04.
32. Find the sum of 1.001 and $8\frac{1}{2}$.
33. Find the sum of 7.5 and $1\frac{1}{2}$.
34. Find the product of 1 and $2\frac{1}{4}$.
35. Subtract 3.35 from the difference of 8 and $\frac{7}{8}$.
36. Find the sum of 30% and 0.05.
37. Find the difference of 125% and 0.075.
38. Find the product of 70% and 220%.
39. Find the difference of the sum of 6.8 and 3.02 and the sum of 5.1 and 1.001.
40. Find the sum of the difference of 7.03 and 5.1 and the difference of 2.18 and 1.
41. Find the quotient of the product of 5 and 2.5 and the product of 4 and 1.5.
42. Find the product of the quotient of 15 and 2.5 and the quotient of 17.5 and 3.5.

Complete each of the following calculations.

43. 20% of 60 equals what?
44. 125% of 10 equals what?
45. 100% of what equals 30?
46. 200% of what equals 70?
47. What % of 60 equals 15?
48. What % of 160 equals 120?
49. 3.5% of 128 equals what?
50. 6.125% of 44 equals what?
51. 0.7% of 248 equals what?
52. 1.5% of 635 equals what?
53. 0.05% of what equals 1?

Perform the indicated operations for Exercises 54–75. A calculator may be helpful.

54. $(0.2)^2 + (0.3)^2$
55. $(0.15)^2 + (0.11)^2$
56. $(1.2)^2 - (1.1)^2$
57. $(1.4)^2 - (0.9)^2$
58. $\sqrt{1.69} + \sqrt{0.25}$
59. $\sqrt{2.25} + \sqrt{0.16}$
60. $\sqrt{2.89} - \sqrt{1.96}$
61. $\sqrt{0.36} - \sqrt{0.04}$
62. $38.83 - 26.07$
63. $56.03 - 47.78$
64. $22.2 \div 11.1 - 1.0$
65. $39.3 \div 13.1 + 2.05$
66. $(2.1)(2.3 + 5.7)$
67. $18.8 \div (5.6 + 3.8)$
68. $(65.0)(6.82 - 1.82)$
69. $1.5 + (2.1)[3.4 + (5.06 - 2.06)]$
70. $3(0.1)^2 + 2(0.2)^2$
71. $(5.2)(0.1) - (3.4)(0.3)^2$
72. Find one-half of the sum of 4.007 and 1.013.
73. Find one-third of the product of 1.25 and 0.3.
74. Subtract $(1.2)^2$ from the sum of 15.25 and 1.18.
75. Subtract $(1.5)^2$ from the difference of $(2.1)^2$ and $(0.4)^2$.
76. The wholesale cost of a curling iron is $12.50. If the markup (the amount added to the cost) is 32% of the wholesale cost, what is the selling price?
77. A textbook costs $21.50 wholesale. If the markup (the amount added to the wholesale cost) is 20% of the cost, what is the selling price?
78. A boom box sells for $58.60 wholesale. If the markup is 30%, what is the retail cost of the boom box?
79. A pad of engineering paper sells for $2.00 wholesale. If the markup is 25%, what is the retail cost of a pad of engineering paper?
80. The regular price of a meal consisting of a hamburger, fries, and a drink is $4.50. If the meal is discounted (lowered) 20%, what is the lower cost of the meal?
81. The price of a shirt is $16.75. If the price of the shirt is discounted (lowered) 20%, what is the new price of the shirt?
82. A new motorcycle sells for $3820. If it depreciates (goes down in value) 18.5% the first year, what is the value of the motorcycle after one year? How much did it depreciate?
83. A new car sells for $12,800. If it depreciates (goes down in value) 17.5% the first year, what is the value of the car after one year? How much did it depreciate?
84. Marlene's base rate of pay is $6.78/hr. She gets time-and-a-half for each hour over 40 hr that she works in one week. If she worked 43.5 hr last week, how much did she earn?
85. Tony earns $5.74/hr during regular working days (Monday through Friday). He earns two and a half times his regular pay if he has to work on the weekend. Last week he worked 40 hr during the week and $5\frac{1}{2}$ hr on Sunday. How much did he earn last week?
86. Wilma received a sales bonus of $7100. If her bonus represents 8% of all she sold, how much did she sell?
87. Mario received a commission of $1200 for selling a house. His commission represents 3.5% of the selling price of the house. How much did the house sell for?
88. Sandi bought a new car for $9850. One year later she sold it for $8274. What was the rate of depreciation on the car (in percent)?

89. Gilbert bought a new algebra textbook for $28.50 and sold it at the end of the semester for $13.11. What was the rate of depreciation (in percent)?

90. Geraldine borrowed $2700 at 9.5% interest. How much interest did she pay?

91. A city collects a sales tax of 1.5%. How much sales tax is collected on a sale of $178.50?

92. John's Discount House sells everything at cost (wholesale) plus 10%. If they sell a TV set for $847, how much profit do they make on the TV?

93. A local credit union charges 1.25% interest for any monthly account with a balance of less than $400. Delores paid $2.80 interest last month on her account. What was the balance in her account?

94. The college bookstore sells a notebook for $4. The store operates on a markup of 25% on its cost. What is the amount the store pays for the notebook?

WRITE IN YOUR OWN WORDS

95. How do you change a decimal fraction to percent form?

96. How do you change a percent to decimal form?

97. How do you change a common fraction to decimal form?

98. How do you determine the number of decimal places in a product of two decimal fractions?

99. How do you change a mixed fraction to decimal form?

100. What does the word "percent" mean?

1.4 Some Concepts from Geometry

HISTORICAL COMMENT   Almost everyone who has ever studied geometric figures is acquainted with the number π (pi), which is used to determine the area and the circumference of a circle. Finding the "true" value of π has been the focus of many mathematicians over the centuries, but it was not until the nineteenth century that it was proven to be a nonrepeating, nonterminating decimal. Ancient Chinese works listed the value of π as 3, which is also the value given in the Old Testament in 1 Kings VII, 23. Here are some other approximations that have been used:

Egypt (c. 1650 B.C.) 3.1605
Greece (225 B.C.) .... between 3.1408 and 3.1428
Rome (20 B.C.) .... $3\frac{1}{8} = 3.1250$
India (A.D. 628) .... $3\frac{1}{7}$ or approximately 3.1428

In the mid-1800s, the value of π was calculated to over 500 decimal places, which was quite a task without the aid of modern computing machinery. Today in nonscientific work, we often use 3.14 as an approximation for π, although it is only correct to the first two decimal places. The value of π to nine decimal places is 3.141592654.

OBJECTIVES   In this section we will learn to
1. compute the area and the perimeter of common geometric figures;
2. compute the volume of common geometric solids;
3. convert from one measure of area or volume to another;
4. use the Pythagorean theorem; and
5. work with the angles of a triangle.

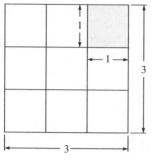

FIGURE 1.2

The **perimeter** of a closed geometric figure is the distance around it. In the case of a square, where all the sides are the same length, the perimeter is obtained by multiplying the length of one side by 4. The **area** of a geometric figure is the number of square units it contains. For example, a square with sides of length 3 has an area of 9 square units. That is, 9 squares with sides one unit in length fit exactly inside its perimeter. See Figure 1.2.

Some common geometric figures and the formulas for determining their perimeters and areas are shown in Figures 1.3(a) through (e). The square, rectangle, triangle, and circle are commonly known. A **trapezoid** is a four-sided plane geometric figure in which two sides, b and B, are parallel. See Figure 1.3(e).

(a) Square
$P = 4s$
$A = s^2$

(b) Rectangle
$P = 2l + 2w$
$A = lw$

(c) Triangle
$P = a + b + c$
$A = \tfrac{1}{2}bh$

(d) Circle
$C = 2\pi r = \pi d$
$A = \pi r^2$

(e) Trapezoid
$P = a + b + B + d$
$A = \tfrac{h}{2}(b + B)$

FIGURE 1.3

EXAMPLE 1 Find the perimeter and the area of a square with 2.5-meter sides; see Figure 1.3(a).

SOLUTION
$P = 4s = 4(2.5) = 10$ meters
$A = s^2 = (2.5)^2 = 6.25$ m^2 m^2 means square meters.

EXAMPLE 2 Find the perimeter and the area of a rectangle of width 5 ft and length 7 ft; see Figure 1.3(b).

SOLUTION
$P = 2w + 2l$
$= 2(5) + 2(7) = 10 + 14$
$= 24$ ft
$A = lw$
$= 5(7) = 35$ ft^2 ft^2 means square feet.

EXAMPLE 3 Find the perimeter of a triangle with sides 4, 7, and 9 inches; see Figure 1.3(c).

SOLUTION
$P = a + b + c = 4 + 7 + 9 = 20$ in.

The formula for the area of a triangle is $A = \tfrac{1}{2}bh$, as in Figure 1.3(c). In this formula, b is the **base**, which is the length of any side, and h is the **height**, which is the distance from the base to the opposite **vertex** (the point where the other two sides meet). If a triangle has two sides that are **perpendicular** (they meet at right angles), then either of the perpendicular sides (the **legs**) is the base and the other is the height. Such a triangle is called a **right triangle.** The perpendicular sides in a right triangle are identified by drawing a small square, □, where they meet. The longest side of a right triangle is called the **hypotenuse.**

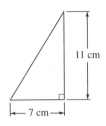

FIGURE 1.4

EXAMPLE 4   Find the area of a triangle with base 7 cm and height 11 cm; see Figure 1.4.

SOLUTION
$$A = \frac{1}{2}bh = \frac{1}{2}(7)(11)$$
$$= \frac{1}{2}(77) = 38.5 \text{ cm}^2$$

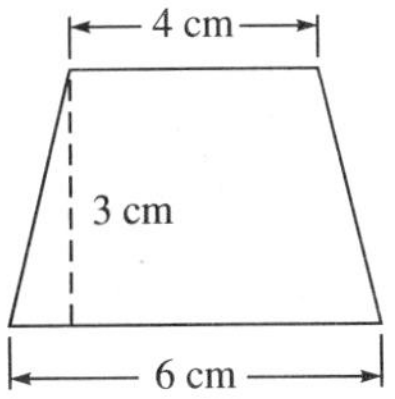

EXAMPLE 5   Find the area of the trapezoid such that the parallel sides are 4 cm and 6 cm and the height is 3 cm; see Figure 1.5.

SOLUTION
$$A = \frac{h}{2}(b + B)$$
$$= \frac{3}{2}(4 + 6) \quad \text{The parallel sides are } b = 4 \text{ and } B = 6.$$
$$= \frac{3}{2}(10)$$
$$= 15 \text{ cm}^2$$

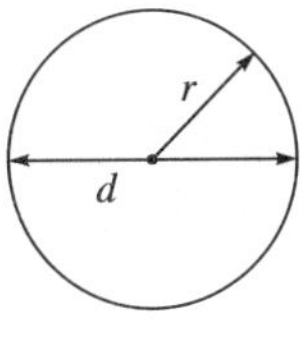

FIGURE 1.5

FIGURE 1.6

The perimeter of a circle is more commonly known as its **circumference.** The **diameter** of a circle is the straight-line from any one point on the circle through the center to the opposite point on the circle, and the **radius** of a circle is the line from the center to any point on the circle. Thus, *the length of the diameter is twice the length of the radius;* see Figure 1.6.

Both the circumference and the area of a circle are related to the radius by the number π. (See the Historical Comment at the beginning of this section.) Many answers in this text will be left in terms of π rather than approximating their value.

EXAMPLE 6   Find the circumference and the area of a circle with a diameter of length 6.5 in. Leave both answers in terms of π.

SOLUTION
$$C = 2\pi r = \pi d \quad \text{The length of the diameter is twice the length of the radius.}$$
$$= 6.5\pi \text{ in.}$$

The area of a circle is given in terms of the radius:
$$r = \frac{1}{2}d = \frac{1}{2}(6.5) = 3.25 \text{ in.}$$
$$A = \pi r^2 = \pi(3.25)^2 = 10.5625\pi \text{ in.}^2$$

36 CHAPTER 1 · A REVIEW OF ARITHMETIC

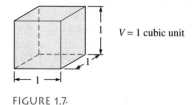

$V = 1$ cubic unit

FIGURE 1.7

We now turn our attention to another geometric concept in which we consider the amount of *space* occupied by a three-dimensional figure. **Volume** is the measure of the number of cubic units contained in a solid, such as a box or a sphere. One cubic unit is the volume of a cube that is one unit on a side; see Figure 1.7.

The volumes for some common geometric solids are listed in Figures 1.8(a) through (e).

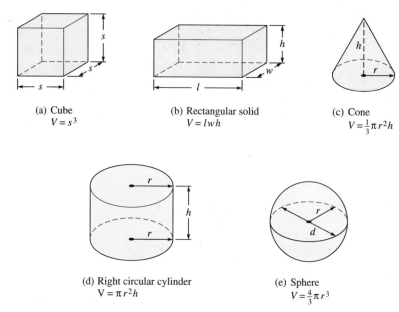

(a) Cube
$V = s^3$

(b) Rectangular solid
$V = lwh$

(c) Cone
$V = \frac{1}{3}\pi r^2 h$

(d) Right circular cylinder
$V = \pi r^2 h$

(e) Sphere
$V = \frac{4}{3}\pi r^3$

FIGURE 1.8

EXAMPLE 7 Find the volume of a cube with sides of 4 m; see Figure 1.9.

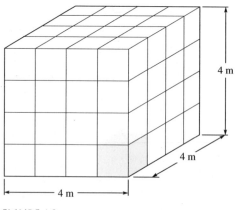

FIGURE 1.9

SOLUTION $V = s^3$
 $= 4^3 = 64$ cubic meters   Each side is 4 meters long.
 $= 64 \text{ m}^3$ m³ means cubic meters.

EXAMPLE 8   Find the volume of a box with length 6 in., width $4\frac{3}{4}$ in., and height $5\frac{1}{2}$ in.; see Figure 1.8(b).

SOLUTION
$$V = lwh = (6)\left(4\frac{3}{4}\right)\left(5\frac{1}{2}\right)$$
$$= (6)(4.75)(5.5) \qquad 4\frac{3}{4} = 4.75 \text{ and } 5\frac{1}{2} = 5.5$$
$$= 156.75 \text{ in}^3$$

Areas and volumes can be measured in many different units. To find how many square inches are contained in a figure with an area given in square feet, we have to multiply by 144, because each square foot is 12 inches on a side and so contains $12 \cdot 12 = 144$ square inches. Some conversions that you should verify follow:

1 square foot = 144 square inches
1 square yard = 9 square feet = 1296 square inches
1 cubic foot = 1728 cubic inches
1 cubic yard = 27 cubic feet = 46,656 cubic inches

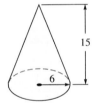

FIGURE 1.10

EXAMPLE 9   Find the volume of a cone with a height of 15 millimeters (mm) and a base with radius 6 mm; see Figure 1.10.

SOLUTION
$$V = \frac{1}{3}\pi r^2 h = \frac{1}{3}\pi(6)^2(15)$$
$$= \frac{1}{3}(36)(15)\pi = 180\pi \text{ mm}^3$$

EXAMPLE 10   Find the volume both in cubic feet and cubic inches of a sphere with diameter 4 ft; see Figure 1.8(e).

SOLUTION
$$V = \frac{4}{3}\pi r^3 = \frac{4}{3}\pi(2)^3 \qquad \text{The radius is } \frac{1}{2} \text{ of the diameter.}$$
$$= \frac{4}{3}\pi(8) = \frac{32}{3}\pi \text{ ft}^3$$

To convert to cubic inches, we multiply by 1728:
$$V = \frac{32}{3}\pi(1728) \text{ in}^3 \qquad 1 \text{ ft}^3 = 1728 \text{ in}^3$$
$$= \frac{55,296}{3}\pi \text{ in}^3$$
$$= 18,432 \text{ in}^3$$

The answer $18{,}432\pi$ in³ is exact. We can use a calculator to approximate the answer to two decimal places by approximating π by 3.14. The symbol "≈" means "approximately equal to."

$$18{,}432\pi \text{ in}^3 \approx 18{,}432(3.14) \text{ in}^3 = 57{,}876.48 \text{ in}^3$$

A special relationship between the sides of a right triangle exists. Its proof is generally attributed to the Pythagoreans, a society of Greek mathematicians founded by Pythagoras (c. 580–500 B.C.), although it was known to many other cultures.

The Pythagorean Theorem

In any right triangle, the square of the length of the hypotenuse is equal to the sum of the squares of the lengths of the other two sides (the legs). For a right triangle with sides a and b and hypotenuse c,

$$a^2 + b^2 = c^2$$

EXAMPLE 11 Find the length of the hypotenuse of a right triangle with legs:

(a) 3 and 4 inches   (b) 5 and 7 inches

SOLUTION (a) $a^2 + b^2 = c^2$

$3^2 + 4^2 = c^2 \qquad a = 3 \text{ and } b = 4$

$9 + 16 = c^2$

$25 = c^2$

$c = 5 \text{ in.}$

(b) $a^2 + b^2 = c^2$

$5^2 + 7^2 = c^2 \qquad a = 5 \text{ and } b = 7$

$25 + 49 = c^2$

$74 = c^2$

$c = \sqrt{74} \text{ in.}$

Calculator Note

A calculator can be used to determine the square root of a number. The sequence of keys to depress varies from calculator to calculator. Two such sequences are as follows:

Scientific Calculator: 74 $\boxed{\sqrt{}}$
   The display shows $\sqrt{74} = 8.602325267$.

Graphics Calculator: $\boxed{\sqrt{}}$ 74 $\boxed{\text{ENTER}}$
   The display shows $\sqrt{74} = 8.602325267$.

The accuracy displayed by a calculator is often more than is necessary for ordinary work. The results can be rounded to any degree of accuracy by using the following rules.

> **To Round a Decimal Number**
>
> 1. Decide how many digits to the right of the decimal point you want to keep. The last digit to be retained is called the **rounding digit.**
> 2. If the digit immediately to the right of the rounding digit is 5 or greater, **round up** by increasing the rounding digit by 1 and dropping the remaining digits.
> 3. If the digit immediately to the right of the rounding digit is 4 or less, **round down** by leaving the rounding digit unchanged and dropping the remaining digits.

EXAMPLE 12   Round each of the following decimals to the indicated number of places.

(a) 8.6023 rounded to two places is 8.60.
(b) 92.1676 rounded to three places is 92.168.
(c) 121.459 rounded to one place to one place is 121.5.

> *In any triangle, the sum of the measures of its angles is 180°.*

An angle that contains less than 90° is an **acute angle,** an angle that contains exactly 90° is a **right angle,** and an angle that contains more than 90° but less than 180° is an **obtuse angle.**

EXAMPLE 13   Two angles of a triangle are 36° and 44°. What kind of angle is the third angle?

SOLUTION   The sum of the two given angles is

$$36° + 44° = 70°$$

Since the sum of the angles of the triangle is 180°, the third angle is

$$180° - 70° = 110°$$

The angle is obtuse.

CHAPTER 1 · A REVIEW OF ARITHMETIC

Do You Remember?

Can you match these?

____ 1. The measure of area
____ 2. The measure of volume
____ 3. The Pythagorean theorem
____ 4. The formula for the area of a rectangle
____ 5. The formula for the area of a square
____ 6. The formula for the area of a circle
____ 7. The formula for the volume of a cone
____ 8. The volume of a cube of side 10 in.
____ 9. The number of cubic inches in a cubic foot
____ 10. 2.654 rounded to two decimal places

a) $A = s^2$
b) 2.66
c) 144 in^3
d) 1728 in^3
e) $A = \pi r^2$
f) square units
g) cubic units
h) $A = lw$
i) $c^2 = a^2 + b^2$
j) $V = \frac{4}{3}\pi r^3$
k) $\frac{1}{2}bh$
l) 2.65
m) $V = \frac{1}{3}\pi r^2 h$
n) 1000 in^3

Answers: 1. f 2. g 3. i 4. h 5. a 6. e 7. m 8. n 9. d 10. l

Exercise Set 1.4

Find (a) the perimeter and (b) the area of each of the figures in Exercises 1–16. Do not change π to decimal form.

1. Square
$4\frac{1}{2}$ in.
$4\frac{1}{2}$ in.

2. Rectangle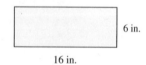
6 in.
16 in.

3. Triangle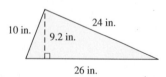
10 in.   24 in.
9.2 in.
26 in.

4. Circle
12 in.

5. Norman window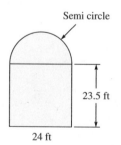
Semi circle
23.5 ft
24 ft

6. Trapezoid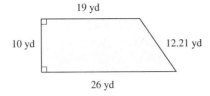

7. A circle with a radius of $1\frac{1}{4}$ yards
8. A square with a side of 1.5 meters
9. A rectangle with length $5\frac{3}{8}$ centimeters and width $2\frac{1}{4}$ centimeters
10. A right triangle with base 1.7 meters, height 1.9 meters, and hypotenuse 3.1 meters
11. A trapezoid with $B = 5\frac{1}{3}$ centimeters, $b = 3\frac{1}{4}$ centimeters, h (height) $= 2\frac{1}{2}$ centimeters, and the fourth side $4\frac{1}{8}$ centimeters
12. A rectangle with length 13.8 inches and width 7.1 inches
13. A square with a side of $\frac{15}{2}$ meters
14. A circle with a radius of $3\frac{2}{3}$ feet
15. A right triangle with base $8\frac{7}{8}$ feet, height $\frac{3}{5}$ foot, and hypotenuse $11\frac{1}{3}$ feet
16. A trapezoid with $B = 5.2$ millimeters, $b = 3.7$ millimeters, h (height) $= 2.1$ millimeters, and the fourth side 4.3 millimeters

Use the Pythagorean theorem to find the missing side of each right triangle. Round your answers to three decimal places if needed. You may find a calculator helpful.

17.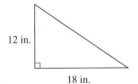

18.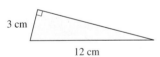

19.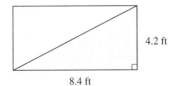

20.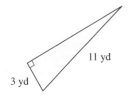

21.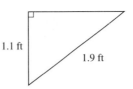

22.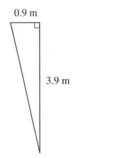

23.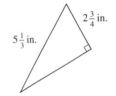

24.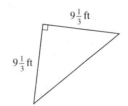

25.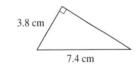

26.

Find the area of the given shape. Round your answers to three decimal places. You may find a calculator helpful.

27. A circle with a diameter of 17 inches
28. A dime with a diameter of 0.5 centimeters
29. A rectangular wall that is 11.5 feet long and $7\frac{1}{2}$ feet high
30. A semicircle with a diameter of $3\frac{7}{8}$ feet
31. A circular mirror with a radius of 11 inches
32. One-half of a rectangle with width $6\frac{2}{9}$ feet and length $5\frac{1}{4}$ feet
33. One-fourth of a trapezoid with $B = 13.7$ inches, $b = 8.9$ inches, and $h = 5.8$ inches.

Find the measure of the missing angle in each of the following triangles.

34.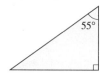

35.

36.

37.

38.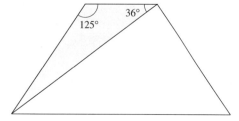

39.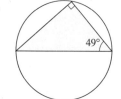

40.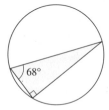

Find the volume of each of the following solids. Do not change π to decimal form.

41. A cube with side $s = 5$ ft
42. A sphere with radius $r = 3$ in.
43. A cone with radius $r = 40$ ft and height $h = 27$ ft
44. A rectangular solid with $l = 21$ cm, $w = 16$ cm, and $h = 7$ cm
45. A right circular cylinder with radius $r = 7\frac{1}{2}$ ft and height $h = 17\frac{1}{4}$ ft
46. A box in the shape of a cube with side $s = 18.2$ in.
47. A basketball with a diameter of 10.5 in.
48. An ice cream cone with $r = 1\frac{1}{8}$ in. and $h = 3\frac{7}{8}$ in.
49. A rectangular box with $l = 13\frac{3}{8}$ in., $w = 12\frac{1}{5}$ in., and $h = 7\frac{1}{4}$ in.
50. A cylindrical oilcan with diameter $d = 4\frac{1}{3}$ in. and height $h = 7\frac{1}{4}$ in.

Work each of the following applied problems. Use 3.1416 for π. You may find a calculator helpful.

51. How many square feet of carpeting will Pauline need to carpet a rectangular room that measures 24 ft by 15 ft? How many square yards?
52. How many gallons of paint will Matt need to paint a rectangular ceiling that is 30 ft by 20 ft if a gallon of paint covers 400 square feet (ft²)?
53. How many square feet of sheathing is needed to cover the gable at the end of the house, as shown in the accompanying figure?

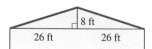

54. Jeannette is making a tablecloth to cover a circular table that has a diameter of 42 in. She wants the tablecloth to hang down 9 in. all around the table (making a diameter of 60 in.). What is the area of the table cloth in square inches? In square feet? In square yards?
55. How many gallons of paint (rounded to the whole gallon) are needed to paint the walls and ceiling of a room $16\frac{1}{2}$ ft long, $12\frac{3}{8}$ ft wide, and $9\frac{1}{2}$ ft high? Figure it as if there were no doors or windows. Assume the paint covers 400 ft²/gal.
56. How many square tiles, each 9 in. on a side, are needed to cover a floor that is 9 ft × 12 ft?
57. How many acres (rounded to the nearest hundredth of an acre) are contained in a rectangular lot 1460 ft wide by 1728.5 ft long? (43,560 ft² equals one acre.)
58. A water tank in the shape of a right circular cylinder is 64 in. in diameter and $3\frac{1}{2}$ ft high. If there are 231 in³ in one gallon, how many gallons of water does the tank hold?
59. How many boxes that are 18 in. long, 12 in. wide, and 8 in. high can be loaded into a closed container that is 21 ft long, 8 ft wide, and 6 ft high?

60. A swimming pool is 30 ft long, 20 ft wide, and filled to a depth of 5 ft. How many cubic feet of water are in the pool? How many gallons of water are in the pool if there are approximately $7\frac{1}{2}$ gal of water in one cubic foot?

61. A basement that is 35 ft long, 24 ft wide, and 9 ft deep is being excavated. A dump truck holds 14 yd³. How many truckloads will it take to haul all of the dirt away?

Write in Your Own Words

62. What is a perimeter?
63. What is the relationship between the area of a triangle and a rectangle with the same base and height?
64. What is the difference between the perimeter and the circumference of a circle?
65. Explain the relationship between a square and a rectangle.
66. What is a triangle?
67. What is a trapezoid?
68. Explain the relationship between the radius and the diameter of a circle?
69. What is a quadrilateral? (Use a dictionary if necessary.)
70. State the Pythagorean theorem.
71. What is the sum of the two non–right angles in a right triangle?
72. What is a semicircle?   73. What is an acute angle?
74. What is a right angle?   75. What is an obtuse angle?
76. What is the hypotenuse of a right triangle?

Summary

1.1 Basic Concepts of Arithmetic and Algebra

In order to use mathematics to solve real-life problems, it is important to be able to translate verbal statements into mathematical symbolism. Success in doing this depends on the correct interpretation of words such as **sum, difference, product, quotient,** or other words with the same meanings. Care must also be taken to interpret punctuation correctly, since the location of something as basic as a comma can change the entire meaning of a statement.

Some of the symbols used, besides those that indicate addition, subtraction, multiplication, and division, are $>$, $\geq$, $<$, $\leq$, $=$, $\neq$, $\sqrt{}$, and **exponents.** When an exponent is a counting number, it indicates the number of times the **base** is to be used as a factor.

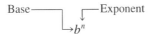

When carrying out the operations within a mathematical sentence, a specified order of operations must be followed. Aids to indicating this order are called **grouping symbols;** they are the parentheses, ( ); brackets, [ ]; braces, { }; the fraction bar, ——; and the radical sign, $\sqrt{}$.

Order of Operations
Simplify a problem from left to right in the following order:

1. Carry out the operations within the grouping symbols working from the innermost symbols outward.
2. Evaluate numbers with exponents.
3. Carry out multiplication and division from left to right.
4. Carry out addition and subtraction from left to right.

A **prime number** is a whole number greater than 1 that is divisible only by itself and 1. The **prime factored form** of a number shows it as a product of all of its **prime factors,** that is, the prime numbers that divide into it evenly. To determine the prime factored form of a number, the following rules for divisibility are used:

1. A number is divisible by 2 if it is even.
2. A number is divisible by 3 if the sum of its digits is divisible by 3.
3. A number is divisible by 5 if it ends in a 0 or 5.

1.2 Common Fractions

A fraction consists of a **numerator** and a **denominator.** In a **proper fraction,** the numerator is less than the denominator. An improper fraction has a numerator that is greater than or equal to its denominator. An improper fraction can also be written as a counting number or a mixed number.

Fractions are simplified or built up by using **the fundamental principle of fractions:**

If $\frac{a}{b}$ represents a fraction and $b \neq 0$, then $\frac{a \cdot c}{b \cdot c} = \frac{a}{b}$, $c \neq 0$.

Fractions are multiplied, divided, added, and subtracted by the following rules:

If $\frac{a}{b}$ and $\frac{c}{d}$ represent fractions, $b \neq 0$ and $d \neq 0$, then $\frac{a}{b} \cdot \frac{c}{d} = \frac{a \cdot c}{b \cdot d}$.

If $\frac{a}{b}$ and $\frac{c}{d}$ represent fractions, b, c, and $d \neq 0$, then $\frac{a}{b} \div \frac{c}{d} = \frac{a}{b} \cdot \frac{d}{c}$.

If $\frac{a}{c}$ and $\frac{b}{c}$ represent fractions, $c \neq 0$, then $\frac{a}{c} + \frac{b}{c} = \frac{a+b}{c}$ and $\frac{a}{c} - \frac{b}{c} = \frac{a-b}{c}$.

When fractions do not share a common denominator, the fundamental principle of fractions can be used to write them as **equivalent fractions** in terms of a **least common denominator,** or **LCD.**

When two fractions have a common denominator, the one with the greater numerator is the greater fraction.

1.3 Decimals and Percents

To convert a fraction to a decimal, divide the numerator by the denominator. Sometimes the division will result in a decimal that either terminates or repeats, while sometimes it will neither terminate nor repeat.

When decimal numbers are added, care must be taken to align the decimal points.

To multiply two decimal numbers, locate the decimal point in the product so that the number of decimal places is equal to the sum of the number of places after the decimal point in the factors.

To divide two decimal numbers, move the decimal point the same number of places to the right in both the divisor and dividend to make the divisor a whole number. Keep the decimal point in the quotient directly above the one in the dividend, and divide as with whole numbers.

To take the square root of a decimal number, particular care must be used to locate the decimal point in the correct position. The square root of a number between 0 and 1 is always greater than the number itself.

One **percent** of a number is **one one-hundredth** of the number. The symbol for percent is %.

$$1\% = \frac{1}{100} = 0.01$$

To change from a percent to a decimal, move the decimal point two places to the left and omit the percent symbol. **To change from a decimal to a percent,** move the decimal point two places to the right and write a percent symbol.

The relationship between a number and a percent of the number is given by the **percent equation:**

Amount = Percent × Base   or $A = P \cdot B$

To find a percent of a number, multiply by the percent expressed as a decimal. If any two of the three parts of the percent equation are known, the third part can be found by using the various forms of the percent equation:

$$A = P \cdot B \qquad B = \frac{A}{P} \qquad \text{and} \qquad P = \frac{A}{B}$$

1.4 Some Concepts from Geometry

The **perimeter** of a geometric figure is the distance around it and is measured in linear units such as feet, inches, or meters. **Areas** of geometric figures are measured in **square units** and **volumes** are measured in **cubic units.** Some areas and volumes require the use of the number π to approximate them.

Figure	Perimeter	Area
Square	$P = 4s$	$A = s^2$
Rectangle	$P = 2l + 2w$	$A = lw$
Triangle	$P = a + b + c$	$A = \frac{1}{2}bh$
Circle	$C = 2\pi r = \pi d$	$A = \pi r^2$
Trapezoid	$P = a + b + c + B$	$A = \frac{h}{2}(b + B)$

The following formulas are used to determine volumes.

Figure	Volume
Cube	$V = s^3$
Rectangular solid	$V = lwh$
Cone	$V = \frac{1}{3}\pi r^2 h$
Right circular cylinder	$V = \pi r^2 h$
Sphere	$V = \frac{4}{3}\pi r^3$

The **Pythagorean theorem** states the relationship between the three sides of a right triangle. For the triangle with legs a and b and hypotenuse c,

$$a^2 + b^2 = c^2$$

The sum of the interior angles of a triangle is 180°.

Decimal numbers can be rounded up or down to the desired degree of accuracy. The last digit to be retained is called the **rounding digit.** If the digit following the rounding digit is 5 or greater, the rounding digit is increased by 1 and the remaining digits are dropped. If the digit following the rounding digit is 4 or less, the rounding digit is unchanged and the remaining digits are dropped.

Cooperative Exercise

Morro College has established a new policy for graduating students, namely, every student must take a course called **Community Services** in order to graduate. In this course, the class is required to do a community service project that requires a minimum number of hours. The class must choose a project, get approval from the Dean of Humanities, and then carry through with the project before they can graduate.

This year the Community Services class has been given permission by the City Council to estimate the cost of materials and provide the labor to put in a flower garden in the center of Cuesta Park. The city will pay for all materials. City equipment and some employees will be available at no charge.

The class is to provide estimates for all materials costs for the garden shown in the accompanying diagram. The requirements are as follows:

1. The concrete area will be 4 in. thick. Two inches of dirt is to be removed first. This dirt is to be spread over the areas where lawn will be planted. Concrete costs $77 per cubic yard delivered.
2. A twelve-inch-high brick wall will surround the center flower area. Brick costs $2.25 per square foot delivered.
3. One inch of mulch will be spread over the lawn areas and rototilled into the soil. Mulch costs $2.79 per bag. A bag holds 2 cubic feet of mulch.
4. The lawn area will be planted with sod, which costs 39.5 cents per square foot delivered.
5. Two inches of mulch will be spread over the flowerbeds and rototilled into the soil. (Same cost as in #3.)
6. Three hundred flowers will be planted in the flowerbeds. The flowers cost $2.29 per pack (four in a pack).

The Community Services class is divided into three groups.

Group I will make estimates of the costs and do all of the concrete and brick work (with some help from city employees).

Group II will make estimates of the costs, rototill, dig in the mulch, and plant the lawn areas.

Group III will make estimates of the costs, rototill in the mulch, and plant the flowerbeds.

The three estimates will be put together into one total estimate for the City Council, along with complete itemizations of the costs and quantities.

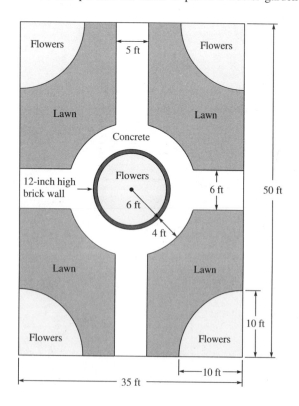

Review Exercises

Simplify, rounding to 3 decimal places as needed.

1. The sum of 15 and 8
2. The difference of 34 and 21
3. The product of 13 and 7
4. The quotient of 63 and 9
5. The sum of 5 squared and 2 cubed
6. The difference of 6 squared and 5 squared
7. The product of 3 squared and 1 cubed
8. The quotient of the square root of 64 and 2 cubed
9. Five more than the product of 7 and 3
10. Six less than the quotient of 28 and 4
11. The product of 6 and the difference of 8 and 3
12. The quotient of 9 squared and the sum of 21 and 6
13. The sum of 5 and 8 multiplied by the difference of 11 and 10
14. Sixty-four divided by 8 divided by the sum of 5 and 3
15. The sum of $\frac{3}{8}$ and $\frac{1}{8}$
16. The difference of $\frac{5}{16}$ and $\frac{1}{16}$
17. The product of $3\frac{1}{2}$ and $4\frac{1}{3}$
18. The quotient of $5\frac{1}{4}$ and $1\frac{3}{4}$
19. Two more than the product of $\frac{5}{8}$ and $\frac{7}{3}$
20. One less than the quotient of $15\frac{1}{3}$ and $7\frac{2}{3}$
21. One-half squared and added to one-fourth squared
22. Three-fourths squared less one-third squared
23. The sum of five and one-fourth and seven and one-third
24. The difference of seven and three-fifths and four and one-eighth
25. Subtract $2\frac{5}{8}$ from the sum of 5 and $2\frac{1}{3}$.
26. Add $9\frac{1}{4}$ to the difference of 11 and $3\frac{1}{7}$.
27. Subtract $4\frac{2}{3}$ from the product of $1\frac{1}{3}$ and 4.
28. Multiply the difference of $4\frac{7}{8}$ and 3 by the square of $1\frac{1}{2}$.
29. Add the square of $1\frac{1}{4}$ to the difference of $9\frac{2}{5}$ and 6.
30. Subtract 11 from the quotient of $25\frac{1}{2}$ and $\frac{1}{4}$.
31. Add 3.75 to 6.2.
32. Subtract 5.09 from 13.6.
33. Find the product of 6.1 and 3.2.
34. Find the quotient of 11 and $2\frac{1}{3}$.
35. Subtract the sum of 8 and 1.75 from the difference of 16 and 3.02.
36. Find the product of 1.24 and 4 divided by the product of 3.1 and 5.2.
37. Subtract 6.01 from the product of 9 and $1\frac{1}{2}$.
38. Divide 16.1 by 4.1.
39. Find the sum of 13.85 and $3\frac{3}{4}$.
40. Multiply the square of $\frac{1}{2}$ by the cube of $\frac{1}{2}$.
41. Add 25% to $1\frac{3}{4}$.
42. Subtract 30% from $4\frac{3}{5}$.
43. Find 40% of 60.
44. Find 200% of $7\frac{1}{3}$.
45. Divide $6\frac{1}{2}$ by 100%.
46. Find 70% of the sum of $3\frac{1}{4}$ and $7\frac{1}{2}$.
47. Multiply 0.7 by 15.
48. Add the square of 1.3 to 125%.
49. Subtract the square root of 0.25 from the square root of 0.64.
50. Add the square root of 4.41 to the square root of 0.81.
51. Multiply the square root of 36 by the square root of 4.
52. Find the quotient of the square root of 100 and the square of 5.

Factor into prime factored form.

53. 12
54. 28
55. 380
56. 720

State whether each number is divisible by (a) 2, (b) 3, and/or (c) 5.

57. 30
58. 120
59. 75
60. 123

Use the fundamental principle of fractions to rewrite each of the following fractions with the given denominator.

61. $\frac{1}{3} = \frac{?}{12}$
62. $\frac{5}{8} = \frac{?}{24}$
63. $\frac{2}{9} = \frac{?}{63}$
64. $\frac{3}{7} = \frac{?}{49}$
65. $\frac{5}{11} = \frac{?}{99}$
66. $\frac{2}{5} = \frac{?}{75}$

Complete each of the following equations.

67. 60% of 20 = what?
68. 90% of what = 27?
69. What % of 50 = 30?
70. 1.6% of 200 = what?

Find the area and perimeter of each of the following shapes. Do not change π to decimal form.

71. A square with a side of $3\frac{5}{8}$ feet

72. A circle with a radius of 1.3 inches

73. A rectangle with length of 6.7 meters and width of $1\frac{1}{2}$ meters

74. A right triangle with base $2\frac{1}{2}$ yards, height 6 yards, and hypotenuse $6\frac{1}{2}$ yards

75. A trapezoid with $B = 3\frac{1}{5}$ feet, $b = 4\frac{1}{3}$ feet, height $h = 2\frac{1}{4}$ feet, and fourth side 2.7 feet

Find the area of each of the following figures. Do not change π to decimal form.

76.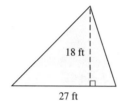

77.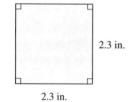

78.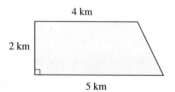

79.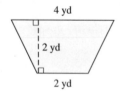

80.

81.

Use the Pythagorean theorem to find the missing side in each of the following right triangles, where a and b are the legs and c is the hypotenuse.

82. $a = 3$, $b = 4$, $c = ?$

83. $a = 5$, $b = 12$, $c = ?$

84. $a = 9$, $b = ?$, $c = 15$

85. $a = ?$, $b = 24$, $c = 26$

86. $a = ?$, $b = 8$, $c = 10$

87. $a = 12$, $b = ?$, $c = 13$

Solve each of the following applied problems and simplify your answer.

88. If 12 of a total of 18 pads of paper have been used, what fractional part of the paper is left?

89. Cindy has a tune-up on her car. Two of the six spark plugs need replacing. What fractional part of the spark plugs does not need replacing?

90. Jim answered 16 of 20 questions correctly on a test. What fractional part did he answer correctly? Rewrite your answer as a percent.

91. Jose's basketball team won 49 of the 56 games they played. What fractional part of the games did the team win? Rewrite your answer as a percent.

92. Joe's favorite cookie recipe calls for $\frac{3}{4}$ c of sugar and $\frac{1}{2}$ c of milk. How much sugar and milk will he need to make a double recipe?

93. A jar holds $\frac{3}{4}$ quart (qt) of water. What part of a quart does it hold when it is $\frac{2}{3}$ full?

94. A gasoline tank holds 18 gal. The tank is filled and, after draining for 3 hr, $\frac{1}{4}$ of a tank of gasoline is left. How many gallons were drained out?

95. Three-eighths of a class of 32 students earned an A or B in an algebra class. How many students earned a grade less than a B?

96. Lisa bought an algebra book for $34.76 and an English book for $29.48. What did she pay for the two books together, including a sales tax of 7%?

97. James ordered a sports jacket for $59.00 through a mail-order catalog. The cost for shipping and handling is 11% of the price of the jacket. What did he pay for the jacket including shipping and handling?

98. Juan earns $4.96 per hour. If he works $30\frac{3}{4}$ hr in one week, how much does he earn?

99. If gasoline costs $1.29 per gallon, how much does Jill pay for 14 gal of gasoline?

100. A car rental agency charges $35.95 per day plus 25 cents per mile. What is the cost to rent a car for 7 days if it is driven 380 mi?

101. Grace plans to upholster her couch. The cost is $14.98 yd² for material and she needs $9\frac{1}{2}$ yd². What does the material cost?

102. If Megan can drive her small car 340.8 mi on 8 gal of gas, how many miles per gallon does she get?

103. A car has depreciated 65% of its original cost. If the value of the car is now $5070, what was its original cost?

104. Harold pays a 15% annual rate of interest on his car loan. If the interest is $75.00 per year, how much is his loan?

105. In a poll of a group of students, 5 ride bicycles to school, 7 ride the bus, 10 drive their own car, and 3 ride in a car pool. What percent of the group drives their own car?

106. Roscoe bought a motorcycle for $1955.00. He made a down payment of 22% of the price. How much was his down payment?

107. Mindy is paid a commission of 7% on everything she sells. If she earned $140.00 in commissions last week, how much did she sell?

108. A state charges a gasoline tax of 9% of the cost. The federal government tax is 4% of the cost. If gasoline is $1.18 per gallon before taxes, what is the price per gallon including both taxes? Round your answer to the nearest cent.

Chapter Test

Simplify each of the following expressions.

1. $56 \div 4 - 3$
2. $(5^2 + 10) \div 7 + 1$
3. $\dfrac{11 + 3^2}{3^2 + 1}$
4. $1\dfrac{5}{8} + 3\dfrac{7}{8}$
5. $\left(\dfrac{1}{5} + \dfrac{1}{2}\right) \cdot \dfrac{7}{10}$
6. $\dfrac{360}{144}$
7. $\left(7 \cdot 2\dfrac{1}{3}\right) \div \left(7\dfrac{1}{2} \cdot 5\right)$
8. 68% of 54
9. $(1.3)(0.2) - 2(0.3)^2$
10. $\left(\dfrac{5}{4}\right)\left(\dfrac{3}{5}\right)\left(\dfrac{20}{9}\right)$
11. $\sqrt{0.25} + \sqrt{(0.2)^2}$
12. $\dfrac{3}{2} + \dfrac{21}{10}[5.06 + (6.4 - 4.38)]$

Work each of the following applied problems.

13. How much does a meal cost if it's discounted 15% from its usual price of $5.50?

14. Subtract $(1.2)^2$ from the sum of $1\frac{1}{2}$ and $2\frac{3}{4}$.

15. How much interest will you pay if you borrow $1200 at 8% interest for one year?

16. What are (a) the area and (b) the circumference of a circle with a radius of length 4 m? Do not change π to decimal form.

17. Find (a) the hypotenuse, (b) the area, and (c) the perimeter of a right triangle with a base of 3 ft and a height of 4 ft.

18. Find (a) the area and (b) the perimeter of a rectangle that is $3\frac{1}{2}$ yd long and $2\frac{1}{5}$ yd wide.

19. Find the sum of $\frac{2}{5}$ and $\frac{1}{4}$ subtracted from the difference of $3\frac{1}{2}$ and $1\frac{1}{4}$.

20. Find 50% of the sum of $5\frac{1}{2}$ and $7\frac{1}{4}$.

21. If you can drive your car for 185.5 mi on 7 gal of gas, how many miles per gallon do you get?

22. In a class of 26 students, 4 earn an A. What percent of the class is this?

23. A recipe calls for $1\frac{3}{4}$ c of water. If you want to make $\frac{2}{3}$ of a recipe, how much water will you need?

24. A new couch costs $598. The store requires a 20% down payment. How much is the down payment?

25. A chain store is discounting all of its merchandise 25%. If a shirt was originally priced at $19, what is the price of the shirt after being discounted? What is the cost after a 7% sales tax is added to the discounted price?

26. Thardis answers 36 out of 50 questions correctly on a test. What percent of the questions does she answer correctly?

27. Celia is paid a salary of $50/week plus a commission of 8% of everything she sells. If she sold $640 last week, what was her total salary?

28. Max earns $5.95/hr. If he works 25 hr and 45 min one week, how much did he earn?

CHAPTER 2
ALGEBRA AND THE REAL NUMBERS

2.1 Sets and Real Numbers
2.2 Symbols, Variables, and Absolute Value
2.3 Addition and Subtraction of Real Numbers
2.4 Multiplication and Division of Real Numbers
2.5 Properties of Real Numbers
2.6 More Applications

*A*PPLICATION

Paul plans to exercise in order to burn the same number of calories as he eats. On Thursday his lunch consists of half an avocado (185 calories), a glass of milk (159 calories), and a slice of cake (225 calories). He then plays a set of tennis (which burns 200 calories) and decides to swim enough to make up for the rest of the calories. If he always swims sets of ten laps at a time (185 calories), how many sets of ten laps must he swim to be sure he has burned off at least as many calories as he has consumed?

See Exercise Set 2.6, Exercise 41.

The higher arithmetic gives us an almost inexhaustible store of exciting truths, which stand in close internal connection, and from which we constantly find new usually unexpected ties. They are usually very simple, and can be discovered by induction, yet so profound that we cannot find how to use them until after many tries. Even if we do succeed the simpler methods may long be concealed.

ALBERT EINSTEIN (1879–1955)

2.1 Sets and Real Numbers

HISTORICAL COMMENT The numerals that are most commonly used in mathematics today are the **Hindu–Arabic numerals:** 1, 2, 3, 4, 5, 6, 7, 8, 9, 0. The name "Hindu" implies origins in India, and "Arabic" reflects the Middle-Eastern culture that eventually reached Europe. It is noteworthy that the zero was invented in India, where contemplating "nothingness" reflects part of the philosophy of the culture. Hindu–Arabic numerals replaced the more cumbersome Roman numerals of the time, although it is not uncommon to see Roman numerals used in certain special circumstances today. For example, see if you can locate the Roman numeral on a one dollar bill. What does it represent?

OBJECTIVES In this section we will learn to
1. identify and use set notation; and
2. identify the set of real numbers and its subsets.

One of the most basic concepts in mathematics is that of a *set*.

DEFINITION Set	A **set** is a collection of objects.

The objects within a set are called its **elements** or **members.** For example, in the set of the days of the week, Sunday is an element. In the set of numbers less than ten, 7 is an element.

A set is indicated by placing its elements within braces, { }. For example,

$$\{1, 2, 3, 4\}$$

is read as "the set containing 1, 2, 3, and 4." To indicate that an object is an element of a set, we use the symbol $\in$, which means "is an element of." To indicate that an object is not an element of a set, $\notin$ is used. Sets are usually designated by using capital letters.

EXAMPLE 1 $A = \{1, 2, 3\}$ is read "A is the set containing 1, 2, and 3."

EXAMPLE 2 $B = \{$Sunday, Monday, Tuesday, Wednesday, Thursday, Friday, Saturday$\}$ is expressed as "B is the set of the days of the week."

EXAMPLE 3   The statement, "Four is an element of the set $C = \{1, 2, 3, 4, 5, 6\}$" is written in symbols as $4 \in C$ or $4 \in \{1, 2, 3, 4, 5, 6\}$, while the statement, "Seven is not an element of the set C" is written as $7 \notin C$.

In Examples 1 and 2 we listed the elements of each set. This method of writing a set is called a **listing** or a **roster**. A second method, called **set-builder notation,** or the **rule method,** is one that gives a description of the common characterisitics of *all* of the elements within the set.

EXAMPLE 4   If B is the set of counting numbers less than 1000, we can write B by using set-builder notation as follows:

$$B = \{x \mid x \text{ is a counting number less than } 1000\}$$

We read B in Example 4 as "the set of all x such that x is a counting number less than 1000." The vertical bar means "such that."

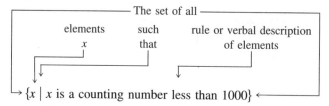

Notice that x can represent *any* element in the set. A letter that represents any element in a set is called a **variable.** A number with a value that cannot change, such as 5, is called a **constant.**

EXAMPLE 5   Write $A = \{4, 5, 6\}$ using set-builder notation.

SOLUTION $A = \{x \mid x \text{ is counting number between 3 and 7}\}$.

EXAMPLE 6   Write $C = \{x \mid x \text{ is a common vowel}\}$ using the listing method.

SOLUTION $C = \{a, e, i, o, u\}$

One method of classifying sets is with respect to the number of elements the set contains. A set is a **finite set** if it contains a limited number of elements. A set that is not finite—does not contain a limited number of elements—is an **infinite set.** Some sets have no members. When this occurs, we call such a set **the empty set** and represent it by the symbol $\emptyset$.

EXAMPLE 7 State whether the following sets are finite, infinite, or empty:

(a) $A = \{2, 4, 6\}$
(b) $B = \{x \mid x \text{ is a counting number}\}$
(c) $C = \{x \mid x \text{ is a person over 11 feet tall}\}$

SOLUTION

(a) $A = \{2, 4, 6\}$ is a finite set, since it contains exactly 3 elements.
(b) $B = \{x \mid x \text{ is a counting number}\}$ is an infinite set, since there is no largest counting number.
(c) $C = \{x \mid x \text{ is a person over 11 feet tall}\}$ is an empty set.

Suppose we state that D is the set of all dogs and that S is the set of all Scottish terriers. Since every Scottish terrier is a dog, every element of S is also an element of D. The set S is a **subset** of D that we indicate as $S \subseteq D$. The symbol $\subseteq$ is read "is a subset of." When a line is drawn through the symbol for a subset, $\not\subseteq$, it is read "is *not* a subset of."

DEFINITION **Subset**	Set A is a **subset** of set B if every element of A is also an element of B. This can be written in set notation as $$A \subseteq B \quad \text{if whenever} \quad x \in A, \quad \text{then} \quad x \in B$$

EXAMPLE 8 Given $A = \{1, 2, 3\}$ and $B = \{1, 2, 3, 4\}$, then $A \subseteq B$.

EXAMPLE 9 $\{1, 5, 8\} \subseteq \{x \mid x \text{ is a counting number}\}$

A set can have many subsets, as shown in the next example.

EXAMPLE 10 The subsets of $\{x, y, z\}$ are

$\{x\}, \{y\}, \{z\}, \{x, y\}, \{x, z\}, \{y, z\}, \{x, y, z\}$, and $\emptyset$

Notice in Example 10 that we included the empty set as a subset of $\{x, y, z\}$. *The empty set is a subset of every set.*

Real Numbers

The set of numbers we use in arithmetic is called **the set of real numbers,** but before we consider that, let's look at the main subsets of the real numbers that we will learn to identify by name. The most basic subset is called *the set of natural numbers* or *the counting numbers,* which we used in Chapter 1.

DEFINITION	$N = \{\text{natural numbers}\} = \{1, 2, 3, 4, \ldots\}$
The Set of Natural Numbers	

The three dots in $\{1, 2, 3, \ldots\}$ indicates that the set continues in the same manner. The set of natural numbers is an infinite set. When the number zero is included in the set, a new set is formed, called *the set of whole numbers.*

DEFINITION	$W = \{\text{whole numbers}\} = \{0, 1, 2, 3, 4, \ldots\}$
The Set of Whole Numbers	

The set of whole numbers is sufficient to carry out many of the operations of arithmetic, but it lacks a means for indicating such everyday things as debt and temperatures below zero. To overcome deficiencies such as this, a new set of numbers was invented. If a number is less than zero, it is called a **negative number.** To indicate that a number is negative, the symbol "−" is placed before it. For example, a number that is 4 less than 0 (negative 4) is written as −4.

We now have the same symbol, −, meaning two different things. We use it to indicate subtraction as well as that a number is negative. The way we use it makes its meaning clear:

$\quad\quad 7 - 5 \quad$ minus; subtract

$\quad\quad -5 \quad$ negative five; 5 less than 0

The sign "+" also has two meanings. Its usual meaning is to indicate addition, but it can also be used to indicate that a number is *greater* than zero. A number that is greater than zero is called a **positive number.** Once again, the way we use it makes its meaning clear:

$\quad\quad 7 + 5 \quad$ plus; add

$\quad\quad +5 \quad$ positive five; 5 more than 0

The introduction of the negative numbers results in another subset of the set of real numbers: *the set of integers.* It consists of the whole numbers together with the negatives of the natural numbers.

DEFINITION	$J = \{\text{integers}\} = \{\ldots, -4, -3, -2, -1, 0, 1, 2, 3, 4, \ldots\}$
The Set of Integers	

A good way to visualize the relationship between the natural numbers, the whole numbers, and the integers is to place them on a **number line.** To do this, we draw a straight line, label a convenient point as zero, and mark off spaces of equal length in both directions from zero. The positive numbers lie to the right of zero and the negative numbers to the left; see Figure 2.1.

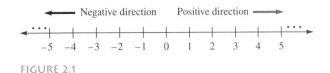

FIGURE 2.1

The point where zero is located is called the **origin.** The number that we use to locate a particular point on the number line is called its **coordinate.** In Figure 2.2, the coordinate of the point P is 2, while the coordinate of the point X is -3.

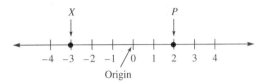

FIGURE 2.2

Notice that the set of integers does not account for points on the number line such as $\frac{1}{2}$, $1\frac{2}{3}$, or $-\frac{15}{16}$. These numbers, called *rational numbers,* represent the division, or quotient, of two integers. Several examples of rational numbers are shown on the number line in Figure 2.3.

FIGURE 2.3

It is easy to list the elements of the set of natural numbers, whole numbers, and integers because a definite pattern between them exists. This is not true, however, for the rational numbers. They are identified by the following definition.

DEFINITION **Rational Number**	A **rational number** is any number that can be expressed as the quotient (ratio) of two integers. If a and b represent integers, and $b \neq 0$, then $\dfrac{a}{b}$ is a rational number.

*C*AUTION

Never divide by zero. Division by zero is not defined.

Notice in the definition that $b \neq 0$. Why must zero be excluded as a divisor? To help answer this question, recall that $\frac{8}{2} = 4$ because $2 \cdot 4 = 8$. Now suppose there is a number q such that $\frac{8}{0} = q$. This would mean that $0 \cdot q = 8$. But we know that $0 \cdot q = 0$, so $0 \cdot q = 8$ can never be true. Therefore q does not exist in this case. Is there a number q such that $\frac{0}{0} = q$? This means that $0 \cdot q = 0$, which is correct. However, $0 \cdot 1 = 0$, $0 \cdot 2 = 0$, and $0 \cdot 17 = 0$. We cannot determine which number q represents. In the first example, $\frac{8}{0} = q$, there was no correct answer, while in the second example, $\frac{0}{0} = q$, we could not determine q, so we say that ***division by zero is not defined.***

EXAMPLE 11 Decide whether the following expressions represent rational numbers:

(a) $\dfrac{2}{5}$ (b) $\dfrac{-3}{8}$ (c) $2\dfrac{1}{2}$ (d) $\dfrac{3}{0}$ (e) 0.3

SOLUTION (a) $\dfrac{2}{5}$ is rational because it is the quotient of two integers, 2 and 5.

(b) $\dfrac{-3}{8}$ is rational because it is the quotient of two integers, -3 and 8.

(c) $2\dfrac{1}{2} = \dfrac{5}{2}$ is rational because 5 and 2 are integers.

(d) $\dfrac{3}{0}$ is *not* rational because division by zero is not defined.

(e) $0.3 = \dfrac{3}{10}$ is rational because 3 and 10 are integers.

EXAMPLE 12 Express 6 as a rational number.

SOLUTION $6 = \dfrac{6}{1}$ because 6 and 1 are integers.

The set of integers is a subset of the set of rational numbers. Every integer can be expressed as a rational number with denominator 1. In fact, a number of subset relationships exists:

natural numbers $\subseteq$ whole numbers $\subseteq$ integers $\subseteq$ rational numbers

Notice that the set of rational numbers as well as the sets of natural numbers, whole numbers, and integers are all infinite sets.

We mentioned in Chapter 1 that some fractions, when expressed as decimal numbers, either terminate or repeat. Those fractions are the ones that form the set of rational numbers.

EXAMPLE 13 Express each of the following rational numbers in decimal form: (a) $\dfrac{1}{2}$ (b) $\dfrac{5}{8}$

SOLUTION (a) To change $\dfrac{1}{2}$ to a decimal we divide the numerator, 1, by the denominator, 2. The division terminates when the remainder is zero.

$$\dfrac{1}{2} = 0.5$$

(b) To change $\dfrac{5}{8}$ to a decimal we divide the numerator, 5, by the denominator, 8. The division terminates when the remainder is zero.

$$\dfrac{5}{8} = 0.625$$

EXAMPLE 14   Express $\frac{5}{27}$ in decimal form.

SOLUTION
$$\begin{array}{r} 0.185 \\ 27\overline{)5.0000} \\ \underline{27} \\ 2\,30 \\ \underline{2\,16} \\ 140 \\ \underline{135} \\ 5 \end{array}$$
←The digits in the quotient will repeat because we started with 5.

We indicate the repetition by placing a bar over the repeating digits. Thus,

$$\frac{5}{27} = 0.\overline{185}$$

Not all decimals repeat or terminate. For example,

   1.01001000100001 . . .   and   2.323323332 . . .

are two such numbers. No repetition of digits occurs for either of them. They belong to another subset of the real numbers called the **irrational numbers.** Common examples of irrational numbers are

$\sqrt{2} \approx 1.414213562$ $\approx$ means approximately equal to.
$\sqrt{3} \approx 1.732050808$

which we will discuss in detail later. Also π, which we use to find the area of a circle, is an irrational number.

To Classify a Real Number by Its Decimal Form
A number expressed in decimal form is
  1. *rational* if it is repeating or terminating;
  2. *irrational* if it is nonrepeating and nonterminating.

We can now answer the question, "What is the set of real numbers?"

DEFINITION
The Set of Real Numbers

The set of real numbers is formed by joining the set of rational numbers and the set of irrational numbers.

Another way to describe a real number is to say that it is the coordinate of some point on a real-number line. The number line in Figure 2.4 shows the location of some sample rational and irrational numbers.

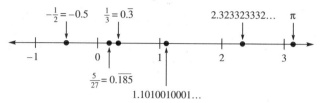

FIGURE 2.4

The diagram of the relationship that exists between the real numbers and its subsets is shown in Figure 2.5. Each rectangle in the figure is to be interpreted as enclosing all of the numbers in the set named inside it.

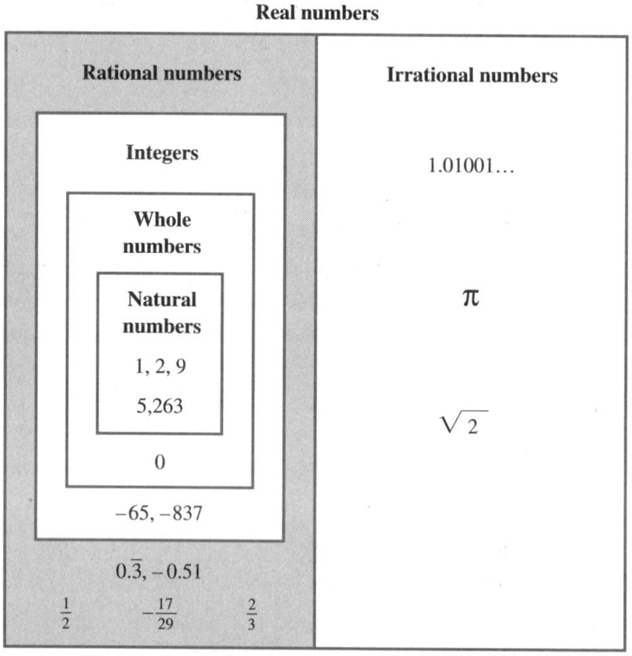

Real numbers

Rational numbers	**Irrational numbers**
Integers **Whole numbers** **Natural numbers** 1, 2, 9 5,263 0 −65, −837 $0.\overline{3}$, −0.51 $\frac{1}{2}$ $-\frac{17}{29}$ $\frac{2}{3}$	1.01001... π $\sqrt{2}$

FIGURE 2.5

Do You Remember?

Can you match these?

_f_   1. Is an element of
_g_   2. Is a subset of
_j_   3. The set of natural numbers
_i_   4. The set of whole numbers
_c_   5. The set of integers
_d & k_   6. An example of a repeating decimal
_l_   7. An example of a nonrepeating, nonterminating decimal
_a_   8. 4
_e_   9. An example of a set-builder notation
_h_   10. Coordinate

a) The number of subsets of {1, 2}
b) The set of real numbers
c) {..., −2, −1, 0, 1, 2, ...}
d) 3.2534
e) {x | x is a whole number}
f) ∈
g) ⊆
h) The name for a point on a number line
i) {0, 1, 2, 3, ...}
j) {1, 2, 3, 4, ...}
k) A rational number
l) π

Answers: 1. f 2. g 3. j 4. i 5. c 6. d, k 7. l 8. a 9. e 10. h

Exercise Set 2.1

Indicate what numbers are indicated by the three dots.

1. {1, 2, 3, ..., 8, 9, 10}
2. {4, 5, 6, ..., 11, 12, 13}
3. {1, 3, 5, ..., 17, 19, 21}
4. {2, 4, 6, ..., 18, 20, 22}
5. {4, 8, 12, ..., 40}
6. {3, 6, 9, ..., 33}
7. {5, 10, 15, ..., 35}
8. {7, 14, 21, ..., 84}

List the elements in each of the following sets.

9. {months of the year}
10. {days of the week}
11. {two-digit numbers less than 50 with both digits the same}
12. {two-digit numbers with digits that add to 4}
13. {the three smallest natural numbers}
14. {the smallest two-digit natural number}
15. {natural numbers between 5 and 10}
16. {natural numbers less than 12}
17. {x | x is the car you drive}
18. {p | p is a class you are taking}
19. {m | m is the twenty-seventh letter of our alphabet}
20. {y | y is the largest natural number}

Fill in the blank with "∈" or "∉" to make each statement true.

21. 3 __ {3, 4, 5}
22. 18 __ {1, 2, 3, ..., 20}
23. 2 __ {even natural numbers}
24. 7 __ {odd natural numbers}
25. π __ {our alphabet}
26. Jim Brown __ {Presidents of the United States}
27. 28 __ {natural numbers less than 30}
28. 5 __ {digits we use for counting}

Fill in the blank with "⊆" or "⊄" to make each statement true.

29. {8, 10} __ {2, 4, 6, ..., 14}
30. {15, 21} __ {3, 6, 9, ..., 33}
31. {48, 56} __ {4, 8, 12, ..., 52}

32. {5, 40} __ {10, 15, 20, ..., 50}
33. {A, B} __ {grades students can earn}
34. {history, math} __ {courses some students like}
35. {spinach} __ {foods some people like}
36. {New York} __ {cities in Canada}

Write each of the following sets using set-builder notation. Answers may vary.

37. {1, 3, 5}
38. {2, 4, 6}
39. {5, 10, 15, ..., 45}
40. {8, 16, 24, ..., 64}
41. {2, 4, 6, ...}
42. {1, 3, 5, ...}
43. {Saturday, Sunday}
44. {January, June, July}
45. $C = \{1, 2, 3, ...\}$
46. $E = \{2, 4, 6, ...\}$
47. $M = \{1, 2, 3, ..., 800\}$
48. $Q = \{3, 6, 9, ..., 54\}$
49. $P = \{May\}$
50. $B = \{June, July\}$

List all of the subsets of each of the following sets.

51. {1, 2}
52. {4, 7}
53. {shoe, sock}
54. {face, cream}
55. {1, 2, 3}
56. {3, 5, 7}

Identify each of the following statements as true *or* false.

57. $5 \in$ {natural numbers}
58. $6 \in$ {integers}
59. $-3 \in$ {whole numbers}
60. $\frac{1}{3} \in$ {integers}
61. {whole numbers} $\subseteq$ {integers}
62. {natural numbers} $\subseteq$ {whole numbers}
63. {real numbers} $\subseteq$ {irrational numbers}
64. {rational numbers} $\subseteq$ {irrational numbers}
65. $\pi \in$ {rational numbers}
66. $\frac{15}{0} \in$ {real numbers}
67. $7 \in$ {rational numbers}
68. $\frac{9}{3} \in$ {rational numbers}

Identify each of the following sets as finite *or* infinite.

69. {natural numbers}
70. {integers}
71. {2, 4, 6, ...}
72. {1, 3, 5, ...}
73. {irrational numbers}
74. {rational numbers}
75. {4, 5, 6}
76. $\left\{\frac{1}{3}, \frac{2}{3}, \frac{3}{4}\right\}$

77. {whole numbers}
78. {real numbers}
79. $\left\{\dfrac{a}{b} \,\middle|\, a, b \text{ are integers}, b \neq 0\right\}$
80. {..., -3, -2, -1, 0, 1, 2, 3, ...}

For Exercises 81–92, list by name all of the common subsets of the real numbers (the integers, the whole numbers, etc.) that each of the following numbers belongs to.

81. $\frac{3}{8}$
82. $\frac{1}{2}$
83. -4
84. -18
85. 0
86. $-\frac{1}{3}$
87. 0.3
88. 0.1
89. 0.675
90. 0.012
91. $\sqrt{15}$
92. $\sqrt{3}$

93. Locate the point represented by each of the following numbers on the same real-number line.

   (a) $\frac{3}{2}$ (b) $-\frac{5}{2}$ (c) $-\frac{3}{4}$
   (d) $\frac{7}{4}$ (e) $4\frac{1}{2}$ (f) $-7\frac{1}{2}$
   (g) -4.75 (h) $\frac{9}{2}$

94. Name the coordinates represented by each of the following points *a* through *j*.

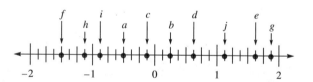

Express each of the following numbers in decimal form. If the decimal form repeats, indicate it with a bar over the repeating part. If the decimal form does not repeat, round it to five decimal places. A calculator may be helpful.

95. $\frac{1}{3}$
96. $\frac{1}{11}$
97. $\frac{1}{9}$
98. $\frac{1}{7}$
99. $\frac{3}{7}$
100. $\frac{2}{3}$
101. $\frac{3}{11}$
102. $\frac{5}{9}$
103. $\frac{1}{13}$
104. $\frac{3}{5}$
105. $\frac{5}{10}$
106. $\frac{3}{12}$

Give an example of each of the following numbers. If it is not possible to do so, write "not possible."

107. A rational number that is also a natural number
108. An integer that is not a whole number
109. A real number that is irrational
110. A whole number that is not a natural number
111. A rational number that is also irrational
112. An integer that is not rational
113. A quotient that is undefined
114. A quotient that is a whole number
115. A real number that is not an integer
116. A real number that is not a whole number

WRITE IN YOUR OWN WORDS

Define each of the following.

117. A set
118. The empty set
119. The set of whole numbers
120. The set of natural (counting) numbers
121. $\in$
122. $\subseteq$
123. $\{x \mid x \in \text{real numbers}\}$
124. A constant
125. A variable
126. A subset

2.2 Symbols, Variables, and Absolute Value

HISTORICAL COMMENT The development of mathematical symbols over the centuries was anything but systematic. Early writers often developed their own symbols to accommodate the concepts they were trying to convey. There is ample evidence that they adopted symbols used by other important writers of their time, often modifying them to serve their own needs. The symbols that have survived to this day sometimes had no better reason for being chosen other than it was easy for the early printers to reproduce them. The symbol "÷" that we use for division was used by the German mathematician Adam Riese in 1525 to indicate subtraction. Leibniz (1646–1716), after experimenting with various symbols for several years, was one of the first actively to pursue standardization of symbols that were both convenient to use and easy to print.

OBJECTIVES In this section we will learn to
1. write algebraic expressions involving variables;
2. simplify expressions involving absolute value;
3. evaluate algebraic expressions for given values of the variables; and
4. translate from verbal expressions into algebraic expressions.

Algebra, unlike arithmetic, often uses letters in place of numbers. The letters represent numbers and are used in the same ways. In the following tables, the variables x and y can represent any numbers, and we use them to show how the operations of arithmetic and algebra are similar.

OPERATION	ARITHMETIC	ALGEBRA
Addition	$15 + 60$	$x + y$
Subtraction	$27 - 15$	$x - y$
Multiplication	60×30	$x \times y$
Division	$50 \div 17$	$x \div y$

Recall that addition results in a **sum**, subtraction in a **difference**, multiplication in a **product**, and division in a **quotient**. In algebra, multiplication and division are often written in several different ways.

Arithmetic		Algebra	
Product	Quotient	Product	Quotient
$6 \times 3 = 18$	$6 \div 3 = 2$	$x \times y$	$x \div y$
$6 \cdot 3 = 18$	$6/3 = 2$	$x \cdot y$	x/y
$6(3) = 18$	$\frac{6}{3} = 2$	$x(y)$	$\frac{x}{y}$
$(6)3 = 18$	$3\overline{)6}$	$(x)y$	$y\overline{)x}$
$(6)(3) = 18$		$(x)(y)$	
		xy	

Certain symbols are less desirable to use than others, since they can be misinterpreted. An "$\times$" for multiplication can be mistaken for the variable x. Using x/y for division sometimes leads to confusion concerning the divisor. Except where there is no chance of confusion, we will avoid these symbols. Graphics calculators, however, will display the symbol "/" on the screen when the division key is pressed.

One of the main reasons for studying algebra is its usefulness in solving applied problems. To be successful in solving applied problems, we must first learn how to translate phrases into algebraic symbols. Here are some common phrases and their algebraic interpretation.

Expression in Words	Expression in Symbols

1. a plus b
    the sum of a and b
    a increased by b
    b more than a
	1. $a + b$

2. a minus b
    the difference of a and b
    b less than a
	2. $a - b$

3. a times b the product of a and b	3. $a \cdot b$ and others
4. a divided by b the quotient of a and b	4. $a \div b$ and others

When we translate verbal statements into algebraic expressions, it is often necessary to use more than one operation.

EXAMPLE 1   **(a)** "Two more than 3 times x" is written

$$3x + 2$$

(b) "y less than the sum of a and b" is written

$$(a + b) - y$$

(c) "Three less than 8 times x" is written

$$8x - 3$$

(d) "One-third of r subtracted from one-fifth of p" is written

$$\frac{1}{5}p - \frac{1}{3}r \quad \text{or} \quad \frac{p}{5} - \frac{r}{3}$$

In the two parts of Example 2 that follow, notice how the use of a comma can alter the meaning of a statement.

EXAMPLE 2 (a) "The product of 3 and x, plus 7" means

$$3x + 7$$

(b) "The product of 3, and x plus 7" means

$$3(x + 7)$$

Example 2 shows how easy it is to misinterpret a statement. We must be careful to read an expression carefully before translating it into symbols. In part (b) the meaning, changed by the placement of the comma, was clarified in its symbolic form by using grouping symbols.

When working with applied problems, there is a real tendency to use the letters x and y almost exclusively to represent numbers. It is quite useful, however, to use letters that remind you of what they represent. For example, d can be used to represent dollars, A to represent area, P to represent perimeter, M to represent Mary's age, and l to represent length.

EXAMPLE 3 Translate each statement into algebraic symbols.

(a) Two times Tom's age
(b) Six years less than Julio's age
(c) Three dollars less than the cost of four oil filters
(d) Sean has $2600 to spend on a computer system. If he spends M dollars on the monitor, how much is left to spend on the other components?

SOLUTION (a) Let $T =$ Tom's age.
$2T =$ two times Tom's age
(b) Let $J =$ Julio's age.
$J - 6 =$ six years less than Julio's age
(c) Let $f =$ the cost of an oil filter.
$4f - 3 =$ three dollars less than the cost of four oil filters
(d) To find the amount left to spend on the other components, we subtract the amount spent from the total funds available:

$$\$2600 - M \text{ dollars}$$

is left for the other components.

EXAMPLE 4 A man earns 6% on the x dollars he has invested in the money market and 4.01% on the y dollars he has invested in a savings account. Write an expression to describe his income from these two investments.

SOLUTION $0.06x + 0.0401y$ Amount = Percent × Base

Now let's look further at the comparison of positive and negative numbers, but this time in relation to their position on the number line. Two numbers that are the same distance from the origin but on opposite sides are said to be **additive inverses** or **opposites** of each other. Consider 6 and −6 on the number line in Figure 2.6.

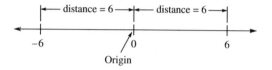

FIGURE 2.6

The two numbers are the same distance, 6 units, from the origin. Thus 6 and −6 are additive inverses or opposites of each other.

The additive inverse of a number can be indicated by writing a negative sign, −, in front of it. Using this notation, the additive inverse of 6 is −6. By the same process, the additive inverse of −6 is −(−6). But since a number can have only one additive inverse, we conclude that

$-(-6) = 6$

This result leads to the following generalization.

The Double Negative Property

For any real number x,

$-(-x) = x$

EXAMPLE 5 Find the additive inverse of each number:

(a) 5   (b) $-\frac{1}{4}$ (c) −7   (d) −2.751   (e) 0

SOLUTION   (a) The additive inverse of 5 is −5.
(b) The additive inverse of $-\frac{1}{4}$ is $-\left(-\frac{1}{4}\right) = \frac{1}{4}$.
(c) The additive inverse of −7 is −(−7) = 7.
(d) The additive inverse of −2.751 is −(−2.751) = 2.751.
(e) The additive inverse of 0 is 0 itself.

Zero is the only number that is its own additive inverse.

Absolute Value

The distance that a number lies from the origin is called its **absolute value.** The absolute value of a number is designated by placing it between vertical bars. Thus,

$$|-6| = 6 \quad \text{and} \quad |6| = 6$$

EXAMPLE 6 Find the absolute value of each number: (a) 3 (b) -11 (c) -0.567

SOLUTION
(a) $|3| = 3$ because 3 is 3 units from the origin.
(b) $|-11| = 11$ because -11 is 11 units from the origin.
(c) $|-0.567| = 0.567$ because -0.567 is 0.567 units from the origin.

Absolute value can be defined in a more formal way. The following definition will be used extensively in later chapters.

DEFINITION
Absolute Value

$$|x| = \begin{cases} x & \text{if } x \geq 0 \\ -x & \text{if } x < 0 \end{cases} \quad \text{or} \quad |x| = \begin{cases} x & \text{if } x \text{ is positive or zero} \\ -x & \text{if } x \text{ is negative} \end{cases}$$

EXAMPLE 7 Use the formal definition to find the absolute value: (a) 7 (b) -9

SOLUTION (a) Seven is greater than zero, so the first part of the definition applies:

$$|7| = 7$$

(b) Negative 9 is less than zero, so the second part of the definition applies:

$$|-9| = -(-9) = 9 \quad \text{Double negative property}$$

EXAMPLE 8 Simplify: (a) $|3 + 7|$ (b) $-|-5|$ (c) $|-13| - |-9|$

SOLUTION
(a) $|3 + 7| = |10| = 10$ Consider the absolute value symbol as a grouping symbol when it contains computation.
(b) $-|-5| = -5 \quad |-5| = 5$
(c) $|-13| - |-9| = 13 - 9$
$ = 4$

EXAMPLE 9 For what values of x is $|x| + 3 > 0$?

SOLUTION The absolute value of any real number is always positive or zero, since the smallest value that $|x|$ can be is zero. Therefore the smallest the sum $|x| + 3$ can be is 3. Thus, $|x| + 3 > 0$ for all real numbers x.

EXAMPLE 10 Evaluate each of the following for $w = -\dfrac{2}{3}$, $x = -11$, $y = 3$, and $z = 5$:

(a) $|x| + |yz|$ **(b)** $|w|^2 + y^3$

SOLUTION **(a)** To *evaluate* means to substitute the given values for w, x, y, and z into the given expression and carry out the indicated operations.

$$|x| + |yz| = |-11| + |3 \cdot 5| \quad \text{\small yz means y times z.}$$
$$= |-11| + |15| \quad \text{\small Carry out the operations within the absolute value symbols first.}$$
$$= 11 + 15$$
$$= 26$$

(b) $|w|^2 + y^3 = \left|-\dfrac{2}{3}\right|^2 + 3^3$ Substitute $w = -\dfrac{2}{3}$ and $y = 3$.

$$= \left(\dfrac{2}{3}\right)^2 + 3^3 \quad \left|-\dfrac{2}{3}\right| = \dfrac{2}{3}$$
$$= \dfrac{4}{9} + 27$$
$$= \dfrac{4}{9} + \dfrac{243}{9} \quad \text{The LCD is 9.}$$
$$= \dfrac{247}{9}$$

Do You Remember?

Can you match these?

___h___ 1. The additive inverse of y a) y
___f___ 2. $|-7|$ b) $-x$
___b___ 3. $|x|$ if $x < 0$ c) x
___g___ 4. $-|7|$ d) 4
___c___ 5. $|x|$ if $x \geq 0$ e) -4
___d___ 6. $-(-4)$ f) 7
___e___ 7. The additive inverse of 4   g) -7
___a___ 8. The opposite of $-y$ h) $-y$

Answers: 1. h 2. f 3. b 4. g 5. c 6. d 7. e 8. a

Exercise Set 2.2

Translate the following words into algebraic expressions using the given variable.

1. The sum of y and 6
2. The sum of p and 11
3. Four more than m
4. Seven added to x
5. Six less than R
6. L minus 7
7. Two times y
8. One-half of B
9. x divided by 24
10. The quotient of P and 16
11. Five percent of Q
12. The cost of x pencils at 10 cents each
13. Two more than twice m
14. Eight added to three times n
15. Twelve less than one half of D
16. Thirty-seven subtracted from one-third of P

Find the absolute value of the following quantities and simplify.

17. $|5|$
18. $|13|$
19. $|-16|$
20. $|-7|$
21. $|0|$
22. $|8-8|$
23. $|-1.24|$
24. $|-5.16|$
25. $\left|\dfrac{19}{0}\right|$
26. $\left|\dfrac{0}{22}\right|$
27. $|6|+|3|$
28. $|-5|+|23|$
29. $|-15|+|-9|$
30. $|3.4|+|-5.2|$
31. $\left|-\dfrac{1}{8}\right|+\left|-\dfrac{1}{4}\right|$
32. $\left|-\dfrac{3}{7}\right|+\left|-\dfrac{2}{5}\right|+|3|$
33. $|-0|+|-5|+|2|$
34. $[|-7.2|+|3.1|]+|5|$
35. $\left[\left|-\dfrac{5}{9}\right|+\left|\dfrac{1}{4}\right|\right]+|4|$
36. $\left[\left|-\dfrac{1}{3}\right|+|0.71|\right]+|3.2|$

Identify each of the following statements as true or false.

37. The absolute value of -3 is greater than 0.
38. The absolute value of -5 is greater than the absolute value of 3.
39. The absolute value of 4 is less than 4.
40. The absolute value of 3 is less than the absolute value of -4.
41. $|-2|+|-4|=6$
42. $5|-3|=15$
43. The absolute value of any negative number is greater than zero.
44. The absolute value of any positive number is less than zero.
45. $|x| \geq 0$
46. $|x|+1 \geq 0$
47. If $|x|+3=8$, then $x=5$ or $x=-5$.
48. If $|x|=0$, then $x=0$.
49. If $|x|-5=2$, then $x=7$ or $x=-7$.
50. $|a+b|=|a-b|$ when $a=19$ and $b=0$.
51. $|a+b|=|a-b|$ when $a=0$ and $b=-7$.
52. Of the two numbers -15 and 14, the one with the larger absolute value is -15.
53. Of the two numbers 13 and -18, the one with the smaller absolute value is -18.
54. $|x|<0$ for some real number x.
55. $|a+b|=|b+a|$ for all real numbers a and b.
56. $|x|=-5$ is true if $x=-5$.

Evaluate each expression when $x=-3$, $y=2$, and $z=0$.

57. $-y$
58. $-(-x)$
59. $|xz|$
60. $|xyz|$
61. $|x|+|y|$
62. $|y|+|z|$
63. $5|x|+3|y|+2|z|$
64. $3|y|^2$
65. $2|y+z|^2$
66. $|x|^2+5|z|^3$
67. $-|x|^2$
68. $\left|\dfrac{y}{z}\right|^3$

Translate the following words into algebraic expressions using variables, constants, and operation symbols.

69. Twice Jim's weight
70. Four pints
71. The cost of two hamburgers
72. One-fifth of the value of a gold necklace
73. Two more than twice Leslie's score
74. The cost of three hamburgers increased by 30¢
75. The price of a car discounted by $600
76. The price of a pair of shoes discounted by $4

77. The price of 100 stamps at 33¢ each
78. The charge for six people to ski
79. Seven less than four times Jill's salary
80. Ten dollars off of the cost of a lawn chair
81. Three times the sum of a number and 9
82. Eleven times the difference of a number and 2
83. The distance traveled at 53 miles per hour (mph) for 6 hr
84. The number of quarts of water in q quarts of snow if each quart of snow is 12% water
85. Sally has $1000 to invest. If she invests x dollars, how much does she have left?
86. Gil has $400. If he spends y dollars, how much money does he have left?
87. The new computer is five times as fast as the older one.
88. A small Coke is two-thirds the size of a large one.
89. Bill studied 2 hr less for the exam than twice as much as Joe.
90. Angelina's recipe calls for 300 less grams (g) of sugar than flour.
91. Indicate the total cost of x books costing $24.75 each and two parking stickers costing $12.00 each.
92. What is the value in dollars of 60 lb of nuts if x pounds are priced at $1.27/lb and the rest at $2.15/lb?
93. The population of the world one particular year was x people. Find the population the following year if it increased by 2%.
94. A good rule of thumb on how to get a good price when purchasing a new car is to take 15% off of the sticker price, s, and then add $200.
95. A business must sell its product at cost, c, plus 40% of the cost.
96. Write the average of the sum of a counting number and the next larger counting number.
97. Don sells lawn mowers. His monthly salary is $600 plus $15 for each lawn mower he sells. Indicate Don's monthly income.
98. Pauline's car gets 28 mpg. If her car holds x gallons, how far can she travel on a full tank of gas?
99. Doug budgets D dollars to buy 300 g of potassium nitrate for his science project. The potassium nitrate sells for 20¢/g. How much money does he have left?
100. In figuring his income tax, Neil has determined his adjusted gross income to be $20,000. He is allowed to deduct $750 from his adjusted gross income for each of his children. If he has x children, what is his taxable income?
101. A man willed his estate as follows: $3593 to each of his two sons, $5000 to each of his three daughters, and the rest to his wife. How much did his wife receive?
102. A famous painting by Henri Matisse was accidentally hung upside down for Q hours before the error was discovered. If 6000 people viewed it during that time, what was the average number of people viewing it per hour?

WRITE IN YOUR OWN WORDS

103. Why is $|-9|$ greater than 5?
104. Why does $|x| = x$ if $x \geq 0$ and $|x| = -x$ if $x < 0$?
105. What does $|-x|$ mean?
106. Why is $|a + b| = |b + a|$ for all real numbers a and b?
107. Why do we call x a variable and 5 a constant?
108. How do we know that if $|x| + 5 = 13$ then $x = 8$ or $x = -8$?

2.3 Addition and Subtraction of Real Numbers

HISTORICAL COMMENT It took many centuries before negative numbers were considered anything but absurd. In the third century, the Greeks refused to accept solutions to equations that were negative. The Hindus, however, discussed negative quantities in the seventh century and gave rules for their operation. In thirteenth-century Europe, Fibonacci (c.1180–c.1250) wrote about a man's profit by saying, "I will show that this question cannot be solved unless it is conceded that a man be in debt." It was not until the close of the sixteenth century that negative numbers were readily accepted everywhere.

2.3 · ADDITION AND SUBTRACTION OF REAL NUMBERS

OBJECTIVES In this section we will learn to
1. add and subtract signed numbers with the same sign; and
2. add and subtract signed numbers with opposite signs.

Addition

A number can be represented by an arrow on a number line. Because the positive direction on a number line is to the right, a positive number is represented by an arrow pointing to the right. Similarly, a negative number is represented by an arrow pointing to the left. The size of the number corresponds to the length of the arrow. See Figure 2.7.

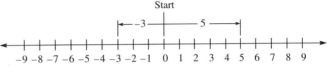

FIGURE 2.7

We can add signed numbers (positive and negative) by using arrows and the number line. For example, to add 3 to 4, we start at the origin and draw an arrow to the point 3. Then, beginning at the tip of the first arrow, we draw a second arrow 4 more units to the right. The sum of the two numbers is the point at the tip of the second arrow, as shown in Figure 2.8.

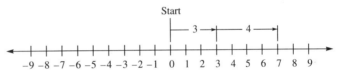

FIGURE 2.8

Thus,

$3 + 4 = 7$

The sum of -5 and -3 can be found using a number line as shown in Figure 2.9.

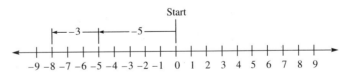

FIGURE 2.9

The arrows indicate that

$(-5) + (-3) = -8$

In summary, the sum of two numbers of the same sign, both positive or both negative, is indicated by two arrows placed end to end and pointing in the same direction, right for positive and left for negative.

When two numbers have opposite signs, the arrows that represent them will point in opposite directions. Figure 2.10 illustrates that $-5 + 4 = -1$.

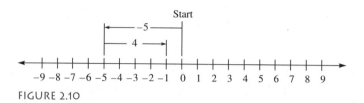

FIGURE 2.10

Finally, we use the number line in Figure 2.11 to show that $-4 + 5 = 1$.

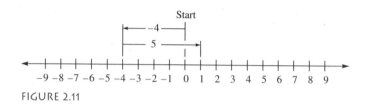

FIGURE 2.11

It is not practical to draw a number line every time we want to add two or more real numbers, so we make the following observations based on the preceding number lines.

Figure 2.8	$3 + 4 = 7$	The sum of two positive numbers is positive.
Figure 2.9	$-3 + (-5) = -8$	The sum of two negative numbers is negative.
Figure 2.10	$-5 + 4 = -1$	The sum of a positive and a negative number may be either positive or negative.
Figure 2.11	$-4 + 5 = 1$	

To Add Two Signed Numbers

1. To find the sum of two numbers with the same signs (both positive or both negative), find the sum of their absolute values and give the sum the same sign.
2. To find the sum of two numbers with opposite signs (one positive and one negative), find the difference of their absolute values and give the answer the sign of the one with the greater absolute value.

EXAMPLE 1 Find the sum $-11 + (-13)$.

SOLUTION Since the numbers have the same sign, we find the sum of their absolute values and give the sum the same sign.

$$|-11| = 11 \quad \text{and} \quad |-13| = 13$$

and

$$11 + 13 = 24$$

The numbers are both negative, so the sum will be negative:

$$-11 + (-13) = -24$$

EXAMPLE 2 Find the sum $-63 + 25$.

SOLUTION The two numbers have opposite signs, so we begin by finding the difference of their absolute values.

$$|-63| = 63 \quad \text{and} \quad |25| = 25$$

and

$$63 - 25 = 38$$

The numbers have opposite signs, so we give the answer the sign of the one with the greater absolute value:

$$(-63) + 25 = -38 \quad \text{-63 has the greater absolute value.}$$

EXAMPLE 3 Find the sum $23 + (-13)$.

SOLUTION The numbers have opposite signs, so we find the difference of their absolute values.

$$|23| - |-13| = 23 - 13 = 10 \quad \text{Find the difference of the absolute values.}$$

Since 23 has the greater absolute value, the result is positive:

$$23 + (-13) = 10$$

EXAMPLE 4 Find each of the following sums.

(a) $2.7 + 6.9 = 9.6$

(b) $\left(-\dfrac{6}{13}\right) + \left(-\dfrac{2}{13}\right) = -\dfrac{8}{13}$ The sum of two negative numbers is negative.

(c) $-3 + (-2) + [5 + (-4)] = -3 + (-2) + [1]$ $5 + (-4) = 1$
$ = -5 + 1$ $-3 + (-2) = -5$
$ = -4$

(d) $-9.76 + |-5.45| = -9.76 + 5.45$
$ = -4.31$

(e) $-3 + 2\{7 + 2[9 + (-8 + 4)]\}$
We follow the order of operations, working from the innermost grouping symbols outward.

$-3 + 2\{7 + 2[9 + (-8 + 4)]\} = -3 + 2\{7 + 2[9 + (-4)]\}$ $-8 + 4 = -4$
$\phantom{-3 + 2\{7 + 2[9 + (-8 + 4)]\}} = -3 + 2\{7 + 2[5]\}$ $9 + (-4) = 5$

$$= -3 + 2\{7 + 10\} \qquad 2[5] = 10$$
$$= -3 + 2\{17\} \qquad 7 + 10 = 17$$
$$= -3 + 34 \qquad 2\{17\} = 34$$
$$= 31$$

EXAMPLE 5   Marilyn balances her checkbook at the end of each week. One week she started with a balance of $76.84 before making a deposit of $189.25. If she wrote checks for $50, $117, and $34.46, what was her balance at the end of the week?

SOLUTION   We can use symbols of grouping to indicate this activity.

Ending balance = (beginning balance + deposit) − (checks written)
$$= (76.84 + 189.25) - (50 + 117 + 34.46)$$
$$= 266.09 - 201.46 = \$64.63$$

Subtraction

One way to consider subtraction of real numbers is to think about how we check a subtraction problem in arithmetic. For example,

$$11 - 7 = 4 \quad \text{because} \quad 4 + 7 = 11$$

Notice also that using the rules for addition yields $11 + (-7) = 4$. Thus,

$$11 - 7 = 11 + (-7) = 4$$

It appears that to subtract one number from another, we can add its additive inverse. Does this work when the numbers are both negative or have opposite signs? Let's consider the following examples to see if that is the way it's done.

SUBTRACTION		CHECK
$8 - (-6) = 8 + 6 = 14$	The additive inverse of -6 is 6.	$14 + (-6) = 8$
$-7 - 8 = -7 + (-8) = -15$	The additive inverse of 8 is -8.	$-15 + 8 = -7$
$12 - (-6) = 12 + 6 = 18$	The additive inverse of -6 is 6.	$18 + (-6) = 12$

We can also view subtraction of signed numbers on a number line. For example, $8 - (-6) = 14$ is illustrated in Figure 2.12.

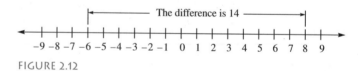

FIGURE 2.12

To Subtract Two Signed Numbers

To subtract a real number b from a real number a, add the additive inverse of b to a:

$$a - b = a + (-b)$$

2.3 · ADDITION AND SUBTRACTION OF REAL NUMBERS 73

EXAMPLE 6 Carry out each of the following subtractions.

(a) $17 - (-18) = 17 + 18$ To subtract -18, add its additive inverse, 18.
$= 35$

(b) $-25 - 9 = -25 + (-9)$ To subtract 9, add its additive inverse, -9.
$= -34$

(c) $-15.2 - (-47.05) = -15.2 + 47.05$
$= 31.85$

In problems that involve more than one operation, we can use the order of operations to simplify them.

EXAMPLE 7 Add or subtract as indicated.

(a) $(-4 + 9) - 13 = (-4 + 9) + (-13)$ Rewrite the subtraction as addition.
$= 5 + (-13)$ Work from left to right: $-4 + 9 = 5$.
$= -8$

(b) $-26 - (-3) - [4 - (-6)]$

$= -26 - (-3) - [4 + 6]$ Do operations within the innermost grouping symbols first: $[4 - (-6)] = 4 + 6$.

$= -26 + 3 - 10$ $4 + 6 = 10, -26 - (-3) = -26 + 3$

$= -23 - 10$ Work from left to right: $-26 + 3 = -23$.

$= -23 + (-10)$ Rewrite the subtraction as addition.

$= -33$

EXAMPLE 8 The temperature in Fargo, North Dakota, was 30° above zero at 4 P.M. on November 26. During the night, a cold front swept down from Canada, lowering the temperature to 6° below zero. How far did the temperature fall?

SOLUTION To find how far the temperature fell, we must find the difference in the two readings. By representing 6° below zero as -6, we have

$$30 - (-6) = 30 + 6 = 36$$

The temperature fell 36°.

EXAMPLE 9 The highest point in California is the top of Mount Whitney, at 14,494 feet above sea level. The lowest point in the state is in Death Valley, at 282 feet below sea level. What is the difference in elevation between these two points?

SOLUTION We represent 282 feet below sea level as -282. Then

$$14,494 - (-282) = 14,494 + 282$$
$$= 14,776$$

The difference in elevation is 14,776 feet.

> **Calculator Note**
>
> A calculator can be used to add and subtract signed numbers. However, it is advisable to **wait until you are thoroughly familiar with the rules that govern the use of signed numbers before trying such calculations.**

Using a Scientific Calculator

To enter a negative number, we first enter its absolute value and then press the $\boxed{\pm}$ key. For example, pressing $\boxed{3}$ followed by $\boxed{\pm}$ will appear as -3 on the display. Pressing $\boxed{\pm}$ again will change the display back to 3. To calculate

$$9 + (-3) - (-16) - 5$$

we depress the keys in the following order:

$\boxed{9}\ \boxed{+}\ \boxed{3}\ \boxed{\pm}\ \boxed{-}\ \boxed{16}\ \boxed{\pm}\ \boxed{-}\ \boxed{5}\ \boxed{=}$

The display shows 17. Thus,

$$9 + (-3) - (-16) - 5 = 17$$

Using a Graphics Calculator

The key for entering a negative number is located away from the key for subtraction; it shows a short dash enclosed in parentheses on its face.

Subtract: $\boxed{-}$ Negative: $\boxed{(-)}$

To calculate $9 + (-3) - (-16) - 5$, we depress the keys in the following order:

$\boxed{9}\ \boxed{+}\ \boxed{(-)}\ \boxed{3}\ \boxed{-}\ \boxed{(-)}\ \boxed{16}\ \boxed{-}\ \boxed{5}\ \boxed{=}$

The display again shows 17.

Do You Remember?

Can you match these?

____ 1. $-7 + (-6)$ a) 13
____ 2. $-7 - (-6)$ b) -13
____ 3. $7 - (-6)$ c) 1
____ 4. $7 - 6$ d) -1
____ 5. $7 + (-6)$
____ 6. $-7 + 6$
____ 7. $-7 - 6$
____ 8. $7 + 6$

Answers: 1. b 2. d 3. a 4. c 5. c 6. d 7. b 8. a

EXERCISE SET 2.3

Add. A number line may be helpful.

1. $17 + 8$
2. $13 + 3$
3. $(+5) + (+12)$
4. $(+8) + (+9)$
5. $8 + (-5)$
6. $13 + (-7)$
7. $(-29) + (3)$
8. $(-36) + (9)$
9. $-44 + 12$
10. $-56 + 15$
11. $(-9) + (-6)$
12. $(-4) + (-7)$
13. $-23 + (-17)$
14. $-19 + (-11)$
15. $81 + (-19) + (-24)$
16. $76 + (-15) + (-22)$
17. $103 + [(-17) + 3]$
18. $57 + [(-18) + 7]$
19. $|49| + |-5|$
20. $|76| + |-8|$
21. $(1.5) + (-1.3)$
22. $(1.8) + (-1.7)$
23. $(0.08) + (-2.1) + (-0.6)$
24. $(0.2) + (-1.5) + (-0.3)$
25. $\frac{3}{5} + \frac{4}{5}$
26. $\left(-\frac{7}{8}\right) + \left(-\frac{3}{8}\right)$
27. $\frac{2}{7} + \frac{4}{7}$
28. $\left(-\frac{4}{11}\right) + \left(-\frac{15}{11}\right)$
29. $(-2) + (-3) + (4) + (1)$
30. $(-7) + (-2) + (6) + (3)$
31. $|1.2| + |-3.4| - 6$
32. $|2.1| + |-2.7| - 7$
33. $\left|-\frac{3}{11}\right| + \frac{8}{11} + \left(-\frac{2}{11}\right)$
34. $\left|-\frac{1}{6}\right| + \frac{5}{6} + \left(-\frac{3}{6}\right)$
35. $-4 + 3\{2 + 4[(-3) + 6]\}$
36. $-5 + 2\{3 + 6[8 + (-7)]\}$

Complete as indicated.

37. Add -13 to 18.
38. Add 24 to -31.
39. Add -56 to -14.
40. Add -7 to -67.
41. Find the sum of 83 and negative 19.
42. Find the sum of 47 and negative 44.
43. Find the sum of negative 31 and negative 3.
44. Find the sum of negative 49 and negative 19.

Subtract and check.

45. $6 - 5$
46. $7 - 3$
47. $21 - 18$
48. $37 - 14$
49. $28 - 15$
50. $51 - 39$
51. $42 - 27$
52. $16 - (-19)$
53. $13 - (-81)$
54. $-6 - (-9)$
55. $-13 - (-18)$
56. $7 - 26$
57. $16 - 81$
58. $-34 - 93$
59. $-6 - 23$
60. $-70 - (-43)$
61. $-50 - (-12)$
62. $8 - (3 - 2)$
63. $9 - (7 - 4)$
64. $-45 - [3 - (-2)]$
65. $-39 - [6 - (-4)]$
66. $27 - (-11) - 5$
67. $16 - (-9) - 4$
68. $-35 - (-18) - (-20)$
69. $-24 - (-8) - (-22)$

Complete as indicated.

70. Subtract 4 from 9.
71. Subtract 8 from 17.
72. Find the difference of 30 and 7.
73. Find the difference of 27 and 11.
74. Subtract negative 63 from negative 38.
75. Subtract negative 57 from negative 23.
76. Subtract 84 from negative 19.
77. Subtract 78 from negative 11.
78. Subtract the absolute value of negative 2.3 from negative 6.7.
79. Subtract the absolute value of negative 5.4 from negative 4.8.
80. Subtract negative fifteen-eighths from negative five-eighths.
81. Subtract negative three-sevenths from negative one-seventh.

Add or subtract, as indicated.

82. $7 + (-5) - 4$
83. $9 + (-2) - 6$
84. $-13 + (-3) + 18$
85. $-17 + (-2) + 22$
86. $|8 + (-3)| - |-5 + 4|$
87. $|14 + (-7)| - |-9 + 5|$
88. $-|-7 + 2| + |-5 + 1|$
89. $-|-9 + 1| + |-8 + 3|$
90. $-38 - [18 - (-14)]$
91. $-27 - [15 - (-11)]$
92. $-50 - (-31) - 15 - (-11)$
93. $-72 - (-48) - 30 - (-51)$
94. $86 - (-50) - 40 - (-20) + 18$
95. $92 - (-70) - 60 - (-12) + 46$

Add or subtract as indicated. A calculator may be helpful.

96. $887.64 - 327.81$
97. $963.84 - 721.89$
98. $121.893 - (-117.641)$
99. $385.773 - (-281.847)$
100. $|-0.125| - |-0.375|$
101. $|-0.215| - |-0.176|$
102. $\left|-\frac{1}{4}\right| + \left|-\frac{1}{8}\right|$
103. $\left|\frac{-1}{5}\right| + \left|\frac{-1}{10}\right|$
104. $-\left|\frac{-1}{3}\right| + \left|\frac{-1}{4}\right|$
105. $-|-3.5| - |-7.1|$
106. $-6.27 - 1.64 + 8.1$
107. $-4.81 - 3.7 + 9.35$

Solve the following applied problems.

108. Ramona has a checkbook balance of $633.64. She deposits $321.38 and then writes checks of $121.81, $178.57, and $83.27. What is her new checkbook balance?

109. Max went on a diet and kept a record of the number of pounds he gained or lost every day for a week: Monday, $+1$; Tuesday, 0; Wednesday, -1; Thursday, $+1$; Friday, -1; Saturday, 0; Sunday, -1. How many pounds did Max gain or lose by the end of the week?

110. A climbing expedition climbs 6732 ft on the first day. On the second day, a snowstorm forces the climbers to retreat 3118 ft to find shelter. The third day they reach the top by climbing 4784 ft. How many feet is it to the top of the mountain?

111. At a football game, the Cougars had the ball on their own 45-yd line. On first down they gained 7 yd; on second down they lost 15 yd; and on third down they gained 4 yd. What yard line were they on on the fourth down?

112. Jose's algebra class has 68 assignments to do. If they have already done 11, how many are left?

113. A plane leaves an airport with an elevation of 542 ft above sea level and climbs to an elevation of 6500 ft above sea level. How many feet does the plane climb?

114. On Wednesday, the temperature in Pocatello, Idaho, reached a high of $+18°F$ and a low of $-9°F$. What is the difference between the high and low temperatures?

115. Tom has a balance of $-\$48.15$ (he's overdrawn) in his checking account. The bank then charges $20.00 for two returned checks. Tom deposits $187.43 and then writes checks for $68.00 and $103.87. What is his new balance?

116. Matt Bowman leaves from Scotty's Castle in Death Valley (252 ft below sea level) and drives to the top of Cuesta Grade (1500 ft above sea level). What is the change in his elevation?

117. The temperature in Juneau, Alaska, was $-6°F$ at 10:00 A.M., rose $12°$ by noon, dropped $10°$ by 4:00 P.M., and dropped $15°$ more by midnight. What was the temperature at midnight?

118. One morning GM stock was $\$64\frac{7}{8}$. That day it rose in value $\$\frac{3}{8}$. The second day it lost $\$\frac{3}{4}$. What was the value of the stock after the second day?

119. A car dealer had a profit of $16,700 in 1990, a loss of $2850 in 1991, a loss of $19,645 in 1992, a profit of $6940 in 1993, and a profit of $12,025 in 1994. Find the profit or loss for this five-year period.

For Extra Thought

120. What number must be added to negative 5 to obtain 0?

121. What number must be added to negative 7 to obtain 3?

122. The sum of what number and 13 is -8?

123. The sum of what number and 4 is -11?

124. What number subtracted from 5 is 8?

125. What number subtracted from -14 is -4?

126. The difference of what number and -6 is 7?

127. The difference of 9 and what number is 13?

128. If x is a positive number, is $-x$ positive or negative?

129. If x is a negative number, is $-x$ positive or negative?

130. If y is a positive number, is $-|y|$ positive or negative?

131. If y is a negative number, is $-|y|$ positive or negative?

132. If x and y are both positive numbers, is $|x| + |y|$ positive or negative?

133. If x and y are both negative numbers, is $|x + y|$ positive or negative?

2.4 Multiplication and Division of Real Numbers

OBJECTIVES In this section we will learn to
1. multiply real numbers; and
2. divide real numbers.

Multiplication is often referred to as "repeated addition." We can use this concept to discover how to multiply pairs of real numbers. Consider the meaning of the product $3 \cdot 2 = 6$.

$3 \cdot 2 = 2 + 2 + 2 = 6$ The sum of three 2s

Now consider the product $2 \cdot 3$, which is equal to $3 \cdot 2$:

$2 \cdot 3 = 3 + 3 = 6$ The sum of two 3s

In both cases we get repeated additions of a positive value, which results in a positive answer. Notice that 2, 3, and 6 are all positive numbers.

The product of any two positive real numbers is positive.

In the same manner,

$3 \cdot 0 = 0 + 0 + 0 = 0$

The product of any real number and zero is zero.

We will now modify the multiplication problem to include the case where one of the numbers being multiplied is negative and show that the product will be negative. To see this, consider the product $3(-2)$, using the concept of repeated addition.

$3(-2) = (-2) + (-2) + (-2) = -6$ The sum of three negative 2s

This case results in the sum of negative values, which produces a negative answer.

The product of any positive real number and any negative real number is negative.

We cannot find the product of two negative numbers by using the concept of repeated addition. For instance, the product $(-2)(-3)$ can't be obtained in this way. What does it mean to add -3 two less than zero times? Fortunately, there are patterns we can analyze to discover the product.

If we study the following patterns of products, we see that the first one leads to our last discovery, that the product of a positive number and a negative number is negative. The second pattern leads to the discovery that the sign of the product of two negative numbers is positive.

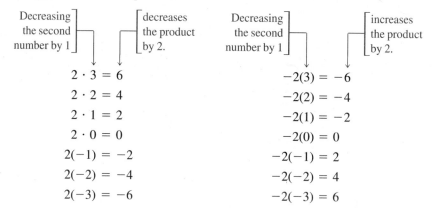

$$2 \cdot 3 = 6 \qquad\qquad -2(3) = -6$$
$$2 \cdot 2 = 4 \qquad\qquad -2(2) = -4$$
$$2 \cdot 1 = 2 \qquad\qquad -2(1) = -2$$
$$2 \cdot 0 = 0 \qquad\qquad -2(0) = 0$$
$$2(-1) = -2 \qquad\qquad -2(-1) = 2$$
$$2(-2) = -4 \qquad\qquad -2(-2) = 4$$
$$2(-3) = -6 \qquad\qquad -2(-3) = 6$$

Assigning a positive sign to the last three products in the second column maintains the pattern of increasing by 2. This last pattern indicates that *the product of two negative numbers is positive.*

To Multiply Two Real Numbers

1. The product of two numbers with like signs (both positive or both negative) is a positive number.
2. The product of two numbers with unlike signs (one positive and one negative) is a negative number.

In general, if a, b, and c represent positive real numbers and $a \cdot b = c$, then

$$a \cdot b = c \qquad (-a)(-b) = c \qquad (-a)b = -c \qquad a(-b) = -c$$

EXAMPLE 1 Find each of the following products.

(a) $3 \cdot 7 = 21$ The product of two numbers of like sign is positive.
(b) $(-8)(-9) = 72$ The product of two numbers of like sign is positive.
(c) $5(-9) = -45$ The product of two numbers of unlike sign is negative.
(d) $(-7)(15) = -105$ The product of two numbers of unlike sign is negative.

Now let's look at the product of more than two signed numbers.

EXAMPLE 2 Find each of the following products:

(a) $(-4)(2)(-5)$ (b) $(-5)(3)(4)$ (c) $(-1)(-2)(-3)$ (d) $(-3)(-2)(-3)(-4)$

SOLUTION (a) $(-4)(2)(-5) = (-8)(-5)$ $(-4)(2) = -8$
$\qquad\qquad\qquad\qquad\quad = 40$

2.4 · MULTIPLICATION AND DIVISION OF REAL NUMBERS

CAUTION

Care must be taken when working with exponents and negative numbers.

$$-a^2 = -1 \cdot a^2$$

while

$$(-a)^2 = (-a)(-a) = a^2$$

Grouping symbols must be used to indicate that an exponent applies to the sign before a number as well as to the number itself.

(b) $(-5)(3)(4) = (-15)(4)$ $(-5)(3) = -15$
$= -60$

(c) $(-1)(-2)(-3) = 2(-3)$ $(-1)(-2) = 2$
$= -6$

(d) $(-3)(-2)(-3)(-4) = 6 \cdot 12$ $(-3)(-2) = 6; (-3)(-4) = 12$
$= 72$

Examples 2(c) and (d) show that *the product of an odd number of negative numbers is negative and the product of an even number of negative numbers is positive.*

When **exponents** are used with signed numbers, we must pay particular attention to their meaning.

EXAMPLE 3 Carry out the indicated operations. See the caution above.

(a) $-2^2 = -1 \cdot 2^2 = -1 \cdot 2 \cdot 2 = -4$
(b) $(-2)^2 = (-2)(-2) = 4$

EXAMPLE 4 Perform the indicated operations: **(a)** $\left(-\frac{3}{5}\right)\left(-\frac{2}{3}\right)^2$ **(b)** $-\left| -4^2(-7)^3 \right|$

SOLUTION   **(a)** $\left(-\frac{3}{5}\right)\left(-\frac{2}{3}\right)^2 = \left(-\frac{3}{5}\right)\left(\frac{4}{9}\right)$ $\left(-\frac{2}{3}\right)^2 = \left(-\frac{2}{3}\right)\left(-\frac{2}{3}\right) = \frac{4}{9}$

$= -\frac{4}{15}$

(b) $-\left| -4^2(-7)^3 \right| = -\left| (-16)(-343) \right|$
$= -\left| 5488 \right| = -5488$ $|5488| = 5488$

EXAMPLE 5 Perform the indicated operations. Be sure to follow the order of operations.

(a) $-4 + 2(-3) + (-7)$
(b) $(-3)4 + (-7)(-5)$
(c) $-9 + (-3)(-5) - (-19) - (-3)^2$

SOLUTION   **(a)** $-4 + 2(-3) + (-7) = -4 + (-6) + (-7)$ Multiplication precedes addition.
$= -10 + (-7)$ Work from left to right.
$= -17$

(b) $(-3)4 + (-7)(-5) = -12 + 35$ Multiplication precedes addition.
$= 23$

(c) $-9 + (-3)(-5) - (-19) - (-3)^2$
$= -9 + (-3)(-5) - (-19) - 9$ Exponents precede multiplication.
$= -9 + 15 - (-19) - 9$ Multiplication precedes addition.

$$= -9 + 15 + 19 + (-9) \qquad \text{Rewrite the subtractions as addition.}$$
$$= 6 + 19 + (-9) \qquad \text{Work from left to right.}$$
$$= 25 + (-9) = 16$$

EXAMPLE 6 Perform the indicated operations:

(a) $(-2)^2(-3)^3 + 2 \cdot 6$ **(b)** $-5(6) - 3[2 - 5(6 - 3)]$

SOLUTION **(a)** $(-2)^2(-3)^3 + 2 \cdot 6 = 4(-27) + 2 \cdot 6 \qquad (-2)^2 = 4;\ (-3)^3 = -27$
$$= -108 + 12$$
$$= -96$$

(b) $-5(6) - 3[2 - 5(6 - 3)] = -5(6) - 3[2 - 5(3)]$
$$= -5(6) - 3[2 - 15]$$
$$= -5(6) - 3[2 + (-15)]$$
$$= -5(6) - 3[-13]$$
$$= -30 - (-39)$$
$$= -30 + 39 = 9$$

Division of any two real numbers follows the same rules that we have always used in arithmetic. The only difference is that we must determine the sign of the quotient. To find a quotient, it is helpful to recall how we check division in arithmetic. For example,

$$\frac{6}{2} = 3 \qquad \text{because} \qquad 2 \cdot 3 = 6$$

and

$$\frac{18}{9} = 2 \qquad \text{because} \qquad 9 \cdot 2 = 18$$

In general, for any real numbers a, b, and c, where $b \neq 0$,

$$\frac{a}{b} = c \qquad \text{if and only if} \qquad a = b \cdot c$$

If the divisor and dividend have the same sign, the quotient will be positive.

For example,

Both positive ⟶ $\frac{8}{2} = 4$ ⟵ The quotient is positive.   because $2 \cdot 4 = 8$

Both negative → $\dfrac{-8}{-2} = 4$ — The quotient is positive.   because $(-2)(4) = -8$

If the divisor and dividend have different signs, the quotient will be negative.

For example,

Different signs → $\dfrac{-8}{2} = -4$ — The quotient is negative.   because $2(-4) = -8$

Different signs → $\dfrac{8}{-2} = -4$ — The quotient is negative.   because $(-2)(-4) = 8$

To Divide Real Numbers

1. The quotient of two numbers with like signs (both positive or both negative) is positive.
2. The quotient of two numbers with unlike signs (one positive and one negative) is negative.

In general, if a, b, and c represent positive numbers and $\dfrac{a}{b} = c$, then

$$\dfrac{a}{b} = c \qquad \dfrac{-a}{-b} = c \qquad \dfrac{a}{-b} = -c \qquad \dfrac{-a}{b} = -c$$

EXAMPLE 7   Find the following quotients and justify your answers using the multiplication rules for real numbers: **(a)** $\dfrac{15}{3}$ **(b)** $\dfrac{-36}{12}$ **(c)** $-45 \div \left(-\dfrac{5}{2}\right)$ **(d)** $\dfrac{-12.1}{1.1}$

SOLUTION   **(a)** $\dfrac{15}{3} = 5$ because $5 \cdot 3 = 15$.

(b) $\dfrac{-36}{12} = -3$ because $12(-3) = -36$.

(c) $-45 \div \left(-\dfrac{5}{2}\right) = (-45)\left(-\dfrac{2}{5}\right) = 18$ because $\left(-\dfrac{5}{2}\right)(18) = -45$.

(d) $\dfrac{-12.1}{1.1} = -11$ because $(1.1)(-11) = -12.1$.

82 CHAPTER 2 · ALGEBRA AND THE REAL NUMBERS

If a quotient isn't evident by inspection, we can find it either by long division or by using a calculator. The sign is determined by the preceding rules.

EXAMPLE 8 Find the quotient of $(-10.9568) \div 2.56$.

SOLUTION Because the quotient of two real numbers of unlike sign is negative, the answer will be negative.

$$
\begin{array}{r}
4.28 \\
2.56{\overline{\smash{\big)}\,10.95\,68}} \\
\underline{10\ 24} \\
71\ 6 \\
\underline{51\ 2} \\
20\ 48 \\
\underline{20\ 48}
\end{array}
$$

Move the decimal point two places to the right and then divide.

The quotient is -4.28.

NOTE The preferred forms for a fraction are $-\dfrac{a}{b}$ and $\dfrac{-a}{b}$. They will be used in this text.

There are three equivalent forms for the quotient of a positive and a negative number:

$$-\frac{a}{b} = \frac{-a}{b} = \frac{a}{-b}$$

For example,

$$-\frac{4}{2} = -2 \qquad \frac{-4}{2} = -2 \qquad \frac{4}{-2} = -2$$

EXAMPLE 9 Perform the indicated operations. Follow the rules for the order of operations.

(a) $\dfrac{7-(-8)}{-7-(-7)} = \dfrac{7+8}{-7+7}$ To subtract, add the additive inverse.

$= \dfrac{15}{0}$

Division by zero is undefined. No quotient exists.

(b) $(-6)(-2) + \dfrac{-6}{-2} = 12 + 3$ Multiplication and division first

$= 15$

(c) $-6\{3 + [8 \div (-2) + 5]\} = -6\{3 + [-4 + 5]\}$ $\quad 8 \div (-2) = -4$

$\phantom{-6\{3 + [8 \div (-2) + 5]\}} = -6\{3 + 1\}$ $\quad -4 + 5 = 1$

$\phantom{-6\{3 + [8 \div (-2) + 5]\}} = -6\{4\} = -24$

(d) $-45 \div (-9)(-2) \div (-5)$

This problem involves only multiplication and division, so we proceed from left to right.

$$(-45) \div (-9)(-2) \div (-5) = 5(-2) \div (-5) \qquad (-45) \div (-9) = 5$$
$$= (-10) \div (-5) \qquad 5(-2) = -10$$
$$= 2$$

In the next example, we must take particular care when we use exponents with negative numbers.

EXAMPLE 10 Perform the indicated operations. Follow the rules for the order of operations.

(a) $-2^4 + (-2)^4 + [2 - 5]^4$

$\qquad = -2^4 + (-2)^4 + [-3]^4 \qquad$ Within grouping symbols first

$\qquad = -16 + 16 + 81 \qquad -2^4 = -(2)(2)(2)(2) = -16;$

$\qquad = 0 + 81 = 81 \qquad (-2)^4 = (-2)(-2)(-2)(-2) = 16$

(b) $\dfrac{(-3)^2 - 2(-4)}{(-5)(-2) - (-2)^3} = \dfrac{9 - 2(-4)}{(-5)(-2) - (-8)}$

$\qquad\qquad\qquad\qquad = \dfrac{9 - (-8)}{10 - (-8)}$

$\qquad\qquad\qquad\qquad = \dfrac{9 + 8}{10 + 8} = \dfrac{17}{18}$

EXAMPLE 11 Divide the difference of negative three cubed and negative sixteen by the sum of negative seven and the fourth power of two.

SOLUTION $\dfrac{(-3)^3 - (-16)}{-7 + 2^4} = \dfrac{(-27) - (-16)}{-7 + 16} \qquad$ The division bar is a grouping symbol; carry out operations above and below it first.

$\qquad\qquad\qquad = \dfrac{-27 + 16}{-7 + 16}$

$\qquad\qquad\qquad = \dfrac{-11}{9}$

Do You Remember?

Can you match these?

____ 1. The product of two numbers of like signs
____ 2. The product of two numbers of unlike signs
____ 3. Division by zero
____ 4. -3^2
____ 5. $(-5)^2$
____ 6. The product of an odd number of negative numbers
____ 7. The quotient of two numbers of unlike signs
____ 8. A negative number raised to an even power
____ 9. The quotient of two numbers of like signs
____ 10. Multiplication is sometimes called repeated ____.

a) positive
b) negative
c) addition
d) -9
e) 9
f) 25
g) -25
h) undefined

Answers: 1. a 2. b 3. h 4. d 5. f 6. b 7. b 8. a 9. a 10. c

Exercise Set 2.4

Find each of the following products.

1. $6 \cdot 9$
2. $8 \cdot 7$
3. $(-12) \cdot 2$
4. $(-10) \cdot 3$
5. $-18(0)$
6. $-23(0)$
7. $-3(-6)(-5)$
8. $-2(-7)(-9)$
9. $14(-2)(-4)$
10. $16(-3)(-1)$
11. $28 \cdot 0 \cdot 33$
12. $25 \cdot 0 \cdot 39$
13. $4 \mid -3 \mid \cdot (-2)$
14. $6 \mid -2 \mid \cdot (-3)$
15. $(-1)(-5)(-2)(3)$
16. $(-1)(-7)(-2)(4)$
17. $8 \mid -2 \mid \cdot \mid -3 \mid$
18. $9 \mid -3 \mid \cdot \mid -4 \mid$
19. $(-2)^2(-3)$
20. $(-3)^2(-4)$
21. $(-1)[(3)^2(-2^2)]$
22. $(-1)[(4)^2(-3^2)]$
23. $(-1)(-2)(-3)(0)(-4)$
24. $(-2)(0)(-5)(-3)(-1)$
25. $\mid (-2)^3(3) \mid$
26. $\mid (-3)^3(2) \mid$
27. $\left(\dfrac{1}{2}\right)\left(-\dfrac{1}{3}\right)$
28. $\left(\dfrac{2}{5}\right)\left(-\dfrac{3}{4}\right)$
29. $\left(-\dfrac{3}{7}\right)\left(-1\dfrac{5}{9}\right)$
30. $\left(-1\dfrac{1}{2}\right)\left(-3\dfrac{1}{3}\right)$

Find the following quotients.

31. $8 \div 4$
32. $12 \div 3$
33. $10 \div (-5)$
34. $6 \div (-3)$
35. $(-24) \div 6$
36. $(-28) \div 7$
37. $(-56) \div (-7)$
38. $(-63) \div (-9)$
39. $(-45) \div 5$
40. $(-27) \div 3$
41. $(-81) \div (-3)$
42. $(-64) \div (-2)$
43. $\dfrac{-180}{5}$
44. $\dfrac{-160}{10}$

45. $\dfrac{-132}{-11}$ 46. $\dfrac{-156}{-12}$

47. $\left(\dfrac{2}{3}\right) \div \left(\dfrac{-1}{4}\right)$ 48. $\left(\dfrac{-5}{9}\right) \div \left(2\dfrac{1}{4}\right)$

49. $\left(-3\dfrac{1}{2}\right) \div \left(-2\dfrac{1}{3}\right)$ 50. $\left(-6\dfrac{2}{3}\right) \div \left(-1\dfrac{1}{9}\right)$

Evaluate each of the following expressions.

51. $3 + 2(-1)$ 52. $5 + 2(-3)$
53. $(-3)(-2 + 5)$ 54. $(-2)(-6 + 8)$
55. $(-7)11 + (-6)(12)$ 56. $(-8)(10) + (-9)(8)$
57. $(-8)[2^3 \cdot 3 - (-6)(-4)]$
58. $(-1)[3^2 \cdot 2 - (-5)(-5)]$
59. $9(-15) - (-10)(-12)$
60. $11(-12) - (-13)(-9)$
61. $|3(-8) - (-5)(-4)|$
62. $|2(-9) - (-7)(-3)|$
63. $(-7)(-5) + 8(-4) \div 2$
64. $(-1)(-4) + 9(-3) \div 9$
65. $(-18) \div 2 \div (-3)$ 66. $(-24) \div 3 \div (-8)$
67. $(-15) \div 0 \cdot (4)$ 68. $(-26) \div 0 \cdot (3)$
69. $88 \div (-11) \div 2$ 70. $96 \div (-4) \div 6$
71. $6[(-3)(-11) - (-15)(4)] \cdot 0$
72. $9[(-1)(-18) - (17)(3)] \cdot 0$
73. $4 + 8 - (-3) \cdot 5 - 2(-6)$
74. $3 + 9 - (-2) \cdot 7 - 5(-6)$
75. $17 - (-3|-2| + 4)$
76. $27 - [7 + (-3)(6)] - 5$
77. $-\{-3[-2 + (-5)(4)] + 2\}$
78. $-\{-2[-7 + (-3)(6)] - 5\}$
79. $-\{-[-5(-3 + 4)] - 1\}$
80. $-\{-[-6(7 - 9)] + 2\}$
81. $18 + [5 - (6 - 5) + 2]$
82. $15 + [4 - (11 - 9) + 3]$
83. $8 \cdot 2 - (5^2 - 3^2 \cdot 2) + (-3)$
84. $7 \cdot 3 - (4^2 - 2^2 \cdot 3) + (-7)$
85. $14(4) \div (-8)(-7)$
86. $(-18)(-5) \div (-3)(-2)$
87. $(-17)(-6) \div (-2) + 5$
88. $(-19)(-9) \div (-3) + 4$
89. $(-15) \div (5)(-3) \div 9$
90. $(-22) \div (11)(-3) \div 2$
91. $-27 \cdot 7 \div (-21)$ 92. $(-48) \cdot 5 \div (-15)$
93. $\dfrac{18}{0}$ 94. $\dfrac{63}{0}$
95. $\dfrac{0}{3} + \dfrac{1}{2}\left(-\dfrac{1}{4}\right)$ 96. $\dfrac{0}{11} - 3\dfrac{1}{2} \div \dfrac{1}{3}$

Write an expression for each of the following statements and simplify.

97. The quotient of 18 and 3
98. Five divided into -40
99. The product of 13 and 7
100. The product of -15 and $-1\dfrac{2}{3}$
101. The result of dividing -54 by -3
102. The result of dividing $-\dfrac{5}{3}$ by $\dfrac{7}{8}$
103. The result of multiplying $5\dfrac{1}{8}$ by $-2\dfrac{2}{3}$
104. The result of multiplying $-1\dfrac{2}{9}$ by $4\dfrac{1}{4}$
105. The sum of 8 and 7 divided by the difference of 11 and 2
106. The difference of 23 and 5 divided by the difference of 11 and 2
107. The product of -8 and -6 divided by the product of -12 and 2
108. The sum of -15 and -12 divided by the sum of -12 and 3
109. The difference of -29 and 7 divided by the difference of 26 and 4
110. The sum of -36 and 17 divided by the difference of 23 and 4

For Extra Thought

Given that both a and b are positive real numbers, indicate whether each of the following is positive, negative, zero, or undefined.

111. $a \cdot b$ 112. $-a(-b)$
113. $-a(b)$ 114. $a(-b)$
115. $\dfrac{0}{a}$ 116. $\dfrac{b}{0}$
117. $\dfrac{a + b}{0}$ 118. $\dfrac{0}{-a + (-b)}$
119. $a + (-b)$, if a is greater than b
120. $a - b$, if a is less than b
121. $-a - (-b)$, if a is less than b
122. $|-a| \cdot 0 + b$
123. $-|-a + (-b)| \cdot 0 - 3$
124. $-(b - a)$, if a is less than b

125. $-\{-[(-b) + a]\}$, if a is less than b

126. $|a - b|$, if b is greater than a

WRITE IN YOUR OWN WORDS

127. How do you add two real numbers with like signs?

128. How do you add two real numbers with unlike signs?

129. How do you subtract two real numbers?

130. How do you multiply or divide two real numbers with like signs?

131. How do you multiply or divide two real numbers with unlike signs?

2.5 Properties of Real Numbers

OBJECTIVES

In this section we will learn to
1. recognize the properties of real numbers; and
2. simplify expressions using the properties of real numbers.

Although we have already used many of the properties of real numbers in this chapter, we did not identify them by name or mention them at the time. As we continue studying mathematics, however, it will become necessary to understand exactly what can or can't be done with the real numbers. What may be obvious when working with known quantities may not be so easy to see when variables are involved.

> **The Commutative Properties**
>
> If a and b represent real numbers, then
>
> $a + b = b + a$ Commutative property for addition
> $a \cdot b = b \cdot a$ Commutative property for multiplication

To *commute* means to exchange or to change places, much as a commuter does in going to and from work. If we change the order in which two or more numbers are added or multiplied, the sum or product will not change.

EXAMPLE 1   Verify the commutative properties for the number 2 and 3.

SOLUTION   **(a)** $2 + 3 = 5$ and $3 + 2 = 5$. Thus,

$2 + 3 = 3 + 2$ Commutative property for addition

(b) $2 \cdot 3 = 6$ and $3 \cdot 2 = 6$. Thus,

$2 \cdot 3 = 3 \cdot 2$ Commutative property for multiplication

EXAMPLE 2   Complete each equation so the commutative properties hold.

(a) $-5 + 7 = \underline{} + (-5)$ **(b)** $\dfrac{1}{2} \cdot \left(-\dfrac{3}{5}\right) = \left(-\dfrac{3}{5}\right) \cdot \underline{}$

SOLUTION (a) By the commutative property for addition, $-5 + 7 = 7 + (-5)$.
(b) By the commutative property for multiplication,
$$\frac{1}{2} \cdot \left(-\frac{3}{5}\right) = \left(-\frac{3}{5}\right) \cdot \frac{1}{2}$$

Note The commutative properties still apply when more than two numbers are involved.

Let's try adding three numbers three different ways:

$2 + 3 + 4 = 2 + 4 + 3$ The 3 and 4 have been commuted.
$2 + 3 + 4 = 3 + 2 + 4$ The 2 and 3 have been commuted.
$2 + 3 + 4 = 4 + 3 + 2$ The 2 and 4 have been commuted.

What are three other orders in which $2 + 3 + 4$ can be written?

In the same manner, the commutative law for multiplication allows us to write the product of 2, 3, and 4 in six different orders. Here are three:

$$2 \cdot 3 \cdot 4 = 2 \cdot 4 \cdot 3 = 3 \cdot 2 \cdot 4$$

What are the other three?

Notice that the commutative property applies only to addition and multiplication. The next two examples show that subtraction and division of real numbers are *not* commutative.

$6 - 2 = 4$ and $2 - 6 = -4$ so $6 - 2 \neq 2 - 6$

$6 \div 3 = 2$ and $3 \div 6 = \dfrac{1}{2}$ so $6 \div 3 \neq 3 \div 6$

The Associative Properties

If *a*, *b*, and *c* represent real numbers, then

$a + (b + c) = (a + b) + c$ Associative property for addition
$a(bc) = (ab)c$ Associative property for multiplication

The associative properties state that the way you *associate*, or *group*, real numbers does not affect their sum or product.

EXAMPLE 3   Verify the associative properties for each of the following.

ADDITION
(a) $2 + (3 + 4) \stackrel{?}{=} (2 + 3) + 4$
$2 + 7 \stackrel{?}{=} 5 + 4$
$9 \stackrel{\checkmark}{=} 9$

MULTIPLICATION
(b) $2(3 \cdot 4) \stackrel{?}{=} (2 \cdot 3)4$
$2(12) \stackrel{?}{=} (6)4$
$24 \stackrel{\checkmark}{=} 24$

Just as subtraction and division are not commutative, they are not associative either.

SUBTRACTION	DIVISION
$12 - (4 - 2) \stackrel{?}{=} (12 - 4) - 2$	$12 \div (6 \div 2) \stackrel{?}{=} (12 \div 6) \div 2$
$12 - 2 \stackrel{?}{=} 8 - 2$	$12 \div 3 \stackrel{?}{=} 2 \div 2$
$10 \neq 6$	$4 \neq 1$

Sometimes an expression will involve either of the commutative or associative properties or both.

EXAMPLE 4 Identify the properties used in each of the following statements.

(a) $2 + (4 + 6) = 2 + (6 + 4)$ (b) $(a + b) + c = (a + c) + b$

SOLUTION (a) *No change in the grouping occurs,* so the associative property was not used. *The order of the numbers changes* from 2, 4, 6 to 2, 6, 4, which means that the commutative property for addition was used.

(b) *The order changes* from a, b, c to a, c, b while *the grouping changes* from $(a + b)$ to $(a + c)$. Therefore both the commutative and associative properties for addition were used.

The Identity Properties

For any real number a,

$a + 0 = a$ Identity property for addition
$a \cdot 1 = a$ Identity property for multiplication

The identity properties state that if we add 0 (zero) to any real number, the sum will be *identical* to the real number we started with, or if we multiply any real number by 1 (one), the product will be *identical* to the original real number. For these reasons, zero is called the **additive identity** and one is called the **multiplicative identity.**

EXAMPLE 5 Find the additive or multiplicative identity in parts (a) through (d).

(a) $5 + 0 = 5$ 0 is the additive identity.
(b) $0 + \dfrac{3}{4} = \dfrac{3}{4}$ 0 is the additive identity.
(c) $-y \cdot 1 = -y$ 1 is the multiplicative identity.
(d) $1 \cdot (2.75) = 2.75$ 1 is the multiplicative identity.

The Inverse Properties

For any real number a,

$a + (-a) = 0$ Inverse property for addition

$a \cdot \dfrac{1}{a} = 1, \quad a \neq 0$ Inverse property for multiplication

The inverse properties state that every real number a has exactly one real number $-a$ that can be added to it resulting in a sum of zero; and every nonzero real number a has exactly one real number $\frac{1}{a}$ that can be multiplied by it to produce a product of 1. The number $-a$ is called the **additive inverse** or **the opposite** of a. The number $\frac{1}{a}$ is called the **multiplicative inverse** or the **reciprocal** of a.

EXAMPLE 6   Find the additive or multiplicative inverses, as indicated.

(a) $5 + (-5) = 0$ 5 and -5 are additive inverses.
(b) $-2.6 + 2.6 = 0$ -2.6 and 2.6 are additive inverses.
(c) $5 \cdot \frac{1}{5} = 1$ 5 and $\frac{1}{5}$ are multiplicative inverses.
(d) $\left(-\frac{1}{m}\right)(-m) = 1$ $-\frac{1}{m}$ and $-m$ are multiplicative inverses.

The Distributive Property

If a, b, and c represent real numbers, then

$$a(b + c) = ab + ac$$

The distributive property suggests that there is an alternative order of operations. We have the option of first finding the sum and then the product, as shown on the left side of the equation, or of first finding the products before adding, as shown on the right side.

EXAMPLE 7   Verify the distributive property for each of the following:

(a) $6(5 + 4)$ (b) $3(9 + 5)$ (c) $4 \cdot 9 + 4 \cdot 6$

SOLUTION   (a) $6(5 + 4) \stackrel{?}{=} 6 \cdot 5 + 6 \cdot 4$
Add first. $\rightarrow 6 \cdot 9 \stackrel{?}{=} 30 + 24 \leftarrow$ Multiply first.
$54 \stackrel{\checkmark}{=} 54$

(b) $3(9 + 5) \stackrel{?}{=} 3 \cdot 9 + 3 \cdot 5$
$3(14) \stackrel{?}{=} 27 + 15$
$42 \stackrel{\checkmark}{=} 42$

(c) $4 \cdot 9 + 4 \cdot 6 \stackrel{?}{=} 4(9 + 6)$
$36 + 24 \stackrel{?}{=} 4(15)$
$60 \stackrel{\checkmark}{=} 60$

The distributive property holds for subtraction as well as for addition, since subtraction is just addition of the additive inverses of the numbers.

EXAMPLE 8 Verify the distributive property for each of the following:

(a) $5(11 - 6)$ (b) $4(9 - 6)$

SOLUTION (a)
$$5(11 - 6) \stackrel{?}{=} 5 \cdot 11 - 5 \cdot 6$$
Subtract first. → $5(5) \stackrel{?}{=} 55 - 30$ ← Multiply first.
$$25 \stackrel{\checkmark}{=} 25$$

(b) $4(9 - 6) \stackrel{?}{=} 4 \cdot 9 - 4 \cdot 6$
$$4(3) \stackrel{?}{=} 36 - 24$$
$$12 \stackrel{\checkmark}{=} 12$$

When more than two numbers are within the symbols of grouping, for example,
$$a(b + c + d + e)$$
the distributive property still applies.

EXAMPLE 9 Verify that the distributive property applies to $9(6 + 3 + 4)$.

SOLUTION
$$9(6 + 3 + 4) \stackrel{?}{=} 9 \cdot 6 + 9 \cdot 3 + 9 \cdot 4$$
$$9(13) \stackrel{?}{=} 54 + 27 + 36$$
$$117 \stackrel{\checkmark}{=} 117$$

EXAMPLE 10 Use the distributive property to write $2(x - 4)$ another way.

SOLUTION
$$2(x - 4) = 2 \cdot x - 2 \cdot 4$$
$$= 2x - 8$$

*C*AUTION

When removing a symbol of grouping preceded by a negative sign, **change the signs before all terms within the grouping symbol.**

An important use of the distributive property is in the removal of grouping symbols preceded by a negative sign. For example, $-(x + y)$ can be thought of as $-1(x + y)$. Then, using the distributive property,

$$-1(x + y) = -1 \cdot x + (-1)y$$
$$= -x + (-y)$$
$$= -x - y \quad \text{To subtract, add the additive inverse.}$$

EXAMPLE 11 Use the distributive property to rewrite and simplify each of the following:

(a) $(x + 12)5$ (b) $3 \cdot x + 3 \cdot y$ (c) $-(2 + a - c)$

SOLUTION (a) $(x + 12)5 = 5(x + 12)$ Commutative property for multiplication
$$= 5 \cdot x + 5 \cdot 12 \quad \text{Distributive property}$$
$$= 5x + 60$$

(b) $3 \cdot x + 3 \cdot y = 3(x + y)$

(c) $-(2 + a - c) = -1(2 + a - c)$
$= -1(2) + (-1)(a) - (-1)(c)$
$= -2 + (-a) - (-c)$
$= -2 - a + c$

EXAMPLE 12 Identify each expression as an example of the use of the commutative, associative, identity, or inverse property of addition or multiplication.

(a) $y + (-y) + z = 0 + z$
(b) $d + \frac{1}{z} \cdot z = d + 1$
(c) $x^3 + x^2y = yx^2 + x^3$

SOLUTION (a) $y + (-y) = 0$ by the inverse property for addition.
(b) $\frac{1}{z} \cdot z = 1$ by inverse property for multiplication.
(c) $x^2y = yx^2$ by the commutative property for multiplication. Also, x^3 appears last in $yx^2 + x^3$ instead of first, as in $x^3 + x^2y$, so we also used the commutative property for addition.

DEFINITION **Properties of Real Numbers**	If a, b, and c are any real numbers, then the following properties are true.

	ADDITION	MULTIPLICATION
Commutative property	$a + b = b + a$	$a \cdot b = b \cdot a$
Associative property	$a + (b + c) = (a + b) + c$	$a(bc) = (ab)c$
Identity property	$a + 0 = a$	$a \cdot 1 = a$
Inverse property	$a + (-a) = 0$	$a \cdot \frac{1}{a} = 1, \quad a \neq 0$

	ADDITION AND MULTIPLICATION
Distributive property	$a(b + c) = ab + ac$

Do You Remember?

Can you match these?

What property justifies each of the following? More than one answer may apply.

_____ 1. $2 + 3 + 5 = 5 + 3 + 2$
_____ 2. $9 + (-9) = 0$
_____ 3. $x + (y + z) = (x + y) + z$
_____ 4. $t + 0 = 0 + t$
_____ 5. $-x + (x) = x + (-x)$
_____ 6. $3(4 + 5) = (5 + 4)3$
_____ 7. $a(b \cdot c) = (b \cdot c)a$
_____ 8. $3 + (7 + 2) = 2 + (3 + 7)$
_____ 9. $8 \cdot \frac{1}{8} = 1$
_____ 10. $3(5 + 2) = 15 + 6$
_____ 11. $5 \cdot (4 \cdot 7) = (5 \cdot 4) \cdot 7$

a) Commutative property for addition
b) Commutative property for multiplication
c) Associative property for addition
d) Associative property for multiplication
e) Identity property for addition
f) Identity property for multiplication
g) Inverse property for addition
h) Inverse property for multiplication
i) Distributive property

Answers: 1. a; 2. g; 3. c; 4. e; 5. a; 6. b; 7. b; 8. a, c; 9. h; 10. i; 11. d

Exercise Set 2.5

Identify each expression as an example of the commutative, associative, identity, or inverse property of addition or multiplication.

1. $8 + 4 = 4 + 8$
2. $-15 + 0 = -15$
3. $a \cdot b = b \cdot a$
4. $(6 + 3) + 2 = 6 + (3 + 2)$
5. $7 + (-7) = 0$
6. $9(1) = 9$
7. $6 + [(-8) + (-3)] = [6 + (-8)] + (-3)$
8. $b + (-b) + 3 = 0 + 3$
9. $-6.4(1) = -6.4$
10. $3\left(\frac{1}{3}\right) = 1$
11. $7[8 \cdot (-6)] = (7 \cdot 8)(-6)$
12. $a(3 \cdot b) = (a \cdot 3)b$
13. $m + (-m) + x = 0 + x$
14. $x\left(\frac{1}{x}\right) + b = 1 + b, \quad x \neq 0$
15. $(m + n + p) + (a + b + c)$
    $= m + n + (p + a) + b + c$
16. $r + s \cdot t + p = r + t \cdot s + p$
17. $0 + 6 = 0$
18. $\frac{2}{5}\left(\frac{5}{2}\right) = 1$
19. $13x \cdot 1 = 13x$
20. $3(4b) = (3 \cdot 4)b$
21. $-16 + 16 = 0$
22. $-3x + 3x = 0$
23. $(3x)y = 3(xy)$
24. $45 + (-11) = (-11) + 45$
25. $6 + (5 + 9) = (6 + 5) + 9$
26. $5(-40) = (-40)5$
27. $5x + 0 = 5x$
28. $y\left(\frac{1}{y}\right) + x = 1 + x$
29. $13x \cdot 3y = 13 \cdot 3xy$
30. $x^2 + xy = x^2 + yx$
31. $17(xy) = (17x)y$
32. $-5x \cdot 4y = -5 \cdot 4xy$

Use the distributive property to rewrite each of the following.

33. $3(4 + 7)$
34. $2(8 + 3)$
35. $2(x + 5)$
36. $9(m + 3)$
37. $8(p - 5)$
38. $4(y - 7)$

39. $(r + 4)12$
40. $(m - 3)11$
41. $5 \cdot 7 + 5 \cdot 2$
42. $3 \cdot 9 + 3 \cdot 4$
43. $7x - 7 \cdot 3$
44. $14x - 14 \cdot 2$
45. $-3(x + 9)$
46. $-4(y + 5)$
47. $-6x - 6 \cdot 3$
48. $-9x - 9 \cdot 4$
49. $13x + 26$
50. $-11p + 33$
51. $-(m - 18)$
52. $-(q - 27)$
53. $-2p + 6$
54. $-6r + 18$
55. $-(x + 3y - 4)$
56. $-(2m - 3n + 5)$

Complete each equation using the specified property of real numbers.

57. Commutative property of multiplication:
    $13(-5) =$ _____
58. Commutative property of addition:
    $y + 8 =$ _____
59. Identity property of multiplication:
    $1(3x + 2) =$ _____
60. Identity property of addition:
    $(2y - 5) + 0 =$ _____
61. Associative property of addition:
    $(-7 + x) + 3 =$ _____
62. Associative property of multiplication:
    $x(yz) =$ _____
63. Inverse property of multiplication:
    $\left(\dfrac{x + 5}{3}\right)\left(\dfrac{3}{x + 5}\right) =$ _____ *reciprocle*
64. Inverse property of addition:
    $(3m) + (-3m) =$ _____
65. Distributive property: $3(a - 4) =$ _____
66. Distributive property: $5x - 20 =$ _____

Use the properties of real numbers to simplify each expression.

67. $3(2x)$
68. $(m + 2) + 7$
69. $\left(\dfrac{1}{2}\right)(2x)$
70. $-7 + (y + 7)$
71. $7(3x - 1)$
72. $-2(y + 3)$
73. $(3x + 8) - 8$
74. $-\dfrac{1}{3}x(9)$
75. $-15y\left(\dfrac{2}{3}\right)$
76. $-(4a + 2b - c)$
77. $-(-x + 3y - 7)$
78. $3\left(\dfrac{1}{3}x - 9\right)$
79. $6x - 24$
80. $-3y + 27$
81. $-8\left(\dfrac{3}{8}x - \dfrac{7}{8}\right)$
82. $-6\left(\dfrac{2}{3}y - \dfrac{1}{2}\right)$
83. $(3p - 1) - 7$
84. $\dfrac{4}{3}\left(\dfrac{3}{4}x - 3\right)$
85. $8x \cdot \dfrac{1}{4}y$
86. $(5y - 4)(-2)$
87. $-4\left(\dfrac{x}{2} - \dfrac{y}{2} + \dfrac{1}{4}\right)$
88. $-9\left(\dfrac{-x}{3} - \dfrac{y}{9} + \dfrac{2}{3}\right)$

Answer each question. Justify any NO answer with an example.

89. What is the multiplicative inverse of $\dfrac{3}{4}$?
90. What is the additive inverse of -8?
91. What is the additive identity?
92. What is the multiplicative identity?
93. Does the commutative property apply to subtraction?
94. Does the commutative property apply to division?
95. Is there an additive inverse of zero?
96. Is there a multiplicative inverse of zero?
97. Is $0 \cdot 1 = 1$?
98. Is $x + (-x) = 0$?
99. Does the associative property apply to subtraction?
100. Does the associative property apply to division?
101. Can the distributive property be used on $8(x - 5)$?
102. Can the distributive property be used on $12x + 6$?
103. What is the additive inverse of y?
104. What is the additive inverse of $|y|$?
105. What is the multiplicative inverse of x: (Assume $x \neq 0$.)
106. What is the multiplicative inverse of $-|x|$: (Assume $x \neq 0$.)

REVIEW EXERCISES

107. Does $|x| + |y| = |x + y|$ for all negative real numbers?
108. Does $|x| - |y| = |x - y|$ for all positive real numbers?
109. Does $|x| + |y| = |x + y|$ for all real numbers?
110. Does $|x| - |y| = |x - y|$ for all real numbers?

Simplify each of the expressions in Exercises 111–116.

111. $-9[-12 + 2(3 \cdot 4 - 15)]$

112. $12[-2(2 \cdot 5 - 13)^2] - 15$

113. $5(-8)^2 + (-7)(-18 - 12)$

114. $4^2(-7 + 16)^2 - 12(5 \cdot 2 - 11)$

115. $3\frac{3}{5} \div 1\frac{1}{5}$

116. $2\frac{1}{7} \cdot 3\frac{1}{5}$

117. If Lea picks $1\frac{1}{2}$ bushels of beans in one hour, how many bushels can she pick in $2\frac{1}{2}$ hr?

118. Lewis uses $\frac{3}{16}$ cup of sugar in a recipe for 12 cookies. How many recipes can he make with $\frac{3}{4}$ cup of sugar?

119. Ida can buy a package of nine stickers for $1.75. How many stickers can she buy for $8.75?

120. On a map, one inch represents 8 miles. How many miles are represented by $3\frac{3}{4}$ in.?

121. Maria uses 8 oz of milk to make a recipe for 12 people. How many ounces of milk are needed to make the same recipe for 18 people?

122. The directions given on a box of plant fertilizer calls for $\frac{1}{4}$ oz of fertilizer for two plants. How much fertilizer is needed for 11 plants?

2.6 More Applications

OBJECTIVES

In this section we will learn to use signed numbers to solve applied problems. One of the most important reasons for studying algebra is to be able to use it as a tool in problem solving. The problems that arise are often stated in words. To help you to be successful we suggest that you follow the steps outlined below.

To Solve an Applied Problem

Step 1 Read the problem from beginning to end.

Step 2 Read the problem through a second time and attempt to answer the question, "What am I asked to find?"

Step 3 Reread the problem another time to gather the information necessary to answer the question in Step 2. You then need to decide what operation or operations are required.

Step 4 Write your answer in a complete sentence.

As you study the following examples, try to follow Steps 1 through 4. The question "What am I asked to find?" is italicized in each case.

EXAMPLE 1 If the difference of 19 and 7 is decreased by the difference between 11 and 5, *what is the result?*

SOLUTION *Difference* implies subtraction:

$$(19 - 7) - (11 - 5) = 12 - 6 = 6$$

The answer is 6.

2.6 · MORE APPLICATIONS   95

EXAMPLE 2   Some of the tallest buildings in North America are the World Trade Center at 1350 feet; the Empire State Building at 1250 feet; the Sears Tower at 1450 feet; and the John Hancock Center at 1127 feet. *How much less than a mile is their combined heights?* (1 mile = 5280 feet)

SOLUTION   The word *combined* indicates addition. The sum of their heights is

$$1350 + 1250 + 1450 + 1127 = 5177 \text{ feet}$$

To find the difference between their combined heights and one mile, we subtract 5177 from 5280:

$$5280 - 5177 = 103 \text{ feet}$$

The combined heights of the buildings is 103 feet less than a mile.

EXAMPLE 3   The average cost of home construction has risen by $25 per square foot over the last six years. *How much more does it cost to construct a 2150-square-foot house now than it did an identical house six years ago?*

SOLUTION   To find the increase in cost, we *multiply* the number of square feet by the increase per square foot:

$$2150 \cdot 25 = \$53{,}750$$

It costs $53,750 more to build the house now than it did six years ago.

EXAMPLE 4   A nurse finds that a bottle holds 1000 cc (cubic centimeters) of a drug. The average dose for that particular drug is 25 cc. *How many doses are in the bottle?*

SOLUTION   The answer to this problem requires *dividing* the total amount of the drug by the amount per dose:

$$1000 \div 25 = 40$$

There are 40 doses in the bottle.

EXAMPLE 5   In one day, doctors in a large metropolitan hospital prescribed doses of phenobarbital in the amounts of 40 mg (milligrams), 30 mg, 40 mg, 20 mg, 25 mg, 40 mg, 30 mg, 40 mg, and 50 mg to nine different patients. *What was the amount of the average dose prescribed?*

SOLUTION   The **average** of a set of quantities is found by dividing the sum of the quantities by the number of quantities. In this case there were nine patients, so

$$\frac{40 + 30 + 40 + 20 + 25 + 40 + 30 + 40 + 50}{9} = \frac{315}{9} = 35$$

The average dose per patient was 35 mg.

EXAMPLE 6   The deepest part of the ocean is the Mariana Trench, at 36,198 feet below sea level. The highest point on earth is Mount Everest, at 29,028 feet above sea level. *What is the difference in elevation between these two points?*

SOLUTION   The word *difference* implies subtraction. Since the deepest point is *below* sea level, we represent it as a negative number. (Sea level is considered to be at zero elevation.)

$$29{,}028 - (-36{,}198) = 29{,}028 + 36{,}198 \quad \text{To subtract, add the inverse.}$$
$$= 65{,}226$$

The difference is 65,226 ft, which is more than 12 miles.

EXAMPLE 7   A student has test scores of 86%, 94%, 74%, 85%, and 94% on five algebra tests. *What is her test average to the nearest percent?*

SOLUTION   To determine her test average, we find the sum of her scores and divide it by the number of scores:

$$\frac{86 + 94 + 74 + 85 + 94}{5} = \frac{433}{5} = 86.6\%$$

Rounded to the nearest percent, her average is 87%.

EXAMPLE 8   A right circular cylinder has a 12-inch-diameter base and is 14 inches high (see Figure 2.13). *Find its volume.* Use $\frac{22}{7}$ to approximate π.

SOLUTION   The formula for the volume of a right circular cylinder is $V = \pi r^2 h$. The radius is $\frac{1}{2}$ of the diameter.

$$V = \pi r^2 h$$
$$= \frac{22}{7}(6)^2 14 \quad r = \frac{1}{2}(12) = 6$$
$$= 22(36)(2) \quad \text{Reduce before multiplying.}$$
$$= 1584 \text{ in}^3$$

The volume is 1584 cubic inches.

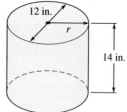

FIGURE 2.13

EXAMPLE 9   By purchasing new machinery, a mining company increases its output from 2635 to 3200 tons daily. *How many additional tons are produced in a 30-day period? What percent increase does this represent over their previous 1-day capacity?*

SOLUTION   In one day the increase in output is

$$3200 - 2635 = 565 \text{ tons}$$

The total increase over a 30-day period is obtained by multiplying the daily increase by 30:

$$30(565) = 16,950 \text{ tons}$$

We calculate the percent increase for one day by using the formula:

Amount = Percent increase × Base

565 = percent increase × 2635

Thus the percent increase is

$$\frac{565}{2635} = 0.2144$$

$$= 21.44\%$$

Exercise Set 2.6

Solve each of the following applied problems.

Money

1. Receipts in a small boutique over a five-day period were $485, $196, $397, $918, and $612. What were the total receipts for the five-day period?

2. A student received a notice that his checking account was overdrawn. He had $450 in his account when he registered in the fall and wrote checks for $70, $45, $23.60, $100, $151, and $80.10. How much was he overdrawn?

3. The student council lost $786.54 on the last band concert. If the council's bank balance was $618.37 before the concert, what is it now?

4. The cost per thousand feet of lumber has gone down by $77.80. What was the cost before, if it now costs $846.30?

5. Ramona buys and sells lots. She purchased a lot for $16,925, but during the year the price of lots went down, and she sold the lot for $14,870. Express her loss as an integer. Find the loss as a percent.

6. In March, 50 shares of Goble Stock sold for $2845. In May, 50 shares of the same stock sold for $2690. Express the loss as an integer. Find the loss as a percent.

7. On September 1, Mauricio's bank account had a balance of $308. On September 15, the account was overdrawn by $68. What was the total value of the checks Mauricio wrote?

8. On June 10, Darlene's bank account had a balance of $285.17. Ten days later her account was overdrawn by $47.88. What was the total value of the checks that Darlene wrote?

9. On a one-day sale, a store sold 15 radios at a loss of $7.25 each. The next six days they sold 13 of the same radios at a profit of $4.85 each. Express their total profit or loss on the radios.

10. During a two-hour sale, a store sold 16 can openers at a loss of $2.25 each. During the rest of the day, they sold 7 of the same can openers at a profit of $5.05 each. Express their profit or loss as an integer.

11. Felix buys a TV for $85 down and $48 per month for 15 months. What is the total price he pays for the TV?

12. Monica buys a chair for $68 down and $13 per month for 18 months. What is the total price she pays for the chair?

Temperature

13. A thermometer indicates −31° for a high and −122° for a low over a period of time. What was the change in the temperature?

14. The surface temperature of a rocket ship was recorded for one day. The highest reading was −75° and the lowest was −149°. What was the change in the temperature?

15. The boiling point of water is 100° Celsius (°C) and that of helium is −269°C. What is the difference in their boiling points?

16. At 4:00 A.M. the temperature is −6° Fahrenheit (°F). By noon the temperature has risen by 36°F. What is the temperature at noon?

Elevation

17. A hawk sitting on a rock on a hillside in Death Valley at 187 ft below sea level sees a mouse scamper into its burrow, which is 233 ft below sea level. What is the difference in their elevations? Express the difference as an integer.

18. A submarine descends 324 m below the surface of the ocean and then descends another 117 m. An airplane is 613 m above the surface of the ocean. How many meters above the submarine is the airplane?

19. What is the difference in elevation between the highest point in the United States (Mt. McKinley at 20,320 ft above sea level) and Death Valley (282 ft below sea level)?

20. What is the difference in elevation between the highest point in the world (Mt. Everest at 29,028 ft above sea level) and a point on the shore of the Dead Sea (1,291 ft below sea level)?

Numbers

21. If the difference of 13 and 5 is decreased by 14, what is the result?

22. If the difference of 7 and -13 is decreased by 18, what is the result?

23. If the quotient of -28 and 7 is decreased by 23, what is the result?

24. If the quotient of 35 and -5 is decreased by 24, what is the result?

25. If the product of negative 13 and 7 is decreased by 5, what is the result?

26. If the product of 23 and negative 3 is decreased by 7, what is the result?

27. If the quotient of -15 and -3 is added to the quotient of -45 and 9, what is the result?

28. If the product of $-\frac{1}{2}$ and $\frac{1}{4}$ is subtracted from the quotient of -3 and 8, what is the result?

Medicine

29. Doses of 150 mg, 70 mg, 125 mg, 225 mg, and 115 mg of a drug were prescribed in a clinic on a Tuesday. How many milligrams were prescribed?

30. Clark's Rule for determining the proper dosage for a child is "the weight of the child times the average adult dose divided by 150." Use Clark's Rule to determine the dose of paregoric for a 75 lb child if the adult dosage is 2 drams.

Averages

31. A student has scores of 75%, 86%, 97%, 68%, 75%, and 85% on six exams. What is his average score? Find the difference in his average and the class average of 80%. Use an integer to indicate this difference.

32. Emma has scores of 84%, 93%, 69%, and 78%. What is her average? Find the difference in her average and the class average of 89%. Express this difference as an integer.

33. Find the average of -36, -15, 24, and -5.

34. Find the average of -11, -15, 43, -5, and -19.

35. Find the average of -86, -78, -14, and -16.

36. Find the average of 0, 55, 37, 21, and 12.

37. Over a period of 8 weeks, Marty lost a total of $2832 in his stock market account. What was his average loss per week? Express this as an integer.

38. A football team lost a total of 52.5 yd in 15 plays. What was the average loss per play? Express this loss as an integer.

Diet/Weight

Use the following calorie information to answer Exercises 39–42.

Calories Contained in Food

1 avocado: 370
1 slice cake: 225
1 glass milk: 159
1 slice pie: 175

Calories Consumed by Exercise

1 set tennis: 200
1 mile jogging: 150
10 laps swimming: 185

39. Sherry eats a slice of pie and drinks a glass of milk after playing a set of tennis. What is the result of these activities when measured in calories?

40. Matt eats an avocado and then jogs for a mile. Still feeling hungry, he eats a piece of cake and drinks a glass of milk. What is the net result in calories of these activities?

41. Paul plans to exercise in order to burn the same number of calories as he eats. On Thursday his lunch consists of half an avocado, a glass of milk, and a slice of cake. He then plays a set of tennis and decides to swim enough to make up for the rest of the calories. If he always swims sets

of 10 laps at a time, how many sets of 10 laps must he swim to be sure he has burned off at least as many calories as he has consumed?

42. Lisa jogs five miles. Has she burned off enough calories so that she can have a slice of cake and a glass of milk and still have burned off more than she will consume?

43. For six consecutive weeks Verle records a weight loss of 2 lb/week. Express his total loss as an integer.

44. For 15 days Laura records a loss of $1\frac{1}{2}$ lb/day. Find her total loss.

GEOMETRY/AREA/VOLUME

45. Find the area of a circle, $A = \pi r^2$, if radius r is $\frac{5}{11}$ in. Use $\frac{22}{7}$ for π.

46. Find the circumference of a circle, $C = 2\pi r$, if $r = \frac{5}{8}$ in. Use $\frac{22}{7}$ for π.

47. Find the surface area S and volume V of a cube with sides of length $s = 4$ in.

48. A triangle has a perimeter of 27 cm. If the formula for the perimeter of a triangle is $P = a + b + c$, where a, b, and c are the sides, and if two of the sides have lengths of $a = 5$ cm and $b = 13$ cm, what is the length of c, the third side?

49. The formula for the perimeter of a *quadrilateral* (a four-sided figure) is $P = a + b + c + d$, where a, b, c, and d are the sides. If a four-sided figure has a perimeter of 344 ft, and if three of the sides measure $a = 87$ ft, $b = 59$ ft, and $c = 73$ ft, what is the length of d, the fourth side?

50. Find the volume of a right circular cylinder, $V = \pi r^2 h$, if $r = \frac{3}{4}$ yd, and $h = 3\frac{1}{5}$ yd. Use $\frac{22}{7}$ for π.

51. Find the volume of a right circular cone, $V = \frac{1}{3}\pi r^2 h$, if $r = \frac{2}{7}$ ft, and $h = 2\frac{1}{3}$ ft. Use $\frac{22}{7}$ for π.

FALLING BODIES

52. The formula for finding the speed of a falling body is approximated by $S = v + gt$, where S is the speed in feet per second (ft/sec). Find S if $t = 4$ sec, $v = 7$, and $g = 32$.

53. Find S if $t = 7$ sec, $v = 2$, and $g = 32$. (Use the formula $S = v + gt$.)

54. Use the formula $S = v + gt$ to find v if $S = 155$, $g = 32$, and $t = 5$ sec.

55. Use the formula $S = v + gt$ to find t if $S = 160$ ft/sec, $v = 0$, and $g = 32$.

MISCELLANEOUS

56. From 1985 to 1990, hundreds of tornadoes occurred in the United States, causing many deaths. The figures are as follows:

YEAR	NUMBER	DEATHS
1985	915	299
1986	570	105
1987	912	116
1988	660	131
1989	604	66
1990	649	73

What was the total number of tornadoes and the total number of lives lost during this period? What was the average number of tornadoes and the average number of lives lost per year? Round all answers to the nearest whole number.

57. Alaska, the largest state in the union, has an area of 586,412 square miles. Rhode Island, the smallest, has an area of 1212 square miles. What is the difference in their areas? How many times as big as Rhode Island is Alaska?

58. In one year, before the energy shortage became critical, Americans consumed 2,165,395,000 barrels of gasoline. The following year consumption increased to 2,242,949,000 barrels. How much did consumption increase? What was the percent of increase?

59. Shasta Dam, near Redding, California, is used for flood control and water storage. If the normal release of 3755 ft^3/sec of water is increased to 6291 ft^3/sec to prepare for heavy spring runoff, how much additional water is released per second? per minute? per hour? What is the percent increase per second?

60. Good-quality barley weighs about 47 lb/bushel. How many bushels of barley does a truck hold if it weighs 14,000 lb empty and 38,600 lb loaded with barley?

61. In a recent pro-am golf tournament, the scores ranged from 10 strokes under par to 15 strokes under par. What is the difference between the high and low scores?

62. A coal mine in Colorado has three shafts radiating from a central elevator. The first shaft is 192 ft below the surface, the second shaft is 80 ft below the first, and the third is 90 ft below the second. Use integers to represent the position of each shaft with respect to the surface.

63. An electric current can be defined as voltage divided by resistance. Find the current in amperes (amps) given a total voltage of 256 volts and a total resistance of 16 ohms.

64. In producing a machine part, the maximum tolerance is given as ±2 millimeters (mm). This means that the part can vary no more than 2 mm more or less than the design size. If the design size is 45 mm, what are the maximum and minimum permissible dimensions?

65. The compression in the engine of Ronaldo's car is 160 lb/in². If it drops to 87 lb/in², how much compression has it lost?

66. Don went north on Interstate 5 a distance of 387 mi and then returned south on Interstate 5 a distance of 439 mi. If north is considered positive and south negative, how far is he from his starting point, and in which direction?

Write in Your Own Words

67. Is it possible for two different integers to have the same absolute value? Why?

68. When is the difference of a positive integer and a negative integer positive, negative, or zero? Explain.

69. When is the product of n negative integers positive? When is it negative? Is it ever zero? Explain.

70. The additive inverses of nonzero integers are sometimes positive and sometimes negative. Explain.

71. Explain how a subtraction problem involving real numbers can be checked.

72. Explain how a division problem involving real numbers can be checked.

73. Explain how to subtract one real number from another real number.

74. Explain how to divide a negative real number by a positive real number.

75. Why is division by zero said to be undefined?

Summary

2.1 Sets and Real Numbers

A **set** is a collection of objects. The objects within a set are called its **elements** or **members.** Sets are indicated by placing the elements within braces, { }. To indicate that an object is a member of a set, the symbol $\in$, which means "is an element of" is used. There are two ways to indicate the elements of a set. In the **listing method** the elements are written out; in the **rule method,** also called **set-builder notation,** the elements are described verbally:

$$\{x \mid x \in \text{whole numbers}\}$$

which is read, "the set of all x such that x is an element of the set of whole numbers." A **variable** is a letter that can represent any element within a set, while a **constant** cannot change its value.

There are two types of sets, **finite** and **infinite.** A set is finite if there is a number that describes exactly how many elements it contains. If a set is not finite, it is infinite. A set that has no elements is called **the empty set.** The symbol for the empty set is $\emptyset$.

Set A is a **subset** of a set B if every element of A is also an element of B. The symbol $\subseteq$ means "is a subset of."

The main subsets of the set of real numbers are the **natural numbers,** $\{1, 2, 3, \ldots\}$; the **whole numbers,** $\{0, 1, 2, 3, \ldots\}$; the **integers,** $\{\ldots, -3, -2, -1, 0, 1, 2, 3, \ldots\}$; the **rational numbers,** each of which can be expressed as the quotient of two integers; and the **irrational numbers,** which cannot be expressed as the quotient of two integers. When rational numbers are expressed as decimals, the decimals will either repeat or terminate, whereas irrational numbers neither repeat nor terminate.

natural numbers $\subseteq$ whole numbers $\subseteq$ integers $\subseteq$ rational numbers $\subseteq$ real numbers

2.2
Symbols, Variables, and Absolute Value

Two numbers that are on opposite sides of the origin and the same distance away are said to be **additive inverses** or **opposites** of each other. The additive inverse of a number is formed by placing a negative sign in front of it.

The **double negative** property asserts that $-(-x) = x$ for any real number x.

The **absolute value** of a number x, symbolized as $|x|$, is its distance from zero on a number line. The formal definition of absolute value is

$$|x| = \begin{cases} x & \text{if } x \geq 0 \\ -x & \text{if } x < 0 \end{cases}$$

2.3
Addition and Subtraction of Real Numbers

To find the sum of two numbers with like signs (both positive or both negative), find the sum of their absolute values and give the sum the same sign. To find the sum of two numbers with opposite signs (one positive and one negative), find the difference of their absolute values and give the answer the sign of the one with the greater absolute value.

Subtraction is defined as adding the opposite or additive inverse of a number. To subtract a real number b from a real number a, add the additive inverse of b to a:

$$a - b = a + (-b)$$

Calculators can be used to carry out operations that involve signed numbers. The scientific calculator and the graphics calculator differ in the way that the numbers are entered, as explained in the user manuals.

2.4
Multiplication and Division of Real Numbers

The **product** or **quotient** of two numbers with **like signs** (both positive or both negative) is **positive**. The **product** or **quotient** of two numbers with **unlike signs** (one positive and one negative) is **negative**. The product of an odd number of negative numbers is negative and the product of an even number of negative numbers is positive. The quotient of a positive and a negative number can be written in three ways:

$$-\frac{a}{b} = \frac{-a}{b} = \frac{a}{-b}$$

Either $-\dfrac{a}{b}$ or $\dfrac{-a}{b}$ are the preferred forms for this text.

2.5
Properties of Real Numbers

The real numbers exhibit several properties that form the basis for arithmetic and algebra. In the following summary, note in particular that the *commutative property* refers to a change in the order in which the numbers are written, while the *associative property* refers to the way that the numbers are grouped.

Properties of Real Numbers
If a, b, and c are any real numbers, then the following properties are true.

	ADDITION	MULTIPLICATION
Commutative property	$a + b = b + a$	$a \cdot b = b \cdot a$
Associative property	$a + (b + c) = (a + b) + c$	$a(bc) = (ab)c$
Identity property	$a + 0 = a$	$a \cdot 1 = a$
Inverse property	$a + (-a) = 0$	$a \cdot \dfrac{1}{a} = 1, \quad a \neq 0$

	ADDITION AND MULTIPLICATION
Distributive property	$a(b + c) = ab + ac$

2.6 More Applications

Many applied problems require the use of signed numbers. Success in solving applied problems is aided by having a definite plan of attack. The following approach is suggested.

To Solve an Applied Problem

Step 1 Read the problem from beginning to end.
Step 2 Read the problem through a second time and attempt to answer the question, "What am I asked to find?"
Step 3 Reread the problem another time to gather the information necessary to answer the question in Step 2. You then need to decide what operation or operations are required.
Step 4 Write your answer in a complete sentence.

Cooperative Exercise

John Venn (1834–1923) introduced the idea of diagrams to illustrate set operations. We now call these **Venn diagrams** and use them in many fields of study that involve sets. Venn diagrams are usually represented by circles or ovals inside of a rectangle. Here is an example of the use of Venn diagrams in the study of blood types.

Human blood can contain either no antigens, the A antigen, the B antigen, the Rh antigen, or combinations of the three antigens. Blood is called type A-positive if the individual has the A and Rh antigens but not the B antigen. An individual with both the A and B antigens but not the Rh antigen has blood type AB-negative. An individual with the Rh antigen only has blood type O-positive. An individual with the B antigen only has blood type B-negative. Other blood types are defined in a similar manner. Use the accompanying Venn diagram and data to answer the following questions.

In a certain laboratory, the following data on individuals and their blood types were recorded.

240 had only the Rh antigen.
96 had none of the antigens.
128 had the A and Rh antigens.
120 had all three antigens.
176 had the B and Rh antigens.
216 had only the B antigen.
136 had the A and B antigens.
200 had only the A antigen.

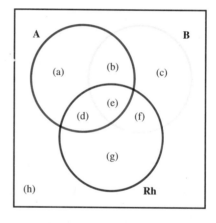

GROUP I

Identify the blood types of the individuals in each of regions (a) through (g) in the Venn diagram. How many regions are represented?

GROUP II

Determine the number of individuals who:

1. are represented here;
2. have no antigens;
3. have exactly one antigen;
4. have exactly two antigens;
5. have all three antigens.

GROUP III
Determine the number of individuals who are:
1. A-positive;
2. B-positive;
3. O-positive;
4. AB-positive;
5. A-negative;
6. B-negative;
7. AB-negative.

Review Exercises

List the elements in each of the following sets.
1. {natural numbers less than 3}
2. {whole numbers less than 4}
3. {integers between 5 and 8}
4. {counting numbers between 6 and 9}
5. {even natural numbers less than 10}
6. {even whole numbers less than 7}
7. {integers between −3 and 2}
8. {natural numbers that are multiples of 4 and less than or equal to 12}
9. {integers between −6 and −2}
10. {integers between 0 and 9 that are multiples of 3}
11. $\{x \mid x < 5, x \in \text{whole numbers}\}$
12. $\{y \mid -3 \leq y < 2, y \in \text{integers}\}$
13. {negative integers larger than negative 4}
14. {positive integers that are multiples of 5}
15. {negative integers that are multiples of 4}
16. {multiples of 6 between 5 and 29}
17. {negative even numbers}
18. {rational numbers that are also irrational numbers}
19. {whole numbers that are also real numbers}
20. {counting or natural numbers}

Write Exercises 21–24 using set-builder notation.
21. {natural numbers}
22. {rational numbers}
23. {2, 4, 6, ...}
24. {3, 6, 9, 12, ...}

Identify Exercises 25–29 as true or false.
25. $3 \in \{\text{whole numbers}\}$
26. $0.25 \in \{\text{rational numbers}\}$
27. $\sqrt{3} \in \{\text{rational numbers}\}$
28. $\{1, 2, 3\} \subseteq \{\text{natual numbers}\}$
29. $\{\text{irrational numbers}\} \subseteq \{\text{real numbers}\}$

30. Draw a real-number line and graph each of the following: 4; −2; 0.5; −0.75; $2\frac{1}{2}$; $-3\frac{1}{4}$; and 0.25.
31. Write without an absolute value symbol: $|6|$; $|-3|$; $-|-4|$; $|0|$; and $|x|$.
32. Evaluate $5x - |2xy| - y$ for $x = -4$ and $y = 0$.

Simplify each of the following expressions.

33. $18 + (+6)$
34. $51 + (-5)$
35. $6 - (-23)$
36. $\frac{1}{3} - \frac{1}{2}$
37. $1\frac{1}{3} + \left(-2\frac{1}{2}\right)$
38. $5\frac{1}{8} - \left(-1\frac{1}{3}\right)$
39. $0.45 - 1.27$
40. $1.81 + (-2.81)$
41. $-3.20 - (-1.80)$
42. $2^2 - 3^2$
43. $3^2 + (-2)^2$
44. $4^2 - 3^3$
45. $\sqrt{25} - 2^3$
46. $\sqrt{121} - (\sqrt{9})^2$
47. $\sqrt{64} - (-\sqrt{49})$
48. $\frac{3}{4} \cdot \frac{5}{3}$
49. $(-8)(-17)$
50. $(-2^2)(3)^2(-5)$
51. $(6 - 8.3)(0)$
52. $(63)(-7)(0)$
53. $\frac{63 - 27.3}{6 - 6}$
54. $\frac{0}{56 - 27}$
55. $\frac{-16}{-2}$
56. $\frac{-91}{-15 - (-8)}$
57. $-|-8 + 3| + |-9 + 6|$
58. $-|7 - 4| - |-11 + 5|$
59. $\frac{2[3 - (2)(5)]}{-13}$
60. $\frac{5[7 + (-4)(6)]}{-42 + 8}$
61. $6 \cdot 3 + 5(8) \div 4$
62. $3 + 8 - (-2)(7) + 15 \div 3$
63. $\frac{-8[2^3 \cdot 5 - (-7)(-2)]}{5^2 + 12^2}$
64. $\frac{-\sqrt{36}(3^3 \cdot 5 - 15)}{\sqrt{64} \cdot 5 - 10}$

65. $5 - 2\{2(6 - 8) + 3[5 + (-7)]\} \div (-5)$

66. $\left(\dfrac{3}{2} - \dfrac{1}{3} \div \dfrac{1}{6}\right)^2$

67. $\dfrac{8.1}{0.09} - (-10)^2$

68. $-\dfrac{1}{5}\left(\dfrac{-1}{4}\right)(-16)(15)$

69. $\left(\dfrac{1}{4} - \dfrac{1}{2} \div 2\right)^2$

70. $6|0 - 5| \div (-5)^2$

71. $16 \div 4(-2) - 2^2$

Write an algebraic expression for each of the following and simplify.

72. The sum of 6 and -3
73. The difference of -8 and -5
74. The quotient of 2.5 and 0.5
75. The product of $\tfrac{3}{4}$ and 25%
76. Negative 16 subtracted from 32
77. The absolute value of -5 added to -11
78. The square root of the sum of $2\tfrac{3}{4}$ and $3\tfrac{1}{2}$
79. The square of the sum of 7 and its additive inverse
80. The difference of 5 and the quotient of 0 and 7

Complete each equation using the given property of real numbers.

81. Associative property: $\quad 6 + (\tfrac{2}{3} - \tfrac{1}{2}) = $ _____
82. Commutative property: $\quad -0.75(8) = $ _____
83. Multiplicative inverse property: $\quad (-13)(\_\_\_\_) = 1$
84. Additive inverse property: $\quad -8.12 + \_\_\_\_ = 0$
85. Distributive property: $\quad 6x + 18 = $ _____
86. Additive identity property: $\quad 6 + \_\_\_\_ = 6$
87. Multiplicative identity property:
    $(-4.96)(\_\_\_\_) = -4.96$
88. Associative property:
    $(-6)[(-7)(2)] = $ _____
89. Commutative property: $\quad 6 + (-5) = $ _____
90. Distributive property: $\quad -5(x - 3) = $ _____

Solve each of the following applied problems.

91. A car is purchased for $625 down and $108 per month for 24 months. How much is paid for the car?

92. A mountain bike costs $480 after the dealer has discounted it by 20%. What was the original price of the bike?

93. If the difference of 56 and -12 is decreased by 20, what is the result?

94. A hawk on Cuesta Ridge, 1520 ft above sea level, flies down the side of the ridge to an altitude of 1275 ft above sea level, catches an upward air current, and sails back up 80 ft in altitude. What is the hawk's final altitude?

95. A student has scores of 81%, 84%, and 75%. What is his average score? If the class average is 84%, find the difference in his average and the class average.

96. The volume of a right circular cylinder is given by $V = \pi r^2 h$. If $h = \dfrac{3}{\pi}$ and $r = \dfrac{1}{2}$, find V.

97. A gallon of water is approximately 0.1368 ft³. What is the measure in cubic feet of a barrel of water containing 55 gal?

98. If $8000 is invested in the stock market and six months later the investment is worth only $7200, how much was lost? Represent the loss as a percent of the investment.

99. A student's bank balance on the 10th of the month was $297.60. On the 20th of the month he received a notice that he was overdrawn by $24.80. What was the total amount of money he had withdrawn from his account?

100. Leah wants to buy a car stereo that is listed for $116.00. When the sales person offers her a discount of 16%, she decides to buy the stereo. After paying a sales tax of 7%, what is the total cost of the stereo?

Chapter Test

Simplify each of the following expressions, if possible.

1. $-8 + 3(5 - 7)$
2. $-9[2 - 3(-1)]$
3. $16 - 2^3(4 \div 2)$
4. $|3^2 - 7^2| \div (-2^3)$
5. $\dfrac{4^2 + 5^2}{42 + (-1)}$
6. $\left(\dfrac{1}{2} + \dfrac{1}{3}\right) \div \sqrt{9}$
7. $-3\{7 - [4 - (-1)]\}$
8. $|-\sqrt{36}| - |-14|$
9. $\left(-\dfrac{3}{4}\right)\left(\dfrac{-8}{3}\right) \div \left(\dfrac{-1}{2}\right)$
10. $(8.25)\left(-\dfrac{1}{4}\right) \div 0$
11. $[-2(-6 + 3) - (5 - 3)](0)$
12. $(-9 + 9) \div \left(2 - \dfrac{1}{2}\right)$

Identify the property of real numbers illustrated by each statement.

13. $-13(x - 2) = -13x + 26$

14. $8xy = 8yx$ **15.** $4\left(\dfrac{1}{4}\right) = 1$

16. $x + (-x) = 0$ **17.** $5x + 0 = 5x$

18. $8 + (6 - 3) = (8 + 6) - 3$

19. $17 + (3 - x) = (3 - x) + 17$

20. $-15y - 30 = -15(y + 2)$

21. $1x = x$ **22.** $(16x)y = 16(xy)$

List the elements in each set.

23. $\{x \mid x \in \text{whole numbers}\}$

24. $\{x \mid x \text{ is an odd natural number}\}$

25. $\{x \mid x \in \text{integers}\}$

26. $\{y \mid y \in \text{whole numbers}, y \leq 7\}$

27. $\{x \mid x \in \text{natural numbers}, x \in \text{irrational numbers}\}$

28. $\{y \mid y \in \text{whole numbers}, y < 4\}$

Solve each of the following applied problems.

29. How many rafters 32 in. long can be cut from a piece of lumber 192 in. long? (Neglect the width of the cut.)

30. The trees in Darlene's orchard are planted 15 ft apart. If there are 63 trees in a row, what is the length of a row in yards?

31. The time sheet for the operations performed on a certain machine list the following: chucking, $\frac{2}{3}$ min; spotting and drilling, $3\frac{1}{3}$ min; facing, $1\frac{2}{3}$ min; grinding, $4\frac{1}{3}$ min; and reaming, $\frac{1}{3}$ min. What is the total time for the operations?

32. The volume of a cone is given by $V = \frac{1}{3}\pi r^2 h$. Find the volume of a cone with radius $r = 6$ ft and height $h = 15$ ft; see the figure. Leave the answer in terms of π.

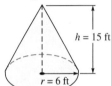

CHAPTER 3
LINEAR EQUATIONS AND INEQUALITIES IN ONE VARIABLE

3.1 Simplifying Algebraic Expressions
3.2 Solving Linear Equations Using the Addition and Subtraction Properties of Equality
3.3 Solving Linear Equations Using the Multiplication and Division Properties of Equality
3.4 Literal Equations and Formulas
3.5 The Addition and Subtraction Properties of Inequality
3.6 The Multiplication and Division Properties of Inequality
3.7 Solving Absolute Value Equations and Inequalities
3.8 More Applications

APPLICATION

The total costs C per month for Mary's Rose Shop are given by the formula $C = \$0.85x + \$20.00$, where x is the number of roses purchased. Find the costs to the store in February if Mary purchases 225 roses.

See Exercise Set 3.8, Exercise 23.

I don't know how I appear to the world; but to myself I feel like a boy playing on the seashore, and diverting myself now and then by finding another pebble or a prettier shell than is originally found, while a great ocean of truths lies undiscovered all around me.

ISAAC NEWTON (1642–1727)

3.1 Simplifying Algebraic Expressions

HISTORICAL COMMENT Maria Agnesi (1718–1799) was one of the first women mathematicians of note in Western culture. By the age of 11 she was able to speak six languages, including Italian (her native tongue), and was already defending the concept of higher education for women. As a teenager she studied many scientific fields and mastered much that was known in her time. Her work in mathematics was creative and her interests ranged from algebra to the calculus, which was only just being developed at that time.

OBJECTIVES In this section we will learn to
1. identify terms, algebraic expressions, and numerical coefficients;
2. recognize and combine like terms; and
3. evaluate algebraic expressions for specified values of the variables.

A **term** can be a single number or the product or quotient of one or more numbers and variables raised to powers. An **algebraic expression** is a sum or difference of such terms. Here are some examples of algebraic expressions and their individual terms:

ALGEBRAIC EXPRESSIONS	TERMS OF THE EXPRESSION
$5 + 4x$	5 and $4x$
$x^2 - 3x + \dfrac{x}{4}$	x^2, $-3x$, and $\dfrac{x}{4}$
$x^4y + 4x^2 - 9$	x^4y, $4x^2$, and -9

One of the first things necessary for solving equations is the ability to simplify algebraic expressions. To **simplify an algebraic expression,** we reduce the number of terms, if possible. We do this by applying the properties of the real numbers. For example, to simplify $6x + 9x$, we apply the distributive property

$$6x + 9x = 6 \cdot x + 9 \cdot x$$
$$= (6 + 9)x \qquad \text{Distributive property: } ba + ca = (b + c)a$$
$$= 15x$$

Notice that only the 6 and the 9 are added; the x remains unchanged. The numbers 6 and 9 are called the **numerical coefficients** of x in the terms $6x$ and $9x$, respectively. In the same manner, the numerical coefficient in $3x^2$ is 3, in $4m^9$ it is 4, and in $2xy$ it is 2.

EXAMPLE 1 Simplify the following algebraic expressions: (a) $3x + 5x$ (b) $5y + 7y - 3y$

SOLUTION
(a) $3x + 5x = (3 + 5)x$ Distributive property
$ = 8x$ Add the numerical coefficients.
(b) $5y + 7y - 3y = (5 + 7 - 3)y$ Distributive property
$ = 9y$ Simplify the numerical coefficient.

In order to simplify an algebraic expression using the distributive property, some of the terms must be **like terms,** which means that *their variable parts have exactly the same base and exponent.*

LIKE TERMS	NOT LIKE TERMS
$3x^3$ and $5x^3$	$7y^5$ and $7y^4$
$2m^4y^3$ and $-4m^4y^3$	$15m^2y^3$ and $13m^3y^2$

EXAMPLE 2 Identify the like terms in each of the following expressions.

(a) $6x^5 + 4x^2 + 2x^5$ The like terms are $6x^5$ and $2x^5$.
(b) $2xy - 5xy^2 + x^2$ There are no like terms.
(c) $4xy + 3yx - 5x^2$ The like terms are $4xy$ and $3yx$ because $xy = yx$.

EXAMPLE 3 Simplify: $4x^2 - 9x + 2x^2$.

SOLUTION
$4x^2 - 9x + 2x^2 = (4x^2 + 2x^2) - 9x$ Commutative and associative properties
$ = (4 + 2)x^2 - 9x$ Distributive property
$ = 6x^2 - 9x$ Add the numerical coefficients.

EXAMPLE 4 Simplify: $2y - 3x + 4y - 7x$.

SOLUTION
$2y - 3x + 4y - 7x = 2y + (-3x) + 4y + (-7x)$ Definition of subtraction
$ = (2y + 4y) + [(-3x) + (-7x)]$ Commutative and associative properties
$ = (2 + 4)y + [(-3) + (-7)]x$ Distributive property
$ = 6y + (-10)x$ Add numerical coefficients.
$ = 6y - 10x$ Definition of subtraction

EXAMPLE 5 Simplify: $\frac{1}{2}x - \frac{1}{3}y + \frac{2}{3}x + \frac{1}{4}y$.

SOLUTION
$\frac{1}{2}x - \frac{1}{3}y + \frac{2}{3}x + \frac{1}{4}y = \frac{1}{2}x + \left(-\frac{1}{3}y\right) + \frac{2}{3}x + \frac{1}{4}y$
$\phantom{\frac{1}{2}x - \frac{1}{3}y + \frac{2}{3}x + \frac{1}{4}y} = \left(\frac{1}{2}x + \frac{2}{3}x\right) + \left(-\frac{1}{3}y + \frac{1}{4}y\right)$ Commutative and associative properties

$$= \left(\frac{1}{2} + \frac{2}{3}\right)x + \left(-\frac{1}{3} + \frac{1}{4}\right)y \quad \text{Distributive property}$$

$$= \frac{7}{6}x + \left(-\frac{1}{12}\right)y \quad \frac{1}{2} + \frac{2}{3} = \frac{7}{6}; \ -\frac{1}{3} + \frac{1}{4} = -\frac{1}{12}$$

$$= \frac{7}{6}x - \frac{1}{12}y \quad \text{Definition of subtraction}$$

Sometimes we must use the distributive law before we can add or subtract like terms.

EXAMPLE 6 Simplify: $(7x - 5y) + 2(2x - 3y)$.

SOLUTION We begin by removing the grouping symbols.

$(7x - 5y) + 2(2x - 3y) = 7x - 5y + 4x - 6y$ Distributive property: $2(2x - 3y) = 4x - 6y$

$\qquad = 7x + 4x + (-5y) + (-6y)$ Commutative property; definition of subtraction

$\qquad = 11x + (-11y)$ Add like terms.

$\qquad = 11x - 11y$ Definition of subtraction

As you gain more experience in algebra, you will be able to eliminate some of the steps shown in the preceding examples. Those that are justified by the distributive property can generally be completed mentally. Another time-saving step can be used when removing grouping symbols preceded by a negative sign. For example, if we think of

$$5x - (3x - 2y) \text{ as } 5x - 1(3x - 2y)$$

we simply follow the rules for multiplication of signed numbers. *When removing a grouping symbol preceded by a negative sign, change the sign of every term within the symbol.* Notice how many fewer steps are necessary to arrive at the same result.

JUSTIFICATION METHOD

$5x - (3x - 2y)$
$\quad = 5x - [3x + (-2y)]$ To subtract, add the inverse.

$\quad = 5x + [-3x + 2y]$ To subtract, add the inverse.

$\quad = [5x + (-3)x] + 2y$ Associative property
$\quad = [5 + (-3)]x + 2y$ Distributive property
$\quad = 2x + 2y$

TIME-SAVER METHOD

$5x - 1(3x - 2y)$
$\quad = 5x - 3x + 2y$ $-1(-2y) = 2y$

$\qquad\qquad\qquad\qquad -1(3x) = -3x$

$\quad = 2x + 2y$

EXAMPLE 7   Simplify: $8 - 3[-2y - (-11y - 8) - 4y] - 2y$.

SOLUTION   We follow the order of operations, working with the innermost grouping symbols first. Recall that removing a grouping symbol preceded by a subtraction sign is the same as multiplying by -1.

$8 - 3[-2y - (-11y - 8) - 4y] - 2y$
$= 8 - 3[-2y - 1(-11y - 8) - 4y] - 2y$
$= 8 - 3[-2y + 11y + 8 - 4y] - 2y$ Multiply by -1.
$= 8 - 3[5y + 8] - 2y$ Combine like terms.
$= 8 - 15y - 24 - 2y$ Multiply by -3.
$= -16 - 17y$ Combine like terms.

EXAMPLE 8   Evaluate $3xy^2 + 2xy^2 + 5x + 3x$ for $x = 3$ and $y = -2$.

SOLUTION   There are two different orders we can use to evaluate the expression: **(a)** we can simplify the expression by combining like terms before we substitute for x and y, or **(b)** we can substitute for x and y first.

(a) We combine like terms first:

$3xy^2 + 2xy^2 + 5x + 3x = 5xy^2 + 8x$ Simplify first.
$= 5(3)(-2)^2 + 8(3)$ $x = 3$ and $y = -2$
$= 15(4) + 24$
$= 60 + 24$
$= 84$

(b) We substitute for x and y first:

$3xy^2 + 2xy^2 + 5x + 3x = 3(3)(-2)^2 + 2(3)(-2)^2 + 5(3) + 3(3)$
$= 9(4) + 6(4) + 15 + 9$
$= 36 + 24 + 24$
$= 84$

EXAMPLE 9   Translate: "Four times the sum of a number and 5 is subtracted from 6 times the sum of the number and 9." Simplify your answer.

SOLUTION   We let $x = $ the number. Then the sum of the number and 5 translates to $x + 5$. Four times this sum is $4(x + 5)$. Next we translate the sum of the number and 9 to $x + 9$. Six times this sum is $6(x + 9)$.

$6(x + 9) - 4(x + 5) = 6x + 54 - 4x - 20$ Distributive property
$= 2x + 34$ Combine like terms.

EXAMPLE 10   A coin collection consists of pennies, nickels, and dimes. There are twice as many nickels as pennies and 4 less dimes than pennies. Write an algebraic expression describing the face value of the collection.

SOLUTION We let p = the number of pennies
$2p$ = the number of nickels
$p - 4$ = the number of dimes

Each penny has a value of 1 cent, each nickel 5 cents, and each dime 10 cents, so the value of the collection is

$$1p + 5(2p) + 10(p - 4) = p + 10p + 10p - 40$$
$$= (21p - 40) \text{ cents}$$

EXAMPLE 11 The length of a rectangle is 6 feet more than twice its width. Find an expression for its perimeter in terms of the width.

SOLUTION We let w = the width of the rectangle
$2w + 6$ = the length of the rectangle

The formula for the perimeter of a rectangle is $P = 2l + 2w$, so

$$P = 2(2w + 6) + 2w$$
$$= 4w + 12 + 2w$$
$$= 6w + 12$$

The perimeter is 12 feet more than 6 times the width.

Exercise Set 3.1

Simplify each algebraic expression.

1. $3x + 8x$
2. $7x + 4x$
3. $14y - 6y$
4. $12p - 5p$
5. $3z + 5z + z$
6. $2m + 7m + m$
7. $10r + r - 7r - 3r$
8. $13r - 6r + r - 2r$
9. $3x - 11y - 8x + 5y$
10. $18q - 3q + 4q - 9q$
11. $x^2 - 3x + 4x^2 + 2x$
12. $y^2 - 5y + 7y^2 + 8y$
13. $5y^3 + 2y^2 - 7y^3 + 4y^2$
14. $8x^3 + x^2 - 9x^3 + 8x^2$
15. $6p - 4 + 7p + 9$
16. $5m - 6 - 3m + 17$
17. $5a - 11b - a + 10b$
18. $-8c + 4d + 5c - 19d$
19. $2x + 3y - 4 + 2y - 2x + 4$
20. $2a - 3b + 6 + 3a + 3b - 6$
21. $\frac{1}{3}x + \frac{1}{4}y + \frac{1}{2}x - \frac{1}{3}y$
22. $3x - y - \frac{3}{4}x + \frac{3}{5}y$
23. $\frac{1}{4}a + 2\frac{3}{7}b - a - 1\frac{1}{4}b$
24. $3\frac{3}{8}a - 2b + c - \frac{1}{4}a - \frac{1}{3}b + \frac{1}{4}c$
25. $0.8x - 1.2y - 1.3x + 2y$
26. $10.7y - 3z - 5y - 1.9z$
27. $1.4a + b - c + 1.8a - 2\frac{1}{3}b - 1\frac{1}{4}c$
28. $1\frac{1}{2}a + 2\frac{1}{4}b - 0.8a - 0.6b$
29. $\frac{3}{5}\left(\frac{10}{3}c - \frac{5}{4}ab\right) + \frac{4}{7}\left(\frac{14}{8}ab - \frac{28}{12}c\right)$
30. $-\frac{12}{5}\left(\frac{10}{12}x^2y + 15xy^2\right) + \frac{4}{3}\left(\frac{15}{2}xy^2 - \frac{21}{4}x^2y\right)$

Evaluate each of the following expressions for $x = 3$ and $y = -2$.

31. $3x + 2$
32. $6x - 7$

33. $7y - 4$
34. $8y + 9$
35. $-3x + 2y$
36. $-x + 4y$
37. $-11x - 20y$
38. $-13x - 15y$
39. $7x - 2y + 10$
40. $3x - 11y + 4$
41. $-17x + 42y - 26$
42. $-13x + 36y - 29$

Evaluate each of the following expressions for the indicated values.

43. $\dfrac{4x - 3y}{3x}$ for $x = \dfrac{1}{2}, y = -\dfrac{1}{3}$

44. $-\dfrac{1}{3}x + x - \dfrac{1}{5}y + \dfrac{1}{6}y$ for $x = 5, y = -\dfrac{1}{2}$

45. $\dfrac{|3x - y|}{|-2x|}$ for $x = -\dfrac{1}{2}, y = -2$

46. $\dfrac{|3xy^2z|}{x}$ for $x = -2, y = 0, z = 3$

47. $5xy^2 - 5x^2y + 3(x - y)$ for $x = -\dfrac{1}{4}, y = 0$

48. $3x^2yz^3 - xyz$ for $x = -\dfrac{1}{2}, y = \dfrac{1}{2}, z = -\dfrac{1}{3}$

Simplify each of the following expressions.

49. $(3x + 2) + (-4x + 7)$
50. $(2y - 5) + (-6y + 1)$
51. $(17x + 5) - (10x + 6)$
52. $(11x - 6) - (8x - 8)$
53. $(m - 3) - (-2m - 5)$
54. $(p - 2) - (-6p - 7)$
55. $(2y^2 + 5) + (3y^2 + y - 2)$
56. $(3x^2 + 1) + (4x^2 + 2x + 3)$
57. $-(q - 9) - (q^2 - q - 3)$
58. $-(m - 8) - (m^2 - m + 4)$
59. $8(y - 6) - 11(2y - 1)$
60. $-13(r - 1) + 9(r + 2)$
61. $-(x + 1) + 2(x - 3) - 4(x + 5)$
62. $-(y - 1) + 4(y - 3) - 2(2y - 5)$
63. $-[6 + (3p - 1) + 2(p + 2)]$
64. $-[5 + 3(x - 2) + (5x + 7)]$
65. $3 + 4[6x - (5x - 7) - x] + 3x$
66. $7 + 2[13y - (4y - 3) - 2y] + 5y$
67. $8 - 3[-2y - (13y - 8) - 5y] - y$
68. $11 - 2[5x - (7x - 9) - 4x] - x$
69. $\dfrac{1}{2}\left[\dfrac{1}{3}a - b + c\right] + \dfrac{1}{3}\left(\dfrac{1}{2}a + 2b - \dfrac{1}{4}c\right)$
70. $-\dfrac{1}{4}\left[\dfrac{-1}{2}a + 2b - c\right] + \dfrac{-1}{2}\left[\dfrac{1}{3}a - 3b + 5c\right]$

Simplify each of the following algebraic expressions. A calculator may be helpful.

71. $3.84x - 5.21y + 8.79x - 1.03y$
72. $6.28a - 3.09b + 7.91a - 4.84b$
73. $0.35(1.28x + 6.08)$
74. $0.29(4.15y + 9.67)$
75. $0.6(3x - 3y) - 0.8(5x - y)$
76. $0.9(3x - 4y) - 0.2(7x - y)$

Translate each of the following phrases or sentences into an algebraic expression and simplify where possible.

77. One-half of a number added to three times the number
78. Eight less than twice a number
79. Four more than three-fourths of a number
80. Julie had $85 for a two-day ski trip. If she spends x dollars the first day, how much does she have left for the second day?
81. Twice the sum of x and negative 5
82. Four times the sum of y and 7
83. Nine times the difference of a and 9
84. Three times the difference of 17 and b
85. A number is subtracted from five times the sum of 3 and the number.
86. Four times the sum of twice a number and 5 is subtracted from five times the number.
87. Add the difference of 13 and a number to the sum of 7 and three times the number.
88. Subtract the difference of 9 and twice a number from the difference of 6 and five times the number.
89. The length of a rectangle is 3 ft more than its width. Find an expression for the perimeter in terms of the width.
90. In a right triangle, the longest side is 8 units more than the shortest side. The middle side is 1 unit less than the longest side. Find an expression for the perimeter in terms of the shortest side.
91. In a triangle, the longest side is 3 units less than twice the shortest side. The middle side is 4 units less than the longest side. Find an expression for the perimeter in terms of the shortest side.
92. The width of a rectangle is $(2x + 8)$ in. The length is 5 in. more than one-half the width. Find an expression for the perimeter in terms of the width.
93. Milt has 3 more dimes than nickels. Find an expression for the value of the nickels and dimes.

94. Sylvia has 6 more dimes than quarters. Find an expression for the value of the dimes and quarters.

95. Jadene has some pennies, nickels, and dimes. She has 6 more pennies than nickels and 3 more nickels than dimes. Find an expression for the value of the pennies, nickels, and dimes.

96. Lester has some nickels, dimes, and quarters. He has 5 less quarters than dimes and 2 less dimes than nickels. Find an expression for the value of the nickels, dimes, and quarters.

WRITE IN YOUR OWN WORDS

97. Define "like terms."
98. Define "algebraic expression."
99. Describe "numerical coefficient."
100. Describe how to add like terms.

101. What property or definition justifies each of the following steps?

$$-5(a - 3) + 4a = -5a + 15 + 4a$$
$$= -5a + 4a + 15$$
$$= a(-5 + 4) + 15$$
$$= -a + 15$$

102. What property or definition justifies each of the following steps?

$$|7 - 11|(3x - 7) + 2^2 x = |-4|(3x - 7) + 2^2 x$$
$$= 4(3x - 7) + 2^2 x$$
$$= 12x - 28 + 2^2 x$$
$$= 12x - 28 + 4x$$
$$= 12x + 4x - 28$$
$$= 16x - 28$$

3.2 Solving Linear Equations Using the Addition and Subtraction Properties of Equality

HISTORICAL COMMENT François Viète (1540–1603) was a French lawyer who studied mathematics for recreation. His works on mathematics, published privately for distribution among friends, were the first that were known to use letters to represent numbers. Viète used consonants to represent known quantities and vowels for unknown quantities.

OBJECTIVES In this section we will learn to
1. identify linear equations;
2. solve linear equations by using the addition property of equality; and
3. solve linear equations by using the subtraction property of equality.

An **equation** is a statement that two quantities are equal. Some examples of simple equations are

$$3 + 5 = 8 \quad 15 - 8 = 7 \quad \text{and} \quad 10x - 3x - 4 = 3$$

Notice that the first two equations are true, but when an equation involves a variable, it can be true or false depending on the value of the variable. Any value of the variable that makes an equation true is called a **solution** or **root** of the equation. Consider, for example,

$$x + 6 = 9$$

When 5 is substituted for x, the equation is false:

$x + 6 = 9$
$5 + 6 \stackrel{?}{=} 9$ Substitute 5 for x.
$11 = 9$ False, 5 is not a solution

When 3 is substituted for x, the equation is true:

$x + 6 = 9$
$3 + 6 \stackrel{?}{=} 9$
$9 \stackrel{\checkmark}{=} 9$ True, 3 is a solution.

To check a potential solution to an equation, we start by writing the equation and then substitute the proposed solution for the variable to see if it makes the equation true.

EXAMPLE 1 Verify that 4 is the solution to $3x + 3 = 15$.

SOLUTION We substitute 4 for x to see if it makes the equation true:

$3x + 3 = 15$ Original equation
$3(4) + 3 \stackrel{?}{=} 15$ Substitute 4 for x.
$12 + 3 \stackrel{?}{=} 15$
$15 \stackrel{\checkmark}{=} 15$ True, the solution is 4.

EXAMPLE 2 Is 7 a solution to $5 - 2x = -5$?

SOLUTION We substitute 7 for x to see if it makes the equation true.

$5 - 2x = -5$ Original equation
$5 - 2(7) \stackrel{?}{=} -5$ Substitute 7 for x.
$5 - 14 \stackrel{?}{=} -5$
$-9 = -5$ False, 7 is not a solution.

Definition	A **linear equation in one variable** is an equation that can be written in the form
Linear Equation in One Variable	$ax + b = c, \quad a \neq 0$
	where x is a variable and a, b, and c represent real numbers.

EXAMPLE 3 Identify the value of a, b, and c in the following linear equations:

(a) $5 + 7x = 2$ (b) $13x - 2 = -18$

SOLUTION (a) $5 + 7x = 2$
$7x + 5 = 2$ Commutative property for addition
Here $a = 7$, $b = 5$, and $c = 2$.

(b) $13x - 2 = -18$

$13x + (-2) = -18$ Definition of subtraction

Here $a = 13$, $b = -2$, and $c = -18$.

Many linear equations are so simple that their solutions can be determined by inspection. For example, it is easy to see that the solution to $x + 9 = 12$ is 3, since $3 + 9 = 12$. Not every equation is as easy to solve, so we must develop a method we can use to simplify it until it takes the form

$x =$ the solution

> *When the solution to an equation has been found, the variable will stand alone on one side of the equation with a coefficient of* 1.

Let's return to the equation $x + 6 = 9$. If we could do something to the left side of the equation to make x stand alone, we would have the solution. We can accomplish this by subtracting 6. If we subtract 6 from the left side, however, we must also subtract it from the right side. If we didn't, the two sides of the equation would no longer be equal.

(a) $x + 6 = 9$ Given equation

$x + 6 - 6 = 9 - 6$ Subtract 6 from each side.

$x = 3$ Solution

To solve the equation $x - 6 = 9$, we *add* 6 to each side:

(b) $x - 6 = 9$ Given equation

$x - 6 + 6 = 9 + 6$ Add 6 to each side.

$x = 15$ Solution

The steps we have used to solve these two equations can be likened to using a balance (a scale) with weights. To keep an equation in balance we must either add the same amount of weight to each side [Figure 3.1(a) on page 116] or subtract the same amount of weight from each side [Figure 3.1(b)].

The properties that justify solving equations in this way are summarized as follows.

Addition Property of Equality

If a, b, and c represent real numbers and $a = b$, then

$a + c = b + c$

Subtraction Property of Equality

If a, b, and c represent real numbers and $a = b$, then

$a - c = b - c$

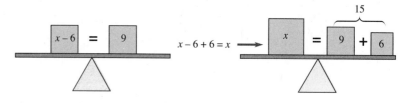

(a) Add 6 to both sides

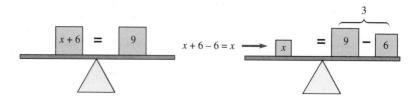

(b) Subtract 6 from both sides

FIGURE 3.1

In the simplest terms, the addition property of equality states that *if equals are added to equals, the results are equal.* The subtraction property of equality states that *if equals are subtracted from equals, the results are equal.*

Linear equations in one variable have a single solution. In later sections we will solve equations with more than one solution; in such cases we generally speak of a **solution set.** In this text we will always refer to a solution or solutions as a solution set and enclose it (them) within braces for identification.

EXAMPLE 4   Solve the equation $x - 13 = 27$.

SOLUTION   When we solve the equation, x will stand alone on one side of the equal sign. Since subtraction is indicated on the left side, we must add 13 to each side to get x alone:

$$x - 13 = 27$$
$$x - 13 + 13 = 27 + 13 \quad \text{Add 13 to each side.}$$
$$x = 40$$

To check the result, we substitute 40 for x in the original equation.

$$x - 13 = 27 \quad \text{Original equation}$$
$$40 - 13 \stackrel{?}{=} 27 \quad \text{Substitute 40 for } x.$$
$$27 \stackrel{\checkmark}{=} 27$$

The solution set is {40}.

EXAMPLE 5   Solve the equation $x + 9 = 12$.

SOLUTION   When we solve the equation, x will be isolated on one side of the equal sign. Since addition is indicated on the left side, we subtract 9 from each side to get x alone:

$$x + 9 = 12$$
$$x + 9 - 9 = 12 - 9 \quad \text{Subtract 9 from each side.}$$
$$x = 3$$

To check the result, we substitute 3 for x in the original equation.

$$x + 9 = 12 \quad \text{Original equation}$$
$$3 + 9 \stackrel{?}{=} 12 \quad \text{Substitute 3 for } x.$$
$$12 \stackrel{\checkmark}{=} 12 \quad \text{True}$$

The solution set is $\{3\}$.

EXAMPLE 6 Solve the equation $5y + 9 = 6y$.

SOLUTION In order to find the solution, we collect all terms involving y on one side of the equation. We can accomplish this by subtracting $5y$ from each side:

$$5y + 9 = 6y$$
$$5y + 9 - 5y = 6y - 5y \quad \text{Subtract } 5y \text{ from each side.}$$
$$9 = y \quad \text{Combine like terms.}$$

CHECK
$$5y + 9 = 6y$$
$$5(9) + 9 \stackrel{?}{=} 6(9) \quad \text{Substitute 9 for } y.$$
$$45 + 9 \stackrel{?}{=} 54$$
$$54 \stackrel{\checkmark}{=} 54$$

The solution set is $\{9\}$.

Notice in Example 6 that the variable was isolated on the *right* side. Isolating it on the left side requires more steps to find the solution, as we'll show in the next example.

EXAMPLE 7 Solve the equation $5y + 9 = 6y$.

SOLUTION To isolate y on the left side of the equation, we begin by subtracting $6y$ from each side:

$$5y + 9 = 6y$$
$$5y + 9 - 6y = 6y - 6y \quad \text{Subtract } 6y \text{ from each side.}$$
$$-y + 9 = 0 \quad \text{Combine like terms.}$$
$$-y + 9 - 9 = 0 - 9 \quad \text{Subtract 9 from each side.}$$
$$-y = -9 \quad \text{Combine like terms.}$$

If $-y = -9$, then y must be 9. The solution set is $\{9\}$.

EXAMPLE 8 Solve the equation $2x + 6 = x + 8$.

SOLUTION In order to get all of the terms involving the variable on one side of the equation, we'll need to use some of the properties more than once.

$$2x + 6 = x + 8$$
$$2x + 6 - 6 = x + 8 - 6 \quad \text{Subtract 6 from each side.}$$

$$2x = x + 2 \quad \text{Combine like terms.}$$
$$2x - x = x + 2 - x \quad \text{Subtract } x \text{ from each side.}$$
$$x = 2 \quad \text{Combine like terms.}$$

We could have completed both steps at once if we wished to do so:

$$2x + 6 = x + 8$$
$$2x + 6 - 6 - x = x + 8 - 6 - x \quad \text{Subtract } x \text{ and 6 from each side.}$$
$$x = 2 \quad \text{Combine like terms.}$$

CHECK
$$2x + 6 = x + 8$$
$$2(2) + 6 \stackrel{?}{=} 2 + 8$$
$$4 + 6 \stackrel{?}{=} 2 + 8$$
$$10 \stackrel{\checkmark}{=} 10$$

The solution set is {2}.

EXAMPLE 9 Solve the equation $2x + 5x = 6x + 3$.

SOLUTION In this equation, there are like terms on the left side. We'll combine these terms first, before applying any of the properties of equality.

$$2x + 5x = 6x + 3$$
$$7x = 6x + 3 \quad \text{Combine like terms.}$$
$$7x - 6x = 6x + 3 - 6x \quad \text{Subtract } 6x \text{ from each side.}$$
$$x = 3 \quad \text{Combine like terms.}$$

The check is left to the reader. The solution set is {3}.

EXAMPLE 10 Solve the equation $-2[x + 4(x - 4)] = 7 - 11x$.

SOLUTION We begin by simplifying the left side of the equation:

$$-2[x + 4(x - 4)] = 7 - 11x$$
$$-2[x + 4x - 16] = 7 - 11x \quad \text{Innermost grouping symbols first}$$
$$-2[5x - 16] = 7 - 11x \quad \text{Combine like terms.}$$
$$-10x + 32 = 7 - 11x \quad \text{Distributive property}$$
$$-10x + 32 + 11x - 32 = 7 - 11x + 11x - 32 \quad \text{Add } 11x \text{ and subtract 32 from each side.}$$
$$x = -25$$

The solution checks. The solution set is {−25}.

All of the equations in the preceding examples are known as **conditional equations.** Such equations are true only for *specific values of the variable.* On the other hand, an equation such as

$$2x + 2x = 4x$$

is called an **identity**. *An identity is true for all meaningful replacements for the variable.* Its solution set is the set of all real numbers.

EXAMPLE 11 Solve the equation $2x + 2 = 3(x - 2) - x + 8$.

SOLUTION
$$2x + 2 = 3(x - 2) - x + 8$$
$$2x + 2 = 3x - 6 - x + 8 \qquad \text{Distributive property}$$
$$2x + 2 = 2x + 2$$

Since the left and right sides are identical, the equation is an identity. The solution set is $\{x \mid x \text{ is a real number}\}$.

EXAMPLE 12 Five times a number is 18 more than four times the number. Find the number.

SOLUTION We let
$$n = \text{the number}$$
$$5n = \text{five times the number}$$
$$4n = \text{four times the number}$$

Five times a number **is** 18 more than 4 times the number.
$$5n = 4n + 18$$
$$5n - 4n = 4n + 18 - 4n \qquad \text{Subtract } 4n \text{ from each side.}$$
$$n = 18 \qquad \text{Combine like terms.}$$

The number is 18.

We can view the relationship between the numbers in Example 12 in other ways. Since five times the number is 18 more than four times the number, we can subtract 18 from the larger number to make it equal to the smaller one:

$$5n - 18 = 4n$$

Adding 18 to each side of this equation yields the first equation.

EXAMPLE 13 Nine subtracted from what number is equal to 20% of the difference between 7 and 3?

SOLUTION We let $x =$ the number. Then 20% of the difference between 7 and 3 is $0.20(7 - 3)$, so

$$x - 9 = 0.20(7 - 3)$$
$$x - 9 = 0.20(4)$$
$$x - 9 = 0.8$$
$$x - 9 + 9 = 0.8 + 9 \qquad \text{Add 9 to each side.}$$
$$x = 9.8$$

The number is 9.8.

Do You Remember?

Can you match these?

_____ 1. An equation that is true for specified values of the variable
_____ 2. If $a = b$, then $a + c = b + c$.
_____ 3. An equation that is true for all real numbers
_____ 4. If $a = b$, then $a - c = b - c$.
_____ 5. The solution set for $x - 3 = 12$
_____ 6. The solution set for $5x - 4x = 11 - 16$
_____ 7. The solution set for $x + 7x = 11x - 3x$
_____ 8. The solution set for $x + 6 = 2x - 8$

a) An identity
b) A conditional equation
c) The addition property of equality
d) The subtraction property of equality.
e) $\{-5\}$
f) $\{15\}$
g) $\{14\}$
h) $\{x \mid x$ is a real number$\}$

Answers: 1. b 2. c 3. a 4. d 5. f 6. e 7. h 8. g

Exercise Set 3.2

In Exercises 1–10, determine whether the given number is a solution of the equation.

1. $3x - 4 = x - 12$; $x = -4$
2. $4x + 7 = 35$; $x = 7$
3. $5y - 1 = -1.5$; $y = -0.1$
4. $2y + 1 = y + 3.5$; $y = 0.5$
5. $\frac{1}{3}x - \frac{1}{6} = \frac{1}{3}x + 0$; $x = 2$
6. $\sqrt{p} - 3 = 1$; $p = 25$
7. $\frac{w-1}{w+8} = \frac{-1}{2}$; $w = -2$
8. $\frac{m+3}{m-5} = 0$; $m = -3$
9. $-4r - 5r = -4r + 5$; $r = -1$
10. $x - 2 + 27x = 5 + (3 \cdot 3)^2$; $x = 3$

Use the addition or subtraction property of equality to solve each equation. Check your answers.

11. $x + 2 = 6$
12. $y + 3 = 7$
13. $y - 5 = 4$
14. $x - 6 = 7$
15. $9 + p = 11$
16. $19 + k = 20$
17. $13 + x = 2$
18. $7 + y = 3$
19. $3p = 2p + 1$
20. $5m = 4m + 3$
21. $7y - 10 = 6y$
22. $8p - 6 = 7p$
23. $-2x + 4 = -3x - 6$
24. $-3y + 5 = -4y + 1$
25. $18m + 4 = -3 + 17m$
26. $6k - 2 = -5 + 5k$
27. $k + 6 = 0$
28. $x - 5 = 0$
29. $0 = -7 + p$
30. $0 = 13 + m$

Combine any like terms first, then solve using the addition or subtraction property of equality. Check your answers.

31. $3 + 2y - 5y + 4y = 10$
32. $2 - 3x + 7 + 4x = 3$
33. $6x - 4 + 6 = 5x + 9$
34. $7p - 5 + 2 = 6p - 1$
35. $-2p = 8 - 3p - 5$
36. $-4y = -5 - 5y + 4$
37. $4m + 2 - 5 + 3 = 3m$
38. $5k + 7 = -4 + 4k + 11 + k$
39. $2y + 7 = y + 11$
40. $3x - 4 = 2x + 10$
41. $-3k - 2 = -4k + 1$
42. $-5m - 1 = -6m + 1$
43. $-8x - 3 = -9x + 1$
44. $-14y + 2 = 5y + 15 - 20y$
45. $-11p + 5 + 8p - 2 = p + 7 - 5p - 1$
46. $-k - 3 + 2k + 5 = 7k - 6 + 4 - 7k$

47. $\frac{1}{4}x - \frac{1}{3} + \frac{3}{4}x + \frac{4}{3} = 2$
48. $\frac{1}{5}x + \frac{3}{7} + \frac{4}{5}x = -\frac{4}{7}$
49. $\frac{1}{2}x = \frac{3}{2}x - \frac{4}{5}$
50. $\frac{1}{3}x = \frac{1}{3} + \frac{4}{3}x$
51. $-\frac{2}{5}y = \frac{4}{5} - \frac{7}{5}y$
52. $\frac{-3}{7}y = \frac{12}{28} - \frac{10}{7}y$
53. $-\frac{1}{2}(2x - 4) - \frac{1}{3}(3x - 9) = -3(x + 5)$
54. $\frac{2}{3}(3y + 6) + \frac{1}{5}(5y - 10) = 2(y - 7)$
55. $-[-3(2y - 3) + 5(y - 7)] = 0$
56. $-2[-7(3r + 2) + (20r - 5)] = 0$
57. $0.5x + 4 = 1.5x$
58. $3.2p - 2.8 = 2.2p - 3.8$
59. $-0.1(10x - 20) + 0.2(10x - 30) = 0$
60. $3(0.4x - 0.8) - 2(0.1x - 0.3) = 0$
61. $\frac{7k}{3} - \frac{1}{2} = \frac{4k}{3} + \frac{1}{3}$
62. $\frac{5m}{8} + \frac{3}{4} = \frac{-3m}{8} + 0.75$

Translate each sentence into an algebraic equation and solve.

63. Three times a number is 6 more than twice the number. Find the number.
64. If six times a number is added to twice a number, the result is 3 less than seven times the number. Find the number.
65. Fifteen times a number is subtracted from eighteen times the number. The result is 2 more than twice the number. Find the number.
66. One-fourth of a number is 4 less than five-fourths of the number. Find the number.

REVIEW EXERCISES

67. Find the perimeter of a rectangle with length 8 in. and width 5 in.
68. The perimeter of a rectangle is 54 in., and it has a length that is 5 in. more than its width. Find the length and width.
69. The perimeter of a triangle is 26 cm. If the longest side is 8 cm more than the shortest side and the middle side is 5 cm less than the longest side, find the length of each side.
70. The perimeter of a square is 32 ft. Find the length of a side.
71. The radius of a circle is 2 in. Find the circumference in terms of π.
72. The area of a rectangle is 38 in². If the width is 1 in., find the length.
73. Find the volume of a right circular cone (in terms of π) if the radius is 5 ft and the height is 3 ft.
74. Find the volume of a right circular cylinder (in terms of π) if the radius is 3 ft and the height is 8 ft.
75. What number is 16 less than 20% of 80?
76. What number is 12 more than 15% of 70?
77. What number less 5 is 7% of the difference of 33 and 13?
78. What number subtracted from 11 is 10% of the sum of 8 and 12?
79. Sixty percent of 80 added to a certain number is twice the number. What is the number?
80. Twenty-five percent of 40 less three times a number is the negative of twice the number. What is the number?

Simplify each of the following quantities.

81. $\left(-2\frac{4}{7}\right) \div \left(-\frac{4}{21}\right)$
82. $\left(\frac{26}{11}\right)\left(\frac{-39}{44}\right)$
83. $12 - 42 \div 6 + 4$
84. $3\left(1\frac{1}{4}\right) + \frac{11}{8}\left(\frac{-3}{22}\right)$
85. $3\frac{1}{2} \div \left(\frac{3}{4}\right)^2 - \frac{1}{6}$
86. $-2 - 78 \div \sqrt{36}$
87. $56 \div (7 - 5)^3$
88. $3\sqrt{7^2 - 24}$
89. -3^2
90. $[3 - 2(2 - 5^2)] \div (-7)$
91. $-40 \div \left(\frac{5}{8}\right)$
92. $-8 \cdot 5^2 + 3 \cdot 5^3$
93. $\frac{5}{7} \div \left(\frac{-15}{49}\right)$
94. $\dfrac{\frac{1}{2} - \frac{3}{4}}{1 - \frac{1}{3}}$
95. $\sqrt{169} - 3^2 \cdot 2$
96. $-\frac{3}{5} + \frac{1}{2}\left(\frac{2}{3} - 4\right)$
97. $-4|-11 + 5| \div -(5 + 9)$
98. $-6\left(\frac{-10}{9}\right)\left(3\frac{3}{4}\right)$
99. $\dfrac{-(-2) + 6(-2)^2}{3(-2)^2 - (-2)}$
100. $\dfrac{3 - 15 + 2(-3)^2}{6 - 3(-11) + 3(-3)^2}$

3.3 Solving Linear Equations Using the Multiplication and Division Properties of Equality

HISTORICAL COMMENT   The equal sign, =, is said to be the invention of Robert Recorde (1510–1575); he first used it in an English text in 1557. He said that he would use two parallel lines to represent two equal quantities because no two things could be more equal. Despite the simplicity of this symbol, it was not readily accepted and did not appear again in print until 1618. Here are some other symbols for equality and the year they appeared:

[1559   ⊔ 1634   ⊓ 1680
∥ 1575   ∞ 1637

OBJECTIVES   In this section we will learn to
1. solve linear equations by using the multiplication property of equality; and
2. solve linear equations by using the division property of equality.

In Section 3.2 we solved equations using the addition and subtraction properties of equality. There are many equations that cannot be solved using only those properties. For example, consider the following equation:

$$\frac{x}{2} = 6$$

The addition and subtraction properties don't apply to its solution. However, since $\frac{12}{2} = 6$, we know that x must be 12. Another way to look at this problem is to see that if we multiply each side by 2, then it takes the form $x =$ a number.

$$\frac{x}{2} = 6$$

$$2 \cdot \frac{x}{2} = 6 \cdot 2 \quad \text{Multiply each side by 2.}$$

$$x = 12 \quad \text{Simplify.}$$

The solution set is indeed {12}.

In the same way, let's consider the equation

$$2x = 6$$

If we divided each side by 2, then x will stand alone:

$$2x = 6$$

$$\frac{2x}{2} = \frac{6}{2} \quad \text{Divide each side by 2.}$$

$$x = 3 \quad \text{Simplify.}$$

The solution set is {3}.

These two examples lead us to the following properties of equality.

> **Multiplication Property of Equality**
> If a, b, and c represent real numbers and $a = b$, then
> $$ac = bc$$
>
> **Division Property of Equality**
> If a, b, and c represent real numbers and $a = b$, $c \neq 0$, then
> $$\frac{a}{c} = \frac{b}{c}$$

In simplest terms, the multiplication and division properties of equality state that *if equals are multiplied or divided by equals (divisor $\neq$ 0), the results are equal.*

EXAMPLE 1 Solve the equation $3x = 18$.

SOLUTION Since x is multiplied by 3, we must *divide* both sides of the equation by 3 in order to get x alone on one side. When x is alone, its coefficient will be 1.

$$3x = 18$$
$$\frac{3x}{3} = \frac{18}{3} \quad \text{Divide each side by 3.}$$
$$x = 6 \quad \text{Simplify.}$$

We check the solution in the original equation.

$$3x = 18 \quad \text{Original equation}$$
$$3 \cdot 6 \stackrel{?}{=} 18 \quad \text{Substitute 6 for } x.$$
$$18 \stackrel{\checkmark}{=} 18$$

The solution set is {6}.

EXAMPLE 2 Solve the equation $\frac{x}{3} = 4$.

SOLUTION Because x is divided by 3, we must *multiply* each side by 3 to get x to stand alone.

$$\frac{x}{3} = 4$$
$$3 \cdot \frac{x}{3} = 3 \cdot 4 \quad \text{Multiply each side by 3.}$$
$$x = 12 \quad \text{Simplify.}$$

CHECK $\quad \frac{x}{3} = 4 \quad \text{Original equation}$
$$\frac{12}{3} \stackrel{?}{=} 4 \quad \text{Substitute 12 for } x.$$
$$4 \stackrel{\checkmark}{=} 4$$

The solution set is {12}.

EXAMPLE 3   Solve the equation $\frac{3}{4}x = 9$.

SOLUTION   In the expression $\frac{3}{4}x$, two things are happening: x is multiplied by 3 as well as divided by 4. To get x to stand alone on one side of the equation, we must *divide* each side by 3 and *multiply* each side by 4. We can accomplish both steps at once by multiplying each side by $\frac{4}{3}$.

$$\frac{3}{4}x = 9$$

$$\frac{4}{3} \cdot \frac{3}{4}x = \frac{4}{3} \cdot 9 \quad \text{Multiply each side by } \frac{4}{3}; \quad \frac{4}{3} \cdot \frac{3}{4}x = 1 \cdot x = x.$$

$$x = 12 \quad \text{Simplify.}$$

CHECK $\frac{3}{4}x = 9$ Original equation

$\frac{3}{4}(12) \stackrel{?}{=} 9$ Substitute 12 for x.

$9 \stackrel{\checkmark}{=} 9$

The solution set is {12}.

Not all equations using multiplication and division in their solutions involve so few operations as we just saw. Solving equations might require combining like terms as well as one or more of the properties of equality.

EXAMPLE 4   Solve the equation $3y + 4y - 2 = 12$.

SOLUTION   First we combine like terms on the left side of the equation:

$$3y + 4y - 2 = 12$$
$$7y - 2 = 12$$

Now we use the addition property of equality to get the term involving the variable alone on one side, then we use the division property of equality.

$7y - 2 + 2 = 12 + 2$ Add 2 to each side.

$7y = 14$

$\frac{7y}{7} = \frac{14}{7}$ Divide each side by 7.

$y = 2$ Simplify.

CHECK $3y + 4y - 2 = 12$ Original equation

$3(2) + 4(2) - 2 \stackrel{?}{=} 12$ Substitute 2 for x.

$$6 + 8 - 2 \stackrel{?}{=} 12$$
$$12 \stackrel{\checkmark}{=} 12$$

The solution set is {2}.

EXAMPLE 5 Solve the equation $6m - 7m = 12$.

SOLUTION
$$6m - 7m = 12$$
$$-m = 12 \quad \text{Combine like terms on the left side.}$$

Since we must solve the equation for m rather than $-m$, we multiply each side by -1. Remember that the product of two negatives is positive.

$$-1(-m) = -1(12)$$
$$m = -12$$

CHECK
$$6m - 7m = 12 \quad \text{Original equation}$$
$$6(-12) - 7(-12) \stackrel{?}{=} 12 \quad \text{Substitute } -12 \text{ for } m.$$
$$-72 + 84 \stackrel{?}{=} 12$$
$$12 \stackrel{\checkmark}{=} 12$$

The solution set is $\{-12\}$.

EXAMPLE 6 Solve the equation $5 - 3(x + 7) + 7 = 6[2 - 3(2x - 1)] - 6$.

SOLUTION We begin by simplifying each side of the equation.

$$5 - 3(x + 7) + 7 = 6[2 - 3(2x - 1)] - 6$$
$$5 - 3x - 21 + 7 = 6[2 - 6x + 3] - 6 \quad \text{Distributive property}$$
$$-9 - 3x = 6[5 - 6x] - 6 \quad \text{Combine like terms.}$$
$$-9 - 3x = 30 - 36x - 6 \quad \text{Distributive property}$$
$$-9 - 3x = 24 - 36x \quad \text{Combine like terms.}$$
$$-9 - 3x + 9 + 36x = 24 - 36x + 9 + 36x \quad \text{Add 9 and } 36x \text{ to each side.}$$
$$33x = 33 \quad \text{Combine like terms.}$$
$$\frac{33}{33}x = \frac{33}{33} \quad \text{Divide each side by 33.}$$
$$x = 1$$

CHECK
$$5 - 3(x + 7) + 7 = 6[2 - 3(2x - 1)] - 6 \quad \text{Original equation}$$
$$5 - 3(1 + 7) + 7 \stackrel{?}{=} 6[2 - 3(2 \cdot 1 - 1)] - 6 \quad \text{Substitute 1 for } x.$$
$$5 - 3(8) + 7 \stackrel{?}{=} 6[2 - 3(1)] - 6$$
$$5 - 24 + 7 \stackrel{?}{=} 6[-1] - 6$$
$$-12 \stackrel{\checkmark}{=} -12$$

The solution set is {1}.

When an equation involves fractions, we use the multiplication property of equality to rewrite the equation without fractions before applying other properties for simplification. For example, to eliminate the fractions from the equation

$$\frac{x}{2} - 4 = \frac{x+1}{3}$$

we multiply each side by the least common denominator, 6:

$$6\left[\frac{x}{2} - 4\right] = 6\left[\frac{x+1}{3}\right]$$

$$6\left(\frac{x}{2}\right) - 6(4) = 6\left(\frac{x+1}{3}\right) \quad \text{Distributive property}$$

$$3x - 24 = 2(x + 1)$$

We can now solve the equation in the manner of Examples 1–6.

$$3x - 24 = 2x + 2$$
$$3x - 2x = 2 + 24 \quad \text{Subtract } 2x \text{ and add 24 to each side.}$$
$$x = 26$$

CHECK
$$\frac{x}{2} - 4 \stackrel{?}{=} \frac{x+1}{3}$$
$$\frac{26}{2} - 4 \stackrel{?}{=} \frac{26+1}{3}$$
$$13 - 4 \stackrel{?}{=} \frac{27}{3}$$
$$9 \stackrel{\checkmark}{=} 9$$

The solution set is {26}.

Solving linear equations in one variable is easier if we follow these steps.

To Solve a Linear Equation

1. Clear the equation of fractions by multiplying each side by the least common denominator, the LCD.
2. Remove grouping symbols, using the distributive property if necessary.
3. Combine like terms on each side of the equation.
4. Isolate the variable term by using the addition or subtraction property of equality.
5. Use the multiplication or division property to solve for the variable.
6. Check the answer in the original equation.

EXAMPLE 7 Solve the equation $\frac{x-1}{4} + 3 = 2 + \frac{x}{2}$.

SOLUTION

$$\frac{x-1}{4} + 3 = 2 + \frac{x}{2}$$

$$4\left[\frac{x-1}{4} + 3\right] = 4\left[2 + \frac{x}{2}\right] \quad \text{Multiply each side by 4, the LCD.}$$

$$4\left(\frac{x-1}{4}\right) + 4(3) = 4(2) + 4\left(\frac{x}{2}\right) \quad \text{Distributive property}$$

$$x - 1 + 12 = 8 + 2x \quad \text{Remove grouping symbols.}$$

$$x + 11 = 8 + 2x \quad \text{Combine like terms.}$$

$$3 = x \quad \text{Isolate the variable}$$

CHECK

$$\frac{x-1}{4} + 3 = 2 + \frac{x}{2}$$

$$\frac{3-1}{4} + 3 \stackrel{?}{=} 2 + \frac{3}{2} \quad \text{Substitute 3 for } x.$$

$$\frac{1}{2} + 3 \stackrel{?}{=} 2 + \frac{3}{2}$$

$$\frac{7}{2} \stackrel{\checkmark}{=} \frac{7}{2}$$

The solution set is {3}.

EXAMPLE 8 Solve the equation $3.1(x - 4) - 22.24 = -1.2(x - 4.1)$. Check the answer with a calculator.

SOLUTION

$$3.1(x - 4) - 22.24 = -1.2(x - 4.1)$$

$$3.1x - 12.4 - 22.24 = -1.2x + 4.92 \quad \text{Remove grouping symbols.}$$

$$3.1x - 34.64 = -1.2x + 4.92 \quad \text{Combine like terms.}$$

$$3.1x + 1.2x = 4.92 + 34.64 \quad \text{Isolate the variable.}$$

$$4.3x = 39.56 \quad \text{Combine like terms.}$$

$$\frac{4.3x}{4.3} = \frac{39.56}{4.3} \quad \text{Divide each side by 4.3}$$

$$x = 9.2$$

CHECK

$$3.1(x - 4) - 22.24 = -1.2(x - 4.1) \quad \text{Original equation}$$

$$3.1(9.2 - 4) - 22.24 \stackrel{?}{=} -1.2(9.2 - 4.1) \quad \text{Substitute 9.2 for } x.$$

Using a Scientific Calculator

Left side 3.1 × (9.2 − 4) − 22.24 =   The display reads −6.12.
Right side 1.2 ± × (9.2 − 4.1) =   The display reads −6.12.

Using a Graphics Calculator

Left side 3.1 [(] 9.2 [−] 4 [)] [−] 22.24 [ENTER]   The display reads −6.12.
Right side [(−)] 1.2 [(] 9.2 [−] 4.1 [)] [ENTER]   The display reads −6.12.
↑
└ Negative key

The solution set is {9.2}.

> If a solution to an equation is a decimal and the decimal is rounded off, the check using the rounded result will not produce exact results.

For example, the solution to $2.3x = 1.95$ rounded to five decimal places is

$x = 0.84783$

CHECK $2.3(0.84783) \stackrel{?}{=} 1.95$
$1.950009 \stackrel{?}{=} 1.95$ Use a calculator to multiply.
$1.950009 \neq 1.95$

We now return to application problems where the first task is to translate the words into the form of an equation.

EXAMPLE 9 When three times a number is added to 12, the sum is 48. *Find the number.*

SOLUTION We let x = the number
$3x$ = three times the number

$$\underbrace{12}_{12} \underbrace{\text{added to}}_{+} \underbrace{\text{three times a number}}_{3x} \underbrace{\text{is}}_{=} \underbrace{48.}_{48}$$

$12 + 3x = 48$
$12 + 3x - 12 = 48 - 12$ Subtract 12 from each side.
$3x = 36$ Combine like terms.
$\dfrac{3x}{3} = \dfrac{36}{3}$ Divide each side by 3.
$x = 12$

The number is 12. Let's check this solution in the words of the problem. Three times 12 added to 12 is 48. The solution checks.

EXAMPLE 10 Twenty percent of the student body at a local community college voted in the last student election. The records indicate that 2978 votes were cast. *How many students attend the college?*

SOLUTION We let x = the number of students at the college. Then 20% of x is 2978.

$0.2x = 2978$ 20% = 0.2; "of" means multiplication.

$\dfrac{0.2x}{0.2} = \dfrac{2978}{0.2}$ Divide each side by 0.2.

$x = 14{,}890$

There are 14,890 students attending the college.

Exercise Set 3.3

Solve the following equations by using the multiplication or division property of equality. Check your answers.

1. $3x = 12$
2. $5x = 15$
3. $7y = -21$
4. $2y = -10$
5. $-5m = 25$
6. $-3p = 24$
7. $\dfrac{1}{3}p = 8$
8. $\dfrac{1}{4}m = 11$
9. $\dfrac{1}{2}a = -20$
10. $\dfrac{1}{3}k = 14$
11. $\dfrac{x}{7} = 4$
12. $\dfrac{y}{3} = 12$
13. $-\dfrac{m}{3} = 2$
14. $-\dfrac{x}{8} = 3$
15. $0.5k = 6$
16. $0.8m = 4$
17. $\dfrac{-3}{5}y = -6$
18. $\dfrac{-4}{7}p = -12$
19. $-25x = 625$
20. $-12x = 156$
21. $\dfrac{5m}{13} = 2$
22. $\dfrac{3k}{11} = 9$
23. $63p = 0$
24. $84y = 0$

Combine like terms, then solve by using the properties of equality.

25. $5p - 3p = 6$
26. $11x - 8x = 9$
27. $2y - 8y = 15 - 9$
28. $3m - 11m = -5 - 11$
29. $17x - 21x = 7 + 6$
30. $9y - 14y = 5 + 8$
31. $-k - 3k = 2 - 5$
32. $-7p - p = 1 - 8$
33. $7y - 18y + 4y = 5 - 11 + 2$
34. $2k - 13k + 5k = -6 + 9$
35. $9 - 6 - 3 = 11p - p$
36. $16 - 12 - 4 = 9m - m$
37. $x + 2x - 4x + 3x = 7 - 11 + 4 - 3 + 2$
38. $5y - y + 2y - 4y = 11 - 9 + 3 - 4$

Solve each of the following equations. Check your answers.

39. $2(2x - 3) = 20$
40. $5(3x + 4) = 30$
41. $7 = 2(5y - 3)$
42. $25 = 4(2y + 5)$
43. $3y - (y - 8) = 15$
44. $7y - (3y - 5) = 14$
45. $3a - 4 = 5(7 - a)$
46. $3b - 10 = 6(4 - b)$
47. $7 - 2(5x - 1) = 3$
48. $11 - 4(x - 3) = 1$
49. $4(x + 2) = 7(2 - x)$
50. $4(2x - 1) = 11(3 - x)$
51. $2(y + 3) = 7(y - 2)$
52. $9(y + 2) = 4(3y + 5)$
53. $7(3 - 4x) = 6(1 + 5x)$
54. $2 + 3(6 - m) = -21 + 4(2 + m)$
55. $9 + 4(k - 3) = 6 + 3(k - 2)$
56. $\dfrac{1}{4}(8m + 4) - 11 = -\dfrac{1}{2}(2m - 4)$
57. $\dfrac{1}{4}(8a - 72) + 20 = \dfrac{1}{3}(3a + 12)$
58. $2[5 - 2(3 - y)] - 1 = 3[2(3y - 2) + 8] - 25$
59. $6[2(6 + 3x) - 4] - 20 = -5[3(7 - x) - 4(8 + 2x)]$
60. $\dfrac{1}{2}\left(3x - \dfrac{1}{4}\right) - \dfrac{2}{3} = \dfrac{3}{4}$
61. $\dfrac{2}{5}\left(\dfrac{7}{3} - 4x\right) - \dfrac{1}{2} = \dfrac{1}{3}$
62. $\dfrac{1}{2}t - \dfrac{7}{2} = \dfrac{5}{2}t + \dfrac{1}{3}$
63. $\dfrac{3}{4}m - \dfrac{1}{3} = \dfrac{5}{8}m + \dfrac{1}{2}$
64. $1.1 - 0.7(5b + 6) = \dfrac{1}{2}(b + 2)$
65. $6.2 - 0.9(2p + 8) = \dfrac{1}{2}(p + 5)$
66. $-0.2(2x - 1) + 0.5 = 0.1(5x - 3) - 7.3$
67. $1.1(3x + 2) - 4.5 = 0.8(2x - 3) + 8.2$

68. $\dfrac{x-1}{3} = \dfrac{2-x}{4}$

69. $\dfrac{2x+3}{2} - \dfrac{x}{3} = \dfrac{3x-1}{6}$

70. $\dfrac{4-x}{2} - \dfrac{x}{5} = \dfrac{2x-5}{5}$

71. $\dfrac{7-3y}{2} - \dfrac{y}{4} = \dfrac{y-3}{4}$

72. $\dfrac{3y}{11} = \dfrac{5}{8}$

73. $\dfrac{2x}{5} = \dfrac{7}{6}$

74. $2.52y = 5.04$

75. $1.38x = 4.14$

76. $\dfrac{2.4}{3} m = 1.44$

77. $\dfrac{1.7}{5} x = 5.1$

Translate each of the following word problems into an algebraic equation and solve.

78. When a number is divided by 5, the result is 17. Find the number.

79. When a number is divided by -2 the result is 6. Find the number.

80. Matt jogs five times as far as Don. If the sum of the distances they jog is 12 kilometers (km), how far does Matt jog?

81. Rich walks 3 mi more than twice the distance Keith walks. If the sum of the distances they walk is 24 mi, how far does Rich walk?

82. Laura earns $156 on her savings account for one year at 4%. How much is in her savings account?

83. Rick earns $218 interest on his savings account for one year at 5%. How much is in his savings account?

84. Arturo's Motorcycle Shop has red and blue motorcycles in stock. If he has 3 more red than blue motorcycles in stock and has 33 motorcycles altogether, how many red motorcycles does he have?

85. Fred's car rental has three times as many compact cars for rent as large cars. If he has 32 cars for rent, how many of these are large cars?

86. Dan answered 76 questions correctly on his algebra test. If his score was 95%, how many questions were on the test?

87. Jessica took a test with 80 questions on it. If she had a score of 87.5%, how many questions did she answer correctly?

88. Lily rents a couch for one week for $82. If the rental agency charges $40 a week plus a daily fee, how much is the daily fee?

89. A plumber charges $38 for a house call plus $35/hr. If Jim's plumbing bill is $125.50, how long did the plumber work?

90. Grace bought a compact disc player for $198 and some CDs for $11 each. If the total bill (before taxes) was $275, how many CDs did she buy?

91. Last month Greg's telephone bill was $25.25. The company charges a flat rate of $15.20 plus 15¢/min. How many minutes did Greg talk on the phone?

92. Seventeen percent of the students in mathematics at a small college take their tests in the testing center. If 153 students take their test in the center, how many students are in mathematics?

93. Forty-eight percent of the students at Mistro College who start out in mathematics drop out before the end of the semester. If 408 students dropped last semester, how many students started in mathematics?

94. A sail in the shape of a triangle requires 12 yd² of material. If the base of the sail measures 4 yd, what is the height of the sail?

95. A rancher has 500 yd of fencing material to enclose a rectangular pasture. He wants the pasture to be 75 yd wide. How long will it be?

96. Artie drove 435 miles in $7\frac{1}{2}$ hr. What was her speed in miles per hour?

97. Max runs 205.2 m in 28.5 sec. What is his running speed in meters per second?

98. Alec measures his rectangular garden and finds the length to be 14.5 ft. If the area of the garden is 166.75 ft², what is the width?

99. A roll of carpet contains 450 ft². If the carpet is 12 ft wide, how long is it?

100. A rectangular garden is 36 ft long and is 8 ft longer than seven times the width. How wide is the garden?

Write in Your Own Words

101. State the addition/subtraction property of equality.
102. State the multiplication/division property of equality.
103. Describe what an "algebraic expression" is.
104. Describe what a "conditional equation" is.

Solve each of the following equations. Show each step and explain in words what you do in each step.

105. $3x - 4 = 8$

106. $5x + 7 = 11$

107. $2(3 - 2x) = x - 5$

108. $-3(4 - x) + 7 = 2x - 8$

3.4 Literal Equations and Formulas

OBJECTIVES

In this section we will learn to
1. solve literal equations for specified values of the variable; and
2. recognize and use many common formulas.

If we know the length and width of a rectangle, can we find its area? If you know the average speed at which you drive, can you determine how far you will travel in four hours? If you invest $50,000 at 12% yearly interest for a period of four years, can you calculate how much interest you will earn? We can find the answers to problems of this type provided that we know the **literal equation,** or formula, that relates their various parts. The formulas we need to answer these questions are as follows:

A = area = length $\times$ width = lw

d = distance = rate $\times$ time = rt

I = Interest = principal $\times$ rate $\times$ time = prt

EXAMPLE 1 Find the area of a rectangle with length 5 cm and width 2 cm.

SOLUTION

area = length $\times$ width = lw

$A = 5 \cdot 2$ Substitute $l = 5$ and $w = 2$.

$= 10 \text{ cm}^2$ Area is measured in square units.

The area is 10 square centimeters = 10 cm^2

EXAMPLE 2 If you average 55 miles per hour (55 mph or 55 miles/hr) on a trip 4 hours long, *how far will you go?*

SOLUTION

distance = rate $\times$ time = rt

$d = 55 \cdot 4$ Substitute $r = 55$ and $t = 4$.

$= 220$ miles

EXAMPLE 3 You have just been given $50,000 by your favorite aunt to finish your education. You decide to invest the money in a second trust deed that pays 12% interest and will be due in four years. The interest income can be used to help defray school expenses and you will still have the $50,000 when you graduate four years later. *How much interest will the deed earn over the four-year period?*

SOLUTION

interest = principal $\times$ rate $\times$ time = prt

$I = (50,000)(0.12)(4)$ Substitute $p = 50,000$, $r = 0.12$, and $t = 4$.

$= \$24,000$

In the previous examples, we knew the values of some of the variables and we found the value of the remaining variable by substitution. In the examples that follow, we will solve for one variable in terms of the other variables.

EXAMPLE 4   Solve $d = rt$ for r.

SOLUTION   To isolate r on one side of the equation, we divide each side by t.

$$d = rt$$

$$\frac{d}{t} = \frac{rt}{t} \quad \textit{Divide each side by } t.$$

$$\frac{d}{t} = r$$

Thus $r = \dfrac{d}{t}$.

In Example 4, if the distance traveled is given in kilometers and the time in days, then the units describing $\dfrac{d}{r}$ are $\dfrac{\text{kilometers}}{\text{day}}$ or kilometers per day (km/day).

EXAMPLE 5   Solve $P = 2l + 2w$ for w.

SOLUTION   Recall that $P = 2l + 2w$ is the formula for the perimeter of a rectangle. When we solve for w, we will have a formula for the width of a rectangle in terms of the perimeter and the length. We begin by isolating the term involving w on one side of the equation:

$$P = 2l + 2w$$

$$P - 2l = 2w \qquad \text{Subtract } 2l \text{ from each side.}$$

$$\frac{P - 2l}{2} = \frac{2w}{2} \qquad \text{Divide each side by 2.}$$

$$\frac{P - 2l}{2} = w$$

or

$$w = \frac{P - 2l}{2}$$

We can obtain an equivalent form of this result by dividing each of the terms on the left side of the equation $P - 2l = 2w$ by 2.

$$P - 2l = 2w$$

$$\frac{P}{2} - \frac{2l}{2} = \frac{2w}{2}$$

$$\frac{P}{2} - l = w \quad \text{or} \quad w = \frac{P}{2} - l$$

EXAMPLE 6 Newton's second law of motion states that force is equal to the product of the mass of a body and its acceleration:

force = mass × acceleration

$F = ma$

Solve this equation for the acceleration, a.

SOLUTION

$F = ma$

$\dfrac{F}{m} = \dfrac{ma}{m}$ Divide each side by m.

$\dfrac{F}{m} = a$

EXAMPLE 7 To change temperature from degrees Celsius to degrees Fahrenheit, we use the formula

$$F = \dfrac{9}{5}C + 32$$

Find a formula to change temperature from degrees Fahrenheit to degrees Celsius.

SOLUTION If we subtract 32 from each side first, we can then get C to stand alone by multiplying each side by $\dfrac{5}{9}$:

$F = \dfrac{9}{5}C + 32$

$F - 32 = \dfrac{9}{5}C$ Subtract 32 from each side.

$\dfrac{5}{9}(F - 32) = \dfrac{5}{9}\left(\dfrac{9}{5}C\right)$ Multiply each side by $\dfrac{5}{9}$.

$\dfrac{5}{9}(F - 32) = C$ Simplify.

The formula for changing temperature from degrees Fahrenheit to degrees Celsius is

$$C = \dfrac{5}{9}(F - 32)$$

EXAMPLE 8 Solve $bx - 2ay^2 = az$ for x.

SOLUTION

$bx - 2ay^2 = az$

$bx = az + 2ay^2$ Add $2ay^2$ to each side.

$x = \dfrac{az + 2ay^2}{b}$ Divide each side by b.

EXAMPLE 9 Solve $\dfrac{4+y}{a} = \dfrac{2-x}{b}$ for y.

SOLUTION We begin by multiplying each side by the least common denominator, ab, to clear of fractions.

$$ab\left(\dfrac{4+y}{a}\right) = ab\left(\dfrac{2-x}{b}\right)$$

$b(4+y) = a(2-x)$ Simplify by dividing out common factors.

$4b + by = 2a - ax$ Distributive property

$4b + by - 4b = 2a - ax - 4b$ Subtract $4b$ from each side to isolate the y-term.

$by = 2a - ax - 4b$ Combine like terms.

$\dfrac{by}{b} = \dfrac{2a - ax - 4b}{b}$ Divide each side by b.

$y = \dfrac{2a - ax - 4b}{b}$

EXAMPLE 10 Solve for h in the formula $V = \dfrac{1}{3}\pi r^2 h$. Find h if $V = 25\pi$ and $r = 3$.

SOLUTION

$V = \dfrac{1}{3}\pi r^2 h$

$3V = \pi r^2 h$ Multiply each side by 3.

$\dfrac{3V}{\pi r^2} = h$ Divide each side by πr^2.

$h = \dfrac{3(25\pi)}{\pi(3^2)}$ Substitute $V = 25\pi$, $r = 3$.

$= \dfrac{3(25\pi)}{\pi(9)} = \dfrac{25}{3}$

EXAMPLE 11 The diameter of a circle is increased from 10,000 inches to 10,002 inches. How much are the circumference and area increased?

SOLUTION

SMALLER CIRCLE   LARGER CIRCLE

$C = \pi d = 10{,}000\pi$ in. $C = \pi d = 10{,}002\pi$ in.

$A = \pi r^2$ $A = \pi r^2$

$= (5{,}000)^2 \pi$ $= (5{,}001)^2 \pi$

$= 25{,}000{,}000\pi$ in². $= 25{,}010{,}001\pi$ in².

We let x = the increase in circumference. Then

x = larger circumference − smaller circumference

$= 10{,}002\pi$ in. − $10{,}000\pi$ in.

$= 2\pi$ in.

The increase in the circumference is 2π in. Since $\pi \approx 3.14$, the circumference increases about 6.28 in. How much would the circumference increase if the original diameter was 1,000,000 inches and it was increased by 2 in.?

Now let's find the increase in the area. We let
y = the increase in area. Then

y = larger area $-$ smaller area
$= 25{,}010{,}001\pi$ in$^2 - 25{,}000{,}000\pi$ in^2
$= 10{,}001\pi$ in^2

The increase in area is $10{,}001\pi$ in^2

Exercise Set 3.4

Solve the following literal equations for the specified variable.

1. $d = rt$ for r (distance formula)
2. $I = prt$ for t (simple interest)
3. $P = 4s$ for s (the perimeter of a square)
4. $A = \frac{1}{2}bh$ for b (the area of a triangle)
5. $F = ma$ for m (Newton's second law)
6. $C = 2\pi r$ for r (the circumference of a circle)
7. $I = \frac{V}{R}$ for V (Ohm's law)
8. $C = \frac{5}{9}(F - 32)$ for F (degrees Fahrenheit to degrees Celsius)
9. $P = \frac{144p}{y}$ for p (pressure–head formula)
10. $R = \frac{CS}{d}$ for C (revolution of a cutting-head)
11. $P = a + b + c$ for b (the perimeter of a triangle)
12. $A = bh$ for h (the area of a parallelogram)
13. $T = m - n$ for n (tolerance)
14. $m = \frac{d}{n}$ for d (mpg, which is distance divided by number of gallons)
15. $R = 2s$ for s (revolution–stroke formula)
16. $w_1 d_1 = w_2 d_2$ for d_1 (lever–balance formula)
17. $F = \frac{lt}{d}$ for l (wedge–force formula)
18. $A = \frac{a+b}{2}$ for b (the average of two numbers)
19. $I = \frac{100M}{C}$ for M (intelligence quotient formula)
20. $C = 19n + 5$ for n (the cost of sending a package)
21. $C = \frac{100w}{L}$ for w (the cephalic index)
22. $0.001\, l = 1$ cc for l (liters to cubic centimeters)
23. 1000 mcg $= 1$ mg for mcg (converting micrograms to milligrams)
24. $R_x = \frac{R_2 R_3}{R_1}$ for R_3 (Wheatstone Bridge)
25. $P = \frac{I}{O}$ for I (percentage of salary)
26. $E = \frac{O}{I}$ for I (efficiency formula)
27. $V = lwh$ for w (the volume of a rectangular solid)
28. $V = \pi r^2 h$ for h (the volume of a cylinder)
29. $P = 2l + 2w$ for l (the perimeter of a rectangle)
30. $y = mx + b$ for x (the slope–intercept form of the equation of a line)
31. $m = \frac{y_2 - y_1}{x_2 - x_1}$ for y_2 (the slope of a line)
32. $ax + by = c$ for y (the general form of the equation of a line)
33. $A = \frac{1}{2}h(B + b)$ for h (the area of a trapezoid)
34. $A = \frac{1}{3}\pi r^2 h$ for h (the volume of a cone)
35. $W = Fd$ for F (work)
36. $A = \frac{a+b+c+d}{4}$ for c (the average of 4 numbers)

37. $A = \pi r^2$ for π (the area of a circle)
38. $V = \frac{4}{3}\pi r^3$ for π (the volume of a sphere)
39. $y - y_1 = m(x - x_1)$ for m (the point–slope form of the equation of a line)
40. $V = \frac{1}{3}Bh$ for "B" (volume of a pyramid)

Find the value of the unknown variable.

41. $I = prt$; $p = 1000$, $r = 0.09$, and $t = 3$
42. $C = \frac{5}{9}(F - 32)$; $F = 36$
43. $P = 2l + 2w$; $l = 15$ and $w = 10$
44. $F = \frac{lt}{d}$; $l = 100$, $t = 2$, and $d = 4$
45. $A = \frac{a + b}{2}$; $a = 86$ and $b = 74$
46. $P = \frac{I}{O}$; $I = 140$ and $O = 800$
47. $A = \frac{1}{2}bh$; $b = 17$ and $h = 14$
48. $E = \frac{O}{I}$; $O = 60$ and $I = 120$
49. $R = 2s$; $R = 40$
50. $C = 19n + 5$; $C = 214$
51. $m = \frac{d}{n}$; $d = 340$ mi and $n = 16$ gal
52. $T = m - n$; $T = 0.001$ and $m = 0.003$
53. $I = prt$; $I = 160$, $p = 2000$, and $t = 1$
54. $A = \frac{1}{2}h(B + b)$; $h = 12$, $B = 10$, and $b = 8$

Use a calculator, if needed, to find the value of the unknown variable in each of the following. Round all answers to three decimal places.

55. $A = \pi r^2$; $\pi = 3.1416$ and $r = 5.2$
56. $F = ma$; $m = 19$ and $a = 864.6$
57. $2.5L = A$; $A = 127.64$
58. $I = prt$; $I = \$186.18$, $p = \$791.77$, and $t = 1.5$
59. $m = \frac{d}{n}$; $d = 825$ and $n = 26.3$
60. $I = \frac{100M}{C}$; $I = 120$ and $M = 17.4$

Solve each of the following equations for the indicated variable.

61. $x + y = 3$ for y
62. $3x + y = 7$ for y
63. $ax + by = c$ for x
64. $y = mx + b$ for x
65. $ax + b = c$ for x
66. $ky + b = c$ for y
67. $\frac{x}{3} + \frac{6}{3} = \frac{1}{3}$ for x
68. $\frac{x}{5} + \frac{y}{5} = \frac{1}{5}$ for y
69. $\frac{y}{7} + c = g$ for y
70. $\frac{y}{p} - q = t$ for y
71. $-5(x - y) = 7$ for x
72. $3(x + 2y) = 4$ for x
73. $6(x + 3z) = -5$ for z
74. $-2(y - 3z) = 11$ for z
75. $\frac{a}{3} + \frac{y}{3} = p$ for y
76. $\frac{-b}{7} + \frac{x}{7} = t$ for x
77. $\frac{1}{2}(r - t) = B$ for t
78. $\frac{1}{3}(p - q) = m$ for q
79. $\frac{2}{3}(x + y) = \frac{1}{3}d$ for y
80. $\frac{3}{4}(2x + y) = \frac{1}{4}p$ for y

Solve each of the following applied problems. Use formulas as needed.

81. Milt drove along a winding mountain road for 91 mi. If the drive took 7 gal of gas, what was his gas mileage (in miles per gallon)?
82. If Jeannette drove 196 mi and her car averages 28 mpg, how many gallons of gas did she use?
83. Mr. Diaz drives 348 mi at an average speed of 58 mph. Assuming he stops once for an hour, how long does the trip take?
84. Mrs. Meissner plans to travel for $4\frac{1}{2}$ hr at an average speed of 63 mph. How far will she go?
85. Two students leave San Luis Obispo at the same time. One travels north at 60 mph and the other travels south at 50 mph. How long will it take them to be 330 mi apart?
86. Rhona and Marci leave at the same time and travel in the same direction. Marci travels at 67 mph and Rhona travels at 59 mph. How long will it be before they are 28 mi apart?
87. If the radius of a circle is 8 in., find the circumference. Leave the answer in terms of π.
88. The perimeter of a triangle is 66 in. If one side is 12 in. and the second side is 29 in., find the length of the third side.
89. Find the interest JoAnn will accumulate if she deposits $600 at 10% annual interest for two years.
90. Sam invests $600 for a year and accumulates $30 in interest. What was the rate of interest.

91. How much should be invested at 7.5% for two years to accumulate $120 in interest?
92. If $500 is invested at 8% annual interest, how long will it take to accumulate $500 in interest?
93. A leopard can run at a speed of 60 mph for 2 min. How far can it run in that time?
94. An antelope can run at a speed of 34 mph for 20 min. How far can it run in that time?
95. A pizza shop sells two small circular pizzas (7-in. diameter) for the same price as one large circular pizza (10-in. diameter). Which is the better buy?
96. A bakery sells a square cake (8 in. on a side and 2 in. high) for the same price as a circular one (9 in. in diameter and 2 in. high). Which is the better buy?
97. How much water (in cubic feet) will a spherical tank hold if the diameter of the tank is 30 ft? See the figure. Use $\frac{22}{7}$ for π.

98. A running track is to be constructed by putting two semicircles on the ends of parallel straight lines. See the figure. Find the distance around the track. Use $\pi = 3.14$.

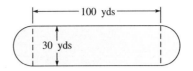

REVIEW EXERCISES

Solve each of the following equations. Check your answers.

99. $5p - 2(p + 3) = p$ 100. $7k - 5(k - 3) = k$
101. $-(y + 2) + 2(y - 3) + 5 = 0$
102. $5(p - 1) - (p + 40) - 11 = 0$
103. $18 - 3(1 - m) + 5 = 0$
104. $12 - 2(1 + x) - 8 = 0$
105. $3(x - 5) = 3x + 4$ 106. $2(y - 8) = 2y + 5$
107. $4k - 2(k + 4) = 3(k - 2) + 2 - k$
108. $5z - 3(z - 4) = 2(z + 4) + 3$
109. $\frac{-4}{3} + \frac{7}{8} = \frac{10}{3}v + \frac{1}{3}v$
110. $\frac{17}{5} - \frac{21}{10} = \frac{11}{10}x + \frac{3}{2}x$
111. $x - \frac{1}{2}x = 5\frac{1}{2} - 7\frac{3}{4}$ 112. $-y + \frac{10}{3}y = -8\frac{1}{3}$
113. $12\frac{2}{3} - 7\frac{5}{6} = \frac{t}{6}$ 114. $-9\frac{1}{3} - 5\frac{3}{4} = \frac{-2s}{3}$

3.5 The Addition and Subtraction Properties of Inequality

HISTORICAL COMMENT The symbols for inequality, $<$ and $>$, date back to the early 1600s and were used by the English mathematician/astronomer Thomas Harriot (1560–1621). Another English mathematician and the inventor of the slide rule, William Oughtred (1574–1660), used the symbols "[" to mean "is greater than" and "]" to mean "is less than." The latter symbols were still in use more than a century later at Harvard University.

OBJECTIVES In this section we will learn to
1. graph intervals on a number line;
2. solve inequalities using the addition property of inequality;
3. solve inequalities using the subtraction property of inequality; and
4. solve compound inequalities.

An inequality is a statement that two quantities are not necessarily equal. If they are not equal, then one must be either greater than or less than the other.

DEFINITION **Less than; Greater than**	A number *a* is **less than** a number *b*, written $a < b$, if *a* is to the left of *b* on the number line. A number *a* is **greater than** a number *b*, written $a > b$, if *a* is to the right of *b* on the number line.

For example, -5 is less than -1 because it is to the left of -1 on the number line; see Figure 3.2

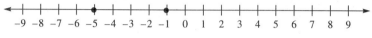

FIGURE 3.2

Let's summarize the meaning of the inequality symbols.

Symbols of Inequality

$<$ means is less than $\qquad$ $>$ means is greater than

$\leq$ means is less than or equal to $\qquad$ $\geq$ means is greater than or equal to

---

The **solution** to an inequality is an **interval** on a number line. To say $x > 2$ means that x can be any real number to the right of 2 on the number line. The **graph** of $x > 2$ is shown in Figure 3.3. The parenthesis at 2 indicates that the number 2 itself is not part of the solution set. The arrow pointing to the right indicates that all numbers to the right of 2 are in the solution set.

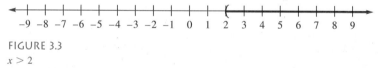

FIGURE 3.3
$x > 2$

If $x \geq 2$, then x can be 2 or any real number to the right of 2 on the number line. To indicate that 2 is in the solution set of this inequality, we put a bracket at 2 (see Figure 3.4). The arrow pointing to the right indicates that all numbers to the right of 2 are also in the solution set.

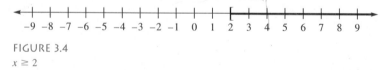

FIGURE 3.4
$x \geq 2$

EXAMPLE 1 Graph the solution set of $x \leq 4$.

SOLUTION The graph of the solution set is shown in Figure 3.5. The bracket at the point 4 indicates that the solution set includes the number 4. The arrow pointing to the left indicates that all numbers to the left of 4 are also in the solution set.

FIGURE 3.5
$x \leq 4$

What is the solution to $2x + 9 < x - 7$? How do we get x to stand alone? Are there properties of inequalities that we can use to isolate x on one side? Let's consider what happens when we add the same number to both sides of an inequality.

$$1 < 4 \quad \text{True}$$
$$1 + 2 \stackrel{?}{<} 4 + 2 \quad \text{Add 2 to each side}$$
$$3 < 6 \quad \text{True}$$

Now let's consider what happens when we subtract the same number from both sides of an inequality.

$$1 < 4 \quad \text{True}$$
$$1 - 5 \stackrel{?}{<} 4 - 5 \quad \text{Subtract 5 from each side.}$$
$$-4 < -1 \quad \text{True}$$

These examples illustrate that adding or subtracting the same quantity to or from both sides of an inequality yields an equivalent inequality. The direction that the inequality symbol points will not change.

Addition Property of Inequality

If a, b, and c represent real numbers and $a < b$, then

$$a + c < b + c$$

Subtraction Property of Inequality

If a, b, and c represent real numbers and $a < b$, then

$$a - c < b - c$$

Although the properties are stated for the symbol $<$, they are also true for $\leq$, $>$, and $\geq$. In words, the properties state that *the same number can be added to or subtracted from each side of an inequality without changing the direction the inequality symbol points*. The solution set is the same as that of the original inequality.

EXAMPLE 2 Solve the inequality $x + 2 \geq 4$ and graph the solution set.

SOLUTION
$$x + 2 \geq 4$$
$$x + 2 - 2 \geq 4 - 2 \quad \text{Subtract 2 from each side.}$$
$$x \geq 2$$

The solution set is $\{x \mid x \geq 2\}$. The graph is shown in Figure 3.6.

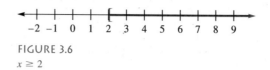

FIGURE 3.6
$x \geq 2$

EXAMPLE 3   Solve the inequality $2x - 3 < x + 4$ and graph the solution set.

SOLUTION   We must use both the addition and subtraction properties.

$$2x - 3 < x + 4$$
$$2x - 3 - x < x + 4 - x \quad \text{Subtract } x \text{ from each side.}$$
$$x - 3 < 4 \quad \text{Combine like terms.}$$
$$x - 3 + 3 < 4 + 3 \quad \text{Add 3 to each side.}$$
$$x < 7 \quad \text{Combine like terms.}$$

The solution set is $\{x \mid x < 7\}$. The graph is shown in Figure 3.7.

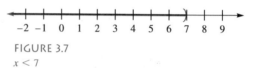

FIGURE 3.7
$x < 7$

EXAMPLE 4   Solve the inequality $3(x - 5) + 2 < 2(x - 2)$.

SOLUTION   We begin by removing the grouping symbols.

$$3(x - 5) + 2 < 2(x - 2)$$
$$3x - 15 + 2 < 2x - 4 \quad \text{Distributive property}$$
$$3x - 13 < 2x - 4 \quad \text{Combine like terms.}$$
$$3x - 13 - 2x < 2x - 4 - 2x \quad \text{Subtract } 2x \text{ from each side.}$$
$$x - 13 < -4 \quad \text{Combine like terms.}$$
$$x - 13 + 13 < -4 + 13 \quad \text{Add 13 to each side.}$$
$$x < 9 \quad \text{Combine like terms.}$$

The solution set is $\{x \mid x < 9\}$.

EXAMPLE 5   Solve the inequality $-5(-x - 1.2) - (-3.4) \geq 3(x - 2.3) + x$.

SOLUTION
$$-5(-x - 1.2) - (-3.4) \geq 3(x - 2.3) + x$$
$$5x + 6 + 3.4 \geq 3x - 6.9 + x \quad \text{Remove grouping symbols.}$$
$$5x + 9.4 \geq 4x - 6.9 \quad \text{Combine like terms.}$$
$$5x - 4x \geq -6.9 - 9.4 \quad \text{Subtract 9.4 and } 4x \text{ from each side.}$$
$$x \geq -16.3 \quad \text{Combine like terms.}$$

The solution set is $\{x \mid x \geq -16.3\}$.

EXAMPLE 6   Graph the solution set for $\{x \mid x > -2 \text{ and } x \text{ is an integer}\}$.

SOLUTION   The integers that are greater than -2 form the set $\{-1, 0, 1, 2, 3, \ldots\}$. Its graph shown in Figure 3.8, consists of separate points, one for each integer greater than -2. The three dots on the right indicate that the pattern continues to the right without end.

FIGURE 3.8

We now turn our attention to **compound inequalities,** which are formed by joining two inequalities involving the same unknown. For example, the compound inequality

$$5 < x + 3 \leq 10$$

which is read, "5 *is less than* $x + 3$ *which is less than or equal to* 10," is made up of the two parts

$$5 < x + 3 \quad \text{and} \quad x + 3 \leq 10$$

It's not necessary to separate a compound inequality into two parts in order to solve it. All that's necessary is to isolate the variable in the middle, as follows:

$$5 < x + 3 \leq 10$$
$$5 - 3 < x + 3 - 3 \leq 10 - 3 \quad \text{Subtract 3 from each part.}$$
$$2 < x \leq 7 \quad \text{Simplify.}$$

The solution set is $\{x \mid 2 < x \leq 7\}$. The graph of the solution set is shown in Figure 3.9. The parenthesis on the left side indicates that 2 *is not* in the solution set, while the bracket on the right side indicates that 7 *is* in the solution set.

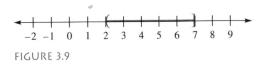

FIGURE 3.9

EXAMPLE 7   Solve and graph the compound inequality $-6 \leq x - 5 \leq -1$.

SOLUTION
$$-6 \leq x - 5 \leq -1$$
$$-6 + 5 \leq x - 5 + 5 \leq -1 + 5 \quad \text{Add 5 to each part.}$$
$$-1 \leq x \leq 4 \quad \text{Combine like terms.}$$

The solution set is $\{x \mid -1 \leq x \leq 4\}$. The graph is shown in Figure 3.10.

FIGURE 3.10

EXAMPLE 8   Solve: $4 + 2y < 3(y - 2) < 2(y + 3)$.

SOLUTION   We begin by removing the grouping symbols.

$$4 + 2y < 3(y - 2) < 2(y + 3)$$
$$4 + 2y < 3y - 6 < 2y + 6 \qquad \text{Distributive property}$$

To get the *y*-term alone in the middle, we subtract $2y$ and add 6 to each part.

$$4 + 2y - 2y + 6 < 3y - 6 - 2y + 6 < 2y + 6 - 2y + 6$$
$$10 < y < 12$$

The solution set is $\{y \mid 10 < y < 12\}$.

To solve applied problems involving inequalities, we must be able to interpret the meaning of certain phrases correctly. Here are a few common phrases

Phrase	Interpretation
x lies between 8 and 12	$8 < x < 12$
x lies between 8 and 12, *inclusive*	$8 \leq x \leq 12$
x is at least 9	$x \geq 9$
x is no more than 5	$x \leq 5$
x is more than 5	$x > 5$
x is less than 12	$x < 12$
x is at most 8	$x \leq 8$
x exceeds 5	$x > 5$

EXAMPLE 9   If 3 is added to five times a number, the result is greater than 6 more than four times the number. *Find the number.*

SOLUTION   We let $x =$ the number
$5x + 3 =$ three added to five times the number,
$4x + 6 =$ six more than four times the number

$$5x + 3 > 4x + 6$$
$$x > 3 \qquad \text{Subtract } 4x \text{ and 3 from each side.}$$

The solution set is any number greater than 3, or $\{x \mid x > 3\}$.

EXAMPLE 10   The length of a rectangle is 3 feet more than the width. If the width lies between 5 and 7, what compound inequality describes the possible length?

SOLUTION   We let $w =$ the width
$w + 3 =$ the length

$5 < w < 7$ The width lies between 5 and 7.

Since the width lies between 5 and 7 and since the length is represented as $w + 3$, we can add 3 to each part of the inequality $5 < w < 7$ to get an inequality that describes the length.

3.5 · THE ADDITION AND SUBTRACTION PROPERTIES OF INEQUALITY   143

$$5 < w < 7$$
$$5 + 3 < w + 3 < 7 + 3 \qquad \text{Add 3 to each part.}$$
$$8 < l < 10 \qquad \text{Simplify.}$$

The length lies between 8 ft and 10 ft.

Do You Remember?

Can you match these?

More than one answer may apply.

____ 1. The addition property of inequality
____ 2. The subtraction property of inequality
____ 3. A compound inequality
____ 4. x is at least 3.
____ 5. x is no more than 4.
____ 6. x lies between 8 and 12.
____ 7. x is at most 4.
____ 8. x exceeds 4.
____ 9. A statement equivalent to $x < 10$
____ 10. An incorrect statement

a) If $x < 3$, then $x + 3 > 3 + 3$.
b) $x \geq 3$
c) $x \leq 4$
d) $x > 4$
e) $8 \leq x \leq 12$
f) If $a > b$, then $a + c > b + c$.
g) $10 > x$
h) If $a < b$, then $a - c < b - c$.
i) $8 < x < 12$
j) If $9 > 5$, then $9 - 11 > 5 - 11$.

Answers: 1. f 2. h, j 3. c, i 4. b 5. c 6. i 7. c 8. d 9. g 10. a

Exercise Set 3.5

Draw a number line and graph each of the following sets. Assume the variable is an element of the set of real numbers unless specified.

1. $\{x \mid x < 3, x \text{ is a natural number}\}$
2. $\{x \mid 3 < x \leq 5, x \text{ is a natural number}\}$
3. $\{x \mid x \leq 5, x \text{ is a whole number}\}$
4. $\{x \mid x \geq 2, x \text{ is a whole number}\}$
5. $\{x \mid -3 \leq x \leq 1, x \text{ is an integer}\}$
6. $\{x \mid 4 < x < 5, x \text{ is an integer}\}$
7. $\{y \mid -4 \leq y < 0, y \text{ is an integer}\}$
8. $\{y \mid 0 \leq y < 1, y \text{ is a whole number}\}$
9. $\{x \mid x \geq 3\}$
10. $\{x \mid x < -2\}$
11. $\{y \mid -1 < y < 1\}$
12. $\{y \mid 4 \leq y \leq 5\}$
13. $\{y \mid 1 \leq y \leq 7\}$
14. $\{x \mid x > -3\}$
15. $\{x \mid x < 7\}$

Solve the following inequalities.

16. $x + 3 > 5$
17. $x + 6 > 4$
18. $y - 5 < 7$
19. $y - 2 < 9$
20. $m + 8 \geq 1$
21. $p - 6 \leq 5$
22. $3x + 4 - 2x < 6$
23. $5x - 2 - 4x > 1$
24. $-x - 5 + 2x \geq -7 + 5$
25. $-3y + y + 3y < 8 - 6$
26. $\frac{7}{2}x \geq \frac{5}{2}x + 5$
27. $\frac{5}{3}x \leq \frac{2}{3}x + 3$
28. $6y + 4 \leq 5y + 2$
29. $7y + 1 \geq 6y + 3$
30. $0 < x - 1 < 0.5$
31. $2 < x + 1 < 4.1$
32. $-4 \leq x + 2 \leq 6$
33. $-3 \leq x - 1 \leq 2$
34. $2(y - 5) > y - 4$
35. $4(x - 2) < 3x - 6$
36. $8x - 11 \leq 7(x - 2)$
37. $7y - 5 \geq 3(2y - 2)$
38. $2 \leq p - 3 \leq 5$
39. $-7 \leq x - 4 \leq -1$

40. $\frac{3}{4} \geq \frac{6y}{5} - \frac{y}{5} \geq 0$ **41.** $\frac{6}{5} \geq \frac{5m}{2} - \frac{3m}{2} \geq -1$

42. $4(m - 2) < 3(m + 5)$ **43.** $7(y + 2) > 2(3y - 6)$

44. $-6(-x + 1) + 5 \geq 5(x - 1) + 2$

45. $-8(-p - 2) - 12 \leq 7(p + 1) - 2$

Solve and graph the following inequalities on a real number line. Assume x is a real number.

46. $x - 4 \geq -3$ **47.** $x + 5 \geq 2$

48. $x - 6 < -5$ **49.** $x + 2 < -5$

50. $0 < x \leq 5$ **51.** $0 \leq x < 7$

52. $-4 \leq x < -1$ **53.** $-6 \leq x \leq 2$

54. $3x - 4 > 2x + 7$ **55.** $7x - 2 > 6x - 9$

56. $2(x - 5) \geq x + 4$ **57.** $4(x - 2) \geq 3x - 1$

Solve each of the following inequalities.

58. $-5 \leq 7 + x \leq 23$ **59.** $-1 \leq x + 7 < 2$

60. $-2 \leq \frac{-2y}{3} + \frac{5}{3}y \leq 3$ **61.** $-5 \leq \frac{2y}{3} + \frac{y}{3} \leq -2$

62. $-7 < 6 + x < 4$ **63.** $-7 < 5 + x < 8$

64. $2x + 3 \leq 3(x - 1) \leq 2x + 7$

65. $x - 2 \leq 2(x + 5) \leq x + 10$

66. $3y - 1 < 4(1 + y) < 3y + 2$

67. $5y + 7 < -6(1 - y) < 8 + 5y$

68. $\frac{1}{3}x - \frac{1}{4} < \frac{4}{3}(2 + x) \leq \frac{1}{3}\left(x - \frac{1}{2}\right)$

69. $\frac{1}{2}x + \frac{1}{3} \leq \frac{-3}{2}(4 - x) < \frac{1}{2}(3 + x)$

A calculator may be helpful to solve the following inequalities.

70. $x - 7.63 \geq 6.18$ **71.** $x - 9.75 \geq 2.35$

72. $0.6x + 1.9 \leq -0.4x + 7.1$

73. $1.9x + 3.2 \leq 0.9x + 4.8$

74. $2.181x - 1.913 > 1.181x + 0.007$

75. $4.091x - 8.671 > 6.199 + 3.091x$

Translate the following applied problems into inequalities and solve.

76. Seven less than a natural number is greater than or equal to 6.

77. The sum of x and 42 is no more than 84.

78. The difference of a whole number and 5 is between 6 and 9. Find all such whole numbers.

79. When 4 is subtracted from an integer, the result is between 14 and 29. Find all such integers.

80. Polly's scores on her first three history tests were 87, 92, and 85. What must she get on her fourth test so that the sum of her test scores is at least 360?

81. Bob earned 94 points on the last test. Jan earned x points on the last test. The sum of their scores was more than 178. What is the lowest score Jan could have received?

82. To earn a B in algebra, Tio's final score on five tests must be at least 400 but less than 450. If his scores were 86, 85, 91, and 88 on the first four tests, what range of scores could he get on the fifth test to earn a B?

83. To earn an A in trigonometry, Jill's final score on five tests must be at least 450. Her scores on the first four tests were 92, 88, 84, and 90. What is the least she must earn on the fifth test to earn an A?

84. Last year's prices at Bennie's Tire Shop for B78-13 tires ranged from a low of $35.23 to a high of $72.84. This can be expressed as $35.23 \leq p \leq 72.84$, where p represents the price. If p is decreased by $7.61, write a new compound inequality that expresses the new price range.

85. Diana sells bouquets of flowers from a low price of $6.50 to a high price of $13.75. This can be expressed as $6.50 \leq p \leq 13.75$. If p is increased by $3.25, write a new compound inequality that expresses this new price range.

86. The side of a square is no more than 14 in. Express the perimeter as an inequality.

87. The side of an equilateral triangle is at least 13 cm. Express the perimeter of the triangle as an inequality.

88. A rectangle's length is 2 ft more than its width, which is between 4 ft and 6 ft. What can the length be? What can the perimeter be?

89. A parallelogram's width is 5 m less than its length. The width is between 4 m and 6 m. What can its length be? What can its perimeter be?

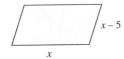

For Extra Thought

90. In order that a certain drug have a beneficial effect, its concentration in the bloodstream must exceed a certain value, called *the minimum therapeutic level*. When this drug is administered, there is a certain time interval t when this occurs, given by the inequality

$$(t - 1)(t - 4) < 0$$

Determine the time interval; write a compound inequality representing it.

91. The speed limit on Highway 37 has been established at 55 mph. Since it is difficult for a motorist to maintain a speed of exactly 55 mph, a "reasonable variance" of the speed s has been established and is given by the inequality

$$(s - 52)(s - 58) < 0$$

Determine the range of speeds that are considered to be within the "reasonable variance" and write it as a compound inequality.

3.6 The Multiplication and Division Properties of Inequality

OBJECTIVES

In this section we will learn to
1. solve inequalities using the multiplication property of inequality;
2. solve inequalities using the division property of inequality;
3. solve applied problems using multiplication and division.

With only the addition and subtraction properties of inequality, we can't solve inequalities such as $\frac{1}{2}x \geq 5$ and $-3x < 7$. Like equations, these require the use of multiplication or division to find the solution. To see whether or not an inequality is changed by multiplying each side by a constant, let's consider the following examples.

MULTIPLICATION BY A POSITIVE NUMBER

$-2 < 4$	True
$2(-2)$? $2 \cdot 4$	Multiply each side by 2.
$-4 < 8$	True

Both the original inequality and the result of multiplying both sides by 2 are true statements. Thus,

> Multiplying each side of an inequality by a positive quantity does not change the direction of the inequality symbol.

Now let's consider what happens when we multiply both sides of an inequality by a *negative* number.

MULTIPLICATION BY A NEGATIVE NUMBER

$-2 < 4$ True

$-2(-2) \; ? \; -2(4)$ Multiply each side by -2.

$4 > -8$ True

> Multiplying each side of an inequality by a negative number reverses the direction of the inequality symbol.

Division of both sides of an inequality by the same quantity has the same effect as multiplication.

DIVISION BY A POSITIVE NUMBER

$4 < 8$ True

$\dfrac{4}{2} \; ? \; \dfrac{8}{2}$ Divide each side by 2.

$2 < 4$ True

> Dividing each side of an inequality by a positive quantity does not change the direction of the inequality symbol.

DIVISION BY A NEGATIVE NUMBER

$4 < 8$ True

$\dfrac{4}{-2} \; ? \; \dfrac{8}{-2}$ Divide each side by -2.

$-2 > -4$ True

> Dividing each side of an inequality by a negative quantity reverses the direction of the inequality symbol.

> **Multiplication Property of Inequality**
>
> If a, b, and c represent real numbers and $a < b$, then
>
> $\quad a \cdot c < b \cdot c$ if c is **positive** $(c > 0)$
>
> and
>
> $\quad a \cdot c > b \cdot c$ if c is **negative** $(c < 0)$

> **Division Property of Inequality**
>
> If a, b, and c represent real numbers and $a < b$, then
>
> $$\frac{a}{c} < \frac{b}{c} \quad \text{if} \quad c \text{ is } \textbf{positive} \ (c > 0)$$
>
> and
>
> $$\frac{a}{c} > \frac{b}{c} \quad \text{if} \quad c \text{ is } \textbf{negative} \ (c < 0)$$

Although the properties are stated for the symbol $<$, they are also true for $\leq$, $>$, and $\geq$.

EXAMPLE 1 Solve and graph the solution set of the inequality $3x + 2 \geq x + 6$.

SOLUTION

$3x + 2 \geq x + 6$	
$3x + 2 - x \geq x + 6 - x$	Subtract x from each side.
$2x + 2 \geq 6$	Combine like terms.
$2x + 2 - 2 \geq 6 - 2$	Subtract 2 from each side.
$2x \geq 4$	Combine like terms.
$\dfrac{2x}{2} \geq \dfrac{4}{2}$	Divide each side by 2.
$x \geq 2$	

The solution set is $\{x \mid x \geq 2\}$. The graph is shown in Figure 3.11.

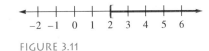

FIGURE 3.11

EXAMPLE 2 Solve the inequality $3(y - 2) - 4(y + 1) < 7$.

SOLUTION

$3(y - 2) - 4(y + 1) < 7$	
$3y - 6 - 4y - 4 < 7$	Distributive property
$-y - 10 < 7$	Combine like terms.
$-y - 10 + 10 < 7 + 10$	Add 10 to each side.
$-y < 17$	Combine like terms.

To solve for y, we multiply each side by -1 and reverse the inequality symbol:

$$-1(-y) > -1(17)$$
$$y > -17$$

The solution set is $\{y \mid y > -17\}$.

It is worth noting that if we had isolated the variable in Example 2 on the right side instead of on the left side of the inequality symbol, the solution would have been $-17 < y$, which is the same as $y > -17$.

In general,

> $x < a$ is equivalent to $a > x$
>
> and
>
> $x > a$ is equivalent to $a < x$

EXAMPLE 3 Solve the inequality $\frac{3}{4}x - \frac{1}{3} < \frac{2}{5}\left(x - \frac{1}{3}\right)$.

SOLUTION

$$\frac{3}{4}x - \frac{1}{3} < \frac{2}{5}\left(x - \frac{1}{3}\right)$$

$$\frac{3}{4}x - \frac{1}{3} < \frac{2}{5}x - \frac{2}{15} \qquad \text{Distributive property}$$

$$60\left(\frac{3}{4}x - \frac{1}{3}\right) < 60\left(\frac{2}{5}x - \frac{2}{15}\right) \qquad \text{Multiply each side by 60, the LCD.}$$

$$60\left(\frac{3}{4}x\right) - 60\left(\frac{1}{3}\right) < 60\left(\frac{2}{5}x\right) - 60\left(\frac{2}{15}\right) \qquad \text{Distributive property}$$

$$45x - 20 < 24x - 8 \qquad \text{Simplify.}$$

$$45x - 20 - 24x < 24x - 8 - 24x \qquad \text{Subtract } 24x \text{ from each side.}$$

$$21x - 20 < -8 \qquad \text{Combine like terms.}$$

$$21x - 20 + 20 < -8 + 20 \qquad \text{Add 20 to each side.}$$

$$21x < 12 \qquad \text{Combine like terms.}$$

$$\frac{21x}{21} < \frac{12}{21} \qquad \text{Divide each side by 21.}$$

$$x < \frac{4}{7} \qquad \text{Simplify.}$$

The solution set is $\left\{x \mid x < \frac{4}{7}\right\}$.

EXAMPLE 4 Solve and graph the compound inequality $11 \geq -2x + 4 \geq 8$.

SOLUTION

$$11 \geq -2x + 4 \geq 8$$

$$11 - 4 \geq -2x + 4 - 4 \geq 8 - 4 \qquad \text{Subtract 4 from each part.}$$

$$7 \geq -2x \geq 4 \qquad \text{Combine like terms.}$$

$$\frac{7}{-2} \le \frac{-2x}{-2} \le \frac{4}{-2}$$ Divide each part by -2 and reverse the inequality symbols.

$$-\frac{7}{2} \le x \le -2$$ Simplify.

The solution set is $\left\{ x \mid -\frac{7}{2} \le x \le -2 \right\}$; the graph is shown in Figure 3.12.

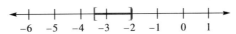

FIGURE 3.12
$-\frac{7}{2} \le x \le -2$

EXAMPLE 5
Use set-builder notation to write the solution set for the inequalities with the following graphs:

(a)    (b)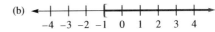

SOLUTION
(a) A bracket appears at -2, so -2 is a solution. A parenthesis appears at 3, so 3 is not a solution. All values between these two numbers are in the solution set, which is $\{x \mid -2 \le x < 3\}$.

(b) A bracket appears at -1, so -1 is a solution. The arrow going to the right indicates that all numbers greater than -1 are also in the solution set. The solution set is $\{x \mid x \ge -1\}$.

EXAMPLE 6
Sarah had scores of 74, 85, 72, and 82 on four algebra tests. What score should she get on her next test to maintain an average of at least 80?

SOLUTION
We let $x =$ the score on the fifth test. To compute her average, we add the test scores and divide by the number of scores.

$$\frac{74 + 85 + 72 + 82 + x}{5} \ge 80$$

$$\frac{313 + x}{5} \ge 80$$ Combine like terms.

$$5\left(\frac{313 + x}{5}\right) \ge 5 \cdot 80$$ Multiply each side by 5.

$$313 + x \ge 400$$ Simplify.

$$x \ge 87$$ Subtract 313 from each side.

Sarah must get at least an 87 on her next test.

EXAMPLE 7 Company A rents a certain motorhome for $250/week plus 20¢/mile. Company B rents the identical motorhome for $300/week but charges only 10¢/mile. If you rented a motorhome from Company B to go on vacation, how far would you have to drive each week before it was cheaper than renting from Company A?

Let's let m = the number of miles to be driven. Then

cost for Company B < cost for Company A

$300 + 0.10m < 250 + 0.20m$

$300 - 250 + 0.1m < 250 + 0.2m - 250$ Subtract 250 from each side; $0.10m = 0.1m$ and $0.20m = 0.2m$.

$50 + 0.1m < 0.2m$ Combine like terms.

$50 + 0.1m - 0.1m < 0.2m - 0.1m$ Subtract $0.1m$ from each side.

$50 < 0.1m$ Combine like terms.

$\dfrac{50}{0.1} < \dfrac{0.1m}{0.1}$ Divide each side by 0.1.

$500 < m$ or $m > 500$

If you plan to average over 500 miles/week, it's cheaper to rent from Company B.

Notice that the properties we use to solve linear inequalities closely parallel those we use to solve linear equations. Here are the only differences:

1. *When both sides of a linear inequality are multiplied by a negative number, the direction of the inequality must be reversed.*
2. *When both sides of a linear inequality are divided by a negative number, the direction of the inequality must be reversed.*

Adding or subtracting the same quantity from both sides of an equation or an inequality creates an equivalent equation or inequality, as does multiplying both sides of an equation or an inequality by a positive quantity.

Do You Remember?

Can you match these?

_____ 1. If $a < b$ and $c < 0$, then ac ___ bc.   a) $<$

_____ 2. If $a > b$ and $c > 0$, then ac ___ bc.   b) $>$

_____ 3. If $a \geq b$ and $c > 0$, then $\dfrac{a}{c}$ ___ $\dfrac{b}{c}$.   c) $\leq$

                                            d) $\geq$

_____ 4. If $a \geq b$ and $c < 0$, then $\dfrac{a}{c}$ ___ $\dfrac{b}{c}$.   e) $y > -17$

                                            f) $y < -17$

_____ 5. If $-2y < 34$, then _____ .

Answers: 1. b 2. b 3. d 4. c 5. e

Exercise Set 3.6

Solve and graph each inequality.

1. $3x < 12$
2. $7x < 35$
3. $-10y \geq 30$
4. $-15y \geq 45$
5. $\dfrac{x}{4} > -1$
6. $\dfrac{x}{5} > -12$
7. $9p \leq -63$
8. $21k \leq -84$
9. $7y + 4 < -10$
10. $2x + 3 \leq 9$
11. $\dfrac{x}{2} - 7 > 11$
12. $\dfrac{p}{3} - 9 < 21$
13. $-14x - 5 \leq 37$
14. $-13y - 3 \geq 40$
15. $-3x + 11 \geq 2$
16. $-11y + 15 > 4$
17. $8y + 5 < 3y + 15$
18. $11x + 2 \leq 6x + 7$

Write the inequality illustrated by each graph.

19.
20.
21.
22.
23.
24.
25.
26.
27.
28.

Find the solution set for each of the following inequalities.

29. $5x - 18 > 2x + 3$
30. $6x - 5 < 2x + 11$
31. $x + 9 \leq 7x - 3$
32. $y + 1 \geq 7y - 5$
33. $2y - 5 \geq 4y + 11$
34. $3k - 7 > 6k + 2$
35. $2(x + 4) > -3(2x)$
36. $-5(p - 1) \leq 2(p + 2)$
37. $\dfrac{p}{-4} - 6 \geq -9$
38. $\dfrac{x}{2} - 8 < 7$
39. $4(-y + 5) - 15 > 0$
40. $6(x - 7) + 30 \leq 0$
41. $-4 \leq 5y + 6 \leq 21$
42. $2 < 3p - 7 < 14$
43. $-3 \leq 3 - 2x < 5$
44. $-5 < 7 - 4x \leq 15$
45. $\dfrac{1}{2}x + 7 \geq \dfrac{2}{3}x$
46. $\dfrac{1}{6}y \leq 1 - \dfrac{3}{4}y$
47. $\dfrac{-5}{6}x < \dfrac{3}{8} - \dfrac{3}{4}x$
48. $\dfrac{-1}{2}x > \dfrac{1}{4} - \dfrac{2x}{3}$
49. $\dfrac{x}{10} > -\dfrac{2}{3} - \dfrac{3x}{5}$
50. $\dfrac{7}{12}y > \dfrac{3}{8} + \dfrac{5y}{4}$
51. $p \geq 2.7 + 1.5p$
52. $x \geq 3.5 + 0.5x$
53. $x - 3284 < -6102$
54. $y - 1011 > -2516$
55. $0.10p < -1.8 + 5.2$
56. $1.41 - 2.9 + 0.2p \leq 5$
57. $7.9x + 8.39 \geq 1.1x - 3$
58. $x - 14.52 > 90$
59. $3(x - 4) + 2(x + 3) < 24$
60. $5(t + 1) - 3(1 - t) > -7$
61. $-5(5 - x) < -11$
62. $-2(x - 5) > -14$
63. $\dfrac{y + 3}{5} + \dfrac{1 - y}{2} > \dfrac{1}{5}$
64. $\dfrac{p + 3}{7} - \dfrac{5 - p}{4} < \dfrac{1}{14}$
65. $\dfrac{-(6 - p)}{8} + \dfrac{-(p + 2)}{7} > -1$
66. $\dfrac{y - 3}{7} + \dfrac{-(2 - y)}{4} \leq \dfrac{3}{14}$
67. $\dfrac{7 - y}{-2} + \dfrac{3 + y}{3} > 5$
68. $\dfrac{-(4 - n)}{4} + \dfrac{n - 2}{3} < 8$
69. $\dfrac{2(x + 3)}{6} + \dfrac{-(-1 - x)}{5} < 0$
70. $\dfrac{-3(6 - x)}{7} - \dfrac{-2(3 + x)}{2} > 0$

Write an inequality for each of the following statements.

71. x is negative.
72. x is positive.
73. y is nonnegative.
74. y is not positive.
75. x is at least 7.
76. x is at most -5.
77. y is no more than 10.
78. y is no less than 6.

79. x is greater than -8 and less than 4.
80. x is at least 0 and at most 8.

Write an inequality for each of the following applied problems and solve it.

81. Five more than three times a number is greater than 29. Find all such numbers for which this is true.
82. Six less than six times a number is less than 31. Find all such numbers for which this true.
83. The perimeter of a rectangle is to be no greater than 81 in. and the length of the rectangle is three more than twice the width. Find the possible values of the width.

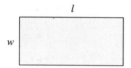

84. One side of a triangle is twice as long as another side. The third side is 12 ft long. If the perimeter is at least 39 ft, find the possible values for the shortest side.

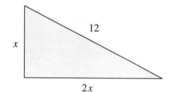

85. Kim has $5000 to invest. If she invests $3000 at 9%, at what rate must she invest the rest ($2000) so that she will earn at least $480 in total interest?
86. Max has $8000 to invest. If he invests $4000 at 8%, at what rate must he invest the rest ($4000) so that he will earn no less than $500 in total interest?
87. Terry bowled 136, 150, and 157 on her first three games. What is the least she can bowl on her fourth game to maintain an average of more than 150?
88. James shot rounds of 88, 87, 79, and 84 on his first four rounds of golf. What can he shoot on his fifth round and maintain an average of 84 or less for the five rounds?
89. In designing a house, Adcock's Architectural Designs decide that the length of one room will be 14 ft and the area of the room will be at least 196 ft². What is the minimum width of the room?
90. The base cost of a mixer is $18. The base cost plus the profit margin must be less than $27 in order for the mixer to sell. Find the profit margin.
91. Keith works at odd jobs while attending college. He has earned $200, $300, and $225 during the last three months. If his average earnings over a four-month period must be at least $250, what must he earn during the fourth month?
92. The number of male geology students is 6 more than the number of female students. If the total number of students in the class is to be no more than 44, find the maximum number of male and female students.
93. The board of trustees of Cedarvale Community College decide to pave a rectangular parking lot. If the length of the parking lot is to be 40 m and the total area they can afford to pave is less than 720 m², find the maximum width (in whole numbers).
94. The cost of a long distance telephone call is $0.49 for the first minute and $0.29 for each additional minute. If the total cost of a call must not exceed $6.29, find the longest the call can be.

REVIEW EXERCISES

95. Solve for B; $A = \frac{1}{2}h(B + b)$
96. Solve for x; $y = mx + b$
97. Solve for h; $A = \frac{1}{2}bh$
98. Solve for t; $d = rt$
99. Solve for r; $C = 2\pi r$
100. Solve for m; $y - y_1 = m(x - x_1)$
101. Solve for x; $3 = \frac{y - x}{x}$
102. Solve for R_1; $\frac{F}{F_1} = \frac{R}{R_1}$

3.7 Solving Absolute Value Equations and Inequalities

OBJECTIVES In this section we will learn to
1. solve absolute value equations;
2. solve absolute value inequalities involving $>$ and $\geq$; and
3. solve absolute value inequalities involving $<$ and $\leq$.

In Chapter 2, we defined the absolute value of a number to be its distance from the origin. This definition turns out to be quite useful in determining the solution set of equations and inequalities involving absolute value. For example, consider the equation

$$|x| = 4$$

There are two numbers that are four units from the origin, 4 and -4. Therefore the solution set of this equation is $\{-4, 4\}$.

EXAMPLE 1 Solve and graph the equation $|x + 5| = 3$.

SOLUTION Two numbers, 3 and -3, have an absolute value of 3. Thus,

$x + 5 = 3$	or	$x + 3 = -3$	
$x + 5 - 5 = 3 - 5$	or	$x + 5 - 5 = -3 - 5$	Subtract 5 from each side.
$x = -2$	or	$x = -8$	

CHECK $|-2 + 5| \stackrel{?}{=} 3$ or $|-8 + 5| \stackrel{?}{=} 3$ $\quad x = -2$ or -8
$|3| \stackrel{?}{=} 3$ or $|-3| \stackrel{?}{=} 3$
$3 \stackrel{\checkmark}{=} 3$ or $3 \stackrel{\checkmark}{=} 3$

The solution set is $\{-2, -8\}$. The graph is shown in Figure 3.13.

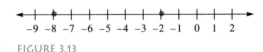

FIGURE 3.13

EXAMPLE 2 Solve: $\left|\dfrac{1}{2}x - 3\right| = 4$.

SOLUTION Two numbers, 4 and -4, have an absolute value of 4.

$\dfrac{1}{2}x - 3 = 4$	or	$\dfrac{1}{2}x - 3 = -4$	
$\dfrac{1}{2}x - 3 + 3 = 4 + 3$	or	$\dfrac{1}{2}x - 3 + 3 = -4 + 3$	Add 3 to each side.
$\dfrac{1}{2}x = 7$	or	$\dfrac{1}{2}x = -1$	
$x = 14$	or	$x = -2$	Multiply each side by 2.

CHECK $\left|\frac{1}{2}(14) - 3\right| \stackrel{?}{=} 4$ or $\left|\frac{1}{2}(-2) - 3\right| \stackrel{?}{=} 4$ $x = 14$ or -2

$|7 - 3| \stackrel{?}{=} 4$ or $|-1 - 3| \stackrel{?}{=} 4$

$|4| \stackrel{?}{=} 4$ or $|-4| \stackrel{?}{=} 4$

$4 \stackrel{?}{=} 4$ or $4 \stackrel{\checkmark}{=} 4$

The solution set is $\{-2, 14\}$.

The Examples 1 and 2 lead us to the following rule for solving absolute value equations.

> **To Solve an Absolute Value Equation**
>
> The solution to $|x| = p$, where p is a positive number, can be found by solving the two equations
>
> $x = p$ or $x = -p$

EXAMPLE 3 Solve: $|2x - 3| - 1 = 6$.

SOLUTION To solve this equation, we must first isolate the term involving the absolute value on one side.

$|2x - 3| - 1 = 6$

$|2x - 3| - 1 + 1 = 6 + 1$ Add 1 to each side.

$|2x - 3| = 7$

We can now solve the equation by applying the preceding rule.

$2x - 3 = 7$ or $2x - 3 = -7$

$2x = 10$ or $2x = -4$ Add 3 to each side.

$x = 5$ or $x = -2$ Divide each side by 2.

The solution set is $\{-2, 5\}$. The check is left to the reader.

*C*AUTION

In order to solve an equation involving absolute value, the term involving the absolute value must be isolated on one side.

EXAMPLE 4 Solve: $\left|\frac{2}{3}x - \frac{1}{2}\right| + \frac{1}{6} = \frac{1}{4}$.

SOLUTION First we isolate the term involving the absolute value.

$\left|\frac{2}{3}x - \frac{1}{2}\right| + \frac{1}{6} = \frac{1}{4}$

$\left|\frac{2}{3}x - \frac{1}{2}\right| = \frac{1}{4} - \frac{1}{6}$ Subtract $\frac{1}{6}$ from each side.

$$\left|\frac{2}{3}x - \frac{1}{2}\right| = \frac{1}{12}$$ Simplify the right side: $\frac{1}{4} - \frac{1}{6} = \frac{1}{12}$.

$\frac{2}{3}x - \frac{1}{2} = \frac{1}{12}$ or $\frac{2}{3}x - \frac{1}{2} = -\frac{1}{12}$

$8x - 6 = 1$ or $8x - 6 = -1$ Multiply each side by 12, the LCD.

$8x = 7$ or $8x = 5$ Add 6 to each side.

$x = \frac{7}{8}$ or $x = \frac{5}{8}$ Divide each side by 8.

The check is left to the reader. The solution set is $\left\{\frac{5}{8}, \frac{7}{8}\right\}$.

Recall that the rule for solving the absolute value equation $|x| = p$ requires that p be a positive number. The next example shows what happens if p is not positive.

EXAMPLE 5 Solve the following absolute value equations:

(a) $|3x + 1| = -4$ (b) $|2x - 1| = 0$

SOLUTION (a) The absolute value of a number cannot be negative, so the solution set is the empty set, $\emptyset$.

(b) The only way that $|2x - 1|$ can be zero is if $2x - 1$ itself is zero, since $|0| = 0$.

$2x - 1 = 0$

$2x = 1$

$x = \frac{1}{2}$

The solution set is $\left\{\frac{1}{2}\right\}$. The check is left to the reader.

We now turn our attention to absolute value inequalities involving the symbols $<$ or $\leq$. Consider the inequality

$|x| < 3$

This inequality refers to all numbers that are less than 3 units from the origin. Such numbers lie between -3 and 3. Therefore the solution set is $\{x \mid -3 < x < 3\}$. The graph is shown in Figure 3.14.

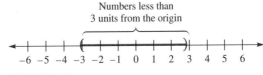

FIGURE 3.14

EXAMPLE 6 Solve and graph: $|x + 2| < 2$.

SOLUTION All of the numbers with absolute value less than 2 lie between -2 and 2. Thus, $|x + 2| < 2$ means that

$$-2 < x + 2 < 2$$
$$-2 - 2 < x + 2 - 2 < 2 - 2 \quad \text{Subtract 2 from each part.}$$
$$-4 < x < 0 \quad \text{Simplify each part.}$$

The solution set is $\{x \mid -4 < x < 0\}$. The graph is shown in Figure 3.15.

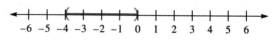

FIGURE 3.15

EXAMPLE 7 Solve: $|3 - 2x| + 2 < 9$.

SOLUTION We must first isolate the absolute value term by subtracting 2 from each side.

$$|3 - 2x| + 2 < 9$$
$$|3 - 2x| < 7 \quad \text{Subtract 2 from each side.}$$

All numbers with absolute value less than 7 lie between -7 and 7. Thus

$$-7 < 3 - 2x < 7$$
$$-7 - 3 < 3 - 2x - 3 < 7 - 3 \quad \text{Subtract 3 from each part.}$$
$$-10 < -2x < 4 \quad \text{Simplify each part.}$$
$$5 > x > -2 \quad \text{Divide each side by } -2; \text{ change the direction of the inequalities.}$$

The solution set is $\{x \mid -2 < x < 5\}$.

When an absolute value inequality involves the symbol $>$ or $\geq$, it is referring to numbers that are *more* than a certain distance from the origin. For example, the inequality

$$|x| \geq 3$$

refers to all numbers that are 3 or more units from the origin. It is satisfied by those numbers for which $x \leq -3$ or $x \geq 3$, as shown in Figure 3.16

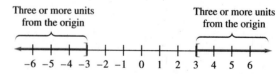

FIGURE 3.16

3.7 · SOLVING ABSOLUTE VALUE EQUATIONS AND INEQUALITIES 157

EXAMPLE 8 Solve the graph: $|2x - 4| \geq 2$.

SOLUTION The inequality $|2x - 4| \geq 2$ means

$2x - 4 \leq -2$ or $2x - 4 \geq 2$
$2x \leq 2$ or $2x \geq 6$ Add 4 to each side.
$x \leq 1$ or $x \geq 3$ Divide each side by 2.

The solution set is $\{x \mid x \leq 1 \text{ or } x \geq 3\}$. The graph is shown in Figure 3.17.

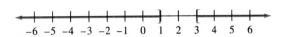

FIGURE 3.17

EXAMPLE 9 Solve: $\left|\dfrac{5 - 2x}{3}\right| \geq 2$.

SOLUTION The inequality $\left|\dfrac{5 - 2x}{3}\right| \geq 2$ means

$\dfrac{5 - 2x}{3} \leq -2$ or $\dfrac{5 - 2x}{3} \geq 2$
$5 - 2x \leq -6$ or $5 - 2x \geq 6$ Multiply each side by 3.
$-2x \leq -11$ or $-2x \geq 1$ Subtract 5 from each side.
$x \geq \dfrac{11}{2}$ or $x \leq -\dfrac{1}{2}$ Divide each side by -2; reverse the inequality.

The solution set is $\left\{x \mid x \leq -\dfrac{1}{2} \text{ or } x \geq \dfrac{11}{2}\right\}$.

To Interpret Absolute Value Inequalities

If p represents a positive number, then

$|x| < p$ means $-p < x < p$

and

$|x| > p$ means $x < -p$ or $x > p$

Notice that these guidelines for interpreting absolute value inequalities are stated in terms of a *positive* number p. What happens when this is not the case?

EXAMPLE 10 Solve: $|x + 5| > -3$.

SOLUTION The absolute value of a number is always positive or zero, so it will be greater than a negative number for any value of *x*. Hence we can conclude that the solution set is $\{x \mid x \text{ is a real number}\}$.

The next example shows a special case where the symbol "≤" is involved.

EXAMPLE 11 Solve: $|x + 2| \leq 0$.

SOLUTION The absolute value cannot be negative. However, $|x + 2| = 0$ if $x = -2$. Therefore the solution set is $\{-2\}$.

EXAMPLE 12 Solve: $-|x + 2| > -2$.

SOLUTION We begin by multiplying each side by -1 so that $|x + 2|$ has a positive coefficient:

$$-1(-|x + 2|) < -1(-2)$$

$$|x + 2| < 2$$

Multiplying by a negative number changes the direction of the inequality.

This problem was solved in Example 6. The solution set is $\{x \mid -4 < x < 0\}$.

CAUTION — The absolute value of a number can be less than or equal to zero only if the number itself is zero.

Do You Remember?

Can you match these?

___ 1. If $|x| = p$ means $x = p$ or $x = -p$, then p is a ___ number.
___ 2. If $|x| \leq 5$, then ___.
___ 3. If $|x| > 5$, then ___.
___ 4. The solution set for $|x| > -3$ is ___.
___ 5. The solution set for $|x| < -3$ is ___.
___ 6. The solution set for $|x| \leq 0$ is ___.
___ 7. The solution set for $|x| < p, p > 0$, is ___.
___ 8. The solution set for $|x - 3| - 1 = 2$ is ___.

a) $-5 \leq x \leq 5$
b) $\emptyset$
c) $\{x \mid x \text{ is a real number}\}$
d) $\{0, 6\}$
e) $x < -5$ or $x > 5$
f) positive
g) negative
h) $\{x \mid -p < x < p\}$
i) $\{0\}$

Answers: 1. f 2. a 3. e 4. c 5. b 6. i 7. h 8. d

Exercise Set 3.7

Solve each absolute value equation. If there is no solution, say so.

1. $|x| = 4$
2. $|x| = 7$
3. $|y| = \frac{1}{2}$
4. $|y| = \frac{1}{4}$
5. $|x| = 0$
6. $|x| = -2$
7. $|y| = -3$
8. $|x| - 2 = -2$
9. $|2x| = 4$
10. $|3x| = 6$
11. $|-y| = 5$
12. $|-y| = -2$
13. $-|x| = -8$
14. $-|x| = -4$
15. $-|-2x| = 11$
16. $-|-5y| = 1$
17. $\left|\frac{1}{2}x\right| = \frac{1}{3}$
18. $\left|\frac{1}{3}x\right| = \frac{1}{2}$
19. $|x| + 2 = 3$
20. $|x| - 3 = 5$
21. $|x + 2| = 4$
22. $|x - 3| = 13$
23. $|2 - y| = 1$
24. $|3 - 2y| = 0$
25. $|2x + 3| = 1$
26. $|3x + 4| = 2$
27. $|4y + 5| = 0$
28. $|3 - 4x| = 2$
29. $|5 - 2x| = \frac{1}{2}$
30. $|7 - 4x| = \frac{2}{3}$
31. $\left|2 - \frac{1}{2}x\right| = \frac{1}{3}$
32. $\left|4 - \frac{3}{8}y\right| = \frac{1}{4}$
33. $|2 - x| + 5 = 7$
34. $|5 - x| - 3 = 11$
35. $|2 - 4x| - 3 = 0$
36. $|x - 8| - 2 = 3$
37. $|x + 5| - 4 = -3$
38. $|2x + 7| - 8 = -7$
39. $\left|\frac{1}{2}x + 2\right| + 3 = 3$
40. $\left|\frac{1}{3}x + 5\right| - 7 = -7$

Solve each absolute value inequality. If there is no solution, say so.

41. $|x| < 3$
42. $|x| < -1$
43. $|x| > 2$
44. $|x| > 0$
45. $|y| > -2$
46. $|p| < 0$
47. $|k| \leq 0$
48. $|s| \geq 0$
49. $-|x| < 2$
50. $-|y| > -3$
51. $-|y| > 7$
52. $-|x| < -4$
53. $|x + 2| \leq 1$
54. $|x - 3| \geq 4$
55. $|x + 7| \geq -5$
56. $|y - 3| \leq 0$
57. $|4y + 5| < 8$
58. $|5x - 1| > 12$
59. $|7x + 2| \geq 0$
60. $-|3y + 5| \geq 0$
61. $|x - 7| + 3 \leq 0$
62. $\left|\frac{1}{2}x + 1\right| \leq \frac{1}{4}$
63. $\left|\frac{2}{3}x - 4\right| \geq \frac{3}{8}$
64. $\left|3 + \frac{5}{8}y\right| > -\frac{1}{2}$
65. $\left|\frac{x}{3} - \frac{1}{3}\right| \leq \frac{5}{8}$
66. $\left|\frac{2x}{3} - \frac{5}{7}\right| < \frac{8}{13}$
67. $\left|\frac{3y}{5} - \frac{1}{3}\right| \geq \frac{5}{11}$
68. $\left|\frac{x}{3}\right| + \frac{1}{2} < 0$
69. $\left|\frac{7x}{4} + \frac{7}{10}\right| < \frac{1}{2}$
70. $\left|\frac{x + 2}{4}\right| \geq 5$
71. $\left|\frac{2x - 1}{3}\right| \leq 15$
72. $\left|\frac{3x + 8}{11}\right| < 0$
73. $\left|\frac{5y - 7}{3}\right| > -1$
74. $|-x - 3| - 2 > 6$
75. $|-y - 4| + 6 \leq 13$
76. $-|3 - y| + 5 \geq 4$
77. $-|2 - x| - 1 > 0$
78. $|5 - 4x| - 1 < 7$
79. $|7 - 3x| + 2 > -5$
80. $|x + 6| < 0$

Review Exercises

Solve each of the following equations or inequalities. If there is no solution, say so.

81. $\frac{3}{4} = \frac{x}{10}$
82. $\frac{m}{3} = \frac{1}{4}$
83. $\frac{5}{9} = \frac{k}{7}$
84. $\frac{x}{3} = 0$
85. $\frac{7p}{5} = \frac{4}{0}$
86. $\frac{3}{4}k = \frac{1}{2}$
87. $\frac{5}{7}m = \frac{3}{8}$
88. $\frac{x}{3} < \frac{4}{7}$
89. $\frac{1}{2}y < \frac{1}{4}$
90. $\frac{k - 2}{7} = \frac{4}{7}$
91. $\frac{p + 1}{5} = \frac{1}{5}$
92. $\frac{x - 3}{3} = \frac{4}{3}$
93. $\frac{2y + 1}{3} = \frac{1}{3}$
94. $\frac{5 - t}{4} = \frac{1}{4}$
95. $\frac{2p - 1}{2} = 2$
96. $\frac{3s + 5}{6} = 2$
97. $\frac{x - 2}{2} = \frac{7}{2}$
98. $\frac{3x - 2}{2} = \frac{1}{2}$
99. $\frac{5m + 7}{3} = \frac{5}{3}$
100. $\frac{3x}{2} = \frac{3}{2}$
101. $\frac{4y}{8} = \frac{2}{4}$
102. $\frac{6p - 7}{3} = 0$
103. $\frac{9x - 5}{3} = 0$

Write each of the following applied problems as an equation and solve it.

104. A lawn mower uses 3 qt of gas to mow 8 acres of lawn. How many quarts are needed to mow 32 acres of lawn?

105. If 2 oz of fertilizer cover 50 ft² of lawn, how many ounces are needed to cover 225 ft² of lawn?

106. If 1 in. on a map represents 16 miles, how many miles are represented by $3\frac{1}{4}$ in.?

107. One-half of an inch on an architectural drawing represents 2 ft on a building. How many feet are represented by $5\frac{1}{4}$ in.?

108. Three peanut candy bars contain 450 calories. How many peanut candy bars contain 1050 calories?

109. If walking 3 mi burns 270 calories, how many miles does Marvin need to walk to burn 490.5 calories?

110. If one-half of an orange provides 45 calories, how many calories do 6 oranges provide?

111. A shop charges $2.25 for a half dozen donuts. What is the cost for 27 donuts?

3.8 More Applications

OBJECTIVES

In this section we will learn to
1. solve applied problems by using the properties of equality; and
2. solve applied problems by using the properties of inequality.

We have gradually developed a process to help us arrive at a method for solving word problems. Some suggestions for planning strategy were given in Section 2.6, but to solve more difficult problems, we need to approach them through a logical sequence of steps. Learning to follow these steps will avoid a lot of frustration. For a particular formula, refer to Exercise Set 3.4, where many of them can be found.

To Solve a Word Problem

1. Read the problem through completely as many times as necessary to understand the meaning of the problem.
2. In reading the problem, ask yourself, "What am I asked to find?"
3. Represent what you are asked to find by a variable.
4. Establish the relationship between the variable and other parts of the problem with an equation or an inequality.
5. Solve the equation or inequality and check your results in the words of the original problem.

The next two examples illustrate these steps.

EXAMPLE 1 Paul is 2 years older than Sean and the sum of their ages is 36. *How old is each?*

SOLUTION **Step 1 Read the problem completely.**
Paul is 2 years older than Sean and the sum of their ages is 36. How old is each?

Step 2 What am I asked to find?
Paul is 2 years older than Sean and the sum of their ages is 36. *How old is each?*

Step 3 **Represent the unknown quantities by a variable.**
We are asked to find the ages of Paul and Sean. We let

x = Sean's age
$x + 2$ = Paul's age

Step 4 **Establish the relationship with an equation.**

Sean's age + Paul's age = 36
$x \quad + \quad x + 2 \quad = 36$

Step 5 **Solve the equation and check it.**

$$x + x + 2 = 36$$
$$2x + 2 = 36$$
$$2x = 34$$
$$x = 17$$

Sean's age is 17 and Paul's age is $x + 2 = 19$. The sum of their ages is indeed 36, so the solution meets the conditions of the original problem.

EXAMPLE 2 The length of a rectangle is twice its width, and the perimeter is 36 in. *Find its dimensions.*

SOLUTION **Step 1** **Read the problem completely.**
The length of a rectangle is twice its width and the perimeter is 36 in. Find its dimensions.

Step 2 **What am I asked to find?**
The length of a rectangle is twice its width and the perimeter is 36 in. *Find its dimensions.*

Step 3 **Represent the unknown quantities by a variable.**
We let $\quad w$ = the width of the rectangle
$\quad\quad 2w$ = the length of the rectangle

Step 4 **Establish the relationship with an equation.**
Recall that $P = 2w + 2l$. Therefore

$$36 = 2w + 2(2w) \quad \text{Substitute } 2w \text{ for } l.$$

Step 5 **Solve and check.**

$$36 = 2w + 4w$$
$$36 = 6w$$
$$6 = w \quad \text{Divide each side by 6.}$$

The width is 6 in. and the length is twice the width, or 12 in. The perimeter is twice the width added to twice the length, so $P = 2(6) + 2(12)$. $P = 12 + 24 = 36$ in. The solution checks.

Try to identify each of the steps in the next examples. Then continue to use the steps to solve the applied problems in the Exercise Set.

EXAMPLE 3   The sum of two consecutive integers is 65. *Find the integers.*

SOLUTION   The difference between any two consecutive integers is 1 and the larger one can be obtained from the smaller by adding 1. Some examples are 5 and 6, or 9 and 10, as well as -11 and -10. With this in mind, we let

x = the first integer

$x + 1$ = the second integer   Consecutive integers differ by 1.

Then

$x + x + 1 = 65$	The sum of the integers is 65.
$2x + 1 = 65$	Combine like terms.
$2x = 64$	Subtract 1 from each side.
$x = 32$	Divide each side by 2.

The two integers are 32 and 33. Their sum is 65, so the solution checks.

EXAMPLE 4   A woman borrowed $20,000 from a finance company for 2 years at simple interest. The total interest charged was $7200. *What was the rate of interest?*

SOLUTION   Recall that

I = interest = principal $\times$ rate $\times$ time = prt

Therefore

$7200 = 20{,}000 \cdot r \cdot 2$	Substitute $I = 7200, p = 20{,}000,$ and $t = 2$.
$7200 = 40{,}000r$	
$\dfrac{7200}{40{,}000} = r$	Divide each side by 40,000.
$0.18 = r$	

The interest rate is 18%.

EXAMPLE 5   Robert invests $40,000, part at 6% and the rest at 8%. His yearly income from the two investments is $2720. *How much does he invest at each rate?*

SOLUTION   We let x = the amount invested at 6%

$40{,}000 - x$ = the amount invested at 8%

Amount Invested	Interest Rate	Interest Earned ($I = prt$)
x	6%	$0.06x$
$40{,}000 - x$	8%	$0.08(40{,}000 - x)$

The yearly interest earned is $2720, so

Interest at 6% + interest at 8% is 2720.

$0.06x \quad + \quad 0.08(40{,}000 - x) = 2720$

We can avoid working with decimals by multiplying each side by 100. Notice that when $0.08(40{,}000 - x)$ is multiplied by 100, only the decimal part changes. However, when using a calculator, one form is just as easy to work with as the other.

$$100[0.06x + 0.08(40{,}000 - x)] = 100(2720)$$
$$6x + 8(40{,}000 - x) = 272{,}000 \quad \text{Distributive property}$$
$$6x + 320{,}000 - 8x = 272{,}000 \quad \text{Distributive property}$$
$$-2x + 320{,}000 = 272{,}000 \quad \text{Combine like terms.}$$
$$-2x = -48{,}000 \quad \text{Subtract 320,000 from each side.}$$
$$x = 24{,}000 \quad \text{Divide each side by } -2.$$

Robert invests $24,000 at 6% and $16,000 at 8%.

EXAMPLE 6 A chemist finds she needs 3 liters of a 20% acid solution. She has 10% and 40% solutions available in the stockroom. *How many liters of each solution should she mix to get 3 liters of a 20% solution?*

SOLUTION A 20% acid solution means that 20% of the solution is acid and 80% is something else. For example, in 10 liters of a 20% acid solution, $0.20(10) = 2$ liters are acid. We let

x = the number of liters of the 40% acid solution

$3 - x$ = the number of liters of the 10% acid solution

When the two acid solutions are mixed, the *total* number of liters of acid in the mixture does not change. This means that 40% of x liters plus 10% of $(3 - x)$ liters equals 20% of 3 liters, as shown in Fig. 3.18.

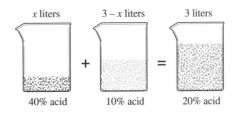

FIGURE 3.18

Hence

$$0.40x + 0.10(3 - x) = 0.20(3)$$

Multiplying each side by 100 to eliminate decimals will simplify the work.

$$40x + 10(3 - x) = 20(3)$$
$$40x + 30 - 10x = 60 \quad \text{Distributive property}$$
$$30x + 30 = 60 \quad \text{Combine like terms.}$$
$$30x = 30 \quad \text{Subtract 30 from each side.}$$
$$x = 1 \quad \text{Divide each side by 30.}$$

Our chemist needs 1 liter of the 40% solution and 2 liters of the 10% solution.

EXAMPLE 7 A coin collection of 60 coins consists of nickels and dimes. The value of the collection does not exceed $4.40. *What is the least number of nickels that are in the collection?*

SOLUTION We let $x =$ the number of nickels in the collection
$60 - x =$ the number of dimes

The value of the total collection is equal to the value of the nickels added to the value of the dimes.

	NUMBER OF COINS	VALUE OF EACH COIN IN CENTS	TOTAL VALUE IN CENTS
NICKELS	x	5	$5x$
DIMES	$60 - x$	10	$10(60 - x)$

The total value of the collection does not exceed $4.40 = 440$ cents. Therefore

$5x + 10(60 - x) \leq 440$
$5x + 600 - 10x \leq 440$ Distributive property
$-5x + 600 \leq 440$ Combine like terms.
$-5x \leq -160$ Subtract 600 from each side.
$x \geq 32$ Divide each side by -5; reverse the inequality.

The collection must have at least 32 nickels in it.

In the coin collection of Example 7, what is the greatest number of nickels it can contain? Read the problem carefully.

EXAMPLE 8 Two cars leave San Luis Obispo at the same time, headed in the same direction. If one car averages 62 mph and the other 55 mph, *how long will it be before they are 35 miles apart?*

SOLUTION Saying that they are 35 miles apart is the same as saying that the difference in the distances they have traveled is 35 miles. The distance formula is

$d =$ distance $=$ rate $\times$ time $= rt$

We let $x =$ the elapsed time until they are 35 miles apart.

	TIME	RATE	DISTANCE $= rt$
FIRST CAR	x	62	$62x$
SECOND CAR	x	55	$55x$

$\begin{bmatrix}\text{Distance for}\\\text{the first car}\end{bmatrix} - \begin{bmatrix}\text{distance for}\\\text{the second car}\end{bmatrix}$ is 35.

$62x - 55x = 35$
$7x = 35$
$x = 5$

It will be 5 hours before they are 35 miles apart.

Exercise Set 3.8

Translate each of the following applied problems into an equation or inequality and solve it.

1. Find three consecutive integers that add to 105.
2. Find four consecutive even integers with a sum of 84.
3. Find the three largest consecutive odd integers with a sum that is less than 100.
4. Find the three smallest consecutive multiples of 4 with a sum that is at least 60.
5. Find the four largest consecutive integers with a sum that is at most 31.
6. Find the four smallest consecutive even integers with a sum that is no more than 80.
7. The Eggleston family traveled 1060 mi in four days. They arranged their driving so that on each day after the first one, they drove 40 mi less than on the previous day. How many miles did they travel each day?
8. David needs to cut a 4-ft-long piece of wood into four pieces so that each piece will be 2.5 in. longer than the previous piece. If there is to be no wasted material, what should be the length of each piece? Neglect the width of the cuts.
9. Nelson is installing five shelves for his stereo equipment and books. The longest shelf is to be the bottom shelf, and each one above it is to be 3 in. shorter than the one immediately below. If the sum of the lengths of the shelves is 120 in., find the length of each shelf.
10. Lisa's collection of 53 dimes and quarters is worth $7.25. Find the number of each kind of coin in her collection.
11. Mary's collection of 36 nickels and quarters is worth at least $2.29. What is the least number of nickels and quarters she can have in her collection?
12. Marty has collected twice as many dimes as quarters. If his coin collection is worth at most $3.10, what is the greatest number of dimes and quarters he can have?
13. Sherry sold 320 tickets for the school play. General admission tickets cost $1.50 each and center section tickets cost $2.50 each. If Sherry collected a total of $560 for the tickets, how many tickets of each kind did she sell?
14. Two bicycle riders, Rich and Keith, are 55 km apart and are riding toward each other. Rich's average speed is 12 km/hr and Keith's is 10 km/hr. How long will it take them to meet?
15. Juan starts walking at 7:00 A.M. at an average rate of 2.5 mph. Two hours later Lonnie starts walking along the same route at an average rate of 4.5 mph. How long will it take Lonnie to overtake Juan?
16. Elena has saved $55 to buy books for this semester. If the cost of the first book is $17 and the second book costs twice as much as the first, how much money will she have left?
17. The state levies a 6% sales tax. If Rodney purchases a car for $5500, what will his total cost be, including the sales tax and a license fee of $78.50?
18. Lil sells real estate. Her income consists of a $6000 base salary plus a 3% commission of her sales. How much real estate must Lil sell in order to have an income of $16,000?
19. What is the price of a suit that normally sells for $169.50 if it is on sale at a 30% discount?
20. Milt's Grading Service estimates that his graders lose 15% of their current value each year. After two years, what would be the value of a grader that originally cost $31,500?
21. The members of the Alpha Gamma Club want to sell 405 boxes of stationery so that they can gross $603.75 to fund a ski trip. One type of stationery sells for $1.25 and another type for $1.75. How many boxes of each will they need to sell?
22. Mrs. Wood needs to earn at least $125/month from interest, and she has $15,000 to invest. She will place part of the money in a passbook account drawing 6% interest so that it will be available if she needs it. She will place the rest in a mutual fund drawing 13% interest. What is the most she can put in the passbook account and satisfy her needs?
23. The total costs C per month for Mary's Rose Shop are given by $C = \$0.85x + \$20.00$, where x is the number of roses purchased. Find the costs to the store in February if Mary purchases 225 roses.
24. Tamra pays $540 interest on a three-year loan of $1800 to purchase a used car. What is the annual rate of interest on the loan?
25. Rex designs his rectangular organic vegetable garden so that the length is twice the width. What are the dimensions of the garden if the perimeter is 36 ft?
26. A rule of thumb useful to auto mechanics is that a car with a pressure radiator cap will boil at a temperature of three times the cap rating plus 212°. What is the cap rating on a Chrysler Newport if it boils at 260°?
27. The Campbell Tool Company receives an order for 14,472 hinges. If they pack two dozen hinges in a box, how many boxes do they need to ship the order?

28. A new stamping machine produces washers at the rate of 960/min. If an order calls for 288,000 washers, how many hours will it take to produce them?

29. A new table at Manuel's Machine Shop is designed to support 8 lathes weighing 228 lb each, 3 drill presses weighing 295 lb each, and 2 milling machines weighing 1920 lb each. How many pounds is the table designed to support?

30. Hydrological engineers often measure large amounts of water in acre-feet. One acre-foot is equivalent to a volume with a base of 4840 yd^2 and a depth of 1 ft. How many acre-feet of water will fit in a rectangular reservoir with a length of 121 yd, a width of 40 yd, and a depth of 6 yd?

31. The cost to residential customers for the first 240 kilowatt-hours (KWH) of electricity is 3.7¢/KWH. The cost of the next 240 KWH is 5.9¢/KWH. If the Dawsons used 420 KWH of electricity last month, how much was their bill?

32. Bob and Celia decide to paint their living room walls and ceiling. The room is 22 ft long, 18 ft wide, and 8 ft high. If Colorglo paint covers 400 ft^2/gal, how many gallons of paint will they need? (This paint is sold only in gallon containers.) How much will it cost if the paint sells for $11.95/gal? (Disregard windows and doors.)

33. Jim's Welding Shop finds that they need 2083 ft^3 of acetylene gas to weld one bracket. How much gas do they need to weld 21 brackets?

34. Auto-Specialists charges a flat rate of $17.50/hr to overhaul a motor plus the cost of parts. If it takes 23.5 hr of labor and $225 in parts to overhaul a motor, what is the total cost of the overhaul?

35. Verle's car averages 21 mpg and uses 18 gal of gas on a given trip. If at the end of the trip the odometer has a reading of 63,275 mi, what was the reading at the beginning of the trip?

36. Janice's car averages 27 mpg. If she wants to travel more than 219 mi before stopping, what is the least amount of gasoline (whole gallons) she must have in the tank?

37. Dennis's car holds 16 gal of gas. His car averages 32 mpg. Can he travel at least 513 mi before stopping for gas?

38. Lewis needs to cut a piece of sheet metal into a triangular shape. He wants the base of the triangle to be 5 in. longer than the shortest side and 7 in. shorter than the longest side. If the perimeter is to be 32 in., find the length of each side.

39. A pharmacist at Economic Drugs uses 800 milliters (ml) of cough syrup containing 5% alcohol and x ml of cough syrup containing 2% alcohol to make a new solution of cough syrup containing 4% alcohol. How much of the 2% strength cough syrup does the pharmacist add?

40. Students in Mr. Beldon's biology class are going to study two different types of bacteria. They need a total of 1092 bacteria, with twice as many of strain C as strain K. How many of each type of bacteria do they need?

41. The doctor tells you to take 600 mg of aspirin every four hours and get plenty of rest. If 1 grain (gr) = 60 mg, how many aspirin tablets should you take every four hours if your bottle of aspirin contains 5 gr tablets?

42. The perimeter of a rectangle is no more than 44 m and it's length is 8 m more than its width. What inequality represents the possible lengths of the rectangle?

43. The perimeter of a rectangle is less than 36 ft and its length is 2 ft more than three times its width. What inequality represents the possible widths of the rectangle?

44. Leah leaves school at 4:00 P.M., traveling at a speed of 42 mph. An hour later Dean leaves the school traveling in the same direction at 56 mph. How much time does Dean need to overtake Leah?

45. George leaves the gym traveling north at a speed of 70 mph. At the same time, Beverly leaves the gym traveling south at a speed of 30 mph. In how many hours will they be at least 120 mi apart?

46. Mom's Honey sells for $4/pint (pt) and Dad's Honey sells for $1.50/pt. How many pints of each type of honey must be mixed to create 25 pt selling at a price of $2.60/pt?

47. A flu epidemic is striking 6 out of 10 students in Morovia. At this rate, how many students will be sick in a school of 12,000 students?

48. A poll indicates that 3 out of 7 voters are voting for Mr. Adams. If there are 19,000 voters in Valencia, how many will vote for Mr. Adams?

49. If a home valued at $85,000 is assessed $900 in real estate taxes, what are the taxes on a $115,000 home?

50. An investment of $1000 earns $45 in a year. How much would have to be invested at this rate to earn $84 in a year?

51. How many liters of a 40% acid solution must be mixed with 80 liters of a 70% acid solution to get a liquid that is 50% acid?

52. How many drums of paint worth $100/drum must be mixed with 30 drums of paint worth $60/drum to get a mixture worth $80/drum?

SUMMARY

3.1
Simplifying Algebraic Expressions

A **term** is a single number or the product or quotient of one or more numbers and variables raised to powers. An **algebraic expression** is the sum or difference of such terms. In the term $3xy$, the number 3 is called the **numerical coefficient**. **Like terms** have variable parts with exactly the same base and exponent.

3.2
Solving Linear Equations Using the Addition and Subtraction Properties of Equality

An **equation** is a statement that two quantities are equal. In an equation involving a variable, any value of the variable that makes the equation true is a **solution** or a **root** of the equation.

A **linear equation in one variable** is an equation that can be written in the form $ax + b = c$, $a \neq 0$, where x is a variable and a, b, and c represent real numbers.

Linear equations in one variable have a single solution; enclosed within braces, it forms the **solution set** of the equation.

Many linear equations can be solved using only the addition and subtraction properties of equality. They state that *if equals are added to or subtracted from equals, the results are equal*. Specifically, the **addition property of equality** says that if a, b, and c represent real numbers and $a = b$, then $a + c = b + c$, while the **subtraction property of equality** says that if a, b, and c represent real numbers and $a = b$, then $a - c = b - c$.

An equation that is true for specific replacements for the variable is called a **conditional equation.** One that is true for all meaningful replacements for the variable is called an **identity.**

3.3
Solving Linear Equations Using the Multiplication and Division Properties of Equality

Some linear equations require the multiplication or division properties of equality to solve for the variable. They state that *if equals are multiplied or divided by equals (divisor $\neq 0$), the results are equal*. Specifically, the **multiplication property of equality** says that if a, b, and c represent real numbers and $a = b$, then $ac = bc$, while **the division property of equality** says that if a, b, and c represent real numbers and $a = b$, $c \neq 0$, then $\frac{a}{c} = \frac{b}{c}$.

3.4
Literal Equations and Formulas

A **literal equation** or **formula** is an equation that involves more than one variable. Literal equations occur frequently in applied problems.

3.5
The Addition and Subtraction Properties of Inequality

An **inequality** is a statement that two quantities are not necessarily equal. Inequalities make use of the symbols $>$, $\geq$, $<$, and $\leq$. A number a is said to be **less than** number b, written $a < b$, if a is to the left of b on the number line, while a is **greater than** b, written $a > b$, if a is to the right of b on the number line.

Inequalities are solved using the properties of inequality. The **addition property of inequality** says that if a, b, and c represent real numbers and $a < b$, then $a + c < b + c$, while the **subtraction property of inequality** says that if a, b, and c represent real numbers and $a < b$, then $a - c < b - c$.

In words, these properties state that *the same number can be added to or subtracted from each side of an inequality without changing the direction the inequality symbol points*. The solution set will remain the same as that of the original inequality.

Compound inequalities such as $-3 < x < 4$ are formed by joining two inequalities involving the same unknown.

3.6 The Multiplication and Division Properties of Inequality

Some inequalities require multiplication and division to find their solution sets. Particular care must be taken when multiplying or dividing by a negative number, since multiplying or dividing each side of an inequality by a positive number does not change the direction of the inequality symbol, while multiplying or dividing by a negative number does. More precisely, the **multiplication property of inequality** says that if a, b, and c represent real numbers and $a < b$, then $\boldsymbol{a \cdot c < b \cdot c}$ if c is **positive** ($c > 0$) and $a \cdot c > b \cdot c$ if c is **negative** ($c < 0$), while the **division property of inequality** says that if a, b, and c represent real numbers and $a < b$, then $\dfrac{a}{c} < \dfrac{b}{c}$ if c is **positive** ($c > 0$) and $\dfrac{a}{c} > \dfrac{b}{c}$ if c is **negative** ($c < 0$).

Although the multiplication and division property of inequality are stated for $<$ they are also true for $\leq$, $>$ and $\geq$.

3.7 Solving Absolute Value Equations and Inequalities

Absolute value equations and inequalities are solved by the following rules: If p is a positive number, then

$|x| = p$ means that $x = p$ or $x = -p$
$|x| < p$ means that $-p < x < p$
$|x| > p$ means that $x < -p$ or $x > p$
$|0| = 0$

Although the rules for solving absolute value inequalities are stated for $<$ and $>$, they are equally true for $\leq$ and $\geq$. Equations of the form $|x| = -p$, where p is positive, have no solution because the absolute value of a number can never be negative. Also, $|x| > -p$, where p is positive, has all real numbers as its solution set.

3.8 More Applications

Consistent success in solving applied problems is enhanced by using the following steps and knowing where to find the formulas necessary for solving the problem.

1. Read the problem through completely as many times as necessary to understand the meaning of the problem.
2. In reading the problem, ask yourself, "What am I asked to find?"
3. Represent what you are asked to find by a variable.
4. Establish the relationship between the variable and other parts of the problem with an equation or an inequality.
5. Solve the equation or inequality and check your results in the words of the original problem.

Cooperative Exercise

In a Business Math class, the students are given a project to calculate and compare the cost of purchasing a car with a three-year loan, purchasing it with a four-year loan, and leasing it for three years. Once the calculations are done, they are to establish reasons for and against each of these methods of purchasing/leasing a car. Last, they are to recommend (as a class) which of the three methods is best in different situations. They are given the following information:

1. The value of the car is $18,850. A tax of 7.25% and a license fee of $364 is added to the purchase price of the car.
2. On the three-year and four-year loans, a 20% down payment of the total cost (including tax and license) is required.
3. The rate of interest is 7.5%/yr on the total amount for the three-year loan and 8.5%/yr on the total amount for the four-year loan.
4. The lease requires an $800 deposit plus payment of the tax and license. The **lease residual value** (the end-of-lease value) of the car is $10,280. The interest on the lease is 6%.
5. The Blue Book resale value after 3 years is $10,680, and after 4 years is $9600.

Group I is to calculate the total cost of the car on the three-year loan.

Group II is to calculate the total cost of the car on the four-year loan.

Group III is to calculate the total cost of the car on the three-year lease.

Each group is to determine:

A. the total paid for the car;
B. the total interest paid;
C. the cost if the car is sold at Blue Book value or is turned in at lease residual value at the end of the time period;
D. the monthly payments; and
E. the reasons that a person should/should not use that method to purchase/lease a car.

The class as a whole is to provide a recommendation about which method they think is the best and why.

Review Exercises

Translate the following phrases into algebraic expressions.

1. The sum of two consecutive integers
2. The sum of two consecutive even integers
3. The product of two consecutive odd whole numbers
4. The product of two consecutive multiples of 3
5. The difference of a number and its multiplicative inverse
6. The sum of a number and its additive inverse
7. The sum of a number and 5 less than the number.
8. The sum of a number and one-half of the number.
9. Four less than three times a number
10. Five more than one-fourth of a number
11. Twice the sum of two consecutive even natural numbers
12. Eight times the sum of two consecutive odd natural numbers
13. The sum of three consecutive odd natural numbers
14. The sum of three consecutive even natural numbers
15. Three more than twice y
16. One-half of the difference of x and 7
17. Twice the sum of 11 and four times a number
18. An $80 radio discounted by y dollars
19. The distance traveled at 58-mph for $5\frac{1}{2}$ hr
20. The amount left after investing x dollars of an $800 fund

Solve the following equations and inequalities.

21. $x - 3 = 13$
22. $x + 5 = 4$
23. $3 - x = 7$
24. $2x - 3 = -5$
25. $3(2 - x) = -3$
26. $\frac{1}{2}x = \frac{1}{4}$
27. $-x > 3$
28. $3y < -7$
29. $-(y - 1) = 4$
30. $4t - 11 \geq 9$
31. $3m + 4 \leq -1$
32. $-4s \geq -5$
33. $15 - 3y \leq -11$
34. $-2(4 - x) > x - 1$
35. $3(x - 4) = -5x$
36. $-7(3 - 2x) = 3x$
37. $-4(5x - 2) = 0$
38. $x - 2 = 2x + 7$
39. $\frac{2}{3}\left(\frac{5}{7}x - \frac{3}{2}\right) = \frac{1}{2}$
40. $-\frac{3}{4}\left(\frac{2}{7}x + \frac{1}{3}\right) = x$
41. $\frac{2}{3}(5 - y) > 2y$
42. $5\left(p + \frac{1}{3}\right) \leq 3p + \frac{2}{3}$
43. $\frac{1}{4}y - \frac{5}{8} \geq 0.375$
44. $0.22x - \frac{1}{2} < 0.12x$

Solve for the indicated variable or constant.

45. $C = 2\pi r$ for π
46. $A = \dfrac{1}{2}bh$ for h
47. $V = \pi r^2 h$ for r^2
48. $V = lwh$ for w
49. $A = \dfrac{1}{2}h(B + b)$ for h
50. $I = prt$ for r

Evaluate each of the following equations for the indicated values.

51. $V = \dfrac{1}{3}Bh$; find V when $h = 10$ and $B = 20$.
52. $y = 2a + ab - 3b$; find y when $a = 2$ and $b = -7$.
53. $3|3x - 2xy| = 5$; find x when $y = 4$ (assume that $x > 0$).
54. $-5x^2 + 6xy - 2y^2$; evaluate for $x = -1$ and $y = 3$.

Solve the following equations for x.

55. $2x - 3y = 5$
56. $-7y + 6x = 1$
57. $7y = -4x$
58. $y = mx + b$
59. $ax + by = c$
60. $\dfrac{1}{4}ay - \dfrac{1}{2}bx = d$
61. $\dfrac{x}{3} - \dfrac{y}{4} = \dfrac{3}{5}$
62. $\dfrac{3x + 2}{5} = \dfrac{2x - 1}{3}$
63. $\dfrac{x - 5}{2} = \dfrac{4}{7}$
64. $\dfrac{-2x}{7} < \dfrac{3x}{2} + 6$
65. $\dfrac{x}{3} + \dfrac{1}{2} = \dfrac{1}{5}$
66. $m \leq \dfrac{2x - 3}{5}$

Solve each absolute value equation or inequality.

67. $|x| = 5$
68. $|y| < 3$
69. $|y| > -2$
70. $|x - 2| = -3$
71. $|2x + 3| < 0$
72. $-|x| = -5$
73. $|-3 - 2x| \geq 4$
74. $|1 - 4y| \leq 0$
75. $|x + 2| - 3 = 8$
76. $|2p - 5| = 0$

Translate each applied problem into an algebraic equation or inequality and solve it.

77. One number is 5 more than a second number. Ten times the smaller number minus the larger number is 6. Find the smaller number.
78. One number is twice a second number. Three times the smaller one added to four times the larger one is 22. Find the larger number.
79. Eunice has at least $2.65 in nickels and dimes. If she has 8 more nickels than dimes, what is the smallest number of nickels that she has?
80. The sum of three odd integers is less than 27. Find the largest three odd integers.
81. After a 30% discount, an item is priced at $154. What was the price before the discount?
82. Willis's yearly salary is $18,000, which is a 15% increase over the previous year. What was his previous year's salary?
83. A salesperson wraps packages at the rate of one every three minutes. A second salesperson wraps packages at the rate of one every two minutes. How many packages can they wrap in an hour working together?
84. The perimeter of a rectangle is less than 56 cm. The width is 6 cm less than the length. Write an inequality for the length.

Chapter Test

Simplify the following expressions.

1. $7x + 4 - 3x - 6$
2. $3(2y - 5) - 4(y + 6)$
3. $2p^2 + 6 - 5p - p^2 + 7 + 3p$
4. $3(r^2 - 2) + 7(r + 5)$

Evaluate each of the following expressions for $x = 3$ and $y = -1$.

5. $3x - \sqrt{9y} + xy$
6. $5|x - 4| - 3y$
7. $\dfrac{x^2 - 2x + 1}{5y}$
8. $-2^2(x^2 - x) + \dfrac{y^2}{3} - \dfrac{5y}{4}$

Solve the following equations.

9. $x - 6 = 5$
10. $2x + 4 = 10$
11. $7x - 2 = 11x + 2$
12. $5x - 7 = 8(x + 2)$
13. $3(4 - x) = 9x + 8 - 2(3 - x)$
14. $-(x + 1) = 3(x - 2) - 5(2 - x)$

Solve and graph the following inequalities. Assume x is a real number unless stated otherwise.

15. $x + 3 < 5$, x is a natural number
16. $2x - 5 \leq 3$, x is an integer
17. $x + 9 \geq 3x + 7$
18. $0 > x + 5$

19. $-3 \leq -3(5 - 2x) < 4$

20. $-3 > -2(x - 3) > -8$

Solve the following absolute value equations and inequalities. If no solution exists, write ϕ.

21. $|x - 5| = 7$
22. $|2x + 3| < -1$
23. $\left|\dfrac{4 - \sqrt{9}x}{2}\right| \geq -2$
24. $\left|\dfrac{7 + 5x}{3}\right| + 4 \leq 7$

Solve each literal equation for the specified variable.

25. $I = prt$ for p
26. $A = 2\pi rh + 2\pi r^2$ for h
27. $V = \dfrac{1}{3}\pi r^2 h$ for h
28. $A = \dfrac{1}{2}h(B + b)$ for h

Translate each word problem into an algebraic equation or inequality and solve it.

29. The difference of two integers is 16. The sum of the two integers is 48. Find the integers.

30. Jadene has more than $1.14 in pennies, nickels, and dimes. She has 6 more nickel than pennies and 6 fewer dimes than pennies. Find the least number of each kind of coin she can have.

31. Two automotive technology students offer to repair their instructor's pickup for $56. They decide to divide up the pay so that the first student makes twice as much as the second one, per hour. If the first student works 3 hr and the second one works 2 hr, how much does each make per hour?

32. Two couples leave camp at the same time, the first traveling west and the second traveling east. Two and one-half hours later, they are $277\frac{1}{2}$ miles apart. If the second couple averages 5 mph slower than the first, what is the first couple's average speed?

CUMULATIVE REVIEW FOR CHAPTERS 1–3

Simplify Exercises 1–10.

1. $(4 - 9)(3 - 7)$
2. $4 \cdot 3^2 - 30$
3. $16 \div 4 \cdot 3$
4. $35 \div (8 - 1)$
5. $2(4^2 - 3^2) \div 7$
6. $(4 - 1)[-2(3 - 5)]$
7. $\dfrac{18 \div (8 - 2)}{4^2 - 7}$
8. $9^2 - 3(3^3)$
9. $\dfrac{7[5 \div (6 - 1) \cdot 2]}{35 \div 5 \cdot 1^2}$
10. $1\dfrac{3}{8} + \dfrac{3}{4}$

11. Rewrite $\dfrac{3}{8}$ in (a) decimal form and (b) percent form.

12. Rewrite 67% (a) in decimal form and (b) as a proper fraction.

13. Rewrite $1\frac{1}{9}$ in decimal form.

14. Factor 210 into prime factored form.

15. Find 28% of 20.

16. Find $\sqrt{64} + \sqrt{0.16}$.

Find the perimeter and area of each of the following shapes.

17. A square with side $1\frac{3}{5}$ in.
18. A rectangle with length $l = 15$ ft and width $w = 7$ ft
19. A right triangle with base $b = 3$ yd and height $h = 4$ yd
20. A right trapezoid with $B = 18$ m, $b = 10$ m, $h = 9$ m, and fourth side 14 m; see figure.

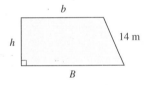

21. A circle with radius $r = 3$ in. (Use 3.14 for π.)

Find the volume of each of the following figures. Do not change π into decimal form.

22. A cube with side $s = 4$ ft
23. A sphere with radius $r = 3$ mm
24. A right circular cylinder with radius $r = \frac{1}{2}$ ft and height $h = 3\frac{1}{4}$ ft
25. A cone with radius $r = 1.2$ in. and height $h = 3$ in.
26. A rectangular box of length $l = 11.1$ ft, width $w = 2.7$ ft, and height $h = 3$ ft

Identify each of the following statements as true *or* false.

27. $2 \in$ {even natural numbers}
28. $\{3, 5\} \subseteq$ {odd whole numbers}
29. $\dfrac{1}{4} \in$ {rational numbers}

30. $\frac{3}{4} \in$ {irrational numbers}
31. $\{x \mid x \text{ is a natural number}, x < 2\} = \{1\}$
32. $\sqrt{4} \in$ {natural numbers}
33. $\{3, 6, 9\} \subseteq \{x \mid x \text{ is a whole number}, x \text{ a multiple of 3}\}$
34. {whole numbers} $\subseteq$ {natural numbers}
35. $|-4| \in$ {natural numbers}
36. If $|x| = 3$, then $x = 3$ or $x = -3$.
37. $|-3.45| + |0.55| = 4$
38. The value of $-|xy|^2 = 4$ when $x = 1$ and $y = -2$.

Simplify each of the following expressions. If undefined, say so.

39. $9 - 11$
40. $-3 - (-7)$
41. $18 - (11 - 1)$
42. $0.35 - 8.45$
43. $6.1 + (-1.8)$
44. $-2[-3(4 - 6) + 1]$
45. $-|-8 + 3| - |-2 + 5|$
46. $\frac{1}{4}\left(\frac{1}{2} - \frac{1}{3}\right)$
47. $16 \div 4 + 6$
48. $-3[5 - (6 - 1)] + 4$
49. $|(-2)^3| - 6.25$
50. $34 \div 0(6)$

State the property illustrated by each equation.

51. $x + 3 = 3 + x$
52. $6(x - 3) = 6x - 18$
53. $6x + 0 = 6x$
54. $3(5y) = (3 \cdot 5)y$
55. $3a\left(\frac{1}{3a}\right) = 1$
56. $7 + (3 + 1) = (7 + 3) + 1$
57. $7y + 14 = 7(y + 2)$
58. $7a + (-7a) = 0$

Simplify each algebraic expression.

59. $6x + 11x$
60. $3a - 4a + 2a$
61. $7y^2 - 5y + 6y - y^2$
62. $0.1a + 0.25b - a - b + 3$
63. $-3(4x - 5y) + 2(-3x + y)$
64. $-1[2a - (3a - 4) + 2]$
65. $6x - 7y - 2(3x)$
66. $-\frac{95}{8}y + 4y + \frac{101}{4}x - 4y$
67. $1.3x + 0.6y - 1.45x - 1.4y$
68. $4xy + y - 2xy - y$

Solve each equation. Check your answers.

69. $x + 3 = 5$
70. $y - 2 = 3$
71. $3p - 4 = 5$
72. $-4y = 3 + y + 2$
73. $0 = 5 - m$
74. $\frac{-3}{5}x = \frac{4}{15} + \frac{2}{5}x$
75. $3.5y - 2 = 1.5y + 6$
76. $3(2 - 5x) = -(x - 1)$
77. $\frac{1}{3}(3a - 12) = \frac{-1}{2}(2a - 6)$
78. $\frac{2x}{5} = \frac{3}{5}$
79. $\frac{3y}{4} - 1 = \frac{y}{2}$
80. $15p + 8 = 2(p + 4)$

Solve for the specified variable.

81. $C = 2\pi r$ for r
82. $y = mx + b$ for x
83. $C = \frac{5}{9}(F - 32)$ for F
84. $I = prt$ for t
85. $A = \frac{1}{2}h(B + b)$ for h
86. $P = 2l + 2w$ for l
87. $\frac{3}{5}(4x - y) = \frac{1}{5}p$ for x

Solve each of the following inequalities.

88. $y - 5 < 4$
89. $x + 2 > 3$
90. $p - 3 \leq -2$
91. $6x + 3 \geq 4x - 2$
92. $4(x - 3) + 2 > 5(1 + x)$
93. $-6 < b + 2 < 1$
94. $-2(-a + 3) - 4 \geq 3(2 - a) + 1$
95. $0.5(p - 2) - 3 \leq 0.1(p - 1)$
96. $\frac{1}{2}y - \frac{1}{4}y > \frac{3}{2}y + 4$

Solve each of the following. If no solution exists, write ϕ.

97. $|x| = 4$
98. $|3y| = 6$
99. $-|2y| = -4$
100. $-|3x| = 0$
101. $|5x| = -10$
102. $|x| > 3$
103. $|2x| \leq 0$
104. $|7y| > -7$
105. $|x - 5| + 4 \leq 0$
106. $|3y - 2| \geq 2$
107. $|4 - x| - 2 = 3$
108. $|3 - 2y| > 3$
109. $|5 - y| + 3 \geq 3$
110. $|2x + 3| > 0$

Translate each of the following applied problems into an equation or an inequality and solve it.

111. Find two consecutive integers that add to -35.
112. Find the two smallest consecutive even whole numbers with a sum of at least 26.

113. Kylee has a collection of nickels and dimes worth $1.05. Find the number of each kind of coin in her collection if she has three less nickels than dimes.

114. A flower garden is two and a half times as long as it is wide. Find the dimensions if the perimeter is 84 ft.

115. Two bicycle riders are 70 mi apart and are riding toward each other. If one averages 12 mph and the other 16 mph, how long will it take them to meet?

116. What is the sale price of a dress that regularly sells for $148 if it is now selling at a 25% discount?

117. Francesca plans to travel nonstop to Newark, a distance of 248 mi. If her car averages 27 mpg, what is the least amount of gas (in whole gallons) she must have in her car?

118. The perimeter of a rectangle is 50 ft. Find the dimensions if the width is 3 ft less than the length.

119. Find the unknown side of the right triangle shown in the accompanying figure.

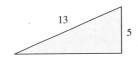

120. How many pounds of peanuts at $3/lb must be mixed with cashews at $5/lb to get a 5-lb mixture worth $20?

CHAPTER 4
LINEAR EQUATIONS AND INEQUALITIES IN TWO VARIABLES; FUNCTIONS

4.1 The Rectangular (Cartesian) Coordinate System
4.2 Graphing Linear Equations in Two Variables
4.3 The Slope of a Line; Equations of a Line
4.4 Graphing Linear Inequalities in Two Variables
4.5 Relations, Functions, and Function Notation

Application

Jones Metals makes and sells chrome bumpers. The company has a fixed cost of $4000/week. If it makes and sells ten identical bumpers in one week, the total income from the sales is $5000. Assume the weekly profit P equals the total income I less the fixed cost, where income is given by the formula $I = 500x$, with x equal to the number of bumpers made and sold in one week. This information is used to determine the company's profitability.

Chapter 4 Review Exercises 66–71.

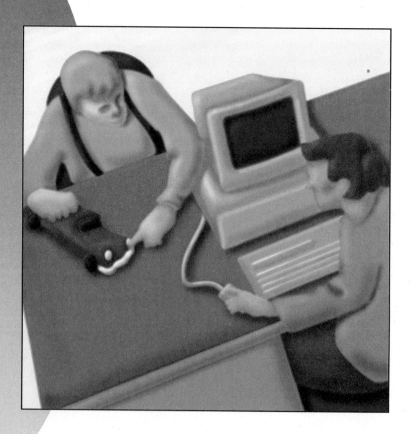

I have so many mathematical ideas that may be of some use in future times if others more penetrating than I will go deeply into them some day and join the beauty of their minds to the labor of mine.

GOTTFRIED WILHELM LEIBNIZ (1646–1716)

4.1 The Rectangular (Cartesian) Coordinate System

HISTORICAL COMMENT René Descartes (1596–1650) was a French mathematician, philosopher, and physician who developed the rectangular coordinate system we will study in this section. It is often called the *Cartesian coordinate system* in honor of his work. There are several stories of how he first thought of the idea. One is that he was watching a fly crawl across the ceiling near the corner of a room and decided to develop a system that would allow the location of the fly to be known in terms of its distance from the walls. Two other stories have him lying in bed. In one, the idea came to him in a dream; in the other, it was a cold day and, since the town was under siege, he remained in bed and thought about mathematics.

OBJECTIVES In this section we will learn to
1. identify the parts of a rectangular coordinate system; and
2. locate points on a rectangular coordinate system.

As we mentioned in the Historical Comment, René Descartes is credited with being the inventor of the rectangular coordinate system. His system allowed algebraic equations in two variables to be represented graphically. For the first time, mathematicians could study equations by looking at graphs (pictures).

The system that Descartes invented was simplicity itself. It allowed him to locate any point on a flat (plane) surface. To do this, he drew two perpendicular number lines, one horizontal and one vertical, that intersected at the zero point on each line. The point of intersection is called the **origin.** The horizontal line is called the *x*-**axis** and the vertical line is called the *y*-**axis.** These two lines divide the plane into four sections called **quadrants,** which are numbered in a counterclockwise direction, as shown in Figure 4.1. The axes themselves are not a part of any quadrant.

Points in a plane are located in terms of their distance to the right or left of the origin and above or below the origin. These *x*- and *y*-distances are known as the **coordinates** of the point. Points to the right of the origin have a **positive x-coordinate** and those to the left of the origin have a **negative x-coordinate.** Points above the origin have a **positive y-coordinate** and those below the origin have a **negative y-coordinate.** Every point on the coordinate system can be located in terms of its *x*-coordinate and *y*-coordinate. These points are described by **ordered pairs,** which are written (x, y):

$$(x, y)$$
x-coordinate ⎯⎯↑ ↑⎯⎯ *y*-coordinate

To **graph** any ordered pair, we use both its *x*-coordinate and its *y*-coordinate. To locate (4, 5), for example, we start at the origin and move 4 units to the right,

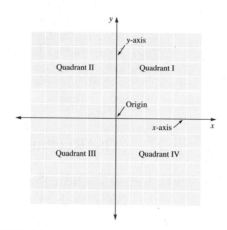

FIGURE 4.1

since the *x*-coordinate is 4. From this position we move up 5 units, because the *y*-coordinate is 5. By placing a dot at our final position, we have **plotted,** or **graphed,** the point (4, 5); see Figure 4.2.

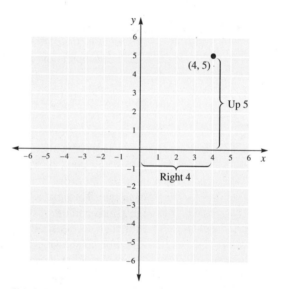

FIGURE 4.2

We use various names to describe the *x*- and *y*-axes. The *x*-axis is often called the **horizontal axis** or the **axis of abscissas.** The *y*-axis is called the **vertical axis** or the **axis of ordinates.** Using the latter terminology in each case suggests that the coordinates of the ordered pair (*x*, *y*) can also be referred to as the abscissa and the ordinate, respectively:

$$(x, y)$$
abscissa ⎯⎯↑ ↑⎯⎯ ordinate

EXAMPLE 1 Plot the following points on a rectangular coordinate system: $(3, -2)$, $(-2, 4)$, $(-2, -2)$, and $(0, -3)$.

SOLUTION To locate the point $(3, -2)$, we start at the origin and move 3 units to the right; then we move 2 units down. This point and the others are shown in Figure 4.3.

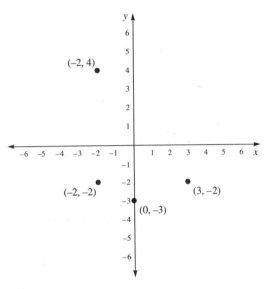

FIGURE 4.3

An equation in two variables, x and y, describes a relationship between the x-coordinates and the related y-coordinates. For example, in the equation $y = 3x + 3$, when $x = 2$ we get $y = 9$.

$y = 3x + 3$
$= 3(2) + 3$ Substitute 2 for x.
$= 9$

By substituting arbitrary values for x, we can find infinitely many ordered pairs that satisfy this equation and therefore are *solutions* to the equation.

EXAMPLE 2 Given the equation $y = 2x + 1$, complete the ordered pairs $(0, \ )$, $(2, \ )$, $(\ , -3)$, and $(\ , 0)$, and plot them on the same coordinate system.

SOLUTION If $x = 0$, then $y = 2(0) + 1 = 1$.   The point is $(0, 1)$.
If $x = 2$, then $y = 2(2) + 1 = 5$.   The point is $(2, 5)$.

If $y = -3$, then $-3 = 2x + 1$
$$-4 = 2x$$
$$-2 = x$$

Subtract 1 from each side then divide by 2.
The point is $(-2, -3)$.

If $y = 0$, then $0 = 2x + 1$. The point is $\left(-\dfrac{1}{2}, 0\right)$.

The graphs of the four points are shown in Figure 4.4.

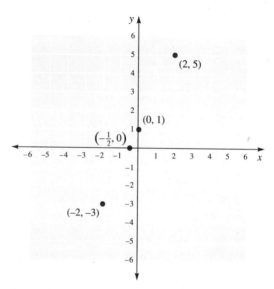

FIGURE 4.4

EXAMPLE 3 Given the equation $2y - 3x = 2$, find the y-coordinate of the point where $x = -1$.

SOLUTION
$$2y - 3x = 2$$
$$2y - 3(-1) = 2 \quad \text{Substitute } -1 \text{ for } x.$$
$$2y + 3 = 2 \quad (-3)(-1) = +3$$
$$2y = -1 \quad \text{Subtract 3 from each side.}$$
$$y = -\dfrac{1}{2} \quad \text{Divide each side by 2.}$$

The y-coordinate is $-\dfrac{1}{2}$.

EXAMPLE 4 Plot seven different points between $x = -4$ and $x = 4$ such that the y-coordinate is twice the x-coordinate.

SOLUTION If the y-coordinate is twice the x-coordinate, then $y = 2x$. To find an ordered pair, we assign a value to x and compute the value of y. Some examples are $(-3, -6)$, $(-2, -4)$, $(-1, -2)$, $(0, 0)$, $(1, 2)$, $(2, 4)$, and $(3, 6)$. These points are shown in Figure 4.5.

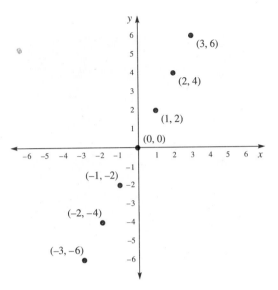

FIGURE 4.5

EXAMPLE 5 Three vertices of a parallelogram are $(-2, 0)$, $(2, 0)$, and $(3, 3)$. Find possible locations for the fourth vertex.

SOLUTION We begin by plotting the given points, as shown in Figure 4.6(a). The opposite sides of a parallelogram are parallel to each other, so one point that completes the parallelogram is $(-1, 3)$, as shown in Figure 4.6(b). Are there any others?

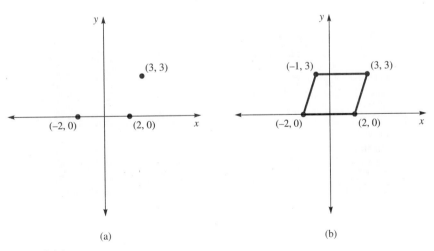

FIGURE 4.6

Do You Remember?

Can you match these?

_____ 1. The first coordinate in an ordered pair
_____ 2. The horizontal axis
_____ 3. The ordinate of a point
_____ 4. The abscissa of a point
_____ 5. The vertical axis
_____ 6. The point where the axes intersect
_____ 7. The direction in which the quadrants are numbered
_____ 8. An ordered pair

a) The coordinates of a point
b) The y-coordinate
c) Counterclockwise
d) The x-coordinate
e) The y-axis
f) The x-axis
g) The origin
h) Clockwise

Answers: 1. d 2. f 3. b 4. d 5. e 6. g 7. c 8. a

Exercise Set 4.1

In which quadrant or quadrants are the following points located?

1. The x-coordinate is positive.
2. The y-coordinate is negative.
3. The x-coordinate is negative.
4. The y-coordinate is positive.
5. The x-coordinate is negative and the y-coordinate is positive.
6. Both the x- and y-coordinates are negative.
7. The x-coordinate is positive and the y-coordinate is negative.
8. The x-coordinate is positive and the y-coordinate is positive.

Find an ordered pair describing each of the following points.

9. It is 3 units above the x-axis and 2 units to the right of the y-axis.
10. It is 2 units below the horizontal axis and 1 unit to the left of the vertical axis.
11. It is 3 units to the left of the vertical axis and 2 units below the horizontal axis.
12. It has a y-value of -3 and an x-value of -2.
13. It has a y-value of 0 and an x-value of 0.
14. It has a y-value of 5 and an x-value of 0.
15. It is 6 units to the left of the y-axis and it is on the x-axis.
16. It is 3 units below the x-axis and it is on the y-axis.
17. It is 5 units to the right of the y-axis and 2 units below the x-axis.
18. It has a y-value of 3 and an x-value of -4.
19. From the origin, go 6 units left and 2 units down.
20. From the origin, go 3 units right and 4 units up.
21. From the point (3, 0), go 4 units horizontally in a positive direction and 1 unit vertically in a positive direction.
22. From the point (0, 4), go 1 unit horizontally in a negative direction and 2 units vertically in a positive direction.
23. From the point $(-2, 3)$, go 2 units left and 4 units down.
24. From the point $(-5, -4)$, go right 1 unit and 3 units up.

Plot these points on the same coordinate system.

25. (1, 2)
26. $(-2, -5)$
27. $(-3, 4)$
28. (2, -4)
29. $(-5, 2)$
30. (0, 3)
31. (0, 0)
32. $(-4, 0)$
33. (5, 0)
34. (0, -5)
35. (3, 5)
36. $(-1, -4)$

4.1 · THE RECTANGULAR (CARTESIAN) COORDINATE SYSTEM   181

In Exercises 37–50, plot all points on the same coordinate system. Let each space on the axes represent two units.

37. (0, 4)
38. (2, 8)
39. (−2, 4)
40. (−4, 12)
41. (−6, −10)
42. (−2, 0)
43. (0, 12)
44. (0, 0)
45. (1, 1)
46. (3, −5)
47. (−1, 3)
48. (−5, −11)
49. (−13, 0)
50. (0, −9)

51. Give the coordinates of each point on the coordinate system shown in the accompanying figure. Each space represents one unit.

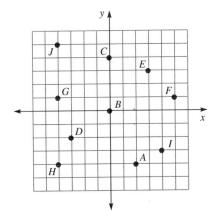

Use a coordinate system to help answer Exercises 52–64. What do you observe about each set of points? (Answers will vary.)

52. Plot six different points that have x-coordinates equal to their y-coordinates.
53. Plot six different points with y-values equal to one-half their x-values.
54. Plot six different points, three in Quadrant I and three in Quadrant III, with y-coordinates 1 greater than their x-coordinates.
55. Plot six different points, three in Quadrant I and three in Quadrant IV, with x-coordinates equal to 2.
56. Plot six different points, three in Quadrant III and three in Quadrant IV, with y-coordinates equal to −3.
57. How many points are there of the form $(2y, y)$? Plot six of them, three with y negative and three with y positive.
58. How many points are there of the form $(x − 1, x)$? Plot six of them, three with x negative and three with x positive.
59. Plot four points such that their y-coordinates are additive inverses of their x-coordinates.

60. Plot four points such that their x-coordinates are multiplicative inverses of their y-coordinates.
61. Plot four points with y-coordinates 1 more than their x-coordinates.
62. Plot four points with x-coordinates 1 less than their y-coordinates.
63. Plot three points with y-coordinates 2 less than three times their x-coordinates.
64. Plot three points with y-coordinates 1 more than twice their x-coordinates.

Plot the following ordered pairs and use them to find a relationship between the x- and y-coordinates in the form of an equation.

65. (−1, −1), (0, 0), (1, 1), (2, 2)
66. (−1, −2), (0, 0), (1, 2), (2, 4)
67. (−1, −4), (0, −4), (1, −4), (2, −4)
68. (−1, −3), (−1, −2), (−1, −1), (−1, 0)
69. (−1, 0), (0, 0), (1, 0), (2, 0)
70. (0, 4), (0, 3), (0, 2), (0, 1)

Use the following equations to complete the ordered pairs.

71. $y = x$; (−2,), (0,), (, 3), (, −1)
72. $y = -x$; (0,), (1,), (−1,), (, 3)
73. $y = -3x$; (−1,), (0,), (, 1), (, −3)
74. $y = 2x$; (−2,), (0,), (2,), (4,)
75. $y = x - 3$; (0,), (1,), (, 0), (2,)
76. $y = 2x - 1$; (0,), (1,), (−1,), (, 0)
77. $y = 3x + 5$; (0,), (1,), (−2,), (, 0)
78. $2y + x = 4$; (, 0), (0,), (2,), (−2,)
79. $3x - 2y = 6$; (0,), (, 3), (, 0), (4,)
80. $y = 1.2x - 4.7$; (0.1,), (1,), (−0.2,), (0,), (, 0)
81. $0.5x + 3.5y = 0.7$; (1,), (, 3), (, 0), (0,)
82. $1.7x - 4y = 11.9$; (0,), (, 0)
83. $1.69x + 1.32y = 2.2308$; (0,), (, 0)
84. $y = \frac{1}{2}x + \frac{1}{3}$; (0,), (, 0), (2,), (4,)
85. $y = \frac{1}{4}x + \frac{1}{2}$; (0,), (, 0), (4,), (8,)
86. $\frac{1}{2}y = \frac{1}{3}x$; (0,), (, 0), (3,), (, 3)
87. $\frac{3}{4}y = \frac{2}{3}x + 1$; (0,), (, 0), $\left(\frac{3}{2},\ \right)$, (3,)
88. $y = \frac{-5}{3}x - 2$; (0,), (, 0), $\left(\frac{3}{5},\ \right)$, $\left(\frac{3}{2},\ \right)$, (3,)

For Extra Thought

89. Three vertices of a rectangle are $A(0, 0)$, $B(4, 0)$, and $C(4, 5)$. Find the coordinates of the fourth vertex.
90. Three vertices of a parallelogram are $A(-3, 2)$, $B(5, -4)$, and $C(-5, -4)$. Find the coordinates of a fourth vertex.
91. Three points on a circle are $A(4, 1)$, $B(0, 5)$, and $C(0, -3)$. Find the coordinates of a possible fourth point, given that BC is a diameter.
92. If the endpoints of the hypotenuse of a right triangle are $(1, 4)$ and $(5, 0)$, find the coordinates of a possible third vertex.
93. Three vertices of a square are $A(-3, 2)$, $B(3, 2)$, and $C(-3, -4)$. Find the coordinates of the fourth vertex.
94. Three vertices of a square are $A(-3, -1)$, $B(4, -1)$, and $C(4, 6)$. Find the coordinates of the fourth vertex.

Review Exercises

Solve each of the following.

95. $|2x - 1| = 5$
96. $|3y + 2| < 5$
97. $|4 - y| \geq 3$
98. $|5 - 2x| \leq 1$
99. $\left|\dfrac{4x + 7}{5}\right| = 0$
100. $\left|\dfrac{6 - 5x}{4}\right| = 1$
101. $\left|\dfrac{3x + 4}{7}\right| = \dfrac{1}{7}$
102. $\left|\dfrac{7y - 1}{8}\right| = \dfrac{3}{4}$

4.2 Graphing Linear Equations in Two Variables

OBJECTIVES

In this section we will learn to
1. graph a linear equation;
2. use check points; and
3. identify the intercepts of an equation.

In the Exercise Set 4.1, you plotted several points satisfying a definite relationship between x and y. For example, the equation describing all ordered pairs where the y-coordinate is 1 greater than the x-coordinate is $y = x + 1$. Six ordered pairs for which this is true are

$$(-3, -2), (-1, 0), (0, 1), (1, 2), (2, 3), (3, 4)$$

When we plot these points on the same coordinate system, we find that they lie on a straight line, as shown in Figure 4.7

We used six points to determine the line in Figure 4.7, but we could have used 60 or 600 if time and space allowed. As long as y is 1 greater than x, the points will all lie on the same straight line. Arrowheads are used at the end of the line to indicate that it extends forever in either direction. Equations such as $y = x + 1$ are called **linear equations** because their graphs are straight lines.

Definition

Linear Equation in Two Variables

A **linear equation in two variables,** x and y, is one that can be put in the form

$$ax + by = c$$

where a, b, and c are real numbers and a and b are not both zero.

4.2 · GRAPHING LINEAR EQUATIONS IN TWO VARIABLES

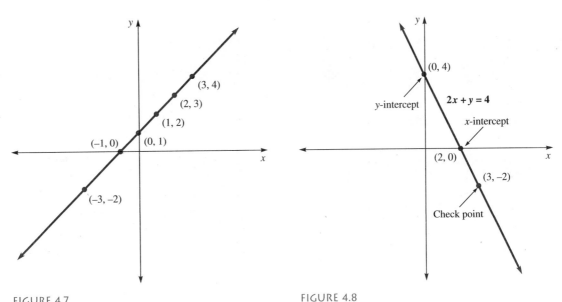

FIGURE 4.7

FIGURE 4.8

EXAMPLE 1 Graph the equation $2x + y = 4$.

SOLUTION To sketch the graph of a straight line, we need only two points. However, obtaining a third point, called a **check point,** provides a check on the accuracy of the other two. To find the coordinates of points that satisfy the equation, we construct a **table of values.** Two of the easiest values to use in constructing the table are $x = 0$ and $y = 0$. The symbol "$\Rightarrow$" means implies.

Equation			Table of Values	
			x	y
$2x + y = 4$				
$2(0) + y = 4 \Rightarrow y = 4$	Let $x = 0$.		0	4
$2x + 0 = 4 \Rightarrow x = 2$	Let $y = 0$.		2	0
$2(3) + y = 4 \Rightarrow y = -2$	←Check point→		3	-2

The three ordered pairs $(0, 4)$, $(2, 0)$, and $(3, -2)$ that satisfy the equation are shown in the table of values. The graph is shown in Figure 4.8.

The point where a line crosses or *intercepts* the x-axis is called the **x-intercept,** and the point at which it crosses the y-axis is called the **y-intercept.** To find the y-intercept, we let $x = 0$, since all points on the y-axis have an x-coordinate of zero. To find the x-intercept, we let $y = 0$, since all points on the x-axis have a y-coordinate of zero.

EXAMPLE 2 Graph the equation $3x + 2y = 4$ by using the x- and y-intercepts and a check point.

SOLUTION

Table of Values		Equation
x	y	$3x + 2y = 4$
0	2	$3(0) + 2y = 4 \Rightarrow y = 2$
$\frac{4}{3}$	0	$3x + 2(0) = 4 \Rightarrow x = \frac{4}{3}$
2	-1	$3(2) + 2y = 4 \Rightarrow y = -1$

Three ordered pairs on the line are the *y*-intercept, (0, 2); the *x*-intercept, $\left(\frac{4}{3}, 0\right)$; and the check point (2, −1). The point $\left(\frac{4}{3}, 0\right) = \left(1\frac{1}{3}, 0\right)$ is $\frac{1}{3}$ of the way between 1 and 2 on the *x*-axis. The graph is shown in Figure 4.9.

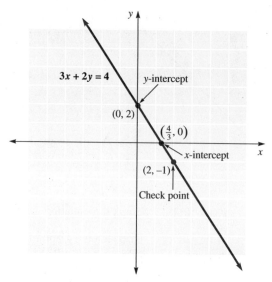

FIGURE 4.9

EXAMPLE 3   Graph the equation $x = 3$.

SOLUTION   The equation $x = 3$ is actually a linear equation of the form $ax + by = c$. To see this, we think of $x = 3$ as $x + 0y = 3$. Since zero times *y* is zero for any value of *y* that we choose, *x* will be 3 for any value of *y* we choose. Three ordered pairs are (3, −2), (3, 0), and (3, 4). The graph does not cross the *y*-axis, so the *y*-intercept does not exist. The graph is shown in Figure 4.10.

The graph of the equation $x = k$, where *k* is any real number, will always be a vertical line. If *k* is positive, the line will be *k* units to the right of the *y*-axis and parallel to it. If *k* is negative, the line will be *k* units to the left of the *y*-axis and parallel to it. Similarly, the line $y = k$ is a line parallel to the *x*-axis. If *k* is positive it will be *k* units above the *x*-axis, and if *k* is negative it will be *k* units below the *x*-axis. The graph of $y = −3$ is shown in Figure 4.11.

EXAMPLE 4   The *y*-coordinate in a linear equation is three times the difference of the *x*-coordinate and 2. Find the equation and sketch its graph.

SOLUTION

The *y*-coordinate	is	three times	the difference of the *x*-coordinate and 2.
↓	↓	↓	↓
y	=	3 ·	(*x* − 2)

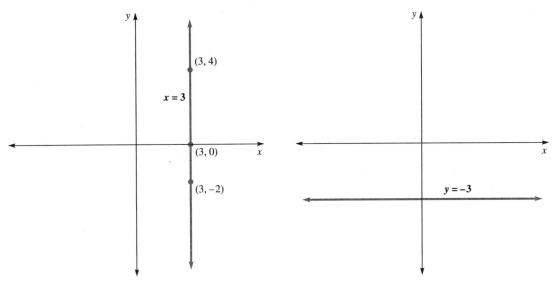

FIGURE 4.10

FIGURE 4.11

$$y = 3(x - 2)$$
$$y = 3x - 6 \qquad \text{Distributive property}$$

x	y
0	-6
2	0
3	3

$y = 3x - 6$
$y = 3(0) - 6 \;\Rightarrow\; y = -6$
$y = 3(2) - 6 \;\Rightarrow\; y = 0$
$y = 3(3) - 6 \;\Rightarrow\; y = 3$

Three ordered pairs are $(0, -6)$, $(2, 0)$, and $(3, 3)$. The graph is shown in Figure 4.12.

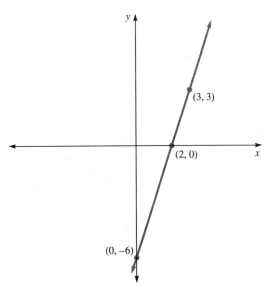

FIGURE 4.12

EXAMPLE 5   Graph the equation $C = 100 - 50T$ for the ordered pairs (0,), (1,), and (2,) if the vertical axis is the C-axis.

SOLUTION   To determine the values of C, we substitute 0, 1, and 2 for T:

T	C
0	100
1	50
2	0

$C = 100 - 50T$
$C = 100 - 50(0) \Rightarrow C = 100$
$C = 100 - 50(1) \Rightarrow C = 50$
$C = 100 - 50(2) \Rightarrow C = 0$

The ordered pairs are (0, 100), (1, 50), and (2, 0). To sketch the graph, we let each mark on the vertical axis represent 25 units and each mark on the horizontal axis 1 unit. The graph is shown in Figure 4.13.

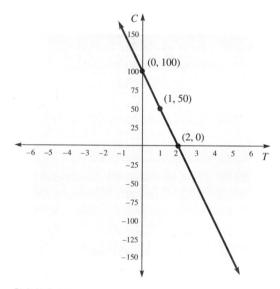

FIGURE 4.13

When graphing equations with variables other than x or y, you are free to decide which variable is represented on the vertical axis and which on the horizontal axis. The graphs will not look the same for the two possible choices, but the relation between the variables will not change.

EXAMPLE 6   Each day a company manufactures and sells n fasteners at a profit of $3 each. Fixed costs associated with the manufacture of these fasteners are $25/day. If P represents the company's daily profit, write an equation relating the profit to the number of fasteners manufactured. Graph the equation.

SOLUTION   Profit is equal to income minus the fixed cost. Income is $3 for each of the n fasteners manufactured, or $3n$. Fixed costs are $25. Thus,

$$P = 3n - 25$$

Three ordered pairs satisfying the equation are (0, −25), (10, 5), and (15, 20). To sketch the graph, we let P be represented by the vertical axis and n by the horizontal axis. We use no negative values of *n* because the company cannot manufacture a negative number of units. The graph, shown in Figure 4.14, assumes that some items are in various stages of completion on any one day by going through fractional values of *n*.

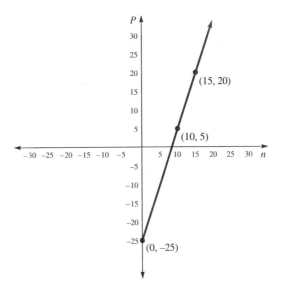

FIGURE 4.14

Note that if a point lies on the graph of a line, its coordinates must satisfy the equation of that line. Conversely, if the coordinates of a point satisfy the equation of a line, the point lies on the graph of that line. For example, the point (2, 4) lies on the line $2x + 3y = 16$ because

$$2 \cdot 2 + 3 \cdot 4 \stackrel{?}{=} 16 \qquad \text{Substitute 2 for } x \text{ and 4 for } y.$$
$$4 + 12 \stackrel{?}{=} 16$$
$$16 \stackrel{\checkmark}{=} 16$$

Observations such as this one allow us to prove that a line passes through a particular point on a coordinate system.

Exercise Set 4.2

Graph the following linear equations using the x- and y-intercepts and a third (check) point.

1. $y = x + 1$
2. $y = 2x + 1$
3. $y = -x - 2$
4. $y = -3x - 1$
5. $y = x + 2$
6. $y = x + 3$
7. $y = -x + 1$
8. $y = -x + 3$
9. $y = -2x + 1$
10. $y = -2x - 3$
11. $x + y = 2$
12. $x + 2y = 4$
13. $x - 3y = 3$
14. $x - y = 5$
15. $2x - y = -1$
16. $x - 3y = -3$
17. $x + 5y = 10$
18. $4x - y = 8$
19. $2x + 3y = 6$
20. $5x - 3y = -15$
21. $y = \frac{1}{2}x + 2$
22. $y = \frac{1}{4}x + 2$

23. $y = \dfrac{1}{4}x - 1$ **24.** $y = \dfrac{1}{2}x - 2$

Graph the following linear equations by finding three ordered pairs.

25. $x = 6$ **26.** $x + 5 = 0$ **27.** $y + 4 = 0$
28. $y = 3$ **29.** $x = y$ **30.** $x - y = 0$
31. $3x + 2y = 0$ **32.** $4x - 3y = 0$
33. $x + y + 1 = 0$ **34.** $x - y + 2 = 0$
35. $5x + 4y - 20 = 0$ **36.** $7x - 3y - 21 = 0$
37. $7x - 4y = 5$ **38.** $5x + 4y = 11$
39. $y = 0$ **40.** $x = 0$
41. $x = -4$ **42.** $y = -2$
43. $5y = -10$ **44.** $3x = -6$

Write an equation from the given information, find three ordered pairs, and graph the line.

45. The *y*-coordinate is 1 more than the *x*-coordinate.
46. The *y*-coordinate is 2 less than the *x*-coordinate.
47. The *y*-value is twice the *x*-value.
48. The *y*-value is one-half the *x*-value.
49. Twice the *y*-coordinate added to the *x*-coordinate is 4.
50. The *y*-coordinate subtracted from the *x*-coordinate is 2.
51. Three times the *x*-value added to twice the *y*-value is 6.
52. The *y*-coordinate is 3.
53. Twice the *x*-coordinate is 8.
54. The sum of the *x*- and *y*-coordinates reduced by 2 is zero.
55. The difference of the *x*- and *y*-coordinates increased by 4 is zero.
56. The *y*-value is twice the difference of the *x*-value and 3.
57. The *y*-value is three times the sum of the *x*-value and 1.
58. Twice the difference of the *x*- and *y*-coordinates less 6 is zero.

Complete the ordered pairs and graph each linear equation using an appropriate vertical and horizontal axis.

59. $d = 60t$; $(0, \ )$, $(3, \ )$, $(6, \ )$
60. $V = 120 + 32t$; $(1, \ )$, $(5, \ )$, $(0, \ )$
61. $I = 1200t$; $(0, \ )$, $\left(\dfrac{1}{2}, \ \right)$, $(2, \ )$
62. $C = 100 + 18t$; $(0, \ )$, $(-20, \ )$, $(16, \ )$

63. $C = \dfrac{1}{2}t$; $(0, \ )$, $(2, \ )$, $(6, \ )$
64. $C = \dfrac{1}{3}t$; $(0, \ )$, $(3, \ )$, $(9, \ )$
65. $C = \dfrac{2}{3}t - \dfrac{1}{2}$; $(0, \ )$, $(2, \ )$, $(6, \ )$
66. $C = 0.5t + 0.5$; $(0, \ )$, $(2, \ )$, $(-1, \ )$
67. $C = 0.1t + 1.75$; $(0, \ )$, $(10, \ )$, $(-10, \ )$
68. $C = 0.75t + 4$; $(0, \ )$, $(10, \ )$, $(-10, \ )$

Solve each applied problem.

69. As a person descends beneath the surface of the ocean, the water pressure P is related to the depth d by the equation $P = \tfrac{5}{11}d + 15$. Find ordered pairs (d, P) for $d = 0, 22, 55,$ and 88.

70. The force F required to lift an object of a given mass m to a given height h is given by the equation $F = mgh$. If an object has a mass of 30 kg, and if $g = 9.8$, find the force required to lift the object 1 m, 5 m, and 20 m.

71. Mrs. Topliff has concluded that her profit P in dollars from the sale of x dresses is determined by the equation $P = 5.5x - 30$. Find her profit for selling 10, 50, and 100 dresses.

72. The normal weight w in pounds of an average man is related to his height h in inches by the equation $w = 5.5h - 220$. Find the weight of men whose heights are 69 in., 6 ft, and 74 in.

73. The cost of making x toys is given by $C = 2.5x + 100$. What is the cost of making 0, 10, and 100 of these toys? Write the answers as ordered pairs.

74. The cost of making x baskets is given by the equation $C = 0.5x + 30$. What is the cost of making 0, 10, and 20 of these baskets? Write the answers as ordered pairs.

75. A linear equation can be written in the form $y = mx + 3$. If the ordered pair $(1, 0)$ is a solution of the equation, find m. Then find y when x is 5, 10, and -2.

76. A linear equation can be written in the form $ax + by = c$. If $a = c = 2$ and the ordered pair $(3, 4)$ is a solution of the equation, find b.

77. Tony's Delivery Service finds that the cost C of making t delivery trips is a linear relationship. The cost of making one trip is $2.50 and the cost of two trips is $5.00.

  (a) Write this information as ordered pairs, (t, C).
  (b) Plot these points on a Cartesian coordinate system.
  (c) Draw a straight line through these points.
  (d) Use the graph to predict the number of trips that can be made for a cost of $12.50.

78. The value of a Bright Lawn Mower depends upon its age. When new, it has a value of $350, and after 3 years its value is $150.
   (a) Write this information as ordered pairs, (a, v), where a = age and v = value.
   (b) Plot these points on a Cartesian coordinate system.
   (c) Draw a straight line through these points.
   (d) Use the graph to estimate the value after 5 years.

WRITE IN YOUR OWN WORDS

79. Describe an x-intercept and how you find it.

80. Describe a linear equation; why do you think it's called that?

81. What is the origin in a Cartesian coordinate system?

82. Describe the Cartesian coordinate system.

83. Describe what the graph of $y = k$ looks like (k is a constant).

84. Why is it a good idea to find three ordered pairs when graphing a linear equation?

85. Describe what the graph of $x = k$ looks like (k is a constant).

86. Describe what is meant by an ordered pair.

4.3 The Slope of a Line; Equations of a Line

OBJECTIVES

In this section we will learn to
1. find the slope of a line using two points;
2. write the equation of a line by using the slope and a point on the line;
3. find the equation of a line from its slope and its y-intercept;
4. use slope to determine if two lines are parallel or perpendicular.

Driving along a highway, you encounter a sign that says "WARNING—6% GRADE AHEAD—TRUCKS USE LOW GEARS." What does this mean? It means that for every 100 feet that the road advances horizontally, it drops 6 feet in elevation. When a couple wants to roof their house with a new type of shingle and they are told that the pitch of the roof must be at least "3 in 12" to use it, what does that mean? It means that the roof must rise three feet for every 12 feet along the side of the house. These are but two of many practical examples of a very important mathematical concept called *slope*.

In Section 4.2 we saw that we can draw the graph of a line by using any two points on it. We can also draw it if we know one point on the line and how rapidly the line rises or falls from the known point. Figure 4.15(a) shows the graph of a line that rises from left to right and Figure 4.15(b) shows the graph of a line that falls from left to right.

The rate at which a line rises or falls as we move along it from left to right is measured by its *slope*, or steepness. If (x_1, y_1) and (x_2, y_2) are any two points on a line, as in Figure 4.16, its *slope* is defined as follows.

DEFINITION

Slope of a Line

If (x_1, y_1) and (x_2, y_2) are any two points on a line, then

$$\text{slope} = m = \frac{\text{change in } y\text{-values}}{\text{change in } x\text{-values}} = \frac{y_2 - y_1}{x_2 - x_1}, \quad x_2 \neq x_1$$

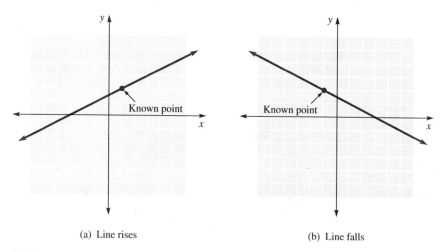

(a) Line rises (b) Line falls

FIGURE 4.15

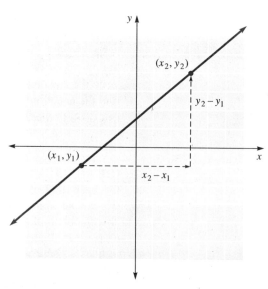

FIGURE 4.16

The ordered pair (x_1, y_1) is read "x sub-1, y sub-1" and (x_2, y_2) is read "x sub-2, y sub-2." The small numbers, called **subscripts,** are used to distinguish one point from the other.

EXAMPLE 1 Graph the line joining (1, 2) and (3, 5), and find its slope by calculating the change in y and the change in x.

SOLUTION The graph is shown in Figure 4.17. Notice that the change in y from (1, 2) to (3, 5) is 3 and the change in x from (1, 2) to (3, 5) is 2. The slope is

$$\frac{\text{change in the } y\text{-values}}{\text{change in the } x\text{-values}} = \frac{3}{2}$$

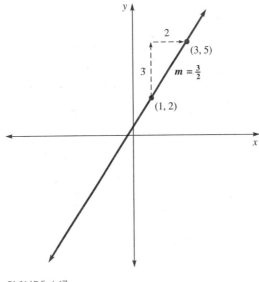

FIGURE 4.17

Notice that the line in Figure 4.17 rises from left to right. In general,

> *A line that rises from left to right has a positive slope.*

EXAMPLE 2   Find the slope of the line joining (1, 2) and (3, 5) in Example 1 using the slope formula $m = \dfrac{y_2 - y_1}{x_2 - x_1}$.

SOLUTION   We let $(x_1, y_1) = (1, 2)$ and $(x_2, y_2) = (3, 5)$. Then

$$\text{slope} = m = \frac{y_2 - y_1}{x_2 - x_1} = \frac{5 - 2}{3 - 1} = \frac{3}{2}$$

The slope of the line is $\dfrac{3}{2}$. The choice of (1, 2) for (x_1, y_1) was arbitrary. We could have let $(x_1, y_1) = (3, 5)$ and $(x_2, y_2) = (1, 2)$, in which case

$$\text{slope} = m = \frac{y_2 - y_1}{x_2 - x_1} = \frac{2 - 5}{1 - 3} = \frac{-3}{-2} = \frac{3}{2}$$

Both calculations indicate that the line rises 3 units for every 2 units of positive change in the horizontal direction; see Figure 4.17.

EXAMPLE 3   Find the slope of the line joining (0, 4) and (5, 1), and graph the line.

SOLUTION We let $(x_1, y_1) = (0, 4)$ and $(x_2, y_2) = (5, 1)$. Then

$$m = \frac{y_2 - y_1}{x_2 - x_1} = \frac{1 - 4}{5 - 0} = \frac{-3}{5}$$

The slope of the line is $-\frac{3}{5}$. The graph is shown in Figure 4.18.

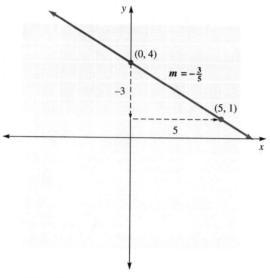

FIGURE 4.18

Notice that the line in Figure 4.18 falls from left to right. In general,

A line that falls from left to right has a negative slope.

EXAMPLE 4 Find the slope of the line joining (0, 2) and (4, 2), and graph the line.

SOLUTION We let $(x_1, y_1) = (0, 2)$ and $(x_2, y_2) = (4, 2)$. Then

$$m = \frac{y_2 - y_1}{x_2 - x_1} = \frac{2 - 2}{4 - 0} = \frac{0}{4} = 0$$

The slope of the line is 0. Since the two y-values, y_2 and y_1, are the same, the line is horizontal, as shown in Figure 4.19.

From Example 4, we conclude that

The slope of a horizontal line is zero.

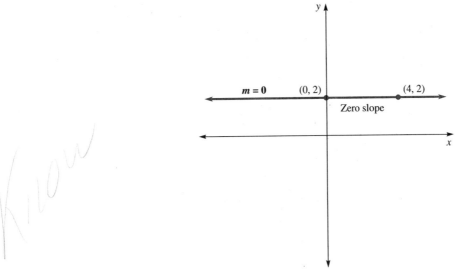

FIGURE 4.19

EXAMPLE 5   Find the slope of the line joining (3, 1) and (3, 5), and graph the line.

SOLUTION   We let $(x_1, y_1) = (3, 1)$ and $(x_2, y_2) = (3, 5)$. Then

$$m = \frac{y_2 - y_1}{x_2 - x_1} = \frac{5 - 1}{3 - 3} = \frac{4}{0}$$

Since division by zero is not defined, the slope of this line is undefined. The two x-coordinates, x_1 and x_2, are the same, so the line is vertical, as shown in Figure 4.20.

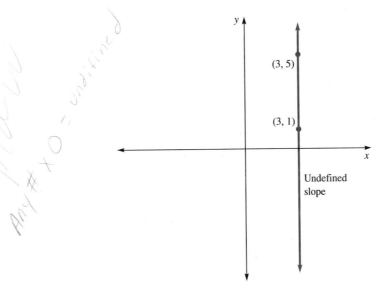

FIGURE 4.20

From Example 5, we see that

> *The slope of a vertical line is undefined.*

It is not always necessary to know two points on a line in order to sketch its graph. If we know one point and the slope, we can graph a line.

EXAMPLE 6   Graph the line that goes through (1, 1) and has slope $\frac{2}{3}$.

SOLUTION   We begin by locating the point (1, 1) on the graph, as shown in Figure 4.21. Then, using (1, 1) as a starting point, we move two units in a vertical direction (the change in y is 2) and 3 units in a horizontal direction (the change in x is 3). The point we have reached, (4, 3), is a second point on the line, and we can use it to sketch the graph.

𝒞AUTION

Care must be taken when using the slope formula to ensure that the values from (x_1, y_1) and (x_2, y_2) are correctly substituted. The order in the numerator must be the same as the order in the denominator, and the change in y must appear in the numerator.

Right: $m = \dfrac{y_2 - y_1}{x_2 - x_1}$

Wrong: $m = \dfrac{y_2 - y_1}{x_1 - x_2}$

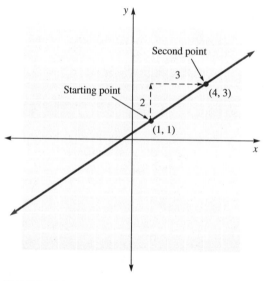

FIGURE 4.21

EXAMPLE 7   Graph the line that goes through $(-1, 3)$ and has slope $-\frac{1}{2}$.

SOLUTION   We begin by locating the point $(-1, 3)$ on the graph, as shown in Figure 4.22. Then, using $(-1, 3)$ as a starting point, we use the slope to find a second point on the line. To do this, we recall that $-\frac{1}{2} = \frac{-1}{2} = \frac{1}{-2}$. If we use $\frac{-1}{2}$, we move 1 unit down from $(-1, 3)$ and 2 units to the right to find the second point. If we use $\frac{1}{-2}$, we move one unit up and 2 units to the left of $(-1, 3)$. Both of the points located in this manner lie on the same line with $(-1, 3)$; see Figure 4.22.

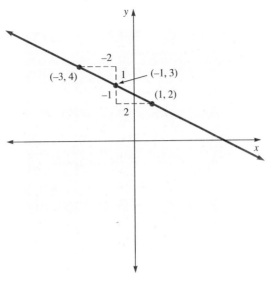

FIGURE 4.22

Equations of a Line

In Section 4.2 we defined a linear equation in two variables as an equation that can be put in the form $ax + by = c$. This form is known as the **general form** of the equation of a line. We can find other forms by using known information about the line. For example, if we know a single point on the line and its slope, we can find its equation. Suppose (x_1, y_1) is a point on a line with slope m. If (x, y) is any other point on the line, then

$$m = \frac{y - y_1}{x - x_1} \qquad \text{Let } (x, y) = (x_2, y_2) \text{ in the slope formula.}$$

Multiplying each side by $x - x_1$ and interchanging the two sides of the equation, we have

$$y - y_1 = m(x - x_1)$$

This is called the *point–slope* form of the equation of a line.

DEFINITION **Point–Slope Form**	The **point–slope form** of the equation of a line through the point (x_1, y_1) with slope m is $$y - y_1 = m(x - x_1)$$

EXAMPLE 8   Find the general form of the equation of the line that goes through $(-1, 2)$ and has slope $\frac{-2}{3}$. Graph the line.

SOLUTION We know one point on the line and we know its slope. The point–slope form of the equation of a line can be used to determine its equation.

$$y - y_1 = m(x - x_1)$$ Point–slope form

$$y - 2 = \frac{-2}{3}[x - (-1)]$$ Substitute $x_1 = -1$, $y_1 = 2$, and $m = \frac{-2}{3}$.

$$y - 2 = \frac{-2}{3}(x + 1)$$

$$3(y - 2) = 3\left[\frac{-2}{3}(x + 1)\right]$$ Clear of fractions; multiply each side by 3.

$$3y - 6 = -2x - 2$$ Distributive property

$$3y + 2x = -2 + 6$$ Add $2x$ and 6 to each side.

$$3y + 2x = 4$$ Simplify.

$$2x + 3y = 4$$ General form

We can graph the line by using the given point, $(-1, 2)$, and the slope, $\frac{-2}{3}$, as shown in Figure 4.23(a). Another method is to find a second point and a check point, using the general form of the equation of the line.

$$2x + 3y = 4$$
$$2x + 3(4) = 4$$ Let $y = 4$.
$$2x + 12 = 4$$ Simplify.
$$2x = -8$$ Subtract 12 from each side.
$$x = -4$$ Divide each side by 2.

Since we let $y = 4$, a second point is $(-4, 4)$. If we let $y = -2$, then we get $x = 5$. The point $(5, -2)$ can serve as a check point. The graph in Figure 4.23(b) shows the line that goes through the two points $(-1, 2)$ and $(-4, 4)$.

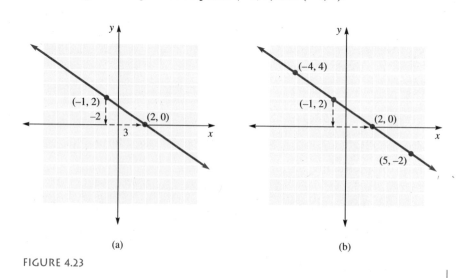

FIGURE 4.23

EXAMPLE 9 Find the equation of the line through $(1, 2)$ and $(2, -1)$.

SOLUTION We begin by finding the slope of the line:

$$m = \frac{y_2 - y_1}{x_2 - x_1} = \frac{-1 - 2}{2 - 1} = \frac{-3}{1} = -3$$

We can now use either of the known points and the slope to write the equation.

$y - y_1 = m(x - x_1)$

$y - 2 = -3(x - 1)$ Let $(x_1, y_1) = (1, 2)$ and $m = -3$.

When we simplify the equation and write it in general form, it becomes

$3x + y = 5$

We now turn our attention to a special case of the point–slope form. When the known point is the y-intercept, the equation of the line can be written in a special form called the *slope–intercept form*. Consider a line with slope m that crosses the y-axis at $(0, b)$. We can write its equation using the point–slope form:

$y - y_1 = m(x - x_1)$	Point–slope form
$y - b = m(x - 0)$	$(x_1, y_1) = (0, b)$
$y - b = mx$	Distributive property
$y = mx + b$	Add b to each side.

Notice the position of the slope m and the point $(0, b)$ in the final equation.

DEFINITION

Slope–Intercept Form

The **slope–intercept form** of the equation of a line with slope m and y-intercept b is

$$y = mx + b$$

slope ↑   ↑ y-intercept

EXAMPLE 10 Find the equation of the line with slope $\frac{2}{3}$ and y-intercept 5.

SOLUTION Substituting $m = \frac{2}{3}$ and $b = 5$ in the slope–intercept form gives us

$$y = \frac{2}{3}x + 5$$

EXAMPLE 11 Write the equation of the line $2x + 3y = 4$ in slope–intercept form. Identify the slope and y-intercept of the line.

SOLUTION The general form of the equation is $2x + 3y = 4$. To write the equation in the slope–intercept form, we solve it for y:

$2x + 3y = 4$ General form
$3y = -2x + 4$ Subtract $2x$ from each side.
$y = -\frac{2}{3}x + \frac{4}{3}$ Divide each side by 3.

When $y = -\frac{2}{3}x + \frac{4}{3}$ is compared to $y = mx + b$, we can see that $m = -\frac{2}{3}$ and $b = \frac{4}{3}$. Therefore the slope is $-\frac{2}{3}$ and the y-intercept is $\frac{4}{3}$.

The last topic we will consider in this section concerns parallel and perpendicular lines. The relationship between their slopes is as follows.

DEFINITION

Parallel and Perpendicular Lines

If two lines are parallel, their slopes are equal.

If two lines are perpendicular, their slopes are negative reciprocals.

CAUTION

The point–slope form and the slope–intercept form of the equation of a line and the information concerning the slopes of perpendicular lines apply only to nonvertical lines.

For example, $y = \frac{2}{3}x + 2$ is parallel to $y = \frac{2}{3}x - 1$ as their slopes are both $\frac{2}{3}$; see Figure 4.24(a). The line $y = \frac{2}{3}x + 2$ is perpendicular to the line $y = -\frac{3}{2}x + 3$ as their slopes, $\frac{2}{3}$ and $-\frac{3}{2}$, are negative reciprocals; see Figure 4.24(b).

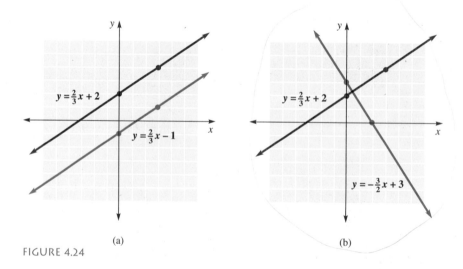

FIGURE 4.24

EXAMPLE 12 Find the equation of the line that goes through (0, 4) and is parallel to the line $5x + 2y = 4$. Write the equation in slope–intercept form.

SOLUTION   We begin by solving $5x + 2y = 4$ for y:

$$5x + 2y = 4$$
$$2y = -5x + 4 \quad \text{Subtract } 5x \text{ from each side.}$$
$$y = -\frac{5}{2}x + 2 \quad \text{Divide each side by 2.}$$

Therefore the slope of the given line is $-\frac{5}{2}$, so the slope of the parallel line is also $-\frac{5}{2}$. The point $(0, 4)$ is the y-intercept, so $b = 4$. The equation of the parallel line is $y = -\frac{5}{2}x + 4$.

EXAMPLE 13   Find the equation of the line that goes through $(1, 3)$ and is perpendicular to the line $y = 2x - 5$. Write the equation in general form.

SOLUTION   The slope of the given line is 2, so the slope of the perpendicular line is $-\frac{1}{2}$. We are looking for the equation of the line through $(1, 3)$ with slope $-\frac{1}{2}$. Using the point–slope form, we have

$$y - 3 = -\frac{1}{2}(x - 1)$$
$$2y - 6 = -x + 1 \quad \text{Multiply each side by 2, the LCD.}$$
$$x + 2y = 7 \quad \text{Add } x \text{ and 6 to each side.}$$

The general form of the equation of the perpendicular line is $x + 2y = 7$.

Summary of the Forms for the Equation of a Line

General Form

$ax + by = c$, a, b, and c real numbers; a and b not both zero

Point–Slope Form

The equation of the line through (x_1, y_1) with slope m is

$y - y_1 = m(x - x_1)$

To use this form, the slope and one point on the line must be known.

Slope–Intercept Form

The equation of a line with slope m and y-intercept b is

$y = mx + b$

To use this form, the y-intercept and the slope of the line must be known.

Computers and Calculators

With the proper software, a computer can aid in graphing straight lines as well as other more complex equations. The graph of $2x - y = 2$ in Figure 4.25 was produced on a Macintosh computer. Graphics calculators also have many of the same capabilities. We will discuss the use of a graphics calculator for graphing in later sections.

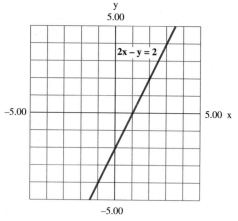

FIGURE 4.25

Do You Remember?

Can you match these?

_____ 1. The general form of the equation of a line

_____ 2. The point–slope form of the equation of a line

_____ 3. The slope–intercept form of the equation of a line

_____ 4. The y-intercept of $y = \frac{2}{7}x + 9$

_____ 5. The slope of a line parallel to $y = \frac{-3}{7}x + 4$

_____ 6. The slope of a line perpendicular to $y = \frac{-3}{7}x + 4$

_____ 7. The formula for the slope of a line

_____ 8. The slope of a horizontal line

_____ 9. The slope of a vertical line

_____ 10. A line with positive slope

_____ 11. A line with negative slope

a) A rising line
b) 0
c) $\frac{-3}{7}$
d) $-\frac{7}{3}$
e) $y = mx + b$
f) $ax + by = c$
g) Undefined
h) $\frac{7}{3}$
i) A falling line
j) $y - y_1 = m(x - x_1)$
k) (0, 9)
l) $\frac{y_2 - y_1}{x_2 - x_1}$

Answers: 1. f 2. j 3. e 4. k 5. c 6. h 7. l 8. b 9. g 10. a 11. i

Exercise Set 4.3

Find the slope of each of the following lines. Each space represents one unit.

1.
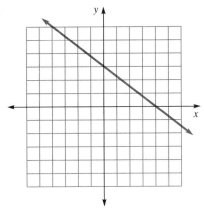

2.
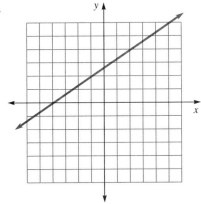

3.
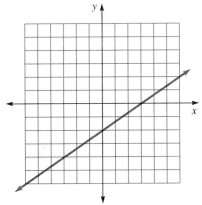

4.
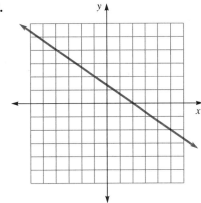

5.
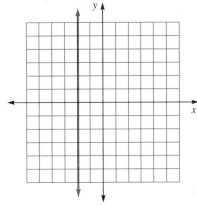

6.
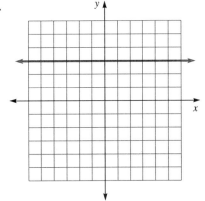

Find the slope of the line through the given pairs of points.

7. (2, 3) and (5, 6) **8.** (3, 1) and (4, 7)
9. (4, 8) and (6, 8) **10.** (5, −3) and (2, −3)
11. (−2, −1) and (4, −6) **12.** (−7, 4) and (3, −5)
13. (0, −2) and (−3, −5) **14.** (−1, −2) and (−3, 0)
15. (3, −5) and (3, 4) **16.** (−2, 7) and (−2, 3)

Graph the line that goes through the given point and has the given slope.

17. (1, 1); $m = \dfrac{2}{3}$ **18.** (2, 2); $m = \dfrac{3}{4}$

19. (0, 0); $m = 4$ **20.** (0, 0); $m = -3$

21. (−3, 1); $m = \dfrac{-3}{2}$ **22.** (4, −3); $m = \dfrac{-5}{4}$

23. (2, 0); $m = 0$ **24.** (0, −5); $m = 0$

25. (3, 1); m undefined **26.** (−2, 4); m undefined

Find the equation of the line that goes through the given point and has the given slope. Write the equation in $y = mx + b$ form and graph the line.

27. (0, 0); $m = 1$ **28.** (0, 0); $m = -1$
29. (2, 3); $m = 2$ **30.** (3, 1); $m = 4$

31. (1, 4); $m = \dfrac{-1}{2}$ **32.** (2, 1); $m = \dfrac{-1}{3}$

33. (6, −4); $m = 0$ **34.** (−3, 5); $m = 0$
35. (−1, 0); m undefined **36.** (4, 0); m undefined

37. (−5, −2); $m = \dfrac{-3}{5}$ **38.** (−1, −7); $m = \dfrac{-2}{5}$

39. (0, −5); $m = \dfrac{-2}{3}$ **40.** (0, 2); $m = \dfrac{3}{4}$

Find the equation in slope–intercept form for each graph pictured in Exercises 41–46. Each space represents one unit.

41.
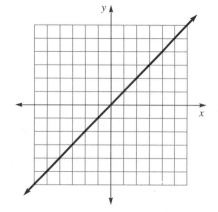

42.
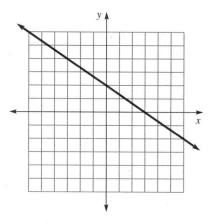

43.
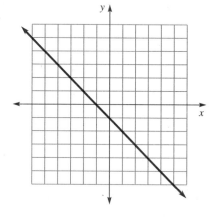

44.
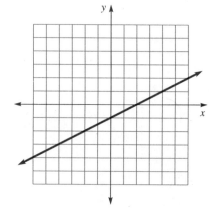

45.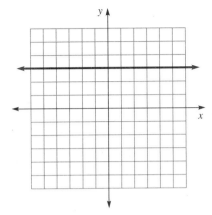

46.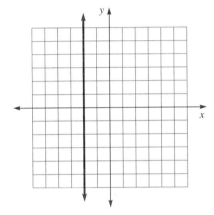

Find the equation of the line going through the two given points. Write your answer in slope–intercept form.

47. (0, 0) and (1, 2)
48. (2, 1) and (0, 0)
49. (0, 0) and (3, 3)
50. (0, 0) and (1, 1)
51. (0, 0) and (2, 3)
52. (3, 2) and (4, 3)
53. (2, 4) and (1, 3)
54. (5, 7) and (1, 4)
55. (0, −3) and (−2, 3)
56. (4, 1) and (4, 5)
57. $\left(\frac{1}{2}, \frac{1}{4}\right)$ and (0, 1)
58. (−2, −3) and $\left(\frac{1}{4}, \frac{1}{2}\right)$
59. (5, −1) and (0, 0)
60. (2, 2) and (5, 3)
61. (0, 0) and (3, 4)
62. (0, 2) and (2, 2)
63. (5, 0) and (0, 0)
64. (2, 1) and (3, 2)
65. (2, −3) and (6, 0)
66. (3, 4) and (4, 6)

Find the slope–intercept form of the equation of a line satisfying each of the following conditions.

67. A vertical line through the origin
68. A vertical line through the point (4, 5)
69. A horizontal line through the point (2, −3)
70. A horizontal line through the origin

71. A line that goes through (2, 3) and is parallel to the line $y = x$
72. A line that has y-intercept (0, 5) and is parallel to the line $y = -2x$
73. A line that has y-intercept (0, 6) and is perpendicular to the line $y = 2x$
74. A line that goes through (−2, 4) and is perpendicular to the line $y = -3x$
75. A line that goes through the origin and is parallel to the line $x + y = 4$
76. A line that goes through the origin and is parallel to the line $x - y = 3$
77. A line that goes through the origin and is perpendicular to the line $2x - 4y = 1$
78. A line that goes through the origin and is perpendicular to the line $3x + 2y = 5$
79. A line that is parallel to the x-axis and has a y-intercept of (0, 6)
80. A line that is parallel to the y-axis and has an x-intercept of (−3, 0)
81. A line that is perpendicular to the x-axis and has an x-intercept of (1, 0)
82. A line that is perpendicular to the y-axis and has a y-intercept of (0, 2)

Solve as indicated.

83. The linear equation $C = \frac{5}{9}(F - 32)$ relates temperature in degrees Celsius (°C) to degrees Fahrenheit (°F). If water boils at 212°F, determine its boiling point in °C. If water freezes at 0°C, determine its freezing point in °F. Plot these two ordered pairs, letting F be the horizontal axis and C the vertical axis. Graph the line, and approximate from the graph the temperature in °F when it is 50°C. What is the slope of the line?

84. Cuesta Grade, which is part of the California highway system, has a 6% grade (meaning the slope is 6% = $\frac{6}{100}$ = 0.06). The elevation h of the grade in feet is related to the horizontal distance d by the equation $h = 0.06d + 650$, where d is also in feet. The highest point on the grade is 1650 ft. Find the horizontal distance to this point from the bottom ($d = 0$) of the grade. Find the elevation of the base of the grade. Find the elevation when $d = 100$ ft. Plot these three ordered pairs and graph the line. Use the graph to find h when $d = 1000$.

WRITE IN YOUR OWN WORDS

85. Define the word *slope*.
86. How do you find the slope of a line through two points?
87. What do the m and b represent in $y = mx + b$?

88. What do m, x_1, and y_1 represent in $y - y_1 = m(x - x_1)$?
89. Why is a vertical line said to have an *undefined* slope?
90. Why does a horizontal line have zero slope?

REVIEW EXERCISES

Solve and graph the following linear inequalities in one variable.

91. $3x + 2 < 5$

92. $-5(2 - x) > 2x + 3$

93. $-5 \leq 2x - 7 < 4$

94. $4 > 1 - 3x \geq -5$

95. $3 - 4(1 - x) + 2(3x - 1) \geq x - 5$

96. $6 + 3(2 - 5x) + 4x < -2(1 - x)$

97. $-2[3 - (4 - 2x) - 5x] \leq 0$

98. $-\dfrac{1}{3}\left[4 + \dfrac{2}{3}\left(\dfrac{1}{2} - \dfrac{x}{4}\right)\right] + \dfrac{1}{2} > 0$

4.4 Graphing Linear Inequalities in Two Variables

OBJECTIVES

In this section we will learn to
1. graph linear inequalities in two variables; and
2. use linear inequalities in two variables to solve applied problems.

In the last two sections we saw that graphs of linear equations such as $x = 1$, $y = 2$, and $2x + y = 4$ are straight lines. By extending these ideas, we can learn how to graph *linear inequalities* such as $x < 1$, $y > 2$, or $2x + y \leq 4$.

Consider the set of ordered pairs for which $x < 1$. There is an infinite number of such pairs, each of which represents a point on a rectangular coordinate system. It's not important what the y-coordinate is as long as the x-coordinate is less than 1. The following ordered pairs are but a few that satisfy the condition $x < 1$:

$$(0, 7) \quad (-1, 7) \quad (-2, 4) \quad \left(-\dfrac{1}{2}, 2\right) \quad (0.1, 5) \quad (0, -7) \quad (-9, 2)$$

The graph of the linear inequality $x < 1$ is shown in Figure 4.26. The graph of the line $x = 1$ is a vertical line one unit to the right of the y-axis. Any point to the left of the line $x = 1$ satisfies the condition $x < 1$. The line $x = 1$ is broken to indicate that it is not a part of the graph. We indicate the region that is the solution set to an inequality by shading it.

The solution set of a linear inequality in two variables is a region in the plane called a **half-plane**, which consists of all points on one side of a line called the **boundary line.** This line is found by replacing the inequality symbol by an equal sign and graphing the resulting line. The half-plane of solutions may or may not include the boundary line. When the inequality symbol is $<$ or $>$, the boundary line is *not* included and it is shown as a **broken line.** When the inequality is $\leq$ or $\geq$, the boundary line *is* included and is shown as a **solid line.**

> When the boundary line of the graph of an inequality in two variables is included in the solution set of an inequality, it is shown as a solid line on the graph. When it is not included, it is shown as a broken line.

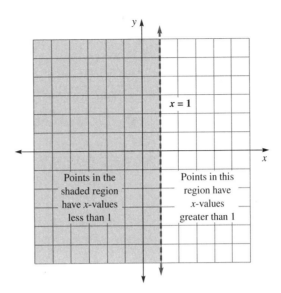

FIGURE 4.26

EXAMPLE 1   Graph the inequality $y \geq 2$.

SOLUTION   The line $y = 2$ is a horizontal line two units above the x-axis. The points represented by the ordered pairs for which the y-coordinate is greater than or equal to 2 lie above or on the line; see Figure 4.27.

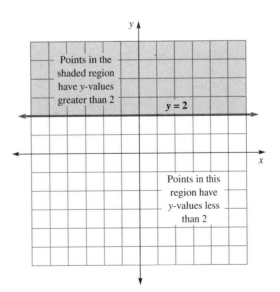

FIGURE 4.27

EXAMPLE 2   Graph the inequality $2x + y \leq 4$.

SOLUTION   First we graph the line $2x + y = 4$ by using the intercepts $(2, 0)$ and $(0, 4)$, and a check point, $(3, -2)$. The boundary line is solid because the "$=$" part of "$\leq$" indicates that it is a part of the solution set; see Figure 4.28. To decide which half-plane is the solution set, we use a **test point,** which is any point in the plane, such as $(0, 0)$, that is not on the boundary line. If the test point satisfies the inequality (makes it true), it is a part of the half-plane that is the solution set. If it does not satisfy the inequality, the desired half-plane is on the other side of the boundary line.

$$2x + y \leq 4 \quad \text{Original inequality}$$
$$2 \cdot 0 + 0 \stackrel{?}{\leq} 4 \quad \text{Test point; let } (x, y) = (0, 0).$$
$$0 \leq 4 \quad \text{True}$$

The half-plane containing the point $(0, 0)$ is therefore the solution set, so that's the side we shade, as shown in Figure 4.28.

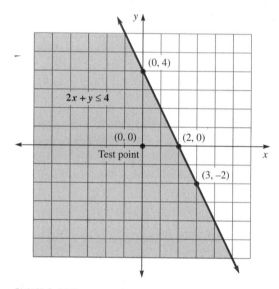

FIGURE 4.28

EXAMPLE 3   Graph the inequality $3x - 4y > 12$.

SOLUTION   We begin by graphing the line $3x - 4y = 12$ as a *broken* line, as shown in Figure 4.29. To determine which half-plane forms the solution set, we'll use the origin, $(0, 0)$, as a test point.

$$3x - 4y > 12 \quad \text{Original inequality}$$
$$3 \cdot 0 - 4 \cdot 0 \stackrel{?}{>} 12 \quad \text{Test point; let } (x, y) = (0, 0).$$
$$0 - 0 \stackrel{?}{>} 12$$
$$0 > 12 \quad \text{False}$$

Since $0 > 12$ is a false statement, the solution set does *not* contain the origin. We shade the half-plane that does not contain the origin, as shown in Figure 4.29.

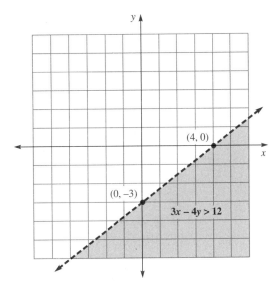

FIGURE 4.29

Another way to determine which half-plane is the solution set is to begin by solving the inequality for y. For example, using the inequality in Example 3,

$$3x - 4y > 12$$
$$-4y > -3x + 12 \quad \text{Subtract } 3x \text{ from each side.}$$
$$y < \frac{3}{4}x - 3 \quad \text{Divide each side by } -4; \text{ reverse the inequality symbol.}$$

Since $y < \frac{3}{4}x - 3$, the solution set contains the region of the plane that is below the line, as we saw in Figure 4.29. If after solving for y the inequality symbol is $>$, the solution set will be above the line.

CAUTION

An inequality must be written in the form $y > mx + b$ or $y < mx + b$ before deciding if the solution set lies above its graph or below it, respectively. A test point can be used when the inequality is in any form.

EXAMPLE 4 Graph the inequality $y > -x$.

SOLUTION The line $y = -x$ can be sketched by using three points or by using its slope, -1, and its y-intercept, $(0, 0)$. The boundary line is a broken line because ">" does not include an equal sign. The region above the line $y = -x$ forms the solution set; see Figure 4.30.

We could also have used a test point to determine the solution set in Example 4. Since the boundary line passes through the origin, we must use a test point other than $(0, 0)$. We'll use $(-2, -2)$, although any point not on the boundary line can be used.

$$y > -x \quad \text{Original equation}$$
$$-2 \overset{?}{>} -(-2) \quad \text{Test point; let } (x, y) = (-2, -2).$$
$$-2 > 2 \quad \text{False}$$

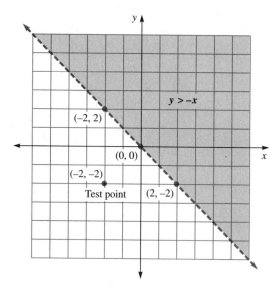

FIGURE 4.30

Since $-2 > 2$ is a false statement, the desired half-plane does not contain the point $(-2, -2)$. We shade the other side of the line, as shown in Figure 4.30.

EXAMPLE 5 Write an inequality that describes all points on a plane that lie between the lines $x = 3$ and $x = 5$, and graph the solution set.

SOLUTION The solution set is $\{(x, y) \mid 3 < x < 5\}$. Its graph is shown in Figure 4.31.

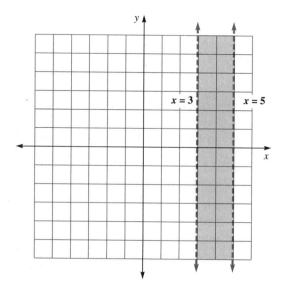

FIGURE 4.31

EXAMPLE 6 Write an inequality that describes the graph in Figure 4.32.

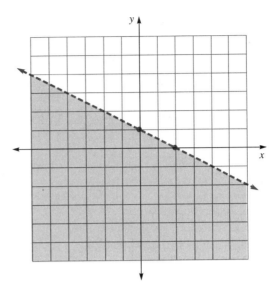

FIGURE 4.32

SOLUTION   We begin by writing the equation of the boundary line. The line passes through the points (0, 1) and (2, 0), so its slope is

$$\frac{0-1}{2-0} = -\frac{1}{2}$$

The y-intercept is the point (0, 1). Using the slope–intercept form to write the equation, we have

$$y = -\frac{1}{2}x + 1$$

The boundary line is broken, so the desired inequality symbol is either $<$ or $>$. Since the shaded region is below the boundary line, we want $<$. Thus

$$y < -\frac{1}{2}x + 1$$

EXAMPLE 7   Write an inequality that illustrates the fact that the perimeter P of a rectangle of length l and width w is at least 340 feet.

SOLUTION   The perimeter of a rectangle is given by the formula $P = 2w + 2l$. To say that the perimeter is at least 340 feet means that P is 340 feet or more:

$$2w + 2l \geq 340$$

EXAMPLE 8   Write an inequality to describe the fact that the difference in the prices of two identical items in different stores is less than twelve dollars.

SOLUTION   We let P_1 be the higher price and P_2 the lower price. The difference in the two prices is less than $12, so

$$P_1 - P_2 < 12$$

Do You Remember?

Can you match these?

_____ 1. A _____ line forms the boundary for $x + y < 3$.

_____ 2. The region satisfying $y < x + 1$ lies _____.

_____ 3. The region satisfying a linear inequality in two variables is a _____.

_____ 4. A _____ line forms the boundary for $y + 3x \geq 2$.

_____ 5. To determine which half-plane satisfies $x + 4y > 7$, a _____ can be used.

a) Test point
b) Half-plane
c) Solid
d) Broken
e) Boundary line
f) Above the line $y = x + 1$
g) Below the line $y = x + 1$

Answers: 1. d 2. g 3. b 4. c 5. a

Exercise Set 4.4

In Exercises 1–6, the boundary line for each inequality has been drawn. Shade the correct half-plane to complete the graph.

1. $y > -2$

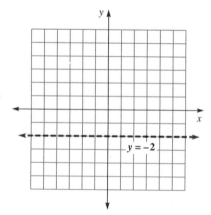

2. $x \geq 1$

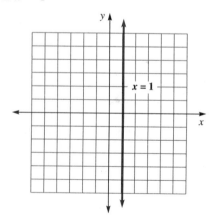

3. $y \geq x + 2$

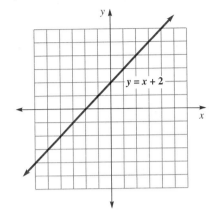

6. $2x + 5y \geq 10$

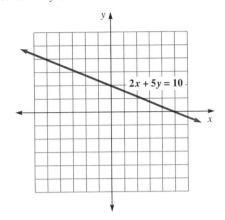

4. $2x + y < 4$

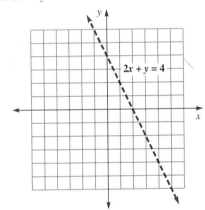

Graph each linear inequality.

7. $y \geq -4$ **8.** $x \leq 3$
9. $x \leq 4$ **10.** $y \geq 2$
11. $x > 2$ **12.** $x > 1$
13. $y > -2$ **14.** $x - y > 2$
15. $y > x + 2$ **16.** $y > -x + 2$
17. $y < x - 3$ **18.** $y < 2x + 5$
19. $x + 2y \geq 4$ **20.** $2x - y < 6$
21. $x + 3y > 6$ **22.** $2x - 3y \geq 12$
23. $4x + 5y \geq 20$ **24.** $6x + 2y < 12$

Write an inequality to describe each of the following regions.

25. All points above the *x*-axis.

26. All points on and to the left of the *y*-axis

27. All points to the right of the line $x = 3$

28. All points below the line $y = -2$

29. All points below the line $y = 3$ and above the line $y = -4$

30. All points to the right of the line $x = -5$ and to the left of the line $x = 1$

31. All points to the right of the line $x = 0$ or to the left of the line $x = -4$

32. All points below the line $y = -3$ or above the line $y = 4$

5. $x > 3y$

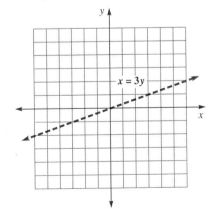

Write an inequality that describes each of the following graphs.

33.
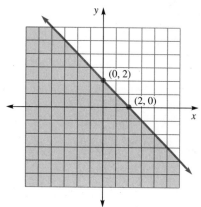

34.
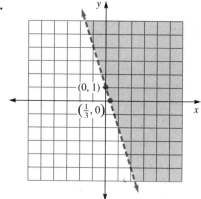

35.
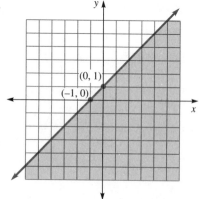

36.
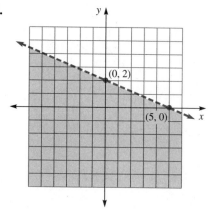

37.
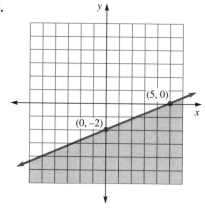

38.
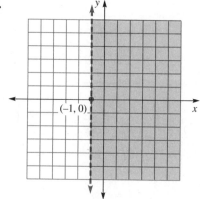

39.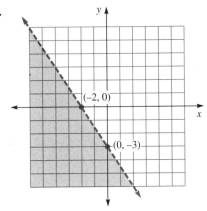

40.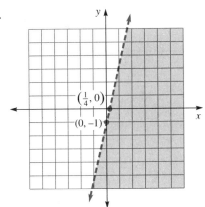

41.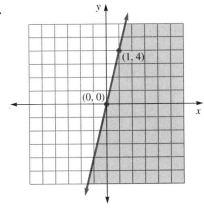

42.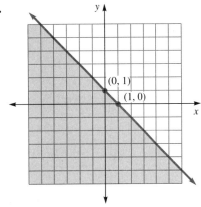

43.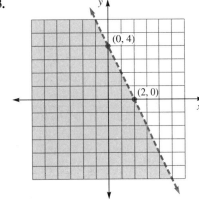

44.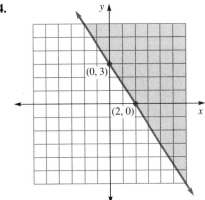

Write a verbal description of each of the following sets.

45. $\{(x, y) \mid x \leq 2\}$ **46.** $\{(x, y) \mid y \geq 3\}$

47. $\{(x, y) \mid y < 0\}$ **48.** $\{(x, y) \mid x > 2\}$

49. $\{(x, y) \mid -3 < x < 4\}$ **50.** $\{(x, y) \mid -2 \leq y \leq 5\}$

51. $\{(x, y) \mid -4 < y < 0\}$ **52.** $\{(x, y) \mid 0 \leq x \leq 7\}$

Write an inequality that expresses each of the following sentences.

53. The difference of the *x*- and *y*-coordinates is no more than 5.
54. The sum of the *x*- and *y*-coordinates exceeds 4.
55. The sum of twice the *x*-coordinate and three times the *y*-coordinate is no more than 6.
56. The perimeter of Dean's rectangular yard is at least 200 ft.
57. The average of two test scores, *x* and *y*, is greater than 84.
58. The difference between two test grades, *x* and *y*, is at most 6.
59. The difference of the *x*-coordinate and twice the *y*-coordinate is less than 4.
60. To keep an airplane in the air, the pilot knows that the speed of the airplane, *s*, minus the speed of the wind, *w*, must be at least 95 mph.
61. The difference of *a* and *b*, the speeds of two cars, is less than 4 mph.

Exercises 62–66 require the use of the inequality $P - 14.25n + 1852.50 \leq 0$, *which relates the profit P in dollars to the number n of tennis racquets produced by Marty's Racquet Company. Note:* $n \geq 0$.

62. What is the greatest possible profit if the company produces 2500 racquets?
63. If it produces no racquets, what is the profit or loss?
64. What is the break-even point ($P = 0$)?
65. What is the least number of racquets the company must produce to realize a $10,000 profit?
66. Graph the inequality.

Use the following information to answer Exercises 67–72. Rhonda makes a silk plant in 3 hr at a cost of $6. She makes a silk flower arrangement in 5 hr at a cost of $9.

67. How many hours does Rhonda need to make *x* silk plants?
68. How many hours does Rhonda need to make *y* silk flower arrangements?
69. The combined time for both jobs (in hours) cannot exceed 40 hr/week. Express this as an inequality.
70. Graph the inequality. *Note:* $x \geq 0$, $y \geq 0$.
71. The total cost of both cannot exceed $54.00. Express this as an inequality.
72. Do the two boundary lines in Exercises 69 and 71 seem to intersect? If so, approximate the ordered pair where this occurs.

REVIEW EXERCISES

Find the slope–intercept form of the equation for each of the following lines.

73. Passes through the points (0, 4) and (2, 5)
74. Has *y*-intercept (0, 4), passes through the point (5, 4)
75. Has *x*-intercept (3, 0), passes through the point (3, −5)
76. Passes through the points (2, 4) and (3, −1)
77. Has *x*-intercept (2, 0), passes through the point (−2, 5)
78. Passes through the origin and is parallel to the line $y = 3x + 2$
79. Passes through the origin and is perpendicular to the line $x + 2y = 6$
80. Passes through the point (−2, −3) and is perpendicular to the line $x + y = 2$
81. Passes through the point (−4, 1) and is parallel to the line $x - y = 0$
82. Has *y*-intercept (0, 5) and *x*-intercept (3, 0)
83. Has *y*-intercept (0, 8) and *x*-intercept (−2, 0)
84. Has *x*-intercept (4, 0) and slope −1
85. Has *x*-intercept (−6, 0) and slope −3
86. The *x*- and *y*-intercepts are both (0, 0) and the slope is −4.

4.5 Relations, Functions, and Function Notation

OBJECTIVES   In this section we will learn to
1. understand the definitions of a relation and a function;
2. determine the domain and range of a relation;
3. recognize a function from its graph; and
4. use function notation.

One of the most fundamental concepts in mathematics involves the relationship between two or more quantities. The distance you travel in a given period of time is directly related to the average speed you maintain. The circumference of a circle is directly related to the length of its radius. The profit a merchant receives when selling an article is related to how much money she has invested in it. Mathematicians refer to such relationships as *relations*.

DEFINITION Relation	A **relation** is a set of ordered pairs.

For example, the set of ordered pairs

$$A = \{(1, 2), (2, 3), (7, 4), (9, -3)\}$$

is a relation. The set of ordered pairs defined by

$$B = \{(x, y) \mid y = x + 2\}$$

is also a relation. Some ordered pairs in B can be found by assigning a value to x and computing the corresponding values of y. A few of the infinitely many ordered pairs in B are $(-2, 0)$, $(-1, 1)$, $(0, 2)$, $(1, 3)$, and $(2, 4)$.

The set of all first elements in a relation is called its **domain** and the set of all second elements is called its **range**. In the set $A = \{(1, 2), (2, 3), (7, 4), (9, -3)\}$, the domain is $\{1, 2, 7, 9\}$ and the range is $\{-3, 2, 3, 4\}$.

EXAMPLE 1   Find the domain and range of $C = \{(-2, 3), (4, 2), (-3, -7), (2, 12), (5, 4)\}$.

SOLUTION   The domain is $\{-3, -2, 2, 4, 5\}$, and the range is $\{-7, 2, 3, 4, 12\}$.

EXAMPLE 2   Find the domain and range of $y = \{(x, y) \mid y = x + 4\}$.

SOLUTION   In the equation $y = x + 4$, the variable x can be replaced by any real number, so the domain is the set of all real numbers. Furthermore, whenever x is replaced by a real number, y will be a real number, so the range is the set of all real numbers.

There is a special type of relation called a *function,* where each element in the domain corresponds to exactly one element in the range. For example, the set of ordered pairs

$$A = \{(1, 2), (3, 4), (-2, 6)\}$$

is a function, while

$$B = \{(3, 4), (5, 2), (11, 7), (3, -7), (9, -3)\}$$

is not a function. The set B contains two ordered pairs, $(3, 4)$ and $(3, -7)$, with the first element (3) corresponding to two different second elements (4 and -7).

DEFINITION	A **function** is a set of ordered pairs in which each first element corresponds to exactly one second element.
Function	

The set $C = \{(1, 2), (3, 4), (5, 6), (7, 8)\}$ describes a function. Each first (domain) element corresponds to exactly one second (range) element, as shown in Figure 4.33. The set $D = \{(1, 2), (1, 4), (6, 2), (9, 3)\}$ does not describe a function, since 1 corresponds to both 2 and 4, as shown in Figure 4.34.

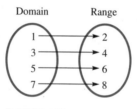
FIGURE 4.33

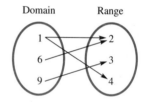
FIGURE 4.34

EXAMPLE 3   Determine whether the equation $y = x + 2$ is a function.

SOLUTION   The equation is a function because there is exactly one value of y for every value of x. In other words, there is no way that we can get two different values of y when we substitute the same value for x.

EXAMPLE 4   Determine whether the equation $x = y^2$ yields a function.

SOLUTION   This equation is not a function, since each value of x (first element) corresponds to two different values of y. For example, if $x = 1$, then y can be either 1 or -1. Therefore two ordered pairs are $(1, 1)$, and $(1, -1)$.

EXAMPLE 5   Determine whether each of the following equations yields a function:
 (a) $x = 4$ **(b)** $y = 2$ **(c)** $y < x + 2$

SOLUTION   **(a)** $x = 4$ is a vertical line, so the first element is 4 for all values of y. Some ordered pairs are $(4, 1), (4, 2), (4, 3)$ and $(4, -8)$. Therefore $x = 4$ is not a function.

(b) $y = 2$ is a horizontal line, so every value of x corresponds to the same value of y. Since it is impossible to use the same first element and get different second elements, the equation $y = 2$ is a function. In general,

> *Every linear equation with a graph that is not a vertical line is a function.*

(c) The graph of an inequality in two variables is a half-plane. Every value of x corresponds to an infinite number of y-values. Therefore an inequality in two variables is not a function. In fact,

> *No inequality in two variables is a function.*

It is important, when assigning values to x, to make sure that they are in the domain. For instance, in the function $y = \dfrac{1}{x}$, zero cannot be assigned to x because division by zero is not defined. Zero is not in the domain of this function.

EXAMPLE 6   Find the domain of the function $y = \dfrac{1}{x - 2}$.

SOLUTION   The only value that x cannot assume is 2, because that would make the denominator zero. Therefore the domain is all real numbers except 2.

EXAMPLE 7   Find the domain and range of $y = |x|$.

SOLUTION   We can substitute any real number for x, so the domain is the set of all real numbers, $\{x \mid x \text{ is a real number}\}$. The absolute value of any real number is always positive or zero, so y cannot be negative. The range is $\{y \mid y \geq 0\}$.

It can be difficult to determine whether an equation describes a relation or a function. If we know the graph of an equation, it is easy to decide whether it represents a function by using a simple test called the **vertical line test.** If a vertical line L intersects the graph at more than one point, the relation cannot be a function. Such an intersection would indicate that there was more than one second element for the same first element; see Figure 4.35(a). Figure 4.35(b) is the graph of a function, where each value of x corresponds to exactly one value of y.

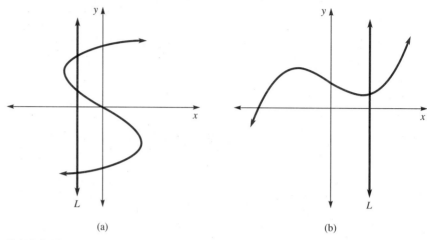

FIGURE 4.35

EXAMPLE 8   Find the domain and range of the relation graphed in Figure 4.36.

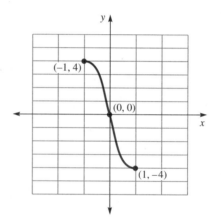

FIGURE 4.36

SOLUTION   First, we note that x goes from -1 to 1, so the domain is $\{x \mid -1 \leq x \leq 1\}$. Since y goes from -4 to 4, the range is $\{y \mid -4 \leq y \leq 4\}$.

Functions are named with letters, such as f, g, F, or P. For example, to indicate that $y = x + 3$ is a function, we often use the notation

$$f(x) = x + 3$$

The notation $f(x)$ is read "f of x" or "the function of x." Likewise, $g(x)$ is read "g of x," and $m(x)$ is read "m of x." These forms of notation all designate a function of x. This type of notation is called **function notation,** and it has the advantage that it calls attention to the fact that a function is being considered.

When a function is written in function notation, it is easy to see the domain and range elements at the same time.

*C*AUTION

The notation $f(x)$ does not mean f times x. It represents the y-value that corresponds to each value of x.

4.5 · RELATIONS, FUNCTIONS, AND FUNCTION NOTATION   219

EXAMPLE 9   In function notation, the equation $y = 2x + 1$ is written $f(x) = 2x + 1$. When

$$x = 3, \quad y = 2(3) + 1 = 7$$

This can be written

$$\underset{\text{Domain element}}{f(3)} = 2(3) + 1 = 6 + \underset{\text{Range element}}{1 = 7}$$

EXAMPLE 10   Given $f(x) = 2x - 4$, find   **(a)** $f(2)$ **(b)** $f(3)$ **(c)** $f(-2)$ **(d)** $f(a)$

SOLUTION
(a) $f(2) = 2(2) - 4 = 4 - 4 = 0$
(b) $f(3) = 2(3) - 4 = 6 - 4 = 2$
(c) $f(-2) = 2(-2) - 4 = -4 - 4 = -8$
(d) $f(a) = 2(a) - 4 = 2a - 4$

EXAMPLE 11   If $P(x) = x^2 + 5x + 1$, find   **(a)** $P(0)$ **(b)** $P(-3)$ **(c)** $P(b)$

SOLUTION
(a) $P(0) = 0^2 + 5(0) + 1 = 0 + 0 + 1 = 1$
(b) $P(-3) = (-3)^2 + 5(-3) + 1 = 9 - 15 + 1 = -5$
(c) $P(b) = b^2 + 5b + 1$

EXAMPLE 12   Graph $f(x) = 3x - 2$.

SOLUTION   We can think of $f(x) = 3x - 2$ as $y = 3x - 2$. Then the y-intercept is -2 and the slope is 3. The graph is shown in Figure 4.37.

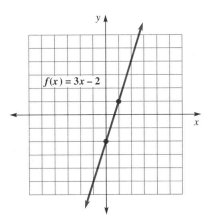

FIGURE 4.37

Do You Remember?

Can you match these?

____ 1. $f(x)$ is read _____.
____ 2. The domain of a relation
____ 3. The range of a relation
____ 4. A relation
____ 5. The domain of $y = \dfrac{3}{x-5}$
____ 6. Is $\{(1, 3), (2, 3), (1, 5)\}$ a function?
____ 7. The range of $f(x) = |x|$

a) A set of ordered pairs
b) Yes
c) A function of x
d) $\{x \mid x \text{ is a real number and } x \neq 5\}$
e) The set of all second elements in a set of ordered pairs
f) $\{y \mid y \geq 0\}$
g) The set of all first elements in a set of ordered pairs
h) No

Answers: 1. c 2. g 3. e 4. a 5. d 6. h 7. f

Exercise Set 4.5

Determine whether each relation is a function.

1. $\{(1, 2), (2, 2), (3, 4), (4, 6)\}$
2. $\{(0, 4), (1, 4), (2, 6), (3, 7)\}$
3. $\{(-2, 2), (-1, 2), (-1, 3), (2, 4)\}$
4. $\{(-4, 1), (-2, 0), (-2, 2), (0, 4)\}$
5. $y = 3x - 4$
6. $y = 5x - 7$
7. $y = |x| + 3$
8. $y = |x| - 5$
9. $y = \dfrac{2}{x-1};\ x \neq 1$
10. $y = \dfrac{3}{x+2};\ x \neq -2$
11. $y = 8$
12. $y = -3$
13. $x = 4$
14. $x = -7$
15. $y = \dfrac{5}{3}x + 3$
16. $y = \dfrac{3}{4}x + 7$
17. $y < x + 1$
18. $y \geq x - 5$
19. $y = |x + 2|$
20. $y = |2x - 3|$
21. $2x + y \geq 3$
22. $3x - 2y \leq 1$

Find the domain and range of each function.

23. $f(x) = x$
24. $f(x) = -x$
25. $f(x) = 3x - 1$
26. $f(x) = 5x + 4$
27. $f(x) = \dfrac{1}{4}x$
28. $f(x) = -\dfrac{1}{2}x$
29. $f(x) = 2(x - 3)$
30. $f(x) = -3(x + 4)$
31. $f(x) = |x|$
32. $f(x) = -|x|$
33. $f(x) = \dfrac{1}{2}x - 3$
34. $f(x) = -\dfrac{1}{4}x + 2$

Find the domain of each function.

35. $f(x) = |x|$
36. $f(x) = |2x|$
37. $f(x) = \dfrac{1}{2}(x + 1)$
38. $f(x) = -\dfrac{2}{3}(x - 2)$
39. $f(x) = \dfrac{2}{x-3}$
40. $f(x) = \dfrac{3}{x+1}$

Find the indicated values. Some may be undefined.

41. $f(x) = 2x - 5;\ f(2), f(0), f(-3)$
42. $f(x) = 3x - 7;\ f(1), f(-2), f(0)$
43. $g(x) = \dfrac{5}{x-3};\ g(4), g(-2), g(3)$
44. $g(x) = \dfrac{5}{x-5};\ g(5), g(0), g(-1)$
45. $f(x) = |2x| - 5;\ f(0), f(3), f(-1)$
46. $g(x) = -|3x| - 1;\ g(0), g\left(\dfrac{1}{4}\right), g(-2)$

47. $g(x) = x^2 + 3x$; $g(0), g(-1), g(2)$
48. $f(x) = x^3 + 5x^2 - 11x + 3$; $f(0), f(-1), f(1)$
49. $f(x) = \sqrt{x}$; $f(0), f(4), f(9)$
50. $f(x) = -\sqrt{x}$; $f(0), f(1), f(16)$

Use the vertical line test to determine which of the following graphs represent functions.

51.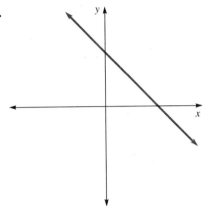

52.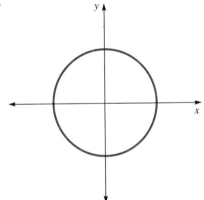

53.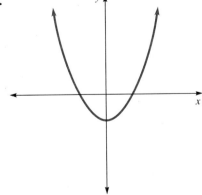

54.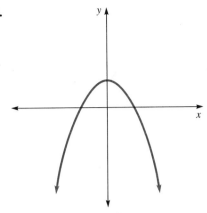

55.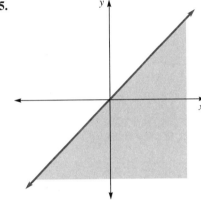

56.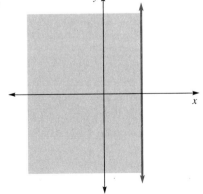

57.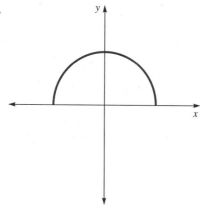

60.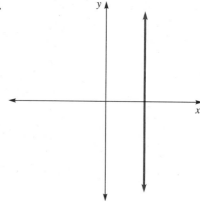

58.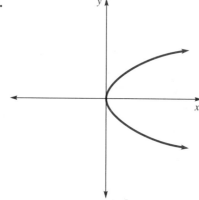

61.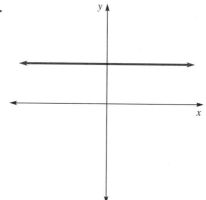

59.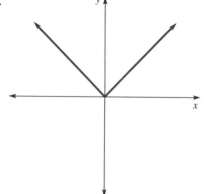

62.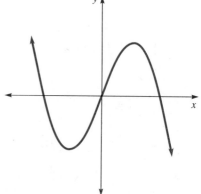

Determine the domain and range of each of the following functions.

63. (a) 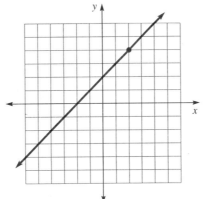 **(b)**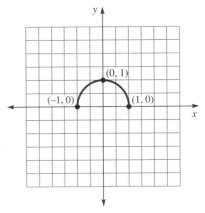

64. (a) 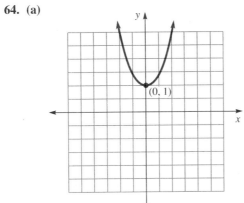 **(b)**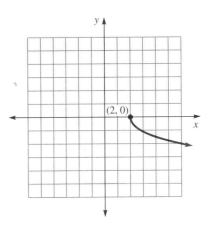

65. (a) 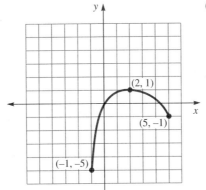 **(b)**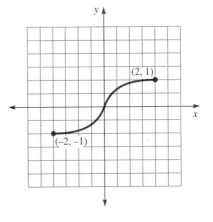

Graph each of the following functions.

66. $f(x) = 3x$
67. $f(x) = 2x + 1$
68. $f(x) = -x + 2$
69. $f(x) = -2x - 3$
70. $f(x) = \frac{1}{3}x + 4$
71. $f(x) = -\frac{3}{4}x + 1$
72. $f(x) = -\frac{2}{3}x + 4$
73. $f(x) = \frac{3}{8}x - 5$
74. $f(x) = 5$
75. $f(x) = -4$

WRITE IN YOUR OWN WORDS

76. Define a relation.
77. Define a function.

78. How do the graphs of $x + y = 8$ and $x + y < 8$ differ?
79. Can you think of an inequality in two variables that is a function? Why or why not?
80. Why is $y = mx + b$ called the slope–intercept form of the equation of a line?
81. The vertical line test is used to determine whether a graph represents a function. Explain why.
82. Are there any linear equations that are not functions? If so, which ones?
83. How do relations and functions differ?

SUMMARY

4.1 The Rectangular (Cartesian) Coordinate System

The rectangular (Cartesian) coordinate system is formed by drawing two perpendicular number lines, one horizontal and one vertical. The two lines are called **axes**, the horizontal one the **x-axis** and the vertical one the **y-axis**. The point of intersection of these lines is called the **origin**. The x- and y-axes divide the plane into four sections called **quadrants**, which are numbered in a counterclockwise direction. Every point in the plane can be represented by an **ordered pair** of numbers called the **coordinates** of the point. The first number in the ordered pair is the **x-coordinate** or the **abscissa** of the point. The second number is the **y-coordinate** or the **ordinate**.

Given an equation in x and y, there is an infinite number of ordered pairs that are solutions. When a value is assigned to either x or y, the missing coordinate in the ordered pair can be computed from the equation.

4.2 Graphing Linear Equations in Two Variables

A **linear equation in two variables,** x and y, is one that can be put in the form

$$ax + by = c$$

where a, b, and c are real numbers and a and b are not both zero.

The graph of a linear equation is a straight line. The point at which the line crosses the x-axis is called the **x-intercept** and is found by letting $y = 0$. The point where the graph crosses the y-axis is the **y-intercept.** It is found by letting $x = 0$.

A minimum of three points should be used to graph a linear equation. Two points are sufficient to graph the line, but the third point serves to check the accuracy of the other two. The third point is called a **check point.**

4.3 The Slope of a Line; Equations of a Line

The **slope** of a line is the measure of the rate at which a line rises or falls as it moves from left to right. A rising line has positive slope and a falling line has negative slope. The slope m of a line through two points (x_1, y_1) and (x_2, y_2) is

$$m = \frac{\text{change in } y\text{-values}}{\text{change in } x\text{-values}} = \frac{y_2 - y_1}{x_2 - x_1}, \qquad x_2 \neq x_1$$

Given a single point (x_1, y_1) on a line and the slope m of the line, the equation of the line can be determined by using the **point–slope form:**

$$y - y_1 = m(x - x_1)$$

If the slope m and the y-intercept b of a line are known, its equation can be determined by using the **slope–intercept form:**

$$y = mx + b$$

A linear equation written in the form $ax + by = c$ is said to be in **general form.**

Parallel lines have equal slopes. The slopes of perpendicular lines, neither of which is vertical, are negative reciprocals.

4.4 Graphing Linear Inequalities in Two Variables

A linear inequality has a solution set that is a region of the coordinate plane called a **half-plane.** The solution set includes the **boundary line** only if the inequality symbol is either $\leq$ or $\geq$. In that case the boundary line is drawn solid.

There are two ways to find which half-plane is the solution set. The first method involves using a **test point** to see if it satisfies the inequality. The second method involves solving the inequality for y.

4.5 Relations, Functions, and Function Notation

A **relation** is a set of ordered pairs. The **domain** of a relation is the set of all first elements in its ordered pairs, and the **range** of a relation is the set of all second elements. A **function** is a special type of relation in which each element in the domain corresponds to exactly one element in the range.

The notation $f(x)$, read "f of x" or "function of x," is called **function notation.** It represents the y-value that corresponds to x.

COOPERATIVE EXERCISE

Many of us have taken up a variety of aerobic exercises, such as swimming, bicycling, jogging, and walking. When doing aerobic exercises, we are advised to monitor our pulse rate to determine whether we are exercising too hard, not hard enough, or the right amount. Studies indicate that we should keep our pulse rate within a range known as the **target zone,** which is dependent upon our age.

The lower limit L of our heartbeats per minute is a function of our age a in years and is given by

$$L(a) \geq -\frac{2}{3}a + 150$$

The upper limit U of our heartbeats per minute is also a function of our age a and is given by

$$U(a) \leq -a + 190$$

These functions are valid in the age range $10 \leq a \leq 70$.

1. Graph these three inequalities on one Cartesian coordinate system and shade the *target zone*. Label the vertical axis as the pulse rate p and the horizontal axis as the age a, using suitable numbers.

2. Would a 20-year-old person exercising with a pulse rate of 140 be in the target zone?

3. Would a 40-year-old person exercising with a pulse rate of 140 be in the target zone?

4. Approximate the range of pulse rates in the target zone of a 21-year-old person.

5. Approximate the range of pulse rates in the target zone of a 30-year-old person.

6. Approximate the range of pulse rates in the target zone of a 40-year-old person.

7. Approximate the range of pulse rates in the target zone of a 69-year-old person.

8. Explain what happens to the target zone as a person ages.

9. Why do you think the target zone is restricted between the ages of 10 and 70, inclusive?

10. Based upon the given inequalities, project what you think the range of pulse rates would be for a person 81 years old. Does this seem reasonable? Why?

11. Based upon the given inequalities, project what you think the range of pulse rates would be for a person 3 years old. Does this seem reasonable? Why?

Review Exercises

Complete the given ordered pairs.

1. $y = 5x + 1$; (0,), (, 0), (1,), (−1,)
2. $y = 3$; (−3,), (0,), (3,)
3. $x + 4 = 0$; (, −2), (, 0), (, 3)

Graph each equation using the x- and y-intercepts and one other point.

4. $y = x - 4$
5. $2x + 3y = 6$
6. $3x - 4y = 12$
7. $y = 3x$
8. $x + 3y = 0$

Write the equation of a line satisfying each of the following conditions.

9. Goes through (3, 0) and has slope $m = 2$
10. Goes through (0, 5) and (1, −1)
11. Goes through (5, −1) and is parallel to the line $y = 2x + 3$
12. Goes through (0, 0) and is perpendicular to the line $3x - 2y = 1$
13. Goes through (−3, −4) and (2, −1)
14. Has slope $m = -\frac{1}{2}$ and y-intercept (0, −4)
15. Has x-intercept (−3, 0) and undefined slope m
16. Has y-intercept (0, −1) and slope $m = 0$
17. Has slope $m = -\frac{3}{4}$ and y-intercept $(0, -\frac{1}{4})$
18. Goes through the origin and is parallel to the y-axis
19. Goes through (2, 0) and is perpendicular to the x-axis
20. The y-coordinate is 1 more than the x-coordinate.
21. Twice the y-coordinate less the x-coordinate is 4.
22. Three times the x-coordinate subtracted from the y-coordinate is 0.

Graph each inequality.

23. $x + y < 2$
24. $y > 3x - 1$
25. $y \leq 1$
26. $x \geq -3$

27. $y \leq 2x + 3$
28. $y \geq -x + 2$
29. $3x - 4y < 0$
30. $2x + 5y > 10$

Determine which of the following are functions.

31. {(−3, 0), (−1, 1), (1, 3), (3, 3)}
32. {(−2, −1), (−1, 0), (−1, 1), (2, 3)}
33. $y = 3$
34. $y = x + 5$
35. $x - y < 5$
36. $2x + y \geq 2$
37. $x = 4$
38. $y = |x|$
39. $y = \sqrt{4x}$
40. $y = 4x + 3$

Find the domain and range of each of the following functions.

41. {(−4, −2), (−2, 0), (0, 2), (2, 4)}
42. {(−2, −1), (−1, 0), (0, 1), (2, 3)}
43. $f(x) = 2$
44. $f(x) = 2x + 3$
45. $f(x) = -2x + 4$
46. $f(x) = |x| + 2$
47. $f(x) = \sqrt{5x}$
48. $f(x) = x - 1$

Use the vertical line test to determine which of the following graphs represent functions.

49.

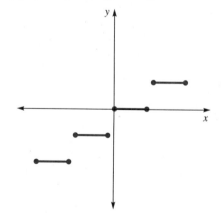

50.

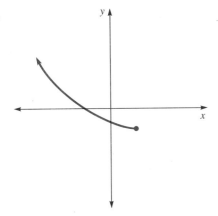

51.

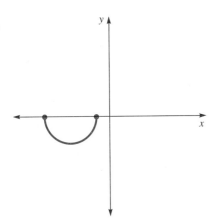

52.

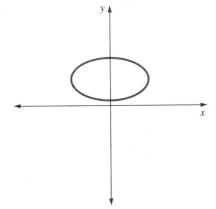

Given $f(x) = 2x + 3$ and $g(x) = 2x^2 - 8$, find the following values.

53. $f(0)$ **54.** $g(0)$ **55.** $f\left(-\dfrac{3}{2}\right)$

56. $f(-2)$ **57.** $g(-2)$ **58.** $g(2)$

Find the domain for Exercises 59 and 60. State any values for which the functions are undefined.

59. $f(x) = \dfrac{4}{x-5}$ **60.** $f(x) = \dfrac{2}{3x+2}$

61. A ski slope ascends 2000 ft in a horizontal distance of 12,000 ft. What is its slope?

62. Old Cuesta Grade ascends 15 ft in a horizontal distance of 200 ft. What is its slope?

63. The *pitch* of a roof (its slope) is determined by the kind of material used and the amount of snowfall in the area. Find the pitch of the roof in the accompanying figure.

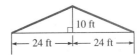

64. Ginger wants the average of her two grades, *x* and *y*, to be no less than 80. Write and graph the inequality. Give one set of possible grades satisfying the conditions.

65. The difference in the cost of two shirts, *x* and *y*, is to be less than $2.00. Write and graph the inequality. Find one set of possible costs satisfying these conditions.

Jones Metals makes and sells chrome bumpers. The company has a fixed cost of $4000/week. If it makes and sells ten identical chrome bumpers in one week, the total income from the sales is $5000. Assume the weekly profit P equals the total income I less the fixed cost, where income is given by the formula $I = 500x$, with x equal to the number of bumpers made and sold in one week. Use this information to answer the following questions:

66. What is the weekly profit equation?

67. What is the profit/loss if ten chrome bumpers are made and sold in one week?

68. What is the profit/loss if the company closes down for one week?

69. What is the break-even point (where $P = 0$)?

70. If the company makes and sells an average of ten bumpers/week for one year, what is its annual profit/loss?

71. If this were your company, would you stay in business if you could make and sell only seven bumpers each week? eight bumpers each week? nine bumpers each week?

Chapter Test

Graph the equation or inequality in Exercises 1–12.

1. $x = 1$
2. $y = 0$
3. $y \geq 3$
4. $x < -2$
5. $y = 3x$
6. $y = -2x$
7. $y - x < 0$
8. $y + x > 2$
9. $2x + y = 4$
10. $x - 4y = 12$
11. $3x - 2y \geq 6$
12. $5x + y \leq 5$
13. Find the slope of the line passing through $(0, 4)$ and $(2, -1)$.
14. Write the equation of the line with slope 0 and y-intercept $(0, 5)$.
15. Graph the line that goes through $(-3, 2)$ and has slope $-\frac{1}{4}$.
16. What is the slope of a vertical line?
17. Write the equation of the line that goes through $(4, -5)$ and has slope 3.
18. Graph $f(x) = -\frac{3}{4}x - 7$.
19. On which side of the line $2x - 3y = 4$ is the graph of $2x - 3y > 4$ shaded?
20. Write the equation of the line that has slope zero and goes through the origin.
21. Find the x- and y-intercepts of the line $3x - 5y = 7$.
22. Write the equation of the line through $(0, 4)$ and $(2, -5)$. Put your answer in slope–intercept form.
23. What is another name for the rectangular coordinate system?
24. Give a verbal description of the set $\{(x, y) \mid y = 2x\}$.
25. Write the equation of the line that passes through the origin and has a slope of 6.
26. What is the slope of a horizontal line?
27. Write the equation of a line with undefined slope that goes through $(7, 9)$.
28. Write the equation of a line that is perpendicular to the line $x - 2y = 5$ and goes through the point $(0, 0)$.
29. Find the domain of $f(x) = \dfrac{7}{x - 1}$.
30. Given $f(x) = -x^2 + 4$, find $f(0)$ and $f(-2)$.
31. Find the domain and range of each of the following functions:
    (a) $f(x) = 2x + 3$
    (b) $\{(-3, 0), (-1, 2), (0, 3), (2, 4)\}$
32. Use the vertical line test to determine which of the following graphs represent functions.

(a)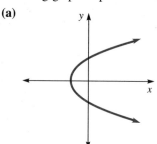

(b)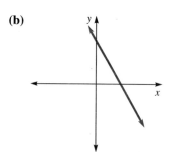

(c)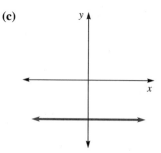

CHAPTER 5
SYSTEMS OF LINEAR EQUATIONS AND INEQUALITIES

5.1 Solving Systems of Linear Equations by the Graphing Method
5.2 Solving Systems of Linear Equations by the Elimination Method
5.3 Solving Systems of Linear Equations by the Substitution Method
5.4 Solving Systems of Linear Inequalities
5.5 More Applications

Application

The Fun Center owner wants admission for a family of three (two adults and a child) to cost no more than $10. He wants adult tickets to cost at least twice children's tickets. Let y be the cost of an adult ticket and x the cost of a child's ticket. A system of inequalities describing these conditions is

$$x + 2y \leq 10$$
$$y \geq 2x$$
$$x \geq 0$$
$$y \geq 0$$

Graph the region that shows the possible costs of adult and children's tickets. Select one possible set of costs.

Exercise Set 5.4, Exercise 46.

The study of nature and natural events is the most productive source of mathematical discoveries. It is the best place to find out what is most important. It provides a practical application for mathematical ideas and exposes an unending challenge for the mathematician.

JEAN BAPTISTE FOURIER (1768–1830)

5.1 Solving Systems of Linear Equations by the Graphing Method

HISTORICAL COMMENT Pierre de Fermat (1601–1665) preceded René Descartes in determining the equation of a straight line. He failed to publish his work, however, and consequently Descartes received much of the credit for the development. This habit of Fermat—failing to publish proofs of important theorems he claimed to have proved—set a task for mathematicians that continues today. In the most famous case, he stated that $x^n + y^n = z^n$ has no solutions if x, y, and z are integers and n is greater than 2. He stated, "I have found a truly wonderful proof for this but the margin is too small to hold it." It was not until 1994 that it was finally proved in a paper over 100 pages long.

OBJECTIVES In this section we will learn to
1. identify systems of equations as consistent and independent, consistent and dependent, or inconsistent and independent; and
2. solve systems of equations by graphing.

Systems of equations are very useful in solving applied problems because they usually make it easier to relate the various parts of the problem to one another. Suppose we are asked to find two numbers that add to 10 and that have a difference of 2. The problem is simple enough that the two numbers, 6 and 4, can be found by trial and error. They can also be found using an equation with one unknown. But many problems of this type are not so easily solved by these means, so other approaches have been developed.

In the preceding problem, finding two numbers that add to 10 and have a difference of 2 can be represented by *two* equations in *two* variables. If we let $x =$ one number and $y =$ the other number, then

$x + y = 10$ Their sum is 10.
$x - y = 2$ Their difference is 2.

When we consider two or more linear equations at the same time, we refer to them as a **system of linear equations.** Recall that the solution set of a linear equation consists of ordered pairs that lie in a straight line. Figure 5.1 shows that when the

231

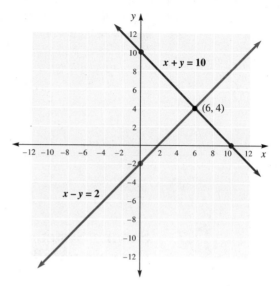

FIGURE 5.1

two preceding equations are graphed on the same coordinate system, they intersect at the point (6, 4). Since (6, 4) lies on both lines, and since it is the only point that does, it is the only point that is a solution to both equations. There are infinitely many points that satisfy one of the equations but not the other. For example, the point (8, 2) satisfies $x + y = 10$, since $8 + 2 = 10$, but it does not satisfy $x - y = 2$, since $8 - 2 \neq 2$. In general,

> If an ordered pair is a solution to every equation in a system of equations, it is a solution to the system.

EXAMPLE 1 Solve the following system of equations by graphing:

$$x + y = 6$$
$$x + 2y = 10$$

SOLUTION First we construct a table of values for each equation by using the x- and y-intercepts and a third check point. The table of values and the graphs of the lines are shown in Figure 5.2. The lines *appear* to intersect at (2, 4), but we must check the solution. We do this by substituting 2 for x and 4 for y in both equations.

$$\text{CHECK} \quad x + y = 6 \qquad x + 2y = 10 \qquad \text{Original equations}$$
$$2 + 4 \stackrel{?}{=} 6 \qquad 2 + 2(4) \stackrel{?}{=} 10$$
$$6 \stackrel{\checkmark}{=} 6 \qquad 10 \stackrel{\checkmark}{=} 10$$

The solution set is {(2, 4)}.

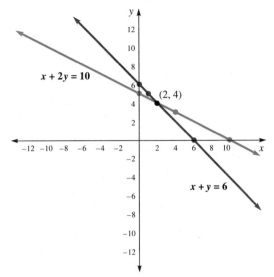

FIGURE 5.2

EXAMPLE 2   Solve the following system of equations by graphing:

$$x + 2y = 1$$
$$2x + y = -1$$

SOLUTION   We begin by constructing tables of values, then graphing the two lines in Figure 5.3. It appears that the solution is the ordered pair $(-1, 1)$, but we must check it. We

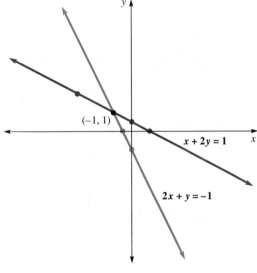

FIGURE 5.3

substitute -1 for x and 1 for y in each equation:

CHECK $\quad x + 2y = 1 \qquad\qquad 2x + y = -1 \qquad$ Original equations
$\quad\quad\quad\;\; -1 + 2 \cdot 1 \stackrel{?}{=} 1 \qquad 2(-1) + 1 \stackrel{?}{=} -1$
$\quad\quad\quad\;\; -1 + 2 \stackrel{?}{=} 1 \qquad\quad -2 + 1 \stackrel{?}{=} -1$
$\quad\quad\quad\;\; 1 \stackrel{\checkmark}{=} 1 \qquad\qquad\quad\;\; -1 \stackrel{\checkmark}{=} -1$

The solution set is $\{(-1, 1)\}$.

In the preceding examples, the lines intersected in a single point. This is not always the case. The lines in the system could be parallel, in which case there would be no intersection and therefore no solution. On the other hand, the lines could coincide (one lies on top of the other), so that any point on either line would be a solution to the other. There are infinitely many solutions in the latter case. A system with a single solution, such as those in Examples 1 and 2, is said to be **consistent and independent**. A system of parallel lines is called **inconsistent and independent**. A system of lines that coincide is **consistent and dependent**. Examples of the three types of systems are shown in Figure 5.4.

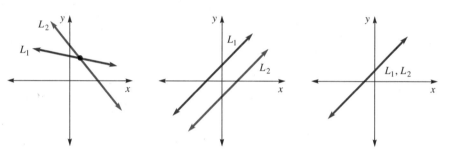

Unique solution; consistent and **independent**

No solution; **inconsistent** and independent

Infinitely many solutions; consistent and **dependent**

FIGURE 5.4

EXAMPLE 3 Solve the following system of equations by graphing:

$x + y = 1$
$x + y = 3$

SOLUTION Figure 5.5 shows that the lines are parallel, so no solution exists.

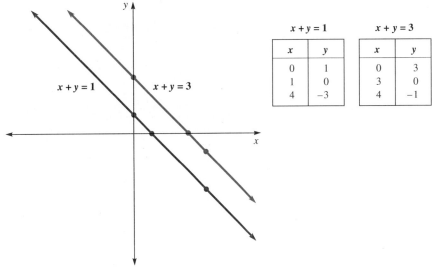

FIGURE 5.5

If we study the equations in Example 3, we can see that it is impossible for the sum of the same two numbers to be both 1 and 3. The statements $x + y = 1$ and $x + y = 3$ show an inconsistency. That's why we say the system is inconsistent and independent. The solution set is the empty set, $\varnothing$. The fact that the lines in Example 3 are parallel can be confirmed by writing them in slope–intercept form, $y = mx + b$:

$x + y = 1 \rightarrow y = -x + 1$ Subtract x from each side.

$x + y = 3 \rightarrow y = -x + 3$ Subtract x from each side.

The slope of both lines is -1; the lines are parallel.

EXAMPLE 4 Solve the following system by graphing:

$$2x - y = 1$$
$$4x - 2y = 2$$

SOLUTION If we write the lines in the slope–intercept form first, we find that they represent the same line.

$2x - y = 1$	Original equation	$4x - 2y = 2$	Original equation
$-y = 1 - 2x$	Subtract $2x$ from each side.	$-2y = 2 - 4x$	Subtract $4x$ from each side
$y = -1 + 2x$	Multiply each side by -1.	$y = -1 + 2x$	Divide each side side by -2.
$y = 2x - 1$		$y = 2x - 1$	

The system is consistent and dependent. The graph of the lines is shown in Figure 5.6.

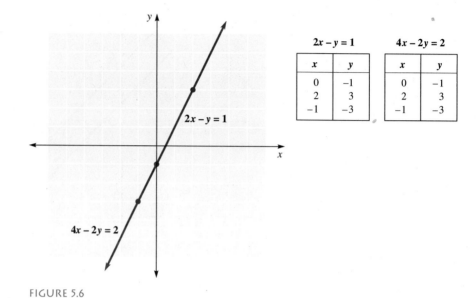

FIGURE 5.6

When the equations in a system of two linear equations are both written in the form $y = mx + b$, they can be classified by the following guidelines.

Slopes	y-Intercepts	Type of System
Slopes not equal	y-intercepts may or may not be equal.	Consistent and independent
Slopes equal	y-intercepts equal	Consistent and dependent
Slopes equal	y-intercepts not equal	Inconsistent and independent

Although systems that have a single solution are both consistent and independent, for the sake of brevity we will call them **independent.** Similarly, systems with no solutions (inconsistent and independent) will be called **inconsistent,** and the systems with an infinite number of solutions (consistent and dependent) will be called **dependent.**

EXAMPLE 5 Classify each of the following systems as independent, dependent, or inconsistent.

(a) $2x - 3y = 6$
$2x + y = 5$

(b) $3x - 2y = 4$
$6x - 4y = 3$

(c) $2x - y = 1$
$4x - 2y = 2$

SOLUTION   We begin by writing each system in slope–intercept form.

(a) $2x - 3y = 6 \rightarrow y = \dfrac{2}{3}x - 2$

$2x + y = 5 \rightarrow y = -2x + 5$

The slopes are not equal; the system is independent.

(b) $3x - 2y = 4 \rightarrow y = \dfrac{3}{2}x - 2$

$6x - 4y = 3 \rightarrow y = \dfrac{3}{2}x - \dfrac{3}{4}$

The slopes are equal but the y-intercepts are not, so the lines are parallel. The system is inconsistent.

(c) $2x - y = 1 \rightarrow y = 2x - 1$

$4x - 2y = 2 \rightarrow y = 2x - 1$

The slopes are the same and so are the y-intercepts. The system is dependent; both equations represent the same line.

EXAMPLE 6   The sum of the ages of two friends is 15 years and the difference of their ages is 3 years. Find their ages by graphing.

SOLUTION   We let $x =$ the age of the older friend

$y =$ the age of the younger friend

Two equations representing the relationship between their ages are

$x + y = 15$ The sum of their ages is 15.

$x - y = 3$ The difference of their ages is 3.

The graphs of the equations are shown in Figure 5.7.

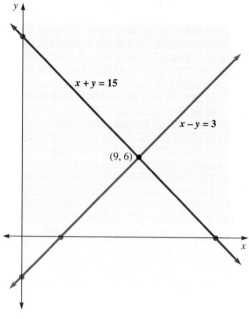

FIGURE 5.7

The lines appear to intersect at (9, 6), which would mean that the older of the two is 9 and the younger one is 6.

CHECK The sum of their ages is $9 + 6 = 15$ and the difference is $9 - 6 = 3$. The solution checks.

Do You Remember?

Can you match these?

Each of the following refers to a system of two linear equations.

_____ 1. A system with an infinite number of solutions is _____.

_____ 2. A system with a single solution is _____.

_____ 3. A system with no solutions is _____.

_____ 4. In a dependent system, the graphs of lines _____.

_____ 5. A system in which both the slopes and the y-intercepts of the equations are equal is said to be _____.

a) coincide
b) dependent
c) inconsistent
d) independent
e) inconsistent and dependent

Answers: 1. b 2. d 3. c 4. a 5. b

Exercise Set 5.1

Determine which ordered pair is the solution to the given independent system of equations.

1. $(0, 0)$, $(1, 1)$
   $x + y = 0$
   $x - y = 0$

2. $(1, 0)$, $(2, 2)$
   $x + y = 1$
   $x - y = 1$

3. $(-1, 6)$, $(0, 4)$
   $2x + y = 4$
   $2x + 2y = 8$

4. $(0, 0)$, $(-6, 3)$
   $x + 2y = 0$
   $3x + 5y = 3$

5. $(4, -3)$, $(4, 3)$
   $x - 4 = 0$
   $y + 3 = 0$

6. $(-7, 2)$, $(-2, 2)$
   $x + y = 0$
   $y - 2 = 0$

7. $(4, -7)$, $(5, -7)$
   $x + 2y + 10 = 0$
   $2x - 3y + 30 = 0$

8. $(-3, 1)$, $\left(-3, \dfrac{11}{4}\right)$
   $3x + y + 8 = 0$
   $2x + 4y - 5 = 0$

9. $(0, -4)$, $\left(\dfrac{20}{27}, -\dfrac{110}{27}\right)$
   $3x + 6y = -20$
   $2x - 5y = 20$

10. $(9, 1)$, $\left(\dfrac{33}{11}, \dfrac{80}{11}\right)$
    $2x + 5y = 23$
    $3x + 2y = 29$

Classify each consistent system of equations as independent or dependent; otherwise classify as inconsistent.

11. $x + y = 3$
    $2x + 2y = 2$

12. $3x - y = 2$
    $6x - 2y = 3$

13. $x - y = 2$
    $x + y = 2$

14. $2x + 3y = 5$
    $3x + 2y = 5$

15. $4x - 2y = 6$
    $2x - y = 3$

16. $y = \dfrac{1}{2}x + 4$
    $2y - x = 8$

17. $y = 3x + 2$
    $3x - y = -2$

18. $4y - 2x = 2$
    $x = 2y - 1$

19. $y = 2x + 6$
    $2x - y = -6$

20. $y = \dfrac{1}{3}x - 1$
    $x = 3y + 3$

21. $y = -\dfrac{1}{4}x + 1$
    $4y - x = 4$

22. $2x - y = 5$
    $x - 2y = 5$

23. $3x + 4y = 8$
    $6x + 4y = 4$

24. $x - 2y = 1$
    $2x - 4y = 1$

Solve each system of equations by the graphing method. If the system is not independent, state whether it is dependent or inconsistent.

25. $x + y = 3$
    $x - y = 1$

26. $x + y = 4$
    $x - y = 3$

27. $2x - y = 4$
    $-2x - y = 2$

28. $2x + y = 6$
    $x - y = 0$

29. $3x - y = 5$
    $6x - 2y = 10$

30. $4x + y = 3$
    $8x + 2y = 6$

31. $2x + y = 6$
    $3x + 2y = 8$

32. $3x - 4y = -8$
    $5x + 2y = -22$

33. $5x - 7y = 35$
    $5x - 7y = 10$

34. $x - 7y = 4$
    $x = 7y + 1$

35. $4y + x = -12$
    $2y = 7x - 6$

36. $x + 5y = -10$
    $2y = x + 3$

37. $x - 2y = -4$
    $2x - y = 1$

38. $3x - y = 3$
    $x - 3y = 1$

39. $3x = 5y + 15$
    $x = \dfrac{5}{3}y + 5$

40. $6x + 2y = 4$
    $3x = 2 - y$

41. $x = 7$
    $y + 3 = 0$

42. $x = 0$
    $y - 4 = 0$

43. $x + y = 2$
    $x - y = 3$

44. $2x + y = 1$
    $x - 2y = 1$

45. $2x + 5y - 20 = 0$
    $x + y - 10 = 0$

46. $x = 2y - y$
    $4x + 3y = -3y$

47. $6x = 5 + 2y$
    $3x - y = 5$

48. $8y = 6x - 16$
    $3x = 4y + 5$

49. $3x + 5y = 15$
    $x + 3y = 3$

50. $2x - 3y = 5$
    $x + 4y = -7$

51. $3x + y = 5$
    $x - y = 3$

52. $y = -x + 2$
    $y = 2x + 1$

53. $x = \dfrac{3}{4}y + 4$
    $y = 0$

54. $x = \dfrac{4}{3}y + 1$
    $x = y + 1$

55. $x - y = 7$
    $2x + y = 2$

56. $x - y = 2$
    $3x - 2y = 6$

57. $2x + y = -10$
    $2y - x = 10$

58. $4x - 18y = -2$
    $3x - 12y = 0$

59. $x - \dfrac{y}{2} = -3$
    $\dfrac{x}{3} + \dfrac{y}{4} = -1$

60. $\dfrac{x}{2} - 2y = 1$
    $y = \dfrac{1}{4}x - \dfrac{1}{2}$

61. $0.5x + 0.5y = 1$
    $0.2x - 0.6y = 2$

62. $0.1x + 0.1y = 0.5$
    $0.4x - 0.2y = 0.2$

Use two variables to set up a system of equations for each of the following applied problems. Solve each system by the graphing method.

63. The sum of two numbers is 60 and their difference is 20. Find the two numbers.

64. The sum of two numbers is 50 and their difference is 10. Find the two numbers.
65. A man is 5 years older than his wife and the sum of their ages is 55. How old is each?
66. The sum of two students' ages is 40; the difference in their ages is 10. Find their ages.
67. Michelle has 10 more dimes than Jeff. The total value of their dimes is $4.00. How many dimes does each have?
68. Frank has 10 more nickels than Tina. The total value of their nickels is $1.50. How many nickels does each have?
69. The length of a rectangle is 110 in. more than the width. If the perimeter of the rectangle is 1300 in., find the length and width.
70. The width of a rectangle is 10 cm less than the length. If the perimeter is 40 cm, find the length and width.
71. A couple cut their lawn in 6 hr working together. If one worked twice as long as the other, how many hours did each work?
72. Dennis and Cathy cleaned their house in 4 hr working together. If Dennis worked 2 hr more than Cathy, how many hours did each work?

Write in Your Own Words

73. Define a consistent and dependent system of equations.
74. Define a consistent and independent system of equations.
75. Define an inconsistent and independent system of equations.
76. Describe the solution set of a consistent and independent system of equations.

5.2 Solving Systems of Linear Equations by the Elimination Method

HISTORICAL COMMENT Systems of equations were known in Egypt, Greece, India, and China as early as A.D. 275. Various methods were used to indicate such systems, and it was the Hindus who first wrote their equations one below another, much as we will do in this section. They used the names of colors to indicate the various unknowns. The elimination method, which we will discuss in this section, was developed in 1559 when various letters were used to represent different unknowns.

OBJECTIVES In this section we will learn to
1. solve systems of linear equations by the elimination method; and
2. write systems of equations to solve various applied problems.

Solving a system of equations by graphing has numerous drawbacks. The most serious one is that the lines may not intersect at a point where x and y can be read directly from the graph. If the solution to a system is $(\frac{2}{17}, \frac{11}{17})$, for example, it would take a keen eye indeed to read the graph this closely. Graphics calculators and computers with graphing software programs can be used to solve systems of equations; an example of their use appears at the end of this section. However, there are other methods of solving systems of linear equations that overcome the difficulty of reading a graph. One is called the **elimination method,** and it is based on a modified form of the addition property of equality.

> **The Modified Addition Property of Equality**
> If A, B, and C represent algebraic expressions such that
> $$A = B \quad \text{and} \quad C = D$$
> then
> $$A + C = B + D$$

EXAMPLE 1 Find the solution to the system of equations

$$x + y = 4$$
$$x - y = 2$$

SOLUTION To solve this system, we use the modified addition property of equality. Adding the left sides and adding the right sides of the two equations, we obtain a new equation in a single variable:

$$\text{Add} \begin{cases} x + y = 4 \\ x - y = 2 \end{cases} \text{Add}$$
$$2x = 6$$
$$x = 3$$

Once we know the value of either x or y, we can obtain the value of the other variable by substituting the known variable into either of the original equations. In this case we know x. Choosing the first equation and substituting $x = 3$ yields

$$x + y = 4$$
$$3 + y = 4 \qquad \text{Substitute } x = 3.$$
$$y = 1 \qquad \text{Subtract 3 from each side.}$$

CHECK $\quad x + y = 4 \quad\quad x - y = 2 \quad\quad$ Original equations
$\quad 3 + 1 \stackrel{?}{=} 4 \quad\quad 3 - 1 \stackrel{?}{=} 2 \quad\quad x = 3, y = 1$
$\quad\quad 4 \stackrel{\checkmark}{=} 4 \quad\quad\quad 2 \stackrel{\checkmark}{=} 2$

The solution checks. The solution set is $\{(3, 1)\}$.

EXAMPLE 2 Solve the following system by the elimination method:

I $\quad x + 2y = 5$
II $\quad x - y = 3$

SOLUTION Before we can eliminate x or y by addition, their coefficients must be additive inverses of each other. If we multiply each side of equation **II** by -1, the coefficients of x will be additive inverses:

$$
\begin{array}{rlrl}
\textbf{I} & x + 2y = 5 & \rightarrow \textbf{I} & x + 2y = 5 \\
\textbf{II} & x - y = 3 & \rightarrow \textbf{III} & -x + y = -3 \quad \text{Multiply each side of \textbf{II} by } -1. \\
& & & 3y = 2 \quad \text{Add \textbf{I} to \textbf{III}.} \\
& & & y = \frac{2}{3} \quad \text{Divide each side by 3.}
\end{array}
$$

To obtain the value of x, we substitute the value of y into either of equations **I** or **II**. Choosing equation **II**, we have

$$
\begin{array}{rl}
\textbf{II} & x - y = 3 \\
& x - \frac{2}{3} = 3 \quad \text{Substitute } \frac{2}{3} \text{ for } y. \\
& x = 3 + \frac{2}{3} \quad \text{Add } \frac{2}{3} \text{ to each side.} \\
& x = \frac{11}{3} \quad \text{Simplify.}
\end{array}
$$

Thus $x = \frac{11}{3}$ and $y = \frac{2}{3}$. We check the solution by substituting $x = \frac{11}{3}$ and $y = \frac{2}{3}$ into *both* of the original equations.

CAUTION

Always check the solution in the original system of equations. It is easy to make an error when multiplying to establish a new system in which the coefficients are additive inverses. Then if no further errors are made, the solution will check in the new incorrect system but not in the original system.

CHECK

$$
\begin{array}{lll}
x + 2y = 5 & x - y = 3 & \text{Original equations} \\
\frac{11}{3} + 2\left(\frac{2}{3}\right) \stackrel{?}{=} 5 & \frac{11}{3} - \frac{2}{3} \stackrel{?}{=} 3 & x = \frac{11}{3}, y = \frac{2}{3} \\
\frac{15}{3} \stackrel{?}{=} 5 & \frac{9}{3} \stackrel{?}{=} 3 & \text{Simplify.} \\
5 \stackrel{\checkmark}{=} 5 & 3 \stackrel{\checkmark}{=} 3 &
\end{array}
$$

The solution set is $\left\{\left(\frac{11}{3}, \frac{2}{3}\right)\right\}$.

EXAMPLE 3 Solve the following system by the elimination method:

$$
\begin{array}{rl}
\textbf{I} & 2x + 3y = 8 \\
\textbf{II} & 3x + 4y = 11
\end{array}
$$

SOLUTION Sometimes we must multiply *both* equations by appropriate numbers to make a pair of coefficients additive inverses of each other. This time we will solve for x first by eliminating y:

$$
\begin{array}{rlrl}
\textbf{I} & 2x + 3y = 8 & \rightarrow \textbf{III} & 8x + 12y = 32 \quad \text{Multiply each side of \textbf{I} by 4.} \\
\textbf{II} & 3x + 4y = 11 & \rightarrow \textbf{IV} & -9x - 12y = -33 \quad \text{Multiply each side of \textbf{II} by } -3. \\
& & & -x = -1 \quad \text{Add \textbf{III} and \textbf{IV}.} \\
& & & x = 1
\end{array}
$$

To solve for y, we substitute $x = 1$ into either of equations **I** or **II**. We choose **I**.

I $2x + 3y = 8$
$2(1) + 3y = 8$
$3y = 6$
$y = 2$

CHECK $2x + 3y = 8$ $3x + 4y = 11$ Original equations
$2(1) + 3(2) \stackrel{?}{=} 8$ $3(1) + 4(2) \stackrel{?}{=} 11$ $x = 1, y = 2$
$2 + 6 \stackrel{?}{=} 8$ $3 + 8 \stackrel{?}{=} 11$
$8 \stackrel{\checkmark}{=} 8$ $11 \stackrel{\checkmark}{=} 11$

The solution checks. The solution set is $\{(1, 2)\}$.

EXAMPLE 4   Solve the following system by the elimination method:

I $\dfrac{x}{2} + \dfrac{y}{3} = 4$

II $\dfrac{x}{4} + \dfrac{y}{3} = 4$

SOLUTION   Systems of equations are generally easier to work with if they are free of fractions. We clear of fractions by multiplying each side of each equation by its LCD:

I $\dfrac{x}{2} + \dfrac{y}{3} = 4$ $\rightarrow$ **III** $3x + 2y = 24$ Multiply each side by 6, the LCD.

II $\dfrac{x}{4} + \dfrac{y}{3} = 4$ $\rightarrow$ **IV** $3x + 4y = 48$ Multiply each side by 12, the LCD.

We can now solve equations **III** and **IV** by the method of Example 3. The solution set is $\{(0, 12)\}$. The check is left to the reader.

We can outline the steps for solving a system of linear equations as follows.

To Solve a System of Linear Equations by the Elimination Method

Step 1 Write both equations in the form $ax + by = c$ if not already in that form.

Step 2 When necessary, multiply one or both equations by appropriate constants so that the coefficients of one of the variables are additive inverses of each other.

Step 3 Add the two equations to eliminate one variable. Solve for the remaining variable.

Step 4 Substitute the result of Step 3 into either of the original equations and solve for the other variable.

Step 5 Check the solution in *both equations* of the original system.

EXAMPLE 5 Solve the system by the elimination method:

$$y = -3x + 5$$
$$x + 2y = 5$$

SOLUTION **Step 1** Rewrite the first equation so that it is in the form $ax + by = c$.

$$y = -3x + 5 \quad \rightarrow \quad 3x + y = 5 \qquad \text{Add } 3x \text{ to each side.}$$

We now have the system

$$3x + y = 5$$
$$x + 2y = 5$$

Step 2 Multiply both sides of the second equation by -3 so that the coefficients of x are additive inverses of each other.

$$x + 2y = 5 \quad \rightarrow \quad -3x - 6y = -15 \qquad \text{Multiply each side by } -3.$$

Step 3 Add the two equations and solve for the remaining variable.

$$3x + y = 5$$
$$\underline{-3x - 6y = -15}$$
$$-5y = -10 \qquad \text{Add the two equations.}$$
$$y = 2$$

Step 4 Substitute the result, $y = 2$, into either of the original equations and solve for x. We choose the second equation.

$$x + 2y = 5$$
$$x + 2(2) = 5 \qquad \text{Substitute 2 for } y.$$
$$x = 1$$

Step 5 Check the solution in the original system.

$y = -3x + 5$	$x + 2y = 5$	Original equation
$2 \stackrel{?}{=} -3(1) + 5$	$1 + 2(2) \stackrel{?}{=} 5$	$x = 1, y = 2$
$2 \stackrel{?}{=} -3 + 5$	$1 + 4 \stackrel{?}{=} 5$	
$2 \stackrel{\checkmark}{=} 2$	$5 \stackrel{\checkmark}{=} 5$	

The solution checks.

The solution set is $\{(1, 2)\}$.

The elimination method will not yield a unique solution if the system is dependent or inconsistent, as is illustrated in the next two examples.

EXAMPLE 6 Solve the following system by the elimination method:

$$x + y = 1$$
$$x + y = 3$$

SOLUTION   These two lines are parallel (see Example 3 of Section 5.1), so there is no solution. If we try to solve the system by the elimination method, a false statement will result:

$$\begin{aligned} x + y = 1 &\rightarrow & x + y &= 1 \\ x + y = 3 &\rightarrow & \underline{-x - y} &= \underline{-3} \quad \text{Multiply each side by } -1. \\ & & 0 &= -2 \end{aligned}$$

If an inconsistent (false) *statement* such as $0 = -2$ *results when solving a system by the elimination method, we know the lines are parallel. The solution set is the empty set.*

EXAMPLE 7   Use the elimination method to solve the following system:

$$\begin{aligned} 2x - y &= 1 \\ 4x - 2y &= 2 \end{aligned}$$

SOLUTION   These two lines coincide (see Example 5 of Section 5.1), so there are infinitely many solutions. When we try to solve this system by the elimination method, the equation $0 = 0$ results, which is a true statement.

$$\begin{aligned} 2x - y = 1 &\rightarrow & -4x + 2y &= -2 \quad \text{Multiply each side by } -2. \\ 4x - 2y = 2 &\rightarrow & \underline{4x - 2y} &= \underline{2} \\ & & 0 &= 0 \end{aligned}$$

If a true statement such as $0 = 0$ *results when solving a system by the elimination method, we know the lines coincide. The solution set consists of all ordered pairs that satisfy either line.* In this case the solution set is $\{(x, y) \mid 2x - y = 1\}$.

We can use systems of equations to solve many different application problems. We will examine two types in Examples 8 and 9 and more in the application section at the end of this chapter.

EXAMPLE 8   Two angles are supplementary. If one is 40° more than the other, find the measure of the angles.

SOLUTION   We let $x =$ the measure of the larger angle
$y =$ the measure of the smaller angle

Since the difference in the size of the two angles is 40°, one equation is

$$x - y = 40$$

If two angles are supplementary, the sum of their measures is 180°, so a second equation is

$$x + y = 180$$

We can solve these two equations by the elimination method.

$$x - y = 40$$
$$x + y = 180$$
$$2x = 220 \quad \text{Add the two equations.}$$
$$x = 110$$

To solve for y, we substitute 110 for x in the second equation.

$$x + y = 180$$
$$110 + y = 180 \quad \text{Substitute 110 for } x.$$
$$y = 70$$

The measures of the two angles are 70° and 110°.

EXAMPLE 9   A collection of quarters and dimes contains 22 coins. The face value of the coins is $3.25. How many quarters are in the collection?

SOLUTION   We let d = the number of dimes
q = the number of quarters

There are 22 coins in the collection. Thus, $d + q = 22$. The value is 325 cents, so a second equation is $10d + 25q = 325$.

$$d + q = 22 \quad \rightarrow \quad -10d - 10q = -220 \quad \text{Multiply each side by } -10.$$
$$10d + 25q = 325 \quad \rightarrow \quad \underline{10d + 25q = 325}$$
$$15q = 105$$
$$q = 7$$

Notice that since we want to know how many quarters there are in the collection, it made sense to solve for q directly. There are 7 quarters.

$\mathcal{E}$XPLORING $\mathcal{G}$RAPHICS

We can also use a computer with graphing software or a graphics calculator to find solutions to systems of equations. In the display shown in Figure 5.8, the system

$$2x + y = 4$$
$$y - x = 1$$

was graphed on the same set of axes using a computer. The screen on a graphics calculator would look essentially the same.

To graph equations using a graphics calculator, we first have to write them as functions. Thus we would write $2x + y = 4$ as $f(x) = -2x + 4$ and $y - x = 1$ as $g(x) = x + 1$. When the graphs are displayed on the screen (Figure 5.8), their intersection appears to be (1, 2), which we can verify as correct in the original equations. Graphics calculators and computer programs incorporate features that

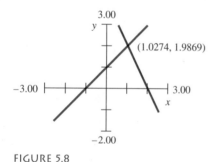

FIGURE 5.8

allow the user to approximate the point of intersection by moving the cursor (tracer) to its apparent location. The coordinates of the cursor are then displayed. At the apparent point of intersection, the cursor reads (1.0274, 1.9869).

Exercise Set 5.2

Solve each system of equations using the elimination method. If the system is not independent, state whether it is dependent or inconsistent. Check your solutions.

1. $x + y = 3$
   $x - y = 1$
2. $x + y = 6$
   $x - y = 2$
3. $x + y = 3$
   $3x - y = 1$
4. $x + y = 2$
   $x + y = 4$
5. $2x - y = 2$
   $x + y = 4$
6. $x + 3y = 10$
   $2x + y = 7$
7. $6x + y = -6$
   $x - 7y = -1$
8. $5x - y = 6$
   $x - y = 2$
9. $x + y = 3$
   $2x + 2y = 8$
10. $3x - y = 13$
    $2x - y = 10$
11. $-x + 4y = -3$
    $-x + 2y = -1$
12. $4x + y = 0$
    $-4x + 2y = 12$
13. $x + y = 4$
    $2x + 2y = 8$
14. $x + y = 1$
    $6x + 2y = 4$
15. $2x + 3y = 2$
    $4x + 9y = 5$

Solve each system of equations by the elimination method after writing each equation in general form, if necessary.

16. $y = 3 - x$
    $x = y + 1$
17. $y = 6 - x$
    $3x + 2y = 10$
18. $2x - 5 = -y$
    $4x + 2y = 10$
19. $x + y = 2$
    $y = 2x - 1$
20. $15x + 2y = 0$
    $-12x - 3y = 24$
21. $6x = 4y + 28$
    $-2 + x = -y$

Solve each system of equations by eliminating the indicated variable.

22. $a - b = 6$
    $a + b = 9$ } b
23. $3a - 4b = -2$
    $5a + 3b = 16$ } b
24. $a - b = 6$
    $a + b = 9$ } a
25. $3m - 4n = -2$
    $5m + 4n = 16$ } m
26. $h + 2k = 1$
    $2h + k = -3$ } k
27. $2s + t = 3$
    $5s - 2t = -15$ } s
28. $h - 2k = 1$
    $2h - k = -3$ } h
29. $2s - t = 3$
    $5s + 2t = -15$ } t
30. $5m - 4n = -1$
    $-7m + 5n = 8$ } n

Solve each system of equations by the elimination method. If the system is not independent, indicate whether it is dependent or inconsistent.

31. $x + y = 10$
    $-x + y = 2$
32. $2x + y = 9$
    $3x - y = 1$
33. $3x - y = 7$
    $10x - 5y = 25$
34. $3x + 2y = 5$
    $-3x + 5y = 25$
35. $x + 3y = 14$
    $4x - y = 4$
36. $x = 3y$
    $y = 5x + 14$
37. $3x - y = 8$
    $x + 2y = 5$
38. $2x + 5y = 9$
    $3x - 2y = 4$
39. $3x - 8y = 11$
    $x + 6y = 8$
40. $3x - 2y = 10$
    $5x + 3y = 4$
41. $2.8x - 4.2y = -7$
    $-2x + 3y = 6.75$
42. $-7.2x + 5.4y = -9$
    $4.8x - 3.6y = 2.4$
43. $0.06x + 0.05y = 0.07$
    $0.4x - 0.3y = 1.1$
44. $1.5x - y = -8.5$
    $0.2x + 0.4y = 1$
45. $\frac{1}{6}x + \frac{1}{2}y = 1$
    $\frac{1}{4}x + \frac{1}{2}y = 1$
46. $\frac{1}{4}x - \frac{1}{8}y = 1$
    $\frac{1}{3}x - \frac{1}{9}y = 1$
47. $\frac{x}{2} + \frac{y}{3} = \frac{2}{3}$
    $\frac{x}{3} + \frac{y}{5} = \frac{1}{3}$
48. $\frac{x}{5} + \frac{y}{3} = \frac{2}{5}$
    $0.2x - 0.7y = 0.1$
49. $\frac{1}{3}x - y = 1.5$
    $0.5x + 0.4y = 0.2$
50. $0.13x - 0.52y = -0.39$
    $0.39x + 0.08y = -2.81$

Use two variables to write a system of equations for each of the following applied problems. Solve by the elimination method.

51. The sum of two numbers is 71 and their difference is 15. Find the two numbers.

52. The sum of two numbers is −100. If one number is three times the other, find the two numbers.

53. The perimeter of a rectangle is 80 cm. If the width is 6 cm less than the length, find the dimensions.

54. The perimeter of a rectangle is 150 in. If the length is three-halves times the width, find the dimensions.

55. Ruth has $6.50 in nickels and dimes. If there are 120 coins altogether, how many of them are nickels?

56. Ida has $2.80 in nickels and quarters. If she has 28 coins altogether, how many of them are quarters?

57. Larry and Bill do lawn jobs. Larry starts a job but has to leave to start another job as Bill arrives. Bill finishes up and writes a bill for $20 for 3 hours' work. If Bill charges $6/hr and Larry charges $8/hr, how many hours did each work?

58. Naomi and Elisa do yard cleanup together. Naomi has to leave a job early, so Elisa finishes the job. Elisa turns in a total bill of $66 for $7\frac{5}{6}$ hr of work. If Naomi charges $9/hr and Elisa charges $8/hr, how much did each make?

59. Nelson wants to cut a chain that is 28 m long into two pieces in such a way that the longer piece is 8 m longer than the shorter one. How long should each piece be?

60. Lester wants to cut a 28-foot-long board in two pieces. If the shorter piece is to be two-fifths the length of the longer piece, find the length of the shorter piece.

61. Two angles have a sum of 180°. (They are supplementary angles.) If one is 30° more than twice the other, find the measure of each angle.

62. Two angles have a sum of 90°. (They are complementary angles.) If one is 10° more than three times the other, find the measure of the angles.

63. Two angles are supplementary and their difference is 52°. Find the measure of the angles.

64. Two angles are complementary and their difference is 24°. Find the measure of the angles.

65. Two angles are complementary (see Exercise 62) and one is twice the other. Find the measure of the angles.

66. Two angles are supplementary and one is three times the other. Find the measure of the angles.

Find the solution to each system of equations by inspecting its graph. Then write the equations of the two lines and solve the system by the elimination method. Compare the solutions found by inspection to those found by the elimination method.

67.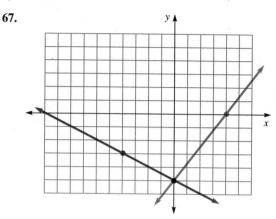

68.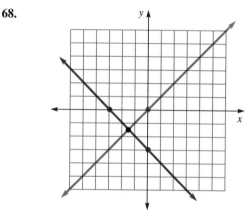

69.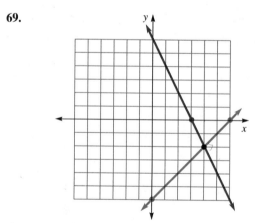

70.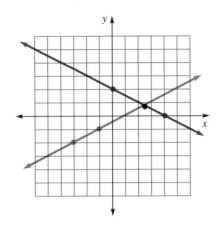

73. $4y + 5\left(\dfrac{2y + 7}{4}\right) = 0$ **74.** $3y + 2\left(\dfrac{1 - 2y}{3}\right) = 6$

75. $-2\left(\dfrac{3 - 2x}{2}\right) + 6x = -2$

76. $-4\left(\dfrac{-2 - x}{5}\right) - 3x = 0$

77. $-\left(\dfrac{5 - 2y}{4}\right) + 3y = 1$ **78.** $-\left(\dfrac{3 - 4y}{3}\right) - y = 2$

79. $3x - 5(3 - x) = 7$ **80.** $-2x + 3(4 - 2x) = 2$

81. $-2p + 3(4 - p) - 2 = 0$ **82.** $r - 5(2 - r) + 4 = 0$

REVIEW EXERCISES

Solve each of the following equations. This practice will help in the next section.

71. $5x - 2(x - 2) = 10$ **72.** $x + 3\left(\dfrac{1 - x}{2}\right) = 5$

5.3 Solving Systems of Linear Equations by the Substitution Method

OBJECTIVES In this section we will learn to
1. solve systems of linear equations by substitution; and
2. determine the easiest method to use for solving a system of linear equations.

In Sections 5.1 and 5.2 we learned how to solve systems of linear equations by the graphing method and by the elimination method. A third method, the **substitution method**, is similar to the elimination method in that it reduces two equations in two variables to a single equation in one variable. However, we need an additional property of equality to solve equations by this method.

> **The Substitution Property of Equality**
>
> Any algebraic expression may be substituted for its equal in an equation. Thus if A, B, and C represent algebraic expressions, and if
>
> $A + B = C$ and $B = D$
>
> then
>
> $A + D = C$

EXAMPLE 1 Use the substitution method to solve the following system:

I $y = 4 - x$
II $x - y = 2$

SOLUTION To solve an equation by substitution, we can solve either equation for either variable and substitute the result into the other equation. Equation **I** is already solved for **y**, so all we have to do is substitute the value of y into equation **II**:

II	$x - y = 2$	
	$x - (4 - x) = 2$	Substitute $4 - x$ for y.
	$x - 4 + x = 2$	Remove grouping symbols.
	$2x - 4 = 2$	Combine like terms.
	$2x = 6$	Add 4 to each side.
	$x = 3$	Divide each side by 2.

By substituting $x = 3$ into $y = 4 - x$, we can obtain the value of y:

$y = 4 - x$
$y = 4 - 3$ Substitute 3 for x.
$y = 1$

To check, we substitute $x = 3$ and $y = 1$ into both of the original equations.

CHECK $y = 4 - x$ $x - y = 2$
          $1 \stackrel{?}{=} 4 - 3$ $3 - 1 \stackrel{?}{=} 2$ $x = 3, y = 1$
          $1 \stackrel{\checkmark}{=} 1$ $2 \stackrel{\checkmark}{=} 2$

The solution checks. The solution set is $\{(3, 1)\}$.

There are five steps we must follow when solving a system of linear equations by the substitution method.

To Solve a System of Linear Equations by the Substitution Method

Step 1   Solve one of the equations for either variable, if necessary.
Step 2   Substitute the result from Step 1 into the other equation.
Step 3   Solve the equation obtained in Step 2.
Step 4   Substitute the solution from Step 3 into the equation found in Step 1, and solve for the variable.
Step 5   Check your results in *both* of the original equations.

We will identify these steps as we work the next example.

EXAMPLE 2 Use the substitution method to solve the system

    I $6x + 3y = 9$
    II $x - 2y = -1$

SOLUTION   **Step 1** **Solve one equation for either variable.**
We choose equation **II** and solve for x. Solving for the variable with a coefficient of 1 avoids working with fractions.

$$x - 2y = -1$$
$$x = -1 + 2y \quad \text{Add } 2y \text{ to each side.}$$

Step 2 **Substitute the result from Step 1 into the other equation.**

    I $6x + 3y = 9$
            $6(-1 + 2y) + 3y = 9$ Substitute $-1 + 2y$ for x.

Step 3 **Solve the equation obtained in Step 2.**

$$6(-1 + 2y) + 3y = 9$$
$$-6 + 12y + 3y = 9 \quad \text{Distributive property}$$
$$-6 + 15y = 9 \quad \text{Combine like terms.}$$
$$15y = 15 \quad \text{Add 6 to each side.}$$
$$y = 1 \quad \text{Divide each side by 15.}$$

Step 4 **Substitute the solution from Step 3 into the equation found in Step 1.**

$$x = -1 + 2y$$
$$x = -1 + 2(1) \quad \text{Substitute 1 for } y.$$
$$x = -1 + 2$$
$$x = 1$$

Step 5 **Check the results** by substituting $x = 1$, $y = 1$ **in both of the original equations.**

$$6x + 3y = 9 \qquad x - 2y = -1$$
$$6(1) + 3(1) \stackrel{?}{=} 9 \qquad 1 - 2(1) \stackrel{?}{=} -1 \qquad x = 1, y = 1$$
$$9 \stackrel{\checkmark}{=} 9 \qquad -1 \stackrel{\checkmark}{=} -1$$

The solution set is $\{(1, 1)\}$.

EXAMPLE 3 Use the substitution method to solve the system

    I $x + 6y = 9$
    II $y = 3x$

SOLUTION Since equation **II** is already solved for *y*, let's substitute this value into equation **I**.

$$x + 6y = 9$$
$$x + 6(3x) = 9 \quad \text{Substitute } 3x \text{ for } y.$$
$$x + 18x = 9$$
$$19x = 9$$
$$x = \frac{9}{19} \quad \text{Divide each side by 19.}$$

Substituting $\frac{9}{19}$ into equation **II** yields

$$y = 3x$$
$$y = 3\left(\frac{9}{19}\right) = \frac{27}{19}$$

CHECK

$$x + 6y = 9 \qquad y = 3x$$
$$\frac{9}{19} + 6\left(\frac{27}{19}\right) \stackrel{?}{=} 9 \qquad \frac{27}{19} \stackrel{?}{=} 3\left(\frac{9}{19}\right) \qquad x = \frac{9}{19}, y = \frac{27}{19}$$
$$\frac{9}{19} + \frac{162}{19} \stackrel{?}{=} 9 \qquad \frac{27}{19} \stackrel{\checkmark}{=} \frac{27}{19}$$
$$\frac{171}{19} \stackrel{?}{=} 9$$
$$9 \stackrel{\checkmark}{=} 9$$

The solution checks. The solution set is $\left\{\left(\frac{9}{19}, \frac{27}{19}\right)\right\}$.

EXAMPLE 4 Solve the following system by the substitution method.

$$\textbf{I} \quad \frac{x}{3} + \frac{y}{4} = -\frac{1}{2}$$
$$\textbf{II} \quad 2x + \frac{y}{2} = \frac{1}{3}$$

SOLUTION We begin by clearing the equations of fractions. To do this, we multiply each side of equation **I** by 12 and each side of equation **II** by 6:

$$\textbf{I} \quad \frac{x}{3} + \frac{y}{4} = -\frac{1}{2} \quad \rightarrow \quad \textbf{III} \quad 4x + 3y = -6 \qquad \text{Multiply each side of } \textbf{I} \text{ by 12.}$$
$$\textbf{II} \quad 2x + \frac{y}{2} = \frac{1}{3} \quad \rightarrow \quad \textbf{IV} \quad 12x + 3y = 2 \qquad \text{Multiply each side of } \textbf{II} \text{ by 6.}$$

We now solve equation **III** for y and substitute the result into equation **IV**.

III $\quad 4x + 3y = -6$

$\qquad\qquad 3y = -6 - 4x \qquad$ Subtract $4x$ from each side.

$\qquad\qquad y = \dfrac{-6 - 4x}{3} \qquad$ Divide each side by 3.

IV $\qquad\qquad 12x + 3y = 2$

$\qquad\qquad 12x + 3\left(\dfrac{-6 - 4x}{3}\right) = 2 \qquad$ Substitute $\dfrac{-6 - 4x}{3}$ for y.

$\qquad\qquad 12x + (-6 - 4x) = 2 \qquad$ Divide out 3.

$\qquad\qquad 12x - 6 - 4x = 2 \qquad$ Remove grouping symbols.

$\qquad\qquad 8x - 6 = 2 \qquad$ Combine like terms.

$\qquad\qquad 8x = 8 \qquad$ Add 6 to each side.

$\qquad\qquad x = 1$

Now we substitute $x = 1$ into $y = \dfrac{-6 - 4x}{3}$:

$y = \dfrac{-6 - 4x}{3}$

$ = \dfrac{-6 - 4(1)}{3} \qquad x = 1$

$ = \dfrac{-10}{3}$

CHECK $\qquad \dfrac{x}{3} + \dfrac{y}{4} = -\dfrac{1}{2} \qquad\qquad 2x + \dfrac{y}{2} = \dfrac{1}{3} \qquad$ Original equations

$\qquad\qquad\quad \dfrac{1}{3} + \dfrac{\frac{-10}{3}}{4} \stackrel{?}{=} -\dfrac{1}{2} \qquad 2(1) + \dfrac{\frac{-10}{3}}{2} \stackrel{?}{=} \dfrac{1}{3} \qquad x = 1, y = \dfrac{-10}{3}$

$\qquad\qquad\quad \dfrac{1}{3} + \left(\dfrac{-10}{12}\right) \stackrel{?}{=} -\dfrac{1}{2} \qquad 2 + \left(\dfrac{-10}{6}\right) \stackrel{?}{=} \dfrac{1}{3}$

$\qquad\qquad\quad \dfrac{2}{6} + \left(-\dfrac{5}{6}\right) \stackrel{?}{=} -\dfrac{1}{2} \qquad \dfrac{6}{3} + \left(\dfrac{-5}{3}\right) \stackrel{?}{=} \dfrac{1}{3}$

$\qquad\qquad\quad \dfrac{-3}{6} \stackrel{?}{=} -\dfrac{1}{2} \qquad\qquad\qquad \dfrac{1}{3} \stackrel{\checkmark}{=} \dfrac{1}{3}$

$\qquad\qquad\quad -\dfrac{1}{2} \stackrel{\checkmark}{=} -\dfrac{1}{2}$

The solution checks. The solution set is $\left\{\left(1, \dfrac{-10}{3}\right)\right\}$.

Notice that Example 4 would be easier to solve by elimination once the original equations are cleared of fractions. In general, the elimination method is preferred to the substitution method whenever one of the variables in one of the equations does not have a coefficient of 1.

What happens when we try to solve a dependent or an inconsistent system by substitution?

EXAMPLE 5   Solve the system

$$x + y = 3$$
$$2x + 2y = 6$$

SOLUTION   This is a dependent system, since the second equation can be obtained from the first one by multiplying each side by 2. Solving the first equation for y yields

$$y = -x + 3$$

Now we substitute $y = -x + 3$ into the second equation.

$2x + 2y = 6$	Second equation
$2x + 2(-x + 3) = 6$	Substitute $-x + 3$ for y.
$2x - 2x + 6 = 6$	Distributive property
$6 = 6$	True

The solution set is $\{(x, y) \mid x + y = 3\}$.

Solving a dependent system by the substitution method results in a statement that is always true. Recall that this is what happened with the elimination method also.

If we try to solve an inconsistent system by substitution, a false statement will result, as it did when we tried to solve such a system by the elimination method.

EXAMPLE 6   Solve the system

$$x + y = 3$$
$$x + y = 4$$

SOLUTION   The lines have the same slope but different y-intercepts. Therefore their graphs are parallel lines, and the system is inconsistent. Solving the first equation for y yields

$$y = 3 - x$$

Substituting this value for y into the second equation, we have

$x + (3 - x) = 4$	Substitute $3 - x$ for y.
$3 = 4$	False

The system is indeed inconsistent and the solution set is the empty set.

As demonstrated in Section 5.2, systems of equations are useful in solving applied problems. In the next example, we will solve a problem involving angles of a triangle by setting up a system of equations and solving it.

EXAMPLE 7 In a particular right triangle, one acute angle is twice as large as the other. Use a system of equations to find the measures of the two acute angles.

SOLUTION In a right triangle, the sum of the measures of the acute angles is 90°. Let

A = the measure of the first angle

B = the measure of the second angle

Then

$A + B = 90$ The sum of the measures of the acute angles is 90°.

$B = 2A$ One angle is twice the other.

Since the second equation is already solved for B, the substitution method is the best choice.

$A + B = 90$

$A + 2A = 90$ Substitute $2A$ for B.

$3A = 90$

$A = 30°$

$B = 2A = 2(30) = 60°$

The measure of one angle is 30° and the other is 60°.

Exercise Set 5.3

Solve each of the following systems of equations by the substitution method.

1. $y = 2x$
   $x + y = 3$

2. $y = 3x$
   $x - y = -2$

3. $x = 2y$
   $x + y = 6$

4. $x = 3y$
   $x - y = -4$

5. $y = -x$
   $2x + y = 8$

6. $y = -\frac{1}{2}x$
   $4x - 2y = 5$

7. $y = -\frac{1}{4}x$
   $2x + 4y = 1$

8. $x = \frac{1}{2}y$
   $3x + y = 0$

9. $x = \frac{1}{4}y$
   $2x + 2y = 5$

10. $x + y = 2$
    $2x - 3y = -1$

Solve each linear system of equations by the substitution method. In each of the following systems, solve the second equation for x first, and substitute that value into the first equation.

11. $2x + 2y = 4$
    $x + 4y = 5$

12. $2x - y = 5$
    $x + y = 1$

13. $x - 2y = -1$
    $x + 2y = 7$

14. $3x - 2y = -7$
    $x = y - 1$

15. $3x - 2y = 1$
    $x - 2y = -5$

16. $4x + y = 5$
    $x + 2y = 4$

Solve each linear system of equations by the substitution method. In each case, solve the first equation for y first, and substitute that value into the second equation.

17. $x + y = 2$
    $2x - y = 1$

18. $6 - y = x$
    $3x + 2y = 10$

19. $x + y = 3$
    $2x + 2y = 8$

20. $x + y = -2$
    $x + y = 4$

21. $2x - y = 2$
    $x + y = 4$

22. $2x + y = 7$
    $x + 3y = 10$

Solve each system of equations by the substitution method. If the system is not independent, indicate whether it is dependent or inconsistent.

23. $x + y = 1$
    $6x + 2y = 4$

24. $3x + 6y = 9$
    $2x + y = 7$

25. $2x - 5 = -y$
    $4x + 2y = 10$

26. $3x - y = 13$
    $2x - 10 = y$

27. $2r + s = 3$
    $5r - 2s = -15$

28. $3x + 2y = -2$
    $x + y = 1$

29. $x + 5y = 1$
    $7x - 15y = -3$

30. $3m + n = -1$
    $n = 2m + 5$

31. $5w - z = 14$
    $z = 4w - 18$

32. $3x = y + 2$
    $3x = y - 4$

33. $x + y = 7$
    $2x + 2y = 14$

34. $6x + 3y = 0$
    $4x + 2y = 6$

35. $y = 2x$
    $\frac{2}{5}x - \frac{1}{10}y = -19$

36. $x = -3 - y$
    $\frac{3}{4}x + \frac{2}{5}y = -4$

37. $y = 1 - x$
    $\frac{1}{2}x - \frac{2}{3}y = 4$

38. $x + \frac{1}{2}y = \frac{-7}{2}$
    $y + 2x = -7$

39. $x - 2y = \frac{3}{2}$
    $\frac{-x}{2} + y = -\frac{3}{4}$

40. $r - 2s = -7$
    $3r - 6s = 10$

41. $x = 3y - 1$
    $-12y + 4x = -3$

42. $0.5x + 0.5y = 15$
    $0.7x - 0.5y = -3$

43. $r + 0.5s = 10$
    $0.8r + 0.3s = 7$

Solve each system of equations by either the elimination method or the substitution method. If the system is not independent, state whether the lines are parallel (inconsistent) or the same line (dependent).

44. $2a + 3b = 8$
    $4a + 3b = 4$

45. $a = -1 - b$
    $2a + 5b = 7$

46. $2r + 4s = 6$
    $s = \frac{5 - r}{2}$

47. $7x + 18y = 17$
    $x = \frac{4y + 19}{3}$

48. $\frac{x}{2} + \frac{y}{3} = \frac{7}{6}$
    $\frac{x}{4} - \frac{3y}{2} = \frac{9}{4}$

49. $\frac{1}{4}x - \frac{3}{2}y = -6$
    $\frac{2x}{3} + \frac{y}{6} = 9$

50. $2a - 5b = -4b - 8 - 4$
    $2b + 12 + 2a = 3b$

51. $3x + 2y = 11$
    $2x - y = 5$

52. $3x - 2y = 0$
    $y = 3$

53. $4y - 5x = 9$
    $x = -2$

54. $\frac{1}{2}a + \frac{2}{5}b = \frac{7}{10}$
    $\frac{1}{3}a + \frac{1}{2}b = -\frac{1}{6}$

55. $\frac{3}{4}x - \frac{1}{2}y = \frac{17}{2}$
    $\frac{1}{2}x - \frac{3}{4}y = \frac{13}{2}$

Use two variables to write a system of equations for each of the following applied problems. Solve by any of the three methods.

56. Find two integers with a sum of -18 and a difference of 6.

57. Find two numbers with a sum of -10 and a difference of -2.

58. Thirty-nine students are enrolled in a ski class. The instructor sees that there are 15 more girls than boys. How many of each are in the class?

59. Mr. Hughes has 50 students in an algebra class. If there are one and a half times as many girls as boys, how many of each are in the class?

60. On opening night of a concert, main floor seats sell for $16 each and balcony seats for $12 each. The following night the seats are $10 and $8, respectively. If both nights are sold out and receipts for the two nights are $22,000 and $14,000, respectively, how many seats are in the balcony?

61. Tickets for the first night of a concert sell at $6 for cushioned seats and $4 for wooden seats. The second-night tickets sell at $8 for cushioned and $6 for wooden seats. If both nights are sold out and receipts are $760 and $1040 for the first and second nights, respectively, how many cushioned and wooden seats does the concert hall have?

62. In one season the Dodgers played 162 games. Twice the number of games won is 6 less than three times the number of games lost. How many games did they win? How many did they lose?

63. If the smaller of two numbers is subtracted from the larger, the result is 8. If the larger number is added to twice the smaller, the result is 53. Find the two numbers.

64. Two angles are supplementary. If the larger is subtracted from twice the smaller, the result is 75°. Find the measures of the two angles.

65. Two angles are complementary. If the larger angle is five times the smaller one, find the measures of the two angles.

66. The perimeter of a parallelogram is $\frac{22}{3}$ ft. The longer side is one foot more than three times the shorter side. Find the dimensions. (See the figure.)

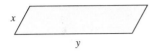

67. The perimeter of a triangle is 15 in. If one side is 1 in. less than twice the second side, find the dimensions. (See the figure.)

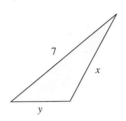

REVIEW EXERCISES

Solve each of the following inequalities by graphing.

68. $y < x + 1$
69. $y > x - 2$
70. $x + y \leq 5$
71. $x + 2y \geq 0$
72. $y \leq 3$
73. $y > -2$
74. $x < -4$
75. $x \geq 1$
76. $3x - 2y \leq -2$
77. $2x + 3y \leq 3$
78. $4x + 3y > 1$
79. $5x - y < 2$
80. $y < |x|$
81. $y > |x|$
82. $y \leq |x| - 2$
83. $y \geq -|x| + 2$

5.4 Solving Systems of Linear Inequalities

OBJECTIVES

In this section we will learn to
1. use graphs to find the solution set of systems of two linear inequalities; and
2. use graphs to find the solution set of systems of three or more linear inequalities.

Before we can graph systems of linear inequalities, we must first review how to graph a single linear inequality. To graph $2x + y \leq 4$ (Example 2 of Section 4.4), we first sketch the line $2x + y = 4$ to establish the boundary of the half-plane represented by the inequality. We draw the line solid to indicate that the points on the line satisfy the inequality. We then choose a test point that does not lie on the line, say, (0, 0), to determine which side of the boundary line contains the solution set.

$$2x + y \leq 4$$
$$2(0) + 0 \stackrel{?}{\leq} 4 \quad x = 0, y = 0$$
$$0 \leq 4 \quad \text{True}$$

Since (0, 0) satisfies the inequality, we shade the region containing (0, 0) to indicate that it, together with the boundary line, forms the solution set. See Figure 5.9.

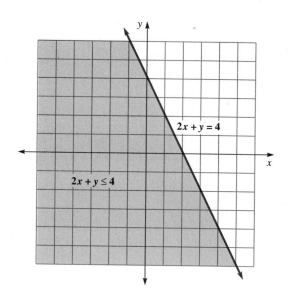

FIGURE 5.9

EXAMPLE 1 Graph the region that is the solution to the system

$$x + y < 2$$
$$x - y > -3$$

SOLUTION We begin by graphing $x + y < 2$. To do this, we sketch the boundary line, $x + y = 2$. The line is broken because the inequality symbol is $<$. Since the test point $(0, 0)$ makes the inequality true, the region on the origin side of the boundary line is the solution set of $x + y < 2$. We shade this region as shown in Figure 5.10.

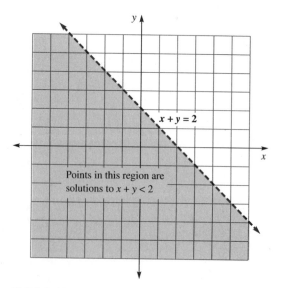

FIGURE 5.10

Next we graph the line $x - y = -3$, which is the boundary line for the half-plane of $x - y > -3$. Since the test point $(0, 0)$ makes the inequality true, we shade the region on the origin side of the boundary, as shown in Figure 5.11.

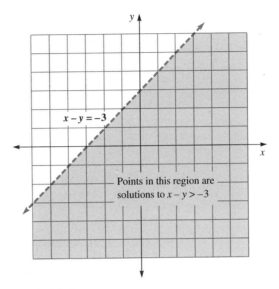

FIGURE 5.11

The solution set of the *system* is the region that is the *overlap* of the shaded regions in Figures 5.10 and 5.11, or the region with the darkest shading in Figure 5.12. It does not include either boundary line.

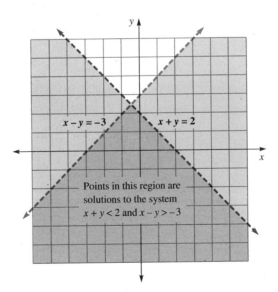

FIGURE 5.12

EXAMPLE 2 Graph the region that is the solution to the system

$$y \leq x + 1$$
$$2y + x \geq 4$$

SOLUTION Here both boundary lines are part of the solution set, so they will both be shown as solid lines. When we use the test point (0, 0), we find that it satisfies one inequality but not the other.

$y \leq x + 1$	$2y + x \geq 4$	
$0 \overset{?}{\leq} 0 + 1$	$0 + 2(0) \overset{?}{\geq} 4$	$x = 0, y = 0$
$0 \leq 1$ True	$0 \geq 4$ False	

SHADE THE ORIGIN SIDE.   SHADE THE SIDE OPPOSITE THE ORIGIN.

The region with the double shading (darkest shading) forms the solution set; see Figure 5.13.

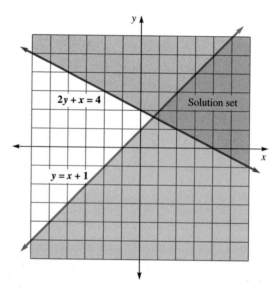

FIGURE 5.13

EXAMPLE 3 Graph the region that is the solution to the system

$$x > 1$$
$$y > 2$$

SOLUTION The line $x = 1$ is vertical and the line $y = 2$ is horizontal. The region satisfying $x > 1$ lies to the right of the line $x = 1$, while the region satisfying $y > 2$ lies above the line $y = 2$. Neither line is part of the solution set. The region with double shading is the solution set; see Figure 5.14.

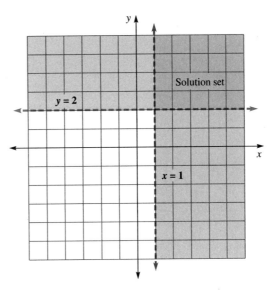

FIGURE 5.14

EXAMPLE 4   Graph the region that is the solution to the system

$$y \leq x + 1$$
$$2y + x = 4$$
$$x \leq 3$$

SOLUTION   The first two inequalities are the same as those in Example 2, so we begin with that region. The half-plane that satisfies the third inequality, $x \leq 3$, is to the left of

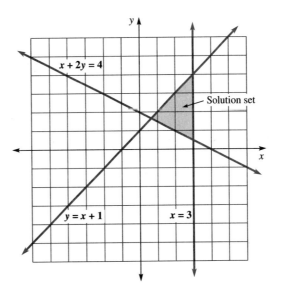

FIGURE 5.15

the vertical line $x = 3$ and includes the line itself. The region that satisfies all three inequalities is shown in Figure 5.15.

EXAMPLE 5   Graph the region that is the solution set for the system

$$x + y \leq 3$$
$$x + y \geq -2$$
$$x \leq 0$$
$$y \leq 4$$

SOLUTION   The solution set of $x \leq 0$ is all points on or to the left of the y-axis, while the solution set of $y \leq 4$ is the set of all points on or below the line $y = 4$. The solution set of $x + y \leq 3$ is the set of all points on or below the line $x + y = 3$, and the solution set of $x + y \geq -2$ is the set of points on or above the line $x + y = -2$. The intersection of the regions that satisfy each of these inequalities is shown in Figure 5.16.

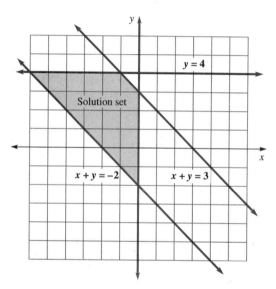

FIGURE 5.16

EXAMPLE 6   The solution set for a system of inequalities is shown in Figure 5.17. Find the inequalities.

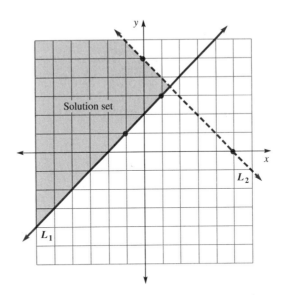

FIGURE 5.17

SOLUTION   We begin by finding the equations of the boundary lines, L_1 and L_2. We see that L_1 passes through $(-1, 1)$ and $(1, 3)$, so its slope is 1. We can find its equation by using the point–slope form of the equation of a line:

$$y - y_1 = m(x - x_1)$$

We let $(x_1, y_1) = (1, 3)$ and $m = 1$. Then

$$y - 3 = 1(x - 1)$$
$$y = x + 2$$

Since L_2 passes through $(0, 5)$ and $(5, 0)$, its slope is -1. Using the slope–intercept form, its equation is $y = -x + 5$. Since the solution set lies above or on L_1, one inequality is $y \geq x + 2$. The solution set lies below L_2. Since L_2 shows as a broken line, the second inequality is $y < -x + 5$.

Exercise Set 5.4

Use the graphing method to find the region that satisfies each system of inequalities.

1. $x \geq 2$
   $y \geq 3$

2. $x > 4$
   $y < 5$

3. $x < -1$
   $y < 2$

4. $y > 0$
   $x < 0$

5. $x \geq 2$
   $y \leq 2$

6. $x > -4$
   $y > -3$

7. $x + y > 1$
   $x > 0$

8. $2x - y \leq 1$
   $y \leq 1$

9. $y > x + 2$
   $x < 1$

10. $y + x < 2$
    $y < -1$

11. $x + y \le 6$
    $x - y \le 1$

12. $x + y \ge 1$
    $y - x \le 1$

13. $x + 2y \le 4$
    $x + 1 \ge y$

14. $x + 3y > 2$
    $x - y < 3$

15. $x + y < 3$
    $x + y > 3$

16. $x + y \ge 3$
    $x - y > 3$

17. $5x + 4y > 7$
    $y > x$

18. $y > 2x$
    $y < 3x$

19. $y \ge 3x$
    $y < 2x$

20. $x > 0$
    $y > 0$
    $y < x + 1$

21. $x < 0$
    $y < 0$
    $y > x - 2$

22. $x > 1$
    $y < 5$
    $y > x - 1$

23. $y < 5$
    $y > x - 1$
    $y > -x - 1$

24. $y > 2x - 1$
    $y > -x$
    $y < 4$

27.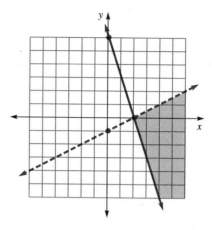

Write the system of inequalities for each of the following graphs.

25.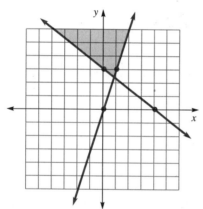

28.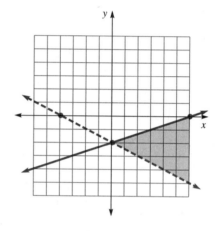

26.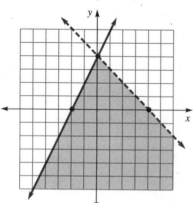

29.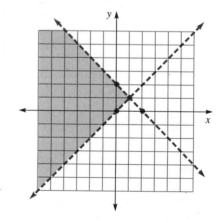

30.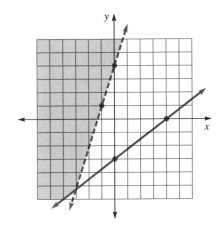

Use the graphing method to find the region that satisfies each system of inequalities.

31. $2x + y \geq 6$
    $y \leq 2(2x - 3)$

32. $3x + y < -2$
    $y > 3(1 - x)$

33. $\dfrac{x}{2} + \dfrac{y}{3} \geq 2$
    $\dfrac{x}{2} - \dfrac{y}{2} < -1$

34. $\dfrac{x}{3} + \dfrac{y}{2} > -1$
    $\dfrac{x}{3} - \dfrac{y}{2} < -3$

35. $3x + 2y < 2$
    $2x + 3y > 1$

36. $x - 0.5y < -0.5$
    $x - 0.5y > -1.5$

37. $0.1x + 0.1y < 0.2$
    $0.01x + 0.01y > 0$

38. $0.5x - 0.75y \geq 2.25$
    $0.5x + 0.25y > 1.5$

39. $0.25x < 0.75y - 0.25$
    $0.5y \geq x - 1.5$

40. $x - 2y \leq 0$
    $x + y \leq 6$
    $-2x + y \leq 2$

41. $x + 3y < 2$
    $x \leq y + 1$
    $y < 1$

42. $x - y \leq 1$
    $y - x \leq 3$
    $x \geq 2$
    $x \leq 4$

43. $x \leq 2y$
    $y \leq 2x + 2$
    $x \leq 3$
    $y \leq 2$

44. A service station has a storeroom that holds at most 200 quarts of oil. Past experience has shown that the station should have at least twice as many quarts of multigrade oil as regular oil on hand. These conditions can be represented by inequalities. Let y be the number of quarts of multigrade oil and x the number of quarts of regular oil. Then

    $x + y \leq 200$
    $y \geq 2x$
    $x \geq 0$
    $y \geq 0$

Graph this system to show the ways the owner could order to stock a new storeroom. Select one possible ordering plan.

45. A farmer has to erect a fence to enclose a chicken yard, and he wants the yard to be in the form of a rectangle with the length at least two times the width. He has money for no more than 100 yards of fencing. Let x = width, y = length. Then

    $2x + 2y \leq 100$
    $y \geq 2x$
    $x > 0$
    $y > 0$

    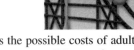

Graph the region that shows the possible dimensions of the rectangle. Select one possible set of dimensions.

46. The Fun Center owner wants admission for a family of three (two adults and one child) to cost no more than $10. He wants adult tickets to cost at least twice children's tickets. Let y be the cost of an adult ticket and x the cost of a child's ticket. A system of inequalities describing these conditions is

    $x + 2y \leq 10$
    $y \geq 2x$
    $x \geq 0$
    $y \geq 0$

Graph the region that shows the possible costs of adult and children's tickets. Select one possible set of costs.

47. The storeroom of a sportswear store holds no more than 300 pairs of shoes. The owner wants to order at least three times as many women's shoes as men's shoes. Let x = number of pairs of women's shoes and y = number of pairs of men's shoes. A system of inequalities describing these conditions is

    $x + y \leq 300$
    $x \geq 3y$
    $x \geq 0$
    $y \geq 0$

Graph the region that shows the possible numbers of pairs of women and mens shoes. Select one possible solution.

REVIEW EXERCISES

Solve each of the following systems of equations by the method indicated.

48. $\left.\begin{array}{r} x + y = 3 \\ x - y = 1 \end{array}\right\}$ graphing

49. $\left.\begin{array}{r}2x + y = 8 \\ x - y = 1\end{array}\right\}$ graphing

50. $\left.\begin{array}{r}x = -3y \\ 3x + y = 6\end{array}\right\}$ substitution

51. $\left.\begin{array}{r}x - y = 5 \\ x + y = 4\end{array}\right\}$ substitution

52. $\left.\begin{array}{r}x - y = 5 \\ y = 3 - x\end{array}\right\}$ elimination

53. $\left.\begin{array}{r}2x + 5y = 10 \\ 5x - 3y = 18\end{array}\right\}$ elimination

5.5 More Applications

HISTORICAL COMMENT Applied problems have been a part of mathematics education as far back as records exist. Although they may not seem too practical while you're studying them, the skills they teach are invaluable in problem analysis later in life. Compare the following problem from 1788 to the one found in Example 4 of this section.

If during an ebb tide, a wherry should set out from Haverhill, to come down the river, and, at the same time, another should set out from Newburyport, to go up the river, allowing the distance to be 18 miles; suppose the current forwards one and retards the other $1\frac{1}{2}$ miles per hour; the boats are equally laden, the rowers equally good, and, in the common way of working in still water, would proceed at a rate of 4 miles per hour; when, in the river, will the two boats meet?

OBJECTIVES In this section we will learn to
1. solve applied problems by using systems of linear equations; and
2. solve applied problems by using systems of linear inequalities.

Many applied problems are easier to solve if two variables are used instead of one. This has been done in the applied problems included in the exercise sets in this chapter. We will continue to use the five-step method for solving word problems from Section 3.8. We will use these steps to outline the solution process in the next example.

EXAMPLE 1   The sum of two numbers is 20 and their difference is 4. Find the numbers.

SOLUTION   **Step 1   Read the problem completely.**
The sum of two numbers is 20 and their difference is 4. Find the numbers.
Step 2   What am I asked to find?
The sum of two numbers is 20 and their difference is 4. *Find the numbers.*
Step 3   Represent the unknown quantities by variables.

We let x = the larger number
y = the smaller number

Step 4 **Establish the relationships with equations.**
The sum of the numbers is 20.

$$x + y = 20$$

The difference of the numbers is 4.

$$x - y = 4$$

Step 5 **Solve the system and check in the words of the original problem.**

$$x + y = 20$$
$$x - y = 4$$

The easiest way to solve these equations is to use the elimination method. When we add the two equations, we have

$$2x = 24$$
$$x = 12$$

Substituting 12 for x in the first equation yields

$$x + y = 20$$
$$12 + y = 20$$
$$y = 8$$

The numbers are 12 and 8. Their sum is 20 and their difference is 4. The solution checks.

EXAMPLE 2   The sum of two consecutive odd integers is 48. *Find the integers.*

SOLUTION   We let $x =$ the larger integer
$y =$ the smaller integer

Consecutive odd integers differ by 2, such as 3 and 5, or 7 and 9. Therefore one equation is

$$x - y = 2$$

The sum of the integers is 48, so a second equation is

$$x + y = 48$$

Once again, the easiest way to solve this system is by the elimination method.

$$x - y = 2$$
$$\underline{x + y = 48}$$
$$2x = 50 \quad \text{Add the two equations.}$$
$$x = 25$$

Substituting $x = 25$ into the second equation yields

$$25 + y = 48$$
$$y = 23 \quad \text{Subtract 25 from each side.}$$

The two integers are 23 and 25. They are both odd and their sum is 48, so the solution checks.

EXAMPLE 3   A chemist has a 20% acid solution and a 50% acid solution in the storeroom. A project she is working on requires 90 milliliters of a 30% acid solution. How many milliliters of each acid solution should she mix to obtain the 30% solution?

SOLUTION   We let x = the number of milliliters of the 20% solution
                     y = the number of milliliters of the 50% solution

She needs a total of 90 milliliters, so

$x + y = 90$

We can obtain the second equation from the fact that the total amount of acid in the mixture does not change. The concentration of acid, however, does change; see Figure 5.18.

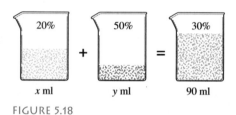

FIGURE 5.18

The picture shows that 20% of x ml plus 50% of y ml equals 30% of 90 ml, or

$0.2x + 0.5y = 0.3(90)$

We can eliminate decimals by multiplying each side by 10:

$2x + 5y = 3(90) = 270$

We now have a system of two equations,

I $x + y = 90$
II $2x + 5y = 270$

To solve the system by substitution, we solve equation **I** for y and substitute the result into equation **II**.

I $y = 90 - x$
II $2x + 5(90 - x) = 270$ Substitute $90 - x$ for y in **II**.
         $2x + 450 - 5x = 270$ Distributive property
         $-3x + 450 = 270$ Combine like terms.
         $-3x = -180$ Subtract 450 from each side.
         $x = 60$ Divide each side by -3.

Substituting $x = 60$ into equation **I** yields $y = 30$. Therefore the mixture will be a mixture of 60 milliliters of the 20% solution and 30 milliliters of the 50% solution.

To check, 20% of 60 plus 50% of 30 is equal to 12 + 15 = 27, which is 30% of 90. The solution checks.

EXAMPLE 4 A boat can go 20 miles upstream in 2 hours or 32 miles downstream in the same amount of time. *Find the speed of the current and the rate the boat travels in still water.*

SOLUTION In going upstream, the boat will be slowed by an amount equal to the rate of the current. When going downstream, it will be sped up by the same amount. Therefore we let

x = the rate of the boat in still water

y = the rate of the current

Direction	Rate (Mph)	Time (Hours)	Distance (Miles)
Upstream	$x - y$	2	20
Downstream	$x + y$	2	32

We have two equations, one for the distance upstream and one for the distance downstream:

Rate · Time = Distance

$2(x - y) = 20$ Distance upstream

$2(x + y) = 32$ Distance downstream

Dividing each side of both equations by 2 yields the system

$x - y = 10$

$x + y = 16$

Adding the two equations to eliminate y gives us

$2x = 26$

$x = 13$

Since $x + y = 16$ and $x = 13$, by substitution $y = 3$. The rate of the boat in still water is 13 mph and the rate of the current is 3 mph.

CHECK If the rate of the boat in still water is 13 mph, it will travel upstream at 10 mph and downstream at 16 mph. This means that it will take 2 hours to go 20 miles upstream or 32 miles downstream. The solution checks.

EXAMPLE 5 The owner of the Sweets Candy Shop find that chocolates priced at $5/lb are not selling well and decides to mix them with some other candy that sells for $2.50/lb. The mixture will sell for $4 lb. *How many pounds of each should she use if she wants to have 20 pounds of the mixture?*

SOLUTION We let $x=$ the number of pounds of the $5 chocolates
$y=$ the number of pounds of the $2.50 candy

There are to be 20 pounds of the mixture, so one equation is

$$x + y = 20$$

She hopes to make the same amount of money for the mixture [20($4) = $80] as she would have had she sold the candies separately. In other words,

(x lb of $5 candy) · price/lb + ($y$ lb of $2.50 candy) · price/lb = $80

or

$$5x + 2.5y = 80$$

The two desired equations are

I $x + y = 20$
II $5x + 2.5y = 80$

We solve the first equation for y and substitute the result into the second one:

I $y = 20 - x$
II $5x + 2.5(20 - x) = 80$ Substitute $20 - x$ for y in **II**.
$5x + 50 - 2.5x = 80$ Distributive property
$2.5x + 50 = 80$ Combine like terms.
$2.5x = 30$ Subtract 50 from each side.
$x = 12$ Divide each side by 2.5.

The mixture should contain 12 lb of the $5 chocolates and 8 lb of the $2.50 candy. The check is left to the reader.

EXAMPLE 6 In business, *equilibrium* is reached when the demand for an item equals the supply. As prices increase, demand usually decreases. Suppose that the demand for an item is given by $D = 8000 - 3p$, where p is the price of the item in dollars, and that the supply is given by $S = 400p - 60$. Since equilibrium means the demand is equal to the supply, we can substitute S for D:

$D = 8000 - 3p$
$400p - 60 = 8000 - 3p$ Substitute $400p - 60$ for D.
$403p = 8060$ Add $3p + 60$ to each side.
$p = 20$ Divide each side by 403.

Thus equilibrium will be reached at a price of $20/item.

EXAMPLE 7 Erin has $30,000 to invest in securities and certificates of deposit (CDs). Because of her concern for the safety of her investments, she plans to invest no more than $10,000 in securities. If x is the amount she will invest in CDs and y represents the amount she will invest in securities, write and graph a system of inequalities that describes her options.

SOLUTION   Four inequalities are

$$x + y \leq 30,000 \quad \text{She has a maximum of \$30,000 to invest.}$$
$$y \leq 10,000 \quad \text{She will invest no more than \$10,000 in securities.}$$
$$x \geq 0 \quad \text{She may or may not invest in CDs.}$$
$$y \geq 0 \quad \text{She may or may not invest in securities.}$$

The graph is shown in Figure 5.19. Units on the axes are in thousands of dollars.

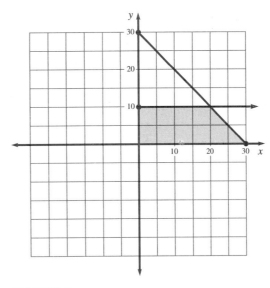

FIGURE 5.19

Exercise Set 5.5

Solve each of the following applied problems using systems of equations or inequalities.

1. Two consecutive odd numbers have a sum of 168. Find the numbers.
2. One number is five times as large as another number, and their difference is 48. Find the numbers.
3. Two motorcyclists start from positions 400 km apart and travel toward each other; they meet in 4 hr. Find the average speed of each if one travels 20 km/hr faster than the other.
4. Two canoeists on a 30-mi trip averaged 5 mph until they tired and slowed to 3 mph. If the trip took 8 hr, how far did they paddle at each speed?
5. Pam's Nut House wants a mixture of 40 lb of nuts to be worth \$1.90/lb. How many pounds of nuts that are worth \$1.80/lb should be mixed with nuts worth \$2.20/lb to accomplish this?
6. Wayne invested part of his money at 8% and the rest at 6%. The income from both investments totaled \$2,000. If he interchanged his investments, his income would increase by \$340. How much did he have in each investment?
7. In a chemistry experiment, pure alcohol and distilled water are mixed to produce 100 centiliters (cl) of solution. If 20 cl more water than alcohol are used, how many centiliters of each are used?
8. During the last student body election, the winner received 85 more votes than his opponent. If there were 877 votes cast, how many votes did each candidate receive?

9. Brian's Energy Store sent two skilled and three unskilled workers to install a solar heating system on a swimming pool. The labor costs were $300. The next day the store sent three skilled and two unskilled workers to install a similar system at a labor cost of $375. What is the cost for each type of worker?

10. If a guitar string 44 cm long is divided into two lengths so that five times one length equals six times the other length, a minor chord will be played when all the strings are plucked. Find the two lengths.

11. How much fencing is needed to enclose a rectangular field if it is three times as long as it is wide and also three times its length exceeds four times its width by 27 m?

12. A radiator holds 12 qt of liquid. How many quarts of pure antifreeze and how many quarts of 40% antifreeze solution must be mixed to fill the radiator with a 50% antifreeze solution?

13. Blake and Nell are 100 km apart and are driving at constant speeds. If they drive toward each other, they will meet in 1 hr. If they drive in the same direction, he will overtake her in 5 hr. Find their speeds.

14. Tamra's Tire Company produces two grades of small tire retreads: the lower grade takes 4 lb of rubber and 2 hr of labor to produce at a production cost of $14, while the higher grade requires 5 lb of rubber and 3 hr of labor at a production cost of $20. What is the cost of a pound of rubber? What is the hourly labor cost?

15. Rachel has a grade point average (GPA) of 2.1, which she has acquired in two semesters of 15 hr each. Grade points are determined as follows: A, 4 points; B, 3 points; C, 2 points; and D, 1 point. During the next two semesters (30 hr of work), Rachel wishes to bring her GPA up to a 2.5. She is sure of an A in two 3-hr courses. How many hours of B and C work does she need in the remaining 24 hr of work to bring her GPA up to a 2.5?

16. Antonio has a GPA of 2.8 for his first semester (15 hr of work). He hopes to raise his GPA to a 3.1 during the second semester (15 hr of work). How many gradepoints must he earn to have a 3.1 GPA for his first two semesters of work?

17. The sum of the voltages in two electric circuits is 120 volts (V). If the polarity of one is reversed (subtracted), the total is 36 V. What is the voltage in each circuit?

18. In an electric circuit consisting of two branches, the current in one branch is 5 amps more than in the other branch. When the two branches are joined, the total current is 18 amps. Find the current in each branch.

19. A tea merchant sells one kind of tea for $1.60/lb and another kind for $2.10/lb. He wishes to make a mixture that will sell for $2/lb. How many pounds of each kind should he use to make 50 lb of the mixture?

20. In 1980 the United States census showed that Las Vegas, Nevada, and Cape Coral, Florida, together grew 163%. The difference in their percentage increase was 25%, Cape Coral having the larger increase. What was the percentage increase for each city?

21. In business, equilibrium has been reached when the supply function S equals the demand function, D. If the supply function for a certain product is given by $S = 2a$ and the demand function for the product is given by $D = 9000 - a$, then for what value of a will equilibrium exist?

22. In business, the break-even point occurs where the revenue function, R, intersects the cost function, C. The cost function for producing a certain kind of tennis shoes is $C(n) = 8n + 9600$, where n is the number of pairs sold. If the selling price for the shoes is $14 a pair, then $R(n) = 14n$. Find the number of pairs of shoes that must be sold in order to break even.

23. The Wests make a substantial down payment on a new house, leaving a balance of $39,000. They borrow some money at 10% and the remaining part at 13%. How much money do they borrow at each rate if they pay $4470 interest the first year?

24. An airplane travels 1500 km in 5 hr with a tailwind. If the return trip against the wind takes 6 hr, find the speed of the plane in still air.

25. Each ounce of a special diet food (A) supplies 5% of the nutrition a person needs. Each ounce of another diet food (B) supplies 12% of the nutrition a person needs. If digestive restrictions requires that the ratio of A to B must be 3 to 5, or $5A = 3B$, how many ounces of each diet food should be used to provide 100% of the required nutrition?

26. From 1960 to 1990, the percentage increase in doctor's fees plus the percentage increase in charges for a semiprivate hospital room was 1463%. The difference in those increases was 781%. Find the percentage increase for each if hospital rooms had the larger increase.

27. The percentage increase in the Consumer Price Index (CPI) plus the percentage increase of health expenditures for the period 1960 to 1990 was 1892%. The difference in the increases was 1488%. Find the percentage increase for each, if the increase was larger for health expenditures. Find the average yearly percentage increase for each.

28. One dozen fritters and 3 dozen donuts cost $8.06. Two dozen donuts and 6 dozen fritters cost $16.52. How much is a dozen donuts?

29. The longest side of a triangle is 18 m, and the shortest side is 4 m less than the middle-sized side. If the perimeter is 38 m, find the length of the two shortest sides.

30. Emilio bought 4 lb of mixed peanuts and cashews for $13.16. If peanuts are $1.19/lb and cashews are $3.99/lb, how many pounds of each did he buy?

31. One tuna sandwich and two chicken sandwiches together cost $10.97. Three tuna sandwiches and one chicken sandwich together cost $12.96. How much is a tuna sandwich?

32. Melanie invested her savings into two accounts that pay 11% and 14%. If her total interest for one year is $7650 and she has invested $60,000, how much has she invested in each account?

33. The perimeter of a rectangle is 54 cm. If the length is 6 cm longer than twice the width, what are the dimensions?

34. Four times a number is 6 more than three times a larger number. If the difference of the numbers is 3, find the numbers.

35. Elena borrowed some money at 16% interest and some more at 12% interest. She paid $104 interest for one year on both loans. If the amount she borrowed at 12% was three times the amount she borrowed at 16%, how much did she borrow at each rate?

36. Mr. Toh makes and sells x tennis rackets at a weekly cost of $C(x) = 25x + 150$. They sell for $R(x) = 50x$. Find the break-even point. Find the profit if 10 rackets are made and sold.

37. The supply and demand equations for x quarts of ice cream in Morro Bay are given by

    Supply: $p = 0.1x + 3.5$
    Demand: $p = -0.8x + 8$

    Find the equilibrium point (where the supply and demand equations intersect).

38. Jim has at most $20,000 available to invest in stocks and bonds, and he wants to invest no more than $8,000 in stocks.
    (a) Write the system of inequalities indicated.
    (b) Graph the system of inequalities.
    (c) From the graph, find one possible set of investments.

39. Geri has up to $10,000 to invest in CDs and notes, and she wants to invest at least $3000 in CDs.
    (a) Write the system of inequalities indicated.
    (b) Graph the system of inequalities.
    (c) From the graph, find one possible set of investments.

40. At Bernard College, students may enroll in no more than five lecture courses and at most three lab courses in a special session. The number of lab courses can be no more than the number of lecture courses.
    (a) Write the system of inequalities indicated.
    (b) Graph the system of inequalities.
    (c) From the graph, find one possible class schedule.

41. Marvin can work at most 10 hr/day. At least four of the hours must be in the morning and at least three of them in the afternoon.
    (a) Write the system of inequalities indicated.
    (b) Graph the system of inequalities.
    (c) From the graph, find one possible work schedule.

SUMMARY

5.1
Solving Systems of Linear Equations by the Graphing Method

Two or more linear equations being considered at the same time are referred to as a **system of linear equations.** A system of linear equations can have a single solution, no solutions, or an infinite number of solutions, depending on whether their graphs (which are lines) intersect at a single point, are parallel, or coincide, respectively. If the two lines intersect at a single point, the solution set consists of a single ordered pair and the system is said to be **independent.** If the two lines are parallel, there are no points that satisfy both equations at the same time. Such a system is said to be **inconsistent** and its solution set is the empty set. In a **dependent system,** the two lines coincide, so any point that lies on one line also lies on the other. The solution set contains an infinite number of points.

When the two equations in a linear system are written in slope–intercept form, $y = mx + b$, it is easy to classify them as independent, dependent, or inconsistent depending on their slopes and y-intercepts.

Slopes not equal:	Independent
Slopes equal and y-intercepts equal:	Dependent
Slopes equal and y-intercepts not equal:	Inconsistent

5.2 Solving Systems of Linear Equations by the Elimination Method

A modified form of the **addition property of equality** is used in solving systems of linear equations. It states:

If A, B, and C represent algebraic expressions such that

$$A = B \quad \text{and} \quad C = D$$

then

$$A + C = B + D$$

The **elimination method** for solving a system of linear equations consists of five steps:

Step 1 Write both equations in the form $ax + by = c$ if not already in that form.
Step 2 If necessary, multiply one or both equations by appropriate constants so that the coefficients of one of the variables are additive inverses of each other.
Step 3 Add the two equations to eliminate one variable. Solve for the remaining variable.
Step 4 Substitute the result of Step 3 into either of the original equations and solve for the remaining variable.
Step 5 Check the solution in *both equations* of the original system.

If the elimination method yields a false statement, such as $0 = 4$, then the system is inconsistent. When an equation like $0 = 0$ results, the system is dependent.

5.3 Solving Systems of Linear Equations by the Substitution Method

The **substitution property of equality,** used for solving a system of linear equations, states:

Any algebraic expression may be substituted for its equal in an equation. Thus if A, B, and C represent algebraic expressions, and if

$$A + B = C \quad \text{and} \quad B = D$$

then

$$A + D = C$$

The **substitution method** for solving a system of linear equations consists of five steps:

Step 1 Solve one of the equations for either variable, if necessary.
Step 2 Substitute the result from Step 1 into the other equation.
Step 3 Solve the equation obtained in Step 2.
Step 4 Substitute the solution from Step 3 into the equation found in Step 1, and solve for the variable.
Step 5 Check your results in *both* of the original equations.

If, when solving a system of equations, a true statement such as 6 = 6 results, the system is dependent. If a false statement such as 0 = 3 results, the system is inconsistent.

5.4 Solving Systems of Linear Inequalities

Systems of linear inequalities are solved by graphical means. Each linear inequality is graphed by shading the area that forms the solution set. The region that satisfies the system is the intersection of the solution sets of each inequality.

5.5 More Applications

Systems of linear equations and inequalities are very useful in solving applied problems. The five-step process outlined in Section 3.8 makes the job easier.

COOPERATIVE EXERCISE

In an introductory architecture class, the students decide to use the given design of a swimming pool to calculate the price (excluding tax) of building the pool and filling it with water. They are given the following information:

1. The dimensions of the outside edge of the pool are given.
2. The soil will have to be excavated and hauled away.
3. A 6-in. layer of concrete is to be spread over the inside of the pool and on a 5-ft walkway surrounding the pool.
4. The inside of the pool and the walkway are to be tiled.
5. The pool is to be filled with water to a depth of 6 in. from the top.

Group I is to calculate the price of excavating and hauling away the soil. The price of excavating the soil is $27/yd³. The price of hauling away the soil is $18/yd³.

Group II is to calculate the price of having all of the concrete work done. The cost is $2.60/ft² (6 in. thick).

Group III is to calculate the price of having all of the tile work done at a cost of $4.90/ft².

Group IV is to calculate the price of filling the pool with water, which costs $2.80/100 ft³. Then they are to combine this with all of the previous prices to find the total cost of building and filling the pool.

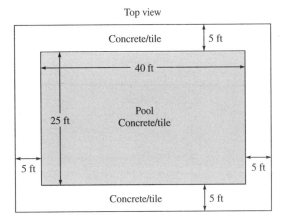

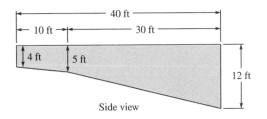

Review Exercises

If possible, solve each system of equations by the method indicated. If the system is not independent, state whether it is dependent or inconsistent.

1. $\left.\begin{array}{r} 3x - y = 1 \\ 2x - 3y = -4 \end{array}\right\}$ graphing

2. $\left.\begin{array}{r} 2x + y = 3 \\ x - 2y = -1 \end{array}\right\}$ graphing

3. $\left.\begin{array}{r} x = -3y \\ 3y = 2x \end{array}\right\}$ graphing

4. $\left.\begin{array}{r} x = 4 \\ y = -2 \end{array}\right\}$ graphing

5. $\left.\begin{array}{r} 3x - 2y = 7 \\ -5y = 4 + 6x \end{array}\right\}$ elimination

6. $\left.\begin{array}{r} 3x = 7 - 5y \\ x - 6y = -13 \end{array}\right\}$ elimination

7. $\left.\begin{array}{r} 5y = 4x + 10 \\ 3y = -4x - 34 \end{array}\right\}$ elimination

8. $\left.\begin{array}{r} y = -1 - \frac{1}{4}x \\ y = -\frac{1}{4}x - 1 \end{array}\right\}$ elimination

9. $\left.\begin{array}{r} 7x - 3y = 4 \\ 14x - 6y = 7 \end{array}\right\}$ elimination

10. $\left.\begin{array}{r} 5(y+1) + 1 = 3(x+1) \\ 6y + 2(3x - 4) = y - 7 \end{array}\right\}$ elimination

11. $\left.\begin{array}{r} x = \frac{2}{3} - 4y \\ \frac{3}{4}x = \frac{9}{4}y - \frac{11}{16} \end{array}\right\}$ substitution

12. $\left.\begin{array}{r} -2x + \frac{9}{2}y = -1 \\ 4x - 9y = -2 \end{array}\right\}$ substitution

13. $\left.\begin{array}{r} x + \frac{3}{2}y = \frac{1}{2} \\ 2x + 3y = \frac{1}{6} \end{array}\right\}$ substitution

If possible, solve each system of equations by any method. If the system is not independent, state whether it is dependent or inconsistent.

14. $\begin{array}{r} 5x + y = -13 \\ y = -2x - 1 \end{array}$

15. $\begin{array}{r} y = \dfrac{x+3}{4} \\ x + y = 7 \end{array}$

16. $\begin{array}{r} 7x - 4y = 13 \\ 7x - 6y = 11 \end{array}$

17. $\begin{array}{r} x - y = -0.1 \\ 3x + 5y = 8.5 \end{array}$

18. $\begin{array}{r} \dfrac{x}{2} - \dfrac{y}{7} = -\dfrac{1}{42} \\ x + y = 9\dfrac{1}{6} \end{array}$

19. $\begin{array}{r} x - 5y = 0.65 \\ -0.9x + y = 0.01 \end{array}$

20. $\begin{array}{r} x = \dfrac{-4 - 3y}{2} \\ 3x - 4y = 11 \end{array}$

21. $\begin{array}{r} -15x + 6y = -7 \\ 15x - 12y = 1 \end{array}$

22. $\begin{array}{r} \dfrac{x}{5} + \dfrac{y}{7} = 1 \\ 0.5x + 0.5y = 5.5 \end{array}$

23. $\begin{array}{r} x - y = -3 \\ 9x - 7y = -63 \end{array}$

24. $\begin{array}{r} 3(x - 1) = 2y - 5 \\ 2(3x - 4) - 3(1 - 2y) = -1 \end{array}$

25. $\begin{array}{r} \dfrac{4x}{19} - \dfrac{9y}{19} = 1 \\ \dfrac{2x}{3} - \dfrac{3y}{2} = 2 \end{array}$

26. $\begin{array}{r} 4.5x - 0.6y = 0.1 \\ \dfrac{5x}{3} - \dfrac{2y}{9} = \dfrac{1}{27} \end{array}$

Solve each of the following systems of inequalities by the graphing method.

27. $\begin{array}{r} x \geq -2 \\ y \geq 0 \end{array}$

28. $\begin{array}{r} x + y > 3 \\ 6x - 5y > 30 \end{array}$

29. $\begin{array}{r} 3x + 2y \geq -6 \\ x - 2y > 2 \end{array}$

30. $\begin{array}{r} x \geq 0 \\ y \geq 0 \end{array}$

31. $\begin{array}{r} x - 3y < 3 \\ -2x + 6y > 6 \end{array}$

32. $\begin{array}{r} -2x + y \leq 4 \\ x \geq 0 \\ y \geq 0 \end{array}$

33. $\begin{array}{r} 3x + 2y \leq 6 \\ y \geq -2 \end{array}$

34. $\begin{array}{r} 2x - 5y \leq -10 \\ 7x + 5y \geq -35 \end{array}$

Find the solution to each system of equations by inspection. Write the equations of the two lines and solve the system by the elimination method or the substitution method. Compare the solutions found by inspection to those found by elimination or substitution.

35.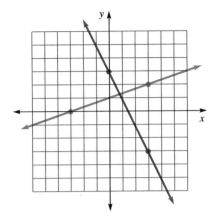

36.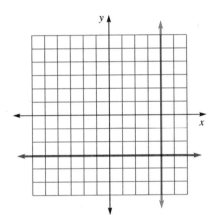

37.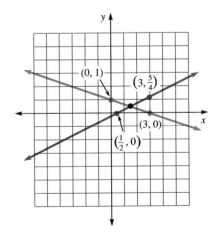

38.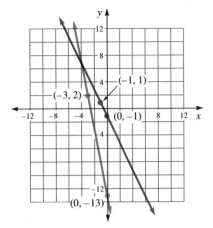

39.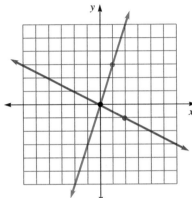

40.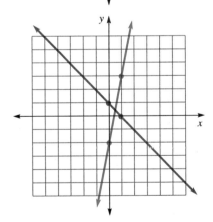

Use two variables to write a system of equations or inequalities and solve by any method.

41. Two negative numbers have a sum of -58 and their difference is -4. Find the numbers.

42. The perimeter of a rectangle is 110 m, and the length is $\frac{3}{2}$ times the width. Find the dimensions.

43. Mario has 39 coins (nickels and quarters) that have a total value of $4.15. Find the number of each coin.

44. Two angles are supplementary. If one angle is 80° more than the other, find the measures of the angles.
45. Two angles are complementary. If one angle is eight times the other, find the measures of the angles.
46. Two numbers have a sum less than 9, while their difference is no more than 3. Solve by graphing and list one possible set of numbers.
47. Carol wants to cut a rope that is 15 ft long into three pieces, and she needs one piece to be 4 ft long. Find the length of the other two pieces if one is 1 ft longer than the other.
48. Two angles have a sum that is less than or equal to 90°. The second angle subtracted from the first one is no more than 30°. Solve by graphing and give one set of possible angles.
49. Rex invested a total of $10,000 in two mutual funds. The first pays 7% interest and the second 12% interest. How much does Rex have invested in each fund if he earns a total of $900 interest in one year?
50. Plain chocolate candy sells for $4.50/lb and nut-filled chocolate candy sells for $6/lb. If Laura spent $26.25 for a 5-lb mixture of the two kinds of chocolate, how many pounds of each kind were in the mixture?
51. In the Atascadero Zoo, certain animals must receive a daily diet that contains at least 20 g of protein and at least 6 g of fat. One food contains 20% protein and 2% fat, while another food contains 10% protein and 6% fat. Write the system of inequalities and solve by graphing. Give one diet that will satisfy the necessary requirements.

Chapter Test

If possible, solve each system of equations by graphing. If the system is not independent, state whether it is dependent or inconsistent.

1. $x + y = 1$
   $x - y = 3$
2. $2x + y = 4$
   $x + y = -2$
3. $x + 3y = 6$
   $3x = 6$
4. $x + y = 0$
   $2x + 2y = 5$
5. $x - 2y = 4$
   $2x - 4y = 8$

If possible, solve each system of equations by the elimination method. If the system is not independent, state whether it is dependent or inconsistent.

6. $3x + y = 2$
   $y = -1$
7. $2x + y = 0$
   $3x - 2y = 7$
8. $2x - 3y = 5$
   $4x - 6y = 10$
9. $4x - y = 6$
   $3x + y = 1$
10. $2x - y = 3$
    $3x + 2y = 8$
11. $4x + 6y = 7$
    $2x + 3y = 1$
12. $3x - 5y = 19$
    $4x + 3y = 6$
13. $15x - 9y = -3$
    $5x - 3y = -1$
14. $3x + 4y = 0$
    $9x + 4y = -12$

Solve each independent system of equations by the substitution method.

15. $x = y + 1$
    $x + y = 2$
16. $x - 2y = 0$
    $x + 2y = 8$
17. $5x - y = 1$
    $3x + y = 7$
18. $2x + 3y = 3$
    $x = 4 - 5y$
19. $2x + 5y = 3$
    $x = 5 - 7y$
20. $5x - 7y = 2$
    $3x - y = 14$

Graph the region that satisfies each system of inequalities.

21. $x < y + 1$
    $y + x > 1$
22. $y \leq 2$
    $x \geq 3$
23. $x \geq 4$
    $x + y < 2$
24. $x - y \geq 2$
    $5x + 3y \leq 2$
25. $3x + 4y > 2$
    $-2x - 5y < -13$
26. $x \geq y$
    $x - 3y < 2$
27. $3x - 5y \leq 15$
    $2x + 3y \geq 12$
    $x < 10$
    $y > 0$
28. $x + 3y \geq 3$
    $3x + y \leq 9$
    $x > 0$
    $y > 0$

Solve the following applied problems by using a system of equations or inequalities.

29. One number is 73 less than another. If their sum is 1899, what are the numbers?
30. The total value of a collection of dimes and quarters is $4.35. If there are 19 more dimes than quarters, how many are there of each?
31. David took a trip of 580 km, part way by bus at an average speed of 40 km/hr and the rest by train at an average speed of 76 km/hr. If his total travel time was 10 hr, how far did he travel by each method?

32. Cindy has $12,000 invested, part at 5% interest and the rest at 7% interest. If she receives the same annual income from each investment, how much is invested at each rate?

33. Susan is planning a meal consisting of a main dish and a fruit plate. She wants the meal to supply at least 1000 mg of calcium and at least 30 mg of protein. Each serving of the main dish supplies 100 mg of calcium and 5 mg of protein, while each serving of the fruit plate supplies 200 mg of calcium and 3 mg of protein.

  (a) Write the system of inequalities indicated.

  (b) Graph the system of inequalities.

  (c) From the graph, find one set of possible servings that would satisfy the requirement.

CHAPTER 6
POLYNOMIALS AND THE LAWS OF EXPONENTS

6.1 Multiplication and Division with Exponents
6.2 More on Exponents; Scientific Notation
6.3 Adding and Subtracting Polynomials
6.4 Multiplying Polynomials
6.5 Dividing Polynomials
6.6 More Applications

APPLICATION

Old Fashioned Chocolates makes and sells x cartons of chocolate candy per month. The monthly cost C (in dollars) is given by the formula $C(x) = 8.5x + 250$. The monthly revenue R (in dollars) is given by $R(x) = -0.10x^2 + 110x$. Find the profit function $P(x)$, where $P(x) = R(x) - C(x)$. Find the profit when $x = 10$; when $x = 20$.

Exercise Set 6.3, Exercise 105

A scientist worthy of the name, especially a mathematician, experiences in his work the same excitement as an artist; his pleasure is as great and even greater when he makes a new discovery or comes up with a new idea.

HENRI POINCARÉ (1854–1912)

6.1 Multiplication and Division with Exponents

HISTORICAL COMMENT   René Descartes, although best known for his development of the Cartesian coordinate system, also contributed significantly to modern mathematical notation. One of his contributions is the notation that we use for writing exponents. Before his time, there were numerous schemes for indicating the power of a number. In 1484 Chuquet used 3^0, 3^1, 3^2 to mean 3, $3x$, $3x^2$, and in 1572 Bombelli used $\overset{0}{3}$, $\overset{1}{3}$, $\overset{2}{3}$ to mean the same thing. Descartes used x, xx, x^3, and x^4 to indicate the first four powers of x. The symbol xx for x^2 was used for many years; Gauss (1777–1855) continued to argue for the notation in about 1800, saying that xx took no more space than x^2 did.

OBJECTIVES   In this section we will learn to

1. use the product rule for exponents;
2. use the quotient rule for exponents; and
3. simplify expressions involving both the product and quotient rules for exponents.

In Chapter 1 we used exponents to write repeated products, such as

$$6^2 = 6 \cdot 6 = 36 \quad \text{and} \quad 2^4 = 2 \cdot 2 \cdot 2 \cdot 2 = 16$$

Recall that in the expression 6^2, the 6 is called the **base** and the 2 is called the **exponent.**

When the base is a variable and the exponent a natural number, the exponent still indicates the number of times the base is used as a factor. Thus the exponent 2 in b^2 indicates that there are two factors of b, and the exponent 3 in b^3 indicates that there are three factors of b. To find the product of b^2 and b^3, we proceed as follows:

$$b^2 \cdot b^3 = \underbrace{(b \cdot b)}_{\text{Two factors}}\underbrace{(b \cdot b \cdot b)}_{\text{Three factors}}$$
$$= \underbrace{(b \cdot b \cdot b \cdot b \cdot b)}_{\text{Five factors of } b} \quad \text{Associative property}$$
$$= b^5$$

This example shows that when the bases are the same, the exponent of the product is the sum of the exponents of the factors, or

$$b^2 \cdot b^3 = b^{2+3} = b^5$$

This generalization is called the *product rule for exponents*.

The Product Rule for Exponents

If m and n represent positive integers, then

$$b^m \cdot b^n = b^{m+n}$$

EXAMPLE 1 Find the product of x^7 and x^{10}.

SOLUTION $x^7 \cdot x^{10} = x^{7+10} = x^{17}$ $\qquad b^m \cdot b^n = b^{m+n}$

EXAMPLE 2 Find the product of a^2, a^3, and a^5.

SOLUTION $a^2 \cdot a^3 \cdot a^5 = a^{2+3+5} = a^{10}$

EXAMPLE 3 $x^2 \cdot x^7 \cdot x^9 \cdot x^{13} = x^{2+7+9+13} = x^{31}$

EXAMPLE 4 Find the product of $2x^3$ and $5x^4$.

SOLUTION
$(2x^3)(5x^4) = (2 \cdot 5)(x^3 \cdot x^4)$ $\qquad$ Commutative and associative properties
$= 10x^{3+4}$ $\qquad b^m \cdot b^n = b^{m+n}$
$= 10x^7$

EXAMPLE 5 Find the product of $(9x^2)(4x^3)(2x^7)$.

SOLUTION
$(9x^2)(4x^3)(2x^7) = (9 \cdot 4 \cdot 2)(x^2 \cdot x^3 \cdot x^7)$
$= 72x^{2+3+7}$
$= 72x^{12}$

CAUTION

The product rule for exponents applies only when the bases are the same. It is impossible to multiply a^2 times b^2, because the bases are not the same. The product of a^2 and b^2 is written simply as a^2b^2.

The rule for division of exponential numbers is obtained in much the same manner as that for multiplication. Consider the quotient $\dfrac{b^6}{b^4}$:

$$\frac{b^6}{b^4} = \frac{\not{b} \cdot \not{b} \cdot \not{b} \cdot \not{b} \cdot b \cdot b}{\not{b} \cdot \not{b} \cdot \not{b} \cdot \not{b}}$$

$$= b \cdot b$$
$$= b^2$$

This example illustrates that the exponent of the quotient is the difference of the exponents in the numerator and the denominator.

$$\frac{b^6}{b^4} = b^{6-4} = b^2$$

EXAMPLE 6 Find each of the following quotients: (a) $\dfrac{x^9}{x^3}$ (b) $\dfrac{y^{11}}{y^6}$ (c) $\dfrac{m^{75}}{m^{16}}$

SOLUTION (a) $\dfrac{x^9}{x^3} = x^{9-3} = x^6$

(b) $\dfrac{y^{11}}{y^6} = y^{11-6} = y^5$

(c) $\dfrac{m^{75}}{m^{16}} = m^{75-16} = m^{59}$

The Quotient Rule for Exponents

If m and n represent positive integers and b is any real number except zero, then

$$\frac{b^m}{b^n} = b^{m-n}$$

In each of the parts of Example 6, notice that the exponent in the numerator is greater than the exponent in the denominator. There are two other possibilities. One is that the exponents in the numerator and denominator are equal; then the quotient is 1.

$$\frac{b^4}{b^4} = \frac{\cancel{b} \cdot \cancel{b} \cdot \cancel{b} \cdot \cancel{b}}{\cancel{b} \cdot \cancel{b} \cdot \cancel{b} \cdot \cancel{b}} = 1$$

If we apply the quotient rule and subtract exponents, we have

$$\frac{b^4}{b^4} = b^{4-4} = b^0$$

For consistency, we must define b^0 as

$$b^0 = 1, \quad b \neq 0$$

The other possibility is that the exponent in the denominator is greater than the one in the numerator. For example,

$$\frac{b^4}{b^6} = \frac{\cancel{b} \cdot \cancel{b} \cdot \cancel{b} \cdot \cancel{b}}{\cancel{b} \cdot \cancel{b} \cdot \cancel{b} \cdot \cancel{b} \cdot b \cdot b} = \frac{1}{b^2}$$

When exponents are subtracted, we have

$$\frac{b^4}{b^6} = b^{4-6} = b^{-2}$$

This means that we must define b^{-2} as

$$b^{-2} = \frac{1}{b^2}$$

DEFINITION	For any nonpositive integer n and any base $b \neq 0$,
Nonpositive Exponents	$b^{-n} = \dfrac{1}{b^n}$ and $b^0 = 1$

EXAMPLE 7 Find each of the following quotients: (a) $\dfrac{x^7}{x^4}$ (b) $\dfrac{x^8}{x^8}$ (c) $\dfrac{x^4}{x^5}$.

(a) $\dfrac{x^7}{x^4} = x^{7-4} = x^3 \qquad \dfrac{b^m}{b^n} = b^{m-n}$

(b) $\dfrac{x^8}{x^8} = x^{8-8} = x^0 = 1$

(c) $\dfrac{x^4}{x^5} = x^{4-5} = x^{-1} = \dfrac{1}{x}$

EXAMPLE 8 Simplify by using the rules for exponents:

(a) 5^0 (b) $\left(\dfrac{1}{2}\right)^{-2}$ (c) $\dfrac{3^2}{3^4}$ (d) $x^2 \cdot x^{-5}$ (e) $(x+y)^2(x+y)^4$

SOLUTION (a) $5^0 = 1$

(b) $\left(\dfrac{1}{2}\right)^{-2} = \dfrac{1}{\left(\dfrac{1}{2}\right)^2} = \dfrac{1}{\dfrac{1}{4}} = 1 \cdot \dfrac{4}{1} = 4$

(c) $\dfrac{3^2}{3^4} = 3^{2-4} = 3^{-2} = \dfrac{1}{3^2} = \dfrac{1}{9}$

(d) $x^2 \cdot x^{-5} = x^{2+(-5)} = x^{-3} = \dfrac{1}{x^3}$

(e) $(x+y)^2(x+y)^4 = (x+y)^{2+4}$ The base is $x + y$.
$= (x+y)^6$

EXAMPLE 9 Simplify: $2^{-2} + 3^{-1}$.

SOLUTION We begin by writing each term as a fraction without negative exponents.

$$2^{-2} + 3^{-1} = \frac{1}{2^2} + \frac{1}{3}$$

$$= \frac{1}{4} + \frac{1}{3}$$

$$= \frac{3}{12} + \frac{4}{12} \quad \text{The LCD is 12.}$$

$$= \frac{7}{12}$$

EXAMPLE 10 Simplify: $(3x^2y^3)\left(\frac{-2x^2y^4}{x^3}\right)$.

SOLUTION

$$(3x^2y^3)\left(\frac{-2x^2y^4}{x^3}\right) = \frac{(3x^2y^3)(-2x^2y^4)}{x^3} \qquad 3x^2y^3 = \frac{3x^2y^3}{1}$$

$$= \frac{(3)(-2)(x^2 \cdot x^2)(y^3 \cdot y^4)}{x^3} \qquad \text{Commutative and associative properties}$$

$$= \frac{-6x^4y^7}{x^3} \qquad b^m \cdot b^n = b^{m+n}$$

$$= -6x^{4-3}y^7 \qquad \frac{b^m}{b^n} = b^{m-n}$$

$$= -6xy^7$$

EXAMPLE 11 Simplify: $5^{-2} \cdot 5^4 \cdot 5^7 \cdot 5^{-9}$.

SOLUTION

$$5^{-2} \cdot 5^4 \cdot 5^7 \cdot 5^{-9} = 5^{-2+4+7+(-9)} \qquad b^m \cdot b^n = b^{m+n}$$

$$= 5^0 = 1$$

EXAMPLE 12 Simplify $\frac{-24y^{-2}}{-6y^{-5}}$ and then evaluate for $y = -4$.

SOLUTION

$$\frac{-24y^{-2}}{-6y^{-5}} = \frac{-24}{-6}y^{-2-(-5)} \qquad \frac{b^m}{b^n} = b^{m-n}$$

$$= 4y^{-2+5}$$

$$= 4y^3$$

When $y = -4$,

$$4y^3 = 4(-4)^3$$

$$= 4(-64) \qquad (-4)^3 = (-4)(-4)(-4) = -64$$

$$= -256$$

An exponential expression is considered to be in **simplest terms** when the final result is free of negative exponents.

EXAMPLE 13 Simplify: $\frac{(a-b)^7}{(a-b)^9}$.

SOLUTION

$$\frac{(a-b)^7}{(a-b)^9} = (a-b)^{7-9} \qquad \frac{b^m}{b^n} = b^{m-n}$$

$$= (a-b)^{-2}$$

$$= \frac{1}{(a-b)^2} \qquad b^{-n} = \frac{1}{b^n}$$

Calculator Note

Calculators can be used to evaluate expressions involving exponents. The sequence of keys depends on the make and model. Many calculators have a $\boxed{y^x}$ or an $\boxed{x^y}$ key for finding powers of numbers, while others use the $\boxed{\wedge}$ key to indicate that an exponent follows. To evaluate $(2.38)^3$, one of the following keystroke sequences applies.

$\boxed{2.38}$ $\boxed{y^x}$ $\boxed{3}$ $\boxed{=}$ The display reads 13.481272.

$\boxed{2.38}$ $\boxed{\wedge}$ $\boxed{3}$ $\boxed{\text{ENTER}}$ The display reads 13.481272.

Exercise Set 6.1

Identify the base and exponent in each of the following powers.

1. x^6
2. 5^{11}
3. -2^5
4. $(-2)^5$
5. $3m^6$
6. $(3m)^6$
7. $-(-2)^7$
8. $-(-a)^7$

Write each expression using exponents.

9. $5 \cdot 5$
10. $7 \cdot a \cdot a \cdot a$
11. $2 \cdot 2 \cdot 2 \cdot 2$
12. $4 \cdot 4 \cdot 4 \cdot 4$
13. $\frac{1}{a \cdot a \cdot a}$
14. $\frac{2}{3 \cdot 3 \cdot 3}$
15. $-(2 \cdot 2 \cdot 2)$
16. $(-3)(-3)(-3)$
17. $x \cdot x \cdot x \cdot x \cdot x$
18. $-y \cdot y \cdot y$
19. $-x \cdot x$
20. $(ab)(ab)$
21. $\frac{r}{s} \cdot \frac{r}{s} \cdot \frac{r}{s}$
22. $-xxxyy$
23. $-x \cdot y + x \cdot x \cdot y$

Find the value of each expression.

24. 6^4
25. 5^3
26. $(-5)^4$
27. -4^7
28. $\left(\frac{1}{2}\right)^0$
29. $\left(\frac{1}{3}\right)^0$
30. $\left(\frac{2}{3}\right)^{-1}$
31. $\left(\frac{1}{4}\right)^{-1}$
32. 2^{-1}
33. 4^{-3}
34. $\left(\frac{1}{2}\right)^{-1}$
35. 2^{-3}
36. 3^{-2}
37. -3^{-3}
38. $\left(\frac{1}{2}\right)^{-2}$
39. $\left(\frac{1}{4}\right)^{-2}$
40. $\frac{1}{4^{-1}}$
41. $\frac{3}{2^{-2}}$
42. $\frac{5}{10^2}$
43. $\frac{2}{4^2}$
44. $(0.1)^{-1}$
45. $(0.2)^{-1}$
46. $(0.01)^{-1}$
47. $(0.25)^{-1}$
48. $\frac{1}{(-5)^{-2}}$
49. $\frac{2}{-(-3)^{-3}}$
50. $\frac{3}{-(-2)^{-3}}$
51. $2 \cdot 4^{-1}$
52. $5 \cdot 10^{-2}$
53. $\frac{-4}{\left(\frac{1}{2}\right)^{-2}}$
54. $\frac{-6}{\left(\frac{1}{3}\right)^{-2}}$

Simplify each expression. Write your answers using only positive exponents. Assume no variable equals zero.

55. $2^2 \cdot 2^6$
56. $3^3 \cdot 3^8$
57. $a^2 \cdot a^3 \cdot a^7$
58. $a^4 \cdot a^3 \cdot a^5$
59. $a^3 \cdot a^{-2} \cdot a^8$
60. $b^3 \cdot b^{-3} \cdot b^6$
61. $x^2 \cdot x^3 \cdot x^{-5} \cdot x^6$
62. $y \cdot y^4 \cdot y^{-6} \cdot y^2$
63. $y^{-5} \cdot y^4 \cdot y^{-3} \cdot y^3$
64. $p^{-7} \cdot p^2 \cdot p^3 \cdot p$

65. $3^{-9} \cdot 3^2 \cdot 3^4 \cdot 3^3$
66. $2^{-11} \cdot 2^7 \cdot 2^3 \cdot 2$
67. $5^{-8} \cdot 5^4 \cdot 5 \cdot 5^2$
68. $8^{-5} \cdot 8 \cdot 8^{-1} \cdot 8^4$
69. $(0.3)^3 (0.3)^{-4} (0.3)^{-1}$
70. $(0.7)^2 (0.7)^{-1} (0.7)^{-3}$
71. $\dfrac{x^7}{x^2}$
72. $\dfrac{x^{10}}{x^3}$
73. $\dfrac{x^3}{x^{11}}$
74. $\dfrac{2^7}{2^4}$
75. $\dfrac{3^8}{3^{10}}$
76. $\dfrac{2^6}{2^{-2}}$
77. $\dfrac{3^{-3}}{3^{-4}}$
78. $\dfrac{a^{-7}}{a^{-4}}$
79. $\dfrac{\left(\frac{1}{2}\right)^3}{\left(\frac{1}{2}\right)^4}$
80. $\dfrac{\left(\frac{1}{3}\right)^5}{\left(\frac{1}{3}\right)^6}$
81. $\dfrac{(0.5)^4}{(0.5)^3}$
82. $\dfrac{(0.1)^5}{(0.1)^4}$
83. $\dfrac{\left(\frac{3}{4}\right)^{-3}}{\left(\frac{3}{4}\right)^{-1}}$
84. $\dfrac{\left(\frac{5}{8}\right)^{-5}}{\left(\frac{5}{8}\right)^{-3}}$
85. $\left(\frac{1}{3}\right)^{-2} + \left(\frac{1}{2}\right)^{-3}$
86. $\left(\frac{3}{4}\right)^{-2} - \left(\frac{3}{5}\right)^{-2}$
87. $(1.5)^{-2} - \left(\frac{2}{3}\right)^2$
88. $(0.5)^{-3} - (2)^3$

Simplify each expression. Write your answers using only positive exponents. Assume no denominator equals zero.

89. $(2x^2)(3x^4)$
90. $(3x^6)(2x^8)$
91. $(2x^2)(5x^4)$
92. $(2x^2 y^2)(3x^3 y)$
93. $(x^2 y^2)(x^3 y^7)$
94. $(x^4 y^2)(x^3 y^7 z)$
95. $(3x^4 y^2)(2x^2 y^3 z^2)$
96. $(xyz^{-2})(x^2 y^2 z^4)$
97. $\dfrac{x^2 y^2}{xy}$
98. $\dfrac{x^3 y^4}{x^2 y^6}$
99. $\dfrac{s^3 t^7}{st^6}$
100. $x^2 y^6 \cdot \dfrac{x^3}{y^4}$
101. $\dfrac{x^2}{2} \cdot \dfrac{8}{y^2} \cdot \dfrac{y^6}{x^4}$
102. $(x+y)^2 (x+y)^7$
103. $(x-y)^3 (x-y)^{-2}$
104. $\dfrac{(a-b)^3}{(a-b)^7}$
105. $\dfrac{(x-2)^5}{(x-2)^5}$
106. $\dfrac{(x-y)^{-2}}{(x-y)^{-3}}$

Simplify each expression first and then evaluate it using the given number.

107. $\dfrac{15y^4}{5y}$; $y = 1$
108. $\dfrac{-30x^5}{6x^4}$; $x = -1$
109. $\dfrac{-27x^6}{3x^3}$; $x = -2$
110. $\dfrac{24y^{-2}}{3y^{-1}}$; $y = 5$
111. $\dfrac{56y^{-4}}{-8y^{-2}}$; $y = -3$
112. $\dfrac{-63x^7}{-9x^5}$; $x = -9$

REVIEW EXERCISES

Find the solution set by the graphing method in Exercises 113–116.

113. $3x + 2y = 6$
114. $x - 4y \geq 8$
115. $x + y = 4$
     $x - y = 6$
116. $2x + y \geq 3$
     $x - 2y \geq -1$

Solve Exercises 117–120 by any method.

117. $3x + 4y = 8$
     $5x - 2y = -4$
118. $y = x - 2$
     $2x + 3y = 11$
119. $5x + 6y = 5.5$
     $4x - 2y = 1.0$
120. $4x - 7y = 10$
     $5x + 3y = -11$

WRITE IN YOUR OWN WORDS

121. How would you simplify $x^4 \cdot x^3$?
122. How would you simplify $\dfrac{y^5}{y^2}$?
123. Why does 7^0 equal 1?
124. Why does x^{-3} equal $\dfrac{1}{x^3}$?

6.2 More on Exponents; Scientific Notation

OBJECTIVES

In this section we will learn to
1. use the power rules to simplify exponential expressions;
2. convert numbers from decimal notation to scientific notation; and
3. convert numbers from scientific notation to decimal notation.

Now let's consider expressions such as $(a^2)^3$, $(ab)^2$, or $\left(\dfrac{a}{b}\right)^2$. We can easily develop the rules for their simplification by using the product and quotient rules, as the following examples show.

EXAMPLE 1 Simplify $(b^2)^3$ by removing the parentheses.

SOLUTION

$$(b^2)^3 = b^2 \cdot b^2 \cdot b^2 \qquad \text{Definition of an exponent}$$
$$= b^{2+2+2} \qquad b^m \cdot b^n = b^{m+n}$$
$$= b^6$$

We could have found the same result by multiplying the exponents:

$$(b^2)^3 = b^{2 \cdot 3} = b^6$$

EXAMPLE 2 Simplify each of the following: **(a)** $(ab)^2$ **(b)** $\left(\dfrac{a}{b}\right)^2$

SOLUTION

(a) $(ab)^2 = (ab)(ab)$
$\qquad\quad = (a \cdot a)(b \cdot b) \qquad$ Commutative and associative properties
$\qquad\quad = a^2 b^2$

(b) $\left(\dfrac{a}{b}\right)^2 = \left(\dfrac{a}{b}\right)\left(\dfrac{a}{b}\right)$
$\qquad\quad\;\; = \dfrac{a \cdot a}{b \cdot b} = \dfrac{a^2}{b^2}$

These examples lead us to the *power rules for exponents*.

The Power Rules for Exponents

If m and n are integers, then

(a) $(a^m)^n = a^{m \cdot n} = a^{mn}$

(b) $(ab)^m = a^m \cdot b^m = a^m b^m$

(c) $\left(\dfrac{a}{b}\right)^m = \dfrac{a^m}{b^m}, \quad b \neq 0$

EXAMPLE 3 Simplify each of the following expressions using the power

(a) $(x^2)^7$ (b) $(2^2)^2$ (c) $\left(\dfrac{2}{3}\right)^2$ (d) $\left(\dfrac{x^2}{y^3}\right)^3$ (e) $(x^2 y^3)$

SOLUTION
(a) $(x^2)^7 = x^{2 \cdot 7} = x^{14}$
(b) $(2^2)^2 = 2^{2 \cdot 2} = 2^4 = 16$
(c) $\left(\dfrac{2}{3}\right)^2 = \dfrac{2^2}{3^2} = \dfrac{4}{9}$
(d) $\left(\dfrac{x^2}{y^3}\right)^3 = \dfrac{(x^2)^3}{(y^3)^3} = \dfrac{x^6}{y^9}$
(e) $(x^2 y^3)^4 = x^{2 \cdot 4} y^{3 \cdot 4} = x^8 y^{12}$

EXAMPLE 4 Simplify: $\left(\dfrac{2x^{-2}}{3y^{-3}}\right)^2$.

SOLUTION
$$\left(\dfrac{2x^{-2}}{3y^{-3}}\right)^2 = \dfrac{2^2 x^{-2 \cdot 2}}{3^2 y^{-3 \cdot 2}} = \dfrac{4x^{-4}}{9y^{-6}}$$

$$= \dfrac{4 \cdot \dfrac{1}{x^4}}{9 \cdot \dfrac{1}{y^6}} = \dfrac{\dfrac{4}{x^4}}{\dfrac{9}{y^6}}$$

$$= \dfrac{4}{x^4} \cdot \dfrac{y^6}{9} = \dfrac{4y^6}{9x^4}$$

Observe that

$$\dfrac{a^{-n}}{b^{-n}} = \dfrac{\dfrac{1}{a^n}}{\dfrac{1}{b^n}} = \dfrac{1}{a^n} \cdot \dfrac{b^n}{1} = \dfrac{b^n}{a^n}$$

When we apply this observation to Example 4, we can eliminate several steps.

$$\left(\dfrac{2x^{-2}}{3y^{-3}}\right)^2 = \dfrac{2^2 x^{-2 \cdot 2}}{3^2 y^{-3 \cdot 2}}$$

$$= \dfrac{4x^{-4}}{9y^{-6}} = \dfrac{4y^6}{9x^4}$$

We now list all of the rules governing exponents.

The Rules of Exponents

Let m and n be integers.

1. Product rule $\quad b^m \cdot b^n = b^{m+n}$

2. Quotient rule $\quad \dfrac{b^m}{b^n} = b^{m-n}, \quad b \neq 0$

3. Power rules $\quad (b^m)^n = b^{m \cdot n}$
   $\quad\quad\quad\quad\quad (ab)^m = a^m b^m$
   $\quad\quad\quad\quad\quad \left(\dfrac{a}{b}\right)^m = \dfrac{a^m}{b^m}, \quad b \neq 0$

4. Definition of a zero exponent $\quad b^0 = 1, \quad b \neq 0$

5. Definition of a negative exponent $\quad b^{-n} = \dfrac{1}{b^n}, \quad b \neq 0$

(Handwritten note: $\sqrt[n]{x^y} = x^{y/n}$; $\sqrt[3]{x^6} = x^{6/3} = x^2$)

EXAMPLE 5 Simplify: $(4x^2y^2x^3y^4)^2$.

SOLUTION

$(4x^2y^2x^3y^4)^2 = (4x^5y^6)^2 \quad\quad x^2 \cdot x^3 = x^5;\ y^2 \cdot y^4 = y^6$

$\quad\quad\quad\quad\quad\quad = 4^2 x^{10} y^{12} \quad\quad (x^5)^2 = x^{10};\ (y^6)^2 = y^{12}$

$\quad\quad\quad\quad\quad\quad = 16 x^{10} y^{12}$

EXAMPLE 6 Simplify: $\dfrac{(3x^2y^2)^{-2}(2xy)^{-3}}{(2x^2y^{-2})^{-1}}$.

SOLUTION

$\dfrac{(3x^2y^2)^{-2}(2xy)^{-3}}{(2x^2y^{-2})^{-1}} = \dfrac{(3^{-2}x^{-4}y^{-4})(2^{-3}x^{-3}y^{-3})}{2^{-1}x^{-2}y^2}$

$\quad\quad\quad\quad\quad\quad\quad\quad = \dfrac{3^{-2} \cdot 2^{-3} x^{-7} y^{-7}}{2^{-1} x^{-2} y^2} \quad\quad x^{-4} \cdot x^{-3} = x^{-7};\ y^{-4} \cdot y^{-3} = y^{-7}$

$\quad\quad\quad\quad\quad\quad\quad\quad = 3^{-2} \cdot 2^{-3-(-1)} x^{-7-(-2)} y^{-7-2} \quad\quad \dfrac{b^m}{b^n} = b^{m-n}$

$\quad\quad\quad\quad\quad\quad\quad\quad = 3^{-2} \cdot 2^{-2} x^{-5} y^{-9}$

$\quad\quad\quad\quad\quad\quad\quad\quad = \dfrac{1}{3^2 \cdot 2^2 x^5 y^9} \quad\quad b^{-n} = \dfrac{1}{b^n}$

$\quad\quad\quad\quad\quad\quad\quad\quad = \dfrac{1}{36 x^5 y^9}$

EXAMPLE 7 Evaluate $\dfrac{(2^{-1}x^4)^{-4}}{4^2(x^{-3})^2(x^{-2})^3}$ for $x = 2$.

SOLUTION We simplify the expression first.

$$\frac{(2^{-1}x^4)^{-4}}{4^2(x^{-3})^2(x^{-2})^3} = \frac{2^4 x^{-16}}{4^2 x^{-6} x^{-6}}$$

$$= \frac{16 x^{-16}}{16 x^{-12}}$$

$$= \frac{1}{x^4} \qquad \frac{x^{-16}}{x^{-12}} = x^{-16-(-12)} = x^{-4} = \frac{1}{x^4}$$

When $x = 2$,

$$\frac{1}{x^4} = \frac{1}{2^4} = \frac{1}{16}$$

Scientific Notation

One of the many practical ways we use exponents is to write very large or very small numbers in a compact form called **scientific notation.** When a number is written in scientific notation, it is written in the form $b \times 10^n$, where $1 \leq b < 10$ and n is an integer. For example, the distance to the sun is about 93,000,000 miles. We write this in scientific notation as

9.3×10^7 miles

Recall that if a number is multiplied by 10, the decimal point shifts one place to the right. We're multiplying 9.3 by 10^7, which is multiplying by 10 seven times, so the decimal point shifts seven places to the right.

$9.3 \times 10^7 = 93{,}000{,}000$
  7 places

We can write very small numbers in the same manner. For example, the mass of a hydrogen atom is approximately

0.0000000000000000000000017 grams

We can write this in scientific notation as

1.7×10^{-24} grams

Dividing a number by 10 moves the decimal point one place to the left. Dividing by 10^{24} moves the decimal point 24 places to the left, and multiplying by 10^{-24} is the same as dividing by 10^{24}, since $10^{-24} = \frac{1}{10^{24}}$. Therefore

$1.7 \times 10^{-24} = 0.0000000000000000000000017$
  24 places

EXAMPLE 8   Write 6,730,000 in scientific notation.

SOLUTION   We start one place to the right of the first nonzero digit, 6, and count the number of places to the decimal point. The result is the power of 10 that we use.

$$6{,}730{,}000 = 6.73 \times 10^6$$

EXAMPLE 9   Write 0.000000712 in scientific notation.

SOLUTION   We start one place to the right of the first nonzero digit, 7, and count the number of places to the decimal point. Since the decimal point is located seven places to the *left*, the power of 10 is *negative*:

$$0.000000712 = 7.12 \times 10^{-7}$$

Calculator Note

Many calculators allow us to enter data directly in scientific notation, and some give answers in this form in certain situations. For example, when 2,356,927 is multiplied by 793,241, the result appears in scientific notation or in different formats, depending on the make and model of the calculator.

$$2{,}356{,}927 \times 793{,}241 = 1.86961113\text{E }12$$

or

$$2{,}356{,}927 \times 793{,}241 = 1.86961113 \times 10^{12}$$

or

$$2{,}356{,}927 \times 793{,}241 = 1.86961113 \quad 12$$

This answer is read $1.86961113 \times 10^{12} = 1{,}896{,}611{,}130{,}000$. It should be noted that this result is not exact. The last four digits in the actual product are not really zero.

EXAMPLE 10   Write each of the following numbers in scientific notation.

(a) $2340 = 2.340 \times 10^3$
(b) $512{,}300 = 5.123 \times 10^5$
(c) $0.00268 = 2.68 \times 10^{-3}$
(d) $125 \times 10^4 = 1.25 \times 10^2 \times 10^4$
$\qquad\quad\; = 1.25 \times 10^6$

EXAMPLE 11   Write each of the following numbers in decimal notation:

(a) 2.35×10^3 (b) 3.42×10^{-4} (c) 5.2×10^{-6}

SOLUTION   (a) $2.35 \times 10^3 = 2350$
(b) $3.42 \times 10^{-4} = 0.000342$
(c) $5.2 \times 10^{-6} = 0.0000052$

EXAMPLE 12 Evaluate $\dfrac{(360{,}000)(0.0093)}{(0.000009)(31{,}000)}$ by using scientific notation.

SOLUTION
$$\dfrac{(360{,}000)(0.0093)}{(0.000009)(31{,}000)} = \dfrac{(3.6 \times 10^5)(9.3 \times 10^{-3})}{(9 \times 10^{-6})(3.1 \times 10^4)}$$
$$= \dfrac{(3.6)(9.3)}{(9)(3.1)} \times 10^{5+(-3)-(-6)-4}$$
$$= (0.4)(3) \times 10^4$$
$$= 1.2 \times 10^4$$
$$= 12{,}000$$

Do You Remember?

Can you match these?

___ 1. $a^4 \cdot a^3$
___ 2. $a^7 \cdot a^{-4}$
___ 3. $(a^5)^3$
___ 4. a^{-3}
___ 5. $\left(\dfrac{a^2}{b^3}\right)^2$
___ 6. $(a^2 b^3)^2$
___ 7. $\dfrac{a^{-2}}{b^{-2}}$
___ 8. 0.00034
___ 9. 34,000
___ 10. 3.4×10^{-6}

a) a^3
b) a^7
c) 0.0000034
d) $a^4 b^6$
e) $\dfrac{b^2}{a^2}$
f) $\dfrac{a^4}{b^6}$
g) a^{15}
h) a^8
i) 3.4×10^4
j) 3.4×10^{-4}
k) $\dfrac{1}{a^3}$

Answers: 1. b 2. a 3. g 4. k 5. f 6. d 7. e 8. j 9. i 10. c

Exercise Set 6.2

Evaluate each of the following powers.

1. 2^0
2. $(-2)^0$
3. -2^0
4. $\left(\dfrac{1}{2}\right)^0$
5. $(0.5)^0$
6. $(-0.25)^0$
7. 8^{-1}
8. -8^{-1}
9. 3^{-2}
10. -3^{-2}
11. $(-3)^{-2}$
12. $-(-3)^{-2}$
13. $\left(\dfrac{2}{3}\right)^{-1}$
14. $\left(\dfrac{5}{8}\right)^{-1}$

15. $(0.5)^{-2}$
16. $(0.1)^{-2}$
17. $(0.2)^{-3}$
18. $(-0.2)^{-3}$
19. $\left(-\dfrac{3}{4}\right)^{-2}$
20. $-\left(-\dfrac{3}{4}\right)^{-2}$
21. $\left[\left(\dfrac{3}{8}\right)^{-1}\right]^{2}$
22. $\left[\left(-\dfrac{7}{8}\right)^{-1}\right]^{2}$
23. $\left[-\left(-\dfrac{1}{4}\right)^{2}\right]^{-1}$
24. $\left[-\left(-\dfrac{1}{2}\right)^{-2}\right]^{-1}$

65. $\dfrac{(4a^2b^2)^{-2}(2ab^{-1})^3}{(3a^2b^{-2})^{-1}}$
66. $\dfrac{[(4a^2b)^{-1}]^{-2}[(3ab)^{-2}]^{-1}}{(a^2b^3)^{-2}}$
67. $\dfrac{(s^6t^{-2})(s^2t^2)^{-4}(st)}{(s^2t^2)^{-3}}$
68. $\dfrac{(3^{-4}a^{-2}b^{-1})^{-2}(5a^4b^{-2})^{-2}}{(3a^{-5}b^{-3})^2}$

Simplify each expression. Write your answers using only positive exponents. Assume no denominator equals zero.

25. $(x^2)^3$
26. $(a^3)^4$
27. $(a^{-2})^5$
28. $(a^{-1})^{-3}$
29. $\left(\dfrac{x}{y}\right)^4$
30. $\left(\dfrac{a}{b}\right)^7$
31. $\left(\dfrac{3}{b}\right)^{10}$
32. $\left(\dfrac{2}{7}\right)^8$
33. $\left(\dfrac{a^0}{b^1}\right)^{-4}$
34. $\left(\dfrac{a^4}{b^3}\right)^3$
35. $(x^2y)^2$
36. $(x^2y^3)^5$
37. $\left(\dfrac{x^{-2}y^2}{z}\right)^4$
38. $\left(\dfrac{x^{-3}y}{z}\right)^3$
39. $\left(\dfrac{3x^2}{y^4}\right)^{-2}$
40. $\left(\dfrac{2^{-1}x}{y^2}\right)^{-1}$
41. $(9^{-2}x^{-3})^{-1}$
42. $(4x^{-6}y)^{-2}$
43. $(7p^{-4})^{-2}$
44. $\left(\dfrac{8p^3}{p^{-3}}\right)^0$
45. $(x^{-1}+y^{-1})^0$
46. $(x^2+xy^{-1})^0$
47. $x^2y^2(x^2)^3$
48. $(x^3y^2)^2$
49. $[(x^2y^3)^2]^2$
50. $[(x^2)^2]^2$
51. $[(y^{-1})^{-1}]^{-2}$
52. $\left(\dfrac{x^2y^{-2}}{x^2y^{-2}}\right)^3$
53. $(5a^2b^2c^2)^2$
54. $(16a^2b^2a^{-3}b^3)^2$
55. $\dfrac{(3x^4)^2}{(6x^2)^3}$
56. $\dfrac{(2x^4)^{-2}}{(3x^2)^{-4}}$
57. $\dfrac{(-3x^2y)^2}{-6(xy^2)^4}$
58. $\dfrac{(-2x^3y^{-2})^2}{-5(x^{-2}y)^{-4}}$
59. $\dfrac{(4x^3y^5)^3}{(2x^4y^2)^4}$
60. $(-3x^2y)^3(2xy^6)^2$
61. $\dfrac{-16x^7y^4z^5}{4x^2y^3z^2}$
62. $\dfrac{5(2x^2y^3)^3}{25(4xy^{-2})^{-1}}$
63. $\dfrac{(4x^2)^3(8x^3)^2}{(16x)^2}$
64. $\dfrac{(m^6n)^{-2}(mn^{-2})^3}{(m^{-2}n^{-2})^{-2}}$

Simplify each expression first and then evaluate using the given number.

69. $\left(\dfrac{27x^5}{9x}\right)^2$; $x=2$
70. $\left(\dfrac{5y^2}{25y^5}\right)^2$; $y=-3$
71. $\left(\dfrac{12p^{-6}}{15p^{-8}}\right)^{-3}$; $p=-1$
72. $\left[\dfrac{80m^{-2}}{16m^3}\right]^2$; $m=5$
73. $\left[\dfrac{(2x^{-4})^2(6x^5)^2}{(3x^{-3})^{-2}(9x^{-4})^2}\right]^1$; $x=1$
74. $\left[\dfrac{(5p^{-3})^2(-p^4)^3}{(25p^6)^1(p^{-1})^2}\right]^{-1}$; $p=-2$

Write each number in scientific notation.

75. 26
76. 0.034
77. 0.002
78. 382
79. 0.00042
80. 0.00231
81. 268,000
82. 387,000
83. 2,856,000
84. 0.000492
85. 0.000000263
86. 7,289,600

Write each number in decimal notation.

87. 2.6×10^3
88. 3.8×10^0
89. 9.23×10^0
90. 4.25×10^7
91. 2.83×10^{-6}
92. 7.312×10^{-5}
93. 4.8×10^{-6}
94. 8.23×10^{-7}
95. 2.95×10^{12}
96. 3.16×10^{-8}

Simplify by first writing each number in scientific notation. Leave your answers in scientific notation form.

97. $(36{,}000)(2000)$
98. $(5000)(0.0007)$
99. $(26{,}000{,}000)(0.00002)$
100. $(0.003)(600{,}000)$
101. $\dfrac{42{,}000}{0.0006}$
102. $\dfrac{210{,}000{,}000}{7000}$
103. $\dfrac{0.00026}{1300}$
104. $\dfrac{(42{,}000)(21{,}000)}{(7000)(90{,}000)}$
105. $\dfrac{(0.000084)(0.0009)}{(0.00012)(0.0000021)}$

Write the number in each statement in scientific notation.

106. The speed of light is 300,000,000 m/sec.

107. The diameter of the sun is approximately 865,000 mi.

108. The diameter of a red corpuscle is approximately 0.000075 cm.

109. The constant of gravitational pull is 0.0000000667 dynes.

110. The mass of an electron is approximately

0.0000000000000000000000000009107 g

111. The number of molecules in a mole of gas is

602,000,000,000,000,000,000,000

Write the number in each statement in decimal notation.

112. The average radius of the earth is 6.37×10^6 m.

113. The number of different three-digit, three-letter license plates possible is approximately 1.76×10^7.

114. Light travels about 5.87×10^{12} mi/yr.

115. The mass of one molecule of water is approximately 3×10^{-23} g.

116. The number of electrons in one coulomb is approximately 6.2×10^{18}.

117. The wavelength of yellow light is approximately 2.3×10^{-5} in.

Evaluate the following using scientific notation and the laws of exponents. Leave your answers in scientific notation form.

118. If the mass of the earth is about 6×10^{29} centigrams (cg) and each centigram is 1.1×10^{-6} tons, find the mass of the earth in tons.

119. If the distance from the sun to the earth is 9.3×10^7 mi and light travels 1.86×10^5 mi/sec, how long does it take light to travel from the sun to the earth?

120. The average life span of an average human is about 2.0×10^9 seconds. How many days is this? How many years is this (assuming 365 days/yr)?

121. Some commercial planes fly at about 3.168×10^4 ft above sea level. How many miles is this? (*Note:* 1 mi = 5.28×10^3 ft.)

6.3 Adding and Subtracting Polynomials

OBJECTIVES

In this section we will learn to
1. identify a term and its degree;
2. identify a polynomial;
3. add and subtract polynomials horizontally;
4. add and subtract polynomials vertically; and
5. use function notation for polynomials.

Recall from Section 3.1 that a **term** is a number or the product of numbers and variables raised to powers. For example,

$$5, \quad 3x^2, \quad 2xy^3, \quad \text{and} \quad \frac{1}{2}x^7$$

are all terms. The **degree** of a term with one variable is the exponent on that variable. If a term contains two or more variables, its degree is the sum of the exponents on the variables.

A **polynomial** is a single term or the sum or difference of a finite number of terms. For example,

$$3, \quad 3x, \quad 2x + y, \quad \text{and} \quad x^2 + 6x - 2$$

are all polynomials. For an expression to be a polynomial, it must meet three conditions:

1. The exponent on any variable in any term must be a whole number. Thus $7x + 2x^{-3}$ is not a polynomial (-3 is not a whole number).
2. When in simplest form, no variable can appear under a radical. Thus $5y^3 + 3\sqrt{x}$ is not a polynomial (x appears under the radical sign, $\sqrt{}$).
3. When in simplest form, no term can contain a variable in the denominator. Thus $5x^2 + \dfrac{6}{x}$ is not a polynomial (x appears in the denominator).

The **degree of a polynomial in one variable** is the largest exponent that appears on that variable in any term.

EXAMPLE 1 (a) $x^6 + 3x + 2$ is a sixth-degree polynomial in x.
(b) $y^4 + 4y + 5$ is a fourth-degree polynomial in y.
(c) 5 is a zero-degree polynomial, since $5 = 5x^0$.
(d) $2y^4 + 3y^3 + y^5$ is a fifth-degree polynomial in y.

A polynomial with a single term is called a **monomial,** one with two terms is called a **binomial,** and one with three terms is called a **trinomial.** Polynomials with more than three terms are usually referred to just as polynomials, although they can be named. One with four terms, for example, is called a quadrinomial.

EXAMPLE 2 (a) $2x$, x^6, and $5x^7$ are monomials.
(b) $x^5 - 1$, $x^7 - 2x$, and $3x^{11} - 19x^2$ are binomials.
(c) $x^4 + 3x^2 - 6$ and $y^9 - 2y^7 + 3y^5$ are trinomials.

When a polynomial is written in **descending powers** of the variable, such as

$$2x^3 + 3x^2 + 2x + 9$$

it is said to be in **standard form.** In this form, we say that 2 is the **leading coefficient.** Since the last term, 9, is a constant, it is called the **constant term.**

EXAMPLE 3 Write $4x^2 - 2x + 6x^3 + 7$ in standard form. Give its degree and identify the leading coefficient and the constant term.

SOLUTION To write the polynomial in standard form, we use the commutative property to rearrange terms from the highest degree to the lowest degree:

$$4x^2 - 2x + 6x^3 + 7 = 6x^3 + 4x^2 - 2x + 7$$

The degree is 3. The leading coefficient is 6 and the constant term is 7.

In Section 3.1 we used the distributive property to combine like terms; for example,

$$6x + 9x = (6 + 9)x = 15x$$

Adding or subtracting polynomials follows a similar process.

EXAMPLE 4   Add $3x^2 + 6x + 7$ and $2x^2 + 5x + 9$.

SOLUTION
$$(3x^2 + 6x + 7) + (2x^2 + 5x + 9)$$
$$= (3x^2 + 2x^2) + (6x + 5x) + (7 + 9) \quad \text{Commutative and associative properties}$$
$$= (3 + 2)x^2 + (6 + 5)x + (7 + 9) \quad \text{Distributive property}$$
$$= 5x^2 + 11x + 16$$

The process we used in Example 4 to add the two polynomials can be completed mentally, as illustrated in the next example.

EXAMPLE 5   Add $6x^2 + 3x + 9$ and $5x^2 + 6x + 4$.

SOLUTION

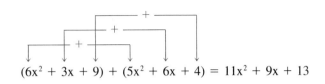

$$(6x^2 + 3x + 9) + (5x^2 + 6x + 4) = 11x^2 + 9x + 13$$

Some students find it easier to add polynomials by writing one below the other. Care must be taken, however, to keep like terms in the same column.

EXAMPLE 6   Add the following polynomials by rewriting them vertically before adding the coefficients of like terms.

(a) $6x^2 + 3x + 9$ and $5x^2 + 6x + 4$
(b) $x^4 + 2x^3 + x^2 - 1$, $x^3 + 3x^2 + 6x + 3$, and $2x^4 - 5x^2 + 9$

SOLUTION   **(a)**
$$\begin{array}{r} 6x^2 + 3x + 9 \\ +\ 5x^2 + 6x + 4 \\ \hline 11x^2 + 9x + 13 \end{array}$$

(b)
$$\begin{array}{r} x^4 + 2x^3 + x^2 - 1 \\ x^3 + 3x^2 + 6x + 3 \\ +\ 2x^4 - 5x^2 + 9 \\ \hline 3x^4 + 3x^3 - x^2 + 6x + 11 \end{array}$$

Subtraction of polynomials also can be performed either horizontally or vertically. Recall that to subtract one quantity from another, we add its additive inverse (its opposite).

When we want to subtract a polynomial, we can use the distributive property by thinking of subtraction as multiplying by -1. For example, we can think of

$$(12x^2 - 3x + 5) - (3x^2 - 9x + 6)$$

as

$$(12x^2 - 3x + 5) - 1(3x^2 - 9x + 6)$$
$$= 12x^2 - 3x + 5 - 3x^2 + 9x - 6 \quad \text{Multiplying by } -1 \text{ changes all signs.}$$
$$= 9x^2 + 6x - 1 \quad \text{Combine like terms.}$$

EXAMPLE 7 Subtract $x^2 - 6x + 5$ from $3x^2 + 9x - 7$.

SOLUTION
$$(3x^2 + 9x - 7) - (x^2 - 6x + 5) = 3x^2 + 9x - 7 - x^2 + 6x - 5 \quad \text{Distributive property}$$
$$= 2x^2 + 15x - 12$$

EXAMPLE 8 Subtract $3x^2 - 9x + 2$ from $6x^4 + 6x^2 + 2x + 7$ vertically.

SOLUTION
$$\begin{array}{r} 6x^4 + 6x^2 + 2x + 7 \\ + \quad -3x^2 + 9x - 2 \\ \hline 6x^4 + 3x^2 + 11x + 5 \end{array} \quad \text{Add the additive inverse.}$$

Examples 7 and 8 lead to the following rule.

To Subtract Two Polynomials

Change the sign of every coefficient in the polynomial being subtracted and add.

EXAMPLE 9 Simplify:

$$(x^4 - 3x^2 - 2x - 1) + (6x^3 - 2x^2 - 5x + 1) - (2x^4 + x^3 - 3x^2 + 9x - 1)$$

SOLUTION We begin by removing grouping symbols, then we combine like terms.

$$(x^4 - 3x^2 - 2x - 1) + (6x^3 - 2x^2 - 5x + 1) - (2x^4 + x^3 - 3x^2 + 9x - 1)$$
$$= x^4 - 3x^2 - 2x - 1 + 6x^3 - 2x^2 - 5x + 1 - 2x^4 - x^3 + 3x^2 - 9x + 1$$
$$= -x^4 + 5x^3 - 2x^2 - 16x + 1$$

Many polynomials involve more than one variable, but addition and subtraction is carried out in the same manner.

EXAMPLE 10 Simplify: $(xy^3z - 3xy^2z^2 + xy^3z^2) - (-2xy^3z + 7xy^2z^2 - 2xy^3z^2)$.

SOLUTION $(xy^3z - 3xy^2z^2 + xy^3z^2) - (-2xy^3z + 7xy^2z^2 - 2xy^3z^2)$
$$= xy^3z - 3xy^2z^2 + xy^3z^2 + 2xy^3z - 7xy^2z^2 + 2xy^3z^2 \quad \text{Distributive property}$$
$$= 3xy^3z - 10xy^2z^2 + 3xy^3z^2 \quad \text{Combine like terms.}$$

EXAMPLE 11 Add $\frac{1}{3}x^3 - \frac{1}{2}x^2 + 3x$ and $\frac{1}{4}x^3 + \frac{1}{5}x^2 - 2$.

SOLUTION
$$\left(\frac{1}{3}x^3 - \frac{1}{2}x^2 + 3x\right) + \left(\frac{1}{4}x^3 + \frac{1}{5}x^2 - 2\right) = \left(\frac{1}{3} + \frac{1}{4}\right)x^3 + \left(-\frac{1}{2} + \frac{1}{5}\right)x^2 + 3x - 2$$
$$= \frac{7}{12}x^3 + \left(-\frac{3}{10}\right)x^2 + 3x - 2$$
$$= \frac{7}{12}x^3 - \frac{3}{10}x^2 + 3x - 2$$

In more advanced courses, the horizontal format for adding and subtracting polynomials is used almost exclusively.

Polynomials in a single variable are often denoted in function notation format. For example, $P(x)$, $R(x)$, and $S(x)$ can be used to represent polynomials in the variable x, for example,

$P(x) = x + 5$,
$R(x) = x^2 + 5x + 6$
$S(x) = x^4 - 3x^3 + 2x^2 - x - 5$

This notation makes it easy to identify the value of a polynomial for a specific quantity.

EXAMPLE 12 If a ball is thrown vertically upward from the ground with an initial velocity of 88 feet per second, its position S above the ground at any time t can be approximated by $S(t) = -16t^2 + 88t$. Find the height of the ball when $t = 1, 2,$ and 3 seconds.

SOLUTION
$S(t) = -16t^2 + 88t$
$S(1) = -16(1)^2 + 88(1) = -16 + 88 = 72$ ft   Substitute 1 for t.
$S(2) = -16(2)^2 + 88(2) = -64 + 176 = 112$ ft   Substitute 2 for t.
$S(3) = -16(3)^2 + 88(3) = -144 + 264 = 120$ ft   Substitute 3 for t.

Do You Remember?

Can you match these?

_____ 1. A polynomial with one term   a) a trinomial
_____ 2. A polynomial with two terms   b) zero
_____ 3. A polynomial with three terms   c) a monomial
_____ 4. The degree of the number 7   d) a binomial
_____ 5. The degree of $5x^{11} - 4x^9$ e) nine
                                             f) eleven

Answers: 1. c 2. d 3. a 4. b 5. f

Exercise Set 6.3

Identify each polynomial as a monomial, a binomial, or a trinomial. State the degree of each polynomial.

1. $3x^2 + 6x + 9$
2. $9x^6 + 4x^2 + 2$
3. $x^7 + 2x$
4. $3x^5 - 2x^4$
5. $x^2 + 1$
6. $3x^9$
7. $2x^2$
8. $2x^8 + 6x$
9. $4x^2 + 6x^6 + 2$
10. 18
11. $7x^0$
12. $3x + 2$

State whether each of the following is a polynomial. If it is not, explain why. If it is a polynomial, state its degree.

13. $3x^2 + 6x + 1$
14. $2x^5 + x^4 - x^2 + 2$
15. $3x^2 + \dfrac{2}{x} + 9$
16. $\dfrac{7}{x+2} + 3$
17. $3x^4 - x^{1/2} + 2x$
18. $y^3 + 2y + 9$
19. 8
20. -26
21. $x^7 + 2x^6 + x^{-3}$
22. $x + 9$
23. $y^6 + \dfrac{y^2}{y^4} - 2y$
24. $\dfrac{3x^7}{4} - x^4 + 3x^2$

Simplify each polynomial by combining all like terms. Then write it in standard form and determine the leading coefficient.

25. $3x + 6 + 2x + 4$
26. $7 + 5y + 3 + 2y$
27. $x + 1 + 3x - 2$
28. $8y - 3 + y - 7$
29. $4 + y - 5 - 2y$
30. $6 + 2x - 7 - 3x$
31. $5 + x + x^2 - 2x - 3$
32. $13 + a - a^2 + 2a - 4$
33. $16p^3 - 2p^2 - p^4 + p^2 - p^3$
34. $17s^2 - 8s^3 + 5s - 16s^2$
35. $0.5y^2 - 0.25y + 0.75y^2 - 0.1y + 0.6$
36. $1.5x^2 - 2.25x - 1.4 - 0.7x^2 + 1.5x + 2$
37. $\dfrac{3}{8}p^2 - \dfrac{5}{8}p + \dfrac{1}{2} - 0.5p + p^2 - 4$
38. $\dfrac{1}{8} + \dfrac{2}{3}a^2 - a + \dfrac{7}{8} - \dfrac{1}{3}a^2 + \dfrac{1}{4}a$
39. $\dfrac{x^2}{2} - \dfrac{5x}{8} - \dfrac{3}{4} + \dfrac{3x^2}{7} - \dfrac{x}{3} + \dfrac{1}{2}$
40. $\dfrac{4s}{3} - s^2 + \dfrac{3}{4} - \dfrac{s}{4} + \dfrac{s^2}{5} - 3$

Add each of the following polynomials.

41. $3x^2 + 2x$
    $\underline{2x^2 + 5x}$

42. $12y^2 + 5y + 3$
    $\underline{y^2 + y + 6}$

43. $6m^2 - 5$
    $\underline{2m^2 + 7}$

44. $2p^2 + 1$
    $\underline{-5p^2 - 6}$

45. $x^2 - 7x + 2$
    $\underline{3x^2 + 2x - 5}$

46. $-4a^2 + 5a - 4$
    $\underline{a^2 - 4a + 2}$

47. $9x^3 + 6x^2 + 4x - 8$
    $\underline{3x^3 - 5x^2 + 8x - 6}$

48. $12p^4 - 6p^3 + 2p^2 - 5p + 1$
    $\underline{9p^4 + 4p^3 - 3p^2 - 8p - 5}$

49. $5x^4 - 7x + 4$
    $x^4 - 3x^2 + 4x - 15$
    $\underline{-3x^4 + 2x^3 + 5x^2 - 19x + 18}$

50. $3y^4 + 2y^3 - 5$
    $ -6y^3 + 4y^2 - 7y$
    $\underline{-8y^4 + y^3 - y^2 + 7y + 5}$

Subtract each of the following polynomials.

51. $3x^2 + 9$
    $\underline{x^2 + 1}$

52. $7y^2 + 2$
    $\underline{3y^2 + 4}$

53. $5p^2 - 5p + 4$
    $\underline{6p^2 + 7p - 1}$

54. $y^2 + 2y - 7$
    $\underline{3y^2 - y + 3}$

55. $6m^2 - 9m - 1$
    $\underline{7m^2 - 8m - 2}$

56. $-2x^2 + x - 11$
    $\underline{11x^2 - 10x + 9}$

57. $8p^4 - 3p^3 + 7p^2 - 9p + 8$
    $\underline{11p^4 + p^3 - 8p^2 + 9p + 8}$

58. $14x^3 - 6x^2 + 4x - 5$
    $\underline{-2x^3 + 7x^2 - 2x - 6}$

59. $5x^4 - 3x^2 + 8$
    $\underline{x^4 + 7x^3 - 3x^2 + 5x + 1}$

60. $m^2 - m + 11$
    $\underline{2m^3 + m^2 + 8m - 29}$

Simplify each of the following expressions by adding or subtracting.

61. $(6y + 2) + (5y + 3)$
62. $(6p + 4) + (5p + 11)$
63. $(-3m + 8) + (7m + 9)$
64. $(-x + 5) + (6x - 7)$
65. $(9y^2 + 2y) + (7y^2 - 4)$
66. $(4x^2 - 7x) + (-20x^2 + 8)$
67. $(10x^2 + 8) - (6x^2 + 4)$
68. $(5p^2 + 4) - (p^2 - 5)$
69. $(6m^2 - 2) - (-3m^2 + 5)$
70. $(y^2 - 5) - (-4y^2 + 1)$

71. $(x^2 + 5x - 7) - (2x^2 - 7x + 1)$
72. $(m^2 + 4m - 3) - (3m^2 - 7m + 5)$
73. $(-7p^2 + 6) + (15p - 4)$
74. $(-5x^2 - 5) + (11x + 9)$
75. $(x^4 + 7x^2 + 3) + (3x^4 - 5x^3 - 8x^2 + 7x - 1)$
76. $(y^4 + y^3 - 7) + (-2y^4 - y^3 - 5y^2 + 8y - 19)$
77. $(m^5 - 5m^4 - 7m^3 - 8) - (-m^4 - 3m^2 + 7)$
78. $(p^5 - p^4 - p^2 + 12) - (-p^4 - p^3 + 2p^2 + 13p - 18)$
79. $(-6y^4 + 12y^2 + 8) - (-8y^4 + 7y + 1)$
80. $6x^3 - (-x^3 + x) + 6x$
81. $12m^3 - (2m^3 - 5m) + m^2 - 4$
82. $5y^3 - (3y^3 + y^2 - 4y) - 2y^2 + 7$
83. $-(-4x^3 + 2x - 7) - 3x^2 + (2x + 4)$
84. $-(-x^3 - 5x^2 + 4) - 2x^3 + x^2 - 8x$
85. $(5y^2 - 7y + 4) - (y^2 + 3) + (-8y^2 - 5y - 11)$
86. $(6y^2 + 11y - 7) - (3y^2 - 5y + 1) + (-2y^2 + 11y + 9)$
87. $(x^6 + 4x^3 - 7x + 19) - (2x^6 - x^5 + 19x^2) + (3x^5 - x^4 - 2x^2 - 13x + 4)$
88. $(x^2 - 7xy^2 + 5x - 9) + (3xy^2 - 9xy + 4xy^2 + 1)$
89. $(-m^2p + mp - 3 + 2mp^2) + (7mp^2 + 4 - 5mp^2 + 6mp)$
90. $(xy^3z - 2xy^2z^2 + 7xyz) - (xy^3z + 3xy^2z^2 + 8x^3yz - 5xyz)$
91. $\left(\dfrac{2}{3}x^2 + \dfrac{1}{2}x - 1\right) + \left(\dfrac{1}{4}x - \dfrac{1}{8} + \dfrac{5}{8}x^2\right)$
92. $\left(-\dfrac{1}{3}y + \dfrac{1}{4} + \dfrac{1}{2}y^2\right) - \left(\dfrac{3}{8} - \dfrac{1}{3}y + \dfrac{1}{4}y^2\right)$
93. $\dfrac{1}{3}(4a + 2) - \dfrac{1}{2}(8a + 5)$
94. $\dfrac{2}{3}(3p + 8) - \dfrac{1}{4}(2p - 3)$
95. $-\dfrac{3}{4}\left(\dfrac{1}{2}m^2 - m + \dfrac{2}{3}\right) - \dfrac{1}{2}\left(\dfrac{2}{3}m^2 + 2m - 1.25\right)$
96. $\dfrac{3}{8}\left(0.5s^2 - 3s + 2\dfrac{2}{3}\right) - \dfrac{1}{4}(0.5s^2 + s - 4)$
97. $(x^2 - 3x + 2) - (2.1x^2 - 5.9x - 11.3)$
98. $(5.1y^2 - 1.4y + 8.1) - (1.3y^2 + 2.3y - 5.9)$
99. $(0.01x^2 - 0.02x - 0.05) - (0.009x^2 + 7.1x - 0.4)$
100. $(4.02p^2 - 1.71p + 0.001) + (0.08p^2 - 8.12p - 0.001)$
101. $(1.2)\left(\dfrac{1}{2}x - 4\right) + (2.4)\left(\dfrac{1}{4}x - 7\right)$
102. $(3.5)(0.5y + 3) - (2.7)(y - 1.1)$

Evaluate each of the following polynomial functions.

103. The cost C (in dollars) of producing x skateboards can be approximated by $C(x) = 38x + 400$. Find the cost when $x = 1$, 4, or 10.

104. The price P (in dollars) of a certain bicycle depends upon the demand x and can be approximated by $P(x) = -8x + 4500$. Find the price when the demand is $x = 100$, 300, or 400.

105. Olde Fashioned Chocolates makes and sells x cartons of chocolate candy per month. The monthly cost C (in dollars) is given by the formula $C(x) = 8.5x + 250$. The monthly revenue R (in dollars) is given by $R(x) = -0.10x^2 + 110x$.
    (a) Find the profit function $P(x)$, where $P(x) = R(x) - C(x)$.
    (b) Find the profit when $x = 10$; when $x = 20$.

106. Jason's Reel Company makes and sells x cartons of fishing reels each month. The cost function C (in dollars) is given by the formula $C(x) = 42x + 380$ and the revenue function R (in dollars) is given by $R(x) = -0.12x^2 + 80x$.
    (a) Find the profit function $P(x)$, where $P(x) = R(x) - C(x)$.
    (b) Find the profit when $x = 10$; when $x = 20$.

WRITE IN YOUR OWN WORDS
107. What is a leading coefficient?
108. What is a monomial?
109. What does it mean to put a polynomial in standard form?
110. What is the degree of a polynomial?
111. What are like terms?
112. What is a polynomial?

FOR EXTRA THOUGHT
Find the perimeter of each of the following figures.
113. Square

$6x + 3$

114. Rectangle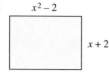

115. Triangle

116. Circle

117. Trapezoid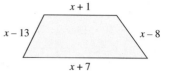

118. Parallelogram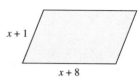

6.4 Multiplying Polynomials

HISTORICAL COMMENT   Gottfried Wilhelm Leibniz (1646–1716) introduced the dot, ·, as a symbol for multiplication in place of ×, arguing that × was too easily confused with the unknown quantity x. He experimented with many others including ⌒, the comma itself, the semicolon, and the asterisk, *. Only the ×, the dot, and the asterisk have survived to this day.

OBJECTIVES   In this section we will learn to
1. multiply a polynomial by a monomial;
2. multiply two polynomials;
3. multiply two polynomials using the "FOIL" method; and
4. find special products of binomials.

In Section 6.1 we learned how to find the product of two monomials, such as $3x^2$ and $2x^5$. Recall that

$$(3x^2)(2x^5) = (3 \cdot 2)(x^2 \cdot x^5) \quad \text{Associative and commutative properties}$$
$$= 6x^{2+5} \quad \text{Product rule for exponents}$$
$$= 6x^7$$

We can find the product of a monomial and a polynomial in a similar manner by applying the distributive property together with the product rule for exponents:

$$2x^2(3x^3 + 4x^2 + 6x - 3) = 2x^2(3x^3) + 2x^2(4x^2) + 2x^2(6x) - 2x^2(3)$$
$$= 6x^5 + 8x^4 + 12x^3 - 6x^2$$

To Multiply a Polynomial by a Monomial

Multiply each term of the polynomial by the monomial using the distributive property and the product rule for exponents.

EXAMPLE 1 Multiply each of the following polynomials:

(a) $3x^2(2x^2 + 6x - 4)$
(b) $-5xy(3x - 2x^2y + 7xy^2)$
(c) $|-3m^3|(4m^3 + 2m - 11)$ for $m > 0$

SOLUTION

(a) $3x^2(2x^2 + 6x - 4) = 3x^2(2x^2) + 3x^2(6x) - 3x^2(4)$
$= 6x^4 + 18x^3 - 12x^2$

(b) $-5xy(3x - 2x^2y + 7xy^2) = -5xy(3x) - (-5xy)(2x^2y) + (-5xy)(7xy^2)$
$= -15x^2y + 10x^3y^2 - 35x^2y^3$

(c) $|-3m^3|(4m^3 + 2m - 11) = (3m^3)(4m^3 + 2m - 11)$ $|-3m^3| = 3m^3$ for $m > 0$
$= 12m^6 + 6m^4 - 33m^3$

EXAMPLE 2 Find the product $\frac{2}{3}a^2b^3\left(\frac{1}{2}ab^3 - \frac{6}{5}ab + \frac{2}{5}ab^4 - \frac{1}{3}a^5\right)$.

SOLUTION

$\frac{2}{3}a^2b^3\left(\frac{1}{2}ab^3 - \frac{6}{5}ab + \frac{2}{5}ab^4 - \frac{1}{3}a^5\right)$

$= \left(\frac{2}{3}\cdot\frac{1}{2}\right)a^3b^6 - \left(\frac{2}{3}\cdot\frac{6}{5}\right)a^3b^4 + \left(\frac{2}{3}\cdot\frac{2}{5}\right)a^3b^7 - \left(\frac{2}{3}\cdot\frac{1}{3}\right)a^7b^3$

$= \frac{1}{3}a^3b^6 - \frac{4}{5}a^3b^4 + \frac{4}{15}a^3b^7 - \frac{2}{9}a^7b^3$

EXAMPLE 3 Find a polynomial that represents the area of a rectangle with length $x + 3$ and width x; see Figure 6.1. Notice that the rectangle is subdivided into two smaller rectangles.

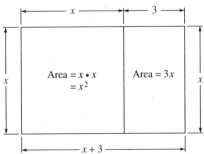

FIGURE 6.1

SOLUTION The area of a rectangle is given by the formula $A = l \cdot w$. Thus, since the length is $x + 3$ and the width is x,

$$A = (x + 3)x = x^2 + 3x$$

The area of the rectangle in Figure 6.1 is also the sum of the two smaller areas, $x \cdot x$ and $3 \cdot x$:

$$x \cdot x + 3 \cdot x = x^2 + 3x$$

which is the polynomial we got before.

The product of two polynomials when neither one is a monomial is found by applying the distributive property more than once. For example,

$$\underset{a\ \cdot\ (b\ +\ c)\ =}{(x + 3)(x + 2)} = \underset{a\ \cdot\ b\ +}{(x + 3)x} + \underset{a\ \cdot\ c}{(x + 3)2}$$

Applying the distributive property a second time yields

$$(x + 3)x + (x + 3)2 = x^2 + 3x + 2x + 6$$
$$= x^2 + 5x + 6$$

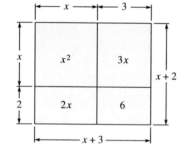

FIGURE 6.2

We can also illustrate the product $(x + 3)(x + 2)$ as the sum of the areas of the subdivisions of a rectangle, as shown in Figure 6.2.

$$A = l \cdot w = (x + 3)(x + 2)$$

The sum of the smaller areas is

$$x^2 + 2x + 3x + 6 = x^2 + 5x + 6$$

Thus

$$(x + 3)(x + 2) = x^2 + 5x + 6$$

Notice that in the product, each term in the first polynomial is multiplied by each term in the second polynomial. We can state this as a general rule.

To Find the Product of Two Polynomials

Multiply each term in one polynomial by each term in the other polynomial. Combine like terms.

EXAMPLE 4 Find each of the following products:

(a) $(6x + 4)(x + 3)$ (b) $(2x + 4)(2x^2 + 3x + 1)$

SOLUTION (a) $(6x + 4)(x + 3) = 6x^2 + 18x + 4x + 12$
$$= 6x^2 + 22x + 12$$

(b) $(2x + 4)(2x^2 + 3x + 1) = 2x(2x^2 + 3x + 1) + 4(2x^2 + 3x + 1)$
$= 4x^3 + 6x^2 + 2x + 8x^2 + 12x + 4$
$= 4x^3 + 14x^2 + 14x + 4$

When we want to multiply two binomials, we can find their product mentally by using the **FOIL method,** which we diagram in the following example.

EXAMPLE 5 Find the product of $2x + 2$ and $3x + 7$.

SOLUTION

F is the product of the *first* terms.
O is the product of the *outer* terms.
I is the product of the *inner* terms.
L is the product of the *last* terms.

$$\text{Product} = \underset{\downarrow}{F} + \underset{\downarrow}{O} + \underset{\downarrow}{I} + \underset{\downarrow}{L}$$

$(2x + 2)(3x + 7) = 2x \cdot 3x + 2x \cdot 7 + 2 \cdot 3x + 2 \cdot 7$
$= 6x^2 + 14x + 6x + 14$
$= 6x^2 + 20x + 14$

EXAMPLE 6 Find the product of $4x - 3$ and $3x - 2$ using the FOIL method.

SOLUTION

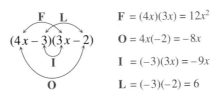

$\mathbf{F} = (4x)(3x) = 12x^2$
$\mathbf{O} = 4x(-2) = -8x$
$\mathbf{I} = (-3)(3x) = -9x$
$\mathbf{L} = (-3)(-2) = 6$

$(4x - 3)(3x - 2) = 12x^2 + [-8x + (-9x)] + 6$
$= 12x^2 - 17x + 6$ Combine like terms.

EXAMPLE 7 Find each of the following products mentally using the FOIL method:

(a) $(7x^2 - 4)(3x^2 + 3)$ **(b)** $(x + 4)(x + 4)$

SOLUTION **(a)** $(7x^2 - 4)(3x^2 + 3) = 21x^4 + 21x^2 - 12x^2 - 12$ FOIL
$= 21x^4 + 9x^2 - 12$ Combine like terms.

(b) $(x + 4)(x + 4) = x^2 + 4x + 4x + 16$ FOIL
$= x^2 + 8x + 16$ Combine like terms.

Example 7(b) is an illustration of the **square of a binomial.** It is one of several types of special products of binomials. Let's consider it again:

$$(x + 4)^2 = (x + 4)(x + 4) = x^2 + 8x + 16$$

First term — Middle term — Last term

Notice that the first and last terms of the product are the squares of the first and last terms of the binomial, respectively. The middle term of the product, $8x$, is twice the product of the two terms of the binomial, as the inner and outer products are identical. If a and b are any two terms, then

$$(a + b)^2 = (a + b)(a + b)$$
$$= a^2 + ab + ab + b^2 \quad \text{FOIL}$$
$$= a^2 + 2ab + b^2 \quad \text{Combine like terms.}$$

Substituting $-b$ for b, we learn that

$$(a - b)^2 = [a + (-b)]^2$$
$$= a^2 - 2a(b) + (-b)^2$$
$$= a^2 - 2ab + b^2$$

The squares of binomials occur so often that it is to your advantage to *memorize* both rules.

*C*AUTION

Note that the following quantities are not equal:

$$(a + b)^2 \neq a^2 + b^2$$
$$(a - b)^2 \neq a^2 - b^2$$

To Square a Binomial

The square of a binomial $a + b$ is equal to the square of the first term plus twice the product of the two terms plus the square of the second term.

$$(a + b)^2 = a^2 + 2ab + b^2$$
↑ ↑
1st term   2nd term

The square of a binomial $a - b$ is equal to the square of the first term minus twice the product of the two terms plus the square of the second term.

$$(a - b)^2 = a^2 - 2ab + b^2$$
↑ ↑
1st term   2nd term

EXAMPLE 8 Find the square of each binomial using the rules for squaring binomials:

(a) $(x + 7)^2$ **(b)** $(2x - 1)^2$ **(c)** $(3x + 2)^2$ **(d)** $(x^2 + 5y)^2$ **(e)** $(4x - 3y^3)^2$

SOLUTION

	First term ↓	Second term ↓	Square of the first term ↓	Twice the product of the two terms ↓	Square of the second term ↓
(a)	$(x + 7)^2$	=	x^2	$+ 14x$	$+ 49$
(b)	$(2x - 1)^2$	=	$4x^2$	$- 4x$	$+ 1$
(c)	$(3x + 2)^2$	=	$9x^2$	$+ 12x$	$+ 4$
(d)	$(x^2 + 5y)^2$	=	$(x^2)^2$	$+ 10x^2y$	$+ (5y)^2$
		=	x^4	$+ 10x^2y$	$+ 25y^2$
(e)	$(4x - 3y^3)^2$	=	$16x^2$	$- 24xy^3$	$+ 9y^6$

EXAMPLE 9 Expand $(x - 2)^3$.

SOLUTION To "expand" means to multiply out.

$(x - 2)^3 = (x - 2)(x - 2)^2$
$= (x - 2)(x^2 - 4x + 4)$ $(x - 2)^2 = x^2 - 4x + 4$
$= x^3 - 4x^2 + 4x - 2x^2 + 8x - 8$ Distributive property
$= x^3 - 6x^2 + 12x - 8$

There is another special type of product; suppose we want to multiply two binomials that differ only by the sign between their two terms. For example, $a - b$ and $a + b$ differ only by the sign between the two terms a and b. We can find their product using the FOIL method:

$(a + b)(a - b) = a^2 - ab + ab - b^2 = a^2 - b^2$

This type of product can be stated as a rule, which should also be memorized.

To Find the Product of the Sum and the Difference of Two Terms

The product of the sum and the difference of the same two terms is equal to the square of the first term minus the square of the second term:

$(a + b)(a - b) = a^2 - b^2$
   ↑   ↑
 1st term 2nd term

EXAMPLE 10 Find each of the following products by using the rule for finding the product of the sum and the difference of two terms:

(a) $(x + 5)(x - 5)$ (b) $(x + 1.1)(x - 1.1)$ (c) $(y^2 + 9)(y^2 - 9)$

(d) $(2a + 3b)(2a - 3b)$ (e) $\left(\dfrac{1}{2}x^3 + \dfrac{2}{3}a^6\right)\left(\dfrac{1}{2}x^3 - \dfrac{2}{3}a^6\right);$

(f) $(5x + 4y^2)(5x - 4y^2).$

SOLUTION

	Sum ↓		Difference ↓		Square of the first term ↓	Square of the second term ↓
(a) $(x$	$+$	$5)(x$	$-$	$5) =$	x^2	$- 25$
(b) $(x$	$+$	$1.1)(x$	$-$	$1.1) =$	x^2	$- 1.21$
(c) $(y^2$	$+$	$9)(y^2$	$-$	$9) =$	y^4	$- 81$
(d) $(2a$	$+$	$3b)(2a$	$-$	$3b) =$	$4a^2$	$- 9b^2$
(e) $\left(\dfrac{1}{2}x^3\right.$	$+$	$\left.\dfrac{2}{3}a^6\right)\left(\dfrac{1}{2}x^3\right.$	$-$	$\left.\dfrac{2}{3}a^6\right) =$	$\dfrac{1}{4}x^6$	$- \dfrac{4}{9}a^{12}$
(f) $(5x$	$+$	$4y^2)(5x$	$-$	$4y^2) =$	$25x^2$	$- 16y^4$

EXAMPLE 11 Simplify $(x - 3)(x + 3) - 2x^2(x^2 - 3x + 4) + (x - 2)^2$. Write the result in standard form.

SOLUTION

$$(x - 3)(x + 3) - 2x^2(x^2 - 3x + 4) + (x - 2)^2$$
$$= x^2 - 9 - 2x^4 + 6x^3 - 8x^2 + x^2 - 4x + 4$$
$$= -2x^4 + 6x^3 - 6x^2 - 4x - 5$$

EXAMPLE 12 Find the area inside the square but outside the circle; see Figure 6.3.

SOLUTION We can obtain the area inside the square but outside the circle by subtracting the area of the circle from the area of the square. The area of the square is $A_s = s^2$:

$$A_s = (2x + 4)^2 = 4x^2 + 16x + 16$$

Since the circle is inscribed in the square, its radius is one-half the length of the side of the square, or $x + 2$. The area of the circle is $A_c = \pi r^2$, so

$$A_c = \pi(x + 2)^2 = \pi(x^2 + 4x + 4)$$

Finally,

$$A = A_s - A_c = 4x^2 + 16x + 16 - \pi(x^2 + 4x + 4)$$
$$= 4x^2 + 16x + 16 - \pi x^2 - 4\pi x - 4\pi$$
$$= (4 - \pi)x^2 + (16 - 4\pi)x + 16 - 4\pi$$

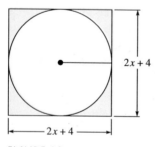

FIGURE 6.3

Do You Remember?

Can you match these?

_____ 1. $x^2(x^2 + 3x + 4)$
_____ 2. $(x + 4)(x - 4)$
_____ 3. $(x + 3)^2$
_____ 4. $(x - 3)^2$
_____ 5. $(x + 2)(x - 3)$
_____ 6. The middle term of $(a + b)^2$
_____ 7. The middle term of $(a + b)(a - b)$
_____ 8. $(2x + 3)(3x - 2)$

a) $x^2 + 6x + 9$
b) $x^2 - x - 6$
c) $2ab$
d) $6x^2 + 5x - 6$
e) 0
f) $x^2 + x - 6$
g) $x^2 - 16$
h) $x^2 - 6x + 9$
i) $a^2 + 2ab + b^2$
j) $a^2 - b^2$
k) $x^4 + 3x^3 + 4x^2$

Answers: 1. k 2. g 3. a 4. h 5. b 6. c 7. e 8. d

Exercise Set 6.4

Find the products. In Exercises 21–26, assume that the variables represent positive real numbers.

1. $x(x + 2)$
2. $y(y + 4)$
3. $x(x - 1)$
4. $y(2y - 3)$
5. $x(x^2 + 6x)$
6. $y(6y^2 - 1)$
7. $x(3x - 2y)$
8. $y(x - xy)$
9. $xy(x^2y + y)$
10. $xy(xy + x^2y)$
11. $xy(x^2y - xy)$
12. $xy(xy^2 - x^2y)$
13. $y^2(y^4 + 6y^2 - 2y + 1)$
14. $x^4(x^5 + 6x^2 + 4x - 1)$
15. $-2y(3y^2 - 2y + 4)$
16. $-2x(-x^2y - 6x + 3)$
17. $-xy^2(2xy^2 - 3x^2y + 2)$
18. $-5ab(-a^2b^3 + ab^2 + 6a)$
19. $3a^2b(6a^2b^3 + 2ab^2 - 6ab)$
20. $-3a^2b(-2ab + ab^2 + 6a + 8)$
21. $\sqrt{-2x}(-12xy + 15y^3 - 8x^2y + 6)$
22. $\sqrt{-pq}(p^2qr^2 - pqr + pq^2r^3)$
23. $\sqrt{9x^2}(-12xy + 15y^3 - 8x^2y + 6)$
24. $\sqrt{3x^4}(17x^3 - 14x^2 + 5x - 11)$
25. $\sqrt{-6m^3}(m^7 - 5m^6 + 11m^2 - 8m - 7)$
26. $\sqrt{-12p^5}(-7p^4 + 12p^3 - 7p^2 + 9p - 8)$
27. $-p^8\left(\dfrac{2}{3}p^5 - \dfrac{6}{7}p^3 + \dfrac{4}{9}p\right)$
28. $-p^8\left(\dfrac{1}{2}p^4 - \dfrac{3}{5}p^2 - 1\right)$
29. $m^5(0.5m^3 - 0.03m - 0.021)$
30. $-b(0.4b^2 - 0.96b + 0.03)$
31. $-12(-2x^4 + 5x^3 - 7x^2 + 6x - 9)$
32. $-17(-2u^4 - 5u^3 + 4u^2 - 7u + 11)$

Find the following products and then simplify by combining like terms.

33. $x(x + 2) - x(x - 3)$
34. $4a(2a - 3) - 7(a + 8)$
35. $3p(5p - 6) + 4p(-2p + 7)$
36. $-7y(2y + 3) + 5y(4y - 8)$
37. $-8(m^2 - 5m) + 6(m^2 + 4m)$
38. $-2t(3 - 7t^3) + 5t^2(8t - 1)$
39. $3y[2y^2 - y(2y + 4)]$
40. $2x[-7x^2 + x(-3x + 9)]$
41. $6t^3 - 2[7 - t(1 - t^2)]$
42. $2a[3a - 4(a^2 - 7) + 6a^2]$

43. $-13s(4s^3 - 6s) - 2s(-12s^3 + 24s)$
44. $4x^2y(6xy^2 - 3x^2y + 5) - 8x^3(3y^3 - 2xy^2 - 1)$
45. $-3z^3(5w^3z + 6) - 2z^2(8w^3z^2 - 9z + 1)$
46. $4x^2[2x^2 - 3(x + 4) - 2x^2(x - 8)]$

Find the following products and simplify where possible.

47. $(x + 2)(x + 3)$
48. $(y + 1)(y + 4)$
49. $(r - 3)(r - 2)$
50. $(p - 2)(p - 5)$
51. $(y + 6)(y - 4)$
52. $(x - 3)(x + 1)$
53. $(m + 5)^2$
54. $(s - 7)^2$
55. $(a + 8)(a - 8)$
56. $(r + 6)(r - 6)$
57. $(2x + 3)(x + 5)$
58. $(3y + 2)(y + 7)$
59. $(3r - 4)^2$
60. $(5m - 1)^2$
61. $(2p - 5)(2p + 5)$
62. $(4x - 7)(4x + 7)$
63. $(3m - 8)(2m + 9)$
64. $(4p + 4)(3p - 9)$
65. $(-2p - 3)(-5p + 8)$
66. $(-5x + 2)(-2x + 7)$
67. $(6 - 5m)(4m + 3)$
68. $(7 - 4a)(3a - 11)$
69. $(7x + 9)^2$
70. $(9y + 8)^2$
71. $(11y - 7)(11y + 7)$
72. $(13m + 8)(13m - 8)$
73. $(3x + 2y)(x - 4y)$
74. $(4p - 3t)(5p + t)$
75. $(2a - 7b)(8a + 5b)$
76. $(7r + 9s)(11r - 3s)$
77. $(9m - n)^2$
78. $(7a - 2b)^2$
79. $(5r + 3s)(5r - 3s)$
80. $(3p - 8q)(3p + 8q)$
81. $(x^2 + 2y)(x^2 - 2y)$
82. $(y^2 + 3x)(y^2 - 3x)$
83. $(x^2 - y^2)(x^2 + y^2)$
84. $(a^2 + b^2)(a^2 - b^2)$
85. $(3x^2 + 4y^2)^2$
86. $(2m^2 - 3n^2)^2$

Simplify each of the following expressions. Write your answers in standard form (descending powers of the variable).

87. $x(x - 1) + (x + 1)(x - 1)$
88. $y(y - 3) + (y + 2)(y - 2)$
89. $2p(3p + 2) + (p - 5)^2$
90. $3r(2r - 3) + (r + 3)^2$
91. $(2m - 1)^2 - 3m(m + 2)$
92. $(5x - 4)^2 - 4x(6x + 5)$
93. $a(a - 4) + 3a(a + 2) - 5a(a + 6)$
94. $p(p - 3) - 7p(p + 2) + 4p(3p + 7)$
95. $y(y - 2)^2 + 2y(y + 3)(y - 3)$
96. $x(x + 7)^2 - 3x(2x + 5)(2x - 5)$
97. $\left(\dfrac{1}{2}x - \dfrac{1}{4}y\right)^2$
98. $\left(\dfrac{3}{4}a + \dfrac{1}{3}b\right)^2$
99. $(0.5r + 0.1s)^2$
100. $(0.2p - 0.3q)^2$

101. $\left(\dfrac{3x}{2} - \dfrac{4y}{3}\right)^2$
102. $\left(\dfrac{5x}{3} + \dfrac{2y}{5}\right)^2$
103. $(x - 2)^3$
104. $(y + 3)^3$
105. $(2x + 1)^3$
106. $(3x - 4)^3$

107. Find the area of the given rectangle.

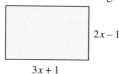

108. Find the area of the given circle.

109. Find the area of the shaded region.

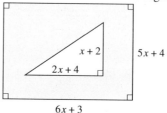

110. Find the area of the shaded region.

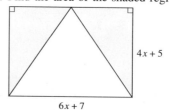

111. For the given rectangular solid: **(a)** find the total surface area and **(b)** find the volume.

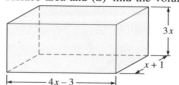

112. For the given sphere: **(a)** find the surface area and **(b)** find the volume.

6.5 Dividing Polynomials

OBJECTIVES In this section we will learn to
1. divide a polynomial by a monomial; and
2. divide a polynomial by a polynomial.

In Sections 6.3 and 6.4 we saw that the sum, the difference, and the product of polynomials is again a polynomial. This is not always the case when we divide two polynomials.

We use the quotient rule for exponents to divide a monomial by a monomial. For example,

$$\frac{8x^7}{2x^3} = 4x^4, \qquad \frac{9x^2y^4}{3x^3y^2} = \frac{3y^2}{x}, \qquad \text{and} \qquad \frac{5a^5b^7}{2ab^9} = \frac{5a^4}{2b^2}$$

To understand how to divide a polynomial by a monomial, let's consider the quotient of 12 and 2.

$$\frac{12}{2} = \frac{10+2}{2} = \frac{10}{2} + \frac{2}{2} = 5 + 1 = 6$$

To Divide a Polynomial by a Monomial

Divide each term of the polynomial by the monomial. Use the quotient rule for exponents to simplify.

EXAMPLE 1 Divide $6x^4 + 3x^3$ by $3x^2$.

SOLUTION
$$\frac{6x^4 + 3x^3}{3x^2} = \frac{\text{A polynomial}}{\text{A monomial}}$$

$$= \frac{6x^4}{3x^2} + \frac{3x^3}{3x^2} \qquad \text{Divide each term of the polynomial by the monomial.}$$

$$= 2x^2 + x \qquad \text{Quotient rule for exponents}$$

EXAMPLE 2 Divide $3y^3 + 6y^2 + 9y$ by $3y$.

SOLUTION
$$\frac{3y^3 + 6y^2 + 9y}{3y} = \frac{3y^3}{3y} + \frac{6y^2}{3y} + \frac{9y}{3y} \qquad \text{Divide each term by the monomial.}$$

$$= y^2 + 2y + 3 \qquad \text{Quotient rule for exponents}$$

EXAMPLE 3 Divide $6a^3 + 4a^2 - 3a$ by $-2a^2$.

SOLUTION
$$\frac{6a^3 + 4a^2 - 3a}{-2a^2} = \left(\frac{6a^3}{-2a^2}\right) + \left(\frac{4a^2}{-2a^2}\right) - \left(\frac{3a}{-2a^2}\right)$$
$$= -3a + (-2) - \left(-\frac{3a}{2a^2}\right)$$
$$= -3a - 2 + \frac{3}{2a}$$

In Example 3, the quotient is not a polynomial because the variable, a, appears in the denominator of the last term.

To divide a polynomial by a polynomial, we use **long division.** This method is similar to the long division process from arithmetic, as illustrated in the following steps. To use long division in algebra, we must write the problem in **standard form,** where both the divisor and the dividend are written in descending powers of the variable. Recall from arithmetic that the divisor is written to the left of the division symbol and the dividend is written under the division symbol.

	Divide 506 by 22.	Divide $x^2 + 5x + 6$ by $x + 2$.
Step 1	**Write the problem in standard form.** $22\overline{)506}$	**Write the problem in standard form.** $x + 2\overline{)x^2 + 5x + 6}$
Step 2	**Divide:** 22 divides into 50 a total of (two) times. $\begin{array}{r} 2 \\ 22\overline{)506} \\ \underline{44} \end{array}$ Multiply: $2(22) = 44$	**Divide:** x divides into x^2 a total of (x) times. $\begin{array}{r} x \\ x+2\overline{)x^2 + 5x + 6} \\ \underline{x^2 + 2x} \end{array}$ Multiply: $x(x+2) = x^2 + 2x$
Step 3	**Subtract and bring down the next term.** $\begin{array}{r} 2 \\ 22\overline{)506} \\ \underline{44} \\ 66 \end{array}$	**Subtract and bring down the next term.** $\begin{array}{r} x \\ x+2\overline{)x^2 + 5x + 6} \\ \underline{x^2 + 2x} \\ 3x + 6 \end{array}$
Step 4	**Divide:** 22 divides into 66 a total of (three) times. $\begin{array}{r} 23 \\ 22\overline{)506} \\ \underline{44} \\ 66 \\ \underline{66} \end{array}$ Multiply: $3(22) = 66$.	**Divide:** x divides into $3x$ a total of (three) times. $\begin{array}{r} x + 3 \\ x+2\overline{)x^2 + 5x + 6} \\ \underline{x^2 + 2x} \\ 3x + 6 \\ \underline{3x + 6} \end{array}$ Multiply: $3(x+2) = 3x + 6.$

$$
\begin{array}{r}
23 \\
22{\overline{\smash{\big)}\,506}} \\
\underline{44} \\
66 \\
\underline{66} \\
0
\end{array}
\quad\text{Subtract: the remainder is zero.}
\qquad
\begin{array}{r}
x + 3 \\
x + 2{\overline{\smash{\big)}\,x^2 + 5x + 6}} \\
\underline{x^2 + 2x} \\
3x + 6 \\
\underline{3x + 6} \\
0
\end{array}
\quad\text{Subtract: the remainder is zero.}
$$

Since there are no more terms to be brought down, the divisions are complete.

Step 5   **To check the result, multiply the divisor by the quotient.**   **To check the result, multiply the divisor by the quotient.**

$$
\begin{array}{r}
23 \\
\underline{\times\, 22} \\
46 \\
\underline{46} \\
506
\end{array}
$$

506 divided by 22 is 23.

$(x + 2)(x + 3) \stackrel{?}{=} x^2 + 3x + 2x + 6$
$\stackrel{\checkmark}{=} x^2 + 5x + 6$

$x^2 + 5x + 6$ divided by $x + 2$ is $x + 3$.

It is interesting to note that when x is 20, the second problem is really the same as the first.

When x is 20, the divisor is $x + 2 = 20 + 2 = 22$.

When x is 20, the dividend is $x^2 + 5x + 6 = (20)^2 + 5(20) + 6 = 506$.

When x is 20, the quotient is $x + 3 = 20 + 3 = 23$.

EXAMPLE 4   Divide $x^2 + x - 6$ by $x - 2$, and check the result.

SOLUTION   **Step 1** $x - 2{\overline{\smash{\big)}\,x^2 + x - 6}}$ Write in standard form.

Step 2 $\begin{array}{r} x \\ x - 2{\overline{\smash{\big)}\,x^2 + x - 6}} \\ \underline{x^2 - 2x} \end{array}$ x divides into x^2 a total of x times; $x(x - 2) = x^2 - 2x$

Step 3 $\begin{array}{r} x \\ x - 2{\overline{\smash{\big)}\,x^2 + x - 6}} \\ \underline{x^2 - 2x} \\ 3x - 6 \end{array}$ Subtract and bring down the next term. Remember: to subtract, add the opposite.

Step 4 $\begin{array}{r} x + 3 \\ x - 2{\overline{\smash{\big)}\,x^2 + x - 6}} \\ \underline{x^2 - 2x} \\ 3x - 6 \\ \underline{3x - 6} \\ 0 \end{array}$

x divides into $3x$ a total of three times; $3(x - 2) = 3x - 6$

Subtract.

The remainder is zero.

Step 5   To **check the result,** multiply the divisor by the quotient. When there is no remainder, the product must equal the dividend

$(x - 2)(x + 3) \stackrel{\checkmark}{=} x^2 + x - 6$

The next example demonstrates how to handle the division problem when the remainder is not zero.

EXAMPLE 5 Divide $4x^3 + 6x^2 + 8x + 2$ by $2x + 1$, and check the result.

SOLUTION **Step 1** $2x + 1 \overline{) 4x^3 + 6x^2 + 8x + 2}$ Write in standard form.

Step 2
$$\begin{array}{r} 2x^2 \\ 2x+1 \overline{) 4x^3 + 6x^2 + 8x + 2} \\ \underline{4x^3 + 2x^2} \end{array}$$

$2x$ divides into $4x^3$ a total of $2x^2$ times.

$2x^2(2x + 1) = 4x^3 + 2x^2$

$$\begin{array}{r} 2x^2 \\ 2x+1 \overline{) 4x^3 + 6x^2 + 8x + 2} \\ \underline{4x^3 + 2x^2} \\ 4x^2 + 8x \end{array}$$

Subtract and bring down the next term.

Step 3
$$\begin{array}{r} 2x^2 + 2x \\ 2x+1 \overline{) 4x^3 + 6x^2 + 8x + 2} \\ \underline{4x^3 + 2x^2} \\ 4x^2 + 8x \\ \underline{4x^2 + 2x} \end{array}$$

$2x$ divides into $4x^2$ a total of $2x$ times.

$2x(2x + 1) = 4x^2 + 2x$

$$\begin{array}{r} 2x^2 + 2x \\ 2x+1 \overline{) 4x^3 + 6x^2 + 8x + 2} \\ \underline{4x^3 + 2x^2} \\ 4x^2 + 8x \\ \underline{4x^2 + 2x} \\ 6x + 2 \end{array}$$

Subtract and bring down the next term.

Step 4
$$\begin{array}{r} 2x^2 + 2x + 3 \\ 2x+1 \overline{) 4x^3 + 6x^2 + 8x + 2} \\ \underline{4x^3 + 2x^2} \\ 4x^2 + 8x \\ \underline{4x^2 + 2x} \\ 6x + 2 \\ \underline{6x + 3} \end{array}$$

$2x$ divides into $6x$ a total of 3 times.

$3(2x + 1) = 6x + 3$

The remainder is $-1. \rightarrow -1$ To subtract, add the opposite.

When the degree of the remainder is less than the degree of the divisor, the division is considered complete. Here the degree of the divisor, $2x + 1$, is 1. The degree of the remainder, -1, is zero. We write the result

$$2x^2 + 2x + 3 + \frac{-1}{2x + 1}$$

Step 5 Check   dividend $\stackrel{?}{=}$ divisor $\times$ quotient $+$ remainder

$$4x^3 + 6x^2 + 8x + 2 \stackrel{?}{=} (2x + 1)(2x^2 + 2x + 3) + (-1)$$
$$\stackrel{?}{=} 4x^3 + 4x^2 + 6x + 2x^2 + 2x + 3 + (-1)$$
$$\stackrel{\checkmark}{=} 4x^3 + 6x^2 + 8x + 2$$

EXAMPLE 6 Divide $27 + y^3$ by $3 + y$.

SOLUTION First we rewrite both the divisor and the dividend in standard form:

$$(y^3 + 27) \div (y + 3)$$

Notice that there is neither a y^2- nor a y-term in the dividend. We get around this by writing those missing terms with zero coefficients.

> **C**AUTION
>
> Both the divisor and the dividend must be in standard form before doing long division.

$$\begin{array}{r} y^2 - 3y + 9 \\ y+3 \overline{\smash{\big)}\, y^3 + 0y^2 + 0y + 27} \\ \underline{y^3 + 3y^2 } \\ -3y^2 + 0y \\ \underline{-3y^2 - 9y } \\ 9y + 27 \\ \underline{9y + 27} \end{array}$$

To subtract, change the sign and add.

The quotient is $y^2 - 3y + 9$.

EXAMPLE 7 Divide $2x^3 + 6x + x^2 - 5$ by $1 + 2x^2 - 3x$.

SOLUTION The dividend and divisor are not in descending powers of the variable, so we must rearrange their terms:

$$\begin{array}{r} x + 2 \\ 2x^2 - 3x + 1 \overline{\smash{\big)}\, 2x^3 + x^2 + 6x - 5} \\ \underline{2x^3 - 3x^2 + x } \\ 4x^2 + 5x - 5 \\ \underline{4x^2 - 6x + 2} \\ 11x - 7 \end{array}$$

The degree of $11x - 7$ is less than the degree of $2x^2 - 3x + 1$, so the division is complete. The result is

$$x + 2 + \frac{11x - 7}{2x^2 - 3x + 1}$$

EXAMPLE 8 What polynomial yields a quotient of $2x - 4$ when divided by $3x + 2$?

SOLUTION The divisor times the quotient must be equal to the dividend. Therefore, since

$$(3x + 2)(2x - 4) = 6x^2 - 8x - 8$$

the polynomial is $6x^2 - 8x - 8$.

EXAMPLE 9 The volume of a right circular cylinder is given by $V = \pi r^2 h$. Find the height of a cylinder with radius $r = (x + 1)$ cm and volume $V = \pi(x^3 + 5x^2 + 7x + 3)$ cm^3, where cm^3 means cubic centimeters.

SOLUTION Substituting the known quantities into the formula for volume gives us

$$\pi(x+1)^2 h = \pi(x^3 + 5x^2 + 7x + 3)$$

To solve for h, we divide each side by $\pi(x+1)^2$:

$$\frac{\pi(x+1)^2 h}{\pi(x+1)^2} = \frac{\pi(x^3 + 5x^2 + 7x + 3)}{\pi(x+1)^2}$$

$$h = \frac{x^3 + 5x^2 + 7x + 3}{x^2 + 2x + 1} \qquad \text{Simplify.}$$

Now we divide to find h.

$$\begin{array}{r} x + 3 \\ x^2 + 2x + 1 \overline{)x^3 + 5x^2 + 7x + 3} \\ \underline{x^3 + 2x^2 + x} \\ 3x^2 + 6x + 3 \\ \underline{3x^2 + 6x + 3} \end{array}$$

The height of the cylinder is $(x + 3)$ cm.

Exercise Set 6.5

Find each of the following quotients. Check your answers.

1. $\dfrac{64}{8}$

2. $\dfrac{34}{17}$

3. $\dfrac{15y^2}{5y}$

4. $\dfrac{48p^3}{6p^2}$

5. $\dfrac{-36a^2}{9a}$

6. $\dfrac{81x^2 y}{-3xy}$

7. $\dfrac{-121a^3 b^9}{-11a^4}$

8. $\dfrac{-91m^5 n^4}{-7mn^5}$

9. $\dfrac{-52a^6 b^3}{-13ab^6}$

10. $\dfrac{84r^3 s^2 t}{56r^2 s^2 t^2}$

11. $\dfrac{-92x^5 y^9 z^2}{46xy^8 z^2}$

12. $\dfrac{51c^8 d^7 e^4}{-34c^7 d^8 e^4}$

13. $\dfrac{9x + 3}{3}$

14. $\dfrac{6x - 2}{2}$

15. $\dfrac{x^2 + 7x}{x}$

16. $\dfrac{3y^2 - 9y}{3y}$

17. $\dfrac{15a^2 - 12a}{3a}$

18. $\dfrac{p^2 + p + 1}{2p}$

19. $\dfrac{8r^3 - 4r^2 + 3r}{2r}$

20. $\dfrac{15s^2 + 7s - 5}{5}$

21. $\dfrac{120t^2 - 80t^5}{30t^3}$

22. $\dfrac{21x^3 - 2x^2}{6x^2}$

23. $\dfrac{17 + 8y + y^2}{y}$

24. $\dfrac{p^7 - p^4}{p^3}$

25. $\dfrac{-24q^3 - 42q^2}{8q}$

26. $\dfrac{-38z^4 - 10z^3}{8z}$

27. $\dfrac{45w^3 + 15w^2 - 9w}{-3w}$

28. $\dfrac{8t^5 - 3t^3 + 2t}{5t^2}$

29. $(-80x^3 y^4 + 32x^2 y^4) \div (-16x^2 y)$

30. $(-100d^5 + 50d^4 - 30d^3 + 20d^2 + 40d) \div (-20d)$

31. $(91x^5 y^5 - 13x^2 y^3) \div (-26x^2 y^3)$

32. $(144a^7 b^6 - 156a^6 b^7 - 84a^3 b^2 - 108a^2 b^3) \div (-12ab)$

Solve each of the following division problems.

33. What polynomial yields a quotient of 24 when divided by 4?

34. What polynomial yields a quotient of $4x^2 - 3x$ when divided by $4x^3$?

35. The quotient of a polynomial divided by $30a^2$ is $6a^8 - 5a^6 + 4a^4 - 3a^2 + 4$. Find the polynomial.

36. The quotient of a polynomial divided by $4y$ is $3y^3 - \dfrac{7}{4}y^2 + \dfrac{1}{4} - \dfrac{1}{y}$. Find the polynomial.

Find each of the following quotients. Check your answers.

37. $1624 \div 29$
38. $3552 \div 37$
39. $(x^2 + 4x + 4) \div (x + 2)$
40. $(y^2 - 6y + 9) \div (y - 3)$
41. $(p^2 + 7p + 6) \div (6 + p)$
42. $(a^2 + a - 6) \div (a - 2)$
43. $(6x^4 + 8x^3 - 10x^2 - 3) \div (x + 3)$
44. $(3m^3 + 8m^2 + 5m + 2) \div (2 + m)$
45. $(a^3 + 1) \div (a + 1)$
46. $(x^5 - 32) \div (x - 2)$
47. $(y^3 - 8) \div (y - 2)$
48. $(c^3 + 64) \div (c + 4)$
49. $(2p^2 - 2p + 5) \div (2p + 4)$
50. $(10x^2 + 31x - 15) \div (2x + 7)$
51. $(2y^3 - y^2 + 3y + 2) \div (2y + 1)$
52. $(54a^3 - 39a^2 - 11a + 1) \div (6a - 5)$
53. $(8x^4 - 12x^3 - 2x^2 + 7x - 6) \div (2x - 3)$
54. $(9y^4 - 6y^3 - 5y^2 - y - 4) \div (4 - 3y)$
55. $(64s^6 - 1) \div (4s^2 - 1)$
56. $(y^6 - 1) \div (y^3 + 1)$
57. $(3a^3 + a^2 + 3a + 1) \div (a^2 + 1)$
58. $(2a^5 + 6a^4 - a^3 + 3a^2 - a) \div (2a^2 - 1)$
59. $(p^5 + 5p^4 + 5p^3 + 1) \div (p^2 + 2p + 1)$
60. $(10y^3 + 17y^2z + 23yz^2 + 4z^3) \div (5y + z)$
61. $(-17y^2 - 3 + 26y + 5y^3) \div (2 - 5y)$
62. $(22p - 17p^2 + 6p^3 - 24) \div (7 - 3p)$
63. $(-14m^3 - 2m^2 + 3m^4 + 19m) \div (-5m + m^2 + 2)$
64. $(3y^2 - 3y^4 - 8y^3 - 4 + 10y) \div (-2y - 3y^2 + 4)$
65. $(13a^3 - 34a + 3a^4 - 11a^2 + 19) \div (-a^2 + 7 - 3a)$
66. $(-23x^2 - 12 - 28x + 6x^4) \div (-4x - 3 + 2x^2)$
67. $(2 + 27y^3) \div (-2 + 3y)$
68. $(6 + 8p^3) \div (1 + 2p)$
69. $(2x^4 - 6x^3 - 3x^2 + 24x - 17) \div (x^2 - x - 5)$
70. $(3y^4 + 19y^3 - 4y^2 - 24y + 13) \div (y^2 + 6y - 2)$

Solve each of the following applied problems.

71. The area of a rectangle is $(x^2 - 5x + 4)$ in^2. If the length is $(x - 1)$ in., find the width.
72. The area of a triangle is $(2x^2 + 2x - 24)$ ft^2. If the base is $(x + 4)$ ft, find the height.
73. The volume of a rectangular box is $(4x^3 + 8x^2)$ cm^3. If the width and height are each $2x$ cm, find the length.
74. The volume of a right circular cylinder is given by $\pi(x^3 - 2x^2 - 4x + 8)$ m^3. If the radius is $(x - 2)$ m, find the height.
75. Three consecutive integers can be represented as x, $x + 1$, and $x + 2$. Show that the average of these three integers is the middle integer. Is this true for five consecutive integers? For seven? For any odd number of consecutive integers?
76. Three consecutive odd (or even) integers can be represented as x, $x + 2$, and $x + 4$. Show that the average of the three consecutive odd integers is the middle integer. Is this true for five consecutive odd integers? For seven? For any odd number of consecutive odd (or even) integers?

REVIEW EXERCISES

Simplify each expression. Write your answers using only positive exponents. Assume no denominator equals zero.

77. $x^5 \cdot x^{-3}$
78. $x^{-8} \cdot x^4$
79. $\dfrac{y^5}{y^3}$
80. $\dfrac{y^{-7}}{y^3}$
81. $\dfrac{x^{-4}}{x^{-3}}$
82. $(x^4 y^{-3})^{-2}$
83. $(3x^{-1})^3$
84. $(x^{-2} y^{-1})^2$
85. $\dfrac{-91y^{-5}}{13y^{-5}}$
86. $\left(\dfrac{x^3}{y^2}\right)^{-2}$
87. $\left(\dfrac{a^{-2}}{b^3}\right)^{-4}$
88. $\left(\dfrac{2x^{-1}}{x^{-3}}\right)^{-2}$
89. $\left(\dfrac{27x^{-3}}{9x^2}\right)^{-1}$
90. $\left(\dfrac{-35y^2}{7y^{-1}}\right)^{-2}$

6.6 More Applications

OBJECTIVES

In this section we will learn to
1. apply the rules of exponents to solve applied problems; and
2. solve applied problems involving polynomials.

Exponents have many applications in everyday problems as well as in many other situations in science and industry. The next few examples show some practical applications.

EXAMPLE 1 The volume of a cube is given by the formula $V = s^3$, where s represents the length of a side. Find the volume of a cube with 3-foot sides. See Figure 6.4.

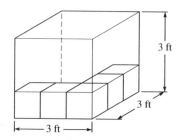

FIGURE 6.4

SOLUTION To find the volume, we substitute 3 feet for s in the volume formula.

$$V = s^3 = (3 \text{ ft})^3 = 3^3 \text{ ft}^3 = 27 \text{ ft}^3$$

The exponent of 3 on feet means the answer is 27 cubic feet. All volumes of objects are given in cubic units.

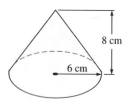

FIGURE 6.5

EXAMPLE 2 The volume of a cone is given by the formula $V = \frac{1}{3}\pi r^2 h$, where r is the radius of the base and h is the height. Find the volume of a cone with a base radius of 6 centimeters and a height of 8 centimeters. Use 3.14 for π. See Figure 6.5.

SOLUTION To find the volume, we substitute 6 cm for r and 8 cm for h in the formula, using 3.14 for π:

$$V = \frac{1}{3}\pi r^2 h$$

$$= \frac{1}{3}(3.14)(6 \text{ cm})^2(8 \text{ cm})$$

$$= \frac{1}{3}(3.14)(36 \text{ cm}^2)(8 \text{ cm})$$

$$= 301.44 \text{ cm}^3 \qquad (\text{cm}^2)(\text{cm}) = \text{cm}^3$$

Note that in Example 2 and in Example 3, which follows, 3.14 is an approximation to π and not its exact value. Therefore the answers to the two examples are approximations to the exact solutions. To be completely correct in the solutions, the equal sign should be replaced by $\approx$, which means "is approximately equal to."

EXAMPLE 3   A spherical water tank is to be coated with rust-resistant paint. How much paint should be ordered for a tank with a 10-foot radius if one gallon covers 300 square feet?

SOLUTION   The formula for the surface area of a sphere is $S = 4\pi r^2$. Substituting 10 ft for r and 3.14 for π gives us

$$S = 4(3.14)(10 \text{ ft})^2 = 4(3.14)(100 \text{ ft}^2) = 1256 \text{ ft}^2$$

To find the number of gallons needed, we divide the surface area, 1256 ft², by the 300 ft² that each gallon covers:

$$\text{number of gallons} = \frac{1256 \text{ ft}^2}{300 \text{ ft}^2/\text{gal}} = \frac{1256}{300} \approx 4.19 \text{ gal}$$

If the paint is available only in gallon containers, 5 gallons must be ordered.

In Examples 1–3, we substituted the dimensions into the formula to show how we should label the final results. This isn't necessary if the units of the final result are apparent.

EXAMPLE 4   A ball is thrown vertically upward from the top of a building that is 100 feet high. If its initial velocity is 66 feet/second and its height at any time t is given by the formula $h = 100 + 66t - 16t^2$, how high will it be after 4 seconds?

SOLUTION   To determine the height of the ball, we substitute 4 for t in the given formula:

$$h = 100 + 66(4) - 16(4^2)$$
$$= 100 + 264 - 256$$
$$= 108$$

The ball is 108 feet high 4 seconds after being released.

EXAMPLE 5   A cube has a side of $2x + 1$ feet. Find its volume and surface area.

SOLUTION
$$\begin{aligned}V = s^3 &= (2x + 1)^3 \qquad \text{Substitute } 2x + 1 \text{ for } s.\\&= (2x + 1)^2(2x + 1)\\&= (4x^2 + 4x + 1)(2x + 1)\\&= 8x^3 + 4x^2 + 8x^2 + 4x + 2x + 1\\&= (8x^3 + 12x^2 + 6x + 1) \text{ ft}^3\end{aligned}$$

The six faces of a cube are squares that are $(2x + 1)$ ft on a side. The surface area is

$$A = 6s^2 = 6(2x + 1)^2 \qquad \text{Substitute } 2x + 1 \text{ for } s.$$
$$= 6(4x^2 + 4x + 1)$$
$$= (24x^2 + 24x + 6) \text{ ft}^2$$

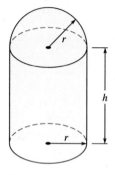

FIGURE 6.6

EXAMPLE 6 The volume of the silo shown in Figure 6.6 is given by $V = \pi r^2 h + \dfrac{2}{3} \pi r^3$. Its dimensions are $r = 2.5 \times 10^1$ ft and $h = 1.7 \times 10^2$ ft. Use $\pi = 3.1416$ to find its volume.

SOLUTION

$$V = \pi r^2 h + \dfrac{2}{3} \pi r^3$$
$$= 3.1416[(2.5 \times 10^1)^2(1.7 \times 10^2)] + \dfrac{2}{3}(3.1416)(2.5 \times 10^1)^3$$
$$= 3.1416(6.25 \times 10^2)(1.7 \times 10^2) + \dfrac{2}{3}(3.1416)(15.625 \times 10^3)$$
$$= 3.1416(10.625 \times 10^4) + \dfrac{2}{3}(49.0875 \times 10^3)$$
$$= 333{,}795 + 32{,}725$$
$$= 366{,}520 \text{ ft}^3$$

Exercise Set 6.6

Solve each applied problem. Use formulas from the book as needed.

1. Find the volume of a sphere if $r = 1$ ft. Use $\pi = 3.14$.
2. If an object is dropped from a tall tower, the distance s in feet it will fall in t seconds is given by the formula $s = 16t^2$. Find the distance the object falls during the first 4 sec.
3. Find the area of a circular enclosure if $r = 10$ m. Use $\pi = 3.14$.
4. In an electrical circuit, the power P equals the electromotive force E multiplied by the current I decreased by the product of the resistance R and the square of the current: $P = EI - RI^2$. Find P for $E = 12$, $I = 3$, and $R = 3$.
5. The population of Corona grew according to the formula $P = 2000t^2 + 100{,}000$ during the last ten years. If we call the time $t = 0$ in 1985, what is the population in 1995?
6. The weight in pounds of a square beam of wood can be found by the formula $W = 0.02lx^2$. Find the weight of a beam of length $l = 60$ in. and width $x = 2$ in. See the accompanying figure.

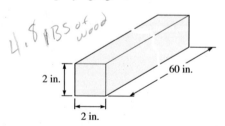

7. Find the power P consumed by a light bulb that has resistance $R = 1.2 \times 10^1$ when the current is $I = 2$. Use the formula $P = I^2R$.
8. The cost C in dollars of producing x hair dryers is given by the formula $C(x) = 360 + 10x + 0.2x^2$. Find the cost of producing 100 hair dryers.

9. Automotive engineers use the formula $H = \dfrac{D^2 N}{0.5}$ to approximate the horsepower rating H of an engine. The letter D represents the diameter of a cylinder in inches and N the number of cylinders. Find the horsepower rating of an eight-cylinder engine with $D = 3.5$ in.

10. Find the volume of a right circular cylinder with radius 2 in. and height 5 in. See the accompanying figure.

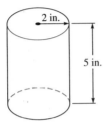

11. The electrical resistance of a length L of wire of diameter D can be found by the formula $R = \dfrac{rL}{D^2}$. Find R given $L = 2.4 \times 10^1$ in., $D = 2.5 \times 10^{-2}$ in., and $r = 1 \times 10^2$ ft.

12. A company's profit P in dollars when x items are manufactured and sold is given by the formula
$$P(x) = -360 + 440x - 0.04x^2$$
What is the profit when $x = 20$?

13. The volume of a football is approximated by the formula $V = \dfrac{\pi L T^2}{6}$, where L is the length and T is the diameter at the widest place. Find the volume of a football with a length of 10 in. and a thickness of 6 in.

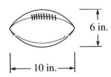

14. Neglecting air resistance, a stone thrown vertically upward will reach a maximum height according to the formula $h = \dfrac{-v^2}{2a}$. Find h when $v = -16$ m/sec and $a = -9.8$ m/sec^2.

15. According to the California State Department of Health and Education, average health costs in recent years can be determined by the formula $C(x) = 2085 + 900x^2$. If x is the number of hospital admittances of a patient in one year, what is the projected cost to a patient who is hospitalized three times in one year?

16. The power P required by a metal punch machine is approximated by the formula $P = \dfrac{t^2 d}{3.78}$, where t is the thickness of the metal being punched and d is the diameter of the hole. Find the power required to punch a 2-in. hole through a $\frac{1}{4}$-in. sheet of metal.

17. The weight in pounds of a circular concrete support column is given by the formula $W = 154\pi r^2 h$. What is the weight of a column 1 ft in diameter ($d = 2r$) and 10 ft high?

18. How much greater in area is the cross section of a pipe with a radius of 4 in. than one with a radius of 2 in.? (The cross section is a circle; see figure.)

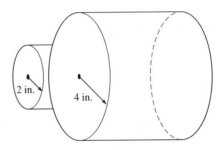

19. How many gallons of water will a conical tank hold if its height is 120 in. and its radius is 40 in.? The volume of one gallon of water is 231 in^3.

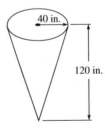

20. The maximum load in tons that a particular cylindrical pillar can carry is given by the formula $L = 4000d^2/h^2$. What weight can a pillar support if $d = 3$ m and $h = 12$ m?

21. The volume of a rectangular box is given by $2x^3 - 11x^2 - 12x + 36$. If the width is $x - 6$ units and the length is $x + 2$ units, find the height.

22. The volume of a conical tank is given by the formula $\frac{1}{3}\pi(3x^3 - 8x^2 + 16x)$. If the radius is $3x - 4$ units, find the height.

23. The volume of a right circular cylinder is given by $\pi(3x^3 - 7x^2 + 5x - 1)$. If the radius is $x - 1$ units, find the height.

24. The radius of a sphere is $x - 2$ units. Find its volume.
25. The edge of a cube is $2x + 5$ units. Find its volume and its surface area.
26. A rectangular box has a length of $4x - 1$ units, a width of $x + 1$ units, and a height of $x - 1$ units. Find its volume and its surface area.

REVIEW EXERCISES

Write a system of equations describing each statement, and solve by any method.

27. A box of apples and a box of pears together weigh a total of 70 lb. The difference between their weights is 20 lb. Find the weight of each box.
28. The sum of two numbers is 24, and one number is 6 more than the other. Find the numbers.
29. The difference of two numbers is 15, and one number is twice the other. Find the numbers.
30. Fred and his daughter Millie together weigh 200 lb. Fred weighs 20 lb more than twice what Millie weighs. Find their weights.

Summary

6.1 Multiplication and Division with Exponents

In the expression b^n, the letter b is called the **base** and n the **exponent.** Two important rules governing operations with exponents are the **product rule for exponents** and the **quotient rule for exponents.** And for consistency with these rules, zero and negative exponents must be defined as follows. If m and n represent integers and $b \neq 0$, then

Product rule $\quad b^m \cdot b^n = b^{m+n}$

Quotient rule $\quad \dfrac{b^m}{b^n} = b^{m-n}$

Definitions $\quad b^0 = 1; \quad b^{-n} = \dfrac{1}{b^n}$

The product and quotient rules for exponents apply only to expressions where the bases are the same.

6.2 More on Exponents; Scientific Notation

Other situations involving exponents are covered by the **power rules.** If m and n are integers, then

$$(b^m)^n = b^{mn}; \quad (ab)^m = a^m b^m; \quad \left(\dfrac{a}{b}\right)^m = \dfrac{a^m}{b^m}, \quad b \neq 0$$

Scientific notation can be used to write any number as a number times a power of 10. When a number is written in scientific notation, it takes the form $b \times 10^n$, where $1 \leq b < 10$ and n is an integer. If the number is less than 1, the power of ten is negative.

6.3 Adding and Subtracting Polynomials

A **term** is a number or the product of numbers and variables raised to powers.

The **degree** of a term in one variable is the exponent on that variable.

A **polynomial** is a single term or the sum or difference of a finite number of terms. The **degree of a polynomial in one variable** is the largest exponent that appears on that variable in any term.

A polynomial with one term is called a **monomial,** a polynomial with two terms is called a **binomial,** and a polynomial with three terms is called a **trinomial.**

When a polynomial is arranged in descending powers of the variable it is said to be in **standard form.**

Polynomials can be added or subtracted by writing them in a horizontal or a vertical format. In more advanced mathematics, the horizontal format is used almost exclusively.

When **subtracting** one polynomial from another, change every sign in the polynomial being subtracted and then follow the rules for addition.

Function notation is often used to indicate a polynomial in one variable. It is a convenient way to indicate the value of the polynomial for a particular quantity.

6.4 Multiplying Polynomials

To multiply a polynomial by a monomial, each term of the polynomial is multiplied by the monomial, using the product rule for exponents.

To find the product of two polynomials, each term in one polynomial is multiplied by each term in the other polynomial and like terms combined. When the two polynomials are binomials, the product can be found by using the **FOIL method:**

$$\left.\begin{array}{l}\textbf{F} \text{ is the product of the first terms.} \\ \textbf{O} \text{ is the product of the outer terms.} \\ \textbf{I} \text{ is the product of the inner terms.} \\ \textbf{L} \text{ is the product of the last terms.}\end{array}\right\} \text{Product} = F + O + I + L$$

Two special products are the **square of a binomial** and the **product of the sum and difference of the same two terms.** These products are used so frequently that the rules that create them should be committed to memory.

To Square a Binomial
The square of a binomial is equal to the square of the first term plus or minus twice the product of the two terms plus the square of the second term:

$$(a + b)^2 = a^2 + 2ab + b^2 \qquad (a - b)^2 = a^2 - 2ab + b^2$$

$\uparrow \quad \uparrow$ $\uparrow \quad \uparrow$
1st term 2nd term   1st term 2nd term

To Find the Product of a Sum and a Difference
The product of the sum and the difference of the same two terms is equal to the square of the first term minus the square of the second term.

$$(a + b)(a - b) = a^2 - b^2$$

$\uparrow \quad \uparrow$
1st term 2nd term

6.5 Dividing Polynomials

To divide a polynomial by a monomial, divide each term of the polynomial by the monomial. The process used to carry out **long division** in algebra is essentially the same as it is in arithmetic. Division of two polynomials does not always yield a polynomial.

6.6 More Applications

There are many applications of exponential expressions and polynomials. Areas, volumes, and scientific notation are but a few of the more common ones.

Cooperative Exercise

The following information pertains to the revenue and costs of the Kim Company over a certain time period.

SALES (5,000 units at $400 per unit): $2,000,000

COSTS	FIXED	VARIABLE	
Material	------	$ 260,000	
Labor	------	410,000	
Overhead	$150,000	520,000	
Administrative	200,000	100,000	
Other expense	190,000	110,000	
TOTALS	$540,000	$1,400,000	$1,940,000
PROFIT			$ 60,000

In general, if x units are sold at s dollars each, with a variable cost of v dollars per unit and a fixed cost of f dollars, then:

The cost function is $C(x) = vx + f$.

The revenue function is given by $R(x) = sx$.

The profit function is given by $P(x) = (s - v)x - f$.

The break-even point is where $x = \dfrac{f}{s - v}$.

For the Kim Company,

(a) Find the revenue function.

(b) Find the cost function.

(c) Find the break-even point in units.

(d) How many units must be sold to generate a profit of $96,000?

(e) What is the break-even point if the management decides to lower fixed costs by $18,000?

(f) What is the break-even point if the management decides to increase fixed costs by $18,000?

(g) What percent of their sales does the profit represent if they sell 5,000 units?

Review Exercises

Simplify. Write your answers using only positive exponents. Assume no denominator equals zero.

1. $x^3 \cdot x^7$
2. $(y^5)^3$
3. $-(2p)^2$
4. $(-3x)^2$
5. $(3x^2y)^2$
6. $2^4 \cdot 2^{-3}$
7. $x^5 \cdot x^{-5}$
8. $(3x^2yz^{-2})^{-2}$
9. $\dfrac{32a^2b^3}{2ab^2}$
10. $\dfrac{28x^2y^2}{14x^2y^3}$
11. $\left(\dfrac{-6m^2n}{3mn^2}\right)^{-1}$
12. $\left(\dfrac{2}{x}\right)^{-3}$
13. $\dfrac{(xy^3)^{-1}}{(xy)^{-2}}$
14. $\left(\dfrac{x^2}{y}\right)^{-2}\left(\dfrac{x^{-1}}{y^{-2}}\right)^{-1}$
15. $2^{-2} + 3^{-1}$
16. $(-2R^2T^{-3})^{-3}$
17. $\dfrac{(27x^2y^2)(x^{-2}y^{-3})}{-6(x^2y^3)^2}$
18. $\dfrac{3^{-2}}{x} - x^{-1}$

19. Write in scientific notation and simplify:

(a) $(48,900)(0.00001)$; (b) $(0.00013)(56)(0.125)$

20. Write in decimal notation:

(a) 1.84×10^{-4}; (b) 7.207×10^5

For each of the following polynomials: (a) simplify; (b) write it in standard form; (c) state its degree; (d) state whether it is a monomial, a binomial, or a trinomial; and (e) state the leading coefficient.

21. $x^2 - 3 - 2x - x^2$
22. $5y - 8y^2 + 3y - 9$
23. $3m^3 - 7m + m^2 - 2m^3$
24. $2(x - 4) + 3(5 - x) + 2$

Simplify each of the following quantities.

25. $(4x^2 - 5x + 1) + (x^2 - 4x + 7)$
26. $(2m^3 - 3m + 4) - (4m^2 - m^3 - 3m)$
27. $5t^3 - 2t^2 - [-t^2 - (7t^3 + 4t^2)]$
28. $(t^2 - 1) - 2(t^2 + 4) - t^2$
29. $(5p^4 + p^2 - 7) - (-p^4 + 7p^2 - 8)$
30. $-[(x + 6) - (5x - 4)] + 3$
31. $-2(x - 7) + 6(2 - 3x) - x$

Carry out the indicated operations.

32. Find the sum of $x^2 - 7x + 8$ and $x^3 + 2x$.

33. Add $4y^3 - 7y^2 + 8y - 1$ to $3y^2 + 8y + 4$.

34. Find the difference of $2x + 18$ and $x^2 - 5x + 17$.
35. Subtract $6x + 3$ from $x^2 + 6x - 3$.
36. Subtract $2y^2 + y + 1$ from the sum of $y^2 + 6$ and $y^2 - y - 1$.
37. Add 15 to the difference of $3s^2 + 8$ and $4s^2 - s - 9$.
38. $\left(\dfrac{1}{2}x - \dfrac{1}{4}y\right) + \left(\dfrac{1}{4}x - \dfrac{1}{2}y\right)$
39. $3p(5p^2 - 6p + 7)$
40. $-2x^2y(xy^2 - 3xy)$
41. $(x - 4)(x + 4)$
42. $(x + 3)^2$
43. $(2x + 3)(x - 5)$
44. $-3rs^2[-2 - (4rs - 2s^2) + s]$
45. $(5m - 7)^2$
46. $(3x + 2y)(3x - 2y)$
47. $(x + 1)^3$
48. $-(2p + 7q)^2$
49. $(0.5m + 0.2n)(0.5m - 0.2n)$
50. $(1.1a - 1.2b)(2a + b)$
51. $\left(\dfrac{1}{2}x + \dfrac{1}{3}y\right)^2$
52. $\left(\dfrac{3}{4}x + \dfrac{7}{8}y\right)\left(\dfrac{3}{4}x - \dfrac{7}{8}y\right)$
53. $(4z + 3)(2z - 1)$
54. $(2q^2 - 3)^2$
55. $\left(\dfrac{5}{8}x - \dfrac{2}{3}y\right)\left(\dfrac{1}{3}x - \dfrac{3}{4}y\right)$
56. $\left(\dfrac{x}{5} + \dfrac{2y}{3}\right)\left(\dfrac{3x}{4} - \dfrac{y}{2}\right)$
57. $\dfrac{3x^2 + 9x}{3x}$
58. $\dfrac{28a^3 + 14a^2 - 21a + 7}{7}$
59. $\dfrac{7y^4 - 8y^3}{2y^3}$
60. $\dfrac{2m^5 - 5m^8}{3m^5}$
61. $(a^2b - ab^2) \div ab$
62. $(5x^4 - 3x^6 - x^3) \div 15x^3$
63. $(x^2 - 7x + 6) \div (x - 1)$
64. $(y^2 + 7y + 12) \div (y + 3)$
65. $(2a^2 + 7a - 1) \div (a + 4)$
66. $(p^3 + 2p^2 + 2p + 1) \div (p + 1)$
67. $(a^3 - 27) \div (a - 3)$
68. $(x^5 - 1) \div (x - 1)$
69. $(-6y^2 + 29y - 4y^3 - 30 + 3y^4) \div (4 - 3y)$
70. $(6 - m + 2m^2 + m^3) \div (3 + m)$
71. $\left(\dfrac{1}{2} + 2y\right)^2$
72. $(0.8x - 0.4)^2$
73. Find the area and perimeter of the given square.

$2x + 3$

74. Find the area and perimeter of the given rectangle.

$4x - 1$
$5x + 4$

75. Find the sum and the product of the integer before n and the one after n.
76. Find the sum and the product of three consecutive even natural numbers if the smallest one is n.
77. Find the area and circumference of the given circle.

$x - 4$

78. Find the surface area and volume of the given cube.

$x + 3$

79. Use the given triangle to find x.

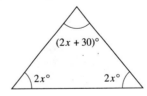
$(2x + 30)°$
$2x°$ $2x°$

80. Jose has grades of $n + 3$, $n - 2$, $n + 4$, and $n - 5$. Find his average grade.

Chapter Test

Simplify each of the following quantities. Write your answers using only positive exponents. Assume no denominator equals zero.

1. x^0
2. $\dfrac{x^3}{x^3}$
3. $(x^4)^3$
4. $\dfrac{x^3}{3} \cdot \dfrac{3^2}{x^5} \cdot \dfrac{x^4}{3}$
5. $-(2x)^2(-3x^2)^2$
6. $(3x^3y)(-2x^2y^4)$
7. $[(x^2y^3)^3]^{-2}$
8. $\left(\dfrac{3x^7}{4x^{11}}\right)^{-1}$
9. $\dfrac{7(3x^2y^3)^3}{27(7xy^{-1}z)^{-1}}$
10. $\dfrac{-64x^8y^{11}z^5}{2^3x^{-3}y^2z^4}$
11. $\left(\dfrac{2^3x^5y^{-4}}{2^2x^{-4}y^5}\right)^{-2}$
12. $\dfrac{(3x-2y)^{-3}}{(3x-2y)^{-2}}$

Write each number in scientific notation.

13. 6300
14. 0.0082

Write each number in decimal notation.

15. 6.21×10^3
16. 2.93×10^{-4}

17. Add:
$$\begin{array}{r} 6x^2 + 5x - 7 \\ 3x^2 - 9x + 2 \end{array}$$

18. Subtract:
$$\begin{array}{r} 17x^3 - 5x + 9 \\ -6x^3 + 2x^2 - 3x - 1 \end{array}$$

Simplify each of the following quantities. Assume no denominator equals zero.

19. $(-x^2 - 9x + 4) + 2(3x - 5)$
20. $(4y^3 + 18y^2 - 9y + 1) - (y^2 + 4y - 5)$
21. $(x^2yz - 5xyz^2) - (xy^2z + 4xyz^2) + (3x^2yz - 8xyz)$
22. $m^3 - (5m^2 + 6) + m^2 - (2m^3 - 5m + 1) + 4m^2$
23. $(x + 5)(x - 5)$
24. $(2y + 3)^2$
25. $(3p + 5)(4p - 3)$
26. $\dfrac{-12x^2 + 24x - 6}{6}$
27. $\dfrac{-56x^2y + 49xy^2 + 9xy}{-7xy}$
28. $\dfrac{x^2 - x - 42}{x + 6}$
29. $(x^3 + 3x^2 + 3x + 1) \div (x + 1)$
30. $(3x^2 + 7x - 11) \div (x - 5)$

Solve each of the following.

31. An object's weight in pounds on the surface of the earth is given by the formula $W = \dfrac{k}{d^2}$. How much does an object weigh if $k = 1.6 \times 10^9$ and $d = 4.4 \times 10^3$?
32. Find the surface area and volume of a cube with an edge of $(x - 4)$ ft.

Cumulative Review for Chapters 4-6

Complete the ordered pairs for each of the following equations.

1. $3x + 4y = 12$; (0,), (, 0), (−4,)
2. $6x - 5y = 30$; (0,), (, 0), (, 6)
3. $3x - y - 2x + y = y - 3$; (0,), (, 0), (−6,)
4. $y = 4x + 5$; (0,), (, 3), (, −1)

Graph each equation by using the x- and y-intercepts and a check point.

5. $3x + 2y = 6$
6. $x - 2y = 4$
7. $2x + 5y = 10$
8. $4x + 3y = 12$

Find the slope of the line through the given points.

9. (2, 0) and (3, −1)
10. (3, 5) and (1, 2)
11. (4, 0) and (4, 3)
12. (5, −1) and (−3, −1)

Write the equation of the line using the given information.

13. Has slope 2 and y-intercept of (0, −4)
14. Has slope $-\dfrac{2}{3}$ and goes through (−1, 2)
15. Goes through (3, 4) and (−1, 0)
16. Goes through (−6, 2) and has undefined slope
17. Goes through (3, −4) and has zero slope

Graph each of the following inequalities.

18. $x + y \leq 2$
19. $3x - y \geq 3$
20. $y > 4$
21. $x < -2$

Find the specified values for each function. Some may be undefined.

22. $f(x) = x - 1$; $f(0), f(-2), f(3)$
23. $f(x) = x^2 - 2$; $f(0), f(3), f(-1)$
24. $P(x) = 5x - 6$; $P(0), P(1), P(-1)$
25. $h(x) = \dfrac{2}{x-3}$; $h(0), h(1), h(3)$

Determine the domain and range of each relation. If it is a function, say so.

26. $y = x - 1$
27. $y = 3$
28. $x = -7$
29. $y = -3(2x - 1)$
30. $y = \dfrac{2}{x}$

Solve each system of linear equations by the method indicated.

31. $\left.\begin{array}{l} x + y = 4 \\ x - y = 2 \end{array}\right\}$ graphing
32. $\left.\begin{array}{l} x = 1 - 2y \\ y = -1 - 2x \end{array}\right\}$ substitution
33. $\left.\begin{array}{l} 3x + 4y = 5 \\ 2x - 3y = 1 \end{array}\right\}$ elimination
34. $\left.\begin{array}{l} x - y = 3 \\ x + 2y = 5 \end{array}\right\}$ substitution
35. $\left.\begin{array}{l} 5x - 7y = -2 \\ 4x + 3y = 7 \end{array}\right\}$ elimination
36. $\left.\begin{array}{l} x + 2y = 5 \\ 3x - y = 1 \end{array}\right\}$ graphing

Find the solution set of each system of linear inequalities by graphing.

37. $\begin{array}{l} x \geq 2 \\ y \leq 3 \end{array}$
38. $\begin{array}{l} y < x \\ x + y \geq 1 \end{array}$
39. $\begin{array}{l} x - y \geq 1 \\ x + 2y \geq 4 \\ x \leq 3 \end{array}$
40. $\begin{array}{l} x < 0 \\ y < 0 \\ y \leq x + 2 \end{array}$

Simplify each of the following quantities using the rules of exponents. Assume no denominator equals zero.

41. 5^2
42. -3^2
43. $2 \cdot 3^3$
44. 4^0
45. 5^{-1}
46. $2^2 \cdot 2^3$
47. $x^3 \cdot x^4$
48. $(2x)^2 \cdot 3x^3$
49. $\left(\dfrac{2}{3}\right)^{-1}$
50. $(x-4)^2(x-4)^6$
51. $b^4 \cdot b^{-2} \cdot b^0$
52. $2a^2 \cdot 3a^3 \cdot 4a^{-4}$
53. $[3^2]^1$
54. $(x^3)^4$
55. $(3x^{-4}y)^{-2}$
56. $\left(\dfrac{17x^3}{9x^{-4}}\right)^0$

Write each number in scientific notation.

57. 8,300
58. 64
59. 0.0012
60. 0.4

Write each number in decimal form.

61. 6.9×10^4
62. 1.23×10^{-3}
63. 8×10^{-2}

Identify each of the following polynomials as a monomial, a binomial, or a trinomial and state its degree.

64. $4x^2 - 7x$
65. $0.03x^3$
66. $x^2 - 5x - 9$

Add or subtract as indicated.

67. $5x - 7 - x + 4$
68. $x^2 - (x^2 + 4)$
69. $(6x^2 + 2x - 7) + (3x^2 - 9x + 2)$
70. $(11x^2 - 10x + 9) - (-2x^2 + x - 8)$
71. $-(-4y^3 + 2y - 7) - 3y^2 + (2y + 4)$
72. $(0.5x^3 - 1.7x + 3.4) - (x^3 + 2.7x^2 + 3.4)$

Multiply and simplify.

73. $3y(2y - 4)$
74. $(x - 8)(x + 8)$
75. $(2x - 3)^2$
76. $4y(2y - 3) - 7(y + 8)$
77. $(3a + 4b)(3a - 4b)$
78. $(2a + 6)(3a + 2)$
79. $(p - 3)(p^2 - 2p + 1)$

Divide each of the following polynomials.

80. $(12x^2 - 15x) \div 3x$
81. $(12y^3 - 8y^2 + 2y) \div (-2y)$
82. $(3x^2 - x - 1) \div (x + 2)$
83. $(a^2 + 6a + 9) \div (a + 3)$
84. $(a^3 + 3a^2 + 3a + 2) \div (a + 1)$
85. $(-6y^3 - 3y^2 + 2y^4 + 24y - 17) \div (y^2 - y - 5)$
86. $(x^3 - 8) \div (x - 2)$

Solve each of the following applied problems by using a system of equations.

87. One number is 7 more than another number and their sum is 11. Find the two numbers.

88. The sum of two numbers is 15 and their difference is 3. Find the two numbers.

89. A boat can go 15 mi upstream in 3 hr and can go 33 mi downstream in the same time. Find the speed of the boat in still water and the speed of the current.

90. How many pounds of candy priced at $1.90/lb should be mixed with candy priced at $2.40/lb to have 20 lb of candy that would sell for $2.20/lb?

91. The sum of two consecutive multiples of 3 is 27. Find the two numbers.

92. Melissa had $10,000 to invest. She invested part at 6% and the rest at 8%. If the interest she received at the end of one year was $740, how much did she invest at each rate?

93. One hundred and forty feet of fencing is required to fence a rectangular yard that has width equal to three-fourths of its length. Find the length and width.

CHAPTER 7
Products and Factors

7.1 Factoring; The Greatest Common Factor
7.2 Factoring Trinomials of the Form $x^2 + bx + c$
7.3 Factoring Trinomials of the Form $ax^2 + bx + c$
7.4 Special Factors; Factoring by Grouping
7.5 A Summary of Factoring
7.6 Solving Factorable Quadratic Equations
7.7 More Applications

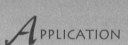

An orchard has 525 fruit trees planted in rows of uniform length. The number of trees in each row is 4 more than the number of rows. Find the number of rows and the number of trees in each row.

Exercise Set 7.7, Exercise 15

7.1 Factoring; The Greatest Common Factor

OBJECTIVES

In this section we will learn to
1. find the greatest common factor of two or more monomials; and
2. use the distributive law to factor polynomials by removing the greatest common factor.

We spent much of Chapter 6 learning how to find the product of two or more polynomials. In this chapter we will learn how to reverse the multiplication process by *factoring*. In multiplication, the product of two or more algebraic quantities is written as a sum of terms. In factoring, a sum of terms is written as the product of two or more quantities.

The number 12 can be expressed as the product of 3 and 4; or, $3 \cdot 4 = 12$. When 12 is expressed as $3 \cdot 4$, we say it is **factored,** and 3 and 4 are called **factors** of 12.

$$\underset{\text{Factors}}{3 \cdot 4} = 12 \leftarrow \text{Product}$$

Twelve can be factored in other ways, such as

$$12 \cdot 1, \quad 6 \cdot 2, \quad 2 \cdot 2 \cdot 3, \quad \text{and} \quad (-2)(-6)$$

to name just a few. However, we usually limit the factors of positive integers to positive integers. Under this condition, the **set of factors** of 12 is

$$\{1, 2, 3, 4, 6, 12\}$$

In general, a factor can be defined as follows.

DEFINITION	
Factor: Exact Divisor	If x and y are two positive integers, then x is a **factor** of y if, when y is divided by x, there is no remainder. When x divides y so that there is no remainder, we say that x is an **exact divisor** of y.

EXAMPLE 1 The positive integer factors (exact divisors) of 18 are

1, 2, 3, 6, 9, and 18

EXAMPLE 2 The positive integer factors of 72 are

1, 2, 3, 4, 6, 8, 9, 12, 18, 24, 36, and 72

In order to factor an expression, it is important to understand the concept of the **Greatest Common Factor,** or **GCF,** which is dependent, in part, on the concept of prime numbers and the prime factorization of a number. Recall from Section 1.1

that a *prime number* is a natural number greater than 1 that is divisible only by itself and 1, and a *composite number* is a natural number greater than 1 that is not prime. Recall also that any composite number can be written in *prime factored form*.

EXAMPLE 3 Write 12 in prime factored form.

SOLUTION $12 = 2 \cdot 2 \cdot 3 = 2^2 \cdot 3$

EXAMPLE 4 Write 540 in prime factored form.

SOLUTION We use the divisibility rules to find the prime factors:

$540 \div 2 = 270;$ $270 \div 2 = 135;$ $135 \div 3 = 45;$ and so on.

We follow the arrows to find the prime factored form.

$2\overline{)540}$
$\downarrow$
$2\overline{)270}$ $540 = 2 \cdot 270$
$\downarrow$
$3\overline{)135}$ $540 = 2 \cdot 2 \cdot 135$
$\downarrow$
$3\overline{)45}$ $540 = 2 \cdot 2 \cdot 3 \cdot 45$
$\downarrow$
$3\overline{)15}$ $540 = 2 \cdot 2 \cdot 3 \cdot 3 \cdot 15$
$\hookrightarrow 5$ $540 = 2 \cdot 2 \cdot 3 \cdot 3 \cdot 3 \cdot 5$

The prime factored form of 540 is $2 \cdot 2 \cdot 3 \cdot 3 \cdot 3 \cdot 5$.

We now return to the concept of the greatest common factor.

DEFINITION

Greatest Common Factor

The **greatest common factor,** abbreviated **GCF,** of two or more integers is the largest integer that is a factor of each.

EXAMPLE 5 Find the greatest common factor of 90 and 72.

SOLUTION We begin by writing each number in prime factored form:

$90 = 2 \cdot 3^2 \cdot 5$ and $72 = 2^3 \cdot 3^2$

To find the greatest common factor, we take each common prime factor the least number of times it appears in either factored form. The least number of times 2 appears in either factored form is one; the least number of times 3 appears is two; 5 is not common to both factored forms. Therefore,

$GCF = 2 \cdot 3^2 = 18$

EXAMPLE 6   Find the greatest common factor of 300 and 450.

SOLUTION $300 = 2^2 \cdot 3 \cdot 5^2$ and $450 = 2 \cdot 3^2 \cdot 5^2$

The least number of times 2 appears is one; the least number of times 3 appears is one; the least number of times 5 appears is two. Therefore

GCF $= 2 \cdot 3 \cdot 5^2 = 150$

EXAMPLE 7   Find the greatest common factor of 252, 294, and 315.

SOLUTION $252 = 2^2 \cdot 3^2 \cdot 7$ $294 = 2 \cdot 3 \cdot 7^2$ $315 = 3^2 \cdot 5 \cdot 7$

The least number of times 3 appears is one; the least number of times 7 appears is one; 2 and 5 are not common to all three numbers. Therefore

GCF $= 3 \cdot 7 = 21$

Notice that the exponent on each prime factor of the GCF in Example 7 is the smallest exponent on that common factor in any of the numbers. Similarly, this reasoning gives us a method of finding the greatest common factor in a series of terms that contain variables. For example, let's consider the GCF of $4y^{10}$, $2y^6$, and $6y^7$:

$4y^{10} = 2^2 \cdot y^4 \cdot y^6$
$2y^6 = 2 \cdot y^6$
$6y^7 = 2 \cdot 3 \cdot y \cdot y^6$

{The greatest common numerical factor is 2. The smallest exponent that appears on the variable in *every* term is 6.

Since the greatest common numerical factor is 2 and the greatest common variable factor is y^6, the GCF is $2y^6$.

To Find the Greatest Common Factor of Monomials

Step 1   Find the prime factorization of each monomial.
Step 2   Determine the least number of times each common prime or variable factor appears in all of the monomials.
Step 3   Find the product of all of the factors to the powers in Step 2.

EXAMPLE 8   Find the greatest common factor of the monomials $21x^2$, $14x^5$, and $-35x^4$.

SOLUTION   **Step 1   Find the prime factorization of each monomial.**

$21x^2 = 3 \cdot 7 \cdot x^2$
$14x^5 = 2 \cdot 7 \cdot x^5$
$35x^4 = 5 \cdot 7 \cdot x^4$

Step 2 **Determine the least number of times each common factor appears in all of the monomial.**

The least number of times 7 appears is once. The other numbers, 2, 3, and 5, are not common factors. The smallest exponent on x is 2.

Step 3 **Find the product of all of the factors to the powers found in Step 2.**

The product of the factors is $7 \cdot x^2$, so the GCF = $7x^2$.

Whenever possible, we try to find the greatest common factor by inspection.

EXAMPLE 9 Find the greatest common factor of the monomials a^2b^3, a^2b^4, and a^3b^2c.

SOLUTION The smallest exponent on a is 2 and the smallest exponent on b is 2, but c is not a common factor, so

GCF = a^2b^2

By using the distributive property, we can extend the greatest common factor concept to polynomials. For example, we can factor $3x^6 + 6x^4$ by removing the greatest common factor of its two terms. The GCF of 3 and 6 is 3. The GCF of x^6 and x^4 is x^4. Therefore the GCF of the two terms is $3x^4$. We remove this factor:

$3x^6 + 6x^4 = 3x^4(x^2 + 2)$ $3x^6 = 3x^4 \cdot x^2;\ 6x^4 = 3x^4 \cdot 2$

To check our work, we use the distributive property to find the product of the factors.

$3x^4(x^2 + 2) = (3x^4)(x^2) + (3x^4)(2)$
$= 3x^{4+2} + 6x^4$
$\stackrel{\checkmark}{=} 3x^6 + 6x^4$

EXAMPLE 10 Find the other factor in $6x^2 - 42x = 6x(\quad)$.

SOLUTION The GCF, which is $6x$, has already been removed. To find the remaining factor, we must determine what quantity will produce each of the terms when we multiply by $6x$.

$6x^2 - 42x = 6x(x - 7)$ $6x^2 = 6x \cdot x;\ 42x = 6x \cdot 7$

The remaining factor is $x - 7$.

EXAMPLE 11 Find the other factor in $63xy^3 - 42x^3y^2 = 21xy^2(\quad)$.

SOLUTION $63xy^3 - 42x^3y^2 = 21xy^2(3y - 2x^2)$ $63xy^3 = 21xy^2(3y);\ 42x^3y^2 = 21xy^2(2x^2)$

The other factor is $(3y - 2x^2)$.

Notice that *when the GCF is removed, the exponents on each variable in the other factor are decreased by amounts equal to the corresponding exponents on the factor that is removed.*

EXAMPLE 12 Factor: $5x^{n+2} - 10x^{n+1}$.

SOLUTION The greatest common numerical factor is 5. To find the greatest common variable factor, we consider that $x^{n+2} = x^{(n+1)+1} = x^{n+1} \cdot x$. So

$$5x^{n+2} - 10x^{n+1} = 5x^{(n+1)+1} - 10x^{n+1}$$
$$= 5x^{n+1}(x - 2)$$

EXAMPLE 13 Factor each polynomial by removing the greatest common factor:

(a) $19a^3b^2 - 38a^2b^3$ (b) $24xy^4 - 6x^2y^3 + 4x^3y^2$
(c) $11x^2 + 100$ (d) $3x^{n+1} + 6x^n$, where n is a positive integer

SOLUTION (a) $19a^3b^2 - 38a^2b^3$
The greatest common numerical factor is 19, and the greatest common variable factor is a^2b^2. Thus,

$$19a^3b^2 - 38a^2b^3 = 19a^2b^2(a - 2b) \quad 19a^3b^2 = 19a^2b^2(a); \; 38a^2b^3 = 19a^2b^2(2b)$$

(b) $24xy^4 - 6x^2y^3 + 4x^3y^2$
The greatest common numerical factor is 2, and the greatest common variable factor is xy^2. Thus,

$$24xy^4 - 6x^2y^3 + 4x^3y^2 = 2xy^2(12y^2 - 3xy + 2x^2)$$

(c) $11x^2 + 100$
Since this polynomial has no common factors other than 1, we say it is prime.

(d) $3x^{n+1} + 6x^n$, where n is a positive integer
The greatest common numerical factor is 3. Since $x^{n+1} = x^n \cdot x$, the greatest common variable factor is x^n. Therefore

$$3x^{n+1} + 6x^n = 3x^n(x + 2)$$

EXAMPLE 14 Factor: $6x^2y^4 + 3xy$.

SOLUTION The GCF is $3xy$. When the GCF is equal to one of the terms of the polynomial, **1** must be put in its place when it is factored out.

$$6x^2y^4 + 3xy = 3xy(2xy^3 + \mathbf{1})$$

To verify that this concept is correct, we use the distributive property:

$$6x^2y^4 + 3xy = 3xy(2xy^3 + \mathbf{1})$$
$$= 3xy(2xy^3) + 3xy(\mathbf{1})$$
$$\stackrel{\checkmark}{=} 6x^2y^4 + 3xy$$

CAUTION

When one of the terms of a polynomial is equal to the greatest common factor of all of the terms, it must be replaced by the number 1 when the GCF is factored out.

If the leading coefficient of the polynomial is negative, we will often factor out the additive inverse of the GCF so that the remaining polynomial has a positive leading coefficient. The next example illustrates this idea.

EXAMPLE 15 Remove the additive inverse of the GCF from $-3x^2y^3 + 6xy^2 - 9x^3y^3$.

SOLUTION Since $3xy^2$ is the greatest common factor, we will remove $-3xy^2$.

$$-3x^2y^3 + 6xy^2 - 9x^3y^3 = -3xy^2(xy - 2 + 3x^2y)$$

EXAMPLE 16 Factor: $-7x^3y^3 - 14x^2y^2 + 21xy^4 + 7xy^2$.

SOLUTION The greatest common numerical factor is 7. The greatest common variable factor is xy^2. Since the lead coefficient is negative, we will remove $-7xy^2$. Notice that $-7xy^2$ is equal to the additive inverse of last term, so we will have to put -1 in its place.

$$-7x^3y^3 - 14x^2y^2 + 21xy^4 + 7xy^2 = -7xy^2(x^2y + 2x - 3y^2 - 1)$$

The reader is encouraged to check the result.

CAUTION — When the greatest common factor is negative, the sign before each term in the remaining factor must be changed.

Do You Remember?

Can you match these?

_____ 1. The GCF of 8, 12, and 20
_____ 2. The GCF of x^4y^6 and x^3y^7
_____ 3. The prime factorization of 54
_____ 4. The factored form of $2xy + 4xy^2$
_____ 5. A word describing a nonfactorable polynomial
_____ 6. $-4x^7y^3 + 12x^5y^3 = -4x^5y^3(\ ?\ )$
_____ 7. A composite integer

a) x^3y^6
b) $2xy(1 + 2y)$
c) Prime
d) 4
e) $2 \cdot 3^3$
f) An integer greater than 1 that is not prime
g) $x^2 + 3$
h) $x^2 - 3$

Answers: 1. d 2. a 3. e 4. b 5. c 6. h 7. f

Exercise Set 7.1

Write each number in prime factored form. If a number is unfactorable, write prime.

1. 6
2. 15
3. 38
4. 84
5. 97
6. 113
7. 264
8. 532
9. 90
10. 56
11. 49
12. 1311
13. 47
14. 43
15. 450
16. 252

Find the greatest common factor (GCF) for each of the following.

17. 24, 36
18. 28, 56
19. 49, 91
20. 72, 88
21. 10, 70, 490
22. 27, 81, 486
23. $3x, 12xy$
24. $18x^2, 30x^5$
25. $15x^3, 35x^2y, 40xy^2$
26. $21xy^2, 42x^2y^3, 49x^3y$
27. $-256r^4s^3, 16r^3t^4, 512s^5$

28. $35x^3y$, $-48y^2z^3$, $36x^2z^5$
29. $-15x^2y$, $17abx$, $35abxy$
30. $-3x^2y^2z$, $7z$, $8xyz$

In Exercises 31–56, the GCF has already been removed. Complete the problem by writing the correct polynomial in the parentheses.

31. $2x + 4 = 2(\quad)$
32. $4x + 8 = 4(\quad)$
33. $3y - 12 = 3(\quad)$
34. $2y - 8 = 2(\quad)$
35. $-2x + 10 = -2(\quad)$
36. $-3y + 12 = -3(\quad)$
37. $p^2 + 7p = p(\quad)$
38. $m^2 - 13m = m(\quad)$
39. $7r^2 - 42r = 7r(\quad)$
40. $5a^2 - 45a = 5a(\quad)$
41. $-7x^2 + 4x = -x(\quad)$
42. $-2y^2 - 5y = -y(\quad)$
43. $28x^2 + 63x = 7x(\quad)$
44. $9b^2 + 12b = 3b(\quad)$
45. $10a^4 - 15a^2 = 5a^2(\quad)$
46. $30x^5 - 42x^3 = 6x^3(\quad)$
47. $6x^3 - 8x^2 - 10x = 2x(\quad)$
48. $10y^3 - 35y^2 + 55y = 5y(\quad)$
49. $-20s^3t^2 + 30s^2t^2 + 10st = -10st(\quad)$
50. $-3a^3b^2 - 18ab^3 + 15a^2b = -3ab(\quad)$
51. $-8y^7z^2 + 24y^9z^5 = -8y^7z^2(\quad)$
52. $-9x^{11}y^5 - 54x^8y^8 = -9x^8y^5(\quad)$
53. $\frac{1}{2}x - \frac{1}{4} = \frac{1}{4}(\quad)$
54. $\frac{3}{8}x + \frac{1}{4} = \frac{3}{8}(\quad)$
55. $0.5y - 1.5 = 0.5(\quad)$
56. $0.1y + 2.3 = 0.1(\quad)$

Factor each of the following polynomials. If the expression is unfactorable, write prime.

57. $2x + 6$
58. $5x + 10$
59. $3y - 13$
60. $2y - 11$
61. $-8p - 12$
62. $-12r - 20$
63. $3x^2 - 9x$
64. $2y^2 + 6y$
65. $ax^2 - ax$
66. $bx^2 - bx$
67. $-5y^2 - 12y$
68. $-9t^2 - 32t$
69. $\pi r^2 - 2\pi r$
70. $\pi r^2 h - \frac{3}{4}\pi r^3$
71. $4x^3 - 8x^2 + 2x$
72. $6y^3 - 12y^2 + 3y$
73. $12x^3y^5 - 18x^2y^6$
74. $9a^4b - 6a^2b^5$
75. $-39y^8 - 52y^6 - 78y^5$
76. $-49x^7y - 56x^6y^2 - 91x^3y^3$
77. $-22x^6y^4 + 40x^4y^3 - 8x^2y^5$
78. $9rs^8 - 24r^4s^6 - 30r^5s^6$
79. $12a^4 - 16a^3 + 24a^2 - 18a$
80. $-15x^4 + 12x^3 - 9x^2 + 6x$
81. $18a^5b^2 - 10a^4b + 4a^2b^3 + 2ab$
82. $-36a^4b^3c^2 - 42a^3b^3c^3 - 54a^3b^3c^4 - 66ab^3c^5$
83. $-60r^5s^2 - 18r^3s^3 - 12r^3s^4$
84. $21x^4y^4z + 35x^8y^3z^2 - 49x^5y^5z^3 - 56x^5y^6z^4$

ℱOR ℰXTRA 𝒯HOUGHT

Factor the GCF from each of the following polynomials.

85. $6x^{n+1} - 3x^n$
86. $14y^{n+2} + 7y^{n+1}$
87. $27a^{3n+2} - 18a^{3n}$
88. $56b^{5n-3} - 16b^{5n}$
89. $91x^{4n+3} + 13x^3$
90. $63r^{2n+5} + 18r^5$
91. $72s^{2n+7} - 24s^{n+1}$
92. $35t^{3n-5} - 14t^{2n-1}$
93. $49x^{n-8} + 14x^{2n+3}$
94. $64y^{n-3} + 24y^{2n+1}$

REVIEW EXERCISES

Multiply each of the following polynomials.

95. $(x + 1)(x - 1)$
96. $(x + 2)(x - 2)$
97. $(y - 3)^2$
98. $(y + 5)^2$
99. $(p - 7)(p + 1)$
100. $(r - 6)(r + 2)$
101. $3(a - 2)(a - 3)$
102. $5(x - 4)(x - 2)$
103. $-2(s + 3)(s - 2)$
104. $-4(y - 8)(y + 3)$
105. $-3x(x - 5)(x - 6)$
106. $-4y(y - 9)(y - 3)$
107. $x^2(x + 3)(x + 7)$
108. $y^2(y + 1)(y + 5)$
109. $-2s^2(s - 6)(s - 9)$
110. $-2s^2(s - 1)(s - 10)$

7.2 Factoring Trinomials of the Form $x^2 + bx + c$

OBJECTIVES In this section we will learn to
1. factor trinomials in which the leading coefficient is 1;
2. factor trinomials after factoring out the GCF; and
3. factor expressions with a common binomial factor.

In Chapter 6 we saw that the product of two binomials is usually a trinomial. In this section we will learn how to reverse this process and *factor* trinomials with leading coefficient 1, such as

$$x^2 + 6x + 8, \quad x^3 + 3x^2 + 2x, \quad \text{and} \quad x^4 + 5x^3 + 4x^2$$

If we try to factor $x^2 + 6x + 8$, it is reasonable to expect that the factors will be two binomials. With that in mind, we want to find two integers, a and b, such that

$$(x + a)(x + b) = x^2 + 6x + 8$$

Now,

$$(x + a)(x + b) = x^2 + ax + bx + ab$$
$$= x^2 + (a + b)x + ab$$

Thus a and b must satisfy the equation

$$x^2 + (a + b)x + ab = x^2 + 6x + 8$$

This means that

$$a + b = 6 \quad \text{and} \quad ab = 8$$

That is, we need two numbers with a product of 8 and a sum of 6. Here are the possibilities:

Product of 8	Sum of 6	
$8 \cdot 1 = 8$	$8 + 1 = 9$	Wrong
$4 \cdot 2 = 8$	$4 + 2 = 6$	Right

The pair of numbers that we seek is 4 and 2. Thus,

$$x^2 + 6x + 8 = (x + 2)(x + 4)$$

To check the result, we use the FOIL method to make sure that the sum of the outer and inner products is $6x$.

$$(x + 2)(x + 4) = x^2 + 6x + 8 \quad 2x + 4x = 6x$$

EXAMPLE 1 Factor $x^2 + 10x + 24$ completely.

SOLUTION   Since the leading coefficient is 1, we need to find two numbers with a product of 24 and a sum of 10. Since all terms in the polynomial are positive, we can look at positive integers only. Here are the possibilities:

Product of 24	Sum of 10	
$24 \cdot 1 = 24$	$24 + 1 = 25$	Wrong
$12 \cdot 2 = 24$	$12 + 2 = 14$	Wrong
$8 \cdot 3 = 24$	$8 + 3 = 11$	Wrong
$6 \cdot 4 = 24$	$6 + 4 = 10$	Right

Only 6 and 4 satisfy the stated conditions. The factorization is

$$x^2 + 10x + 24 = (x + 4)(x + 6)$$

Since multiplication is commutative, we can also write the factorization

$$x^2 + 10x + 24 = (x + 6)(x + 4)$$

EXAMPLE 2   Factor $x^2 - 6x - 16$ completely.

SOLUTION   To factor $x^2 - 6x - 16$, we need two numbers with a product of -16 and a sum of -6. To achieve a product of -16, one of the two integers must be negative and the other one positive.

Product of -16	Sum of -6	
$-16 \cdot 1 = -16$	$-16 + 1 = -15$	Wrong
$-8 \cdot 2 = -16$	$-8 + 2 = -6$	Right

The numbers -8 and 2 satisfy the requirements, so

$$x^2 - 6x - 16 = (x - 8)(x + 2)$$

The list of number pairs that produce the proper product in Example 2 is not complete. Other combinations such as 16 and -1, 8 and -2, and -4 and 4 all produce the correct product but incorrect sums. Once the correct combination is found, there is no need to try other possibilities.

EXAMPLE 3   Factor $x^2 - 9x + 18$ completely.

SOLUTION   The last term, 18, is positive, so both of the numbers in the factors must have the same sign. The middle term, $-9x$, is negative so both numbers must be negative.

Product of 18	Sum of -9	
$(-18)(-1) = 18$	$-18 + (-1) = -19$	Wrong
$(-9)(-2) = 18$	$-9 + (-2) = -11$	Wrong
$(-6)(-3) = 18$	$-6 + (-3) = -9$	Right

The numbers are -6 and -3. Thus,

$$x^2 - 9x + 18 = (x - 6)(x - 3)$$

EXAMPLE 4   Factor $x^2 - 4x + 8$ completely.

SOLUTION   We must find two numbers with a product of 8 and a sum of -4. Since the middle term is negative and the last term is positive, the two numbers we seek must both be negative.

Product of 8	Sum of -4	
$(-8)(-1) = 8$	$-8 + (-1) = -9$	Wrong
$(-4)(-2) = 8$	$-4 + (-2) = -6$	Wrong

No other possibilities exist, so $x^2 - 4x + 8$ is not factorable using only integers.

DEFINITION **Prime Polynomial**	If a polynomial with coefficients that are all integers is not factorable using only integers, it is a **prime polynomial.**

When factoring any polynomial, we always remove common factors first. Once we do this, the work of completing the factoring is easier.

EXAMPLE 5   Factor $2x^4 + 6x^3 + 4x^2$ completely.

SOLUTION   We first remove the common factor $2x^2$:

$$2x^4 + 6x^3 + 4x^2 = 2x^2(x^2 + 3x + 2)$$

To factor $x^2 + 3x + 2$, we need to find two numbers with a product of 2 and a sum of 3. The numbers are 1 and 2:

$$2x^4 + 6x^3 + 4x^2 = 2x^2(x^2 + 3x + 2)$$
$$= 2x^2(x + 1)(x + 2)$$

With practice you will be able to factor most trinomials mentally.

EXAMPLE 6   Factor $x^2 - 6xy - 7y^2$ completely.

SOLUTION   We are looking for two numbers with a product of -7 and a sum of -6. The numbers are -7 and 1. Since the last term, $7y^2$, contains y^2, we must not forget y in our factorization:

$$x^2 - 6xy - 7y^2 = (x - 7y)(x + y)$$

To Factor Trinomials with Leading Coefficient 1

1. Remove any common factors.
2. If the last term of the trinomial is positive, both terms in the factors will have the same sign as the middle term in the trinomial.

$$x^2 + 5x + 6 = (x + 2)(x + 3) \qquad x^2 - 5x + 6 = (x - 2)(x - 3)$$

(Sign positive — Same sign)   (Sign positive — Same sign)

3. If the last term of the trinomial is negative, the terms in the factors will have opposite signs.

$$x^2 - 5x - 6 = (x - 6)(x + 1) \qquad x^2 + 5x - 6 = (x + 6)(x - 1)$$

(Sign negative — Opposite signs)   (Sign negative — Opposite signs)

EXAMPLE 7   Factor $-3x^2 - 12x - 12$ completely.

SOLUTION   First we remove the common factor of -3.

$$-3x^2 - 12x - 12 = -3(x^2 + 4x + 4)$$

Now we need two numbers with a product of 4 and a sum of 4. The numbers are 2 and 2:

$$-3x^2 - 12x - 12 = -3(x^2 + 4x + 4)$$
$$= -3(x + 2)(x + 2) = -3(x + 2)^2$$

EXAMPLE 8   Factor $x(x + 3) - 2(x + 3)$ completely.

SOLUTION   When we multiply $x(x + 3) - 2(x + 3)$ out and simplify it, we get $x^2 + x - 6$, which is a trinomial with leading coefficient 1. Factoring,

$$x^2 + x - 6 = (x + 3)(x - 2)$$

It's not necessary to do this, however, since the original expression already has $x + 3$ as a common factor. We can factor it by using the GCF method from Section 7.1.

$$x(x + 3) - 2(x + 3) = (x + 3)(x - 2)$$

EXAMPLE 9 Factor: $(x^2 + 6x + 5) + 2(x + 1)$.

SOLUTION Like the problem in Example 8, we can complete this one in two different ways.

Method 1 First we simplify the expression by combining like terms.

$$(x^2 + 6x + 5) + 2(x + 1) = x^2 + 6x + 5 + 2x + 2$$
$$= x^2 + 8x + 7$$
$$= (x + 1)(x + 7)$$

Method 2 First we factor the trinomial $x^2 + 6x + 5$.

$$(x^2 + 6x + 5) + 2(x + 1) = (x + 1)(x + 5) + 2(x + 1)$$

Now we remove the common factor of $x + 1$.

$$(x + 1)(x + 5) + 2(x + 1) = (x + 1)[(x + 5) + 2]$$
$$= (x + 1)(x + 7)$$

Method 2 of Example 9 is an illustration of a method of factoring called **factoring by grouping**. This type of factoring is helpful when the polynomial to be factored has four or more terms. The next two examples illustrate this.

EXAMPLE 10 Factor $x^3 + 6x^2 + 5x + 30$ completely by grouping.

SOLUTION Grouping the first two terms and the last two terms causes a common binomial factor to emerge:

$$x^3 + 6x^2 + 5x + 30 = (x^3 + 6x^2) + (5x + 30)$$
$$= x^2(x + 6) + 5(x + 6)$$
$$= (x + 6)(x^2 + 5)$$

Since $x^2 + 5$ is prime, the factorization is complete.

EXAMPLE 11 Write $x^2 + 6x + 5$ as a polynomial with four terms and use grouping to factor it completely.

SOLUTION To use grouping, we choose the groups so that they contain a common factor:

$$x^2 + 6x + 5 = x^2 + 5x + x + 5 \qquad 6x = 5x + x$$
$$= x(x + 5) + 1(x + 5)$$
$$= (x + 5)(x + 1)$$

This can be checked by multiplication.

We will pay more attention to factoring by grouping in Section 7.4, along with other special methods of factoring.

Do You Remember?

Can you match these?

___ 1. $x^2 + 5x + 6$
___ 2. $x^2 - 5x - 6$
___ 3. $x^2 + 5x - 6$
___ 4. $x^2 - 5x + 6$
___ 5. $-2x^4 - 10x^3 + 12x^2$
___ 6. A prime polynomial

a) $(x - 2)(x - 3)$
b) $(x + 6)(x - 1)$
c) $(x - 6)(x + 1)$
d) $(x + 2)(x + 3)$
e) $-2x^2(x - 3)(x - 2)$
f) $-2x^2(x + 6)(x - 1)$
g) A factorable polynomial
h) A nonfactorable polynomial

Answers: 1. d 2. c 3. b 4. a 5. f 6. h

Exercise Set 7.2

Fill in the blank spaces with the correct sign, numbers, or letters.

1. $x^2 + 4x + 4 = (x + 2)(x +)$
2. $y^2 + 3y + 2 = (y + 2)(y +)$
3. $n^2 + 8n + 15 = ()(n + 5)$
4. $p^2 + 7p + 12 = ()(p + 4)$
5. $v^2 - v - 6 = (v 3)(v 2)$
6. $d^2 - 7d - 8 = (d 8)(d 1)$
7. $y^2 - 11y + 18 = (y)(y)$
8. $a^2 - 12a + 20 = (a)(a)$
9. $2x^2 - 6x - 80 = 2(x + 5)()$
10. $3m^2 + 12m - 63 = 3()(m + 7)$
11. $x^3 + 11x^2 + 10x = x()()$
12. $x^4 + 5x^3 + 4x^2 = x^2()()$

Factor each polynomial; be sure to factor out the GCF first. If a polynomial is not factorable, write prime.

13. $x^2 + 3x + 2$
14. $n^2 + 7n + 6$
15. $p^2 - p - 2$
16. $a^2 + 8a + 12$
17. $v^2 - 8v + 15$
18. $b^2 - 12b + 20$
19. $x^2 + x - 72$
20. $y^2 - 11y + 24$
21. $z^2 - 14z + 49$
22. $r^2 - 3r - 40$
23. $3m^2 + 12m + 15$
24. $2a^2 + 8a + 4$
25. $5b^2 + 55b + 60$
26. $4x^2 - 48x - 140$
27. $x^3 - 7x^2 - 30x$
28. $y^3 - 7y^2 - 18y$
29. $a^2 - 2a - 63$
30. $n^2 - 16n + 39$
31. $-2p^2 - 30p - 52$
32. $-2b^2 + 40b - 150$
33. $n^2 - 10n - 48$
34. $t^2 - 19t + 17$
35. $3s^5 - 18s^4 + 15s^3$
36. $x^7 - 5x^6 - 14x^5$
37. $a^2 + 17a + 42$
38. $p^2 + 8p - 33$
39. $-3x^2 + 21x - 30$
40. $-5a^2 - 60a + 140$
41. $p^2 + 4p - 14$
42. $n^2 - 5n + 12$
43. $x^2 - 3xy + 2y^2$
44. $y^2 + 9yz + 8z^2$
45. $p^2 - 14pq + 24q^2$
46. $m^2 - 11mn + 24n^2$
47. $a^2 + 2ab - 15b^2$
48. $b^2 + bc - 30c^2$
49. $3m^4 - 3m^3 - 90m^2$
50. $4p^4 - 16p^3 - 20p^2$
51. $2x^4 - 2x^3 - 144x^2$
52. $2x^3 + 34x^2 + 144x$
53. $10a^3 + 110a^2b + 300ab^2$
54. $11c^3 - 132c^2 + 297c$
55. $-3x^2 - 138x - 135$
56. $-6a^3 + 12a^2 + 48a$
57. $r^2 + 18rs + s^2$
58. $a^2 - 12ab + b^2$
59. $9x^3 + 72x^2 + 108x$
60. $5b^3 + 85b^2 + 80b$
61. $7t^2 - 42st + 56s^2$
62. $9y^2 - 114yz + 192z^2$
63. $x(x - 2) + 3(x - 2)$
64. $y(y - 4) + 2(y - 4)$

65. $m(p + 3) - n(p + 3)$ 66. $r(s - 1) - t(s - 1)$
67. $y(y - 4) + (y - 4)$ 68. $y(y + 5) + (y + 5)$
69. $x^3(x + 1) + x^2(x + 1)$ 70. $x^3(x - 1) - x^2(x - 1)$
71. $m^2(m - 7) - 7m(m - 7)$
72. $2p^3(p - 5) - 10p^2(p - 5)$
73. $-6y^3(y + 2) - 12y^2(y + 2)$
74. $-3r^3(r - 3) + 9r^2(r - 3)$
75. $(x + y)^2 - (x + y)$
76. $(a + b)^2 - (a + b)$
77. $x(x - y) - y(x - y)$
78. $s(s - t) - t(s - t)$
79. $(m + n)(m - n) - (m + n)^2$
80. $(x + y)(x - y) - (x - y)^2$
81. $(x^2 + 8x + 7) + (x + 7)$
82. $(y^2 - 3y + 2) + (y - 2)$
83. $(p^3 + 7p^2 + 6p) - (p^2 + 6p)$
84. $(r^3 - 5r^2 - 14r) - (r^2 - 7r)$
85. $(x^3 + 3x^2 + 2x) + (-2x - 4)$
86. $(y^3 + 9y^2 + 20y) + (-y^2 - 5y)$
87. $\left(\frac{1}{2}x^2 - \frac{1}{4}x\right) - \left(x - \frac{1}{2}\right)$
88. $\left(\frac{1}{3}y^2 - \frac{1}{6}y\right) - \left(y - \frac{1}{2}\right)$
89. $\left(\frac{3}{5}r^2 + \frac{3}{10}r\right) + \left(r + \frac{1}{2}\right)$
90. $\left(\frac{1}{7}p^2 + \frac{1}{21}p\right) + \left(p + \frac{1}{3}\right)$
91. $(0.5s^2 + s) - (s + 2)$
92. $(0.25t^2 + t) - (t + 4)$
93. $(x^2 + 0.3x + 0.02) + (x + 0.2)$
94. $(y^2 + 0.8y + 0.12) + (y + 0.6)$

For Extra Thought

Find two different integers b that allow the trinomial to be factored. Answers may vary.

95. $x^2 + bx + 6$ 96. $x^2 + bx + 4$
97. $x^2 - bx - 9$ 98. $x^2 - bx - 10$

Find two different integers c that allow the trinomial to be factored. Answers may vary.

99. $y^2 - 5y + c$ 100. $y^2 + 7y + c$
101. $y^2 - 8y - c$ 102. $y^2 - y - c$

REVIEW EXERCISES

Multiply out each of the following factors.

103. $(2x + 1)(x - 2)$ 104. $(3y + 5)(y - 1)$
105. $(3p + 2)(2p + 3)$ 106. $(2m - 7)(m + 1)$
107. $2(2y - 5)(y + 4)$ 108. $3(2t + 7)(3t - 1)$
109. $-x(4x - 1)(2x + 5)$ 110. $-y(3y + 4)(2y - 5)$
111. $-4y^2(3y + 8)(2y - 3)$ 112. $-5x^2(5x + 8)(7x - 2)$

7.3 Factoring Trinomials of the Form $ax^2 + bx + c$

OBJECTIVES

In this section we will learn to
1. factor trinomials whose leading coefficient is not 1;
2. evaluate second-degree polynomial functions; and
3. plot the graphs of selected trinomial functions.

In Section 7.2 we factored trinomials with a leading coefficient of 1. In this section we will learn to factor trinomials with leading coefficients other than 1. For example,

$3x^2 + 7x + 2$

We know from past experience that if $3x^2 + 7x + 2$ is factorable, it might be the product of two binomials, so we start as follows:

$3x^2 + 7x + 2 = ()()$

The only possible factors of $3x^2$ are $3x$ and x, so we use these two numbers as the first terms of each factor.

$$3x^2 + 7x + 2 = (3x\quad)(x\quad)$$

Now we examine the last term, 2, in the trinomial. Its only possible factors are 1 and 2. We know that they are both positive, because all of the terms in the trinomial are positive. All we have to do is to use 1 and 2 as the second terms in each potential factor and multiply to see if we get the correct middle term, $7x$.

$$3x^2 + 7x + 2 \stackrel{?}{=} (3x + 2)(x + 1) \quad \text{Wrong; } 2x + 3x = 5x$$

Let's exchange the 2 and the 1 to see how the middle term is affected.

$$3x^2 + 7x + 2 \stackrel{\checkmark}{=} (3x + 1)(x + 2) \quad \text{Right; } x + 6x = 7x$$

Thus $3x^2 + 7x + 2 = (3x + 1)(x + 2)$.

As the number of factors of the leading coefficient and the constant term increases, there will be more ways to arrange their factors when seeking the factors of the polynomial. Only one, however, can be correct.

EXAMPLE 1   Factor $3x^2 + 11x + 10$ completely.

SOLUTION   The only possible factors of the first term are $3x$ and x. The pairs of factors of the last term are 10 and 1 or 5 and 2. Since all the terms of $3x^2 + 11x + 10$ are positive, the signs between the two terms of the factors must be positive. Here are the possible combinations:

Possible Factors	Middle Term	Result
$(3x + 1)(x + 10)$	$30x + x = 31x$	Wrong
$(3x + 10)(x + 1)$	$3x + 10x = 13x$	Wrong
$(3x + 2)(x + 5)$	$15x + 2x = 17x$	Wrong
$(3x + 5)(x + 2)$	$6x + 5x = 11x$	Right

Thus

$$3x^2 + 11x + 10 = (3x + 5)(x + 2)$$

EXAMPLE 2   Factor $2x^2 - 7x + 6$ completely.

SOLUTION   The factors of $2x^2$ are $2x$ and x. Since the last term, 6, is positive and the sign of the middle term, $-7x$, is negative, we must consider negative factors of 6. The choices are -6 and -1 or -2 and -3.

Possible Factors	Middle Term	Result
$(2x - 6)(x - 1)$	$-2x - 6x = -8x$	Wrong
$(2x - 1)(x - 6)$	$-12x - x = -13x$	Wrong

Possible Factors	Middle Term	Result
$(2x - 2)(x - 3)$	$-6x - 2x = -8x$	Wrong
$(2x - 3)(x - 2)$	$-4x - 3x = -7x$	Right

In this example, we could have avoided the first and the third trials. In the first trial, $2x - 6$ has a common factor of 2, and in the third trial, $2x - 2$ also has a common factor of 2. Since the original trinomial, $2x^2 - 7x + 6$, does not have a common factor of 2, none of its binomial factors can have a common factor of 2. The trinomial is factored as

$$2x^2 - 7x + 6 = (2x - 3)(x - 2)$$

EXAMPLE 3   Factor $12x^2 - 8x - 15$ completely.

SOLUTION   The possible pairs of factors of the first term are $12x$ and x, $6x$ and $2x$, or $4x$ and $3x$. Since the last term is negative, we need to consider pairs of factors of 15 where one is positive and one is negative. These are -15 and 1, 15 and -1, -5 and 3, or 5 and -3. One possible combination yields

$(6x - 5)(2x + 3)$ Wrong; $18x + (-10x) = 8x$
$-10x$
$18x$

Notice that in the first trial, the only thing wrong with the middle term is its sign. If we interchange the middle signs of the two factors, we get

$(6x + 5)(2x - 3)$ Right; $-18x + 10x = -8x$
$10x$
$-18x$

Therefore $12x^2 - 8x - 15 = (6x + 5)(2x - 3)$.

EXAMPLE 4   Factor $2x^3 + 10x^2 + 12x$ completely.

SOLUTION   We factor out the greatest common factor, $2x$, first.

$$2x^3 + 10x^2 + 12x = 2x(x^2 + 5x + 6)$$
$$= 2x(x + 2)(x + 3)$$

EXAMPLE 5   Factor $30x^5 + 2x^4 - 56x^3$ completely.

SOLUTION   We factor out the greatest common factor, $2x^3$, first.

$$30x^5 + 2x^4 - 56x^3 = 2x^3(15x^2 + x - 28)$$

Now we factor $15x^2 + x - 28$. Because the middle term, x, has a small coefficient, we start with factors of the first and last terms that are close in size. As it turns out, the factors of $15x^2 + x - 28$ are $5x + 7$ and $3x - 4$. Thus,

$$30x^5 + 2x^4 - 56x^3 = 2x^3(5x + 7)(3x - 4)$$

When factoring is a trial-and-error process, it can become tedious if not approached in an organized manner. For instance, $24x^2 + 34x - 45$ can result in 48 different trials if you are unlucky enough to choose the right combination last. To increase your skill, keep the following steps in mind.

To Factor Trinomials with Leading Coefficient Other Than 1

1. Remove common factors first so that the leading coefficient is positive.
2. If the last term of the trinomial is positive, both terms in the binomial factors will have the same sign as the middle term in the trinomial.

   Sign positive
   ↓
   $6x^2 + 17x + 12 = (2x + 3)(3x + 4)$

   Same sign

   Sign positive
   ↓
   $6x^2 - 17x + 12 = (2x - 3)(3x - 4)$

   Same sign

3. If the last term of the trinomial is negative, the terms in the binomial factors will have opposite signs.

   $6x^2 - x - 12 = (2x - 3)(3x + 4)$
   ↑
   Sign negative   Opposite sign

   $6x^2 + x - 12 = (2x + 3)(3x - 4)$
   ↑
   Sign negative   Opposite sign

4. Neither of the binomial factors of a polynomial can have a common factor if the polynomial has none.

EXAMPLE 6   Factor: $-16x^3 - 4x^2 + 30x$.

SOLUTION   The polynomial $-16x^3 - 4x^2 + 30x$ has a common factor and a negative leading coefficient; we remove it first.

$$-16x^3 - 4x^2 + 30x = -2x(8x^2 + 2x - 15) \qquad \text{$-2x$ is a common factor.}$$
$$= -2x(2x + 3)(4x - 5)$$

Before leaving our work with factoring trinomials, it is appropriate that we review two previous concepts: polynomial functions and the Cartesian coordinate system. Recall that to find a given value of $f(x)$, we replace x with the given value everywhere x occurs in the formula for $f(x)$.

EXAMPLE 7   Given $f(x) = x^2 + 3x + 2$, find seven ordered pairs of the form $(x, f(x))$, for the values $-4, -3, -2, -\dfrac{3}{2}, -1, 0,$ and 1.

SOLUTION

FUNCTION VALUES	ORDERED PAIR
$f(-4) = (-4)^2 + 3(-4) + 2 = 16 - 12 + 2 = 6$	$(-4, 6)$
$f(-3) = (-3)^2 + 3(-3) + 2 = 9 - 9 + 2 = 2$	$(-3, 2)$
$f(-2) = (-2)^2 + 3(-2) + 2 = 4 - 6 + 2 = 0$	$(-2, 0)$
$f\left(-\dfrac{3}{2}\right) = \left(-\dfrac{3}{2}\right)^2 + 3\left(-\dfrac{3}{2}\right) + 2 = \dfrac{9}{4} - \dfrac{9}{2} + 2 = \dfrac{-1}{4}$	$\left(-\dfrac{3}{2}, -\dfrac{1}{4}\right)$
$f(-1) = (-1)^2 + 3(-1) + 2 = 0$	$(-1, 0)$
$f(0) = (0)^2 + 3(0) + 2 = 2$	$(0, 2)$
$f(1) = (1)^2 + 3(1) + 2 = 6$	$(1, 6)$

We can find the graph of this polynomial by plotting each of these points on a Cartesian coordinate system and drawing a smooth curve through them, as shown in Figure 7.1.

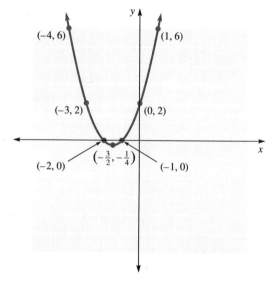

FIGURE 7.1

Do You Remember?

Can you match these?

More than one answer may apply.

___ 1. When factoring, always remove any ___ first.
___ 2. The factors of $x^2 + 8x + 12$ will have all ___ sign(s).
___ 3. The factors of $x^2 - 9x - 10$ will have ___ sign(s).
___ 4. The factors of $2x^2 - x - 28$ are ___.
___ 5. The factors of $3x^2 - 17x + 10$ will have ___ sign(s).
___ 6. To begin factoring a trinomial, the lead term should be ___.
___ 7. The factors of $x^2 - 12x + 20$ are ___.
___ 8. A nonfactorable expression is said to be ___.

a) The same
b) Different
c) Common factors
d) Positive
e) $(2x + 7)$ and $(x - 4)$
f) Prime
g) $(x - 10)$ and $(x + 2)$
h) $(x - 10)$ and $(x - 2)$
i) Negative

Answers: 1. c 2. a, d 3. b 4. e 5. a, i 6. d 7. h 8. f

Exercise Set 7.3

Factor each polynomial completely. If a polynomial is not factorable, write prime.

1. $2y^2 + 5y + 2$
2. $2a^2 + 7a + 6$
3. $5x^2 - 3x - 8$
4. $4p^2 - 12p + 5$
5. $3x^2 - 8x + 4$
6. $11t^2 - 9t - 2$
7. $2s^2 + s - 6$
8. $6x^2 + 5x - 6$
9. $12m^2 - 7m - 11$
10. $6d^2 + 5d - 12$
11. $8y^2 + 47y - 6$
12. $12n^2 - 7n - 12$
13. $6 - 11x + 4x^2$
14. $5 - 23y - 12y^2$
15. $3a^2 + 2a + 5$
16. $7s^2 + 5s + 3$
17. $11b^2 - 9b - 2$
18. $15x^2 - 23x + 6$
19. $4n^2 + 12n + 5$
20. $4z^2 - 12z + 9$

Factor each polynomial completely. Remember to factor out the GCF first. If a polynomial is not factorable, write prime.

21. $-2y^2 - 5y - 3$ (*Hint:* Factor out -1 first.)
22. $-10x^2 + 6x + 16$ (*Hint:* Factor out -2 first.)
23. $-2p^2 + 12p - 10$
24. $-22a^2 - 8a + 14$
25. $625m^2 - 25m^3$
26. $400t^2 - 165t$
27. $-42r^2 + 50r + 68$
28. $-3m^2 - 5m + 12$
29. $225x^2 + 400x^3$
30. $12m^3 - 3m^2$
31. $8y^2 + 40y + 50$
32. $7a^2b - 7ab - 35b^2$
33. $-60x^3y + 2x^2y + 2xy$
34. $-8x^2 + 34x - 33$
35. $28 + 40d - 32d^2$
36. $30 + p - 20p^2$
37. $4y^2 - 28y + 49$
38. $25a^2 + 60ab + 36b^2$
39. $-4x^3 - 12x^2 - 8x$
40. $-5x^3 - 10x^2 - 5x$
41. $-30a^2 - a - 20$
42. $-18d^2 + 19d + 12$
43. $16a^2 - 4a - 56$
44. $32x^4 + 64x^3 + 32x^2$
45. $8y^3 + 26y^2 + 6y$
46. $27p^4 + 42p^3 - 24p^2$
47. $30x^3 - 28x^2y - 130xy^2$
48. $72a^3 + 216a^2b + 112ab^2$
49. $64y^2 + 80yz + 25z^2$
50. $81z^2 - 144zw + 64w^2$

51. $26m^5n^4 + 104m^4n^5$
52. $-14x^4y^4 - 189x^3y^5$
53. $80s^3t - 26s^2t^2 - 72st^3$
54. $120r^3 - 39r^2s - 108rs^2$
55. $-48x^2 - 55xy + 24y^2$
56. $-84y^3 - 82xy^2 - 36x^2y$
57. $-110x^5y + 140x^4y - 30x^3y$
58. $-120x^4y - 72x^3y + 48x^2y$
59. $6x^2 - 7xy - 3y^2$
60. $7x^2 + 33xy - 10y^2$
61. $15y^2 - 41y + 14$
62. $3r^2 - 35r + 50$
63. $4p^2 + 6p + 9$
64. $3t^2 + 14t + 8$
65. $8x^2 - 6xy - 35y^2$
66. $20y^2 - 7xy - 6x^2$
67. $10a^2 - 39ab - 4b^2$
68. $12r^2 - 31rs - 15s^2$
69. $24x^2 + 2xy - 15y^2$
70. $15x^3y + 34x^2y^2 + 15xy^3$
71. $32x^3y - 16x^2y^2 - 30xy^3$
72. $75r^3s - 60r^2s^2 - 36rs^3$
73. $36a^3b - 48a^2b^2 + 16ab^3$
74. $-18x^3 - 30x^2 + 28x$
75. $-10y^4 + 66y^3 + 28y^2$
76. $-15y^4 + 10y^3 + 40y^2$
77. $-40x^5 - 4x^4 + 20x^3$
78. $-14x^2 + 60xy - 16y^2$

For each polynomial function, find ordered pairs $(x, f(x))$ for the given values of x. Plot these ordered pairs on a Cartesian coordinate system and draw a smooth curve to connect the points.

79. $f(x) = x^2 - 3x - 4$; $x = -1, 0, 1, \frac{3}{2}, 2, 3, 4$
80. $f(x) = x^2 - x - 2$; $x = -2, -1, 0, \frac{1}{2}, 1, 2, 3$
81. $f(x) = x^2 + 3x + 2$; $x = -3, -2, -1, -\frac{3}{2}, 0, 1$
82. $f(x) = (x - 1)^2$; $x = -2, -1, 0, 1, 2, 3, 4$
83. $f(x) = -(x + 2)^2$; $x = -5, -4, -3, -2, -1, 0, 1$
84. $f(x) = -x^2 - 4x + 5$; $x = -5, -4, -3, -2, 0, 1, 2$

REVIEW EXERCISES

Multiply each of the following special products.

85. $(x + 1)^2$
86. $(x - 3)^2$
87. $(x + 1)(x - 1)$
88. $(x + 4)(x - 4)$
89. $(y - 5)^2$
90. $(y + 4)^2$
91. $(2x - y)^2$
92. $(3x + y)^2$
93. $(5x + 2y)(5x - 2y)$
94. $(3r + 4s)(3r - 4s)$
95. $(6m - n)(6m + n)$
96. $(7p - 2q)(7p + 2q)$
97. $-(4s + 5t)^2$
98. $-(8x - 3y)^2$
99. $(x + 1)^3$
100. $(x + 2)^3$
101. $(r - 3)^3$
102. $(x - 2)^3$
103. $(x + y)^3$
104. $(s - t)^3$

7.4 Special Factors; Factoring by Grouping

OBJECTIVES In this section we will learn to
1. factor perfect-square trinomials;
2. factor the difference of squares;
3. factor the sum or difference of two cubes; and
4. factor by grouping.

In Section 6.4 we found the following special products:

1. The square of a binomial:

$$(a + b)^2 = a^2 + 2ab + b^2$$
$$(a - b)^2 = a^2 - 2ab + b^2$$

2. The product of the sum and the difference of the same two quantities:

$$(a + b)(a - b) = a^2 - b^2$$

Using these products as a guide, we can see how to factor special polynomials. The expressions $a^2 + 2ab + b^2$ and $a^2 - 2ab + b^2$ are called **perfect-square trinomials** because they are the squares of the binomials $a + b$ and $a - b$, respectively. The key to recognizing a perfect-square trinomial is noticing that two of its terms are perfect squares. The middle term must then be twice the product of the two terms of the binomial. For example,

$$x^2 + 6x + 9 = (x + 3)^2$$

is a perfect-square trinomial. The square of x is x^2 and the square of 3 is 9. The middle term, $6x$, is twice the product of 3 and x.

EXAMPLE 1 Factor each of the following perfect-square trinomials.

(a) $x^2 + 4x + 4 = (x + 2)^2$
(b) $x^2 + 14x + 49 = (x + 7)^2$
(c) $4x^2 - 4x + 1 = (2x - 1)^2$
(d) $9x^2 + 12x + 4 = (3x + 2)^2$
(e) $x^4 + 10x^2y + 25y^2 = (x^2 + 5y)^2$ $x^4 = (x^2)^2$
(f) $a^2 - 2.2a + 1.21 = (a - 1.1)^2$ $1.21 = (1.1)^2$

Not all trinomials with two squared terms are perfect-square trinomials. It is important to check the middle term. For example, in $x^2 + 13x + 36$, the first term, x^2, and the last term, 36, are perfect squares. However, $x^2 + 13x + 36 \neq (x + 6)^2$. In fact,

$$x^2 + 13x + 36 = (x + 4)(x + 9)$$

In the equation

$$(a + b)(a - b) = a^2 - b^2$$

the expression $a^2 - b^2$ is called **the difference of two squares.** It serves as a pattern for factoring the difference of two quantities, each of which is a perfect square. For example,

$$4x^2 - 9y^2 = (2x - 3y)(2x + 3y)$$

The square of $2x$ ↑   ↑ The square of $3y$

EXAMPLE 2 Factor each of the following binomials as the difference of two squares.

(a) $x^2 - 25 = (x - 5)(x + 5)$
(b) $y^2 - 81 = (y - 9)(y + 9)$
(c) $4a^2 - 9b^2 = (2a - 3b)(2a + 3b)$
(d) $25x^2 - 36y^2 = (5x - 6y)(5x + 6y)$
(e) $121 - t^2 = (11 - t)(11 + t)$

Although the sum of two squares cannot be factored, the sum of two cubes and the difference of two cubes both can.

> **CAUTION**
>
> The sum of the squares of two quantities cannot be factored using only real numbers; it is prime. In particular,
>
> $$a^2 + b^2 \neq (a + b)^2$$
>
> since
>
> $$(a + b)^2 = a^2 + 2ab + b^2$$
>
> and
>
> $$a^2 - b^2 \neq (a - b)^2$$
>
> since
>
> $$(a - b)^2 = a^2 - 2ab + b^2$$

To Factor the Sum and the Difference of Two Cubes

$$a^3 + b^3 = (a + b)(a^2 - ab + b^2)$$
$$a^3 - b^3 = (a - b)(a^2 + ab + b^2)$$

To see that this pattern is the correct factorization, we multiply out the right sides.

$$(a + b)(a^2 - ab + b^2) = a^3 - a^2b + ab^2 + a^2b - ab^2 + b^3 = a^3 + b^3$$
$$(a - b)(a^2 + ab + b^2) = a^3 + a^2b + ab^2 - a^2b - ab^2 - b^3 = a^3 - b^3$$

The patterns for the sum and difference of two cubes **must be memorized** because of the potential for confusion with the square of a binomial.

EXAMPLE 3 Factor: $x^3 - 8$.

SOLUTION To factor $x^3 - 8$, we let $x = a$ and $2 = b$ in $a^3 - b^3 = (a - b)(a^2 + ab + b^2)$.

$$x^3 - 8 = x^3 - 2^3 = (x - 2)(x^2 + 2x + 2^2) = (x - 2)(x^2 + 2x + 4)$$
$$a^3 - b^3 = (a - b)(a^2 + a \cdot b + b^2)$$

Thus,

$$x^3 - 8 = (x - 2)(x^2 + 2x + 4)$$

EXAMPLE 4 Factor $x^3 + 125y^3$.

SOLUTION We let $a = x$ and $b = 5y$. Then

$$(a + b)(a^2 - ab + b^2)$$
$$x^3 + 125y^3 = x^3 + (5y)^3 = (x + 5y)(x^2 - 5xy + 25y^2)$$

EXAMPLE 5 Factor each of the following polynomials:

(a) $64x^3 - 27y^3$ (b) $3m^3 - 24n^3$

SOLUTION (a) $64x^3 - 27y^3 = (4x)^3 - (3y)^3 = (4x - 3y)[(4x)^2 + (4x)(3y) + (3y)^2]$
$\qquad\qquad\qquad = (4x - 3y)[16x^2 + 12xy + 9y^2]$

(b) $3m^3 - 24n^3 = 3(m^3 - 8n^3)$ Remove the common factor first.
$\qquad\qquad\quad = 3(m - 2n)(m^2 + 2mn + 4n^2)$

Some expressions require the use of more than one technique in order to factor them completely.

EXAMPLE 6 Factor $a^4 - 18a^2 + 81$ completely.

SOLUTION We begin by factoring $a^4 - 18a^2 + 81$ as a perfect-square trinomial.

$$a^4 - 18a^2 + 81 = (a^2)^2 - 18a^2 + 9^2$$
$$= (a^2 - 9)^2$$

$$= [(a-3)(a+3)]^2 \quad a^2 - 9 \text{ is the difference of two squares.}$$
$$= (a-3)^2(a+3)^2 \quad (ab)^m = a^m b^m$$

EXAMPLE 7 Factor $x^4 - 16$ completely.

SOLUTION $x^4 - 16 = (x^2 - 4)(x^2 + 4)$

Now we notice that $x^2 - 4$ is the difference of two squares and can be factored further.

$$x^4 - 16 = (x^2 - 4)(x^2 + 4)$$
$$= (x-2)(x+2)(x^2+4)$$

The binomial $x^2 + 4$ cannot be factored further.

> **CAUTION**
>
> Do not confuse
>
> $a^2 - ab + b^2$
>
> with
>
> $a^2 - 2ab + b^2 = (a-b)^2$
>
> or
>
> $a^2 + ab + b^2$
>
> with
>
> $a^2 + 2ab + b^2 = (a+b)^2$
>
> The expressions $a^2 - ab + b^2$ and $a^2 + ab + b^2$ cannot be factored further.

We once again turn our attention to polynomials that contain more than three terms. In this situation we should try **factoring by grouping.** To review how it works, consider the polynomial

$$ax + bx + ay + by$$

The first two terms have a common factor of x, and the second two terms have a common factor of y.

$$ax + bx + ay + by = x(a+b) + y(a+b)$$

By grouping the terms this way, we have isolated the common factor $a + b$.

$$ax + bx + ay + by = (a+b)x + (a+b)y$$
$$= (a+b)(x+y) \quad \text{The GCF is } a+b.$$

There is often more than one way to group a polynomial in order to find its factors. Another way to group the preceding example is

$$ax + bx + ay + by = (ax + ay) + (bx + by) \quad \text{Group the first and third terms and the second and fourth terms.}$$
$$= a(x+y) + b(x+y)$$
$$= (x+y)(a+b) \quad \text{The common factor is } x+y.$$

Except for order, the factorization is the same.

EXAMPLE 8 Factor $x^2y + x^2z - y - z$ completely.

SOLUTION
$$x^2y + x^2z - y - z = x^2(y+z) - 1(y+z)$$
$$= (y+z)(x^2-1) \quad y+z \text{ is a common factor.}$$
$$= (y+z)(x+1)(x-1) \quad \text{Factor } x^2-1 \text{ as the difference of two squares.}$$

Factoring by grouping does not always require that the terms be grouped two at a time. Notice in the next example that the grouping involves three terms.

EXAMPLE 9 Factor $x^2 + 10x + 25 - y^2$ completely.

SOLUTION $x^2 + 10x + 25 - y^2 = (x + 5)^2 - y^2$

When rewritten this way, we can see that $x^2 + 10x + 25 - y^2$ is the difference of two squares. We let $a = x + 5$ and $b = y$. Then

$$x^2 + 10x + 25 - y^2 = \underbrace{(x + 5)^2}_{a^2} - \underbrace{y^2}_{b^2}$$

$$= \underbrace{[(x + 5)}_{a} + \underbrace{y]}_{+b}\underbrace{[(x + 5)}_{a} - \underbrace{y]}_{-b}$$

$$= [x + 5 + y][x + 5 - y]$$

Do You Remember?

Can you match these?

_____ 1. $a^2 - b^2$ a) $(a - b)(a^2 + 2ab + b^2)$
_____ 2. $a^3 - b^3$ b) $(a + b)(a^2 - ab + b^2)$
_____ 3. $a^3 + b^3$ c) $(a - b)(a + b)$
_____ 4. $a^2 + b^2$ d) $(a + b)^2$
_____ 5. $a^2 + 2ab + b^2$ e) Prime
_____ 6. $a^2 - 2ab + b^2$ f) $a^2 + ab + b^2$
_____ 7. A factor of $a^3 - b^3$ g) $a^2 - ab + b^2$
_____ 8. A factor of $a^3 + b^3$ h) $(a - b)(a^2 + ab + b^2)$
                                         i) $(a - b)^2$

Answers: 1. c 2. h 3. b 4. e 5. d 6. i 7. f 8. g

Exercise Set 7.4

Factor completely. If not factorable, write prime.

1. $x^2 - 1$
2. $x^2 - 9$
3. $y^2 - 16$
4. $b^2 - 4$
5. $t^2 - 25$
6. $m^2 - 36$
7. $r^2 - 64$
8. $s^2 - 100$
9. $p^2 - 121$
10. $2x^2 - 18$
11. $3y^2 - 27$
12. $12 - 3x^2$
13. $50 - 2t^2$
14. $x^3 - 4x$
15. $5y^3 - 5y$
16. $x^2 - y^2$
17. $a^2 - b^2$
18. $3m^2 - 12n^2$
19. $5a^2 - 45b^2$
20. $x^4 - 1$
21. $y^4 - 16$
22. $a^4 - b^4$
23. $x^4 - y^4$
24. $81 - t^4$
25. $16 - s^4$
26. $3xy^3 - 12x^3y$
27. $16xy^3 - 26x^3y$
28. $x^2 + 4$
29. $y^2 + 9$

Factor completely. If not factorable, write prime.

30. $x^2 + 2x + 1$
31. $y^2 + 4y + 4$
32. $a^2 - 6a + 9$
33. $b^2 - 8b + 16$
34. $t^2 + 10t + 25$
35. $r^2 + 14r + 49$
36. $m^2 - 12m + 36$
37. $s^2 - 18s + 81$
38. $100 + 20x + x^2$
39. $144 + 24p + p^2$
40. $121 - 22y + y^2$
41. $64 - 16q + q^2$
42. $x^2 - 14x + 64$
43. $a^2 - 12a + 49$
44. $2x^2 + 8x + 8$
45. $3y^2 + 24y + 48$
46. $2a^2 - 24a + 72$
47. $3b^2 + 30b + 75$
48. $x^2 + 6xy + 9y^2$
49. $m^2 + 8mn + 16n^2$
50. $4x^2 + 12x + 9$
51. $9p^2 + 12p + 4$
52. $9a^2 - 24ab + 16b^2$
53. $25p^2 - 60pq + 36q^2$
54. $x^4 - 8x^2 + 16$
55. $y^4 - 18y^2 + 81$
56. $4t^4 - 4t^2 + 1$
57. $9x^4 - 6x^2 + 1$
58. $16a^4 - 40a^2b^2 + 25b^4$
59. $36x^4 + 84x^2y^2 + 49y^4$

Factor each of the following polynomials completely.

60. $ax + ay + bx + by$
61. $cx - cy + dx - dy$
62. $x^3 + x^2 - 2x - 2$
63. $y^3 - 3y^2 - 4y + 12$
64. $p^3 - 5p^2 - 4p + 20$
65. $x^3 + 7x^2 - x - 7$
66. $ab + 5a + bc + 5c$
67. $xy + 2x + cy + 2c$
68. $cx - cy - dy + dx$
69. $rx - rs - tx + ts$
70. $x^2y + xy^2 - x - y$
71. $r^2z + r^2x - z - x$
72. $2bc + 2bx + cy + xy$
73. $3abcd - 3cd - ab + 1$
74. $3(x - 4) + 4x(x - 4)$
75. $5y(2x - 1) - (2x - 1)$

Factor each of the following sums or differences of cubes completely.

76. $x^3 - 1$
77. $y^3 + 1$
78. $x^3 + 8$
79. $y^3 - 27$
80. $a^3 - 64$
81. $b^3 + 125$
82. $8x^3 + 27$
83. $64p^3 - q^3$
84. $2x^3 - 16y^3$
85. $3a^3 - 24b^3$
86. $-2x^3 + 16y^3$
87. $-24p^3 - 81q^3$

Factor each of the following differences of squares completely.

88. $(x - 1)^2 - (x + 2)^2$
89. $(x - 3)^2 - (x + 4)^2$
90. $(p - 2)^2 - (p^2 + 4p + 4)$
91. $(a + 3)^2 - (a^2 - 6a + 9)$
92. $x^2 - (x - 2)^2$
93. $y^2 - (y + 1)^2$
94. $(x^2 - 8x + 16) - (x + 4)^2$
95. $(x^2 - 10x + 25) - (x + 5)^2$

WRITE IN YOUR OWN WORDS

96. What are the steps in factoring a binomial?
97. What are the steps in factoring a trinomial?
98. Tell how you can be sure that
$$a^3 + b^3 = (a + b)(a^2 - ab + b^2)$$
99. How can you be sure that $a^2 - b^2 = (a + b)(a - b)$?
100. Why is the sum of squares not factorable?
101. Why is $x^2 - 5x + 3$ prime?
102. How are the operations of multiplication and factoring related?

7.5 A Summary of Factoring

OBJECTIVES

In this section we will review
1. the factoring of general trinomials;
2. factoring the difference of squares;
3. factoring the sum and difference of cubes; and
4. factoring by grouping.

In Sections 7.1 through 7.4, we learned to factor many different polynomials by a variety of methods. However, the problems were grouped according to type, which allowed us to predict the proper approach in factoring them. The Exercise Set at the end of this section contains a variety of problems and it is the task of the reader to decide the best method for factoring each one of them. Take notice of the following caution before applying the suggestions for factoring: In each of the steps for factoring, new expressions can emerge that are still factorable. Factoring must continue until **all factors are prime.**

Suggestions for Successful Factoring

1. If the expression contains any common factors including -1, remove them first.
2. If the expression contains only two terms, check to see if it is
   (a) the difference of two squares: $a^2 - b^2 = (a - b)(a + b)$
   (b) the difference of two cubes: $a^3 - b^3 = (a - b)(a^2 + ab + b^2)$
   (c) the sum of two cubes: $a^3 + b^3 = (a + b)(a^2 - ab + b^2)$
3. If the expression is a trinomial and the first and third terms are perfect squares, check to see if it is perfect-square trinomial:

   $a^2 + 2ab + b^2 = (a + b)^2$ or $a^2 - 2ab + b^2 = (a - b)^2$

   ↑   ↑   ↑   ↑
   Perfect squares   Perfect squares

4. If the leading coefficient of a trinomial is 1, as in $x^2 + bx + c$, try to find two numbers with a product of c and a sum of b. These numbers will become the second terms of the two binomial factors. When the sign before c is positive, the signs in the binomial factors will be the same as the sign before b. If the sign before c is negative, the signs in the binomial factors will differ.
5. If the leading coefficient of a trinomial is not 1, try factoring it as a **general trinomial.** If the sign before the last term in the trinomial is positive, the signs of the second terms of its binomial factors will be the same as the sign before the middle term of the trinomial. Remember that *neither of the binomial factors of a polynomial can have a common factor if the trinomial has none.*
6. If the expression contains more than three terms that don't combine, try **factoring by grouping.**
7. Check all results by multiplication.

EXAMPLE 1   Factor: $96x^4z^2 - 486y^4z^2$.

SOLUTION   First we remove the common factor, $6z^2$:

$$96x^4z^2 - 486y^4z^2 = 6z^2(16x^4 - 81y^4)$$

Since $16x^4 - 81y^4$ is the difference of two squares, we can factor it further.

$$96x^4z^2 - 486y^4z^2 = 6z^2(16x^4 - 81y^4)$$
$$= 6z^2(4x^2 - 9y^2)(4x^2 + 9y^2)$$

Since $4x^2 - 9y^2$ is the difference of two squares, we can factor *it* further.

$$96x^4z^2 - 486y^4z^2 = 6z^2(4x^2 - 9y^2)(4x^2 + 9y^2)$$
$$= 6z^2(2x - 3y)(2x + 3y)(4x^2 + 9y^2)$$

The binomial $4x^2 + 9y^2$ is prime and cannot be factored further. Thus,

$$96x^4z^2 - 486y^4z^2 = 6z^2(2x - 3y)(2x + 3y)(4x^2 + 9y^2)$$

EXAMPLE 2   Factor: $6x^3 + 3x^2 - 3x$.

SOLUTION   We remove the common factor, $3x$, first.

$$6x^3 + 3x^2 - 3x = 3x(2x^2 + x - 1)$$

The leading coefficient of $2x^2 + x - 1$ is not 1, so we factor it as a general trinomial.

$$6x^3 + 3x^2 - 3x = 3x(2x^2 + x - 1)$$
$$= 3x(2x - 1)(x + 1)$$

EXAMPLE 3   Factor $9x^2 + 30xy + 25y^2$ completely.

SOLUTION   The first term, $9x^2$, and last term, $25y^2$, are perfect squares. We'll try factoring $9x^2 + 30xy + 25y^2$ as a perfect-square trinomial.

$$9x^2 + 30xy + 25y^2 \stackrel{?}{=} (3x + 5y)^2$$

The middle term is $2(3x)(5y) = 30xy$. The factorization is correct.

EXAMPLE 4   Factor $a^3(a - b) - b^3(a - b)$ completely.

SOLUTION   We remove the common factor, $a - b$, first.

$$a^3(a - b) - b^3(a - b) = (a - b)(a^3 - b^3)$$

Now we factor $a^3 - b^3$ as a difference of two cubes.

$$a^3(a - b) - b^3(a - b) = (a - b)(a^3 - b^3)$$
$$= (a - b)(a - b)(a^2 + ab + b^2)$$
$$= (a - b)^2(a^2 + ab + b^2)$$

Notice that in Example 4, either $(a - b)(a - b)(a^2 + ab + b^2)$ or $(a - b)^2(a^2 + ab + b^2)$ are acceptable as the factored form of $a^3(a - b) - b^3(a - b)$.

EXAMPLE 5   Factor $x^2 - 2xy + y^2 - z^2$ completely.

SOLUTION   This expression involves more than three terms, so we'll use factoring by grouping. The first three terms form a perfect-square trinomial.

$$x^2 - 2xy + y^2 - z^2 = (x^2 - 2xy + y^2) - z^2$$
$$= (x - y)^2 - z^2$$

We now have the difference of two squares.

$$x^2 - 2xy + y^2 - z^2 = [(x-y)+z][(x-y)-z] \qquad a^2 - b^2 = (a+b)(a-b)$$
$$= (x - y + z)(x - y - z)$$

EXAMPLE 6 Factor $\frac{1}{8}x^3 - \frac{1}{27}y^3$ completely.

SOLUTION The polynomial $\frac{1}{8}x^3 - \frac{1}{27}y^3$ is the difference of two cubes:

$$\frac{1}{8}x^3 - \frac{1}{27}y^3 = \left(\frac{1}{2}x\right)^3 - \left(\frac{1}{3}y\right)^3$$
$$= \left(\frac{1}{2}x - \frac{1}{3}y\right)\left[\left(\frac{1}{2}x\right)^2 + \left(\frac{1}{2}x\right)\left(\frac{1}{3}y\right) + \left(\frac{1}{3}y\right)^2\right]$$
$$= \left(\frac{1}{2}x - \frac{1}{3}y\right)\left(\frac{1}{4}x^2 + \frac{1}{6}xy + \frac{1}{9}y^2\right)$$

Do You Remember?

Can you match these?

Match each of the following expressions to the method you would use to factor it.

____ 1. $x^3 + x^2 + x + 1$
____ 2. $36x^2 - 25y^2$
____ 3. $27m^3 + 125y^3$
____ 4. $x^2 + 6x + 9$
____ 5. $a^3 - 64c^3$
____ 6. $2x^2 + 7x + 6$
____ 7. $2x^3 - 4x^2 - 12x$

a) Difference of two squares
b) Difference of two cubes
c) Perfect-square trinomial
d) Sum of two cubes
e) Factor by grouping
f) Remove a common factor
g) Trial and error

Answers: 1. e 2. a 3. d 4. c 5. b 6. g 7. f

Exercise Set 7.5

Factor each polynomial completely. If not factorable, write prime.

1. $3x^2 - 3$
2. $7y^2 + 7$
3. $xy^3 - 9x^3y$
4. $x^2 + 5x + 4$
5. $y^2 - 3y - 4$
6. $y^2 + 8y + 16$
7. $a^3 + 8b^3$
8. $16m^2 - 25n^2$
9. $36 - 49k^2$
10. $5x^2 - 5x - 60$
11. $b^3 - 8$
12. $16p^2 - 40pq + 25q^2$

13. $y^3 - y^2$
14. $48x - 64$
15. $x^2 + 2x - 15$
16. $3y^2 + 4y - 4$
17. $a^2 + 25$
18. $16p^2 + 46p - 6$
19. $10 - 24y - 18y^2$
20. $48 + 10m - 2m^2$
21. $x^4 - 81$
22. $t^4 - 16$
23. $125 + 8a^3$
24. $t(t - 6) + 7(t - 6)$
25. $a^2 - b^2 + 5a - 5b$
26. $2xy - 5y + 5 - 2x$
27. $9x^2 + 3x - 30$
28. $-42p^2 - 64p - 24$
29. $x^3y^3 + 1$
30. $64x^2 - 121y^2$
31. $ab + 7b + 3a + 21$
32. $3x^2 - 27$
33. $6 - 4x - 32x^2$
34. $m^2 + 14m + 49$
35. $8q^2 + 24q + 18$
36. $24 - 6x^2$
37. $5p^2 + 5p - 5$
38. $5t^2 + 30t + 45$
39. $a^3 + 3a^2 - 10a - 30$
40. $16 + 24y + 9y^2$
41. $x^3 - 25x$
42. $m^2 + 10m - 25$
43. $5b^3 - 5c^3$
44. $6a^3 + 3a^2 - 3a$
45. $18a^2b + 12ab^2 + 30ab$
46. $m^3 + 16m$
47. $8p^2 - 25p + 3$
48. $200 + 10y - y^2$
49. $30 - 11p + p^2$
50. $a^2 + ab + b^2$
51. $-2c^3 - 8c^2 + 10c$
52. $9x^2 + 24xy + 16y^2$
53. $xy^3 - 4xy$
54. $(x - 1)^2 - (x + 3)^2$
55. $-2t^2 - 2t + 144$
56. $16x^5 - 54x^2$
57. $60 - 2a - 6a^2$
58. $a^2(1 - b) - 4(1 - b)$
59. $rs(a - b) + t(a - b)$
60. $d^2 - 81$
61. $x^2 - 169$
62. $-4 + 34x - 42x^2$
63. $4y^2 + 2y + 1$
64. $a^6 + a^5 + a^4 + a^3$
65. $-3t^2 - 3t + 60$
66. $x^4 - 81$
67. $16p^2 + 56pq + 49q^2$
68. $b^2x + b^2y - x - y$
69. $8 - 8m^2$
70. $ab + ab^3$
71. $r^3 - 27$
72. $3x^3 + 3x^2 - 60x$
73. $y^4 - 8y$
74. $15y^2 - xy - 6x^2$
75. $3a^2 + a + 1$
76. $9x^4 - 21x^3 - 8x^2$
77. $54x - 60x^2 + 6x^3$
78. $2ab + 2ax - by - xy$
79. $a^3(x^3 + 1) + b^3(x^3 + 1)$
80. $x^3(t^3 - 1) - y^3(t^3 - 1)$
81. $9y^2 - 12xy + 4x^2$
82. $900 - 9y^2 + x^2y^2 - 100x^2$
83. $x^2 - \dfrac{1}{9}$
84. $y^3 - \dfrac{1}{8}$
85. $\dfrac{1}{8}a^3 + \dfrac{64}{27}b^3$
86. $\dfrac{1}{4}a^2 - \dfrac{9}{49}b^2$
87. $0.16x^2 - 0.25y^2$
88. $0.09a^2 - 0.36$
89. $0.49 - 0.81t^2$
90. $1.21 - 1.69m^2$

For Extra Thought

Find b so that each of the following polynomials factor as shown.

91. $x^2 - bx + 9 = (x - 3)^2$
92. $y^2 + by + 16 = (y + 4)^2$
93. $a^2 - ba + 36 = (a - 6)^2$
94. $t^2 - bt + 121 = (t - 11)^2$
95. $x^b - 4 = (x + 2)(x - 2)$
96. $y^2 - 0.5y + b = \left(y - \dfrac{1}{4}\right)^2$
97. $m^3 - b = (m - 2)(m^2 + 2m + 4)$
98. $p^2 - 8p + b = (p - 4)^2$
99. $bx^2 + 24xy + 16y^2 = (3x + 4)^2$
100. $25x^2 - bxy + 36y^2 = (5x - 6y)^2$

7.6 Solving Factorable Quadratic Equations

HISTORICAL COMMENT Many attempts were made to solve quadratic equations in early times. The Egyptians attempted to solve them using arithmetic only, while Euclid and his followers used very complex geometric methods. It is interesting to note that it was not until 1631 that a work by the astronomer Thomas Harriot appeared and introduced the methods that we will employ in this section. Even so, the author ignored negative solutions, which, as we shall see, are just as valid as positive solutions.

OBJECTIVES

In this section we will learn to
1. solve factorable quadratic equations; and
2. solve selected factorable equations of degree greater than 2.

> **DEFINITION**
>
> **Quadratic Equation**
>
> A **quadratic equation** is an equation that can be put into the form
>
> $$ax^2 + bx + c = 0$$
>
> where a, b, and c are real numbers and $a \neq 0$. When a quadratic equation is written in the form $ax^2 + bx + c = 0$, it is said to be in *standard form*.

Simply stated a quadratic equation is a second-degree equation in one variable. The equations

$$x^2 + 2x - 3 = 0, \quad 2x^2 + 3x + 1 = 0, \quad \text{and} \quad 4x^2 - 9 = 0$$

are all quadratic equations in standard form.

Many quadratic equations can be solved by factoring and applying the **zero-factor property**:

> **Zero-Factor Property**
>
> If a and b represent any real number and if $a \cdot b = 0$, then either $a = 0$ or $b = 0$.

EXAMPLE 1 Solve the equation $(x + 3)(x + 5) = 0$.

SOLUTION Since $(x + 3)(x + 5) = 0$, either $x + 3 = 0$ or $x + 5 = 0$, by the zero-factor property. We now solve each of these first-degree equations by the methods covered in Chapter 3.

$$x + 3 = 0 \quad \text{or} \quad x + 5 = 0$$
$$x = -3 \quad \text{or} \quad x = -5$$

To check the solutions, we substitute them into the original equation.

$(x + 3)(x + 5) = 0$ or $(x + 3)(x + 5) = 0$ Original equation
$(-3 + 3)(-3 + 5) \stackrel{?}{=} 0$ Let $x = -3$ $(-5 + 3)(-5 + 5) \stackrel{?}{=} 0$ Let $x = -5$
$(0)(2) \stackrel{?}{=} 0$ $(-2)(0) \stackrel{?}{=} 0$
$0 \stackrel{\checkmark}{=} 0$ $0 \stackrel{\checkmark}{=} 0$

The solution set is $\{-5, -3\}$.

EXAMPLE 2 Solve the equation $x^2 - 7x - 18 = 0$.

SOLUTION We begin by factoring the equation.
$$x^2 - 7x - 18 = 0$$
$$(x - 9)(x + 2) = 0$$

Now,

$x - 9 = 0$ or $x + 2 = 0$ Zero-factor property
$x = 9$ or $x = -2$

To check, we substitute each of the solutions in the original equation.

$$x^2 - 7x - 18 = 0 \qquad\qquad x^2 - 7x - 18 = 0$$
$$9^2 - 7(9) - 18 \stackrel{?}{=} 0 \quad \text{or} \quad (-2)^2 - 7(-2) - 18 \stackrel{?}{=} 0$$
$$81 - 63 - 18 \stackrel{?}{=} 0 \qquad\qquad 4 + 14 - 18 \stackrel{?}{=} 0$$
$$0 \stackrel{\checkmark}{=} 0 \qquad\qquad\qquad\qquad 0 \stackrel{\checkmark}{=} 0$$

The solution set is $\{-2, 9\}$.

EXAMPLE 3 Solve the equation $2y^2 + y = 3$.

SOLUTION *Note:* Writing the equation as $y(2y + 1) = 3$ **will not help,** since the product of the factors must equal zero. Instead we begin by writing the equation in standard form:

$$2y^2 + y = 3$$
$$2y^2 + y - 3 = 0 \qquad \text{Subtract 3 from each side.}$$

Now we factor the equation and apply the zero-factor property:

$$2y^2 + y - 3 = 0$$
$$(2y + 3)(y - 1) = 0$$

Thus,

$2y + 3 = 0$ or $y - 1 = 0$
$2y = -3$ or $y = 1$
$y = \dfrac{-3}{2}$

CHECK
$$2y^2 + y = 3 \qquad\qquad 2y^2 + y = 3$$
$$2\left(\dfrac{-3}{2}\right)^2 + \left(\dfrac{-3}{2}\right) \stackrel{?}{=} 3 \quad \text{or} \quad 2(1)^2 + 1 \stackrel{?}{=} 3$$
$$2\left(\dfrac{9}{4}\right) - \dfrac{3}{2} \stackrel{?}{=} 3 \qquad\qquad 3 \stackrel{\checkmark}{=} 3$$
$$\dfrac{6}{2} \stackrel{?}{=} 3$$
$$3 \stackrel{\checkmark}{=} 3$$

The solution set is $\left\{\dfrac{-3}{2}, 1\right\}$.

> **Caution**
>
> Be sure to check all solutions in the *original* equation, not in one of the altered forms.

Suppose that we make an error rewriting the original equation in standard form but make no further errors solving the equation. The solutions we obtain will be solutions to the altered equation but not the original equation.

To Solve a Quadratic Equation by Factoring

Step 1 Write the equation in standard form: $ax^2 + bx + c = 0$.
Step 2 Factor the equation completely.
Step 3 Set each factor containing a variable equal to zero, and solve the resulting equation(s).
Step 4 Check each solution in the *original* equation.

EXAMPLE 4 Solve $2x^2 = 18$ by factoring.

SOLUTION **Step 1** Write the equation in standard form.

$$2x^2 = 18 \quad \text{Original equation}$$
$$2x^2 - 18 = 0$$

Step 2 Factor the equation completely.

$$2(x - 3)(x + 3) = 0$$

Step 3 Set each factor containing a variable equal to zero and solve the resulting equation(s).

$$x - 3 = 0 \quad \text{or} \quad x + 3 = 0$$
$$x = 3 \quad \text{or} \quad x = -3$$

We ignore the numerical factor 2, since it does not contain a variable and 2 certainly cannot be equal to zero.

Step 4 Check each solution in the original equation.

$$2x^2 = 18 \qquad 2x^2 = 18$$
$$2(-3)^2 \stackrel{?}{=} 18 \quad \text{or} \quad 2(3)^2 \stackrel{?}{=} 18$$
$$2(9) \stackrel{?}{=} 18 \qquad 2(9) \stackrel{?}{=} 18$$
$$18 \stackrel{\checkmark}{=} 18 \qquad 18 \stackrel{\checkmark}{=} 18$$

The solution set is $\{-3, 3\}$.

EXAMPLE 5 Solve $2x^2 = -6x$ by factoring.

SOLUTION

$$2x^2 = -6x \qquad \text{Original equation}$$
$$2x^2 + 6x = 0 \qquad \text{Write in standard form.}$$

$$2x(x+3) = 0 \qquad \text{Factor the equation completely.}$$
$$2x = 0 \quad \text{or} \quad x + 3 = 0 \qquad \text{Set each factor equal to zero.}$$
$$x = 0 \quad \text{or} \qquad x = -3$$

The solution set is $\{-3, 0\}$. The check is left to the reader.

The zero-factor property can be extended to more than two numbers. If the product of several numbers is zero then at least one of them must be zero. Of course, all of them could be zero. The following examples show how we can use this property to solve factorable equations.

EXAMPLE 6 Solve $a^3 - 4a = 0$ by factoring.

SOLUTION
$$a^3 - 4a = 0$$
$$a(a^2 - 4) = 0 \qquad \text{Remove the common factor first.}$$
$$a(a-2)(a+2) = 0$$

We set each of the three factors equal to zero:
$$a = 0 \quad \text{or} \quad a - 2 = 0 \quad \text{or} \quad a + 2 = 0$$
$$a = 0 \quad \text{or} \qquad a = 2 \quad \text{or} \qquad a = -2$$

CHECK
$$a^3 - 4a = 0 \qquad\qquad a^3 - 4a = 0 \qquad\qquad a^3 - 4a = 0$$
$$0^3 - 4(0) \stackrel{?}{=} 0 \quad \text{or} \quad 2^3 - 4(2) \stackrel{?}{=} 0 \quad \text{or} \quad (-2)^3 - 4(-2) \stackrel{?}{=} 0$$
$$0 - 0 \stackrel{?}{=} 0 \qquad\qquad 8 - 8 \stackrel{?}{=} 0 \qquad\qquad -8 + 8 \stackrel{?}{=} 0$$
$$0 \stackrel{\checkmark}{=} 0 \qquad\qquad 0 \stackrel{\checkmark}{=} 0 \qquad\qquad 0 \stackrel{\checkmark}{=} 0$$

The solution set is $\{-2, 0, 2\}$.

EXAMPLE 7 Solve the equation $(x - 2)(x + 1) = 3x(x + 1)$ by factoring.

SOLUTION We can approach this problem in two ways.

Method 1 Write the equation in standard form and factor.
$$(x-2)(x+1) = 3x(x+1)$$
$$x^2 - x - 2 = 3x^2 + 3x$$
$$-2x^2 - 4x - 2 = 0$$
$$-2(x^2 + 2x + 1) = 0$$
$$-2(x+1)^2 = -2(x+1)(x+1) = 0$$

Both factors containing the variable are the same, so the equation has a single solution:
$$x + 1 = 0$$
$$x = -1$$

Method 2 Notice that the factor $x + 1$ appears on both sides of the equation.

$$(x - 2)(x + 1) = 3x(x + 1)$$
$$(x - 2)(x + 1) - 3x(x + 1) = 0$$
$$(x + 1)[(x - 2) - 3x] = 0 \quad \text{Remove the common factor.}$$
$$(x + 1)(-2x - 2) = 0 \quad \text{Simplify; combine like terms.}$$
$$-2(x + 1)(x + 1) = 0$$

Again both factors containing the variable are the same. The equation has a single solution.

$$x + 1 = 0$$
$$x = -1$$

CHECK
$$(x - 2)(x + 1) = 3x(x + 1) \quad \text{Original equation}$$
$$(-1 - 2)(-1 + 1) \stackrel{?}{=} 3(-1)(-1 + 1) \quad \text{Substitute } x = -1.$$
$$(-3)(0) \stackrel{?}{=} -3(0)$$
$$0 \stackrel{\checkmark}{=} 0$$

The solution set is $\{-1\}$.

EXAMPLE 8 Solve the equation $(x + 2)(x^2 - 4x + 3) = 0$.

SOLUTION We begin by factoring $x^2 - 4x + 3$ as $(x - 1)(x - 3)$.

$$(x + 2)(x^2 - 4x + 3) = 0$$
$$(x + 2)(x - 1)(x - 3) = 0$$

Now we set each factor equal to zero.

$$x + 2 = 0 \quad \text{or} \quad x - 1 = 0 \quad \text{or} \quad x - 3 = 0$$
$$x = -2 \quad \text{or} \quad x = 1 \quad \text{or} \quad x = 3$$

All solutions check. The solution set is $\{-2, 1, 3\}$.

EXAMPLE 9 Solve: $(2x + 1)(x - 3) = 9$.

SOLUTION Here is an example of a common error made by beginning algebra students. It must be studied carefully.

$$2x + 1 = 9 \quad \text{or} \quad x - 3 = 9 \quad \text{FALSE}$$

But if $2x + 1 = 9$ and $x - 3 = 9$, then

$$(2x + 1)(x - 3) = (9)(9) = 81$$

However, the original equation said that $(2x + 1)(x - 3) = 9 \neq 81$.

*C*AUTION

The product of the factors must be equal to zero to solve an equation by factoring.

There are many ways we can use quadratic equations to solve applied problems. The next two examples are but two of them.

EXAMPLE 10 The lengths of the sides of a right triangle are shown in Figure 7.2. Find the numerical length of each side.

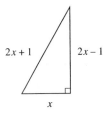

FIGURE 7.2

SOLUTION The lengths of the sides of a right triangle are related by the Pythagorean theorem:

$$a^2 + b^2 = c^2$$

where c is the length of the hypotenuse and a and b are the lengths of the legs. We substitute x, $2x - 1$, and $2x + 1$ for a, b, and c, respectively:

$$x^2 + (2x - 1)^2 = (2x + 1)^2$$
$$x^2 + 4x^2 - 4x + 1 = 4x^2 + 4x + 1$$
$$5x^2 - 4x + 1 = 4x^2 + 4x + 1$$
$$x^2 - 8x = 0$$
$$x(x - 8) = 0$$

We set each factor equal to zero:

$$x = 0 \quad \text{or} \quad x - 8 = 0$$
$$x = 0 \quad \text{or} \quad x = 8$$

Since the length of the side of a triangle cannot be zero, the sides are

$$x = 8$$
$$2x - 1 = 15$$
$$2x + 1 = 17$$

CHECK The length of the three sides must satisfy the Pythagorean theorem.

$$a^2 + b^2 = c^2$$
$$8^2 + 15^2 \stackrel{?}{=} 17^2$$
$$64 + 225 \stackrel{?}{=} 289$$
$$289 \stackrel{\checkmark}{=} 289$$

EXAMPLE 11 Find two consecutive odd numbers with a product of 63.

SOLUTION We let x = the first number
         $x + 2$ = the second number

The product of the two numbers is 63, so

$$x(x + 2) = 63$$
$$x^2 + 2x = 63$$

$$x^2 + 2x - 63 = 0$$
$$(x + 9)(x - 7) = 0$$
$$x + 9 = 0 \quad \text{or} \quad x - 7 = 0$$
$$x = -9 \quad \text{or} \quad x = 7$$

There are two solutions to this problem. If $x = -9$, then $x + 2 = -7$; if $x = 7$, then $x + 2 = 9$. Therefore the pair of odd numbers is either -9 and -7 or 7 and 9.

Do You Remember?

Can you match these?

_____ 1. The standard form for $2x^2 - 2 = 5x$
_____ 2. The _____ property can be used to solve factorable second-degree equations.
_____ 3. The solution set of $(x - 4)(x - 3) = 0$
_____ 4. $x^2 + 4x + 4 = 0$ has __ solution(s).
_____ 5. $x^3 - 6x^2 - 7x = 0$ has __ solution(s).
_____ 6. The solution set of $x^2 - 6x - 7 = 0$
_____ 7. The solution set of $x^2 - 7x = 0$
_____ 8. The solution set of $6x^2 + 13x + 6 = 0$

a) Pythagorean
b) $2x^2 - 5x - 2 = 0$
c) $2x^2 = 5x + 2$
d) One
e) Two
f) Three
g) Zero-factor
h) $\{-7, 1\}$
i) $\left\{-\dfrac{2}{3}, -\dfrac{3}{2}\right\}$
j) $\{0, 7\}$
k) $\{-1, 7\}$
l) $\{3, 4\}$

Answers: 1. b 2. g 3. l 4. d 5. f 6. k 7. j 8. i

Exercise Set 7.6

Solve each of the following factored equations.

1. $(x - 1)(x + 3) = 0$
2. $(x + 4)(x + 5) = 0$
3. $(y + 6)(y - 2) = 0$
4. $(p + 11)(p - 7) = 0$
5. $d(d - 5) = 0$
6. $m(m + 4) = 0$
7. $(2a - 3)(5a + 6) = 0$
8. $(7s - 1)(2s + 6) = 0$
9. $m(m + 7)(2m - 1) = 0$
10. $s(3s - 4)(s + 9) = 0$
11. $b(8b - 3)(5b - 9) = 0$
12. $t(2t + 13)(5t - 1) = 0$
13. $4x(x - 5)(2x + 3) = 0$
14. $7a(13a - 1)(7a + 4) = 0$
15. $(2y - 3)(y + 4)(7y - 1) = 0$
16. $(x - 11)(5x + 7)(9x - 2) = 0$

Solve each of the following equations.

17. $r(r + 7) = 8$
18. $p(p + 3) = 4$
19. $x(x - 4) = 12$
20. $y(y - 6) = 27$
21. $x^2 + 3x + 2 = 0$
22. $x^2 - 3x + 2 = 0$
23. $y^2 - 5y + 6 = 0$
24. $y^2 + 5y + 6 = 0$
25. $x^2 - x - 6 = 0$
26. $x^2 + x - 6 = 0$
27. $a^2 + 5a - 24 = 0$
28. $a^2 - 2a - 15 = 0$
29. $r^2 - 2r - 3 = 0$
30. $r^2 - 3r - 4 = 0$
31. $(x - 1)^2 - 9 = 0$
32. $(y + 2)^2 - 4 = 0$
33. $(t + 3)^2 - 1 = 0$
34. $(m - 1)^2 - 16 = 0$
35. $(x - 6)^2 = 25$
36. $(t + 9)^2 = 49$
37. $6p^2 = p + 2$
38. $2p^2 = p + 10$
39. $6t^2 - 7t = 5$
40. $25t^2 = -4 - 20t$
41. $x^2 - 9 = 0$
42. $x^2 - 25 = 0$
43. $a^3 - a = 0$
44. $m^3 - 4m = 0$
45. $(y - 1)(5y^2 + 11y - 12) = 0$
46. $(2y + 3)(2y^2 - y - 1) = 0$
47. $3p^2 - 6p = 2p^2 - 4p - 1$
48. $s^2 - 7s = 5s^2 - 2s - 6$
49. $x^3 - 2x^2 = 8x$
50. $x^3 - 6x^2 = -8x$
51. $4k^3 - 49k = 0$
52. $4m^3 - 9m = 0$
53. $13s = -2(s^2 - 66)$
54. $-20k = -3(k^2 + 4)$
55. $10 - 2y = (y - 5)^2$
56. $13 - 4y = (y + 2)^2$
57. $m^3 = 10m^2 - 25m$
58. $3p^2 = 20p - 2p^3$
59. $(y - 5)^2 = (y + 5)^2$
60. $(m + 3)(m - 3) = 8m$
61. $(x + 1)(x - 2) = 4$
62. $(a + 2)^2 + 4a = 13$
63. $(2x + 3)(x + 1) = 3x(x + 1)$
64. $2y(y + 3) = (3y + 1)(y + 1)$
65. $9x^2 - 49 = 0$
66. $36y^2 - 81 = 0$
67. $9t^3 + 30t^2 + 25t = 0$
68. $12s^3 + 26s^2 + 10s = 0$
69. $3(2a^2 + 1) = 7a^2 - a - 27$
70. $4(2b^2 + 3) = 10b^2 - 2b - 100$
71. $5x(12x^2 + 7x + 1) = 0$
72. $3y(y^2 + 7y + 12) = 0$
73. $(p^2 - 2p - 15)(p^2 + 4p - 12) = 0$
74. $(x^2 - 6x - 7)(x^2 + 7x - 8) = 0$

For each of the following applied problems **(a)** *write the equation,* **(b)** *simplify it and put it in standard form,* **(c)** *factor and solve it, and* **(d)** *check the solution(s).*

75. Find two consecutive whole numbers with a product of 72.
76. Find two consecutive odd whole numbers with a product of 143.
77. Find two consecutive multiples of 5 with a product of zero.
78. Find two consecutive multiples of 3 with a product of zero.
79. The sum of two numbers is 4, and the sum of their squares is $\frac{25}{2}$. Find the numbers.
80. One negative number is 3 less than another negative number. If their product is 40, find the negative numbers.
81. The area of a rectangle is 24 in.² If the length is 2 in. more than the width, find the dimensions. See the accompanying figure.

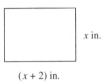

82. Use the given figure of the right triangle to find the lengths of the three sides.

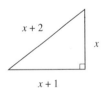

83. An object is dropped from the top of a tower 176 ft above the ground. The height h, in feet, of the object at time t, in seconds, is given by

$$h = -16t^2 + 144$$

Find the time for the object to reach the ground.

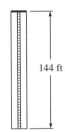

84. A diver jumps from a diving board 48 ft above the water. The height h in feet of the diver is given by

$$h = -16t^2 + 32t + 48$$

Find the time t in seconds it takes the diver to reach the water.

7.7 More Applications

OBJECTIVES

In this section we will learn to solve a variety of applied problems by factoring. To solve applied problems, we developed a five-step process, which we review in the first example.

EXAMPLE 1

Step 1 **Read the problem completely.**
The product of two consecutive integers is 4 more than 4 times their sum. Find the integers.

Step 2 **What am I asked to find?**
The product of two consecutive integers is 4 more than four times their sum. *Find the integers.*

Step 3 **Represent the unknown quantities by variables.**

We let x = the first integer
$x + 1$ = the second integer   Consecutive integers differ by 1.

Step 4 **Establish the relationship with an equation.**
The product is 4 more than four times their sum.

$$x(x + 1) = 4(x + x + 1) + 4$$

Step 5 **Solve the equation and check in the words of the original problems.**

$$x(x + 1) = 4(x + x + 1) + 4$$
$$x(x + 1) = 4(2x + 1) + 4 \quad \text{Combine like terms.}$$
$$x^2 + x = 8x + 4 + 4 \quad \text{Distributive property}$$
$$x^2 + x = 8x + 8 \quad \text{Combine like terms.}$$
$$x^2 - 7x - 8 = 0 \quad \text{Subtract } 8x \text{ and } 8 \text{ from each side.}$$
$$(x - 8)(x + 1) = 0$$
$$x - 8 = 0 \quad \text{or} \quad x + 1 = 0 \quad \text{Zero-factor property}$$
$$x = 8 \quad \text{or} \quad x = -1$$

If $x = 8$, the other number is $x + 1 = 9$. The product of these two numbers is 72 and their sum is 17. The product is 4 more than four times their sum, which is 68. This solution checks. If $x = -1$, then $x + 1 = 0$. This solution also checks.

The two pairs of numbers are 8 and 9, and -1 and 0.

EXAMPLE 2 The longest side of a right triangle is 8 in. longer than the shortest side. The third side is 7 in. longer than the shortest side. Find the lengths of the three sides.

SOLUTION   Since the triangle is a right triangle, we can use the Pythagorean theorem (see Figure 7.3). We let

$x =$ the length of the shortest side
$x + 7 =$ the length of the third side
$x + 8 =$ the length of the longest side

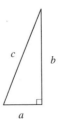

FIGURE 7.3

Then

$(x + 8)^2 = x^2 + (x + 7)^2$	Substitute for c, a, and b.
$x^2 + 16x + 64 = x^2 + x^2 + 14x + 49$	Square the binomials.
$x^2 + 16x + 64 = 2x^2 + 14x + 49$	Combine like terms.
$0 = x^2 - 2x - 15$	Write in standard form.
$0 = (x - 5)(x + 3)$	Factor.

Setting each factor equal to zero gives us

$x - 5 = 0$ or $x + 3 = 0$
$x = 5$ or $x = -3$

Since the side of a triangle cannot have a negative length, we reject -3 as a solution. Hence $x = 5$ is the length of the shortest side, $x + 7 = 12$ is the length of the third side, and $x + 8 = 13$ is the length of the longest side (the hypotenuse). Thus the three sides are 5, 12, and 13. Since $5^2 + 12^2 = 13^2$, the triangle is a right triangle. The solution checks.

EXAMPLE 3   The length of a rectangle is 3 feet more than its width. If the area is 108 square feet, find its dimensions.

SOLUTION   We let $w =$ width
$w + 3 =$ length

(See Figure 7.4.) The area of a rectangle is $A = lw$, so

$w(w + 3) = 108$	Substitute $A = 108$ and $l = w + 3$.
$w^2 + 3w = 108$	Distributive property
$w^2 + 3w - 108 = 0$	Write in standard form.
$(w - 9)(w + 12) = 0$	Factor.

$w - 9 = 0$ or $w + 12 = 0$
$w = 9$ or $w = -12$

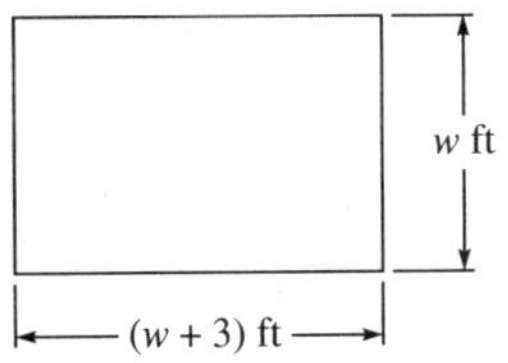

FIGURE 7.4

We reject -12; the width is 9 feet and the length is $w + 3 = 9 + 3 = 12$ feet.

EXAMPLE 4   A commodity sells for a unit price p and generates revenue R according to the formula $R = p(2p + 50)$. What unit price will generate a revenue of $300?

SOLUTION   We substitute 300 for R and solve for p:

$$300 = p(2p + 50)$$
$$300 = 2p^2 + 50p$$
$$0 = 2p^2 + 50p - 300$$
$$0 = 2(p^2 + 25p - 150)$$
$$0 = 2(p - 5)(p + 30)$$
$$p - 5 = 0 \quad \text{or} \quad p + 30 = 0 \qquad \text{The factor 2 cannot be zero.}$$
$$p = 5 \quad \text{or} \quad p = -30$$

We reject $p = -30$; the unit price is $5.

EXAMPLE 5   A vineyard is to be planted in rows. The number of vines in each row is 10 more than the number of rows. If the vineyard is to contain 504 vines, find the number of rows and the number of vines in each row.

SOLUTION   We let $x =$ the number of rows to be planted
$x + 10 =$ the number of vines in each row

The number of vines to be planted can be found by multiplying the number of rows by the number of vines in each row:

$$x(x + 10) = 504$$
$$x^2 + 10x = 504$$
$$x^2 + 10x - 504 = 0$$
$$(x - 18)(x + 28) = 0$$
$$x - 18 = 0 \quad \text{or} \quad x + 28 = 0$$
$$x = 18 \quad \text{or} \quad x = -28$$

We reject -28; there will be $x = 18$ rows with $x + 10 = 28$ vines in each row.

EXAMPLE 6   A box 3 inches high is made from a square piece of sheet metal. To construct the box, a three-inch square is cut from each corner of the sheet of metal, and the sides are folded up. If the box holds 147 cubic inches, what are its dimensions?

SOLUTION   We let $x =$ the length of a side of the square piece of sheet metal
$V = 147 =$ the volume of the box

Since $V = lwh$, we need to determine the width and length of the box. Figure 7.5(a) shows how we can determine these dimensions; Figure 7.5(b) shows the completed box. The dotted lines indicate folds. The length and width of the box are $x - 6$ and the height is 3 inches, so

$$V = lwh$$
$$147 = (x - 6)(x - 6)3 \qquad l = w = x - 6 \text{ and } h = 3$$

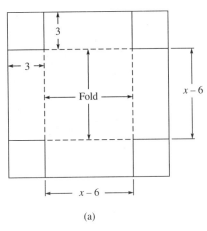
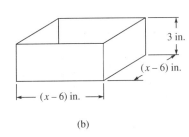

(a)   (b)

FIGURE 7.5

Dividing both sides by 3 gives us

$$(x - 6)^2 = 49$$
$$x^2 - 12x + 36 = 49$$
$$x^2 - 12x - 13 = 0$$
$$x - 13 = 0 \quad \text{or} \quad x + 1 = 0$$
$$x = 13 \quad \text{or} \quad x = -1$$

We reject -1, because the box cannot have a side that is negative in length. The sheet of metal is therefore 13 inches on each side. The box's dimensions will be 7 inches by 7 inches by 3 inches high.

EXERCISE SET 7.7

Write each applied problem as a quadratic equation, factor it, solve it, and then check it.

1. Find two consecutive odd integers with a product of 63.

2. Find two consecutive even integers with a product of 48.

3. Find two consecutive even natural numbers with a product that is 60 more than twice the larger natural number.

4. Find two consecutive odd natural numbers with a product that is 1 less than five times their sum.

5. The sum of the squares of two consecutive odd numbers is 290; find the numbers.

6. The sum of a number and its square is 42; find the number.

7. The length of a rectangle is 4 m greater than its width, and the area is 60 m². Find the width, length, and perimeter.

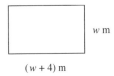

8. The length of a rectangle is 3 cm less than twice the width. The area is 104 cm². Find the length, width, and perimeter.

9. One square has a side that is 10 yd greater than the side of another square. The total area of the two squares

is 1850 yd². Find the length of a side and the area of each square.

10. A neighbor has two square gardens. The larger garden has a side that is 8 m greater than a side of the smaller garden. The total area of the two gardens is 424 m². Find the length of a side and the area of each of the gardens.

11. If the unit price of a commodity is p dollars, the total revenue R is given by $R = p(200 - 5p)$. What price p will yield a total revenue of $2000?

12. If p dollars is invested at r percent *compounded annually*, at the end of two years it will grow to an amount $A = p(1 + r)^2$. What interest rate r is needed for an investment of $100 to grow to $144 in two years?

13. A variable electric current I (in amps) is given by the formula $I = t^2 - 7t + 12$, where t is the number of seconds elapsed from the time a switch is turned on. At what time will the electric current be 2 amps?

14. In an electrical circuit, $P = EI - RI^2$, where P = power in watts (W), E = electromotive force in volts (V), I = current in amps, and R = resistance in ohms. What is the current in a circuit with $R = 2$ ohms, $E = 12$ V, and $P = 18$ W?

15. An orchard has 525 fruit trees planted in rows of uniform length. The number of trees in each row is 4 more than the number of rows. Find the number of rows and the number of trees in each row.

16. A total of 192 tomato plants are planted in rows of uniform length. The number of plants in each row is 3 times the number of rows. Find the number of rows and the number of plants in each row.

17. Polly has a garden plot that is 30 ft by 50 ft. She wants to plant a strip of grass around the garden (see figure). She has enough grass seed for 900 ft². How wide a strip can she plant?

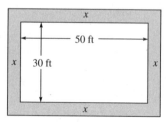

18. A rectangular lawn has an area of 108 ft², and the length is three times the width. A uniformly wide concrete walkway with an area of 180 ft² is poured around the lawn. Find the dimensions of the lawn and the width of the walkway.

19. A tank 4 cm high (see the figure) is made from a square piece of sheet metal by cutting a square out of each corner and folding up the sides. The volume of the tank is 400 cm³. Find the dimensions of the tank and those of the piece of sheet metal.

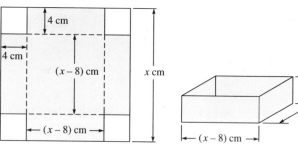

20. A box has a height of x in. Its width is 3 in. less than the height. Its length is 5 in. more than the height. If the area of the base is 84 in.², find the height of the box. See the figure.

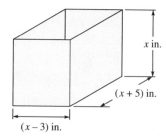

21. A right circular cylindrical tube is made from a rectangular piece of tin with an area of 32π in.² If the height of the cylinder is the same as the radius, find the height. The formula for the surface area is $S = 2\pi rh$, where r is the radius of the tube and h is the height.

22. A right circular cylindrical tube is made from a rectangular piece of tin with an area of 18π in.² If the height of the cylinder is one-half of the diameter, find the diameter. See formula in Exercise 21.

23. Find the length of each side of a right triangle if side b is 1 m longer than side a, and side c is 2 m longer than side a. (See the figure.)

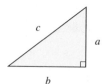

24. Find the length of each side of a right triangle if the two shorter sides are 1 in. and 2 in. shorter than the longest side.

25. Two runners leave the same point and travel at right angles to each other for 1 hr. They are then 13 mi apart. If one runner travels 7 mph faster than the other, what is the rate of the slower runner?

26. Two hikers leave the same point and travel at right angles to each other. At the end of 1 hr they are 5 mi apart. If one walks 1 mph faster than the other, what is the speed of the slower hiker?

27. The lengths of the three sides of a right triangle are three consecutive integers. Find the lengths of the three sides.

28. Use the accompanying figure to find the lengths of the three sides of the right triangle.

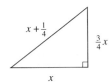

29. Use the given figure to find the lengths of the three sides of the right triangle.

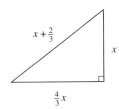

30. The product of two rational numbers is -3. One number is 8 more than four times the other number. Find the rational numbers.

31. One rational number is 11 less than another rational number. If their product is 12, find the rational numbers.

32. The length of a rectangle is 1 m less than three times the width. If the area of the rectangle is 2 m^2, find the length and width.

33. The length of a rectangle is 2 ft more than one-third its width. If the area of the rectangle is 9 ft^2, find the length and width.

34. The length of a rectangle is 2 cm more than three times its width. If the area of the rectangle is 33 cm^2, find the length and width.

35. Use the given figure to find x if the total area of the trapezoid is 160 m^2.

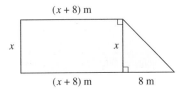

36. Use the accompanying figure to find x if the total area of the rectangle is x^2 ft^2.

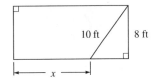

37. A 25 ft ladder leans up against a wall. How high up the wall does it reach if the base is 7 ft from the wall?

38. A cable is attached to the top of a 40 ft antenna from a stake 30 ft from the base of the antenna. How long is the cable?

39. Use the given figure of a trapezoid to find the lengths of the sides if the area of the trapezoid is 10 cm^2.

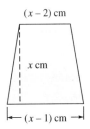

40. Use the given figure of a trapezoid to find the lengths of the three sides if the area is 27 yd^2.

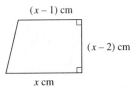

Summary

7.1 Factoring; The Greatest Common Factor

To say that one quantity is a **factor** of another means that it divides it exactly (no remainder). The **greatest common factor**, or **GCF**, of two or more integers is the largest integer that is a factor of each.

The **greatest common factor of two or more monomials** is found as follows:

Step 1 Find the prime factorization of each monomial.
Step 2 Determine the least number of times each common prime or variable factor appears in each monomial.
Step 3 Find the product of all of the factors to the powers found in Step 2.

When factoring, the first thing to look for is a greatest common factor. When the GCF is removed, the exponents on each variable in the remaining factor are decreased by an amount equal to the exponent on the term that is removed. If the additive inverse of the greatest comon factor is used, the sign before each term in the remaining factor must be changed.

7.2 Factoring Trinomials of the Form $x^2 + bx + c$

To factor a polynomial, always remove the greatest common factor first.

When the polynomial is a trinomial of the form $x^2 + bx + c$, the process of factoring can be made easier by observing the following relationship between its terms.

1. If the sign before the last term of the trinomial is positive, both terms in the factors will have the same sign as the middle term in the trinomial. We wish to find two numbers with a product of c and a sum of b.
2. If the sign before the last term of the trinomial is negative, the terms in the factors will have opposite signs. We wish to find two numbers with a product of c and a *difference* of b.

7.3 Factoring Trinomials of the Form $ax^2 + bx + c$

When the leading coefficient of a trinomial to be factored is not 1, it can take many trials to determine the proper factorization. The number of trials can be lessened by observing that neither of the binomial factors of a polynomial can have a common factor if the polynomial has none.

7.4 Special Factors; Factoring by Grouping

Several special formats for factoring are useful.

1. Perfect-square trinomials

$$a^2 + 2ab + b^2 = (a + b)^2 \quad \text{and} \quad a^2 - 2ab + b^2 = (a - b)^2$$

2. The difference of two squares

$$a^2 - b^2 = (a + b)(a - b)$$

3. The sum and difference of cubes

$$a^3 + b^3 = (a + b)(a^2 - ab + b^2)$$
$$a^3 - b^3 = (a - b)(a^2 + ab + b^2)$$

4. Factoring by grouping

   To factor by grouping, the terms of a polynomial are grouped in such a way that a common factor is evident. Factoring by grouping is often indicated when the polynomial contains more than three terms.

7.5 A Summary of Factoring

Factoring is generally easier if there is a plan of attack. Suggestions for successful factoring are as follows:

1. If the expression contains any common factors, remove them first.
2. If the expression contains only two terms, check to see if it is
   (a) the difference of two squares: $a^2 - b^2 = (a - b)(a + b)$
   (b) the difference of two cubes: $a^3 - b^3 = (a - b)(a^2 + ab + b^2)$
   (c) the sum of two cubes: $a^3 + b^3 = (a + b)(a^2 - ab + b^2)$
3. If the expression is a trinomial and the first and third terms are perfect squares, check to see if it is a perfect-square trinomial:

   $$a^2 + 2ab + b^2 = (a + b)^2 \quad \text{or} \quad a^2 - 2ab + b^2 = (a - b)^2$$

4. If the leading coefficient of a trinomial is 1, as in $x^2 + bx + c$, try to find two numbers with a product of c and a sum of b. These numbers will be the second terms of the two binomial factors. When the sign before c is positive, the signs in the binomial factors will be the same as the sign before b. If the sign before c is negative, the signs in the binomial factors will differ.
5. If the leading coefficient of a trinomial is not 1, try factoring it as a **general trinomial.** If the sign before the last term in the trinomial is positive, the signs of the second terms of its binomial factors will be the same as the sign before the middle term of the trinomial. Remember that *if the trinomial does not contain any common factors, neither of the binomial factors can contain a common factor.*
6. If the expression contains more than three terms, try **factoring by grouping.**
7. Always check the factorization by multiplying out the factors.

7.6 Solving Quadratic Equations by Factoring

A **quadratic equation** is one that can be put in the form $ax^2 + bx + c = 0$, where a, b, and c are real numbers and $a \neq 0$.

Equations of degree 2 or higher often can be solved by factoring and then using the **zero-factor property,** which states that

If a and b represent any real numbers and if a · b = 0, then either a = 0 or b = 0.

To Solve a Quadratic Equation by Factoring

Step 1 Write the equation in standard form: $ax^2 + bx + c = 0$.
Step 2 Factor the equation completely.
Step 3 Set each factor with a variable equal to zero and solve the resulting equation(s).
Step 4 Check the solutions in the *original* equation.

Always check the solution to a quadratic equation in the original equation, not in an altered form that may have been used to solve the equation.

7.7 More Applied Problems

Many applied problems involve equations of degree 2 or higher and can be solved by the methods of this chapter.

Cooperative Exercise

A common practice among lending institutions is to *discount* a loan. This is done by calculating the amount of *simple interest* ($I = prt$) the borrower will have to pay over the particular time period and then by deducting it in advance from the loan. The money deducted in advance is called the *lenders' discount*, and the money received by the borrower is called the *proceeds*. The **discount formula** is

$$p = A(1 - rt) \quad \text{or} \quad A = \frac{p}{1 - rt}$$

where

p = proceeds (amount received by borrower)
r = discount rate (rate of interest)
t = time (number of years of the loan)
A = amount (amount of loan before discounting)

I. If the Dodson family borrows $1400 at $6\frac{1}{2}$% for 2 years, how much will they actually receive (their proceeds)? How much is the loan discounted? How much will they actually pay back? What will their monthly payments be?

II. If the Dodson family actually needs the full $1400 now, how much should they borrow from the lending institution for 2 years at $6\frac{1}{2}$% discount? How much is the loan discounted? How much will they pay back? What will their monthly payments be in this case?

III. The *effective rate of interest* paid by the Dodson family is found by the formula

$$\text{effective rate of interest} = \frac{r}{1 - rt}$$

where the effective rate of interest is the rate of interest from the viewpoint of the borrower. Find the effective rate of interest if the Dodsons borrow $1400 at $6\frac{1}{2}$% for 2 years. Round the decimal form of the answer to three decimal places before rewriting it in percent form. How much does the effective rate of interest differ from the discount rate?

Review Exercises

Write each number in prime factored form. If a number is not factorable, write prime.

1. 68
2. 41
3. 2115
4. 91

Find the GCF for each of the following groups of expressions.

5. 49, 84
6. $5x$, $20xy$
7. $6x^2$, $-2xy$, $4y^2$
8. $-15x^2y^3$, $63xy^4$, $-24x^3y^2$

Factor out the GCF. If a trinomial is not factorable, write prime.

9. $18x^2 - 36x + 60$
10. $-27a^2b + 63ab^2 - 81ab$
11. $45s^5 - 15s^4 + 30s^3$
12. $-67x^2y + 53ab^2 - 17abx$

Factor completely. If an expression is not factorable, write prime.

13. $x^2 - 81$
14. $y^2 + 4$
15. $16m^2 - 36n^2$
16. $x^2 - 6x + 9$
17. $y^2 + 14y + 49$
18. $t^4 - 1$
19. $-16 + 8s - s^2$
20. $3y(9y^2 - 4) - 2(9y^2 - 4)$
21. $ax + ay - bx - by$
22. $-x(x^2 - 1) + (x^2 - 1)$
23. $m^4 - 81$
24. $x^2 - 16x + 64$
25. $t^2 + 9$
26. $a^2 + 5a + 3$
27. $x^2 + 5x - 6$
28. $y^2 - 10y + 24$

29. $3t^2 + 12t - 63$
30. $2z^2 - 12z + 8$
31. $2x^2 - 10xy + 12y^2$
32. $y^2(y - 2) - 4(y - 2)$
33. $6p^4 - 18p^3 - 60p^2$
34. $6x^3 - 22x^2y + 20xy^2$
35. $z^3 - 8$
36. $m^3 + 27$
37. $(x + 2y)x + (x + 2y)b$
38. $ab^2 - ac^2 + db^2 - dc^2$
39. $64x^2 - \dfrac{1}{25}$
40. $36y^2 - \dfrac{1}{49}z^2$
41. $0.25x^2 - 0.04$
42. $0.81m^2 - 0.16n^2$
43. $x^7 - 100x^5$
44. $27x^2 + 35x + 9$
45. $9a^2 - 33ab + 30b^2$
46. $8p^3 + 27q^3$
47. $-9 - 12x - 16x^2$
48. $(x + y)^2$
49. $(a^2 - b^2)^3$
50. $121x^2 - 144y^2$

Solve each of the following equations. Check all your answers.

51. $x(x - 2) = 0$
52. $(x + 3)(x - 4) = 0$
53. $(2a - 5)(3a - 2) = 0$
54. $x^2 + 6x = 0$
55. $x(x - 5)(2x + 7) = 0$
56. $5(4x + 1)(2x - 3) = 0$
57. $x^2 = 7x$
58. $6s^2 - 96 = 0$
59. $p^2 + 8p + 16 = 0$
60. $y^2 - 3y = 10$
61. $3a^3 - 6a^2 - 24a = 0$
62. $2x(x + 4) + 3(x + 4) = 0$
63. $x^2 - 4x - 21 = 0$
64. $-2 = a - 10a^2$
65. $(4x^2 + 5x - 6)\left(x - \dfrac{1}{2}\right) = 0$
66. $x^2 - \dfrac{1}{100} = 0$
67. $x^4 - 49x^2 = 0$
68. $21x^2 + 23x - 20 = 0$
69. $7x = 3x^2 - 4(x + 1)$
70. $\dfrac{x^2}{4} - \dfrac{25}{36} = 0$

Write each of the following applied problems as an equation, simplify it, put it in standard form, factor and solve it, and check your answer.

71. Find two consecutive whole numbers with a product of 156.
72. Find two consecutive even whole numbers with a product of 224.
73. Find two consecutive negative odd integers with a product of 143.
74. The product of a number and its additive inverse is -36. Find the number.
75. The rectangle in the accompanying figure has an area of 176 m². Find the dimensions and the perimeter.

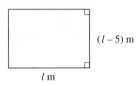

76. Use the right triangle in the figure to find the length of the hypotenuse in feet. Find the area and perimeter.

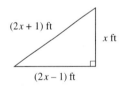

Chapter Test

Factor each of the following expressions completely. If not factorable, write prime.

1. $x^2 - 3x$
2. $5x^4 + 30x^3$
3. $-9y^4 + 12y^3 - 21y^2$
4. $-21a^3b^6 + 63a^4b^4 - 77a^5b^4 + 84a^4b^5$
5. $9x^2 - 30x + 25$
6. $p^2 - 4$
7. $49y^2 + 112yz + 64z^2$
8. $x^3 - 64$
9. $3y^3 + 24z^3$
10. $5x(x^2 - 1) - 3(x^2 - 1)$
11. $8a^2 + 12a - 10$
12. $3x^2 - 27$
13. $2m^2 + 24m + 72$
14. $7x^2 + 11x - 1$
15. $36x^3 + 84x^2y + 49xy^2$
16. $5y^2 - 70yz + 120z^2$
17. $8y^3 - 5xy^2 + 13x^2y$
18. $x^4 - 16$
19. $-3a^2 - 138a - 135$
20. $16b^2 - 400c^2$
21. $-12 - 3m + 20m^2$
22. $-15x^2 + 60$
23. $28s^2t + 130st^2 - 30s^3$
24. $-69xy - 26y^2 - 10x^2$

Solve each equation by factoring and check all your answers.

25. $(x^2 - x)(2x + 3) = 0$
26. $16x^2 - 4x - 56 = 0$

27. $a^3 - 6a^2 + 8a = 0$ 28. $x^4 - 13x^2 + 36 = 0$
29. $3a^2 + 4a = 2a^2 + 6a - 1$

Write the equation for each applied problem, simplify it, put it in standard form, factor and solve it, and check all your answers.

30. Find two consecutive negative integers with a product of 72.

31. A computer switchboard with n outputs can handle C connections according to the formula $C = n(n - 1)$. How many outputs n are there when $C = 506$?

32. The length of a rectangle is twice the width, and the area is 450 m². Find the length and width.

33. Use the given figure of a right triangle to find the lengths of the three sides. Find the perimeter and the area.

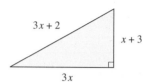

CHAPTER 8
RATIONAL EXPRESSIONS AND FRACTIONS

8.1 Simplifying Rational Expressions
8.2 Multiplying and Dividing Rational Expressions
8.3 Adding and Subtracting Rational Expressions; The Least Common Denominator
8.4 Complex Fractions
8.5 Solving Rational Equations
8.6 Ratio and Proportion
8.7 Variation

Application

On a calm day, Marlene can fly her new miniplane at a speed of 74 kilometers per hour. When the wind is blowing, it takes her as long to go 50 kilometers against the wind as it does to go 75 kilometers with the wind. What is the speed of the wind?

Exercise Set 8.5, Exercise 48

Leonhard Euler (pronounced "oiler") was a great mathematician known for his ingenious algorithms. (An algorithm is a systematic method for solving a problem.) Besides being a mathematical genius, Euler was also a family man with thirteen children. It has been said that he could make discoveries in mere days that were forever beyond the reach of other mathematicians.

LEONHARD EULER (1707–1783)

8.1 Simplifying Rational Expressions

OBJECTIVES

In this section we will learn to
1. recognize rational expressions;
2. simplify rational expressions; and
3. determine the values for which a rational expression is undefined.

In arithmetic, the quotient of two numbers is called a *fraction* or a *ratio*. In algebra, when the numerator and the denominator of a fraction are polynomials, it is called a *rational expression*. For example,

$$\frac{3}{x}, \quad \frac{2+x}{1+x^2}, \quad \text{and} \quad \frac{a^2+3a+9}{a^3+2a}$$

are all rational expressions.

DEFINITION	A **rational expression** is the quotient of two polynomials such that the denominator cannot equal zero. The numerator can have any value.
Rational Expression	

As the definition of a rational expression suggests, there may be some values of the variable for which the denominator is zero. When this occurs, the rational expression fails to exist or is *undefined*. For example, the rational expression $\frac{x+3}{x-2}$ is not defined at $x = 2$ because $2 - 2 = 0$ and division by zero is undefined. Recall that if there were some number a such that $\frac{x+3}{x-2} = a$ when $x = 2$, then $\frac{2+3}{2-2} = a$. This means that $\frac{5}{0} = a$, or $5 = 0 \cdot a$. Since $0 \cdot a = 0$, this is impossible.

In this chapter, except where noted otherwise, we will assume that all denominators are nonzero.

Working with algebraic fractions employs the same rules and operations that we use in arithmetic. We use the same techniques to add, subtract, multiply, and divide rational expressions as we use with numerical fractions.

Rational expressions, like arithmetic fractions, are simplified by *dividing out* common factors from the numerator and the denominator.

ARITHMETIC

$$\frac{4}{6} = \frac{\overset{1}{\cancel{2}} \cdot 2}{\cancel{2} \cdot 3} = \frac{2}{3}$$

$$\frac{30}{42} = \frac{\overset{1}{\cancel{2}} \cdot \overset{1}{\cancel{3}} \cdot 5}{\cancel{2} \cdot \cancel{3} \cdot 7} = \frac{5}{7}$$

ALGEBRA

$$\frac{2x+4}{3x+6} = \frac{2\overset{1}{\cancel{(x+2)}}}{3\cancel{(x+2)}} = \frac{2}{3}$$

$$\frac{3x^2+9x}{6x^3+12x^2} = \frac{\overset{1}{\cancel{3x}}(x+3)}{\underset{2}{\cancel{6x}} \cdot x(x+2)} = \frac{x+3}{2x(x+2)}$$

These illustrations are examples of the **fundamental property of rational expressions,** which we use to simplify the rational expressions.

The Fundamental Property of Rational Expressions

If A, B, and C are polynomials, $B \neq 0$ and $C \neq 0$, then

$$\frac{A \cdot C}{B \cdot C} = \frac{A}{B} \quad \text{and} \quad \frac{A}{B} = \frac{A \cdot C}{B \cdot C}$$

Here $\frac{C}{C} = 1$.

In simplest terms, the fundamental property says the following:

1. Multiplying both the numerator and the denominator of a rational expression by the same nonzero polynomial does not change its value.
2. Dividing a common factor out of the numerator and the denominator of a rational expression does not change its value.

EXAMPLE 1   Simplify: $\frac{6x}{3x^2}$.

SOLUTION $\frac{6x}{3x^2} = \frac{2 \cdot \overset{1}{\cancel{3}} \cdot \overset{1}{\cancel{x}}}{\cancel{3} \cdot \cancel{x} \cdot x}$ Factor numerator and denominator.

$= \frac{2}{x}$ Use the fundamental property to divide out the common factors 3 and x.

EXAMPLE 2   Simplify: $\frac{3x-6}{2x-4}$.

SOLUTION $\frac{3x-6}{2x-4} = \frac{3(x-2)}{2(x-2)}$ Factor numerator and denominator.

$= \frac{3}{2}$ Use the fundamental property to divide out the common factor $x-2$.

EXAMPLE 3 Simplify: $\dfrac{x^2 + 6x + 5}{x^2 + 4x + 3}$.

SOLUTION

$\dfrac{x^2 + 6x + 5}{x^2 + 4x + 3} = \dfrac{(x + 5)(x + 1)}{(x + 3)(x + 1)}$ Factor numerator and denominator.

$= \dfrac{x + 5}{x + 3}$ Use the fundamental property to divide out the common factor $x + 1$.

*C*AUTION

The fundamental property can be used to simplify fractions only when **products** are involved.

If we ignore the caution concerning the fundamental property, we can show *incorrectly* that $\dfrac{8}{2} = 7$, as follows:

$\dfrac{8}{2} = \dfrac{6 + \cancel{2}}{\cancel{2}} = 6 + 1 = 7$, which we know is wrong. A *sum* is involved.

$\dfrac{8}{2} = \dfrac{4 \cdot \cancel{2}}{\cancel{2}} = 4$, which we know is right. A *product* is involved.

EXAMPLE 4 Simplify: $\dfrac{x^2 - 36}{x^3 - 12x^2 + 36x}$.

SOLUTION

$\dfrac{x^2 - 36}{x^3 - 12x^2 + 36x} = \dfrac{(x - 6)(x + 6)}{x(x^2 - 12x + 36)}$ Factor numerator and denominator; remove common factors first.

$= \dfrac{(x - 6)(x + 6)}{x(x - 6)^2}$ $x^2 - 12x + 36 = (x - 6)^2$

$= \dfrac{x + 6}{x(x - 6)}$ Use the fundamental property to divide out the common factor $x - 6$.

EXAMPLE 5 Simplify: $\dfrac{x^3 - 64}{x^2 - 4x}$.

SOLUTION

$\dfrac{x^3 - 64}{x^2 - 4x} = \dfrac{(x - 4)(x^2 + 4x + 16)}{x(x - 4)}$ Factor numerator and denominator.

$= \dfrac{x^2 + 4x + 16}{x}$ Use the fundamental property to divide out the common factor $x - 4$.

EXAMPLE 6 Simplify: $\dfrac{4x - 4y + bx - by}{2b + 8 + bx + 4x}$.

SOLUTION

$\dfrac{4x - 4y + bx - by}{2b + 8 + bx + 4x} = \dfrac{4(x - y) + b(x - y)}{2(b + 4) + x(b + 4)}$ Factor numerator and denominator by grouping.

$= \dfrac{(x - y)(4 + b)}{(b + 4)(2 + x)}$

$= \dfrac{x - y}{2 + x}$ Divide out the common factor $4 + b = b + 4$.

EXAMPLE 7 Simplify: $\dfrac{b - a}{a - b}$.

SOLUTION

$\dfrac{b - a}{a - b} = \dfrac{-1(-b + a)}{a - b}$ Factor -1 from the numerator to create a common factor with the denominator.

$= \dfrac{-1(a - b)}{a - b}$

$= -1$

Notice in Example 7 that each term in the numerator has the opposite sign of the corresponding term in the denominator. Whenever this is the case, the quotient is -1.

EXAMPLE 8 Simplify: $\dfrac{(x^2 - 3x + 6)(x + 4)^2}{(x^2 - 16)(-x^2 + 3x - 6)}$.

SOLUTION The factors $x^2 - 3x + 6$ and $-x^2 + 3x - 6$ differ term by term by sign only. Therefore their quotient is -1:

$\dfrac{(x^2 - 3x + 6)(x + 4)^2}{(x^2 - 16)(-x^2 + 3x - 6)}$

$= \dfrac{(x^2 - 3x + 6)(x + 4)^2}{(x - 4)(x + 4)(-x^2 + 3x - 6)}$ Factor the denominator.

$= \dfrac{(x^2 - 3x + 6)(x + 4)(x + 4)}{(x - 4)(x + 4)(-1)(x^2 - 3x + 6)}$ Factor -1 from $-x^2 + 3x - 6$.

$= \dfrac{x + 4}{-1(x - 4)}$ The fundamental property of rational expressions.

$= \dfrac{-(x + 4)}{x - 4}$

The simplified form of Example 8 can be written in several different ways, depending on where we locate the negative sign:

$$\frac{-1(x+4)}{x-4} = -\frac{x+4}{x-4} = \frac{-x-4}{x-4} = \frac{x+4}{4-x}$$

> **To Change the Signs of a Fraction**
>
> Any two of the three signs of a fraction can be changed without changing its value. In other words, if a and b represent positive quantities, then
>
> $$\frac{-a}{b} = \frac{a}{-b} = -\frac{a}{b}, \quad b \neq 0$$
>
> and
>
> $$\frac{a}{b} = \frac{-a}{-b} = -\frac{-a}{b} = -\frac{a}{-b}, \quad b \neq 0$$

We must remember that when we work with rational expressions, there are some values of the variables for which the expression is undefined. We cannot use any value of the variable that will make the denominator zero. Such values must be restricted.

EXAMPLE 9 Determine the restricted values for $\frac{x+3}{x^2 - 3x + 2}$.

SOLUTION The expression is undefined whenever the denominator, $x^2 - 3x + 2$, is equal to zero. To determine these values, we factor the denominator and determine the values that make the factors zero.

$$\frac{x+3}{x^2 - 3x + 2} = \frac{x+3}{(x-1)(x-2)}$$

The factor $x - 1$ is zero when $x = 1$, and $x - 2$ is zero when $x = 2$. Therefore the restricted values are 1 and 2.

EXAMPLE 10 Are there any restrictions on x in $\frac{x+2}{x^2+9}$?

SOLUTION No: Since x^2 is either positive or zero for any value of x, the sum of x^2 and any positive number is never zero.

Do You Remember?

Can you match these?

____ 1. The restrictions on x in $\dfrac{x^2 - 16}{x^2 - 4}$

____ 2. The reduced form of $\dfrac{x^2 - 81}{x^2 - 18x + 81}$

____ 3. $\dfrac{A \cdot C}{B \cdot C} = \dfrac{A}{B}$ and $\dfrac{A}{B} = \dfrac{A \cdot C}{B \cdot C}$

____ 4. The simplified form of $\dfrac{x - y}{y - x}$

____ 5. The definition of a rational expression

a) the quotient of two polynomials
b) a fraction
c) the fundamental principal of rational expressions
d) 1
e) $\dfrac{x - 9}{x + 9}$
f) $x \neq 2, -2$
g) 2
h) $\dfrac{x + 9}{x - 9}$
i) -1

Answers: 1. f 2. h 3. c 4. i 5. a

Exercise Set 8.1

Factor and simplify (if possible). State any restrictions on the variables in Exercises 11–32.

1. $\dfrac{5}{10}$
2. $\dfrac{7}{21}$
3. $\dfrac{26}{42}$
4. $\dfrac{56}{64}$
5. $\dfrac{-72}{96}$
6. $\dfrac{-54}{63}$
7. $\dfrac{49}{96}$
8. $\dfrac{101}{350}$
9. $\dfrac{-252}{-288}$
10. $\dfrac{-216}{-264}$
11. $\dfrac{9x}{6x}$
12. $\dfrac{27p}{24p}$
13. $\dfrac{-12s^2 t}{3st^2}$
14. $\dfrac{-2ab^2}{8a^2 b}$
15. $\dfrac{-21m^2 n^3}{-70m^3 n}$
16. $\dfrac{-49x^3 y^2}{-91x^2 y^3}$
17. $-\dfrac{-2x^3 y}{-8x^3 y^4}$
18. $-\dfrac{-16m^3 n^2}{-16n^2}$
19. $\dfrac{(-8a^2)^2}{(-4a)^3}$
20. $\dfrac{(-3s^3)^3}{(-9s^4)^2}$
21. $\dfrac{(540x^5 y)^0}{|-9|}$
22. $\dfrac{6(-33a^4)^0}{|-9|}$
23. $\dfrac{4x + 4}{2x + 2}$
24. $\dfrac{9x - 9}{3x - 3}$
25. $\dfrac{(x + 1)^2 (x - 1)}{(x + 1)(x - 1)^2}$
26. $\dfrac{(2m - 3)^2 (m + 2)}{(2m - 3)(m + 2)^2}$
27. $\dfrac{-5x^2 + 10x}{4x^2 - 8x}$
28. $\dfrac{-13a^2 - 39a}{26a^2 + 78a}$
29. $\dfrac{a - b}{b + a}$
30. $\dfrac{m - n}{n + m}$
31. $\dfrac{5x^2 - 5x}{10x - 10}$
32. $\dfrac{3a^2 - 3a}{2a - 2}$

Factor and simplify (if possible).

33. $\dfrac{x^3 - y^3}{y - x}$
34. $\dfrac{p^3 - 1}{1 - p}$
35. $\dfrac{a^2 - 6a + 5}{a^2 - 25}$
36. $\dfrac{x^2 - 3x + 2}{x^2 - 4}$
37. $\dfrac{4y^2 - 4y + 1}{4y - 2}$
38. $\dfrac{p^2 - 16}{p^2 + 3p - 4}$
39. $\dfrac{-3x - 15}{x^2 + 9x + 20}$
40. $\dfrac{11y^2 - 22y}{6 - 12y}$
41. $\dfrac{-8p^2 - 32p - 32}{-6p - 12}$
42. $\dfrac{x^2 + x - 12}{x^2 - x - 6}$
43. $\dfrac{a^2 - b^2}{b^3 - a^3}$
44. $\dfrac{m^2 - n^2}{-(n^3 - m^3)}$

45. $\dfrac{3x^2 - 12x}{12x - 3x^2}$

46. $\dfrac{-(9x^2 + 36x + 36)}{25x^2 + 20x + 20}$

47. $\dfrac{m^2 + n^2}{m^2 - n^2}$

48. $\dfrac{x^2 + 4x + 4}{x^2 - 4x + 4}$

49. $\dfrac{a^4 + a^3}{a^4 - a^3}$

50. $\dfrac{6x^4 + 6y^3}{3x^4 - 3y^3}$

51. $\dfrac{2a^2 + 19a + 24}{2a^2 + 19a + 35}$

52. $\dfrac{2x^2 + 3x + 2}{6x^2 - 11x + 4}$

53. $\dfrac{x^2 - 7xy + 12y^2}{x^2 - 4xy + 3y^2}$

54. $\dfrac{x^2y - 8xy + 15y}{xy - 3y}$

55. $\dfrac{8s^2 - 16t^2}{8t^2 - 4s^2}$

56. $\dfrac{18a^2b + 24ab^2 + 8b^3}{-b(3a + 2b)^2}$

57. $\dfrac{(2x - 5)^2(x + 1)^3(3x + 2)^3}{(x + 1)^2(2x - 5)^2(3x + 2)^3}$

58. $\dfrac{-(7y - 2)^2(3y + 1)^3(y - 4)^3}{(3y + 1)^2(y - 4)^4(7y - 2)^2}$

59. $\dfrac{x^2 + 6x + 8}{x^2 + 5x + 6}$

60. $\dfrac{y^2 - y - 2}{y^2 - y - 12}$

61. $\dfrac{6 - a}{a^2 - 11a + 30}$

62. $\dfrac{-16 + 6m + m^2}{2 - m}$

63. $\dfrac{s^2 + 3s + 2}{3s^4 + 9s^3 + 6s^2}$

64. $\dfrac{bx - by - ax + ay}{7b - 7a}$

65. $\dfrac{ax + bx - ay - by}{2ax - ay - 2bx + by}$

66. $\dfrac{y^2 + 4y + 4}{y^2 + 5y + 6}$

REVIEW EXERCISES

Multiply or divide as indicated and simplify.

67. $\dfrac{4}{5} \cdot \dfrac{5}{8}$

68. $\dfrac{12}{9} \cdot \dfrac{3}{4} \cdot 7$

69. $\dfrac{64}{30} \cdot \dfrac{18}{24}$

70. $\dfrac{7}{24} \cdot \dfrac{5}{48} \cdot 16 \cdot \dfrac{3}{25}$

71. $\dfrac{6}{15} \div \dfrac{3}{5}$

72. $\dfrac{56}{63} \div \dfrac{72}{81}$

73. $16 \div \dfrac{1}{4} \div 48$

74. $56 \div 14 \cdot \dfrac{1}{3} \div 28$

Write each of the following as a ratio.

75. What fractional part of a dollar is a dime?

76. What fractional part of a dollar is a quarter?

77. There are eight buttons on a card, and six are sewn on a blouse. What fractional part of the buttons is used?

78. A bottle of soda contains 32 oz. Seventeen ounces of the pop are consumed. What fractional part of the pop is left in the bottle?

79. There are 35 students in a class. Twelve of them are men, and the rest are women. What fractional part of the class is women?

80. Out of 126 orange and tangerine trees in an orchard, 23 are tangerine trees. What fractional part of the orchard is made up of orange trees?

Evaluate each of the following expressions for the given value of x. Note: Some may be undefined.

81. $\dfrac{3x + 2}{x - 1}; \; x = 0$

82. $\dfrac{2x + 5}{x + 3}; \; x = 0$

83. $\dfrac{5x + 3}{x - 4}; \; x = -1$

84. $\dfrac{2x - 3}{3x + 1}; \; x = \dfrac{3}{2}$

85. $\dfrac{x^2 - 4}{x + 1}; \; x = -1$

86. $\dfrac{x^2 + 9}{x^2 - 9}; \; x = -3$

87. $\dfrac{x^2 + 3x + 2}{x^2 + 3x - 4}; \; x = -1$

88. $\dfrac{x^2 - 8x + 7}{x^2 + 3x + 2}; \; x = 3$

WRITE IN YOUR OWN WORDS

89. What is a rational expression?

90. Explain how a rational *number* differs from a rational *expression*.

91. Explain what it means to ''simplify'' a rational expression.

92. What does the fundamental principle of rational expressions state?

8.2 Multiplying and Dividing Rational Expressions

OBJECTIVES

In this section we will learn to
1. multiply rational expressions; and
2. divide rational expressions.

In arithmetic, the product of two or more fractions is found by multiplying the numerators and writing this product over the product of the denominators. For example,

$$\frac{2}{3} \cdot \frac{5}{7} = \frac{2 \cdot 5}{3 \cdot 7} = \frac{10}{21}$$

If any of the fractions have factors in the numerators that are present in any of the denominators, these factors should be divided out before finding the product:

$$\frac{6}{25} \cdot \frac{5}{14} = \frac{\overset{1}{\cancel{2}} \cdot 3}{5 \cdot \cancel{5}} \cdot \frac{\overset{1}{\cancel{5}}}{\cancel{2} \cdot 7}$$

$$= \frac{3}{5 \cdot 7} \qquad \text{Divide out the common factors 2 and 5.}$$

$$= \frac{3}{35}$$

We use the same process to multiply rational expressions.

To Multiply Rational Expressions

The product of two rational expressions $\frac{A}{B}$ and $\frac{C}{D}$ is

$$\frac{A}{B} \cdot \frac{C}{D} = \frac{A \cdot C}{B \cdot D}, \qquad B \neq 0, D \neq 0$$

EXAMPLE 1 Find each product and reduce it to lowest terms:

(a) $\dfrac{9}{11} \cdot \dfrac{22}{3} \cdot \dfrac{6}{7}$ (b) $-\dfrac{2}{x} \cdot \dfrac{x^2}{12}$ (c) $\dfrac{7a^3}{5b^2} \cdot \dfrac{15bc}{28a^2c^2}$

SOLUTION (a) $\dfrac{9}{11} \cdot \dfrac{22}{3} \cdot \dfrac{6}{7} = \dfrac{\overset{1}{\cancel{3}} \cdot 3}{\cancel{11}} \cdot \dfrac{2 \cdot \overset{1}{\cancel{11}}}{\cancel{3}} \cdot \dfrac{2 \cdot 3}{7}$

$\qquad = \dfrac{3 \cdot 2 \cdot 2 \cdot 3}{7} \qquad$ Divide out the common factors 3 and 11.

$\qquad = \dfrac{36}{7}$

(b) $\dfrac{-2}{x} \cdot \dfrac{x^2}{12} = \dfrac{-1 \cdot \cancel{2}}{\cancel{x}} \cdot \dfrac{\cancel{x} \cdot x}{\cancel{2} \cdot 2 \cdot 3}$

$= \dfrac{-1 \cdot x}{2 \cdot 3}$ Divide out the common factors 2 and x.

$= \dfrac{-x}{6}$ or $-\dfrac{x}{6}$

(c) $\dfrac{7a^3}{5b^2} \cdot \dfrac{15bc}{28a^2c^2} = \dfrac{\cancel{7a^3}}{\cancel{5}b^2} \cdot \dfrac{3 \cdot \cancel{5}bc}{4 \cdot \cancel{7}a^2c^2}$

$= \dfrac{3a^3bc}{4a^2b^2c^2}$ Divide out the common factors 5 and 7.

$= \dfrac{3a}{4bc}$ Use rules for exponents to simplify.

EXAMPLE 2   Find the product $\dfrac{x}{x+y} \cdot \dfrac{(x+y)^2}{y^2}$.

SOLUTION

$\dfrac{x}{x+y} \cdot \dfrac{(x+y)^2}{y^2} = \dfrac{x}{\cancel{x+y}} \cdot \dfrac{\cancel{(x+y)}(x+y)}{y^2}$ Factor the numerator.

$= \dfrac{x(x+y)}{y^2}$ Divide out the common factor $x + y$.

We can also find the product by using the rules for exponents.

$\dfrac{x}{x+y} \cdot \dfrac{(x+y)^2}{y^2} = \dfrac{x(x+y)^{2-1}}{y^2}$

$= \dfrac{x(x+y)}{y^2}$

In the remainder of the text, we will use the rules for exponents to simplify rational expressions involving multiplication and division; we will not continue to strike out like terms from the numerator and denominator.

EXAMPLE 3   Simplify: $\left(\dfrac{6x^2y^3}{5xy^2}\right)^2 \cdot \left(\dfrac{25x^3y^7}{12xy^4}\right)$.

SOLUTION   We simplify the expression first and then carry out the indicated operations.

$\left(\dfrac{6x^2y^3}{5xy^2}\right)^2 \cdot \left(\dfrac{25x^3y^7}{12xy^4}\right) = \left(\dfrac{6xy}{5}\right)^2 \left(\dfrac{25x^2y^3}{12}\right)$

$= \dfrac{36x^2y^2}{25} \cdot \dfrac{25x^2y^3}{12}$

$= 3x^4y^5$ Divide out the common numerical factors; then use the rules for exponents.

EXAMPLE 4 Find the product $\dfrac{x^2-4}{x^2+5x+6} \cdot \dfrac{x+3}{x^2-7x+10}$.

SOLUTION We begin by factoring each rational expression completely.

$$\dfrac{x^2-4}{x^2+5x+6} \cdot \dfrac{x+3}{x^2-7x+10}$$

$$= \dfrac{(x+2)(x-2)}{(x+3)(x+2)} \cdot \dfrac{x+3}{(x-2)(x-5)}$$

$$= \dfrac{1}{x-5} \qquad \text{Divide out the common factors } x+3, x+2, \text{ and } x-2.$$

EXAMPLE 5 Find the product $\dfrac{ax+ay+bx+by}{a-b} \cdot \dfrac{a^2-b^2}{x^2+2xy+y^2}$.

SOLUTION We first factor each expression completely. We use factoring by grouping where necessary.

$$\dfrac{ax+ay+bx+by}{a-b} \cdot \dfrac{a^2-b^2}{x^2+2xy+y^2}$$

$$= \dfrac{a(x+y)+b(x+y)}{a-b} \cdot \dfrac{(a-b)(a+b)}{(x+y)^2}$$

$$= \dfrac{(x+y)(a+b)}{a-b} \cdot \dfrac{(a-b)(a+b)}{(x+y)^2}$$

$$= \dfrac{(a+b)^2}{x+y}$$

CAUTION

Notice that when all of the terms in the numerator are divided out, the numerator of the result is 1, not zero. Consider, for example,

RIGHT $\dfrac{2(x+1)}{6(x+1)(x-1)}$

$= \dfrac{\overset{1}{\cancel{2}}(\cancel{x+1})}{\underset{3}{\cancel{6}}(\cancel{x+1})(x-1)}$

$= \dfrac{1}{3(x-1)}$

WRONG $\dfrac{2(x+1)}{6(x+1)(x-1)}$

$= \dfrac{\cancel{2}(\cancel{x+1})}{\underset{3}{\cancel{6}}(\cancel{x+1})(x-1)}$

$= \dfrac{0}{3(x-1)} = 0$

Recall that to divide two fractions in arithmetic, we multiply the numerator by the reciprocal of the denominator. For example,

$$\dfrac{2}{3} \div \dfrac{5}{7} = \dfrac{\frac{2}{3}}{\frac{5}{7}} = \dfrac{2}{3} \cdot \dfrac{7}{5} = \dfrac{14}{15}$$

We use the same procedure to divide rational expressions.

> **To Divide Rational Expressions**
>
> The quotient of two rational expressions $\dfrac{A}{B}$ and $\dfrac{C}{D}$ is
>
> $$\dfrac{A}{B} \div \dfrac{C}{D} = \dfrac{A}{B} \cdot \dfrac{D}{C} = \dfrac{A \cdot D}{B \cdot C}, \qquad B, C, D \neq 0$$

EXAMPLE 6
Find the quotient $\dfrac{6a^2}{2b} \div \dfrac{a^3}{4b^2}$.

SOLUTION

$\dfrac{6a^2}{2b} \div \dfrac{a^3}{4b^2} = \dfrac{6a^2}{2b} \cdot \dfrac{4b^2}{a^3}$ Multiply the numerator by the reciprocal of the denominator.

$= \dfrac{12b}{a}$ Divide out the common factors 2, a^2, and b; then multiply.

EXAMPLE 7
Find the quotient $\dfrac{3a + 6b}{a - b} \div \dfrac{2a^2 + 8ab + 8b^2}{a^2 - b^2}$.

SOLUTION

$\dfrac{3a + 6b}{a - b} \div \dfrac{2a^2 + 8ab + 8b^2}{a^2 - b^2}$

$= \dfrac{3a + 6b}{a - b} \cdot \dfrac{a^2 - b^2}{2a^2 + 8ab + 8b^2}$ Multiply the numerator by the reciprocal of the denominator.

$= \dfrac{3(a + 2b)}{a - b} \cdot \dfrac{(a - b)(a + b)}{2(a + 2b)^2}$ Factor each rational expression completely.

$= \dfrac{3(a + b)}{2(a + 2b)}$ Divide out the common factors $a + 2b$ and $a - b$; then multiply.

EXAMPLE 8
Simplify: $\dfrac{x^2 + 5x + 4}{x^2 + 4x + 3} \cdot \dfrac{x^2 + 10x + 21}{4 - x} \div \dfrac{x^2 + 11x + 28}{x - 4}$.

SOLUTION First we rewrite the division as multiplication, then we factor all terms completely:

$\dfrac{x^2 + 5x + 4}{x^2 + 4x + 3} \cdot \dfrac{x^2 + 10x + 21}{4 - x} \cdot \dfrac{x - 4}{x^2 + 11x + 28}$

$= \dfrac{(x + 4)(x + 1)}{(x + 3)(x + 1)} \cdot \dfrac{(x + 7)(x + 3)}{4 - x} \cdot \dfrac{x - 4}{(x + 7)(x + 4)}$

$= \dfrac{x - 4}{4 - x}$ Divide out the common factors $x + 1$, $x + 3$, $x + 4$, and $x + 7$.

$= \dfrac{-1(4 - x)}{4 - x}$

$= -1$ Divide out the common factor $4 - x$.

EXAMPLE 9
Find the quotient $\dfrac{4x^2 - 12x + 9}{x + 1} \div (3 - 2x)$.

SOLUTION If we think of $3 - 2x$ as $\dfrac{3 - 2x}{1}$, we can write

$\dfrac{4x^2 - 12x + 9}{x + 1} \div \dfrac{(3 - 2x)}{1}$

$$= \frac{4x^2 - 12x + 9}{x + 1} \cdot \frac{1}{3 - 2x}$$ Multiply the numerator by the reciprocal of the denominator.

$$= \frac{(2x - 3)(2x - 3)}{x + 1} \cdot \frac{1}{3 - 2x}$$

$$= \frac{-1(3 - 2x)(2x - 3)}{x + 1} \cdot \frac{1}{3 - 2x}$$

$$= \frac{-1(2x - 3)}{x + 1}$$ Divide out the common factor $3 - 2x$.

$$= \frac{-(2x - 3)}{x + 1} \quad \text{or} \quad \frac{-2x + 3}{x + 1}$$

EXAMPLE 10 Evaluate $\dfrac{x^3 - 8}{x^2 - 4x + 4} \cdot \dfrac{x^2 - 4}{2x^3 + 8x^2 + 8x} \div \dfrac{x^2 + 2x + 4}{x + 2}$ for $x = -280$.

SOLUTION We simplify the expression first.

$$\frac{x^3 - 8}{x^2 - 4x + 4} \cdot \frac{x^2 - 4}{2x^3 + 8x^2 + 8x} \cdot \frac{x + 2}{x^2 + 2x + 4}$$

$$= \frac{(x - 2)(x^2 + 2x + 4)}{(x - 2)^2} \cdot \frac{(x - 2)(x + 2)}{2x(x + 2)^2} \cdot \frac{x + 2}{x^2 + 2x + 4}$$

$$= \frac{1}{2x} = \frac{-1}{560} \quad \text{Substitute } -280 \text{ for } x.$$

Note that performing the algebraic simplification first made it easier to evaluate the original expression when $x = -280$.

Would Example 10 have been easier to evaluate unsimplified with a calculator or simplified without a calculator? The reader is encouraged to carry out the evaluation with a calculator and then answer the question.

Exercise Set 8.2

Find each product and simplify.

1. $\dfrac{8}{15} \cdot \dfrac{6}{10}$

2. $\dfrac{56}{49} \cdot \dfrac{42}{63}$

3. $\left(\dfrac{-84}{26}\right) \cdot \left(\dfrac{91}{28}\right) \cdot \left(\dfrac{-156}{39}\right)$

4. $\left(\dfrac{-121}{9}\right) \cdot \left(\dfrac{18}{-11}\right)$

5. $\dfrac{3}{x} \cdot \dfrac{x^2}{12}$

6. $\dfrac{5a}{7b} \cdot \dfrac{35b^2}{25a}$

7. $\left(\dfrac{-8y^2}{27x}\right) \cdot \left(\dfrac{-45x}{49y}\right)$

8. $\left(\dfrac{-63m^2}{81n^3}\right) \cdot \left(\dfrac{54n^4}{-42m^5}\right)$

Find each quotient and simplify.

9. $\dfrac{143}{121} \div \dfrac{117}{99}$

10. $\dfrac{75}{63} \div \dfrac{150}{105}$

11. $\left(\dfrac{-8}{15} \div \dfrac{-24}{30}\right) \div \left(\dfrac{-16}{40}\right)$

12. $\left(\dfrac{42}{-54} \div \dfrac{-63}{56}\right) \div \left(\dfrac{280}{-54}\right)$

13. $\dfrac{9a}{8b} \div \dfrac{3a}{4b}$

14. $\dfrac{m}{3n} \div 3n$

15. $\dfrac{-18s}{10t} \div \dfrac{6s^3}{5t^2}$

16. $\dfrac{-5a^3b}{35ab^3} \div \dfrac{45ab^2}{63a^2b}$

Multiply or divide as indicated and simplify.

17. $\dfrac{5a^3}{8b^2} \cdot \dfrac{16bc}{25a^2}$

18. $\dfrac{14x^2}{15y} \cdot \dfrac{20y^2z}{21xz}$

19. $\dfrac{27a^2x}{20by^2} \div \dfrac{18ax^2}{25by^3}$

20. $\dfrac{33abc}{56x^2y} \div \dfrac{11abc^2}{28x^2y^2}$

21. $\dfrac{80x^3y^4}{-33a^2b^3} \cdot \dfrac{-11ab^3}{-40x^3}$

22. $\dfrac{40uv}{-21x^3y} \cdot \dfrac{-42x^3}{-65u^2v^2}$

23. $\dfrac{22xy}{21abc^3} \cdot \dfrac{14a^2bc}{33x^3} \div \dfrac{8y}{9bc}$

24. $\dfrac{3bc}{8x^2y} \cdot \dfrac{25ac}{6by^2} \div \dfrac{5ac^2}{16xy^3}$

25. $\dfrac{-4a}{3cy} \cdot \dfrac{5xy^2}{6ac^2} \cdot \dfrac{9c^3}{10xy}$

26. $\dfrac{-10a}{21b^2y} \cdot \dfrac{9abx}{4y} \cdot \dfrac{14by^2}{15a^2x}$

27. $\left(\dfrac{9x^7}{10a^5} \div \dfrac{4x^6}{6a^4b}\right) \cdot \dfrac{3x}{5a}$

28. $\left(\dfrac{18u^4v}{15ab^2} \div \dfrac{21uv^3}{-10a^3b}\right) \cdot \dfrac{3ab^2}{-v^4}$

29. $\left(\dfrac{3a^2}{8b^3} \cdot \dfrac{-5b}{6a}\right) \div \left(\dfrac{-8b}{9a} \cdot \dfrac{4a^3}{3b^2}\right)$

30. $\left(\dfrac{6m^2}{7n^3} \cdot \dfrac{2n}{3m^2}\right) \div \left(\dfrac{4m}{9n} \cdot \dfrac{3n^3}{4m^2}\right)$

31. $\dfrac{x}{(x+2)^2} \cdot \dfrac{(x+2)}{5x} \cdot \dfrac{10(x-3)}{(x+2)}$

32. $\dfrac{y}{(2y-5)^2} \cdot \dfrac{(2y-5)}{3y} \cdot \dfrac{12(2y-3)}{(2y-5)}$

33. $\dfrac{a-b}{a^2-b^2} \cdot \dfrac{a+b}{b} \div \dfrac{a-b}{b}$

34. $\dfrac{a+b}{a-b} \cdot \dfrac{b-a}{a+b} \div \dfrac{1}{a+b}$

35. $\dfrac{-2(x-y)^2}{(x+y)^3} \cdot \dfrac{(y+x)^2}{(y-x)} \cdot \dfrac{(x+y)}{(x-y)}$

36. $\dfrac{-5(a+2b)^3}{(a-2b)} \cdot \dfrac{(2b-a)}{(2b+a)^2} \div \left(\dfrac{a-2b}{2b+a}\right)$

37. $\dfrac{18r^2s}{35x^2y} \div \dfrac{9rs}{8ax^2} \div \dfrac{16s^2}{15y^2}$

38. $\dfrac{8y}{9bc} \div \dfrac{33x^3}{14a^2bc} \div \dfrac{22xy}{21abc^3}$

39. $\left(\dfrac{2x^3}{a^3+a^2}\right) \div \left(\dfrac{6x^5}{a^2}\right)$

40. $\dfrac{2p^3r}{(y-x)^2} \div \left(\dfrac{p^2r}{x-y}\right)$

41. $\dfrac{x^5-x^3y^2}{4x^2-9y^2} \div \dfrac{3x^3-3x^2y}{8x-12y}$

42. $\dfrac{x^3y-25xy}{5x^2+10x} \div \dfrac{x^2-6x+5}{15x^3+30x^2}$

43. $\left(\dfrac{-6a^2b^3c^4}{2m^3np^2}\right)^2 \div \left(\dfrac{12a^3bc^2}{-4mn^2p^3}\right)^2$

44. $\left(\dfrac{4a^2bc^3}{3xy^2z^4}\right)^2 \div \left(\dfrac{-2ab^3c}{x^2yz^2}\right)^3$

45. $\dfrac{x^2-x-20}{x^2-12x+35} \div \dfrac{x^2+x-12}{x^2-10x+21}$

46. $\dfrac{ax+ay-bx-by}{a^2-b^2} \div \dfrac{x+y}{a+b}$

47. $\dfrac{x^2+4x+16}{x+3} \div \dfrac{ax^3-64a-bx^3+64b}{ax-3a-bx+3b}$

48. $\dfrac{9x^2-3x+1}{6x} \div \dfrac{(27x^3+1)}{x^2}$

49. $(15x+20) \cdot \dfrac{x+4}{45x^2+36}$

50. $(3a-6) \cdot \dfrac{a+4}{a^2+6a-16}$

51. $\dfrac{2ax+6a-bx-3b}{ax+3a+2bx+6b} \cdot \dfrac{a+2b}{2a-b}$

52. $\dfrac{ax+2a+bx+2b}{ax+6a-bx-6b} \cdot \dfrac{a-b}{a+b}$

53. $\dfrac{2t^4-20t^3+18t^2}{t^3+t^2-20t} \cdot \dfrac{2t^2-2t-24}{3t^3-18t^2+15t} \cdot \dfrac{6t^4-150t^2}{4t^2-36}$

Evaluate each of the following rational expressions. Hint: It may save time to simplify first.

54. $\dfrac{a^3-6a^2+9a}{a^2+5a+6} \cdot \dfrac{a^4+8a^3+16a^2}{a^5-6a^4+5a^3}$

$\div \dfrac{a^4-16a^2}{6a^3-12a^2-90a}; \; a=1$

55. $\dfrac{a^5+5a^4-14a^3}{16a^3-16a^2-96a} \div \dfrac{2a^3+16a^2+14a}{8a^3-72a^2+144a}$

$\div \dfrac{a^4-8a^3+12a^2}{3a^2+7a+2}; \; a=1$

56. $\dfrac{x^2-9}{2x+4} \div \dfrac{3-x}{4x+8}; \; x=3$

57. $\dfrac{y^2-3y-10}{3y^2-y-2} \div \dfrac{y^2+3y+2}{6y^2+y-2}; \; y=-2$

58. $\dfrac{p^2 + 8p + 15}{3p^2 + 13p - 10} \cdot \dfrac{3p^2 + p - 2}{p^2 + 2p - 3}$; $p = -1$

59. $\dfrac{a^2 - 5a - 14}{3a^2 - 2a - 1} \div \dfrac{a^2 - 9a + 14}{1 + 2a - 3a^2}$; $a = -2$

REVIEW EXERCISES

Add or subtract as indicated and simplify.

60. $\dfrac{1}{3} + \dfrac{3}{5}$

61. $\dfrac{5}{6} - \dfrac{5}{8}$

62. $3 - \dfrac{4}{5}$

63. $2\dfrac{1}{5} + 1\dfrac{3}{5}$

64. $4\dfrac{1}{7} + 1\dfrac{1}{2}$

65. $5\dfrac{2}{7} - 3\dfrac{3}{7}$

66. $5\dfrac{3}{4} - 3\dfrac{1}{2}$

67. $\dfrac{5}{6} - \dfrac{5}{9}$

68. $1\dfrac{5}{8} + \dfrac{5}{24}$

69. $\dfrac{7}{12} - \dfrac{8}{15}$

70. $6\dfrac{1}{3} + \left(\dfrac{2}{3}\right)^2$

71. $\left(7\dfrac{2}{3}\right)^2 - 58$

72. $6.5 + 12 + 2\dfrac{1}{2} + 0.5$

73. $1\dfrac{4}{5} + 0.75 - 2\dfrac{1}{2}$

74. $\left(\dfrac{2}{3} - \dfrac{1}{6}\right)^3$

75. $\left(1 - \dfrac{3}{2}\right)^3$

76. $1\dfrac{1}{3} - 2\dfrac{1}{4}$

77. $1\dfrac{1}{5} + 0.75 + 2\dfrac{1}{4}$

78. $5\dfrac{1}{7} - 2.5 - 1\dfrac{1}{4}$

79. $\dfrac{a}{b} + \dfrac{c}{d}$, $b, d \neq 0$

80. $\dfrac{a}{b} - \dfrac{c}{d}$, $b, d \neq 0$

Divide using long division.

81. $\dfrac{x^2 + 4x + 3}{x + 2}$

82. $\dfrac{x^2 - x - 6}{x - 1}$

83. $\dfrac{x^5 - 1}{x - 1}$

84. $\dfrac{y^3 + 8}{y + 2}$

85. $\dfrac{y^3 + 3y^2 + 3y + 2}{y - 1}$

86. $\dfrac{9a^4 - 6a^3 - 5a^2 - a - 4}{-3a + 4}$

WRITE IN YOUR OWN WORDS

87. Explain how to multiply two rational expressions.

88. Explain how to divide two rational expressions.

89. Was it better for you to use the hint given for Exercises 54–59? If so, why? If not, why not?

90. If you have both multiplication and division in one problem, what is the correct order of operation?

8.3 Adding and Subtracting Rational Expressions; The Least Common Denominator

OBJECTIVES In this section we will learn to
1. add and subtract rational expressions with a common denominator;
2. find the least common denominator of several rational expressions; and
3. add and subtract rational expressions having different denominators.

We can add or subtract two or more numerical fractions if they have the same denominator. For example, we find the sum of $\dfrac{2}{9}$ and $\dfrac{5}{9}$ by adding the numerators and then writing the sum over the same (common) denominator:

$$\dfrac{2}{9} + \dfrac{5}{9} = \dfrac{2 + 5}{9} = \dfrac{7}{9}$$

Similarly, we find the difference of $\dfrac{2}{9}$ and $\dfrac{5}{9}$ by subtracting the numerators and writing the difference over the same (common) denominator.

8.3 · ADDING AND SUBTRACTING RATIONAL EXPRESSIONS; THE LEAST COMMON DENOMINATOR

The same methods apply to rational expressions with the same denominators:

$$\frac{x}{x^2+2} + \frac{1}{x^2+2} = \frac{x+1}{x^2+2}$$ Add the numerators; the denominator remains the same.

> **To Add or Subtract Two Rational Expressions**
>
> If $\frac{A}{B}$ and $\frac{C}{B}$ are rational expressions, then
>
> $$\frac{A}{B} + \frac{C}{B} = \frac{A+C}{B} \quad \text{and} \quad \frac{A}{B} - \frac{C}{B} = \frac{A-C}{B}, \quad B \neq 0$$

EXAMPLE 1 Find the following sums or differences: (a) $\frac{1}{3x} + \frac{7}{3x}$ (b) $\frac{5x-2}{x+1} - \frac{4}{x+1}$

SOLUTION (a) $\frac{1}{3x} + \frac{7}{3x} = \frac{1+7}{3x} = \frac{8}{3x}$ The denominators are the same; add the numerators.

(b) $\frac{5x-2}{x+1} - \frac{4}{x+1} = \frac{5x-2-4}{x+1}$ The denominators are the same; subtract the numerators.

$$= \frac{5x-6}{x+1}$$

EXAMPLE 2 Find the sum $\frac{1}{x+2} + \frac{2x+3}{x+2}$.

SOLUTION
$$\frac{1}{x+2} + \frac{2x+3}{x+2} = \frac{1+2x+3}{x+2}$$ The denominators are the same; add the numerators

$$= \frac{2x+4}{x+2}$$

$$= \frac{2(x+2)}{x+2}$$ Factor the numerator.

$$= 2$$ Divide out the common factor $x+2$.

To find the sum of two fractions with *different* denominators, we must first write each fraction in terms of the **least common denominator**, abbreviated **LCD**. For example, we can add $\frac{1}{6}$ and $\frac{3}{5}$ by expressing each fraction in terms of the least common denominator, 30. We then use the fundamental principle to rewrite each fraction with the same (common) denominator:

$$\frac{1}{6} = \frac{1 \cdot 5}{6 \cdot 5} = \frac{5}{30} \quad \text{and} \quad \frac{3}{5} = \frac{3 \cdot 6}{5 \cdot 6} = \frac{18}{30}$$

Now we can find their sum.

$$\frac{1}{6} + \frac{3}{5} = \frac{5}{30} + \frac{18}{30}$$

$$= \frac{5 + 18}{30}$$

$$= \frac{23}{30}$$

The question naturally arises, "How do we find the least common denominator of two or more rational expressions?" The answer is, "We use the same method as in arithmetic."

> **To Find the Least Common Denominator of Two or More Rational Expressions**
>
> **Step 1** Factor each denominator completely.
> **Step 2** The LCD is the product of each distinct prime factor the greatest number of times it occurs in any of the denominators. Leave algebraic expressions in the denominator in factored form.

EXAMPLE 3 Find the sum $\dfrac{1}{6} + \dfrac{1}{4}$.

SOLUTION **Step 1** **Factor each denominator completely.**

ARITHMETIC

$$\frac{1}{6} + \frac{1}{4} = \frac{1}{2 \cdot 3} + \frac{1}{2 \cdot 2}$$

ALGEBRA

$$\frac{1}{x^2 - 4} - \frac{2}{x - 2} = \frac{1}{(x + 2)(x - 2)} - \frac{2}{x - 2}$$

Step 2 **The LCD is the product of each distinct prime factor the greatest number of times it occurs in any of the denominators.**

$6 = 2 \cdot 3$
$4 = 2 \cdot 2$

$x^2 - 4 = (x + 2)(x - 2)$
$x - 2 = x - 2$

2 is a factor at most twice.
3 is a factor at most once.
The LCD is $2^2 \cdot 3 = 12$.

$x + 2$ is a factor at most once.
$x - 2$ is a factor at most once.
The LCD is $(x + 2)(x - 2)$.

EXAMPLE 4 Find the sum $\dfrac{1}{x^2 - 9} + \dfrac{2}{x + 3}$.

SOLUTION To find the LCD, we first factor the denominators.

$x^2 - 9 = (x + 3)(x - 3)$
$x + 3 = x + 3$

The greatest number of times $x + 3$ appears as a factor is one.
The greatest number of times $x - 3$ appears as a factor is one.

Therefore the LCD is $(x + 3)(x - 3)$. Now we use the fundamental principle of rational expressions, if necessary, to write each fraction in terms of the LCD, $(x + 3)(x - 3)$.

$$\frac{1}{x^2 - 9} = \frac{1}{(x + 3)(x - 3)}$$

and

$$\frac{2}{x + 3} = \frac{2}{x + 3} \cdot \frac{x - 3}{x - 3} = \frac{2(x - 3)}{(x + 3)(x - 3)}$$

The denominators are now the same. We add the numerators and simplify.

$$\frac{1}{x^2 - 9} + \frac{2}{x + 3} = \frac{1}{(x + 3)(x - 3)} + \frac{2(x - 3)}{(x + 3)(x - 3)}$$

$$= \frac{1 + 2(x - 3)}{(x + 3)(x - 3)}$$

$$= \frac{1 + 2x - 6}{(x + 3)(x - 3)}$$

$$= \frac{2x - 5}{(x + 3)(x - 3)}$$

Once we find the LCD, we can write the fractions over it and add or subtract. The steps are outlined as follows.

To Find the Sum or Difference of Rational Expressions

Step 1 If the denominators are not the same, find the LCD by the method previously described.

Step 2 Write each rational expression as an equivalent one over the common denominator.

Step 3 Find the sum or difference of the numerators and write the result over the common denominator.

Step 4 Simplify if possible.

EXAMPLE 5 Find the sum $\dfrac{x}{x^2 + 3x + 2} + \dfrac{1}{x^2 - 2x - 3}$.

SOLUTION **Step 1 To find the LCD, factor the denominators.**

$$x^2 + 3x + 2 = (x + 1)(x + 2)$$
$$x^2 - 2x - 3 = (x + 1)(x - 3)$$

$$\text{LCD} = (x + 1)(x + 2)(x - 3) \qquad \text{The LCD is the product of the distinct prime factors.}$$

Step 2 Write each expression over the LCD.

$$\frac{x}{x^2 + 3x + 2} + \frac{1}{x^2 - 2x - 3}$$

$$= \frac{x}{(x + 1)(x + 2)} + \frac{1}{(x + 1)(x - 3)}$$

$$= \frac{x(x - 3)}{(x + 1)(x + 2)(x - 3)} + \frac{1(x + 2)}{(x + 1)(x + 2)(x - 3)}$$

Step 3 Add the numerators.

$$\frac{x(x - 3)}{(x + 1)(x + 2)(x - 3)} + \frac{1(x + 2)}{(x + 1)(x + 2)(x - 3)} = \frac{x(x - 3) + 1(x + 2)}{(x + 1)(x + 2)(x - 3)}$$

$$= \frac{x^2 - 2x + 2}{(x + 1)(x + 2)(x - 3)}$$

Since $x^2 - 2x + 2$ is not factorable, no simplification is possible.

EXAMPLE 6 Add: $\dfrac{3}{x - 2} + \dfrac{2}{2 - x}$.

SOLUTION At first it appears that the LCD is $(x - 2)(2 - x)$. However, by multiplying the numerator and denominator of the second expression by -1, the denominator becomes $-1(2 - x) = -2 + x = x - 2$, so the LCD is, in fact, $x - 2$.

$$\frac{3}{x - 2} + \frac{2}{2 - x} = \frac{3}{x - 2} + \frac{-2}{x - 2} \qquad \frac{-1}{-1} = 1; \text{ multiplying the fraction}$$

$$= \frac{3 + (-2)}{x - 2} \qquad \text{by 1 doesn't change its value.}$$

$$= \frac{1}{x - 2}$$

We could also have multiplied the numerator and denominator of the first expression by -1 to get an LCD of $2 - x$. In this case the final result would be $\dfrac{-1}{2 - x}$, which is equivalent to the first result.

The reader should rework Example 6 using $(x - 2)(2 - x)$ as the common denominator to discover how many steps we saved by making the changes we did.

EXAMPLE 7 Find the sum $\dfrac{7}{4 - x} + \dfrac{2}{x^2 - 16}$.

SOLUTION First we factor the denominators to find the LCD.

$$\frac{7}{4 - x} + \frac{2}{x^2 - 16} = \frac{7}{4 - x} + \frac{2}{(x - 4)(x + 4)}$$

We can write the first term, $\dfrac{7}{4-x}$, as $\dfrac{-7}{x-4}$ by multiplying the numerator and denominator by -1. Once we do this, we can see that the LCD is $(x-4)(x+4)$.

$$\dfrac{-7}{x-4} + \dfrac{2}{(x-4)(x+4)}$$

$$= \dfrac{-7(x+4)}{(x-4)(x+4)} + \dfrac{2}{(x-4)(x+4)} \qquad \text{Rewrite each fraction in terms of the LCD, } (x-4)(x+4).$$

$$= \dfrac{-7(x+4)+2}{(x-4)(x+4)}$$

$$= \dfrac{-7x-28+2}{(x-4)(x+4)} \qquad \text{Distributive property}$$

$$= \dfrac{-7x-26}{(x-4)(x+4)} \qquad \text{Combine like terms.}$$

EXAMPLE 8 Simplify: $\dfrac{3}{a-6} + \dfrac{a}{2a+2} - \dfrac{21}{a^2-5a-6}$.

SOLUTION

$$\dfrac{3}{a-6} + \dfrac{a}{2a+2} - \dfrac{21}{a^2-5a-6}$$

$$= \dfrac{3}{a-6} + \dfrac{a}{2(a+1)} - \dfrac{21}{(a-6)(a+1)} \qquad \text{The LCD is } 2(a+1)(a-6).$$

$$= \dfrac{3[2(a+1)]}{2(a+1)(a-6)} + \dfrac{a(a-6)}{2(a+1)(a-6)} - \dfrac{21(2)}{2(a+1)(a-6)}$$

$$= \dfrac{6a+6+a^2-6a-42}{2(a+1)(a-6)} \qquad \text{Distributive property}$$

$$= \dfrac{a^2-36}{2(a+1)(a-6)} \qquad \text{Combine like terms.}$$

$$= \dfrac{(a-6)(a+6)}{2(a+1)(a-6)} \qquad \text{Factor the numerator.}$$

$$= \dfrac{a+6}{2(a+1)} \qquad \text{Divide out the common factor } a-6.$$

CAUTION

When subtracting rational expressions with more than one term in the numerator, be sure to make all the necessary sign changes.

EXAMPLE 9 Simplify: $\dfrac{x-4}{x-5} - \dfrac{x+9}{x^2-7x+10}$.

SOLUTION

$$\dfrac{x-4}{x-5} - \dfrac{x+9}{(x-5)(x-2)} = \dfrac{(x-4)(x-2)}{(x-5)(x-2)} - \dfrac{x+9}{(x-5)(x-2)}$$

$$= \dfrac{x^2-6x+8-(x+9)}{(x-5)(x-2)}$$

—Be careful with this sign.

$$= \frac{x^2 - 6x + 8 - x - 9}{(x - 5)(x - 2)}$$

$$= \frac{x^2 - 7x - 1}{(x - 5)(x - 2)}$$

There are no common factors in the numerator and the denominator, so the result cannot be simplified.

EXAMPLE 10 Simplify: $\frac{x + 1}{x - 1} + \frac{3}{x + 1} - (x + 1)$.

SOLUTION When a term is not written over a denominator, we think of its denominator as 1.

$$\frac{x + 1}{x - 1} + \frac{3}{x + 1} - (x + 1) = \frac{x + 1}{x - 1} + \frac{3}{x + 1} - \frac{x + 1}{1}$$

$$= \frac{(x + 1)(x + 1)}{(x + 1)(x - 1)} + \frac{3(x - 1)}{(x + 1)(x - 1)} - \frac{(x + 1)(x + 1)(x - 1)}{(x + 1)(x - 1)}$$

$$= \frac{(x + 1)^2 + 3(x - 1) - (x^2 + 2x + 1)(x - 1)}{(x + 1)(x - 1)}$$

$$= \frac{(x^2 + 2x + 1) + (3x - 3) - (x^3 - x^2 + 2x^2 - 2x + x - 1)}{(x + 1)(x - 1)}$$

$$= \frac{(x^2 + 2x + 1) + (3x - 3) - (x^3 + x^2 - x - 1)}{(x + 1)(x - 1)}$$

$$= \frac{x^2 + 2x + 1 + 3x - 3 - x^3 - x^2 + x + 1}{(x + 1)(x - 1)}$$

$$= \frac{-x^3 + 6x - 1}{(x + 1)(x - 1)}$$

EXERCISE SET 8.3

Find the least common denominator (the LCD) for each of the following lists of denominators.

1. 4, 8
2. 7, 14
3. 5, 15
4. 3, 12
5. $9x$, $10x$
6. $8x$, $15x$
7. $4x^2$, $12xy$, $9y^2$
8. $5a^2$, $30ab$, $12b^2$
9. $56xy$, $63x$, $72y$
10. $54x$, $48y$, $24xy$
11. $2x + 2$, $x + 1$
12. $5x - 10$, $x - 2$
13. $y^2 - 4$, $y^2 + 3y + 2$
14. $m^2 - 9$, $m^2 + 4m + 3$
15. $a^3 - 5a^2 + 6a$, $a^2 - a - 6$
16. $y^3 - 7y^2 + 12y$, $y^2 - y - 12$

Find the sum or difference and simplify.

17. $\frac{1}{4} + \frac{3}{4} + \frac{5}{4}$
18. $\frac{2}{7} + \frac{5}{7} + \frac{1}{7}$
19. $\frac{3x}{4} + \frac{7x}{4} - \frac{5x}{4}$
20. $\frac{2y}{5} + \frac{4y}{5} - \frac{3y}{5}$
21. $\frac{11}{5n} - \frac{4}{5n} - \frac{2}{5n}$
22. $\frac{12}{9p} - \frac{2}{9p} - \frac{1}{9p}$

23. $\dfrac{s+3}{2} + \dfrac{s-5}{2}$

24. $\dfrac{t-7}{3} + \dfrac{2t+4}{3}$

25. $\dfrac{4m-5}{7st} - \dfrac{3m-8}{7st}$

26. $\dfrac{5r+2}{9xy} - \dfrac{2r-1}{9xy}$

27. $\dfrac{3ab}{8a^2b^2} - \dfrac{11ab}{8a^2b^2}$

28. $\dfrac{18x^2}{10x^2y^2} - \dfrac{8x^2}{10x^2y^2}$

29. $\dfrac{7}{x+9} + \dfrac{x+2}{x+9}$

30. $\dfrac{-3}{2x-3} + \dfrac{2x+3}{2x-3}$

31. $\dfrac{-14}{3a-5} - \dfrac{6a-24}{3a-5}$

32. $\dfrac{s^2}{s-t} - \dfrac{t^2}{s-t}$

33. $\dfrac{5}{x-2} - \dfrac{2}{2x-4}$

34. $\dfrac{3x}{x+2} - \dfrac{x}{2x+4}$

35. $\dfrac{-1}{y-3} + \dfrac{2}{3-y}$

36. $\dfrac{6}{y^2-9} + \dfrac{3}{3-y}$

37. $\dfrac{4}{3x+12} - \dfrac{x}{x+4}$

38. $\dfrac{2}{5p-25} - \dfrac{p}{p-5}$

39. $\dfrac{1}{a} + \dfrac{a+2}{a+1}$

40. $\dfrac{6}{x} + \dfrac{x+1}{x-3}$

41. $\dfrac{2}{4y^2-16} + \dfrac{3}{4+2y}$

42. $\dfrac{3}{a+3} + \dfrac{3a+1}{(a-1)(a+3)}$

43. $\dfrac{1}{x^2-1} - \dfrac{1}{x^2-2x+1}$

44. $\dfrac{1}{y^2-9} - \dfrac{1}{y^2-6y+9}$

45. $\dfrac{5}{p^2+17p+16} + \dfrac{3}{p^2+9p+8}$

46. $\dfrac{10}{r^2+r-6} + \dfrac{3r}{r^2-4r+4}$

47. $\dfrac{3x+1}{x^2+2x+1} + \dfrac{x-4}{x^2+3x+2}$

48. $\dfrac{3x-1}{x^2-3x-4} - \dfrac{x+4}{x+1}$

49. $\dfrac{4n+3}{n+2} - \dfrac{2n-5}{n^2-2n-8}$

50. $\dfrac{3a-2}{a+3} - \dfrac{2a+3}{a^2+a-6}$

51. $\dfrac{4}{y+3} - \dfrac{3y-1}{y^2+3y} - \dfrac{2}{y}$

52. $\dfrac{3}{y-2} - \dfrac{2y-1}{y^2-2y} - \dfrac{4}{y}$

53. $\dfrac{4}{a-3} - \dfrac{3}{3-a} + \dfrac{a}{2}$

54. $\dfrac{7}{t-5} + \dfrac{2}{5-t} + \dfrac{t}{3}$

55. $\dfrac{n+1}{n-1} - 1$

56. $\dfrac{n-1}{n-3} - 2$

57. $\dfrac{4}{2n-3} - \dfrac{2n+1}{(n+2)(2n-3)}$

58. $\dfrac{3}{s+3} - \dfrac{3s+1}{(s-1)(s+3)}$

59. $\dfrac{2p}{p^2-9} + \dfrac{2}{p+3} - \dfrac{1}{p-3}$

60. $\dfrac{2a}{a^2-16} - \dfrac{1}{a-4} + \dfrac{1}{a+4}$

61. $\dfrac{1}{5t-5} - \dfrac{1}{1-t}$

62. $\dfrac{-3}{3r-3} + \dfrac{-3}{1-r}$

63. $\dfrac{2}{x^2} + \dfrac{9}{x-5} - \dfrac{7}{x^2-2x-15}$

64. $\dfrac{2}{a-1} - \dfrac{a+3}{a^2-1} + \dfrac{3}{a^2+a}$

65. $\dfrac{2a-7}{a-5} - \dfrac{3}{a} - 4a - 2$

66. $\dfrac{3n-5}{n+3} - \dfrac{4}{n} - 2n - 3$

67. $\dfrac{n+1}{n-2} - \dfrac{n+2}{n+3} - \dfrac{3n-4}{n^2+n-6}$

68. $\dfrac{2x}{x-1} - \dfrac{x+1}{x-2} - \dfrac{4x-5}{x^2-3x+2}$

REVIEW EXERCISES

Simplify each of the following quotients.

69. $\dfrac{3}{4} \div \dfrac{5}{8}$

70. $1\dfrac{1}{4} \div 2\dfrac{3}{7}$

71. $\dfrac{5}{8} \div \dfrac{3}{4} \div 1\dfrac{1}{2}$

72. $4\dfrac{6}{7} \div 1\dfrac{4}{9} \div 2\dfrac{1}{2}$

73. $\left(\dfrac{1}{2} + \dfrac{1}{3}\right) \div \left(\dfrac{1}{2} - \dfrac{1}{3}\right)$

74. $\left(\dfrac{3}{4} - \dfrac{2}{5}\right) \div \left(\dfrac{2}{3} + \dfrac{1}{4}\right)$

75. $\dfrac{\dfrac{1}{4} - \dfrac{1}{3}}{\dfrac{1}{2} - \dfrac{2}{3}}$

76. $\dfrac{1\dfrac{1}{8} + 2\dfrac{3}{5}}{1\dfrac{1}{5} + 4\dfrac{1}{2}}$

77. $\dfrac{x-3}{x+2} \div \dfrac{x^2-2x-3}{x^2-4}$

78. $\dfrac{x^2-25}{x^2-36} \div \dfrac{x^2+10x+25}{x^2+12x+36}$

79. $\dfrac{x^3-8}{x^3+27} \div \dfrac{x^2+2x+4}{x^2-3x+9} \div \dfrac{x-2}{x+3}$

WRITE IN YOUR OWN WORDS

80. Explain how to find a least common denominator (an LCD).

81. Explain how to add two rational expressions that have the same denominator.

82. Explain how to add two rational expressions that have different denominators.

83. Explain how you would find the LCD of the denominators $56x$, $49xy$, and $63x^2y$.

8.4 Complex Fractions

OBJECTIVES In this section we will learn to
1. simplify complex fractions by treating them as division problems; and
2. simplify complex fractions by multiplying the numerator and the denominator by the same quantity.

A **complex fraction,** sometimes called a **compound fraction,** is a fraction in which the numerator or the denominator (or both) contains fractions. For example,

$$\frac{\frac{1}{2}}{\frac{1}{3}}, \quad \frac{1+\frac{1}{3}}{3-\frac{2}{5}}, \quad \text{and} \quad \frac{\frac{1}{a}+\frac{1}{b}}{\frac{1}{a}-\frac{1}{b}}$$

are complex fractions. We can simplify complex fractions by two different methods. We will use the three preceding examples to illustrate both methods.

To Simplify a Complex Fraction: Method 1

1. If needed, simplify the numerator and the denominator into single fractions.
2. Divide by multiplying the numerator by the reciprocal of the denominator.
3. Simplify.

EXAMPLE 1 Simplify: $\dfrac{\frac{1}{2}}{\frac{1}{3}}$.

SOLUTION To divide, we multiply the numerator by the reciprocal of the denominator.

$$\frac{\frac{1}{2}}{\frac{1}{3}} = \frac{1}{2} \cdot \frac{3}{1} = \frac{3}{2}$$

EXAMPLE 2 Simplify: $\dfrac{1+\frac{1}{3}}{3-\frac{2}{5}}$.

SOLUTION   We want to write the numerator and the denominator as single fractions:

$$\frac{1+\frac{1}{3}}{3-\frac{2}{5}} = \frac{\frac{3}{3}+\frac{1}{3}}{\frac{15}{5}-\frac{2}{5}} \qquad \text{Write numerator and denominator in terms of their LCDs.}$$

$$= \frac{\frac{4}{3}}{\frac{13}{5}} \qquad \text{Combine terms.}$$

$$= \frac{4}{3} \cdot \frac{5}{13} \qquad \text{Multiply the numerator by the reciprocal of the denominator.}$$

$$= \frac{20}{39}$$

EXAMPLE 3   Simplify: $\dfrac{\frac{1}{a}+\frac{1}{b}}{\frac{1}{a}-\frac{1}{b}}$.

SOLUTION   We want to write the numerator and the denominator as single fractions:

$$\frac{\frac{1}{a}+\frac{1}{b}}{\frac{1}{a}-\frac{1}{b}} = \frac{\frac{1}{a}\cdot\frac{b}{b}+\frac{1}{b}\cdot\frac{a}{a}}{\frac{1}{a}\cdot\frac{b}{b}-\frac{1}{b}\cdot\frac{a}{a}} \qquad \text{Write numerator and denominator in terms of } ab, \text{ their LCDs.}$$

$$= \frac{\frac{b+a}{ab}}{\frac{b-a}{ab}}$$

$$= \frac{b+a}{ab} \cdot \frac{ab}{b-a} \qquad \text{Multiply the numerator by the reciprocal of the denominator.}$$

$$= \frac{b+a}{b-a} \qquad \text{Divide out the common factor } ab \text{ and multiply.}$$

To Simplify a Complex Fraction: Method 2

1. Find the LCD of all of the fractions in the complex fraction.
2. Multiply the numerator and the denominator by the LCD found in Step 1.
3. Simplify.

EXAMPLE 4 Simplify: $\dfrac{\frac{1}{2}}{\frac{1}{3}}$.

SOLUTION The denominators of the fractions making up the complex fractions are 2 and 3. The LCD of these two fractions is 6. We multiply both numerator and denominator by 6:

$$\dfrac{\frac{1}{2} \cdot 6}{\frac{1}{3} \cdot 6} = \dfrac{\frac{6}{2}}{\frac{6}{3}} = \dfrac{3}{2} \qquad \text{Simplify numerator and denominator.}$$

EXAMPLE 5 Simplify: $\dfrac{1 + \frac{1}{3}}{3 - \frac{2}{5}}$.

SOLUTION The denominators are 3 and 5, which have an LCD $3 \cdot 5 = 15$. We multiply both the numerator and the denominator by 15:

$$\dfrac{1 + \frac{1}{3}}{3 - \frac{2}{5}} = \dfrac{\left(1 + \frac{1}{3}\right) \cdot 15}{\left(3 - \frac{2}{5}\right) \cdot 15} = \dfrac{15 \cdot 1 + 15 \cdot \frac{1}{3}}{15 \cdot 3 - 15 \cdot \frac{2}{5}} \qquad \text{Distributive property}$$

$$= \dfrac{15 + 5}{45 - 6}$$

$$= \dfrac{20}{39}$$

EXAMPLE 6 Simplify: $\dfrac{\frac{1}{a} + \frac{1}{b}}{\frac{1}{a} - \frac{1}{b}}$.

SOLUTION The denominators are a and b, which have an LCD of ab.

$$\dfrac{\frac{1}{a} + \frac{1}{b}}{\frac{1}{a} - \frac{1}{b}} = \dfrac{\left(\frac{1}{a} + \frac{1}{b}\right) \cdot ab}{\left(\frac{1}{a} - \frac{1}{b}\right) \cdot ab} \qquad \text{Multiply the numerator and the denominator by } ab, \text{ the LCD.}$$

$$= \dfrac{\dfrac{1}{a} \cdot ab + \dfrac{1}{b} \cdot ab}{\dfrac{1}{a} \cdot ab - \dfrac{1}{b} \cdot ab} \quad \text{Distributive property}$$

$$= \dfrac{b + a}{b - a} \quad \text{Simplify numerator and denominator.}$$

The choice of which method to use is left to the reader. When there are single fractions in the numerator and the denominator, it is generally easier to use Method 1. In more complex situations, such as Example 6, Method 2 is usually the method of choice. Since both are equally valid, it is up to you to decide which one you prefer. We will use Method 2 to complete the remaining examples.

EXAMPLE 7 Simplify: $\dfrac{4 - \dfrac{1}{x^2}}{2 - \dfrac{1}{x}}$.

SOLUTION The LCD of the denominators x and x^2 is x^2.

$$\dfrac{4 - \dfrac{1}{x^2}}{2 - \dfrac{1}{x}} = \dfrac{\left(4 - \dfrac{1}{x^2}\right) \cdot x^2}{\left(2 - \dfrac{1}{x}\right) \cdot x^2} \quad \begin{array}{l}\text{Multiply numerator and} \\ \text{denominator by } x^2, \text{ the LCD.}\end{array}$$

$$= \dfrac{4 \cdot x^2 - \dfrac{1}{x^2} \cdot x^2}{2 \cdot x^2 - \dfrac{1}{x} \cdot x^2} \quad \text{Distributive property}$$

$$= \dfrac{4x^2 - 1}{2x^2 - x} \quad \text{Simplify numerator and denominator.}$$

$$= \dfrac{(2x + 1)(2x - 1)}{x(2x - 1)} \quad \text{Factor numerator and denominator.}$$

$$= \dfrac{2x + 1}{x} \quad \text{Divide out the common factor } 2x - 1.$$

EXAMPLE 8 Simplify: $\dfrac{\dfrac{1}{x} + \dfrac{1}{y}}{\dfrac{1}{x + y}}$.

SOLUTION The LCD is $xy(x + y)$.

$$\dfrac{\dfrac{1}{x} + \dfrac{1}{y}}{\dfrac{1}{x + y}} = \dfrac{\left(\dfrac{1}{x} + \dfrac{1}{y}\right) \cdot xy(x + y)}{\left(\dfrac{1}{x + y}\right) \cdot xy(x + y)}$$

$$= \frac{\left(\frac{1}{x}\right) \cdot xy(x+y) + \left(\frac{1}{y}\right) \cdot xy(x+y)}{\left(\frac{1}{x+y}\right) \cdot xy(x+y)}$$ Distributive property

$$= \frac{y(x+y) + x(x+y)}{xy}$$ Simplify numerator and denominator.

$$= \frac{xy + y^2 + x^2 + xy}{xy}$$ Distributive property

$$= \frac{x^2 + 2xy + y^2}{xy}$$ Combine like terms.

$$= \frac{(x+y)^2}{xy}$$

EXAMPLE 9 Simplify: $\dfrac{3 - \dfrac{1}{x+2}}{\dfrac{2}{x-2} + 3}$.

SOLUTION The LCD is $(x+2)(x-2)$, so we multiply the numerator and the denominator by it.

$$\frac{3 - \dfrac{1}{x+2}}{\dfrac{2}{x-2} + 3} = \frac{\left(3 - \dfrac{1}{x+2}\right)(x+2)(x-2)}{\left(\dfrac{2}{x-2} + 3\right)(x+2)(x-2)}$$

$$= \frac{3(x+2)(x-2) - \dfrac{1}{x+2}(x+2)(x-2)}{\dfrac{2}{x-2}(x+2)(x-2) + 3(x+2)(x-2)}$$

$$= \frac{3(x^2-4) - (x-2)}{2(x+2) + 3(x^2-4)}$$ Simplify numerator and denominator.

$$= \frac{3x^2 - 12 - x + 2}{2x + 4 + 3x^2 - 12}$$ Distributive property

$$= \frac{3x^2 - x - 10}{3x^2 + 2x - 8}$$ Combine like terms.

$$= \frac{(3x+5)(x-2)}{(3x-4)(x+2)}$$

There are no common factors in the numerator and denominator, so no further simplification is possible. Either

$$\frac{3x^2 - x - 10}{3x^2 + 2x - 8} \quad \text{or} \quad \frac{(3x+5)(x-2)}{(3x-4)(x+2)}$$

is acceptable as the answer.

Exercise Set 8.4

Simplify each complex fraction.

1. $\dfrac{\frac{5}{8}}{\frac{15}{16}}$

2. $\dfrac{\frac{7}{9}}{\frac{14}{3}}$

3. $\dfrac{\frac{1}{2}}{\frac{3}{4}}$

4. $\dfrac{\frac{7}{8}}{\frac{1}{4}}$

5. $\dfrac{\frac{-2}{3}}{\frac{1}{6}}$

6. $\dfrac{\frac{-3}{5}}{\frac{7}{10}}$

7. $\dfrac{\frac{7}{12}}{\frac{-8}{15}}$

8. $\dfrac{\frac{3}{4}}{\frac{-11}{16}}$

9. $\dfrac{\frac{1}{x}}{\frac{2}{y}}$

10. $\dfrac{\frac{3}{a}}{\frac{2}{b}}$

11. $\dfrac{\frac{5}{x}}{\frac{3}{x^2}}$

12. $\dfrac{\frac{-7}{y}}{\frac{2}{y^2}}$

13. $\dfrac{\frac{-35}{x^2}}{\frac{-20}{x^4}}$

14. $\dfrac{\frac{7}{t^2}}{\frac{-1}{t^3}}$

15. $\dfrac{\frac{-7}{a^3}}{\frac{14}{a^2}}$

16. $\dfrac{\frac{7}{y^4}}{\frac{-27}{y^4}}$

17. $\dfrac{\frac{13}{s^2}}{\frac{2}{s}}$

18. $\dfrac{\frac{46}{y^4}}{\frac{32}{y^5}}$

19. $\dfrac{\frac{56}{xy}}{\frac{49}{x^2}}$

20. $\dfrac{\frac{81}{xy^2}}{\frac{27}{x^2 y}}$

21. $\dfrac{\frac{121}{a^2 b^2}}{\frac{44}{a^2 b}}$

22. $\dfrac{\frac{12a}{x^2 y}}{\frac{18a}{xy}}$

23. $\dfrac{\frac{2}{5}+2}{3+\frac{9}{5}}$

24. $\dfrac{1+\frac{6}{4}}{\frac{3}{4}-7}$

25. $\dfrac{\frac{1}{3}+1}{\frac{5}{27}-5}$

26. $\dfrac{1-\frac{3}{8}}{1+\frac{9}{16}}$

27. $\dfrac{\frac{5}{7}-\frac{10}{63}}{\frac{14}{18}+8}$

28. $\dfrac{4-\frac{1}{4}}{\frac{7}{8}-\frac{5}{3}}$

29. $\dfrac{\frac{1}{x}-5}{\frac{1}{x}+3}$

30. $\dfrac{a+\frac{1}{a}}{\frac{3}{a}-a}$

31. $\dfrac{\frac{m+n}{m}}{\frac{1}{m}+\frac{1}{n}}$

32. $\dfrac{\frac{1}{b}+\frac{1}{c}}{\frac{1}{b+c}}$

33. $\dfrac{x-1}{x-\frac{1}{x}}$

34. $\dfrac{y-5}{y-\frac{25}{y}}$

35. $\dfrac{a+\frac{3}{a}}{\frac{a^2+2}{7}}$

36. $\dfrac{\frac{1}{x}-x}{x+\frac{1}{x}}$

37. $\dfrac{9-\frac{1}{p^2}}{3-\frac{1}{p}}$

38. $\dfrac{36-\frac{1}{t^2}}{6+\frac{1}{t}}$

39. $\dfrac{\frac{1}{x}-\frac{1}{y}}{\frac{x-y}{3}}$

40. $\dfrac{\frac{1}{r}+\frac{1}{s}}{\frac{r+s}{7}}$

41. $\dfrac{\frac{2}{m^2}+\frac{5}{mn}}{\frac{7m}{n^2}-\frac{2n}{m}}$

42. $\dfrac{\frac{a}{b}-\frac{3}{a}}{3+\frac{2}{b^2}}$

43. $\dfrac{\frac{3}{s}+\frac{4}{s^2}}{\frac{4}{s}-1}$

44. $\dfrac{\frac{x}{x-y}}{\frac{x^2}{x^2-y^2}}$

45. $\dfrac{a-3+\frac{2}{a}}{a-4+\frac{3}{a}}$

46. $\dfrac{2-\frac{5}{r}+\frac{2}{r^2}}{6+\frac{7}{r}-\frac{5}{r^2}}$

47. $\dfrac{2+\frac{15}{t-7}}{4+\frac{17}{t-5}}$

48. $\dfrac{2+\frac{5}{2p-1}}{3+\frac{-2}{2p-1}}$

49. $\dfrac{1+\frac{x}{y-x}}{\frac{x}{x+y}-1}$

50. $\dfrac{\frac{x}{x-y}}{\frac{x^2}{x^2+y^2}}$

51. $\dfrac{a+1-\frac{12}{a-2}}{\frac{-2}{a-2}+a-1}$

52. $\dfrac{y+1-\frac{1}{y-1}}{y-1-\frac{1}{y-1}}$

53. $\dfrac{\frac{x}{x-y}-\frac{y}{x+y}}{\frac{y}{x-y}+\frac{x}{x+y}}$

54. $\dfrac{\frac{5}{y^2-y}+\frac{10}{y}}{\frac{-15}{y-1}+\frac{5}{y^2-y}}$

For Extra Thought

55. When two resistors with resistance R and r are connected in parallel, the total resistance of the circuit is given by $\dfrac{1}{\dfrac{1}{R}+\dfrac{1}{r}}$. Simplify as much as possible.

56. The approximate yield to maturity of a bond is given by

$$\dfrac{I+\left(\dfrac{P-C}{n}\right)}{\dfrac{P+C}{2}}$$

Simplify as much as possible.

Review Exercises

Solve each of the following equations. Check your answers.

57. $3x + 2 = 7$
58. $6(y - 3) - 2(3y - 4) = 9$
59. $-5(3 - 2x) + 4(2x - 5) = 7 - x$
60. $(x + 4)(x + 1) = 0$
61. $y(y - 2)(y + 3) = 0$
62. $x^2 - 3x - 4 = 0$
63. $p^2 - 49 = 0$
64. $a^2 = 8a - 16$
65. $x^3 - x^2 = 12x$
66. $x^4 + 7x^3 = -12x^2$
67. $x(x^2 - 4) - 3(x^2 - 4) = 0$
68. $a^2(a^2 - 9) - 4(a^2 - 9) = 0$

8.5 Solving Rational Equations

Historical Comment

In this section we will solve problems involving time, rate, and distance, as well as problems that involve filling and draining tanks. Problems such as these date back to the history of the Roman Legions. The marching stride of the Roman soldiers was so standardized that the time it took them to go a given distance could be easily figured. Therefore problems involving their traveling time were common. Another early favorite was the time it took a man and his wife to drink a given quantity of beer when they drank alone or when they drank together. The *Greek Anthology* (c. 500) contains the following problem concerning the filling of a tank.

> I am a bronze lion; my spouts are my two eyes, my mouth, and the flat of my right foot. My right eye fills a jar in two days, my left in three, and my foot in four. My mouth is capable of filling the jar in six hours. Tell me how long all four working together will take to fill it.

The solution to this problem is given at the end of this section.

Objectives

In this section we will learn to
1. solve equations involving rational expressions; and
2. solve selected applied problems involving rational expressions.

The equation

$$\dfrac{1}{2} + \dfrac{x}{3} = \dfrac{5}{6}$$

involves rational expressions, and it is not difficult to solve if we approach it logically. Since it contains fractions, the first thing we will do is eliminate them by using the multiplicative property of equality and multiplying each side by the least common denominator, 6.

$$\frac{1}{2} + \frac{x}{3} = \frac{5}{6}$$

$$6\left(\frac{1}{2} + \frac{x}{3}\right) = 6\left(\frac{5}{6}\right) \quad \text{Multiply each side by 6, the LCD.}$$

$$6 \cdot \frac{1}{2} + 6 \cdot \frac{x}{3} = 6 \cdot \frac{5}{6} \quad \text{Distributive property}$$

We simplify each part and multiply, as shown in the next step.

$$\overset{3}{\cancel{6}} \cdot \frac{1}{\cancel{2}} + \overset{2}{\cancel{6}} \cdot \frac{x}{\cancel{3}} = \overset{1}{\cancel{6}} \cdot \frac{5}{\cancel{6}}$$

$$3 + 2x = 5 \quad \text{Simplify and multiply.}$$
$$2x = 2 \quad \text{Subtract 3 from each side.}$$
$$x = 1 \quad \text{Divide each side by 2.}$$

We check by substituting 1 for x in the original equation.

$$\frac{1}{2} + \frac{x}{3} = \frac{5}{6} \quad \text{Original equation}$$

$$\frac{1}{2} + \frac{1}{3} \overset{?}{=} \frac{5}{6} \quad \text{Substitute 1 for } x.$$

We simplify the left side of the equation by adding $\frac{1}{2}$ and $\frac{1}{3}$. The LCD is 6.

$$\frac{3}{6} + \frac{2}{6} \overset{?}{=} \frac{5}{6}$$

$$\frac{5}{6} \overset{\checkmark}{=} \frac{5}{6}$$

The solution checks. The solution set is {1}.

In Section 3.3, we outlined a plan for solving linear equations in one variable. We can use a similar plan to solve equations involving rational expressions:

To Solve Rational Equations

Step 1 Note any restrictions on the variable, then clear the equation of fractions by multiplying each side by the least common denominator (the LCD).

Step 2 Remove grouping symbols by using the distributive property.

Step 3 Combine like terms on each side of the equation.

Step 4 If the equation is linear (first degree), isolate the variable to find the solution. If the equation is not linear, set it equal to zero and solve it by factoring if possible.

Step 5 Check the solution(s) in the original equation.

Not all steps are needed to solve every equation, so we show only those that we need in the following examples.

EXAMPLE 1   Solve: $\dfrac{1}{x} + \dfrac{2}{3} = \dfrac{5}{x}$.

SOLUTION   **Step 1**   **Note any restrictions on the variable and clear of fractions.**
Note that $x \neq 0$, *since division by zero is undefined.* To clear of fractions, we multiply each side by $3x$, the LCD.

$$3x\left(\dfrac{1}{x} + \dfrac{2}{3}\right) = 3x\left(\dfrac{5}{x}\right)$$

Step 2   **Remove grouping symbols by using the distributive property.**

$$3x \cdot \dfrac{1}{x} + 3x \cdot \dfrac{2}{3} = 3x \cdot \dfrac{5}{x} \qquad \text{Distributive property}$$
$$3 + 2x = 15 \qquad \text{Simplify each term and multiply.}$$

Step 3   This step is not needed.
Step 4   **Isolate the variable and solve.**

$$2x = 12 \qquad \text{Subtract 3 from each side.}$$
$$x = 6$$

Step 5   **Check in the original equation.**

$$\dfrac{1}{x} + \dfrac{2}{3} = \dfrac{5}{x} \qquad \text{Original equation}$$
$$\dfrac{1}{6} + \dfrac{2}{3} \stackrel{?}{=} \dfrac{5}{6} \qquad \text{Substitute 6 for } x.$$
$$\dfrac{1}{6} + \dfrac{4}{6} \stackrel{?}{=} \dfrac{5}{6} \qquad \text{The LCD is 6.}$$
$$\dfrac{5}{6} \stackrel{\checkmark}{=} \dfrac{5}{6}$$

The solution checks. The solution set is {6}.

EXAMPLE 2   Solve: $\dfrac{6}{x-2} = \dfrac{3}{x+2}$.

SOLUTION   **Step 1**   **Note any restrictions on the variable and clear of fractions.**
The denominators will be zero if $x = 2$ or -2. Thus $x \neq -2$ or 2.

$$(x-2)(x+2)\left(\dfrac{6}{x-2}\right) = (x-2)(x+2)\left(\dfrac{3}{x+2}\right) \qquad \text{The LCD is } (x-2)(x+2).$$
$$(x+2)(6) = (x-2)(3) \qquad \text{Divide out common factors.}$$

Step 2   **Remove grouping symbols.**

$$6x + 12 = 3x - 6$$

Step 4 **Isolate the variable and solve.**

$$6x - 3x = -6 - 12 \quad \text{Subtract } 3x \text{ and } 12 \text{ from each side.}$$
$$3x = -18$$
$$x = -6$$

Step 5 **Check in the original equation.**

$$\frac{6}{x-2} = \frac{3}{x+2} \quad \text{Original equation}$$

$$\frac{6}{-6-2} \stackrel{?}{=} \frac{3}{-6+2} \quad \text{Substitute } -6 \text{ for } x.$$

$$\frac{6}{-8} \stackrel{?}{=} \frac{3}{-4}$$

$$\frac{-3}{4} \stackrel{\checkmark}{=} \frac{-3}{4}$$

The solution set is $\{-6\}$.

Notice in Example 2, as well as in other examples, that not all steps are necessary to find a solution to an equation. If a step is not required, skip it and go on to the next one.

EXAMPLE 3 Solve: $\dfrac{8}{3-x} + 2 = 1 - \dfrac{7}{x-3}$.

SOLUTION Since both denominators $x - 3$ and $3 - x$, are zero when $x = 3$, we must restrict x by saying $x \neq 3$. Note that $x - 3$ and $3 - x$ differ in sign only. Therefore we multiply the numerator and the denominator of $\dfrac{8}{3-x}$ by -1.

$$\frac{8}{3-x} = \frac{8(-1)}{(3-x)(-1)} = \frac{-8}{x-3}$$

Our equation is now

$$\frac{-8}{x-3} + 2 = 1 - \frac{7}{x-3}$$

Now we multiply each side by $x - 3$, the LCD.

$$(x-3)\left(\frac{-8}{x-3} + 2\right) = (x-3)\left(1 - \frac{7}{x-3}\right)$$

$$(x-3)\left(\frac{-8}{x-3}\right) + 2(x-3) = (x-3)1 - (x-3)\left(\frac{7}{x-3}\right)$$

$$-8 + 2x - 6 = x - 3 - 7 \quad \text{Simplify and multiply.}$$
$$-14 + 2x = x - 10 \quad \text{Combine like terms.}$$
$$x = 4$$

CHECK $\dfrac{8}{3-x} + 2 = 1 - \dfrac{7}{x-3}$ Original equation

$\dfrac{8}{3-4} + 2 \stackrel{?}{=} 1 - \dfrac{7}{4-3}$ Substitute 4 for x.

$\dfrac{8}{-1} + 2 \stackrel{?}{=} 1 - \dfrac{7}{1}$

$-8 + 2 \stackrel{?}{=} 1 - 7$

$-6 \stackrel{\checkmark}{=} -6$

The solution set is {4}.

EXAMPLE 4 Solve: $\dfrac{x-3}{x-2} + 1 = \dfrac{-1}{x-2} + 1$.

SOLUTION Since the denominator, $x - 2$, is zero when $x = 2$, we must restrict x so that $x \neq 2$. We multiply each side of the equation by $x - 2$, the LCD.

$(x-2)\left(\dfrac{x-3}{x-2} + 1\right) = (x-2)\left(\dfrac{-1}{x-2} + 1\right)$ Distributive property

$(x-2)\left(\dfrac{x-3}{x-2}\right) + (x-2)(1) = (x-2)\left(\dfrac{-1}{x-2}\right) + 1(x-2)$

$x - 3 + x - 2 = -1 + x - 2$ Simplify and multiply.

$2x - 5 = x - 3$

$x = 2$

Since $x \neq 2$, the equation has no solution; the solution set is $\varnothing$. When a proposed solution to an equation does not check, it is said to be **extraneous.**

EXAMPLE 5 Solve: $\dfrac{-4}{x} + x = 3$.

SOLUTION Since the denominator is zero when $x = 0$, we must restrict x to be nonzero. We multiply each side by x, the LCD.

$x\left(\dfrac{-4}{x} + x\right) = 3(x)$

$x \cdot \dfrac{-4}{x} + x \cdot x = 3x$ Distributive property

$-4 + x^2 = 3x$

The equation is quadratic. We write it in standard form and factor it.

$x^2 - 3x - 4 = 0$

$(x-4)(x+1) = 0$

$x - 4 = 0 \quad \text{or} \quad x + 1 = 0$ Set each factor equal to zero.

$x = 4 \quad \text{or} \quad x = -1$

We check both solutions in the original equation.

$$\frac{-4}{4} + 4 \stackrel{?}{=} 3 \quad \text{or} \quad \frac{-4}{-1} + (-1) \stackrel{?}{=} 3$$
$$-1 + 4 \stackrel{?}{=} 3 \quad\quad\quad 4 - 1 \stackrel{?}{=} 3$$
$$3 \stackrel{\checkmark}{=} 3 \quad\quad\quad 3 \stackrel{\checkmark}{=} 3$$

Both solutions check. The solution set is $\{-1, 4\}$.

EXAMPLE 6   Solve: $\dfrac{4}{x-2} - \dfrac{2}{x+1} = 1$.

SOLUTION   Since the first denominator is zero when $x = 2$ and the second one is zero when $x = -1$, we must restrict x so that $x \neq 2$ or -1.

We multiply each side by $(x + 1)(x - 2)$, the LCD.

$$(x-2)(x+1)\left(\frac{4}{x-2}\right) - (x-2)(x+1)\left(\frac{2}{x+1}\right) = (x-2)(x+1) \cdot 1$$

$(x + 1)(4) - (x - 2)(2) = (x - 2)(x + 1) \cdot 1$ Simplify.
$4x + 4 - 2x + 4 = x^2 - x - 2$ Distributive property
$2x + 8 = x^2 - x - 2$ Combine like terms.
$0 = x^2 - 3x - 10$ Write in standard form.

The equation $x^2 - 3x - 10 = 0$ is quadratic, so we solve it by factoring.

$(x - 5)(x + 2) = 0$
$x - 5 = 0 \quad \text{or} \quad x + 2 = 0$ Set each factor equal to zero.
$x = 5 \quad \text{or} \quad x = -2$

We check both solutions in the original equation.

$$\frac{4}{5-2} - \frac{2}{5+1} \stackrel{?}{=} 1 \quad \text{or} \quad \frac{4}{-2-2} - \frac{2}{-2+1} \stackrel{?}{=} 1 \quad\quad x = 5 \text{ or } -2$$
$$\frac{4}{3} - \frac{2}{6} \stackrel{?}{=} 1 \quad\quad\quad \frac{4}{-4} - \frac{2}{-1} \stackrel{?}{=} 1$$
$$\frac{8}{6} - \frac{2}{6} \stackrel{?}{=} 1 \quad\quad\quad -1 + 2 \stackrel{?}{=} 1$$
$$1 \stackrel{\checkmark}{=} 1 \quad\quad\quad 1 \stackrel{\checkmark}{=} 1$$

Both solutions check, so the solution set is $\{-2, 5\}$.

Many applications result in equations involving rational expressions, as we show in the next four examples.

EXAMPLE 7   One-third of a number is 8 less than one-half of that number. Find the number.

SOLUTION   We let n = the number

$$\frac{1}{3}n = \text{one-third of the number}$$

$$\frac{1}{2}n = \text{one-half of the number}$$

$$\frac{1}{3}n = \frac{1}{2}n - 8$$

$$6\left(\frac{1}{3}n\right) = 6\left(\frac{1}{2}n - 8\right) \qquad \text{Multiply each side by 6, the LCD.}$$

$$2n = 3n - 48 \qquad \text{Distributive property}$$

$$n = 48$$

CHECK   One-half of 48 is 24 and one-third of 48 is 16. The difference between 24 and 16 is 8. The solution checks; the number is 48.

EXAMPLE 8   Due to an uphill climb, it took Alexandra 3 hours to ride her road bicycle from Saratoga to Berring and 2 hours to return. Her average speed on the return trip was 5 mph faster than her speed going. How far apart are the two towns?

SOLUTION   The key to solving this problem is to remember that $d = rt$ and $r = d/t$. We let d = the distance between the two towns.

Direction	Distance	Rate	Time
Saratoga to Berring	d	$\frac{d}{3}$	3
Berring to Saratoga	d	$\frac{d}{2}$	2

The average speed on the return trip is 5 mph faster than the average speed going to Saratoga, so we write

$$\frac{d}{2} = \frac{d}{3} + 5$$

$$6\left(\frac{d}{2}\right) = 6\left(\frac{d}{3} + 5\right) \qquad \text{Multiply each side by 6, the LCD.}$$

$$3d = 2d + 30 \qquad \text{Distributive property}$$

$$d = 30$$

The reader should check the solution in the words of the problem. The two towns are 30 miles apart.

8.5 · SOLVING RATIONAL EQUATIONS

EXAMPLE 9 The numerator of a certain fraction is 4 less than the denominator. If the numerator is increased by 9 and the denominator is decreased by 1, the fraction has a new value of 2. Find the original fraction.

SOLUTION We let x = the numerator of the original fraction

$x + 4$ = the denominator of the original fraction

$\dfrac{x}{x+4}$ = the original fraction

$\dfrac{x+9}{x+4-1} = 2$ The numerator is increased by 9; the denominator is decreased by 1.

$\dfrac{x+9}{x+3} = 2$ Simplify.

$x + 9 = 2(x + 3)$ Multiply each side by $x + 3$, the LCD.

$x + 9 = 2x + 6$ Distributive property

$3 = x$

Now we substitute 3 for x in the original fraction.

$$\dfrac{x}{x+4} = \dfrac{3}{3+4} = \dfrac{3}{7}$$

The original fraction is $\dfrac{3}{7}$.   The check is left to the reader.

EXAMPLE 10 One pipe can fill a tank in 8 hours. The drain can empty the same tank in 12 hours. If the drain is left open by mistake, how long will it take to fill the tank?

SOLUTION We let h denote the number of hours it takes to fill the tank with the pipe if the drain is open. To solve this problem, we determine how much of the tank the pipe fills in 1 hour with the drain open. If the pipe alone fills it in 8 hours, it fills $\dfrac{1}{8}$ of the tank in 1 hour. If the drain empties the tank in 12 hours, it drains $\dfrac{1}{12}$ of the tank in 1 hour. Finally, if it takes h hours for the pipe to fill the tank with the drain open, then $\dfrac{1}{h}$ of the tank is filled in 1 hour.

	PIPE ALONE	DRAIN ALONE	BOTH OPEN
AMOUNT PER HOUR	$\dfrac{1}{8}$	$\dfrac{1}{12}$	$\dfrac{1}{h}$

$$\text{THE AMOUNT FILLED BY THE FILL PIPE} - \text{THE AMOUNT EMPTIED BY THE DRAIN} = \text{THE AMOUNT FILLED WITH BOTH OPEN.}$$

$$\dfrac{1}{8} - \dfrac{1}{12} = \dfrac{1}{h}$$

We multiply each side by 24h, the LCD.

$$3h - 2h = 24$$
$$h = 24$$

It takes 24 hours for the pipe to fill the tank with the drain open. The check is left to the reader.

Solution to the Historical Problem at the Beginning of This Section

I am a bronze lion; my spouts are my two eyes, my mouth, and the flat of my right foot. My right eye fills a jar in two days, my left in three, and my foot in four. My mouth is capable of filling the jar in six hours. Tell me how long all four working together will take to fill it.

SOLUTION

Spout	Time to Fill in Days	Amount Filled in One Day
Right eye	2	$\frac{1}{2}$
Left eye	3	$\frac{1}{3}$
Foot	4	$\frac{1}{4}$
Mouth	6 hr = $\frac{1}{4}$ day	$\frac{1}{\frac{1}{4}} = 4$
Together	d	$\frac{1}{d}$

The sum of the amounts each part of the lion can fill in one day is equal to the amount they can fill together in one day, so

$$\frac{1}{2} + \frac{1}{3} + \frac{1}{4} + 4 = \frac{1}{d}$$

We multiply each side by 12d, the LCD.

$$6d + 4d + 3d + 48d = 12$$
$$61d = 12$$
$$d = \frac{12}{61} \text{ days}$$

The time in hours is $\frac{12}{61} \cdot 24 = \frac{288}{61} \approx 4.72$ hours.

EXERCISE SET 8.5

Solve each rational equation and check your solution. Be careful to note those cases where values of the variable cause it to be undefined.

1. $\dfrac{x}{3} = \dfrac{1}{2}$
2. $\dfrac{y}{5} = \dfrac{1}{3}$
3. $\dfrac{k}{8} = \dfrac{-5}{8}$
4. $\dfrac{t}{7} = \dfrac{-3}{7}$
5. $\dfrac{1}{3} - \dfrac{1}{2} = \dfrac{p}{6}$
6. $\dfrac{1}{4} - \dfrac{1}{2} = \dfrac{m}{8}$
7. $\dfrac{3}{5} + \dfrac{y}{4} = 1$
8. $\dfrac{2}{9} - \dfrac{x}{6} = 1$
9. $3n - \dfrac{1}{4} = \dfrac{11}{4}$
10. $5k - \dfrac{1}{3} = \dfrac{14}{3}$
11. $\dfrac{7}{x} = \dfrac{3}{5}$
12. $\dfrac{4}{y} = \dfrac{2}{9}$
13. $\dfrac{-4}{p} = \dfrac{5}{11}$
14. $\dfrac{-3}{t} = \dfrac{7}{13}$
15. $\dfrac{13}{10} = \dfrac{19}{2k} - \dfrac{3}{k}$
16. $\dfrac{1}{2} = \dfrac{3}{m} + \dfrac{2}{m}$
17. $\dfrac{4}{3} - \dfrac{2}{3a} = \dfrac{4}{a} - \dfrac{1}{2}$
18. $\dfrac{4}{9} - \dfrac{1}{x} = \dfrac{1}{9} - \dfrac{2}{3x}$
19. $\dfrac{1}{2x} - \dfrac{1}{2} = \dfrac{5}{10x} + \dfrac{1}{5}$
20. $\dfrac{3}{2y} - \dfrac{1}{4} = \dfrac{1}{5} + \dfrac{6}{4y}$
21. $\dfrac{5}{a-1} = \dfrac{3}{a+1}$
22. $\dfrac{5}{3m-4} = \dfrac{3}{m-3}$
23. $\dfrac{3p}{p-2} = \dfrac{14-4p}{p-2} + 4$
24. $\dfrac{5t}{t+5} - 2 = \dfrac{-35}{t+5}$
25. $\dfrac{3}{2k-1} = \dfrac{6k}{2k-1} - 4$
26. $\dfrac{6}{L-3} - 5 = \dfrac{2L}{L-3}$
27. $\dfrac{8}{5-x} = \dfrac{6}{x-1}$
28. $\dfrac{3}{2x-1} = \dfrac{7}{3x+1}$
29. $\dfrac{3}{4} = \dfrac{8a-1}{6a+8}$
30. $\dfrac{6y+9}{5y+10} = \dfrac{3}{5}$
31. $\dfrac{m+8}{m-1} + 2 = 0$
32. $\dfrac{k-3}{k+6} + 8 = 0$
33. $-3 + \dfrac{2}{t-2} = -\dfrac{5}{2-t}$
34. $\dfrac{3}{4-m} + 2 = \dfrac{3m}{m-4}$
35. $\dfrac{x}{x+1} - \dfrac{2x}{4x+4} = \dfrac{x}{2x+2}$
36. $\dfrac{5y-5}{5y+5} - \dfrac{5y+1}{3y+3} = -\dfrac{3y-1}{y+1}$
37. $\dfrac{5}{k-3} = \dfrac{33-k}{k^2-6k+9}$
38. $\dfrac{7}{2k+2} = \dfrac{-3}{k+2} + \dfrac{20k-1}{2k^2+6k+4}$

Write each of the following applied problems as a rational equation and solve it. Check your answers.

39. Two less than one-half of a number is $\frac{1}{4}$.
40. Five more than one-third of a number is 2.
41. One-third of a number is 3 more than one-fourth of the number.
42. One less than one-sixth of a number is 2 more than one-fourth of the number.
43. The width of a rectangle is one-sixth of its length. If its perimeter is 98, find its dimensions.
44. The width of a rectangle is two-fifths of its length. If its perimeter is 140, find its dimensions.
45. David knows that he can mow, edge, water, and clean the yard in 7 hr. If his wife Ruth helps him, they can do the job in 4 hr. How long does it take Ruth working alone to do the job?
46. A swimming pool can be filled by either of two pipes. The larger pipe fills it in 10 hr and the smaller pipe in 15 hr. How long does it take to fill the pool using both pipes?
47. Two families travel in opposite directions from a given house at a given time. After 5 hr they are 440 km apart. If one family travels 40 km/hr faster than the other, determine the rate at which each family travels.
48. On a calm day, Marlene can fly her new miniplane at a speed of 74 km/hr. When the wind is blowing, it takes her as long to go 50 km against the wind as it does to go 75 km with the wind. What is the speed of the wind?
49. When the numerator of $\frac{5}{8}$ is increased by a certain number and the denominator is multiplied by the same number, the resulting fraction is $\frac{1}{3}$. Find the number.
50. If Pam can fold 50 handbills per minute and Cindy can fold 75 per minute, how long does it take both of them working together to fold 1800 handbills?
51. A pipe can fill a pool in 20 hr and a drain can empty it in 32 hr. After draining and cleaning the pool, Lee forgets to close the drain before turning on the pipe. If

he doesn't catch his mistake, how long does it take to fill the pool?

52. The numerator of a fraction is $\frac{1}{12}$ of the denominator. When 3 is subtracted from the denominator and 4 is added to the numerator, the new fraction is $\frac{5}{9}$. What was the original fraction?

53. The Tretts leave San Luis Obispo and head north at 8:00 A.M., traveling at an average rate of 40 mph. At 11:00 A.M., their daughter Susanne leaves San Luis Obispo to catch up with them. If she travels at an average rate of 55 mph, at what time does she overtake them?

54. Eighty-eight ft/sec is equivalent to 60 mph. What is the average speed in mph of a cyclist who covers 275 ft in 5 sec?

The following equations will become quadratic equations when they are cleared of fractions. Note any restrictions on the variables, solve by factoring, and check all solutions.

55. $\dfrac{8a + 3}{a} = 3a$ **56.** $\dfrac{3}{k} + k = 4$

57. $m + 8 = \dfrac{m - 10}{m}$ **58.** $y + 1 = \dfrac{y + 1}{y}$

59. $\dfrac{-4}{y} + 2 = \dfrac{y + 1}{5}$ **60.** $\dfrac{a + 4}{a + 7} - \dfrac{3}{8} = \dfrac{a}{a + 3}$

REVIEW EXERCISES

Solve each equation for the indicated variable.

61. $\dfrac{V_1}{V_2} = \dfrac{P_1}{P_2}$; P_2 **62.** $\dfrac{1}{f} = \dfrac{1}{a} + \dfrac{1}{b}$; b

63. $F = G\dfrac{M_1 M_2}{d^2}$; M_1 **64.** $\dfrac{2A}{h} - b_1 = b_2$; h

65. $b = \dfrac{a - 3}{3a - 2}$; a **66.** $2rt - t = 3r + 1$; r

67. $\dfrac{x - h}{a} = \dfrac{y - k}{b}$; y **68.** $y = \dfrac{2x + 5}{x - 3}$; x

69. $y = \dfrac{3x - 1}{x + 2}$; x **70.** $m = \dfrac{y - y_1}{x - x_1}$; y

FOR EXTRA THOUGHT

71. If an object is placed p units from a lens with a focal length of f, the image distance is q. The relationship between these three variables is given by the rational equation

$$\dfrac{1}{f} = \dfrac{1}{p} + \dfrac{1}{q}$$

(a) Solve for p.

(b) If an object has an image distance of 4 cm from a lens with a focal length of 1.4 cm, use part **(a)** to find the distance of the object from the lens.

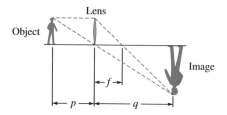

72. If three resistors with resistances of R_1, R_2, and R_3 ohms are placed in parallel, the total resistance R of the circuit is given by the rational equation

$$\dfrac{1}{R} = \dfrac{1}{R_1} + \dfrac{1}{R_2} + \dfrac{1}{R_3}$$

Three resistors are placed in parallel, where the first two resistors have the same resistance and the third resistor has four times the resistance of either of the other two. If the total resistance of the circuit is 10 ohms, find the resistance of each of the three resistors.

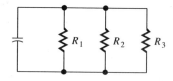

8.6 Ratio and Proportion

OBJECTIVES In this section we will learn to
1. write the ratio between two quantities;
2. write proportions; and
3. use ratios and proportions to solve applied problems.

Modern living requires us to make intelligent comparisons in a variety of situations. For example, we might want to compare the price per ounce of two similar products in the supermarket, or the output of two workers based on the hours they have worked, or the quality control in a company's production based on product sampling. We can make such comparisons using *ratios*.

DEFINITION **Ratio**	The **ratio** of two quantities a and b is their comparison represented by division, written $$\frac{a}{b} \quad \text{or} \quad a:b, \quad b \neq 0$$

For example, suppose that during a sales promotion an automobile dealer sells 20 of a particular model of a compact car while a competing dealer sells 12 of the same model without a promotion. The sales ratio of the cars between the two dealerships is 20 to 12, or

$$\frac{20}{12} = \frac{5}{3}$$

This means that for every 5 cars the first dealer sells, the second dealer sells 3. In this example we are comparing compact car sales to compact car sales. We can also use ratios to compare two different objects as well. Suppose the dealer that sold 20 compact models also sold 10 full-size sedans. Then the ratio of compact sales to full-size sales is 20 to 10, or

$$\frac{20}{10} = \frac{2}{1}$$

This means that two compact cars were sold for every full-size sedan sold.

> *Ratios can be used to compare two types of quantities:*
> **1.** those that are expressed in terms of the same units; and
> **2.** those that are expressed in terms of two different units.

When ratios compare the same units, or can be rewritten so as to compare the same units, they represent pure numbers. Example 1 illustrates this concept.

EXAMPLE 1 (a) The ratio of 1 foot to 5 feet is written

$$\frac{1 \text{ ft}}{5 \text{ ft}} = \frac{1}{5}$$

(b) The ratio of 3 inches to 1 foot is

$$\frac{3 \text{ in.}}{1 \text{ ft}} = \frac{3 \text{ in.}}{12 \text{ in.}} = \frac{1}{4} \qquad \text{Change 1 foot to 12 inches.}$$

(c) The ratio of 7 ounces to 2 pounds is

$$\frac{7 \text{ oz}}{2 \text{ lb}} = \frac{7 \text{ oz}}{32 \text{ oz}} = \frac{7}{32} \qquad \text{Change 2 pounds to 32 ounces (1 lb = 16 oz).}$$

The final ratios in Example 1 are not identified by units such as inches, feet, or pounds. Some ratios, however, have units that cannot be made to be the same. They are, nevertheless, meaningful.

EXAMPLE 2 A motorist drove 200 miles in 4 hours. Find the ratio of the distance traveled to the time it took.

SOLUTION $\quad \dfrac{200 \text{ mi}}{4 \text{ hr}} = 50 \text{ mph} \qquad$ The units are different.

The average speed of the motorist was 50 miles per hour.

An idea closely related to a ratio is that of a *proportion*.

DEFINITION **Proportion**	A **proportion** is a statement that two ratios are equal. $$\frac{a}{b} = \frac{c}{d}, \quad b \neq 0, \, d \neq 0$$ In the proportion, the quantities a, b, c, and d are called the **terms.** In particular, a and d are called the **extremes,** and b and c are called the **means.**

For example,

$$\frac{1}{5} = \frac{2}{10}$$

is a proportion because the ratios $\dfrac{1}{5}$ and $\dfrac{2}{10}$ are equal.

If we clear the proportion

$$\frac{a}{b} = \frac{c}{d}$$

of fractions by multiplying each side by bd, the LCD, we have

$$bd \cdot \frac{a}{b} = bd \cdot \frac{c}{d}$$

$$ad = bc \qquad \text{Simplify and multiply.}$$

This leads to the following important property of proportions.

In any proportion, the product of the extremes is equal to the product of the means:

$$\frac{a}{b} = \frac{c}{d} \quad \text{implies} \quad ad = bc, \quad b \neq 0, d \neq 0$$

EXAMPLE 3   Is $\frac{2}{5} = \frac{8}{20}$ a true proportion?

SOLUTION   To answer this question, we must see whether the product of the extremes is equal to the product of the means.

$$\frac{2}{5} \stackrel{?}{=} \frac{8}{20}$$

Extremes $2 \cdot 20 \stackrel{?}{=} 8 \cdot 5$ Means

$$40 \stackrel{\checkmark}{=} 40$$

Since the products are equal, the proportion is a true one.

EXAMPLE 4   Solve the proportion $\frac{x}{6} = \frac{5}{3}$ for x.

SOLUTION

$$\frac{x}{6} = \frac{5}{3}$$

$3 \cdot x = 6 \cdot 5$ Product of extremes = product of means

$3x = 30$

$x = 10$

CHECK $\frac{x}{6} \stackrel{?}{=} \frac{5}{3}$ Original equation

$\frac{10}{6} \stackrel{?}{=} \frac{5}{3}$ Substitute 10 for x.

$\frac{5}{3} \stackrel{\checkmark}{=} \frac{5}{3}$

The solution set is $\{10\}$.

EXAMPLE 5 Solve the proportion $\dfrac{2}{y+4} = \dfrac{1}{y-3}$ for y. Assume that $y \neq -4, 3$.

SOLUTION
$$\dfrac{2}{y+4} = \dfrac{1}{y-3}$$
$2(y - 3) = y + 4$ Product of extremes = product of means
$2y - 6 = y + 4$ Distributive property
$y = 10$ Subtract y and add 6 to each side.

CHECK $\dfrac{2}{y+4} = \dfrac{1}{y-3}$ Original equation

$\dfrac{2}{10+4} \stackrel{?}{=} \dfrac{1}{10-3}$ Substitute 10 for y.

$\dfrac{2}{14} \stackrel{?}{=} \dfrac{1}{7}$

$\dfrac{1}{7} \stackrel{\checkmark}{=} \dfrac{1}{7}$

The solution set is {10}.

EXAMPLE 6 Solve the proportion $\dfrac{x+6}{x} = \dfrac{x+2}{4}$ for x. Assume $x \neq 0$.

SOLUTION
$4(x + 6) = x(x + 2)$ Product of extremes = product of means
$4x + 24 = x^2 + 2x$
$0 = x^2 - 2x - 24$
$0 = (x - 6)(x + 4)$
$x - 6 = 0$ or $x + 4 = 0$
$x = 6$ or $x = -4$

The two solutions, 6 and -4, check. The solution set is $\{-4, 6\}$.

The concepts of ratio and proportion have many applications.

EXAMPLE 7 An advertisement claims that 8 out of ever 10 doctors surveyed recommend a particular brand of aspirin. If the survey is supposedly representative of all physicians, how many doctors in a group of 130,000 recommend that brand?

SOLUTION We let x = the number of doctors that recommend the brand. Then

$$\dfrac{8}{10} = \dfrac{x}{130{,}000}$$

$8(130{,}000) = 10x$ Product of extremes = product of means

$$1{,}040{,}000 = 10x$$
$$104{,}000 = x$$

We conclude that 104,000 doctors make the recommendation.

Similar triangles are triangles that have the same shape. This means that the angles of one triangle are the same size as the angles of the other. It has been proven that

Corresponding sides of similar triangles are proportional.

In two similar triangles, **corresponding sides** are those opposite the angles with the same measure. (See Figure 8.1.)

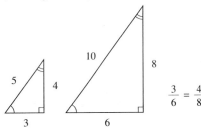

FIGURE 8.1

EXAMPLE 8   To determine the height of a flagpole, a student measures the shadow of a friend who is 6 feet tall. She also measures the length of the shadow of the flagpole at the same time of day. The friend's shadow is 4 feet long; the pole's shadow is 24 feet long. What is the height of the pole?

SOLUTION   The ratio of the two heights and the ratio of the lengths of the shadows must be equal. The triangle with the man as one side is similar to the one with the flagpole as a side. We let $x =$ the height of the pole. (See Figure 8.2.) We can now establish the proportion.

$$\frac{x}{6} = \frac{24}{4}$$
$$4x = 6 \cdot 24$$
$$4x = 144$$
$$x = 36$$

The flagpole is 36 feet high.

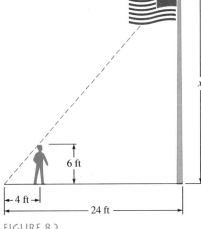

FIGURE 8.2

EXAMPLE 9 Find the lengths of the unknown sides of the similar triangles *ABC* and *ADE*. (See Figure 8.3.)

SOLUTION Since the corresponding sides are proportional, we can set up proportions between the ratio of two known sides and a ratio involving an unknown side.

To Find x
$$\frac{7}{x} = \frac{4}{8}$$
$$4x = 56$$
$$x = 14 \text{ units}$$

To Find y
$$\frac{5}{y} = \frac{4}{8}$$
$$4y = 40$$
$$y = 10 \text{ units}$$

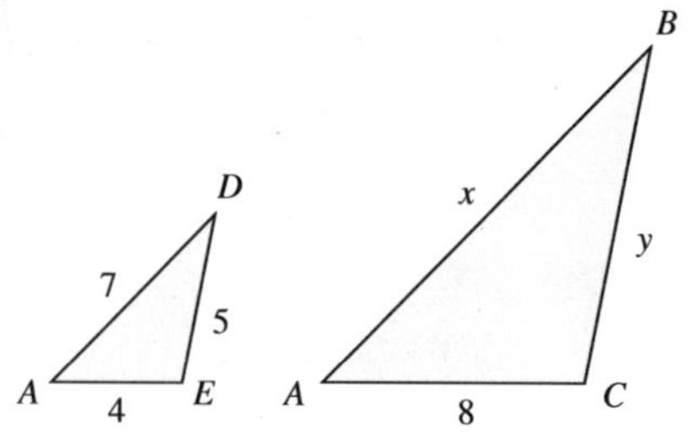

FIGURE 8.3

EXAMPLE 10 A 3-inch by 5-inch photograph is enlarged so that its shortest dimension is 4 inches. What is its longest dimension?

SOLUTION We let x = the longest dimension. Then the ratios between the short and long sides of the two pictures must be the same.

Ratio of the short sides → $\dfrac{3}{4} = \dfrac{5}{x}$ ← Ratio of the long sides

$$3x = 20 \qquad \text{Product of extremes = product of means}$$
$$x = \frac{20}{3} = 6\frac{2}{3}$$

The longest dimension is $6\frac{2}{3}$ inches.

Caution

In setting up a proportion, always be careful that the ratio on the right is comparing the same quantities in the same order as the ratio on the left.

There are other ways we could have compared the dimensions in Example 10:

$$\frac{3}{5} = \frac{4}{x} \implies 3x = 20$$
$$\frac{4}{3} = \frac{x}{5} \implies 20 = 3x$$
$$\frac{x}{4} = \frac{5}{3} \implies 3x = 20$$

EXAMPLE 11 Two business partners agree to share their business profits in the same ratio as their investments in the business, which is 6 to 4. If their profit for the year is $50,000, how much does each get?

SOLUTION We let x = the share of one partner; then the share of the other partner is $50,000 − x$.

Ratio of profits → $\dfrac{x}{50,000 - x} = \dfrac{6}{4}$ ← Ratio of investments

$$6(50,000 - x) = 4x$$

$$300{,}000 - 6x = 4x$$
$$300{,}000 = 10x \quad \text{Add } 6x \text{ to each side.}$$
$$30{,}000 = x$$

The first partner's share is $30,000 and the second's share is $20,000.

Do You Remember?

Can you match these?

More than one answer may apply.

_____ 1. An example of a ratio
_____ 2. An example of a proportion
_____ 3. A statement that two ratios are equal
_____ 4. Corresponding sides of similar triangles are ____.
_____ 5. The solution to $\dfrac{x}{9} = \dfrac{5}{3}$
_____ 6. The means of the proportion $\dfrac{2}{3} = \dfrac{6}{9}$
_____ 7. The extremes of the proportion $\dfrac{2}{3} = \dfrac{6}{9}$

a) 2 and 9
b) 15
c) a proportion
d) 3 and 6
e) proportional
f) $\dfrac{7}{8}$
g) $\dfrac{a}{b} = \dfrac{c}{d}$

Answers: 1. f 2. g 3. c, g 4. e 5. b 6. d 7. a

Exercise Set 8.6

Write each ratio as a common fraction in simplest form.

1. 8 to 4
2. 9 to 3
3. 12 to 56
4. 88 to 99
5. 60 to 90
6. 25 to 85
7. 14 to 91
8. 39 to 65
9. $7\dfrac{1}{8}$ to $2\dfrac{1}{4}$
10. $12\dfrac{1}{2}$ to $5\dfrac{5}{8}$
11. $1\dfrac{2}{7}$ to $2\dfrac{1}{7}$
12. $3\dfrac{1}{2}$ to $6\dfrac{1}{8}$

13. The ratio of men to women, if there are 54 people in a room and 30 of them are men
14. The ratio of 224 miles to 4 hours
15. The ratio of wins to losses (in dollars), if you spend $120 on lottery tickets and win $25 in prizes
16. The ratio of David's overtime wage to his basic wage, if his basic hourly wage is $6.70 and his overtime hourly wage is $10.05

17. Find the ratio of the length to the width in the accompanying rectangle.

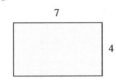

18. Find the ratio of the longest side to the shortest side in the given triangle.

19. A baseball player hit 54 times in 270 times at bat. What is the ratio of the number of hits to the number of times at bat?

20. A car radiator with a capacity of 16 qt is completely full of a mixture of antifreeze and water. If it contains 8 qt of antifreeze, what is the ratio of antifreeze to water?

21. What is the ratio of men to women in a certain medical school if there are 200 men and 160 women?

22. A particular model outboard motor requires 1 qt of oil for every 2 gal of gasoline used for fuel. What is the ratio of oil to gasoline? Of gasoline to oil?

Solve each proportion.

23. $\dfrac{x}{4} = \dfrac{5}{10}$ 24. $\dfrac{x}{7} = \dfrac{140}{5}$ 25. $\dfrac{8}{x} = \dfrac{1}{9}$

26. $\dfrac{7}{2} = \dfrac{11}{x}$ 27. $\dfrac{2x}{7} = \dfrac{14}{5}$ 28. $\dfrac{22}{10} = \dfrac{100}{x}$

29. $\dfrac{x-3}{x} = \dfrac{7}{9}$ 30. $\dfrac{x+2}{x} = \dfrac{4}{5}$

31. $\dfrac{5}{x+9} = \dfrac{7}{2x-1}$ 32. $\dfrac{3}{5x+4} = \dfrac{1}{2x-3}$

33. $\dfrac{7x+5}{6x-1} = \dfrac{-1}{2}$ 34. $\dfrac{9x+11}{5x-8} = \dfrac{-3}{4}$

35. $\dfrac{x+2}{x} = \dfrac{x+7}{2-x}$ 36. $\dfrac{1-x}{x} = \dfrac{5x+1}{x+1}$

37. $\dfrac{x+3}{4} = \dfrac{2x}{x-3}$ 38. $\dfrac{1}{2x+8} = -\dfrac{4-x}{11x-47}$

39. $\dfrac{x}{2} = \dfrac{1}{x+1}$ 40. $\dfrac{x}{1} = \dfrac{x^2+6}{2x+1}$

41. $\dfrac{4x}{x+2} = \dfrac{x-2}{x+2}$

42. $\dfrac{4x^2-11x-4}{4x+3} = -\dfrac{1}{4x+3}$

Write each statement as a proportion and solve it.

43. If you can drive 200 mi on 8 gal of gasoline, how far can you drive on 20 gal of gasoline?

44. If 1 km is approximately 0.6 mi, how many kilometers is it to a town 25 mi away?

45. If 1 cup is equivalent to 240 ml, how many milliliters are in $3\tfrac{3}{4}$ cups?

46. If the scale on a road map reads 1 in. = 500 mi and two cities are $1\tfrac{3}{4}$ in. apart on the map, how far apart are they in miles?

47. For your trip to Canada, you need to change currency. If one U.S. dollar is worth $1.18 in Canadian dollars, how many Canadian dollars will you receive for 400 U.S. dollars?

48. A survey shows that 3 out of 4 people watch television every night. If 12,418 people are surveyed, how many of them will say they watch television every night?

49. Dennis and Dean invest $12,872 in a sailboat. The ratio of Dennis's investment to Dean's is 3 to 5. How much did each invest in the boat?

50. A rectangle is said to be most pleasing to the eye when the ratio of the width to the length is 5 to 8. Using this ratio, what should be the length of a rectangular painting if its width is 12 in.?

51. If 24 lb of fertilizer feeds 4000 ft^2 of lawn, how much fertilizer is needed to feed 10,000 ft^2?

52. Two pancakes contain 120 calories. Ron ate a stack of 8 pancakes for breakfast. How many calories did he consume?

53. Managers at a shirt factory find that out of every 1000 shirts they produce, 17 are defective. How many defective shirts can they expect in a production run of 40,000 shirts?

54. A survey discloses that 2 out of 9 college freshmen never finish their junior year. If the freshmen class at Greenfield College has 693 students enrolled this year, how many will be expected to finish their junior year?

55. Polly's waffle recipe calls for 3 eggs, $1\tfrac{3}{4}$ cups of flour, 2 tsp of baking powder, 1 tbsp of sugar, 6 tbsp of butter, and $1\tfrac{1}{2}$ cups of milk. She discovers she has only 2 eggs on hand. How much of each of the other ingredients should she use?

56. Robert needs 65 yd^2 to carpet his living room, hall, and dining room. If the carpet costs $16.95/yd^2, what is the total cost?

57. The volume of a gas is proportional to its temperature, as given by the formula
$$\dfrac{V_1}{V_2} = \dfrac{T_1}{T_2}$$
If the volume of a gas is 288 cm^3 at 176°, what is its volume at 264°?

58. A 5-ft-tall woman has a shadow 3 ft long. How tall is a tree with a shadow 27 ft long?

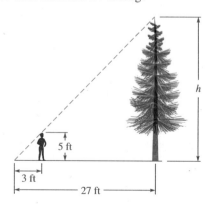

59. A picture 2 in. × 3 in. is enlarged so that the shortest dimension is 4 in. What is the other dimension?

60. A 3 in. × 5 in. picture is enlarged so that the longest dimension is 10 in. What is the shortest dimension?

61. If antifreeze is mixed in a ratio of 1 : 4 with water, how many quarts of antifreeze are needed in a radiator that holds 24 qt?

ℱOR ℰXTRA 𝒯HOUGHT

Use proportions to find the unknown sides of the similar triangles.

62.
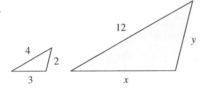

63.
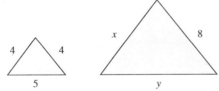

64.
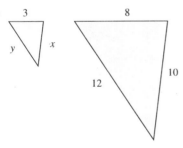

65.
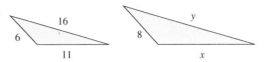

66.
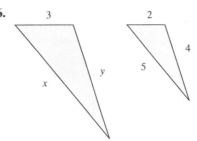

67.
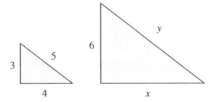

WRITE IN YOUR OWN WORDS

68. In the proportion $\frac{a}{b} = \frac{c}{d}$, the quantities b and c are called the *means,* and a and d are called the *extremes.* If the means and extremes are exchanged, what happens to the proportion?

69. The proportion $\frac{a}{b} = \frac{c}{d}$ can also be written $a:b = c:d$. Explain why it seems reasonable to call a and d the extremes and b and c the means in this form.

8.7 Variation

OBJECTIVES

In this section we will learn to
1. recognize and use direct variation;
2. recognize and use inverse variation; and
3. recognize and use joint variation.

Many jobs in science, engineering, and other fields involve working with formulas that express relationships between two or more quantities. In this section we will discover how to write and work with them.

DEFINITION	**Direct variation** means that *y* varies *directly* with *x*, or
Direct Variation	$$y = kx$$ where *k* is called the **constant of variation.**

For example, the formula for the area of a circle, $C = \pi d$, is an example of direct variation; the constant of variation is the number π (pi). In direct variation, as one quantity increases, the other quantity also increases.

EXAMPLE 1 Translate each example of direct variation into words.

(a) $y = kt$ means *y* varies directly with *t*.
(b) $y = ks^3$ means *y* varies directly with the cube of *s*.
(c) $A = \pi r^2$ means that *A* varies directly with the square of *r*.

Not all variation is direct. There is also *inverse* and *joint variation*.

DEFINITION	**Inverse variation** means that *y* varies *inversely* with *x*, or
Inverse Variation	$$y = \frac{k}{x}$$ where *k* is called the **constant of variation.**

In inverse variation, as one quantity incresaes the other decreases.

EXAMPLE 2 Translate each equation into words.

(a) $y = \dfrac{k}{x^2}$ means *y* varies inversely with the square of *x*.
(b) $y = \dfrac{k}{\sqrt{x}}$ means *y* varies inversely with the square root of *x*.

DEFINITION	**Joint variation** means that one variable varies directly with the *product* of two or more variables, such as
Joint Variation	$$y = kxz \quad \text{or} \quad y = kxzw$$ where it is said to *vary jointly* with *xz* or with *xzw*, respectively.

EXAMPLE 3 Translate each equation into words.

(a) $y = kxz$ means *y* varies jointly with *x* and *z*.
(b) $y = k\dfrac{sp}{t^2}$ means *y* varies jointly with *s* and *p* and inversely with the square of *t*.

EXAMPLE 4 Write a variation equation for each of the following statements.

(a) S varies directly with the square root of t.
(b) P varies directly with the fifth power of h.
(c) f varies inversely with b.
(d) A varies jointly with the product of x and y.

SOLUTION (a) $S = k\sqrt{t}$
(b) $P = kh^5$
(c) $f = \dfrac{k}{b}$
(d) $A = kxy$

The *constant of variation,* as the name implies, remains constant for all values of the variable within the given restrictions. For this reason, it can be readily determined.

EXAMPLE 5 Find k if y varies directly with x, and if $y = 8$ when $x = 2$.

SOLUTION To say that y varies directly as x means that $y = kx$.

$y = kx$
$8 = k \cdot 2$ Substitute 8 for y and 2 for x.
$4 = k$ Divide each side by 2.

EXAMPLE 6 Find k if y varies inversely with the square root of x, and if $y = 7$ when $x = 1.21$.

SOLUTION To say that y varies inversely with the square root of x means that $y = \dfrac{k}{\sqrt{x}}$.

$y = \dfrac{k}{\sqrt{x}}$

$7 = \dfrac{k}{\sqrt{1.21}}$ Substitute 7 for y and 1.21 for x.

$7 = \dfrac{k}{1.1}$ $\sqrt{1.21} = 1.1$

$7.7 = k$ Multiply each side by 1.1.

EXAMPLE 7 If y varies directly with the square of x, and if $y = 16$ when $x = 2$, find y when $x = 3$.

SOLUTION To say that y varies directly with the square of x means $y = kx^2$.

$y = kx^2$
$16 = k \cdot 2^2$ Substitute 16 for y and 2 for x.
$16 = 4k$
$4 = k$ Divide each side by 4.

Now we replace k with 4 in the variation equation.

$y = kx^2$ Variation equation
$y = 4x^2$ Substitute 4 for k.

To find y when $x = 3$, we replace x by 3.

$y = 4 \cdot 3^2 = 36$

EXAMPLE 8 If y varies jointly with x^2 and z and inversely with t, find k when $y = 5$, $x = 2$, $z = 3$, and $t = 9$. Then find y when $x = 4$, $z = 9$, and $t = 18$.

SOLUTION

$y = k\dfrac{x^2 z}{t}$ Variation equation

$5 = k\dfrac{2^2 \cdot 3}{9}$ Substitute 5 for y, 2 for x, 3 for z, and 9 for t.

$5 = \dfrac{4}{3}k$ Simplify.

$\dfrac{15}{4} = k$ Multiply each side by $\dfrac{3}{4}$.

Now we replace k by $\dfrac{15}{4}$ in the variation equation.

$y = \dfrac{15}{4} \cdot \dfrac{x^2 z}{t}$

$y = \dfrac{15}{4}\left(\dfrac{4^2 \cdot 9}{18}\right)$ Substitute 4 for x, 9 for z, and 18 for t.

$= 30$

EXAMPLE 9 The weight of an object varies inversely with the square of the distance of the object from the center of the earth. How much would a man who weighs 200 pounds on the earth's surface weigh 2000 miles above the earth's surface?

SOLUTION The variation equation is

$$W = \dfrac{k}{d^2}$$

We first need to find k. The radius of the earth is approximately 4000 miles, so a person standing on the surface of the earth is 4000 miles from its center. We substitute 4000 for d.

$$200 = \dfrac{k}{(4000)^2}$$

$$200(4000)^2 = k$$

Without multiplying, we substitute $200(4000)^2$ for k in the variation equation:

$$W = \frac{200(4000)^2}{d^2}$$

At 2000 miles above the surface of the earth, a person is 6000 miles from the center. We substitute 6000 for d in the variation equation.

$$W = \frac{200(4000)^2}{(6000)^2}$$
$$= \frac{200(4^2 \cdot 1000^2)}{6^2 \cdot 1000^2}$$
$$= \frac{200 \cdot 4^2}{6^2}$$
$$= \frac{800}{9} = 88\frac{8}{9}$$

The man would weigh $88\frac{8}{9}$ pounds 2000 miles from the earth.

EXAMPLE 10 The interest earned on an investment varies jointly with the rate of interest paid and the length of time for which the money is invested. If $200 is earned on an investment at 5% for 4 years, how long will it take to earn $300?

SOLUTION We use the interest formula for the variation equation:

$$I = prt$$

where I = interest, p = the amount invested (the principal), r = rate of interest paid, and t = the time in years. Since the amount invested is fixed, p is the constant of variation.

$I = prt$	Variation equation
$200 = p(0.05)4$	Substitute 200 for I, 0.05 for r, and 4 for t.
$200 = 0.2p$	
$1000 = p$	

Now we substitute 1000 for p in the variation equation.

$$I = 1000rt$$

To find how long it will take to earn $300, we let $I = 300$ and $r = 0.05$.

$$300 = 1000(0.05)t$$
$$300 = 50t$$
$$6 = t$$

It will take 6 years to earn $300 on a $1000 investment at 5% interest.

Do You Remember?

Can you match these?

_____ 1. An example of joint variation only
_____ 2. An example of inverse variation only
_____ 3. An example of direct variation only
_____ 4. The constant of variation in $F = \dfrac{\pi d}{2}$
_____ 5. An example of both direct and inverse variation

a) $x = k\dfrac{y}{t}$
b) π
c) $S = kt$
d) $S = \dfrac{k}{t^2}$
e) $x = kyt$
f) $\dfrac{\pi}{2}$

Answers: 1. e 2. d 3. c 4. f 5. a

Exercise Set 8.7

Write a variation equation that represents each of the following statements. Use k for the constant of variation.

1. The circumference C of a circle varies directly with the diameter d.
2. The perimeter P of a square varies directly with the length s of a side.
3. Distance traveled, d, varies jointly with the rate r and the time t.
4. The area A of a rectangle varies jointly with the length l and the width w.
5. y varies inversely with x.
6. P varies inversely with t.
7. V varies directly with the square of s.
8. M varies directly with the cube of t.
9. T varies inversely with the square root of n.
10. Q varies inversely with the cube root of p.
11. y varies directly with x and inversely with the square of z.
12. x varies directly with the square root of a and inversely with b.

Write an equation that expresses the indicated variation and solve for k.

13. y varies directly with x, and $y = 12$ when $x = 3$.
14. y varies directly with the square of x, and $y = 16$ when $x = 2$.
15. M varies jointly with a and b, and $M = 38$ when $a = 19$ and $b = \frac{1}{2}$.
16. W varies jointly with p and q, and $W = 84$ when $p = 12$ and $q = 3$.
17. S varies inversely with the square of t, and $S = 2$ when $t = 5$.
18. P varies inversely with the square root of c, and $P = 7$ when $c = \frac{1}{4}$.
19. y varies directly with u and inversely with v, and $y = 18$ when $u = 9$ and $v = 7$.
20. V varies directly with the square of r, and $V = \dfrac{3}{\pi}$ when $r = 1$.

Write a variation equation for each of the following statements; find k and solve as indicated.

21. If y varies directly with x, and if $y = 14$ when $x = 2$, find y when $x = 4$.
22. If R varies directly with x, and if $R = 39$ when $s = 13$, find R when $s = 8$.
23. If S varies inversely with t, and if $S = 8$ when $t = 3$, find S when $t = 8$.

24. If Q varies inversely with w, and if $Q = 56$ when $w = 16$, find Q when $w = 2$.
25. If A varies directly with b^2, and if $A = 2$ when $b = \frac{1}{3}$, find A when $b = 4$.
26. If D varies directly with e^3, and if $D = 8$ when $e = 2$, find D when $e = 5$.
27. If y varies jointly with x and z, and if $y = 18$ when $x = 6$ and $z = 3$, find y when $x = 4$ and $z = 7$.
28. If y varies jointly with x^2 and z, and if $y = 68$ when $x = 4$ and $z = 2$, find y when $x = 1$ and $z = 7$.
29. If y varies jointly with x and z and inversely with w^2, and if $y = 56$ when $x = 5$, $z = 4$, and $w = 1$, find y when $x = 4$, $z = 5$, and $w = 2$.
30. If y varies jointly with x^2 and z^3 and inversely with the square root of w, and if $y = 81$ when $x = 3$, $z = 1$, and $w = 4$, find y when $x = 1$, $z = 2$, and $w = 16$.

Write a variation equation for each of the following statements; find k and solve as indicated.

31. The tension T of a spring varies directly with the distance d the spring is stretched. If the tension is 38 lb when the spring is stretched 4 in., find the tension when the spring is stretched 12 in.
32. The distance d a car travels, at a constant speed, varies directly with the time t. If a car travels 102 mi in 2 hr, how far will it travel in 5 hr?
33. The distance d an object falls from an airplane varies directly with the square of the time t it falls. If an object falls 128 ft in 2 sec, how far will it fall in 4 sec?
34. The weight W of a body varies inversely with the square of the distance d of the body from the center of the earth. If a woman weighs 120 lb 4000 miles from the center of the earth, what would she weigh 5000 miles from the center of the earth?
35. The weight of a certain wire varies directly with its length. If 500 ft of the wire weighs 40 lb, what does one mi of the wire weigh?
36. The amount of income a person receives varies jointly with the number of hours worked and the hourly wage. If Linda works 6 hr at $9/hr and receives $72, how much will she receive if she works 8 hr at $8/hr?
37. The length of a rectangle varies directly with the area and inversely with the width. If a rectangle with a length of 8 ft and a width of 4 ft has an area of 32 ft^2, find the length of a rectangle with a width of 5 ft and an area of 144 ft^2.
38. The width of a rectangle varies inversely with the length when the area is fixed. If the width is 9 cm when the length is 15 cm, find the width when the length is 25 cm.
39. The height of a triangle varies directly with its area and inversely with its base. If the height is 8 in. and the base is 6 in. when the area is 24 in^2, find the height when the base is 8 in. and the area is 26 in^2.
40. The area of a circle varies directly with the square of the radius. If the area of a circle is 49π m^2 when the radius is 7 m, find the area when the radius is $\frac{8}{\pi}$ m.
41. The volume of a sphere varies directly with the cube of the radius. If the volume of a sphere is $\frac{32\pi}{3}$ ft^3 when the radius is 2 ft, find the volume when the radius is 3 ft.
42. The time it takes for a car to travel a certain distance varies inversely with its speed. If a particular trip takes 4 hr when the car travels at 55 mph, how long will the same trip take when the car travels at 60 mph?
43. The rate of interest varies directly with the amount of interest earned and inversely with the principal (the amount invested). If $1000 invested for one year at 5% interest earns $80, what is the interest rate if $1000 invested for one year earns $72?
44. The volume of a gas varies inversely with the pressure on the gas. If the volume is 15 cm^3 when the pressure is 12 kg/cm^2, what is the volume when the pressure is 20 kg/cm^2?

Write each of the following variation equations in words.

45. $P = kmn$
46. $I = kprt$
47. $V = \dfrac{kx^2}{y^3 z}$
48. $T = \dfrac{kp^2 q}{v w^2}$
49. $V = kr^2 h$
50. $V = kab^2 c$

SUMMARY

Simplifying Rational Expressions

A **rational expression** is the quotient of two polynomials such that the denominator cannot equal zero. Before working with a rational expression, any value that makes the denominator zero should be excluded.

Rational expressions are simplified by applying the **fundamental property of rational expressions:**

If A, B, and C are polynomials, $B \neq 0$ and $C \neq 0$, then

$$\frac{A \cdot C}{B \cdot C} = \frac{A}{B} \quad \text{and} \quad \frac{A}{B} = \frac{A \cdot C}{B \cdot C}$$

The final simpification often results in equivalent forms as far as negative signs are involved:

$$\frac{-a}{b} = \frac{a}{-b} = -\frac{a}{b}, \quad b \neq 0$$

Any two of the three signs of a fraction can be changed without changing its value:

$$\frac{a}{b} = \frac{-a}{-b} = -\frac{-a}{b} = -\frac{a}{-b}, \quad b \neq 0$$

When a factor in the numerator is exactly the same as one in the denominator except for the signs of their terms, the quotient of the two factors is -1.

8.2 Multiplying and Dividing Rational Expressions

Multiplication and division of rational expressions is carried out in the same manner as multiplication and division of numerical fractions.

The **product of two rational expressions** $\frac{A}{B}$ and $\frac{C}{D}$ is

$$\frac{A}{B} \cdot \frac{C}{D} = \frac{A \cdot C}{B \cdot D}, \quad B, D \neq 0$$

The **quotient of two rational expressions** $\frac{A}{B}$ and $\frac{C}{D}$ is

$$\frac{A}{B} \div \frac{C}{D} = \frac{A}{B} \cdot \frac{D}{C} = \frac{A \cdot D}{B \cdot C}, \quad B, C, D \neq 0$$

Before multiplying rational expressions, they should be factored and any common factors divided out. If *all* terms divide out, the product is 1, not zero.

8.3 Adding and Subtracting Rational Expressions; The Least Common Denominator

Rational expressions are added and subtracted in the same manner as numerical fractions. If $\frac{A}{B}$ and $\frac{C}{B}$ are rational expressions with the same denominator, then

$$\frac{A}{B} + \frac{C}{B} = \frac{A + C}{B}, \quad B \neq 0$$

If two rational expressions do *not* have the same denominator, they must be written in terms of the **least common denominator,** or **LCD,** before addition can take place.

To find the least common denominator:

1. Factor each denominator completely.
2. The LCD is the product of each distinct prime factor the greatest number of times it occurs in any of the denominators.

Algebraic expressions in the least common denominator should be left in factored form.

To find the sum or difference of rational expressions:

1. If the denominators are not the same, find the LCD by factoring.
2. Write each rational expression as an equivalent one over the common denominator.
3. Find the sum or difference of the numerators and write the result over the common denominator.
4. Simplify if possible.

Finding a common denominator can sometimes be simplified when it is observed that two factors are additive inverses of each other, such as $x - a$ and $a - x = -1(x - a)$.

8.4 Complex Fractions

A **complex fraction** is one in which the numerator or denominator (or both) contains fractions.

There are two ways to simplify complex fractions. One involves writing both the numerator and the denominator as single fractions (if necessary) and then following the rules for dividing fractions. The second involves multiplying both the numerator and the denominator of the complex fraction by the LCD of all of the denominators. Which method is used is a matter of personal preference.

8.5 Solving Rational Equations

Rational equations are solved by a five-step process similar to the one used to solve linear equations. It is important to note any restrictions on the variable before solving.

To solve rational equations:

1. Note any restrictions on the variable, then clear the equation of fractions by multiplying each side by the least common denominator.
2. Remove grouping symbols using the distributive property.
3. Combine like terms on each side of the equation.
4. If the equation is linear (first degree), isolate the variable to find the solution. If the equation is not linear, set it equal to zero and solve it by factoring.
5. Check the answer(s) in the original equation.

When a proposed solution to an equation does not check, it is said to be **extraneous**.

8.6 Ratio and Proportion

A **ratio** is the comparison of two quantities by division. A **proportion** is a statement that two ratios are equal. In the proportion

$$\frac{a}{b} = \frac{c}{d}$$

a, b, c, and d are the **terms**, a and d are the **extremes**, and b and c are the **means**. In any proportion, *the product of the extremes is equal to the product of the means:*

$$ad = bc$$

It is important when working with proportions to make sure that the ratios on both sides of the equal sign compare the same quantities in the same order.

Similar triangles are those in which the measures of the corresponding angles are equal. *Corresponding sides of similar triangles are proportional.*

8.7 Variation

There are three types of variation:

Direct variation: $y = kx$

Inverse variation: $y = \dfrac{k}{x}$

Joint variation: $y = kxz$

The quantity k is called the **constant of variation** because its value never changes in a given equation.

Cooperative Exercise

The Mixter family has found a house they want to buy. The price of the house is $126,000 and requires a 15% down payment. You are the loan officer of their local bank and you're asked to furnish them with information concerning the following options for a loan. Your bank charges two points to make a loan and this is included in the amount of the loan. Use a computer or a calculator with an amortization program to do the following calculations.

OPTION I   A 30-YEAR FIXED-RATE LOAN AT 8.25% INTEREST

1. The monthly payment
2. Total payments over the 30-year period
3. Total interest paid over the 30-year period
4. Total principal paid over the 30-year period
5. What percentage of the total payments is paid out in interest?

OPTION II   A 15-YEAR FIXED-RATE LOAN AT 7.75% INTEREST

1. The monthly payment
2. Total payments over the 15-year period
3. Total interest paid over the 15-year period
4. Total principal paid over the 15-year period
5. What percentage of the total payments is paid out in interest?

OPTION III A 15-YEAR VARIABLE LOAN STARTING AT 6.00% WITH A MAXIMUM OF A 0.5% INCREASE PER YEAR; INTEREST TO BE INCREASED NOT MORE THAN SEVEN TIMES IN THE 15-YEAR PERIOD.

Assume the worst case: an interest increase of 0.5% each year for the first seven years.

1. The monthly payment (each year over the 15-year loan)
2. Total payments over the 15-year period
3. Total interest paid over the 15-year period
4. Total principal paid over the 15-year period
5. What percentage of the total payments is paid out in interest?

IV SUMMARY AND COMPARISONS

1. Compare the payments for the three different loans (each year).
2. Compare the total interest paid for each of the three loans.
3. Compare the total payments paid for each of the three loans.
4. Based upon this information, make comparisons of the three loans.
5. Which of the three loans would you suggest for the family?

Review Exercises

Simplify each of the following rational expressions.

1. $\dfrac{280}{-56}$

2. $\dfrac{5}{x} \cdot \dfrac{x^3}{30}$

3. $\dfrac{-91}{26} \cdot \dfrac{39}{15} \cdot \dfrac{45}{65}$

4. $\left(\dfrac{-28}{5x^2} \div \dfrac{7}{20x} \right) \cdot \dfrac{15x^3}{35}$

5. $\dfrac{2a^2}{a^2+a^3} \div \dfrac{7a^2}{1+a}$

6. $\dfrac{20x^2y}{-ab^2} \div \dfrac{15xy^2}{-a^3b^2}$

7. $\dfrac{y}{(y+3)^2} \cdot \dfrac{(y+2)}{(y-4)} \cdot \dfrac{(y+3)^3}{(y+2)}$

8. $\dfrac{-6(a-b)^2}{(a+b)^2} \div \dfrac{(a-b)^3}{-8(a^2-b^2)}$

9. $\dfrac{x^2+11x+18}{x^2-81} \cdot \dfrac{x^2-7x-18}{x^2-4}$

10. $(30y-40) \div \dfrac{3y-4}{20(y^2-16)}$

11. $\dfrac{x^3-8}{x^2+2x+4} \cdot \dfrac{-1}{x-2}$

12. $\dfrac{(3x-4)^2(x+2)^3(5x-1)^4}{(5x-1)^3(3x-4)^3(x+2)^2}$

13. $\dfrac{x^2-4}{-(4-x^2)(x+2)}$

14. $\left(\dfrac{3x^2}{8y^3} \cdot \dfrac{-5y}{6x}\right) \div \left(\dfrac{-8y}{9x} \cdot \dfrac{4x^3}{3y^2}\right)$

15. $\dfrac{x^2-3x-4}{x^2-16} \div \dfrac{x+1}{x^2-5x+4}$

16. $\dfrac{cx+dx-cy-dy}{2cx-cy+2dx-dy}$

17. $\dfrac{x^3+3x^2+2x}{3x^4+9x^3+6x^2}$

18. $\dfrac{x^2-y^2}{x-y} \cdot \dfrac{x}{x+y} \cdot \dfrac{x-y}{x}$

Find the sum or difference and simplify.

19. $\dfrac{1}{5}+\dfrac{2}{5}+\dfrac{4}{5}$

20. $\dfrac{x-3}{5x}+\dfrac{x+4}{5x}$

21. $\dfrac{-2}{3y-4}+\dfrac{4x-7}{3y-4}$

22. $\dfrac{5}{x^2-9}+\dfrac{2}{3-x}$

23. $\dfrac{3}{x}-\dfrac{1}{4x-x^2}$

24. $\dfrac{5}{x-3}+\dfrac{3x-2}{(x-3)(x+2)}$

25. $\dfrac{1}{x^2-16}-\dfrac{1}{x^2-8x+16}$

26. $\dfrac{4}{x-7}-\dfrac{2x-1}{x^2-7x}-\dfrac{3}{x}$

27. $4x-3-\dfrac{3x+1}{x}-\dfrac{x^2}{5}$

28. $\dfrac{x+1}{x-3}-\dfrac{x+2}{x+3}-\dfrac{5x-2}{x^2-9}$

Simplify each complex fraction.

29. $\dfrac{\dfrac{3}{4}}{\dfrac{9}{24}}$

30. $\dfrac{\dfrac{-1}{5}}{\dfrac{7}{-21}}$

31. $\dfrac{\dfrac{13}{y^3}}{\dfrac{39}{y^2}}$

32. $\dfrac{\dfrac{1}{5}+2}{\dfrac{7}{10}+2}$

33. $\dfrac{\dfrac{1}{y}-6}{\dfrac{2}{y}+2}$

34. $\dfrac{\dfrac{x-y}{x}}{\dfrac{1}{x}-\dfrac{1}{y}}$

35. $\dfrac{x^2-7x}{x-\dfrac{49}{x}}$

36. $\dfrac{25-\dfrac{1}{m^2}}{5-\dfrac{1}{m}}$

37. $\dfrac{\dfrac{1}{x}+\dfrac{1}{y}}{\dfrac{x+y}{11}}$

38. $\dfrac{\dfrac{a}{a-b}}{\dfrac{a^2}{a^2-b^2}}$

39. $\dfrac{t+1-\dfrac{1}{t-1}}{t-1-\dfrac{1}{t-1}}$

40. $\dfrac{\dfrac{m}{m-n}-\dfrac{n}{m+n}}{\dfrac{n}{m-n}+\dfrac{m}{m+n}}$

Solve each rational equation and check all solutions.

41. $\dfrac{x}{5}=\dfrac{1}{2}$

42. $\dfrac{1}{4}-\dfrac{2}{5}=\dfrac{y}{3}$

43. $\dfrac{2}{x}=\dfrac{4}{7}$

44. $\dfrac{7}{2y}=0$

45. $\dfrac{15}{3x}-\dfrac{2}{7}=\dfrac{5}{x}$

46. $\dfrac{x-3}{x+6}+5=0$

47. $\dfrac{3y}{y-4}-2=\dfrac{3}{4-y}$

48. $\dfrac{1}{5}-\dfrac{1}{2y}+\dfrac{1}{2}=\dfrac{5}{10y}$

49. $2=\dfrac{8x-5}{4x+3}$

50. $\dfrac{3}{t+2}-\dfrac{20t-1}{2t^2+6t+4}+\dfrac{7}{2t+2}=0$

51. $\dfrac{6}{m-3}-9=\dfrac{-9m}{m-3}$

Write each ratio as a common fraction and simplify it.

52. 38 to 49

53. $3\tfrac{1}{2}$ to $5\tfrac{1}{2}$

54. 286 miles to 8 hours

Solve each proportion.

55. $\dfrac{y}{5}=\dfrac{7}{9}$

56. $\dfrac{x-1}{x}=\dfrac{3}{8}$

57. $\dfrac{3y}{22} = \dfrac{-1}{7}$

58. $\dfrac{7p + 5}{3p - 2} = \dfrac{-2}{11}$

59. $\dfrac{1}{y + 1} = \dfrac{y}{2}$

60. $\dfrac{2t}{t - 3} = \dfrac{t + 3}{4}$

Write each of the following statements as a variation equation; find k and solve as indicated.

61. If y varies directly with the square of x, and if $y = 15$ when $x = 4$, find y when $x = 2$.

62. The area of a circle varies directly with the square of the radius. If the area is 121π ft² when the radius is 11 ft, find the area when the radius is 13 ft.

63. The height of a triangle with a fixed area varies inversely with the base. If the height is 12 m when the base is 2 m, find the height when the base is $5\tfrac{1}{2}$ m.

Write each of the following statements as a rational equation and solve it. Check your answers.

64. Six more than one-third of a number is 5.

65. When the numerator of $\tfrac{3}{4}$ is increased by a certain number and the denominator is multiplied by the same number, the resulting fraction is $\tfrac{2}{3}$. Find the number.

66. Dave can mow, edge, and trim his yard in 5 hr. If Celeste helps him, the job takes $2\tfrac{1}{2}$ hr. How long does it take Celeste to do the job alone?

67. Matt left Pismo Beach at 7:00 A.M. traveling north at 45 mph. At 9:00 A.M. Jill left Pismo Beach to catch up with him. If she travels at 60 mph, at what time will she overtake him?

Chapter Test

Simplify each fraction.

1. $\dfrac{162}{270}$

2. $\dfrac{-18x^3y^2}{48x^2y^2}$

3. $\dfrac{6a + 8}{9a^2 - 16}$

4. $\dfrac{3x^2 + 18x - 48}{2x^2 + 20x + 32}$

5. $\dfrac{m^2 - 7m}{49m - m^3}$

6. $\dfrac{6x^2 + 5xy - 6y^2}{6x^2 + 17xy + 12y^2}$

7. $\dfrac{9p - 3}{3}$

8. $\dfrac{15a^2b - 24ab^2 - 3ab}{3ab}$

9. $\dfrac{y^2 + 4y + 3}{y + 3}$

10. $\dfrac{4x^2 + 2x - 11}{2x - 3}$

Multiply or divide as indicated.

11. $\dfrac{2y}{x} \cdot \dfrac{x}{4y}$

12. $\dfrac{x}{x + 1} \div \dfrac{x + 2}{2x + 4}$

13. $\dfrac{ab - 2b^2}{ab - b^2} \cdot \dfrac{a^2 - ab}{a^2 - 2ab}$

14. $\dfrac{y^2 - y - 6}{y^2 - 4} \div \dfrac{y^2 - 2y - 3}{y^2 - y - 2}$

Add or subtract as indicated.

15. $\dfrac{8}{3y} - \dfrac{x}{4y}$

16. $\dfrac{9}{x + 3} + \dfrac{3x}{x + 3}$

17. $\dfrac{m + 1}{m} + 2 + \dfrac{m - 1}{m}$

18. $\dfrac{3x}{x^2 - 7x + 6} - \dfrac{3x}{x^2 - 4x - 12}$

Simplify.

19. $\dfrac{\dfrac{2}{x} + x}{\dfrac{3}{x^2} + 5}$

20. $\dfrac{1 - \dfrac{5}{x} - \dfrac{6}{x^2}}{\dfrac{1}{x} - \dfrac{6}{x^2}}$

Solve each equation and check your answers. Indicate any values of the variable that must be excluded.

21. $\dfrac{7}{3x} = \dfrac{2x - 1}{12x} - \dfrac{5}{4x}$

22. $\dfrac{2}{a + 5} + \dfrac{3}{a - 4} = \dfrac{-18}{a^2 + a - 20}$

Write each of the following applied problems as a rational equation, solve it, and check it.

23. David and Ruth can do the dinner dishes together in 20 min. If Ruth can do them alone in 30 min, how long does it take David working alone?

24. Ken can fly his airplane 525 km against the wind in the same time that it takes him to fly 600 km with the wind. If the speed of the wind is 20 km/hr, what is the speed of the airplane in still air?

Write each ratio as a common fraction and simplify it.

25. The ratio of 1450 women to 870 men

26. The ratio of 13.2 lb to 6 kg

27. The ratio of 700 mi to 12 hr

28. The price ratio of items selling for 40¢ to items selling for $1

Use proportions to solve each of the following applied problems. Check your answers.

29. If Emma's ratio of grade points to the total number of units she is carrying is 7 to 2, and if she has completed 60 units, how many grade points does she have?

30. If the price-earnings ratio for a common stock is 5 to 2 and the stock earns $36/share, what is the price of the stock?

Write each of the following applied problems as a variation equation; find the constant of variation and solve as indicated.

31. The volume of a right circular cylinder varies jointly with the height and the square of the radius. If the volume is 64π in^2 when the height is 4 in. and the radius is 4 in., find the volume when the height is 3 in. and the radius is 3 in.

32. The base of a triangle varies directly with the area and inversely with the height. If the base is 8 m when the area is 16 m^2 and the height is 4 m, find the base when the area is 24 m^2 and the height is 2 m.

CHAPTER 9
ROOTS AND RADICALS

9.1 Roots and Radicals
9.2 Simplified Forms of Radicals
9.3 Addition and Subtraction of Radicals
9.4 Multiplication and Division of Radicals
9.5 Solving Equations Containing Radicals
9.6 Rational Exponents and Radicals

Application

A softball diamond is in the shape of a square with 60-foot sides. The catcher stands 3 feet behind home plate and throws to second base. Give both an exact and an approximate answer, rounded to three decimal places, of the distance the ball is thrown.

Exercise Set 9.6, Exercise 110

Mathematics, rightly viewed, possesses not only truth but supreme beauty, a beauty cold and austere, like that of a sculpture. It has no appeal to our weaker nature, not gorgeous like a painting or sculpture, yet sublimely pure, and capable of requiring a stern perfection as only a great artist can show.

BERTRAND RUSSELL (1872–1970), Mathematician and Philosopher

9.1 Roots

HISTORICAL NOTE One of the reasons that history does not credit many women mathematicians is because it was very difficult or even impossible for a woman to attend a university, let alone obtain the support of her family if she wanted to study mathematics. What women learned, they usually had to learn on their own. They often had to read mathematics books on the sly (when they could obtain them), waiting until their families went to bed and then studying in their rooms, often without heat, using candles they had hidden for this purpose. Sophie Germain (1776–1831), a French mathematician, was forced to use a man's name to get her works published. Some of them were so high-quality that she won the Grand Prix from the French Acadamy for her paper, *Memoir of the Vibrations of Elastic Plates*. Other notables are the Englishwoman Mary Somerville (1780–1872) and the Russian mathematician Sofya Kovalevskaya (1850–1891), who moved to Germany to get an education and eventually became the first woman to teach in a European university.

OBJECTIVES In this section we will learn to
1. recognize the parts of a radical expression;
2. determine whether a root is rational, irrational, or not a real number;
3. use a calculator to find the square root of a number; and
4. find roots of expressions containing variables.

In the earlier chapters of this book we worked with many types of algebraic expressions and used them in a variety of applications. The solutions to the problems were always rational numbers. By learning how to work with roots and radicals, we will be able to solve a wide variety of equations that we have previously avoided.

Recall that the square of a number is the product of the number with itself. For example,

$$3^2 = 3 \cdot 3 = 9, \qquad (-4)^2 = (-4)(-4) = 16, \qquad \text{and} \qquad \left(\frac{2}{3}\right)^2 = \frac{2}{3} \cdot \frac{2}{3} = \frac{4}{9}$$

The opposite of squaring a number is finding the *square root*.

DEFINITION **Principal Square Root**	The **principal square root** of any nonnegative number a, written $\sqrt{a}$, is that nonnegative number b such that $b^2 = a$.

EXAMPLE 1   Find the square roots: (a) $\sqrt{25}$ (b) $\sqrt{16}$ (c) $\sqrt{9}$

SOLUTION   (a) $\sqrt{25} = 5$ because $5^2 = 25$.
(b) $\sqrt{16} = 4$ because $4^2 = 16$.
(c) $\sqrt{9} = 3$ because $3^2 = 9$.

The symbol "$\sqrt{}$" is called a **radical sign;** it always indicates a *positive* square root of a number. The number under the radical sign is called the **radicand,** and the entire expression is called a **radical expression.**

Radical sign → $\underbrace{\sqrt{49}}_{\text{Radical expression}}$ ← Radicand

Every positive number has two real square roots, one positive and one negative. Zero has only one square root: itself. The positive (principal) square root of a number is indicated simply by the symbol $\sqrt{}$. Thus,

$\sqrt{49} = 7$ because $7^2 = 49$

To indicate the *negative* square root of a number, we use the symbol $-\sqrt{}$. For example, to indicate the negative square root of 49, we write $-\sqrt{49}$, and

$-\sqrt{49} = -7$

EXAMPLE 2   Find each square root: (a) $\sqrt{81}$ (b) $-\sqrt{36}$ (c) $\sqrt{400}$ (d) $-\sqrt{\dfrac{9}{16}}$

SOLUTION   (a) The radical expression $\sqrt{81}$ indicates the positive square root, so

$\sqrt{81} = 9$ because $9 \cdot 9 = 81$

(b) The radical expression $-\sqrt{36}$ indicates the negative square root, so

$-\sqrt{36} = -6$

(c) $\sqrt{400} = 20$

(d) $-\sqrt{\dfrac{9}{16}} = -\dfrac{3}{4}$

Numbers such as 4 and $\dfrac{9}{25}$ are called **perfect squares** because they are the squares of rational numbers. Numbers such as 2, 3, and 11 are not perfect squares, since they are not the squares of rational numbers. The square roots of such numbers are **irrational.** Recall from Chapter 2 that irrational numbers, when written in decimal form, neither terminate nor repeat. In general,

The square root of any non-negative number that is not a perfect square is an irrational number.

EXAMPLE 3   Classify each of the following expressions as rational or irrational:

(a) $\sqrt{144}$ (b) $\sqrt{23}$ (c) $\sqrt{35}$ (d) $-\sqrt{\dfrac{4}{25}}$

SOLUTION   (a) $\sqrt{144} = 12$ because $12^2 = 144$, so $\sqrt{144}$ is rational.
(b) $\sqrt{23}$ is irrational because 23 is not a perfect square.
(c) $\sqrt{35}$ is irrational because 35 is not a perfect square.
(d) $-\sqrt{\dfrac{4}{25}}$ is rational because $\left(-\dfrac{2}{5}\right)^2 = \dfrac{4}{25}$.

Calculator Note

Modern technology has eliminated the need to use tables to find square roots or to go through the laborious process of calculating them by hand. It is also now possible to achieve much greater accuracy with very little effort. The keystrokes necessary to find the square root of a number depend on the type and model of calculator you have. The "$\sqrt{}$" symbol may be found on the face of a key itself or above or below one of the keys. If it is above or below the key, depress the $\boxed{\text{Inv}}$ or $\boxed{\text{2nd}}$ key before depressing the $\boxed{\sqrt{}}$ key. The Hewlett-Packard calculators have color-coded keys that are used to obtain second functions.

To find $\sqrt{5}$, use the following keystrokes.

Using a Scientific Calculator

Many general brands

5   $\boxed{\sqrt{}}$ The display reads 2.236067978.

Texas Instruments TI-36

5   $\boxed{\text{Inv}}$ $\boxed{\sqrt{}}$ The display reads 2.236067978.

Hewlett-Packard series

5   $\boxed{\text{ENTER}}$ $\boxed{\text{f}}$ $\boxed{\sqrt{}}$ The display reads 2.236067977.

Using a Graphics Calculator

Texas Instruments TI-81

$\boxed{\text{2nd}}$ $\boxed{\sqrt{}}$ 5 $\boxed{\text{ENTER}}$ The display reads 2.236067978.

Sharp EL-5200

$\boxed{\sqrt{}}$ 5 $\boxed{=}$ The display reads 2.236067977.

Consult your owner's manual for special instructions for your calculator.

EXAMPLE 4 Use a calculator to find each of the following square roots, rounding the answers to three decimal places: **(a)** $\sqrt{17}$ **(b)** $\sqrt{14.59}$ **(c)** $\sqrt{2984.859}$

SOLUTION (a) $\sqrt{17} \approx 4.123$ The symbol "$\approx$" means "is approximately equal to."
(b) $\sqrt{14.59} \approx 3.820$
(c) $\sqrt{2984.859} \approx 54.634$

Recall that in our definition of square root, we required that the radicand be a nonnegative number. The reason is that not every number has a square root that is a real number. For example, $\sqrt{-9}$ does not have a real-number square root, since

$$\sqrt{-9} \neq 3 \quad \text{because} \quad 3 \cdot 3 \neq -9$$

and

$$\sqrt{-9} \neq -3 \quad \text{because} \quad (-3)(-3) \neq -9$$

We will discuss square roots of negative numbers in Section 10.4.

There are roots other than square roots. For example, the *cube root* of a number a is written $\sqrt[3]{a}$, the *fourth root* of a is written $\sqrt[4]{a}$, and the *n*th root of a is written $\sqrt[n]{a}$. The small number on the radical symbol is called the **index**, and it indicates which root we are finding.

Index ⟶ $\sqrt[n]{a}$ ⟵ Radicand
         ⟵ Radical sign

> The **nth root** of a number b, written $\sqrt[n]{b}$, is that number a such that $a^n = b$.

In simplest terms, the cube root of a number is a number that, when cubed, is the given number. The fourth root of a number is a number that, when raised to the fourth power, is the given number. And the *n*th root of a number is a number that, when raised to the *n*th power, is the given number.

Using this notation, we could write $\sqrt{b}$ as $\sqrt[2]{b}$. However, in all future work, when no index is shown it is understood to be 2.

EXAMPLE 5 Find $\sqrt[3]{8}$.

SOLUTION $\sqrt[3]{8} = 2 \quad \text{because} \quad 2 \cdot 2 \cdot 2 = 2^3 = 8$

EXAMPLE 6 Find $\sqrt[3]{-27}$.

SOLUTION $\sqrt[3]{-27} = -3 \quad \text{because} \quad (-3)(-3)(-3) = (-3)^3 = -27$

Notice that the *cube* root of a negative number is a negative real number and the cube root of a positive number is a positive real number. In fact,

> Any odd root of a negative number is negative, and an even root of a negative number is not a real number.

EXAMPLE 7   Find each of the following roots:
(a) $\sqrt{4}$ and $-\sqrt{4}$ (b) $\sqrt[3]{8}$ and $\sqrt[3]{-8}$;   (c) $\sqrt[4]{16}$ and $-\sqrt[4]{16}$
(d) $\sqrt[3]{\frac{8}{27}}$ and $-\sqrt[3]{\frac{8}{27}}$ and $\sqrt[3]{-\frac{8}{27}}$

SOLUTION   (a) $\sqrt{4} = 2$ and $-\sqrt{4} = -2$
(b) $\sqrt[3]{8} = 2$ and $\sqrt[3]{-8} = -2$
(c) $\sqrt[4]{16} = 2$ and $-\sqrt[4]{16} = -2$
(d) $\sqrt[3]{\frac{8}{27}} = \frac{2}{3}$ and $-\sqrt[3]{\frac{8}{27}} = -\frac{2}{3}$ and $\sqrt[3]{-\frac{8}{27}} = -\frac{2}{3}$

Calculator Note

Calculators can also be used to find higher roots of numbers, such as cube roots, fourth roots, and so on. Some calculators have a special key devoted to finding cube roots but that is generally not the case. We will show how it is done on a scientific calculator. Your calculator may vary. Consult your owner's manual.

To find $\sqrt[3]{12}$ on most scientific calculators, the keystroke sequence is

12 [y^x] [(] 1 [÷] 3 [)] [=]   The display reads 2.289428485.

To find $\sqrt[4]{123}$, the necessary keystroke sequence is

123 [y^x] [(] 1 [÷] 4 [)] [=]   The display reads 3.330245713.

In general, to find any root of a number, the keystroke sequence is

The number [y^x] [(] 1 [÷] n [)] [=]
↑ The index of the desired root

EXAMPLE 8   Use a calculator to find each of the following roots using one of the sequences of keystrokes just given:
(a) $\sqrt[5]{17.38}$ (b) $\sqrt[3]{-37.92}$ to the nearest ten-thousandth   (c) $\sqrt{64}$

SOLUTION   (a) $\sqrt[5]{17.38}$: the number is 17.38 and $n = 5$.

$\sqrt[5]{17.38} \approx 1.770149531$

(b) $\sqrt[3]{-37.92}$: the number is -37.92 and $n = 3$.

$\sqrt[3]{-37.92} \approx -3.359614468$

≈ -3.3596 To the nearest ten-thousandth

(c) $\sqrt{64}$: the number is 64 and n is 2.

$\sqrt{64} = 8$

In parts (a), (b), and (c) of Example 7, each of the radicands was a perfect power of some integer. This observation provides us with a means of simplifying radicals containing variables. For example,

$\sqrt[2]{4} = \sqrt[2]{2^2} = 2$ $\sqrt[3]{-8} = \sqrt[3]{(-2)^3} = -2$

$\sqrt[4]{16} = \sqrt[4]{2^4} = 2$ $\sqrt[3]{\dfrac{8}{27}} = \sqrt[3]{\left(\dfrac{2}{3}\right)^3} = \dfrac{2}{3}$

> When the index and the exponent on the radicand are the same, the radical simplifies to the base in the radicand, provided the base is nonnegative when the index is even.

The index and the exponent are the same.

$\sqrt[n]{b^n} = b$ ← base number; $b \geq 0$ when n even

*C*AUTION

The square root of a sum of two numbers is **not** the sum of the square roots of the two numbers. In other words,

$\sqrt{a^2 + b^2} \neq a + b$

If it were true, then

$\sqrt{25} = \sqrt{9 + 16}$
$= \sqrt{3^2 + 4^2}$
$= 3 + 4 = 7$

But we know that

$\sqrt{25} = 5 \neq 7$.

EXAMPLE 9 Find each of the indicated roots. Assume all variables represent nonnegative numbers.

(a) $\sqrt[2]{x^2}$ (b) $\sqrt[3]{x^6}$ (c) $\sqrt{9a^2b^4}$ (d) $\sqrt[3]{\dfrac{27a^3}{64b^3}}$

SOLUTION (a) $\sqrt{x^2} = x$ The index is understood to be 2. The index and the exponent are the same; the base is x.

(b) $\sqrt[3]{x^6} = \sqrt[3]{(x^2)^3}$ Think of $x^6 = (x^2)^3$.

$= x^2$ The index and the exponent are the same; the base is x^2

(c) $\sqrt{9a^2b^4} = \sqrt{(3ab^2)^2}$

$= 3ab^2$ The index and the exponent are the same; the base is $3ab^2$.

(d) $\sqrt[3]{\dfrac{27a^3}{64b^6}} = \sqrt[3]{\left(\dfrac{3a}{4b^2}\right)^3} = \dfrac{3a}{4b^2}$ The index and the exponent are the same; the base is $\dfrac{3a}{4b^2}$.

EXAMPLE 10 Simplify: $\sqrt{x^2 + 10x + 25}$, $x \geq 0$.

SOLUTION Since $x^2 + 10x + 25$ is the square of $x + 5$, we can write

$\sqrt{x^2 + 10x + 25} = \sqrt{(x + 5)^2}$ The index and the exponent are the same; the base is $x + 5$.

$= x + 5$

9.1 · ROOTS   447

EXAMPLE 11 A 17-foot ladder is leaned against the side of a two-story building. The bottom of the ladder is 8 feet from the wall. How high does the ladder reach up the wall? See Figure 9.1.

SOLUTION The ladder and the wall together with the ground form a right triangle such that the ladder is the hypotenuse. The Pythagorean theorem applies in this situation. We let x = the height the ladder reaches up the wall. Then

$$17^2 = x^2 + 8^2$$
$$289 = x^2 + 64$$
$$225 = x^2$$

Then $x = 15$ since $15^2 = 225$. The ladder will reach 15 feet up the wall.

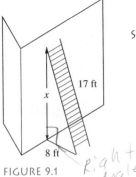
FIGURE 9.1

Do You Remember?

Can you match these?

____ 1. $\sqrt{}$
____ 2. The index of $\sqrt[4]{a}$
____ 3. The term under the radical symbol
____ 4. $\sqrt[4]{16}$
____ 5. $\sqrt[3]{27} = 3$ because ____.
____ 6. $\sqrt[3]{-125}$
____ 7. $\sqrt{8}$
____ 8. $\sqrt{-4}$

a) not a real number
b) -4
c) 2
d) $\sqrt[3]{8}$
e) 5
f) the radicand
g) radical symbol
h) $3^3 = 27$
i) -1
j) 4
k) -5

Answers: 1. g 2. j 3. f 4. c 5. h 6. k 7. d 8. a

Exercise Set 9.1

Find the indicated root. If no real-number root exists, write Not a real number.

1. $\sqrt{4}$
2. $\sqrt{9}$
3. $\sqrt{16}$
4. $\sqrt{25}$
5. $\sqrt{1}$
6. $\sqrt{36}$
7. $\sqrt{81}$
8. $\sqrt{64}$
9. $\sqrt{49}$
10. $-\sqrt{1}$
11. $-\sqrt{9}$
12. $-\sqrt{121}$
13. $-\sqrt{100}$
14. $\sqrt[3]{1}$
15. $\sqrt[3]{8}$
16. $\sqrt[3]{27}$
17. $\sqrt[3]{64}$
18. $\sqrt[3]{-8}$
19. $\sqrt[3]{-1}$
20. $\sqrt[3]{-64}$
21. $\sqrt[3]{-27}$
22. $\sqrt[4]{16}$
23. $-\sqrt[4]{1}$
24. $-\sqrt[4]{81}$
25. $\sqrt[4]{256}$
26. $\sqrt[4]{-1}$
27. $\sqrt[4]{-4}$
28. $\sqrt{-9}$
29. $\sqrt{-16}$
30. $\sqrt[4]{-1}$
31. $\sqrt[4]{-16}$
32. $\sqrt{\dfrac{1}{4}}$
33. $\sqrt{\dfrac{4}{9}}$
34. $\sqrt{\dfrac{9}{25}}$
35. $-\sqrt{\dfrac{16}{25}}$
36. $-\sqrt{\dfrac{25}{49}}$

37. $\sqrt[3]{\dfrac{1}{8}}$ 38. $\sqrt[3]{\dfrac{8}{27}}$ 39. $\sqrt[4]{\dfrac{1}{16}}$

40. $\sqrt[4]{\dfrac{1}{81}}$

Find the indicated root. Assume that all variables represent positive real numbers.

41. $\sqrt{x^2}$ 42. $\sqrt{y^2}$ 43. $\sqrt{x^4}$
44. $\sqrt{y^6}$ 45. $\sqrt{t^{12}}$ 46. $\sqrt{p^8}$
47. $\sqrt[3]{x^3}$ 48. $\sqrt[3]{y^3}$ 49. $-\sqrt{x^6}$
50. $-\sqrt{y^4}$ 51. $\sqrt{x^{12}}$ 52. $\sqrt{y^{20}}$
53. $-\sqrt{m^{16}}$ 54. $-\sqrt{p^{36}}$ 55. $\sqrt[3]{x^6}$
56. $\sqrt[3]{y^{12}}$ 57. $\sqrt[3]{-m^{18}}$ 58. $\sqrt[3]{-t^{24}}$
59. $\sqrt[4]{y^8}$ 60. $\sqrt[4]{x^{20}}$ 61. $\sqrt{81p^6}$
62. $\sqrt{144m^{28}}$ 63. $\sqrt[3]{27a^{51}}$ 64. $\sqrt[3]{125b^{63}}$
65. $\sqrt{49y^2}$ 66. $\sqrt{64x^2}$ 67. $-\sqrt{121t^4}$
68. $-\sqrt{36m^6}$ 69. $\sqrt{4x^2y^2}$ 70. $\sqrt{9a^4b^2}$

71. $\sqrt{81m^4n^4p^2}$ 72. $\sqrt{x^2y^4z^{10}}$ 73. $\sqrt{\dfrac{9x^2}{16y^2}}$

74. $\sqrt{\dfrac{25a^4}{36b^2}}$ 75. $\sqrt[3]{\dfrac{8x^3}{y^3}}$ 76. $\sqrt[3]{\dfrac{a^6}{27b^9}}$

77. $\sqrt[4]{\dfrac{16t^8}{s^4}}$ 78. $\sqrt[4]{\dfrac{x^4}{81y^8}}$

79. $\sqrt{x^2 + 2x + 1}$ 80. $\sqrt{x^2 - 4x + 4}$
81. $\sqrt{x^2 + 4x + 4}$ 82. $\sqrt{x^2 + 6x + 9}$
83. $\sqrt{x^2 + 10x + 25}$ 84. $\sqrt{x^2 + 12x + 36}$

 *Use a calculator to approximate the indicated root. Round your answers to three decimal places.*

85. $\sqrt{3}$ 86. $\sqrt{2}$ 87. $\sqrt{12.5}$
88. $\sqrt{15.8}$ 89. $-\sqrt{2}$ 90. $-\sqrt{3}$
91. $\sqrt{8.6}$ 92. $\sqrt{5.4}$ 93. $\sqrt{4.75}$
94. $\sqrt{17.1}$ 95. $\sqrt{84.69}$ 96. $\sqrt{57.02}$

Find x in each of the given figures.

97. Right triangle

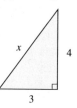

98. Right triangle

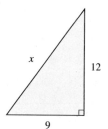

99. Square with area 144 ft²

100. Square with area 225 m²

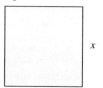

WRITE IN YOUR OWN WORDS

101. Describe $\sqrt[3]{8}$ in words.
102. Describe $\sqrt[n]{a}$ in words.
103. Define the principal square root.
104. In $\sqrt[n]{a}$, the n is called the _____ and the a is called the _____.
105. The square root of a positive real number that is not a perfect square is _____.
106. The square root of a negative real number is not a _____ number.

9.2 Simplified Forms of Radicals

HISTORICAL COMMENT   The notation used to indicate roots has evolved over a period of hundreds of years and was not always as convenient to write as the notation that we use today. A few of the early methods of indicating roots that bear definite similarities to modern notation are shown here.

Year	Square Root	Cube Root	Year	Square Root	Cube Root
1521	√	∿	1585	√	√③
1553	↙	≳	1707	√	√3

Before the bar at the top of the radical sign was added, various methods were used to show what terms the radical affected. In 1585, Simon Steven (Belgium, 1546–1620) used √9 2 to mean $\sqrt{9x^2}$ and √9)2 to mean $\sqrt{9}x^2$.

OBJECTIVES In this section we will learn to
1. multiply and divide radicals;
2. simplify radicals by writing them as products;
3. simplify fractional expressions involving radicals in the denominator; and
4. simplify radicals involving fractions in the radicand.

In order to simplify radicals, we must first introduce two new properties: *the multiplication property of radicals* and *the division property of radicals*.

The Multiplication Property of Radicals

If a and b are nonnegative real numbers, then

$$\sqrt[n]{a \cdot b} = \sqrt[n]{a} \cdot \sqrt[n]{b}$$

In words, *the multiplication property of radicals states that the nth root of the product of two (or more) numbers is the product of the nth roots of the numbers.*

As an illustration of this property, consider $\sqrt{144}$, which is the product of two perfect squares:

$\sqrt{144} = \sqrt{9 \cdot 16}$ $144 = 9 \cdot 16$

$12 = \sqrt{9 \cdot 16}$ The *principal* square root of 144 is 12.

$12 = \sqrt{9} \cdot \sqrt{16}$ The multiplication property of radicals

$12 = 3 \cdot 4$ The principal square root of 9 is 3; the principal square root of 16 is 4.

$12 = 12$

EXAMPLE 1 Use the multiplication property of radicals to simplify each of the following radicals: **(a)** $\sqrt{36}$ **(b)** $\sqrt{100}$ **(c)** $\sqrt[3]{216}$

SOLUTION   **(a)** $\sqrt{36} = \sqrt{9 \cdot 4} = \sqrt{9} \cdot \sqrt{4} = 3 \cdot 2 = 6$
(b) $\sqrt{100} = \sqrt{25 \cdot 4} = \sqrt{25} \cdot \sqrt{4} = 5 \cdot 2 = 10$
(c) $\sqrt[3]{216} = \sqrt[3]{8 \cdot 27} = \sqrt[3]{8} \cdot \sqrt[3]{27} = 2 \cdot 3 = 6$

> **The Division Property of Radicals**
>
> If a and b are nonnegative real numbers, then
>
> $$\sqrt[n]{\frac{a}{b}} = \frac{\sqrt[n]{a}}{\sqrt[n]{b}}, \quad b \neq 0$$

In words, *the division property of radicals states that the nth root of the quotient of two numbers is the quotient of the nth root of the numbers.*

As an illustration of the division property, consider $\sqrt{16}$, which is the quotient of two perfect squares:

$\sqrt{16} = \sqrt{\dfrac{144}{9}} \qquad 16 = \dfrac{144}{9}$

$4 = \sqrt{\dfrac{144}{9}} \qquad$ The principal square root of 16 is 4.

$4 = \dfrac{\sqrt{144}}{\sqrt{9}} \qquad$ The division property of radicals

$4 = \dfrac{12}{3} \qquad$ The principal square root of 144 is 12; the principal square root of 9 is 3.

$4 = 4$

EXAMPLE 2 Use the division property of equality to simplify: (a) $\sqrt{\dfrac{36}{9}}$ (b) $\sqrt{\dfrac{16}{9}}$

SOLUTION (a) $\sqrt{\dfrac{36}{9}} = \dfrac{\sqrt{36}}{\sqrt{9}} = \dfrac{6}{3} = 2$

Also, $\sqrt{\dfrac{36}{9}} = \sqrt{4} = 2$

(b) $\sqrt{\dfrac{16}{9}} = \dfrac{\sqrt{16}}{\sqrt{9}} = \dfrac{4}{3}$

We are now ready to use the properties of radicals to write them in *simplest form*. We will use illustrations with square roots before applying the properties to higher roots. **For a radical to be in simplest form, three conditions must be met.**

CONDITION 1	A square root radical is in **simplest form** if the radicand does not contain any perfect-square factors.

9.2 · SIMPLIFIED FORMS OF RADICALS

EXAMPLE 3 Simplify: $\sqrt{12}$.

SOLUTION Since $\sqrt{12}$ can be written as $\sqrt{4 \cdot 3}$ and since $4 = 2^2$ is a perfect square, the radical is not in simplest form.

$$\sqrt{12} = \sqrt{4 \cdot 3}$$
$$= \sqrt{4}\,\sqrt{3} \qquad \text{Multipication property of radicals}$$
$$= 2\sqrt{3} \qquad \sqrt{4} = 2$$

EXAMPLE 4 Simplify each of the following square roots.

(a) $\sqrt{45} = \sqrt{9 \cdot 5} = \sqrt{9}\,\sqrt{5} = 3\sqrt{5}$ $\quad 9 = 3^2$, a perfect square.
(b) $\sqrt{75} = \sqrt{25 \cdot 3} = \sqrt{25}\,\sqrt{3} = 5\sqrt{3}$ $\quad 25 = 5^2$, a perfect square.
(c) $\sqrt{76} = \sqrt{4 \cdot 19} = \sqrt{4}\,\sqrt{19} = 2\sqrt{19}$ $\quad 4 = 2^2$, a perfect square.
(d) $\sqrt{0.0025} = 0.05$ $\quad 0.0025 = (0.05)^2$, a perfect square.
(e) $\sqrt{x^2y^2} = \sqrt{x^2} \cdot \sqrt{y^2} = xy$ $\quad$ Assume $x \geq 0, y \geq 0$.
(f) $\sqrt{x^6} = \sqrt{(x^3)^2} = x^3$ $\quad$ Assume $x \geq 0$.

A cube root is in simplest form if its radicand does not contain any perfect cubes as factors. Similarly, a fourth root is in simplest form if its radicand does not contain any perfect fourth powers as factors.

EXAMPLE 5 Simplify: $\sqrt[3]{54}$.

SOLUTION Since $\sqrt[3]{54}$ can be written as $\sqrt[3]{27 \cdot 2}$, and since $27 = 3^3$ is a perfect cube, the radical is not in simplest form.

$$\sqrt[3]{54} = \sqrt[3]{27 \cdot 2}$$
$$= \sqrt[3]{27}\,\sqrt[3]{2} \qquad \text{Multiplication property of radicals}$$
$$= 3\sqrt[3]{2} \qquad \sqrt[3]{27} = 3$$

EXAMPLE 6 Simplify each of the following radicals.

(a) $\sqrt[3]{72} = \sqrt[3]{8 \cdot 9} = \sqrt[3]{8}\,\sqrt[3]{9} = 2\sqrt[3]{9}$ $\quad 8 = 2^3$, a perfect cube.
(b) $\sqrt[4]{48} = \sqrt[4]{16 \cdot 3} = \sqrt[4]{16}\,\sqrt[4]{3} = 2\sqrt[4]{3}$ $\quad 16 = 2^4$, a perfect fourth power.
(c) $2\sqrt[3]{2000} = 2\sqrt[3]{1000 \cdot 2}$
$\qquad\qquad\quad = 2\sqrt[3]{1000}\,\sqrt[3]{2}$
$\qquad\qquad\quad = 2 \cdot 10\sqrt[3]{2} = 20\sqrt[3]{2}$ $\quad 1000 = 10^3$, a perfect cube.
(d) $\sqrt[3]{x^7y^3} = \sqrt[3]{(x^2)^3 xy^3}$
$\qquad\qquad = \sqrt[3]{(x^2)^3 y^3}\,\sqrt[3]{x} \qquad \text{Multiplication property of radicals}$
$\qquad\qquad = x^2y\sqrt[3]{x} \qquad (x^2)^3$ and y^3 are perfect cubes.

(e) $\sqrt[5]{32x^7y^6z^{10}} = \sqrt[5]{2^5x^5x^2y^5y(z^2)^5}$
$= \sqrt[5]{2^5x^5y^5(z^2)^5} \cdot \sqrt[5]{x^2y}$
$= 2xyz^2\sqrt[5]{x^2y}$

A second condition for a radical to be in simplest form involves fractions.

CONDITION 2 No fractions can appear under the radical sign.

EXAMPLE 7 Simplify: $\sqrt{\dfrac{5}{16}}$.

SOLUTION Because $\sqrt{\dfrac{5}{16}}$ contains a fraction, it is not in simplest form.

$\sqrt{\dfrac{5}{16}} = \dfrac{\sqrt{5}}{\sqrt{16}}$ Division property of radicals.

$= \dfrac{\sqrt{5}}{4}$ $\sqrt{16} = 4$, simplest form

EXAMPLE 8 Simplify each of the following radicals.

(a) $\sqrt{\dfrac{16}{49}} = \dfrac{\sqrt{16}}{\sqrt{49}} = \dfrac{4}{7}$ $16 = 4^2$ and $49 = 7^2$ are perfect squares.

(b) $\sqrt[3]{\dfrac{5}{27}} = \dfrac{\sqrt[3]{5}}{\sqrt[3]{27}} = \dfrac{\sqrt[3]{5}}{3}$ $27 = 3^3$ is a perfect cube.

(c) $\sqrt[4]{\dfrac{32}{81}} = \dfrac{\sqrt[4]{32}}{\sqrt[4]{81}}$ Division property of radicals

$= \dfrac{\sqrt[4]{16}\sqrt[4]{2}}{\sqrt[4]{81}}$ Multiplication property of radicals

$= \dfrac{2\sqrt[4]{2}}{3}$ $16 = 2^4$ and $81 = 3^4$ are perfect fourth powers.

(d) $\sqrt[3]{\dfrac{2x^3}{27y^6z^9}} = \dfrac{\sqrt[3]{2}\sqrt[3]{x^3}}{\sqrt[3]{(3y^2z^3)^3}}$ Product and quotient rules for radicals

$= \dfrac{\sqrt[3]{2}x}{3y^2z^3}$ $(3y^2z^3)^3$ is a perfect cube.

The third condition for simplifying a radical also involves fractions.

| CONDITION 3 | No fraction can have a radical in the denominator. |

For example, $\dfrac{\sqrt{10}}{\sqrt{2}}$ is not in simplest form because there is a radical in the denominator. To eliminate the radical, we use the division property.

$$\dfrac{\sqrt{10}}{\sqrt{2}} = \sqrt{\dfrac{10}{2}} = \sqrt{5}$$

EXAMPLE 9 Simplify the following expressions.

(a) $\dfrac{\sqrt{51}}{\sqrt{3}} = \sqrt{\dfrac{51}{3}} = \sqrt{17}$ Division property of radicals

(b) $\dfrac{\sqrt{63}}{\sqrt{7}} = \sqrt{\dfrac{63}{7}} = \sqrt{9} = 3$ Division property of radicals

(c) $4\dfrac{\sqrt[3]{81}}{\sqrt[3]{3}} = 4\sqrt[3]{\dfrac{81}{3}} = 4\sqrt[3]{27} = 4 \cdot 3 = 12$

EXAMPLE 10 Simplify: $\sqrt{18} \cdot \sqrt{54}$.

SOLUTION Each radical contains a perfect square:

$$\begin{aligned}\sqrt{18} \cdot \sqrt{54} &= \sqrt{9 \cdot 2} \cdot \sqrt{9 \cdot 6} \\ &= (3\sqrt{2})(3\sqrt{6}) = 9\sqrt{12} \quad 3 \cdot 3 = 9;\ \sqrt{2} \cdot \sqrt{6} = \sqrt{12} \\ &= 9\sqrt{4 \cdot 3} = 9 \cdot 2\sqrt{3} \\ &= 18\sqrt{3}\end{aligned}$$

As you continue to work with radicals, you will discover that you can do the simplification steps in many different orders. All, however, should arrive at the same result if carried out correctly.

We now summarize the results of this section.

Simplified Forms for Radicals

A radical expression is in simplest form if:

1. the radicand contains no perfect squares if the radical is a square root, no perfect cubes if the radical is a cube root, and, in general, no perfect nth powers if the radical is an nth root;
2. no fractions appear in the radicand; and
3. no radical appears in the denominator of an expression involving fractions.

EXERCISE SET 9.2

Simplify each of the following radical expressions.

1. $\sqrt{8}$
2. $\sqrt{12}$
3. $\sqrt[3]{16}$
4. $\sqrt[3]{24}$
5. $-\sqrt{28}$
6. $-\sqrt{54}$
7. $-\sqrt[3]{54}$
8. $-\sqrt[3]{250}$
9. $\sqrt{45}$
10. $\sqrt{72}$
11. $\sqrt[4]{32}$
12. $\sqrt[4]{162}$
13. $\sqrt{\dfrac{6}{25}}$
14. $\sqrt{\dfrac{7}{36}}$
15. $-\sqrt{\dfrac{16}{49}}$
16. $-\sqrt{\dfrac{81}{64}}$
17. $3\sqrt{32}$
18. $2\sqrt{300}$
19. $2\sqrt[3]{40}$
20. $5\sqrt[3]{81}$
21. $5\sqrt{50}\,\sqrt{4}$
22. $4\sqrt{72}\,\sqrt{49}$
23. $3\sqrt{32}\,\sqrt{128}$
24. $6\sqrt{12}\,\sqrt{18}$
25. $\sqrt{\dfrac{121}{64}}$
26. $\sqrt{\dfrac{100}{25}}$
27. $-\sqrt[3]{\dfrac{27}{64}}$
28. $-\sqrt[3]{\dfrac{81}{125}}$
29. $\sqrt[5]{64}$
30. $\sqrt[5]{486}$
31. $\sqrt{\dfrac{288}{25}}$
32. $\sqrt{\dfrac{128}{49}}$
33. $-8\sqrt{288}$
34. $-7\sqrt{450}$
35. $\sqrt{7}\,\sqrt{21}$
36. $\sqrt{27}\,\sqrt{48}$
37. $\sqrt{75}\,\sqrt{27}$
38. $\sqrt{20}\,\sqrt{45}$
39. $\sqrt{\dfrac{1}{5}}\cdot\sqrt{\dfrac{4}{5}}$
40. $\sqrt{\dfrac{2}{3}}\cdot\sqrt{\dfrac{2}{27}}$
41. $\sqrt{\dfrac{3}{7}}\cdot\sqrt{\dfrac{27}{7}}$
42. $\sqrt{\dfrac{5}{9}}\cdot\sqrt{\dfrac{5}{9}}$
43. $\sqrt{\dfrac{11}{16}}$
44. $\sqrt{\dfrac{7}{36}}$
45. $\dfrac{5\sqrt{75}}{\sqrt{5}}$
46. $\dfrac{3\sqrt{200}}{\sqrt{2}}$
47. $\dfrac{18\sqrt{20}}{9\sqrt{10}}$
48. $\dfrac{26\sqrt{10}}{13\sqrt{5}}$

70. $\sqrt{108}$
71. $\sqrt{\dfrac{12}{9}}$
72. $\sqrt{\dfrac{32}{36}}$
73. $\sqrt[3]{72}$
74. $-\sqrt[3]{108}$
75. $\sqrt{0.16}$
76. $\sqrt{0.0049}$
77. $\sqrt[4]{0.0016}$
78. $\sqrt[4]{0.0001}$
79. $\sqrt{\dfrac{72}{2}}$
80. $\sqrt{\dfrac{9}{16}}$
81. $\dfrac{\sqrt{40}}{\sqrt{10}}$
82. $\dfrac{\sqrt{80a^4}}{\sqrt{5a^2}}$
83. $\sqrt[3]{\dfrac{2}{125}}$
84. $\sqrt[3]{\dfrac{3}{8}}$
85. $\sqrt[3]{\dfrac{-5}{64}}$
86. $\sqrt[3]{\dfrac{-1}{27}}$
87. $\sqrt[4]{\dfrac{32}{81}}$
88. $\sqrt[4]{\dfrac{1}{81}}$
89. $\sqrt{\dfrac{3}{4x^2}}$
90. $\sqrt{\dfrac{11}{9a^4}}$
91. $\sqrt[4]{\dfrac{16a}{81b^8 c^4}}$
92. $\sqrt[4]{\dfrac{17x}{16x^5 y^4}}$
93. $\sqrt[5]{\dfrac{-1}{32a^{10}}}$
94. $\sqrt[5]{\dfrac{64x^6}{y^{30} z^5}}$

Simplify first, if possible, then use a calculator to find a decimal approximation for each of the following expressions. Round each answer to three decimal places.

95. $\dfrac{1}{\sqrt{3}}$
96. $\dfrac{\sqrt{5}}{\sqrt{2}}$
97. $\dfrac{\sqrt{3}}{\sqrt{7}}$
98. $\dfrac{\sqrt{2}}{2}$
99. $\dfrac{\sqrt{30}}{\sqrt{5}}$
100. $\dfrac{\sqrt{49}}{\sqrt{7}}$
101. $\dfrac{\sqrt{56}}{\sqrt{8}}$
102. $\dfrac{\sqrt{54}}{\sqrt{3}}$

REVIEW EXERCISES

Add or subtract as indicated.

103. $(x^2 - 5x + 4) + (2x^3 + 7x - 1)$
104. $(7y^3 - 8y) - (y^2 + 5y - 2)$
105. $(x^3 + x^2 - x + 4) + (x^2 - 8) - (x^3 + x^2 + x - 5)$
106. $(7a^2 + 8a - 11) - (a^2 + 4) - (5a^2 - 11a - 9)$
107. $(-11y^2 + 8y + 4) - (y^2 - 9) - (4y^2 - 8y + 7)$
108. $(-4b^2 + 13b - 1) - (-3b + 4) - (-b^2 - 2b - 5)$

The hypotenuse c of a right triangle with height h and base b is found by using the formula $c = \sqrt{b^2 + h^2}$. Find c for each of the following triangles. Round your answers to three decimal places (if needed). See the figure.

109. $b = 5$ ft; $h = 7$ ft

Simplify the following radical expressions. Assume that all variables are positive real numbers.

49. $\sqrt{x^2}$
50. $\sqrt{a^4}$
51. $\sqrt{y^3}$
52. $\sqrt{t^5}$
53. $\sqrt{y}\,\sqrt{y^3}$
54. $\sqrt{x}\,\sqrt{x}$
55. $\sqrt[3]{x^7}$
56. $\sqrt[3]{p^4}$
57. $\sqrt{p}\,\sqrt{p}$
58. $\sqrt{a^2}\,\sqrt{a}$
59. $\dfrac{\sqrt{m^3}}{\sqrt{m}}$
60. $\dfrac{\sqrt{x^5}}{\sqrt{x}}$
61. $\sqrt{8c^3}$
62. $\sqrt{200y^5}$
63. $-\sqrt[3]{a^3 b^2}$
64. $-\sqrt[3]{m^6 n^4}$
65. $\sqrt{\dfrac{54x^3 y^7}{6xy^2}}$
66. $\sqrt{\dfrac{162a^9 b^{15}}{9a^4 b^4}}$
67. $\sqrt[3]{\dfrac{162x^4 y^5}{3xy^2}}$
68. $\sqrt[4]{\dfrac{40t^{11}}{8t^7}}$
69. $\sqrt{72}$

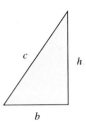

110. $b = 1$ m; $h = 1$ m
111. $b = 3$ cm; $h = 4$ cm
112. $b = 12$ in.; $h = 9$ in.
113. $b = 4$ yd; $h = 5$ yd
114. $b = 8$ ft; $h = 11$ ft
115. $b = 2\frac{1}{2}$ m; $h = 3\frac{3}{4}$ m
116. $b = 4\frac{1}{8}$ ft; $h = 7\frac{1}{4}$ ft

Solve each of the following applied problems.

117. Two cars leave Pismo Beach at the same time, one heading north at 50 mph and the other heading south at 55 mph. How long will it take them to be 315 mi apart?

118. Patricia decides to invest equal amounts in each of three accounts paying 6%, 7%, and 8% annual interest. How much should she invest at each rate if she wants the annual income (her interest) to be $2520?

9.3 Addition and Subtraction of Radicals

OBJECTIVES   In this section we will learn to
1. add and subtract radicals; and
2. simplify expressions involving radicals

To add or subtract radicals, we use the distributive property. For example,

$$2\sqrt{3} + 7\sqrt{3} = (2 + 7)\sqrt{3} = 9\sqrt{3}$$

and

$$7\sqrt{6} - 9\sqrt{6} = (7 - 9)\sqrt{6} = -2\sqrt{6}$$

Only *like* radicals can be added or subtracted. **Like radicals** have the **same index** and the **same radicand**. Their sum is found by adding their coefficients; the radical portion does not change.

EXAMPLE 1   Add: $2\sqrt{6} + 3\sqrt{6} + 5\sqrt{6}$.

SOLUTION $2\sqrt{6} + 3\sqrt{6} + 5\sqrt{6} = (2 + 3 + 5)\sqrt{6}$ Distributive property
$= 10\sqrt{6}$

EXAMPLE 2   Subtract: $15\sqrt{3} - 4\sqrt{3}$.

SOLUTION $15\sqrt{3} - 4\sqrt{3} = (15 - 4)\sqrt{3}$ Distributive property
$= 11\sqrt{3}$

EXAMPLE 3   Add or subtract as indicated.

(a) $3\sqrt{11} + 13\sqrt{11} = 16\sqrt{11}$
(b) $19\sqrt[3]{3} - 4\sqrt[3]{3} = 15\sqrt[3]{3}$
(c) $18\sqrt[4]{7} - 13\sqrt[4]{7} + 5\sqrt[4]{7} = 10\sqrt[4]{7}$
(d) $9\sqrt{13} - 2\sqrt{13} - 12\sqrt{13} = -5\sqrt{13}$

Often radicals to be added or subtracted do not have the same radicand. When this occurs, it is sometimes possible to simplify the radicals first and then to add or subtract them.

EXAMPLE 4   Add: $7\sqrt{2} + \sqrt{8}$.

SOLUTION   The radicands are not the same, but $\sqrt{8}$ can be simplified so that it is expressed in terms of $\sqrt{2}$:

$$7\sqrt{2} + \sqrt{8} = 7\sqrt{2} + \sqrt{4 \cdot 2}$$
$$= 7\sqrt{2} + 2\sqrt{2} = 9\sqrt{2}$$

EXAMPLE 5   Subtract: $5\sqrt{12} - \sqrt{27}$.

SOLUTION   Since neither $\sqrt{12}$ nor $\sqrt{27}$ are in simplest form, we begin by simplifying them.

$$5\sqrt{12} - \sqrt{27} = 5\sqrt{4 \cdot 3} - \sqrt{9 \cdot 3}$$
$$= 5\sqrt{4} \cdot \sqrt{3} - \sqrt{9} \cdot \sqrt{3} \qquad \sqrt[n]{a \cdot b} = \sqrt[n]{a} \cdot \sqrt[n]{b}$$
$$= 5 \cdot 2\sqrt{3} - 3\sqrt{3}$$
$$= 10\sqrt{3} - 3\sqrt{3}$$
$$= 7\sqrt{3}$$

EXAMPLE 6   Simplify: $9\sqrt{75} - 4\sqrt{12} + \sqrt{28}$.

SOLUTION

$$9\sqrt{75} - 4\sqrt{12} + \sqrt{28} = 9\sqrt{25 \cdot 3} - 4\sqrt{4 \cdot 3} + \sqrt{4 \cdot 7}$$
$$= 9\sqrt{25} \cdot \sqrt{3} - 4\sqrt{4} \cdot \sqrt{3} + \sqrt{4} \cdot \sqrt{7}$$
$$= 9 \cdot 5\sqrt{3} - 4 \cdot 2\sqrt{3} + 2\sqrt{7}$$
$$= 45\sqrt{3} - 8\sqrt{3} + 2\sqrt{7}$$
$$= 37\sqrt{3} + 2\sqrt{7}$$

Since $\sqrt{3}$ and $\sqrt{7}$ are not like radicals, no further simplification is possible.

9.3 • ADDITION AND SUBTRACTION OF RADICALS

We now turn our attention to simplifying radical expressions with variables in the radicands.

EXAMPLE 7 Simplify: $\sqrt{x^3} + 2\sqrt{x}$, $x \geq 0$.

SOLUTION If we think of x^3 as $x^2 \cdot x$, then since x^2 is a perfect square, we can remove it from under the radical.

$$\begin{aligned}\sqrt{x^3} + 2\sqrt{x} &= \sqrt{x^2 \cdot x} + 2\sqrt{x} \\ &= \sqrt{x^2}\sqrt{x} + 2\sqrt{x} \qquad \sqrt[n]{a \cdot b} = \sqrt[n]{a} \cdot \sqrt[n]{b} \\ &= x\sqrt{x} + 2\sqrt{x} \\ &= (x + 2)\sqrt{x} \qquad \text{Distributive property}\end{aligned}$$

EXAMPLE 8 Simplify $\sqrt{9m^2n} + \sqrt{16m^4n^3}$, where m and n are positive numbers.

SOLUTION Since both radicals are square roots, we rewrite the radicands so that any perfect squares within them can be identified.

$$\begin{aligned}\sqrt{9m^2n} + \sqrt{16m^4n^3} &= \sqrt{3^2 m^2 n} + \sqrt{4^2(m^2)^2 n^2 n} \\ &= \sqrt{3^2} \cdot \sqrt{m^2} \cdot \sqrt{n} + \sqrt{4^2} \cdot \sqrt{(m^2)^2} \cdot \sqrt{n^2} \cdot \sqrt{n}\end{aligned}$$

We can now remove all perfect-square terms from the radicals.

$$\begin{aligned}\sqrt{9m^2n} + \sqrt{16m^4n^3} &= 3m\sqrt{n} + 4m^2n\sqrt{n} \\ &= (3m + 4m^2n)\sqrt{n} \qquad \text{Distributive property}\end{aligned}$$

EXAMPLE 9 Simplify: $\sqrt[3]{16x^3} + \sqrt[3]{54x^3} - 9x\sqrt[3]{2}$.

SOLUTION
$$\begin{aligned}\sqrt[3]{16x^3} + \sqrt[3]{54x^3} - 9x\sqrt[3]{2} &= \sqrt[3]{8 \cdot 2 \cdot x^3} + \sqrt[3]{27 \cdot 2 \cdot x^3} - 9x\sqrt[3]{2} \\ &= 2x\sqrt[3]{2} + 3x\sqrt[3]{2} - 9x\sqrt[3]{2} \\ &= -4x\sqrt[3]{2}\end{aligned}$$

EXAMPLE 10 Simplify $10\sqrt{150x^3y^2} - 6x\sqrt{54xy^2}$, given that x and y are both positive.

SOLUTION
$$\begin{aligned}10\sqrt{150x^3y^2} - 6x\sqrt{54xy^2} &= 10\sqrt{25 \cdot 6 \cdot x^2 \cdot x \cdot y^2} - 6x\sqrt{9 \cdot 6 \cdot x \cdot y^2} \\ &= 10 \cdot 5xy\sqrt{6x} - 6x \cdot 3y\sqrt{6x} \\ &= 50xy\sqrt{6x} - 18xy\sqrt{6x} \\ &= 32xy\sqrt{6x}\end{aligned}$$

EXAMPLE 11 Simplify: $\dfrac{12 - \sqrt{54}}{3}$.

SOLUTION
$$\frac{12-\sqrt{54}}{3} = \frac{12-\sqrt{9\cdot 6}}{3}$$
$$= \frac{12-3\sqrt{6}}{3}$$
$$= \frac{3(4-\sqrt{6})}{3} = 4-\sqrt{6}$$

EXAMPLE 12 Simplify: $\dfrac{4-\sqrt{8}}{2} - \dfrac{6-\sqrt{32}}{4}$.

SOLUTION $\dfrac{4-\sqrt{8}}{2} - \dfrac{6-\sqrt{32}}{4} = \dfrac{2(4-\sqrt{8})-(6-\sqrt{32})}{4}$ The LCD is 4.
$$= \frac{8-2\sqrt{8}-6+\sqrt{32}}{4}$$
$$= \frac{8-2\cdot 2\sqrt{2}-6+4\sqrt{2}}{4} \qquad \sqrt{8}=2\sqrt{2};\ \sqrt{32}=4\sqrt{2}$$
$$= \frac{8-4\sqrt{2}-6+4\sqrt{2}}{4}$$
$$= \frac{2}{4} = \frac{1}{2}$$

EXAMPLE 13 Use a calculator to evaluate $\dfrac{\sqrt{7}+\sqrt{11}}{\sqrt{3}}$.

SOLUTION We think of $\dfrac{\sqrt{7}+\sqrt{11}}{\sqrt{3}}$ as $(\sqrt{7}+\sqrt{11}) \div \sqrt{3}$. Then, using a scientific calculator, we use the following sequence of keys:

$\boxed{(}$ 7 $\boxed{\sqrt{}}$ $\boxed{+}$ 11 $\boxed{\sqrt{}}$ $\boxed{)}$ $\boxed{\div}$ 3 $\boxed{\sqrt{}}$ $\boxed{=}$

The display reads 3.442379447. With the TI-81 graphics calculator, the following sequence of keys is used:

$\boxed{(}$ $\boxed{\sqrt{}}$ 7 $\boxed{+}$ $\boxed{\sqrt{}}$ 11 $\boxed{)}$ $\boxed{\div}$ $\boxed{\sqrt{}}$ 3 $\boxed{\text{ENTER}}$

The display reads 3.442379447.

Exercise Set 9.3

Add or subtract (if possible) and simplify. Assume all variables represent positive real numbers.

1. $7\sqrt{2} + 8\sqrt{2}$
2. $5\sqrt{3} + 4\sqrt{3}$
3. $4\sqrt{7} - 2\sqrt{7}$
4. $3\sqrt{6} - \sqrt{6}$
5. $\sqrt{11} - 3\sqrt{11}$
6. $4\sqrt{x} + 2\sqrt{x}$
7. $9\sqrt{y} + 7\sqrt{y}$
8. $6\sqrt{a} - 7\sqrt{a}$

9. $\sqrt{m} - 2\sqrt{m}$
10. $3\sqrt{2} - 5\sqrt{3}$
11. $4\sqrt{7} + 6\sqrt{9}$
12. $3\sqrt{5} - 3\sqrt{16}$
13. $2\sqrt{5} - 7\sqrt{5} + 3\sqrt{5}$
14. $15\sqrt{13} + 19\sqrt{13} - 18\sqrt{13}$
15. $2\sqrt{y} - \sqrt{y} + 5\sqrt{y}$
16. $\sqrt{m} + \sqrt{m} - 3\sqrt{m}$
17. $4\sqrt{7} + 3\sqrt{5} - 6\sqrt{7}$
18. $7\sqrt{11} - 5\sqrt{13} + 2\sqrt{11}$
19. $3\sqrt{a} + 2\sqrt{b} - \sqrt{a}$
20. $7\sqrt{n} - 5\sqrt{m} - \sqrt{n}$
21. $6\sqrt{3} - 4\sqrt{2} + \sqrt{7}$
22. $2\sqrt{15} + 3\sqrt{20} - \sqrt{7}$
23. $\sqrt{25x} - \sqrt{9x}$
24. $\sqrt{8m} + 2\sqrt{18m}$
25. $\sqrt{28} - 3\sqrt{63} + 4\sqrt{7}$
26. $2\sqrt{24} - \sqrt{54} - \sqrt{150}$
27. $\sqrt{50} - 7\sqrt{12} + 3\sqrt{32}$
28. $\dfrac{\sqrt{2}}{3} + \dfrac{3\sqrt{2}}{3}$
29. $\dfrac{3\sqrt{5}}{2} - \dfrac{\sqrt{5}}{2}$
30. $\dfrac{\sqrt{7}}{\sqrt{4}} - \dfrac{3\sqrt{7}}{\sqrt{4}}$
31. $\dfrac{\sqrt{3}}{\sqrt{9}} + \dfrac{2\sqrt{3}}{\sqrt{9}}$
32. $\dfrac{5\sqrt{6}}{\sqrt{16}} + \dfrac{3\sqrt{6}}{\sqrt{4}}$
33. $\dfrac{7\sqrt{11}}{\sqrt{25}} - \dfrac{\sqrt{11}}{\sqrt{36}}$
34. $3\sqrt[3]{2} + 5\sqrt[3]{2}$
35. $-7\sqrt[3]{6} + 5\sqrt[3]{6}$
36. $\sqrt[4]{29} - 2\sqrt[4]{29}$
37. $6\sqrt[4]{3} - 7\sqrt[4]{3}$
38. $3\sqrt[3]{5} - 2\sqrt[3]{5}$
39. $2\sqrt[3]{3} - 5\sqrt{3}$
40. $\sqrt[4]{2} + \sqrt[3]{2}$
41. $2\sqrt{7} - \sqrt[4]{7}$
42. $3\sqrt{x^3} - 6x\sqrt{x}$
43. $-y\sqrt{2y} + 3\sqrt{2y^3}$
44. $8\sqrt{a} - \sqrt{a^2} - 5\sqrt{a}$
45. $2\sqrt{p} - 3\sqrt{pt} + 2\sqrt{pt}$
46. $3m\sqrt{5m} - 7\sqrt{5m} + m\sqrt{5m}$
47. $\sqrt{0.16} + \sqrt{0.25}$
48. $\sqrt{0.81} - \sqrt{0.0049}$
49. $\sqrt{0.0121} - \sqrt[3]{0.008}$
50. $\sqrt[3]{0.027} + \sqrt{0.0064}$
51. $\sqrt{176} - 5\sqrt{99} + 3\sqrt{44}$
52. $2\sqrt{54} - \sqrt{8} - 5\sqrt{32} + \sqrt{48} - 3\sqrt{27}$
53. $2\sqrt{243} - 2\sqrt{45} - 3\sqrt{192} + 5\sqrt{40}$
54. $-2\sqrt{125} - 6\sqrt{32} + 3\sqrt{20} - \sqrt{200}$
55. $8\sqrt{27y^3} - 4\sqrt{48y^3}$
56. $7\sqrt{128m^3} - 10\sqrt{72m^3}$
57. $-6\sqrt{18x^3} + 5\sqrt{32x^3}$
58. $\sqrt{\dfrac{x}{4}} + 5\sqrt{\dfrac{x}{9}}$
59. $\sqrt{\dfrac{8y}{25}} + 2\sqrt{\dfrac{2y}{81}}$
60. $7\sqrt{x^3y^2} - 4x\sqrt{xy^2}$
61. $13\sqrt{150a^3b^2} - 4a\sqrt{54ab^2}$
62. $-3\sqrt{125m} + 2\sqrt{98m} - \sqrt{162m}$
63. $5\sqrt{16t} - 2\sqrt{75t} + \sqrt{9t} - \sqrt{12t}$
64. $\dfrac{\sqrt{3}+1}{2} - \dfrac{\sqrt{3}-1}{2}$
65. $\dfrac{\sqrt{6}+1}{5} - \dfrac{\sqrt{6}-1}{3}$
66. $\dfrac{\sqrt[3]{3}+\sqrt[3]{2}}{7} - \dfrac{\sqrt[3]{3}-\sqrt[3]{2}}{5}$
67. $\dfrac{\sqrt[3]{5}-\sqrt[3]{3}}{-2} + \dfrac{\sqrt[3]{5}+\sqrt[3]{3}}{3}$
68. $a\sqrt{a} + 5a\sqrt{25a} - 14\sqrt{16a^4} + \sqrt{4a^3}$
69. $5\sqrt[3]{56y} + \sqrt[3]{7y}$
70. $2\sqrt[3]{135x} + 8\sqrt[3]{54x} - 5\sqrt[3]{40x}$
71. $\sqrt[3]{320y^3} + 6\sqrt[3]{40y^3} - 5\sqrt[3]{32y^2} + 2\sqrt[3]{256y^2}$

Simplify each of the following fractions.

72. $\dfrac{2+\sqrt{8}}{2}$
73. $\dfrac{3+\sqrt{27}}{3}$
74. $\dfrac{12-3\sqrt{8}}{6}$
75. $\dfrac{10-\sqrt{50}}{5}$
76. $\dfrac{6-\sqrt{27}}{3}$
77. $\dfrac{15-4\sqrt{27}}{3}$
78. $\dfrac{2+\sqrt{2^2-4(-1)}}{2}$
79. $\dfrac{3-\sqrt{3^2-4(1)(2)}}{2}$
80. $\dfrac{7-\sqrt{(-7)^2-4(6)(2)}}{4}$
81. $\dfrac{-4-\sqrt{(-4)^2-4(1)(2)}}{2}$

REVIEW EXERCISES

Carry out the indicated operation.

82. $3x^2(5x^2 - 7x + 4)$
83. $4xy^2(-7x^2 + 8xy - 9y^2)$
84. $(x-4)^2$
85. $(2x+7)^2$
86. $(2x+y)(2x-y)$
87. $(4y-3)(4y+3)$
88. $-2x(3x-8)(3x+8)$
89. $-5y(2y-7)(2y+7)$
90. $(3x^2-4y^2)^2$
91. $(4x^2y+5xy^2)^2$
92. $(x^2-3x-4) \div (x-1)$
93. $(4x^2+12x+9) \div (2x+3)$

Use a calculator to find the approximate value of each expression. Round your answers to three decimal places.

94. $\dfrac{1+\sqrt{2}}{2}$
95. $\dfrac{2+\sqrt{3}}{2}$
96. $\dfrac{-3+\sqrt{5}}{4}$
97. $\dfrac{-5+\sqrt{3}}{\sqrt{7}}$

98. $\dfrac{\sqrt{3}-4}{\sqrt{2}}$

99. $\dfrac{\sqrt{5}-\sqrt{3}}{\sqrt{7}}$

100. $\dfrac{\sqrt{11}+\sqrt{13}}{\sqrt{3}}$

101. $\dfrac{-\sqrt{17}+\sqrt{23}}{\sqrt{5}-\sqrt{3}}$

102. $\dfrac{\sqrt{32}-\sqrt{29}}{\sqrt{7}+\sqrt{2}}$

Solve each of the following applied problems.

103. A man has $3.00 in nickels, dimes, and quarters. If he has 4 more nickels than quarters, and twice as many dimes as nickels, how many quarters does he have?

104. A 20-ft chain is to be cut into 2 pieces, one 3 times as long as the other. How long is each piece of chain?

105. A passenger train leaves Portland for Kansas City traveling at 40 mph. Two hours later a second train leaves Portland for Kansas City traveling at 60 mph. How long before the second train overtakes the first train?

106. Two planes leave Dallas Airport at the same time, one headed east traveling at 600 mph, and the other traveling west at 150 mph. How long will it be before they are 900 miles apart? How far will each have traveled?

107. A plane travels 5 hr against the wind and then returns (with the wind) in $4\tfrac{1}{11}$ hours. If the wind speed is 50 mph, what is the airspeed of the plane (its speed in still air)?

108. A man can go downstream in 1 hr less than he can go the same distance upstream. If the current is 5 mph, how fast can he travel in still water if he takes 2 hr to travel upstream?

9.4 Multiplication and Division of Radicals

OBJECTIVES

In this section we will learn to
1. multiply radicals;
2. simplify fractions with a square root in the denominator;
3. simplify fractions with a cube root in the denominator; and
4. simplify fractions with a sum or difference of radicals in the denominator.

We multiply expressions involving radicals in the same way that we multiply polynomials. After finding the product, we must simplify all radicals as much as possible, using the rule $\sqrt[n]{a} \cdot \sqrt[n]{b} = \sqrt[n]{ab}$.

EXAMPLE 1 Multiply each of the following:

(a) $\sqrt{5}\sqrt{2}$ (b) $(-2\sqrt{3})(4\sqrt{3})$ (c) $(-5\sqrt{6})(2\sqrt{3})$.

SOLUTION
(a) $\sqrt{5}\sqrt{2} = \sqrt{10}$

(b) $(-2\sqrt{3})(4\sqrt{3}) = -8(\sqrt{3}\sqrt{3})$
$= -8(\sqrt{9})$ $\sqrt[n]{a} \cdot \sqrt[n]{b} = \sqrt[n]{ab}$
$= -8(3)$ $\sqrt{9} = 3$
$= -24$

(c) $(-5\sqrt{6})(2\sqrt{3}) = (-5 \cdot 2)(\sqrt{6} \cdot \sqrt{3})$ Commutative and associative properties
$= -10\sqrt{18}$ $\sqrt[n]{a} \cdot \sqrt[n]{b} = \sqrt[n]{ab}$
$= -10\sqrt{9 \cdot 2}$
$= -10\sqrt{9} \cdot \sqrt{2}$ $\sqrt[n]{ab} = \sqrt[n]{a} \cdot \sqrt[n]{b}$
$= -30\sqrt{2}$

9.4 · MULTIPLICATION AND DIVISION OF RADICALS

EXAMPLE 2 Multiply $\sqrt{3}(\sqrt{2} - \sqrt{5})$ using the distributive property.

SOLUTION
$$\sqrt{3}(\sqrt{2} - \sqrt{5}) = \sqrt{3} \cdot \sqrt{2} - \sqrt{3} \cdot \sqrt{5} \quad \text{Distributive property}$$
$$= \sqrt{6} - \sqrt{15}$$

Neither 6 nor 15 contains perfect-square factors, so no further simplification is possible.

EXAMPLE 3 Multiply $2\sqrt{3}(\sqrt{3} + 3\sqrt{2})$ using the distributive property.

SOLUTION
$$2\sqrt{3}(\sqrt{3} + 3\sqrt{2}) = 2\sqrt{3} \cdot \sqrt{3} + 2\sqrt{3} \cdot 3\sqrt{2} \quad \text{Distributive property}$$
$$= 2 \cdot 3 + 6\sqrt{6} \quad \sqrt{3} \cdot \sqrt{3} = 3; \sqrt{3} \cdot \sqrt{2} = \sqrt{6}$$
$$= 6 + 6\sqrt{6}$$

EXAMPLE 4 Find the product $(\sqrt{3} + \sqrt{2})(\sqrt{3} + \sqrt{5})$.

SOLUTION This is the product of two binomials, so we will use the FOIL method.

$$(\sqrt{3} + \sqrt{2})(\sqrt{3} + \sqrt{5}) = \overbrace{\sqrt{3}\sqrt{3}}^{F} + \overbrace{\sqrt{3}\sqrt{5}}^{O} + \overbrace{\sqrt{2}\sqrt{3}}^{I} + \overbrace{\sqrt{2}\sqrt{5}}^{L}$$
$$= 3 + \sqrt{15} + \sqrt{6} + \sqrt{10}$$

No radicals contain perfect-square factors and no two radicals are the same, so no further simplification is possible.

EXAMPLE 5 Find the product $(3\sqrt{2} - 5)(\sqrt{6} - 2)$ using the FOIL method.

SOLUTION
$$(3\sqrt{2} - 5)(\sqrt{6} - 2) = \overbrace{3\sqrt{2}\sqrt{6}}^{F} - \overbrace{3\sqrt{2} \cdot 2}^{O} - \overbrace{5\sqrt{6}}^{I} + \overbrace{5 \cdot 2}^{L}$$
$$= 3\sqrt{12} - 6\sqrt{2} - 5\sqrt{6} + 10$$
$$= 3\sqrt{4 \cdot 3} - 6\sqrt{2} - 5\sqrt{6} + 10$$
$$= 3\sqrt{4}\sqrt{3} - 6\sqrt{2} - 5\sqrt{6} + 10$$
$$= 6\sqrt{3} - 6\sqrt{2} - 5\sqrt{6} + 10$$

No radicals contain perfect-square factors and no two radicals are the same, so no further simplification is possible.

EXAMPLE 6   Find the product $(\sqrt{2} - \sqrt{3})^2$.

SOLUTION   This is the square of a binomial; we use the pattern $(a - b)^2 = a^2 - 2ab + b^2$.

$$(a - b)^2 = a^2 - 2 \cdot a \cdot b + b^2$$

$$(\sqrt{2} - \sqrt{3})^2 = (\sqrt{2})^2 - 2\sqrt{2}\sqrt{3} + (\sqrt{3})^2$$
$$= 2 - 2\sqrt{6} + 3$$
$$= 5 - 2\sqrt{6}$$

When two binomials differ only by the sign between the terms, they are said to be **conjugates** of each other. For example, $a + b$ and $a - b$ are conjugates of each other:

$$a + b \quad \text{and} \quad a - b$$

The two binomials differ only by sign.

Recall from Chapter 7 that $(a + b)(a - b) = a^2 - b^2$.

EXAMPLE 7   Simplify: $(\sqrt{5} + \sqrt{6})(\sqrt{5} - \sqrt{6})$.

SOLUTION   These are conjugate pairs; we use the pattern $(a + b)(a - b) = a^2 - b^2$ to find the product.

$$(\sqrt{5} + \sqrt{6})(\sqrt{5} - \sqrt{6}) = (\sqrt{5})^2 - (\sqrt{6})^2$$
$$= 5 - 6$$
$$= -1$$

EXAMPLE 8   Simplify: $(2\sqrt{3} - 3\sqrt{2})(2\sqrt{3} + 3\sqrt{2})$.

SOLUTION   The two binomials are conjugates of each other.

$$(2\sqrt{3} - 3\sqrt{2})(2\sqrt{3} + 3\sqrt{2}) = (2\sqrt{3})^2 - (3\sqrt{2})^2$$
$$= 2^2(\sqrt{3})^2 - 3^2(\sqrt{2})^2$$
$$= 4 \cdot 3 - 9 \cdot 2$$
$$= 12 - 18$$
$$= -6$$

Notice in Examples 7 and 8 that the product of two conjugates involving square root does not contain any square roots in the answer.

In Section 9.2 we stated that a fraction is not in simplest form if the denominator contains a radical. We are now ready to consider what to do when this situation arises. For example, $\dfrac{1}{\sqrt{2}}$ is not in simplest form because the radical $\sqrt{2}$ is in the denominator. To eliminate the radical, we multiply both the numerator and the denominator by $\sqrt{2}$. Because $\dfrac{\sqrt{2}}{\sqrt{2}} = 1$, this doesn't change the value of the fraction.

$$\frac{1}{\sqrt{2}} \cdot \frac{\sqrt{2}}{\sqrt{2}} = \frac{\sqrt{2}}{(\sqrt{2})^2} = \frac{\sqrt{2}}{2}$$

In the original expression, the denominator was the irrational number $\sqrt{2}$. Now that we have multiplied the numerator and the denominator by $\sqrt{2}$, the denominator is the rational number 2. The process of changing an irrational number in the denominator to a rational number is called **rationalizing the denominator.**

EXAMPLE 9   Simplify: $\dfrac{\sqrt{7}}{\sqrt{15}}$.

SOLUTION   To eliminate the radical from the denominator, we multiply numerator and denominator by $\sqrt{15}$.

$$\frac{\sqrt{7}}{\sqrt{15}} \cdot \frac{\sqrt{15}}{\sqrt{15}} = \frac{\sqrt{105}}{15}$$

EXAMPLE 10   Simplify: $\dfrac{\sqrt{2}}{\sqrt{6}}$.

SOLUTION   There are two ways to simplify this expression. The first involves rationalizing the denominator directly:

$$\frac{\sqrt{2}}{\sqrt{6}} = \frac{\sqrt{2}}{\sqrt{6}} \cdot \frac{\sqrt{6}}{\sqrt{6}} = \frac{\sqrt{12}}{6} \qquad \text{Multiply numerator and denominator by } \sqrt{6}.$$

$$= \frac{\sqrt{4 \cdot 3}}{6} = \frac{2\sqrt{3}}{6}$$

$$= \frac{\sqrt{3}}{3}$$

The second way is based on the observation that

$$\frac{\sqrt{2}}{\sqrt{6}} = \sqrt{\frac{2}{6}} = \sqrt{\frac{1}{3}}$$

We can simplify $\sqrt{\dfrac{1}{3}}$ by multiplying the numerator and the denominator of $\sqrt{\dfrac{1}{3}}$ by $\sqrt{3}$:

$$\sqrt{\frac{1}{3}} = \frac{\sqrt{1}}{\sqrt{3}}$$

$$= \frac{\sqrt{1}}{\sqrt{3}} \cdot \frac{\sqrt{3}}{\sqrt{3}}$$

$$= \frac{\sqrt{3}}{3}$$

EXAMPLE 11 Simplify: $\dfrac{3}{\sqrt{12}}$.

SOLUTION The tendency here is to multiply the numerator and the denominator by $\sqrt{12}$ immediately in order to rationalize the denominator. However, if we think of 12 as $\sqrt{4 \cdot 3} = 2\sqrt{3}$, we can see that multiplying the numerator and denominator by $\sqrt{3}$ will accomplish the same thing with fewer steps. We show the two choices for comparison.

$$\dfrac{3}{\sqrt{12}} = \dfrac{3}{\sqrt{12}} \cdot \dfrac{\sqrt{3}}{\sqrt{3}} \quad \text{or} \quad \dfrac{3}{\sqrt{12}} = \dfrac{3}{\sqrt{12}} \cdot \dfrac{\sqrt{12}}{\sqrt{12}}$$

$$= \dfrac{3\sqrt{3}}{\sqrt{36}} \qquad\qquad\qquad = \dfrac{3\sqrt{12}}{12}$$

$$= \dfrac{3\sqrt{3}}{6} \qquad\qquad\qquad = \dfrac{3\sqrt{4 \cdot 3}}{12}$$

$$= \dfrac{\sqrt{3}}{2} \qquad\qquad\qquad = \dfrac{3 \cdot 2\sqrt{3}}{12}$$

$$\qquad\qquad\qquad\qquad\qquad = \dfrac{6\sqrt{3}}{12}$$

$$\qquad\qquad\qquad\qquad\qquad = \dfrac{\sqrt{3}}{2}$$

Example 11 illustrates how, when rationalizing a denominator that is a square root, both the numerator and the denominator should be multiplied by the *smallest* quantity that makes the radicand a perfect square. The next example illustrates what to do when the denominator is a cube root.

EXAMPLE 12 Simplify: $\dfrac{1}{\sqrt[3]{4}}$.

SOLUTION The radical in the denominator is a cube root. To rationalize it, we have to make the radicand into a perfect cube. Since $\sqrt[3]{4} = \sqrt[3]{2^2}$, if we multiply numerator and denominator by $\sqrt[3]{2}$, it will make the denominator a perfect cube.

$$\dfrac{1}{\sqrt[3]{4}} = \dfrac{1}{\sqrt[3]{2^2}} \cdot \dfrac{\sqrt[3]{2}}{\sqrt[3]{2}}$$

$$= \dfrac{\sqrt[3]{2}}{\sqrt[3]{2^3}} = \dfrac{\sqrt[3]{2}}{2} \qquad \sqrt[3]{2^2} \cdot \sqrt[3]{2} = \sqrt[3]{2^3} = 2$$

We also could have rationalized the denominator in Example 12 by multiplying numerator and denominator by $\sqrt[3]{4^2}$. This would have rationalized the denominator, but then we would have had to make further simplifications. The reader is encouraged to try it and compare the effort necessary to get the same results.

EXAMPLE 13 Simplify: $\dfrac{\sqrt{2} + \sqrt{5}}{\sqrt{2}}$.

SOLUTION
$$\dfrac{\sqrt{2} + \sqrt{5}}{\sqrt{2}} = \dfrac{\sqrt{2} + \sqrt{5}}{\sqrt{2}} \cdot \dfrac{\sqrt{2}}{\sqrt{2}}$$
$$= \dfrac{\sqrt{2}\sqrt{2} + \sqrt{2}\sqrt{5}}{2}$$
$$= \dfrac{2 + \sqrt{10}}{2}$$

EXAMPLE 14 Simplify: $\dfrac{2}{\sqrt{3} - 2}$.

SOLUTION To rationalize the denominator, we multiply the numerator and the denominator by the *conjugate* of the denominator, $\sqrt{3} + 2$:

$$\dfrac{2}{\sqrt{3} - 2} = \dfrac{2}{\sqrt{3} - 2} \cdot \dfrac{\sqrt{3} + 2}{\sqrt{3} + 2}$$
$$= \dfrac{2\sqrt{3} + 4}{(\sqrt{3})^2 - 2^2} \qquad (a - b)(a + b) = a^2 - b^2$$
$$= \dfrac{2\sqrt{3} + 4}{3 - 4}$$
$$= \dfrac{2\sqrt{3} + 4}{-1}$$
$$= -(2\sqrt{3} + 4)$$

EXAMPLE 15 Simplify: $\dfrac{3\sqrt{x} - 4}{3\sqrt{x} + 4}$, $x > 0$.

SOLUTION To rationalize the denominator, we multiply numerator and denominator by $3\sqrt{x} - 4$.

$$\dfrac{3\sqrt{x} - 4}{3\sqrt{x} + 4} = \dfrac{3\sqrt{x} - 4}{3\sqrt{x} + 4} \cdot \dfrac{3\sqrt{x} - 4}{3\sqrt{x} - 4}$$
$$= \dfrac{(3\sqrt{x})^2 - 2(3\sqrt{x}) \cdot 4 + 4^2}{(3\sqrt{x})^2 - 4^2} \qquad \begin{array}{l}(a - b)^2 = a^2 - 2ab + b^2 \\ (a + b)(a - b) = a^2 - b^2\end{array}$$
$$= \dfrac{9x - 24\sqrt{x} + 16}{9x - 16}$$

EXAMPLE 16 Determine whether $2 + \sqrt{3}$ is a solution to the equation $x^2 - 4x + 1 = 0$.

SOLUTION To determine whether $2 + \sqrt{3}$ is a solution, we substitute it for x in the given equation:

$$x^2 - 4x + 1 = 0$$
$$(2 + \sqrt{3})^2 - 4(2 + \sqrt{3}) + 1 \stackrel{?}{=} 0$$
$$4 + 4\sqrt{3} + 3 - 8 - 4\sqrt{3} + 1 \stackrel{?}{=} 0$$
$$0 \stackrel{\checkmark}{=} 0$$

Therefore $2 + \sqrt{3}$ is a solution to $x^2 - 4x + 1 = 0$.

Exercise Set 9.4

Perform the indicated operations and simplify.

1. $\sqrt{3}\sqrt{2}$
2. $\sqrt{5}\sqrt{7}$
3. $\sqrt{6}\sqrt{2}$
4. $(3\sqrt{2})(5\sqrt{3})$
5. $(4\sqrt{8})(2\sqrt{10})$
6. $(3\sqrt{14})(2\sqrt{21})$
7. $(6\sqrt{3})(-\sqrt{3})$
8. $(-7\sqrt{2})(\sqrt{2})$
9. $(5\sqrt{32})(-\sqrt{50})$
10. $(-5\sqrt{6})(-2\sqrt{3})$
11. $2(\sqrt{3} - 5)$
12. $4(\sqrt{2} + 1)$
13. $\sqrt{3}(\sqrt{5} + 2)$
14. $\sqrt{2}(\sqrt{3} + 4)$
15. $\sqrt{7}(\sqrt{10} - 3)$
16. $\sqrt{11}(\sqrt{3} - 5)$
17. $\sqrt{6}(\sqrt{5} - \sqrt{7})$
18. $\sqrt{13}(\sqrt{2} + \sqrt{3})$
19. $3\sqrt{2}(2\sqrt{3} + 5)$
20. $5\sqrt{3}(7\sqrt{2} - 3)$
21. $-3\sqrt{7}(2\sqrt{7} - 5\sqrt{3})$
22. $8\sqrt{6}(3\sqrt{6} - \sqrt{24})$
23. $7\sqrt{3}(-2\sqrt{3} + \sqrt{12})$
24. $(\sqrt{2} + 1)(\sqrt{2} - 1)$
25. $(\sqrt{3} - 5)(\sqrt{3} + 5)$
26. $(\sqrt{3} + 7)^2$
27. $(\sqrt{11} - 4)^2$
28. $(2\sqrt{5} - \sqrt{3})(2\sqrt{5} + \sqrt{3})$
29. $(3\sqrt{6} + 2\sqrt{3})(3\sqrt{6} - 2\sqrt{3})$
30. $(7\sqrt{2} - \sqrt{3})^2$
31. $(3\sqrt{6} - \sqrt{2})^2$
32. $(\sqrt{7} + \sqrt{3})(\sqrt{7} - \sqrt{2})$
33. $(\sqrt{5} + \sqrt{2})(\sqrt{6} - \sqrt{3})$
34. $\dfrac{1}{\sqrt{2}}$
35. $\dfrac{1}{\sqrt{3}}$
36. $\dfrac{2}{\sqrt{2}}$
37. $\dfrac{3}{\sqrt{3}}$
38. $\dfrac{2}{\sqrt{8}}$
39. $\dfrac{3}{\sqrt{27}}$
40. $\dfrac{4}{\sqrt{28}}$
41. $\dfrac{5}{\sqrt{50}}$
42. $\dfrac{\sqrt{3}}{\sqrt{2}}$
43. $\dfrac{\sqrt{5}}{\sqrt{3}}$
44. $\dfrac{2 - \sqrt{3}}{\sqrt{2}}$
45. $\dfrac{\sqrt{5} + 2}{\sqrt{3}}$
46. $\dfrac{2 + \sqrt{3}}{\sqrt{12}}$
47. $\dfrac{1 - \sqrt{5}}{\sqrt{45}}$
48. $\dfrac{8 + \sqrt{5}}{8 - \sqrt{5}}$
49. $\dfrac{\sqrt{6} + 1}{\sqrt{12} - 3}$
50. $\dfrac{\sqrt{7} - \sqrt{5}}{\sqrt{7} + \sqrt{5}}$
51. $\dfrac{\sqrt{2}}{\sqrt{3} - \sqrt{5}}$
52. $\dfrac{\sqrt{3}}{\sqrt{3} - \sqrt{6}}$
53. $\dfrac{2\sqrt{5} + 3}{2\sqrt{5} - 3}$
54. $\dfrac{3\sqrt{2} + \sqrt{3}}{3\sqrt{2} - \sqrt{3}}$
55. $\dfrac{7\sqrt{3} - 6\sqrt{5}}{7\sqrt{3} + 6\sqrt{5}}$
56. $\dfrac{2\sqrt{2} - 5\sqrt{3}}{\sqrt{3} - \sqrt{2}}$

Find each of the following products and simplify. Assume all variables represent positive real numbers.

57. $\sqrt{x}\sqrt{x^3}$
58. $\sqrt{y}\sqrt{y}$
59. $\sqrt{y^3}\sqrt{y^2}$
60. $\sqrt{x^2}\sqrt{x^3}$
61. $\sqrt{3x^2}\sqrt{3x}$
62. $\sqrt{2a}\sqrt{2a^3}$
63. $(2\sqrt{y})(-3\sqrt{y})$
64. $(-3\sqrt{x^3})(\sqrt{x})$
65. $(-6\sqrt{a^3})(-2\sqrt{a^4})$
66. $(-\sqrt{x^2y})(\sqrt{xy^2})$
67. $3(\sqrt{x} + 2)$
68. $5(\sqrt{y} - 3)$
69. $x(\sqrt{x} - 7)$
70. $y(\sqrt{y} - 11)$
71. $2\sqrt{x}(3\sqrt{x} - \sqrt{x^2})$
72. $5\sqrt{t}(2\sqrt{t} - \sqrt{t^2})$
73. $(\sqrt{x} + \sqrt{y})(\sqrt{x} - \sqrt{y})$

74. $(\sqrt{a} + 2\sqrt{b})(\sqrt{a} - 2\sqrt{b})$
75. $(3\sqrt{m} - 2\sqrt{n})^2$
76. $(7\sqrt{p} + 5\sqrt{q})^2$
77. $(11\sqrt{s} - 5\sqrt{t})(11\sqrt{s} + 5\sqrt{t})$
78. $(\sqrt{xy} - x\sqrt{y})^2$

Simplify each of the following. Assume all variables represent positive real numbers.

79. $\dfrac{1}{\sqrt{x}}$
80. $\dfrac{3}{\sqrt{y}}$
81. $\dfrac{y}{\sqrt{y}}$
82. $\dfrac{x}{\sqrt{x}}$
83. $\dfrac{3x}{\sqrt{3x}}$
84. $\dfrac{2y}{\sqrt{8y}}$
85. $\dfrac{x+y}{\sqrt{x+y}}$
86. $\dfrac{x-y}{\sqrt{x-y}}$
87. $\dfrac{\sqrt{x}-1}{\sqrt{x}}$
88. $\dfrac{\sqrt{y}-2}{\sqrt{y}}$
89. $\dfrac{1+\sqrt{2x}}{\sqrt{2x}}$
90. $\dfrac{\sqrt{x}+3}{\sqrt{x}}$
91. $\dfrac{2+\sqrt{2p}}{\sqrt{p}}$
92. $\dfrac{8+\sqrt{a}}{8-\sqrt{a}}$
93. $\dfrac{2\sqrt{y}+3}{2\sqrt{y}-3}$
94. $\dfrac{5\sqrt{m}-\sqrt{n}}{5\sqrt{m}+\sqrt{n}}$
95. $\dfrac{2\sqrt{t}-5\sqrt{z}}{3\sqrt{t}+4\sqrt{z}}$
96. $\dfrac{7\sqrt{a}-5\sqrt{b}}{11\sqrt{a}+9\sqrt{b}}$

Determine if the given number is a solution of the equation.

97. $x^2 - 4x + 1 = 0;\ 2 - \sqrt{3}$
98. $2x^2 - 3 = 0;\ \dfrac{\sqrt{6}}{2}$
99. $(x+3)^2 = 25;\ \dfrac{-1+\sqrt{5}}{2}$
100. $x^2 + 6x - 2 = 0;\ 3 - \sqrt{11}$
101. $m^2 + 7m - 12 = 0;\ \dfrac{-7+\sqrt{97}}{2}$
102. $x^2 - x - 1 = 0;\ 1 - \sqrt{2}$

REVIEW EXERCISES

Solve each of the following equations by factoring.

103. $x^2 - 4 = 0$
104. $x^2 - 9 = 0$
105. $y^2 - 4y - 5 = 0$
106. $y^2 + 6y + 9 = 0$
107. $m^2 - 8m + 12 = 0$
108. $a^2 - 4a - 12 = 0$
109. $9x^2 - 12x + 4 = 0$
110. $25x^2 + 20x + 4 = 0$
111. $16x^3 - 9x = 0$
112. $36x^3 - 49x = 0$

Write each applied problem as a quadratic equation and solve it.

113. The square of a number added to 8 times the number is equal to -16. Find the number.
114. Four times the square of a number added to 9 is equal to 12 times the number. Find the number.
115. Nine times the square of a number added to 25 is equal to -30 times the number. Find the number.
116. The square of a number less 12 times the number is equal to -36. Find the number.
117. A dairy wants to mix milk containing 3% butterfat with milk containing 30% butterfat to obtain 900 gal of milk that is 8% butterfat. How much of each must be used in the mixture?
118. How much pure alcohol must be added to 10 cc of a 60% alcohol solution to strengthen it to a 90% solution?

9.5 Solving Equations Containing Radicals

OBJECTIVES

In this section we will learn to
1. solve equations involving square roots;
2. determine the distance between two points in a plane; and
3. identify extraneous solutions of equations involving radicals.

Equations involving radicals are solved by using the *power property of equality*.

Power Property of Equality
If A and B are two algebraic expressions and if
$$A = B \quad \text{then} \quad A^n = B^n$$

When $n = 2$, the power property is called the **squaring property**. For example, since

$$4 = 4$$

then

$$4^2 = 4^2 \quad \text{Squaring property}$$

that is,

$$16 = 16$$

We must be careful when we use the squaring property for solving an equation, because sometimes it can result in an **extraneous** (false) **solution**. To illustrate how this can occur, we start with the equation $x = 2$.

$$x = 2$$
$$x^2 = 2^2 \quad \text{Squaring property}$$
$$x^2 = 4$$

We can solve the new equation, $x^2 = 4$, by factoring after writing it in standard form.

$$x^2 = 4$$
$$x^2 - 4 = 0 \quad \text{Standard form}$$
$$(x - 2)(x + 2) = 0$$
$$x - 2 = 0 \quad \text{or} \quad x + 2 = 0$$
$$x = 2 \quad \text{or} \quad x = -2$$

Notice that squaring has introduced the extraneous solution -2, which does not satisfy the original equation, $x = 2$. For this reason we must *always check all solutions in the **original** equation*.

EXAMPLE 1 Solve the equation $\sqrt{x + 2} = 4$.

SOLUTION   To eliminate the radical, we use the squaring property.

$$\sqrt{x+2} = 4$$
$$(\sqrt{x+2})^2 = 4^2$$
$$x + 2 = 16$$
$$x = 14 \quad \text{Subtract 2 from each side.}$$

We check the answer in the *original* equation.

$$\sqrt{x+2} = 4 \quad \text{Original equation}$$
$$\sqrt{14+2} \stackrel{?}{=} 4 \quad \text{Substitute 14 for } x.$$
$$\sqrt{16} \stackrel{?}{=} 4$$
$$4 \stackrel{\checkmark}{=} 4$$

The solution set is {14}.

EXAMPLE 2   Solve the equation $2\sqrt{x} = 3\sqrt{x-5}$.

SOLUTION
$$2\sqrt{x} = 3\sqrt{x-5}$$
$$(2\sqrt{x})^2 = (3\sqrt{x-5})^2 \quad \text{Squaring property}$$
$$2^2(\sqrt{x})^2 = 3^2(\sqrt{x-5})^2$$
$$4x = 9(x-5)$$
$$4x = 9x - 45$$
$$-5x = -45 \quad \text{Subtract } 9x \text{ from each side.}$$
$$x = 9 \quad \text{Divide each side by } -5.$$

CHECK
$$2\sqrt{x} = 3\sqrt{x-5} \quad \text{Original equation}$$
$$2\sqrt{9} \stackrel{?}{=} 3\sqrt{9-5} \quad \text{Substitute 9 for } x.$$
$$2\sqrt{9} \stackrel{?}{=} 3\sqrt{4}$$
$$2 \cdot 3 \stackrel{?}{=} 3 \cdot 2$$
$$6 \stackrel{\checkmark}{=} 6$$

The solution set is {9}.

EXAMPLE 3   Solve the equation $\sqrt{3x+2} = \sqrt{4x+2}$.

SOLUTION
$$\sqrt{3x+2} = \sqrt{4x+2}$$
$$(\sqrt{3x+2})^2 = (\sqrt{4x+2})^2 \quad \text{Squaring property}$$
$$3x + 2 = 4x + 2$$
$$0 = x \quad \text{Subtract } 3x \text{ and 2 from each side.}$$

CHECK
$$\sqrt{3x+2} = \sqrt{4x+2}$$
$$\sqrt{3 \cdot 0 + 2} \stackrel{?}{=} \sqrt{4 \cdot 0 + 2} \quad \text{Substitute 0 for } x.$$
$$\sqrt{2} \stackrel{\checkmark}{=} \sqrt{2}$$

The solution set is {0}.

EXAMPLE 4 Solve the equation $\sqrt{2x^2 - 5x + 6} = \sqrt{x^2 - 7x + 14}$.

SOLUTION

$$\sqrt{2x^2 - 5x + 6} = \sqrt{x^2 - 7x + 14}$$
$$(\sqrt{2x^2 - 5x + 6})^2 = (\sqrt{x^2 - 7x + 14})^2 \quad \text{Squaring property}$$
$$2x^2 - 5x + 6 = x^2 - 7x + 14$$
$$x^2 + 2x - 8 = 0 \quad \text{Write in standard form.}$$
$$(x - 2)(x + 4) = 0$$
$$x - 2 = 0 \quad \text{or} \quad x + 4 = 0$$
$$x = 2 \quad \text{or} \quad x = -4$$

CHECK
$$\sqrt{2x^2 - 5x + 6} = \sqrt{x^2 - 7x + 14}$$
$$\sqrt{2(2^2) - 5(2) + 6} \stackrel{?}{=} \sqrt{(2)^2 - 7(2) + 14} \quad \text{Substitute 2 for } x.$$
$$\sqrt{8 - 10 + 6} \stackrel{?}{=} \sqrt{4 - 14 + 14}$$
$$\sqrt{4} \stackrel{\checkmark}{=} \sqrt{4}$$

The check of -4 is left to the reader. The solution set is $\{-4, 2\}$.

If we use the squaring property to solve equations with radicals, we must keep the following caution in mind.

CAUTION

When squaring a binomial, do not forget the middle term:

$(a + b)^2 = a^2 + 2ab + b^2$ not $a^2 + b^2$
$(a - b)^2 = a^2 - 2ab + b^2$ not $a^2 - b^2$

EXAMPLE 5 Solve the equation $\sqrt{3x + 3} + 5 = x$.

SOLUTION We want to eliminate the radical by using the squaring property, but first it must be isolated on one side of the equation.

$$\sqrt{3x + 3} + 5 = x$$
$$\sqrt{3x + 3} = x - 5$$
$$(\sqrt{3x + 3})^2 = (x - 5)^2 \quad \text{Squaring property}$$
$$3x + 3 = x^2 - 10x + 25 \quad (a - b)^2 = a^2 - 2ab + b^2.$$
$$0 = x^2 - 13x + 22 \quad \text{Subtract } 3x \text{ and 3 from each side.}$$
$$0 = (x - 11)(x - 2)$$
$$x - 11 = 0 \quad \text{or} \quad x - 2 = 0$$
$$x = 11 \quad \text{or} \quad x = 2$$

CHECK $\sqrt{3x + 3} + 5 = x$ $\sqrt{3x + 3} + 5 = x$ Original equation
$\sqrt{3 \cdot 11 + 3} + 5 \stackrel{?}{=} 11$ or $\sqrt{3 \cdot 2 + 3} + 5 \stackrel{?}{=} 2$ $x = 11$ or 2

$$\sqrt{36}+5 \stackrel{?}{=} 11 \qquad\qquad \sqrt{9}+5 \stackrel{?}{=} 2$$
$$11 \stackrel{\checkmark}{=} 11 \qquad\qquad 8 \ne 2$$

Thus $x = 2$ is extraneous. The solution set is $\{11\}$.

In some cases, we must apply the squaring property of equality more than once to eliminate the radicals.

EXAMPLE 6   Solve the equation $\sqrt{8x+7} = 4\sqrt{x} + 1$.

SOLUTION   To eliminate the radical on the left side of the equation, we square both sides:

$$(\sqrt{8x+7})^2 = (4\sqrt{x}+1)^2$$
$$8x + 7 = (4\sqrt{x})^2 + 2(4\sqrt{x})(1) + 1^2 \qquad (a+b)^2 = a^2 + 2ab + b^2$$
$$8x + 7 = 16x + 8\sqrt{x} + 1$$

Since one radical still remains, we must isolate it on one side of the equation and square again.

$$-8x + 6 = 8\sqrt{x} \qquad \text{Subtract } 16x \text{ and } 1 \text{ from each side.}$$

Since 2 is a factor of each side, we divide it out before squaring again.

$$(-4x+3)^2 = (4\sqrt{x})^2$$
$$16x^2 - 24x + 9 = 16x$$
$$16x^2 - 40x + 9 = 0$$
$$(4x-9)(4x-1) = 0$$
$$4x - 9 = 0 \quad \text{or} \quad 4x - 1 = 0$$
$$4x = 9 \quad \text{or} \quad 4x = 1$$
$$x = \frac{9}{4} \quad \text{or} \quad x = \frac{1}{4}$$

CHECK $\sqrt{8\left(\frac{9}{4}\right)+7} \stackrel{?}{=} 4\sqrt{\frac{9}{4}}+1$ or $\sqrt{8\left(\frac{1}{4}\right)+7} \stackrel{?}{=} 4\sqrt{\frac{1}{4}}+1 \qquad x = \frac{9}{4} \text{ or } \frac{1}{4}$

$$\sqrt{18+7} \stackrel{?}{=} 4\left(\frac{3}{2}\right)+1 \qquad\qquad \sqrt{2+7} \stackrel{?}{=} 4\left(\frac{1}{2}\right)+1$$
$$\sqrt{25} \stackrel{?}{=} 6 + 1 \qquad\qquad\qquad \sqrt{9} \stackrel{?}{=} 2 + 1$$
$$5 \ne 7 \qquad\qquad\qquad\qquad 3 \stackrel{\checkmark}{=} 3$$

The solution set is $\left\{\frac{1}{4}\right\}$.

The distance between any two points on a plane can be found by using a formula that involves a radical. If (x_1, y_1) and (x_2, y_2) are any two points on a plane, we can form a right triangle by using the two points as the end points of the hypotenuse, as shown in Figure 9.2. The lengths of the legs are $x_2 - x_1$ and $y_2 - y_1$. If we let d represent the length of the hypotenuse, then by the Pythagorean theorem, the square of the hypotenuse is equal to the sum of the squares of the two legs, or

$$d^2 = (x_2 - x_1)^2 + (y_2 - y_1)^2$$

Since $\sqrt{d^2} = d$,

$$d = \sqrt{(x_2 - x_1)^2 + (y_2 - y_1)^2}$$

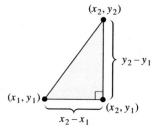

FIGURE 9.2

The Distance Formula

The distance between two points in a plane with coordinates (x_1, y_1) and (x_2, y_2) is

$$d = \sqrt{(x_2 - x_1)^2 + (y_2 - y_1)^2}$$

EXAMPLE 7 Find the distance between each pair of points:

(a) $(5, 2)$ and $(8, 6)$ **(b)** $(-7, -9)$ and $(0, 6)$

SOLUTION **(a)** We let $(x_1, y_1) = (5, 2)$ and $(x_2, y_2) = (8, 6)$. Then

$$\begin{aligned}
d &= \sqrt{(x_2 - x_1)^2 + (y_2 - y_1)^2} \\
&= \sqrt{(8 - 5)^2 + (6 - 2)^2} \\
&= \sqrt{3^2 + 4^2} \\
&= \sqrt{9 + 16} \\
&= \sqrt{25} = 5 \text{ units}
\end{aligned}$$

(b) We let $(x_1, y_1) = (-7, -9)$ and $(x_2, y_2) = (0, 6)$. Then

$$\begin{aligned}
d &= \sqrt{[0 - (-7)]^2 + [6 - (-9)]^2} \\
&= \sqrt{7^2 + (6 + 9)^2} \\
&= \sqrt{49 + 225} \\
&= \sqrt{274}
\end{aligned}$$

When using the distance formula, the choice of which point to call (x_1, y_1) and which to call (x_2, y_2) is arbitrary; the distance will be the same for either choice. The reader should verify this by interchanging the values of x_1 and x_2 and those of y_1 and y_2 in Example 7 and recalculating d.

EXAMPLE 8 The length of time t it takes a pendulum of length l to make a single swing is given by $t = 2\pi\sqrt{\dfrac{l}{32}}$. Solve the equation for l.

SOLUTION   We begin by dividing each side by 2π:

$$\frac{t}{2\pi} = \frac{2\pi\sqrt{\frac{l}{32}}}{2\pi}$$

$$\frac{t}{2\pi} = \sqrt{\frac{l}{32}}$$

To eliminate the radical, we use the squaring property:

$$\left(\frac{t}{2\pi}\right)^2 = \left(\sqrt{\frac{l}{32}}\right)^2$$

$$\frac{t^2}{4\pi^2} = \frac{l}{32}$$

$$\frac{8t^2}{\pi^2} = l \qquad \text{Multiply each side by 32.}$$

EXAMPLE 9   The square root of the sum of two consecutive odd integers is 1 more than the square root of the larger number. Find the two integers.

SOLUTION   We let x = the smaller integer
$x + 2$ = the larger integer

Since the square root of the sum of the two integers is 1 greater than the square root of the larger number, we add 1 to the square root of the larger integer to make the two equal.

$$\sqrt{x + x + 2} = \sqrt{x + 2} + 1$$
$$(\sqrt{2x + 2})^2 = (\sqrt{x + 2} + 1)^2 \qquad \text{Square both sides to eliminate one radical.}$$
$$2x + 2 = x + 2 + 2\sqrt{x + 2} + 1 \qquad (a + b)^2 = a^2 + 2ab + b^2$$
$$x - 1 = 2\sqrt{x + 2} \qquad \text{Isolate the radical.}$$
$$(x - 1)^2 = (2\sqrt{x + 2})^2 \qquad \text{Squaring property}$$
$$x^2 - 2x + 1 = 4(x + 2)$$
$$x^2 - 2x + 1 = 4x + 8$$
$$x^2 - 6x - 7 = 0$$
$$(x - 7)(x + 1) = 0$$
$$x - 7 = 0 \quad \text{or} \quad x + 1 = 0$$
$$x = 7 \quad \text{or} \quad x = -1$$

If $x = 7$, then $x + 2 = 9$, and if $x = -1$, then $x + 2 = 1$. Let's check these number pairs in the words of the problem. The square root of the sum $7 + 9 = 16$ is 4. The square root of the larger integer, 9, is 3. Since 4 is 1 greater than 3, the solution checks.

The square root of the sum $-1 + 1 = 0$ is zero. The square root of the larger integer, 1, is 1. Since 0 is not 1 greater than 1, this solution does not check and is extraneous. The two numbers are 7 and 9.

EXERCISE SET 9.5

Solve each equation. Simplify all solutions and check each one in the original equation.

1. $\sqrt{x} = 9$
2. $\sqrt{y} = 1$
3. $\sqrt{2p} = 8$
4. $\sqrt{3t} = 6$
5. $\sqrt{7y} = -7$
6. $-\sqrt{8a} = 4$
7. $-3\sqrt{x} = 1$
8. $-\sqrt{3x} = 2$
9. $2\sqrt{3y} = 5$
10. $5\sqrt{2y} = 4$
11. $-8\sqrt{3x} = -7$
12. $-11\sqrt{5y} = -1$
13. $\sqrt{t-2} = 3$
14. $\sqrt{x+2} = 1$
15. $\sqrt{9p+7} = 4$
16. $\sqrt{3m+1} = 5$
17. $-\sqrt{s+1} = 3$
18. $-\sqrt{x+1} = 5$
19. $\sqrt{5-y} = -2$
20. $\sqrt{4-x} = -3$
21. $\sqrt{a-1} - 4 = 0$
22. $\sqrt{5p+1} - 1 = 0$
23. $7 - \sqrt{x-3} = 4$
24. $\sqrt{2p+1} + 5 = 7$
25. $\sqrt{3y+1} + 4 = 9$
26. $7 - 3\sqrt{4m} = 0$
27. $\sqrt{7x-4} = -3$
28. $2\sqrt{3a+1} + 5 = 0$
29. $2\sqrt{p-1} + 5 = 0$
30. $2 + 5\sqrt{3x} = 3\sqrt{3x} + 12$
31. $9 + \sqrt{2a} = 11 - 2\sqrt{2a}$
32. $\sqrt{a}\sqrt{a-9} = 0$
33. $\sqrt{2y}\sqrt{y-18} = 0$
34. $\sqrt{x}\sqrt{x+2} = \sqrt{3}$
35. $\sqrt{a}\sqrt{a-2} = \sqrt{8}$
36. $\sqrt{2m+1} = 3\sqrt{m}$
37. $\sqrt{4y+1} = 2\sqrt{y}$
38. $\sqrt{2a-1} = \sqrt{a+3}$
39. $\sqrt{3x-1} = \sqrt{x+2}$
40. $3\sqrt{5p+1} = -2\sqrt{p+1}$
41. $-7\sqrt{2t-9} = 3\sqrt{5t+4}$
42. $\sqrt{8x+1} = \sqrt{4x+1}$
43. $\sqrt{5y+1} = \sqrt{3y+1}$
44. $2a - 7 = \sqrt{2a-1}$
45. $\sqrt{2c-3} = c - 3$
46. $\sqrt{3m+7} = 0$
47. $\sqrt{2a-3} = 0$
48. $\sqrt{5t-2} = 0$
49. $\sqrt{7a+9} = 0$
50. $\sqrt{m} = 2m - 1$
51. $x - 3 - \sqrt{4x} = 0$
52. $\sqrt{y+1} = y + 1$
53. $\sqrt{5x+5} = 5x + 5$
54. $m = \sqrt{m}$
55. $x = \sqrt{2x}$
56. $p + \sqrt{p-2} = 2$
57. $x = \sqrt{x+2}$
58. $\sqrt{5t+1} = t + 1$
59. $y = 4 + \sqrt{2y}$
60. $x = 3 + \sqrt{4x}$
61. $\sqrt{x-5} = \sqrt{x} - 1$
62. $\sqrt{y+8} = \sqrt{y} + 2$
63. $\sqrt{4a+3} = 2\sqrt{a} + 1$
64. $\sqrt{2p+7} = \sqrt{p} + 2$
65. $2\sqrt{m-5} + 5 = 3\sqrt{m}$
66. $\sqrt{8y+7} - 1 = 4\sqrt{y}$
67. $\sqrt{y^2+7y+26} = 4$
68. $\sqrt{m^2-8} = m + 4$
69. $5a + \sqrt{a^2+2} = 4a - 3$
70. $3p - \sqrt{p^2+1} = 2p$
71. $\sqrt{x^2+5x+4} = \sqrt{2x^2+5x+3}$
72. $\sqrt{y^2+2y-4} = \sqrt{2y^2+3y-4}$
73. $\sqrt{3x^2-x+8} = \sqrt{5x^2-x}$
74. $\sqrt{4x^2+7x-1} = \sqrt{x^2+7x}$
75. $\sqrt{2x^2+9} = \sqrt{3x^2+4x+9}$
76. $\sqrt{x^2-4x+8} = \sqrt{2x^2-2x+9}$

Find the distance between the given points.

77. (5, 6) and (2, 2)
78. (4, 6) and (1, 2)
79. (10, −1) and (−5, 7)
80. (4, 16) and (−1, 4)
81. (0, 0) and (1, 2)
82. (0, 0) and (−2, 3)
83. $\left(-\frac{1}{2}, 0\right)$ and $\left(\frac{1}{4}, -2\right)$
84. $\left(\frac{2}{3}, 0\right)$ and $\left(-\frac{1}{2}, 3\right)$
85. $(\sqrt{3}, 0)$ and $(-2\sqrt{3}, \sqrt{3})$
86. $(-\sqrt{2}, 0)$ and $(2\sqrt{2}, \sqrt{5})$

Solve each literal equation for the specified letter. Assume that all variables represent positive real numbers.

87. $C = \sqrt{a^2+b^2}$; a^2
88. $r = \sqrt{\dfrac{A}{\pi}}$; A
89. $T = \sqrt{\dfrac{2s}{g}}$; s
90. $X = \sqrt{\dfrac{mn}{2}}$; n
91. $E = \dfrac{s}{\sqrt{n}}$; n
92. $s = \sqrt{\dfrac{m_1 m_2}{F}}$; F
93. $C = \sqrt{\dfrac{E}{m}}$; E
94. $M = \sqrt{1 - \dfrac{a^2}{b^2}}$; b^2

Solve each of the following applied problems. Check all solutions.

95. The square root of the sum of a number and 4 is 7. Find the number.
96. The square root of the difference of a number and 1 is 15. Find the number.

97. A number less its square root is zero. Find the number.

98. The square root of twice a number is equal to the square root of the sum of 10 and the number. Find the number.

99. Two times the square root of 1 more than a number is equal to 1 more than the number. Find the number.

100. Three times the square root of one-third of a number is equal to one-third of the number. Find the number.

101. The square root of the sum of two consecutive natural numbers is equal to 2 less than the larger number. Find the numbers.

102. The square root of the sum of the squares of two consecutive whole numbers is 2 more than the smaller number. Find the numbers.

For Extra Thought

Use a calculator to evaluate each radical equation. Round your answers to three decimal places.

103. Find
$$t = 2\pi \sqrt{\frac{L}{g}}$$
when $L = 200$ cm and $g = 25$ cm/sec^2; use 3.14 for π.

104. As illustrated in the figure below, if you are h m above

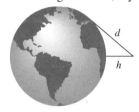

the earth's surface, the distance d in kilometers to the horizon is given by $d = 3.6\sqrt{h}$. Find h when $d = 228$ km.

105. The three sides of a triangle have lengths $a = 3.46$ in., $b = 2.13$ in., and $c = 4.17$ in. Find the area of the triangle if the area can be found using the formula
$$A = \sqrt{s(s - a)(s - b)(s - c)}$$
where $s = \frac{1}{2}(a + b + c)$.

106. The approximate circumference C of an ellipse with axes A and B (see figure) is given by the formula (due to Peano, 1887)

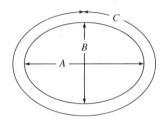

$$C = \pi[A + B + \frac{1}{2}(\sqrt{A} - \sqrt{B})^2]$$

Find C for $A = 18.8$ ft and $B = 12.2$ ft.

Write in Your Own Words

107. Can the sum of 5 and $\sqrt{x}$ equal zero? Explain.

108. Can the difference of 3 and $\sqrt{x}$ equal zero? Explain.

109. Can $\sqrt[3]{x^3} = x$? Explain.

110. Can $\sqrt{x^2}$ equal $-x$? Explain.

111. What is an extraneous solution?

112. Explain the *power property of equality.*

9.6 Rational Exponents and Radicals

OBJECTIVES

In this section we will learn to
1. interpret the expressions $b^{1/n}$ and $b^{m/n}$;
2. convert from radical form to rational exponent form;
3. convert from rational exponent form to radical form; and
4. use rational exponents to simplify radicals.

Until this point we have discussed algebraic expressions with integer exponents only. What meaning should we give to $4^{1/2}$, or $8^{2/3}$, or $32^{-2/5}$? To examine the meaning of $4^{1/2}$, we will look at examples that use the previous rules of exponents that we learned for use with integers.

We begin by assuming that $(b^m)^n = b^{mn}$ and $b^m b^n = b^{m+n}$ for all rational numbers m and n. Suppose we let $b = 4$. Then

$$4^{1/2} = (2^2)^{1/2} = 2^{2(1/2)} = 2 \quad \text{and} \quad \sqrt{4} = 2$$

Also,

$$4^{1/2} \cdot 4^{1/2} = 4^{(1/2)+(1/2)} = 4 \quad \text{and} \quad \sqrt{4} \cdot \sqrt{4} = \sqrt{16} = 4$$

Since both $4^{1/2} \cdot 4^{1/2} = 4$ and $\sqrt{4} \cdot \sqrt{4} = 4$, it is logical to define $4^{1/2} = \sqrt{4}$. In the same manner,

$$(\sqrt[3]{5})(\sqrt[3]{5})(\sqrt[3]{5}) = \sqrt[3]{125} = 5 \quad \text{and} \quad 5^{1/3} \cdot 5^{1/3} \cdot 5^{1/3} = 5^{(1/3)+(1/3)+(1/3)} = 5$$

These examples lead to the following definition.

DEFINITION $b^{1/n}$	If b is a positive number and n is a positive integer greater than 1, then $$b^{1/n} = \sqrt[n]{b}$$

EXAMPLE 1 Write each of the following exponential quantities in radical notation and simplify: **(a)** $25^{1/2}$ **(b)** $32^{1/5}$ **(c)** $8^{1/3}$ **(d)** $256^{1/4}$

SOLUTION **(a)** $25^{1/2} = \sqrt[2]{25} = \sqrt{25} = 5$
(b) $32^{1/5} = \sqrt[5]{32} = 2$
(c) $8^{1/3} = \sqrt[3]{8} = 2$
(d) $256^{1/4} = \sqrt[4]{256} = 4$

EXAMPLE 2 Write each of the following exponential quantities in radical notation:

(a) $5y^{1/3}$ **(b)** $(5y)^{1/3}$ **(c)** $-7x^{1/4}$ **(d)** $(7x)^{1/4}$

SOLUTION **(a)** $5y^{1/3} = 5\sqrt[3]{y}$
(b) $(5y)^{1/3} = \sqrt[3]{5y}$
(c) $-7x^{1/4} = -7\sqrt[4]{x}$
(d) $(7x)^{1/4} = \sqrt[4]{7x}$

How do we define an expression such as $8^{2/3}$? Consider:

$$8^{2/3} = (8^{1/3})^2 = (\sqrt[3]{8})^2 = 2^2 = 4$$

and

$$8^{2/3} = (8^2)^{1/3} = (64)^{1/3} = \sqrt[3]{64} = 4$$

These two examples lead us to the following definition.

9.6 · RATIONAL EXPONENTS AND RADICALS

DEFINITION $b^{m/n}$	If b is a positive number and m and n are positive integers, $n > 1$, then $$b^{m/n} = \sqrt[n]{b^m} = (\sqrt[n]{b})^m$$

EXAMPLE 3 Evaluate each of the following expressions:

(a) $4^{3/2}$ (b) $27^{2/3}$ (c) $32^{3/5}$ (d) $-8^{4/3}$ (e) $(0.027)^{4/3}$

SOLUTION
(a) $4^{3/2} = (4^{1/2})^3 = (\sqrt{4})^3 = 2^3 = 8$
(b) $27^{2/3} = (27^{1/3})^2 = (\sqrt[3]{27})^2 = (3)^2 = 9$
(c) $32^{3/5} = (32^{1/5})^3 = (\sqrt[5]{32})^3 = 2^3 = 8$
(d) $-8^{4/3} = -(8^{1/3})^4 = -(\sqrt[3]{8})^4 = -(2)^4 = -16$
(e) $(0.027)^{4/3} = [(0.027)^{1/3}]^4 = [\sqrt[3]{0.027}]^4 = [0.3]^4 = 0.0081$

EXAMPLE 4 Assume all variables represent positive numbers, and write each of the following radicals in rational exponent form: (a) $\sqrt[3]{y}$ (b) $\sqrt{x^3}$ (c) $-4\sqrt[5]{x^2y^3}$

SOLUTION
(a) $\sqrt[3]{y} = y^{1/3}$
(b) $\sqrt{x^3} = (x^3)^{1/2} = x^{3/2}$
(c) $-4\sqrt[5]{x^2y^3} = -4(x^2y^3)^{1/5} = -4x^{2/5}y^{3/5}$

We can also use the concept of *negative* exponents with rational exponents.

EXAMPLE 5 Evaluate each of the following: (a) $16^{-1/4}$ (b) $64^{-1/3}$ (c) $27^{-2/3}$

SOLUTION
(a) $16^{-1/4} = \dfrac{1}{16^{1/4}} = \dfrac{1}{\sqrt[4]{16}} = \dfrac{1}{2}$
(b) $64^{-1/3} = \dfrac{1}{64^{1/3}} = \dfrac{1}{\sqrt[3]{64}} = \dfrac{1}{4}$
(c) $27^{-2/3} = \dfrac{1}{27^{2/3}} = \dfrac{1}{(27^{1/3})^2} = \dfrac{1}{(\sqrt[3]{27})^2} = \dfrac{1}{3^2} = \dfrac{1}{9}$

As we work with rational exponents, it should be kept in mind that all the rules of exponents that were stated earlier for integers still hold.

EXAMPLE 6 Use the rules of exponents to simplify each of the following quantities.

(a) $2^{1/2} \cdot 2^{1/2} = 2^{(1/2)+(1/2)} = 2$ $b^m \cdot b^n = b^{m+n}$
(b) $(5^{3/4})^4 = 5^{(3/4) \cdot 4} = 5^3 = 125$ $(b^m)^n = b^{mn}$

(c) $\dfrac{3^{1/2}}{3^{1/4}} = 3^{(1/2)-(1/4)}$ $\qquad \dfrac{b^m}{b^n} = b^{m-n}$

$\qquad \qquad = 3^{(2/4)-(1/4)} = 3^{1/4}$

(d) $\left(\dfrac{9}{4}\right)^{1/2} = \dfrac{9^{1/2}}{4^{1/2}}$ $\qquad \left(\dfrac{a}{b}\right)^m = \dfrac{a^m}{b^m}$

$\qquad \qquad = \dfrac{\sqrt{9}}{\sqrt{4}} = \dfrac{3}{2}$

(e) $\dfrac{a^{1/4} \cdot a^{-2}}{a^{2/3} \cdot a^{-3/4}} = a^{(1/4)+(-2)-(2/3)-(-3/4)}$

$\qquad \qquad = a^{(1/4)-2-(2/3)+(3/4)}$

$\qquad \qquad = a^{(3/12)-(24/12)-(8/12)+(9/12)}$ The LCD is 12.

$\qquad \qquad = a^{-20/12}$

$\qquad \qquad = a^{-5/3} = \dfrac{1}{a^{5/3}}$ $\qquad b^{-n} = \dfrac{1}{b^n}$

Rational exponents also provide a means to write some radicals in simpler form by reducing their index.

EXAMPLE 7 Assume all variables represent positive numbers, and simplify each of the following radicals by using rational exponents: **(a)** $\sqrt[4]{4}$ **(b)** $(\sqrt[9]{x})^3$ **(c)** $\sqrt[6]{x^8}$ **(d)** $\sqrt{x} \cdot \sqrt[3]{x^2}$

SOLUTION **(a)** $\sqrt[4]{4} = 4^{1/4} = (2^2)^{1/4} = 2^{2/4} = 2^{1/2} = \sqrt{2}$

(b) $(\sqrt[9]{x})^3 = (x^{1/9})^3 = x^{3/9} = x^{1/3} = \sqrt[3]{x}$

(c) $\sqrt[6]{x^8} = (x^8)^{1/6} = x^{8/6} = x^{4/3} = \sqrt[3]{x^4} = \sqrt[3]{x^3 \cdot x} = x\sqrt[3]{x}$

(d) $\sqrt{x} \cdot \sqrt[3]{x^2} = x^{1/2} \cdot x^{2/3}$

$\qquad \qquad = x^{(3/6)+(4/6)} = x^{7/6}$

$\qquad \qquad = \sqrt[6]{x^7} = \sqrt[6]{x^6 \cdot x} = x\sqrt[6]{x}$

Do You Remember?

Can you match these?

Assume all variables represent positive numbers.

___ 1. $y^{2/3}$ in radical form
___ 2. $\sqrt[4]{x^3}$ in exponential form
___ 3. $(\sqrt[4]{x})^5$ in exponential form
___ 4. $8^{4/3} =$ ___
___ 5. $4^{1/2} + 8^{2/3} =$ ___
___ 6. A simpler radical form of $\sqrt[4]{9}$

a) $\sqrt{3}$
b) 16
c) $\sqrt[3]{y^2}$
d) 6
e) $x^{5/4}$
f) $x^{4/3}$
g) $x^{4/5}$
h) $x^{3/4}$

Answers: 1. c 2. h 3. e 4. b 5. d 6. a

Exercise Set 9.6

Write each of the following exponential quantities in radical form. Do not simplify.

1. $4^{1/2}$
2. $9^{1/2}$
3. $16^{1/2}$
4. $25^{1/2}$
5. $3^{1/4}$
6. $1^{1/4}$
7. $-8^{2/3}$
8. $-12^{3/4}$
9. $x^{1/2}$
10. $y^{1/3}$
11. $4x^{1/2}$
12. $6y^{1/3}$

Write each of the following radicals in rational exponent form. Do not simplify.

13. $\sqrt{x}$
14. $\sqrt{11}$
15. $\sqrt[3]{7}$
16. $\sqrt[3]{13}$
17. $-\sqrt[3]{8^2}$
18. $-\sqrt[3]{4^2}$
19. $\sqrt[4]{27^3}$
20. $\sqrt[5]{32^3}$
21. $-\sqrt{x}$
22. $-\sqrt{y}$
23. $\sqrt[3]{x^2}$
24. $\sqrt[3]{y^2}$
25. $4\sqrt[5]{y^3}$
26. $6\sqrt[7]{x^4}$
27. $-8\sqrt{xy^2}$
28. $-5\sqrt{xy}$
29. $\sqrt[4]{x^3y^7}$
30. $\sqrt[5]{x^2y^8}$

Simplify each of the following expressions.

31. $4^{1/2}$
32. $25^{1/2}$
33. $8^{1/3}$
34. $27^{1/3}$
35. $-8^{2/3}$
36. $-27^{2/3}$
37. $32^{3/5}$
38. $81^{3/4}$
39. $-4^{3/2}$
40. $-9^{3/2}$
41. $(-8)^{2/3}$
42. $(-27)^{2/3}$
43. $(0.25)^{1/2}$
44. $(0.16)^{1/2}$
45. $(0.0016)^{3/4}$
46. $(0.0081)^{3/4}$
47. $\left(\dfrac{1}{4}\right)^{1/2}$
48. $\left(\dfrac{1}{25}\right)^{1/2}$
49. $\left(\dfrac{1}{8}\right)^{2/3}$
50. $\left(\dfrac{1}{64}\right)^{2/3}$
51. $\left(\dfrac{-27}{64}\right)^{1/3}$
52. $\left(\dfrac{-64}{125}\right)^{1/3}$
53. $-\left(\dfrac{16}{81}\right)^{3/4}$
54. $-\left(\dfrac{16}{625}\right)^{3/4}$
55. $(8x^3)^{1/3}$
56. $(27x^6)^{1/3}$
57. $(256x^8)^{1/4}$
58. $(81x^{16})^{1/4}$

Use the product and quotient rules of exponents to simplify each of the following products or quotients.

59. $2^{1/2} \cdot 2^{1/2}$
60. $3^{1/2} \cdot 3^{1/2}$
61. $2^{1/3} \cdot 4^{1/3}$
62. $3^{1/3} \cdot 9^{1/3}$
63. $\dfrac{8^{2/3}}{8^{1/2}}$
64. $\dfrac{9^{4/3}}{9^{1/3}}$
65. $\dfrac{64^{1/4}}{4^{1/4}}$
66. $\dfrac{16^{2/3}}{2^{2/3}}$
67. $3^{2/3} \cdot 9^{2/3}$
68. $2^{2/3} \cdot 4^{2/3}$
69. $\left(\dfrac{2}{3}\right)^{2/3}\left(\dfrac{4}{9}\right)^{2/3}$
70. $\left(\dfrac{9}{8}\right)^{3/4}\left(\dfrac{9}{2}\right)^{3/4}$
71. $(0.2)^{2/3}(0.04)^{2/3}$
72. $(0.03)^{3/4}(0.09)^{3/4}$
73. $(0.5)^{3/2}(0.5)^{1/2}$
74. $(0.16)^{5/4}(0.16)^{3/4}$
75. $\left(\dfrac{3}{4}\right)^{5/4}\left(\dfrac{3}{4}\right)^{3/4}$
76. $\left(\dfrac{5}{8}\right)^{3/2}\left(\dfrac{5}{8}\right)^{1/2}$

Use the exponent rules $x^0 = 1$ and $x^{-n} = \dfrac{1}{x^n}$ (for $x \neq 0$) to simplify each of the following expressions.

77. 8^0
78. 5^0
79. -11^0
80. -15^0
81. $8^{-1/3}$
82. $27^{-1/3}$
83. $4^{-1/2}$
84. $64^{-1/2}$
85. $16^{-3/4}$
86. $81^{-3/4}$
87. $(0.0016)^{-1/4}$
88. $(0.0081)^{-1/4}$
89. $(-27)^{-2/3}$
90. $(-125)^{-2/3}$
91. $(-32)^{-4/5}$
92. $(-1)^{-4/5}$
93. $\left(\dfrac{1}{2}\right)^{-1}$
94. $\left(\dfrac{1}{3}\right)^{-1}$
95. $\left(\dfrac{1}{8}\right)^{-2/3}$
96. $\left(\dfrac{8}{27}\right)^{-2/3}$

Simplify each of the following quantities. Assume all variables represent positive real numbers.

97. $(x^{2/3})^{3/2}$
98. $(y^3)^{1/3}$
99. $(y^8)^{1/4}$
100. $(x^{1/7})^{14}$
101. $\dfrac{y^{1/3}}{y^{4/3}}$
102. $\dfrac{x^{6/5}}{x^{1/5}}$

103. $\left(\dfrac{x^{1/2}}{x^{1/3}}\right)^{12}$
104. $\left(\dfrac{a^{1/4}}{a^{1/3}}\right)^{24}$
105. $\left(\dfrac{x^{7/6}x^{5/6}}{x^{1/3}x}\right)^3$
106. $\left(\dfrac{y^{5/9}y^{4/9}}{y^{2/9}y^{1/9}}\right)^{18}$
107. $\left(\dfrac{a^{1/2}a^{1/3}}{a^{1/4}a^{2/3}}\right)^{12}$
108. $\left(\dfrac{t^{3/4}t^{1/2}}{t^{1/6}t^{1/3}}\right)^{24}$

REVIEW EXERCISES

Solve each of the following applied problems.

109. A 10-ft ladder is leaning against a building. The bottom of the ladder is 3 ft from the building. How high is the top of the ladder? Give an exact answer and an approximate answer rounded to three decimal places.

110. A softball diamond is in the shape of a square with 60-ft sides. The catcher stands 3 ft behind home plate and throws to second base. Give both an exact and an approximate answer rounded to three decimal places, of the distance the ball is thrown.

Summary

9.1 Roots and Radicals

The opposite of squaring a number is finding the **square root**.

The **principal square root** of any positive number a, written $\sqrt{a}$, is that positive number b such that $b^2 = a$.

The symbol "$\sqrt{}$" is called a **radical sign,** and it always indicates a *positive* square root. The number under the radical sign is called the **radicand** and the entire expression is called a **radical expression.**

Radical sign → $\sqrt{b}$ ← Radicand
↑
Radical expression

The square of a rational number is called a **perfect square.** The square root of a number that is not a perfect square is an **irrational number.**

Square roots of numbers can be found by using a calculator. The owner's manual should be consulted to determine the method for each particular brand of calculator.

Roots other than square roots are indicated by the **index number** that appears on the radical symbol:

Index ⌐   ⌐ Radical sign
$\sqrt[n]{a}$ ← Radicand

CAUTION

The square root of a sum of two or more numbers is *not* equal to the sum of the square roots of the numbers:

$\sqrt{a^2 + b^2} \neq \sqrt{a^2} + \sqrt{b^2}$

The **cube root** of a number a is that number b such that $b^3 = a$. The **fourth root** of a number a is that number b such that $b^4 = a$. In general, the **nth root** of a number a, written $\sqrt[n]{a}$, is that number b such that $b^n = a$.

Any odd root of a negative number is a negative number, but an even root of a negative number is not a real number.

9.2 Simplified Forms of Radicals

Radicals are simplified by using the multiplication and division properties of radicals. If a and b are nonnegative real numbers, then

MULTIPLICATION PROPERTY

$$\sqrt[n]{a \cdot b} = \sqrt[n]{a} \cdot \sqrt[n]{b}$$

DIVISION PROPERTY

$$\sqrt[n]{\frac{a}{b}} = \frac{\sqrt[n]{a}}{\sqrt[n]{b}}, \quad b \neq 0$$

Three conditions must be met for a radical expression to be in **simplest form:**

1. The radicand can contain no factors that are perfect squares if the radical is a square root, no factors that are perfect cubes if it is a cube root, or, in general, no factors that are perfect nth powers if the radical is an nth root.
2. No fractions can appear under the radical.
3. No radical can appear in the denominator of an expression involving fractions.

9.3 Addition and Subtraction of Radicals

In order to add or subtract radicals, they must have the same radicand and the same index. When radicals are to be added or subtracted but they do *not* have the same radicand, it is sometimes possible to simplify them first so that they do have the same radicand and then to add or subtract them.

9.4 Multiplication and Division of Radicals

Expressions containing radicals are multiplied using the same methods that are used to multiply polynomials. After the product is found, all radicals should be simplified as much as possible, following the rules in Section 9.2.

An expression with a radical in the denominator is not in simplest form. **Rationalizing the denominator** will simplify such an expression. To rationalize a denominator such as $\sqrt{a} + \sqrt{b}$, multiply both numerator and denominator by its **conjugate,** $\sqrt{a} - \sqrt{b}$.

9.5 Solving Equations Containing Radicals

Equations involving radicals are solved by using the **power property of equality,** which states that if A and B are two algebraic expressions and if

$$A = B \quad \text{then} \quad A^n = B^n$$

Care must be taken when using the power property because sometimes it can produce an **extraneous** (false) **solution.** *All solutions to equations obtained by using the power property must be checked in the* **original** *equation.*

The distance between two points (x_1, y_1) and (x_2, y_2) in a plane is found by using a formula that involves a square root. The formula is derived by using the Pythagorean theorem.

The Distance Formula

$$d = \sqrt{(x_2 - x_1)^2 + (y_2 - y_1)^2}$$

9.6 Rational Exponents and Radicals

Expressions involving rational exponents are evaluated using the same rules that apply to expressions involving integer exponents. Given that m is an *integer*, n is a *positive integer greater than* 1, and b is a *positive number*, the relationship between rational exponents and radicals is as follows:

$$b^{1/n} = \sqrt[n]{b} \quad \text{and} \quad b^{m/n} = \sqrt[n]{b^m} = (\sqrt[n]{b})^m$$

Changing a radical into exponential form is often helpful in simplifying the radical.

Cooperative Exercise

In a right triangle with sides a, b, and c, where c is the hypotenuse, by the Pythagorean theorem,

$$a^2 + b^2 = c^2 \quad \text{or} \quad c = \sqrt{a^2 + b^2}$$

A. Construct a right triangle with sides 3, 4, and 5. Construct a square on each side. Calculate the area of each square. What is the relationship between these squares? How does this relate to the Pythagorean theorem?

B. Construct a right triangle as in part A, but construct a semicircle on each side. Calculate the area of each semicircle. What is the relationship between these three areas? How does this relate to the Pythagorean theorem?

C. Construct a right triangle as in part A, but construct an equilateral triangle on each side. Calculate the area of each equilateral triangle. What is the relationship between these three areas? How does this relate to the Pythagorean theorem?

D. Use parts A, B, and C to generalize a rule regarding the areas of similar figures constructed on the sides of a right triangle.

Review Exercises

Find the indicated root. If no real-number root exists, write Not a real number. *Assume all variables represent positive real numbers.*

1. $\sqrt{49}$
2. $-\sqrt{36}$
3. $\sqrt{\dfrac{9}{64}}$
4. $\sqrt{\dfrac{16}{121}}$
5. $\sqrt{-9}$
6. $-\sqrt{-16}$
7. $-\sqrt[3]{27}$
8. $\sqrt[3]{64}$
9. $\sqrt[4]{16}$
10. $\sqrt[4]{\dfrac{1}{81}}$
11. $\sqrt[3]{0.008}$
12. $\sqrt[4]{0.0016}$
13. $\sqrt{x^2}$
14. $\sqrt{y^4}$
15. $\sqrt{x^2y^2}$
16. $\sqrt{\dfrac{y^4}{x^8}}$
17. $-\sqrt[3]{x^3}$
18. $-\sqrt[3]{-x^3}$
19. $-\sqrt[4]{y^4}$
20. $\sqrt{81x^2y^8z^6}$

Simplify each of the following radical expressions. Assume all variables represent positive real numbers.

21. $\sqrt{8}$
22. $\sqrt{56}$
23. $\sqrt[3]{16x^3}$
24. $\sqrt[3]{-81xy^6}$
25. $\sqrt[3]{54x^7}$
26. $\sqrt[4]{32x^5y^4}$
27. $-\sqrt[3]{108x}$
28. $\sqrt[3]{0.064x^4}$
29. $\sqrt{\dfrac{18x^3y^7}{3xy^2}}$
30. $\sqrt{\dfrac{11a}{9a^3}}$
31. $\sqrt[3]{\dfrac{2a^2b}{125ab^2}}$
32. $\sqrt[4]{\dfrac{32x^3}{162x^7y^4}}$

Add or subtract (if possible) as indicated, and simplify. Assume all variables represent positive real numbers.

33. $5\sqrt{2} + 3\sqrt{2}$
34. $7\sqrt{11} - 8\sqrt{11}$
35. $6\sqrt{x} - 2\sqrt{x}$
36. $5\sqrt{y} - \sqrt{y}$
37. $2\sqrt{3} + 4\sqrt{5}$
38. $5\sqrt{x} - 3\sqrt{y}$
39. $3\sqrt[3]{4} + 5\sqrt[3]{4}$
40. $-8\sqrt[3]{6} + \sqrt[3]{6}$

41. $3\sqrt{x^3} - 5x\sqrt{x}$
42. $3\sqrt{44} - 5\sqrt{99} + \sqrt{176}$
43. $\sqrt{48} - 2\sqrt{27} + 6\sqrt{32} - \sqrt{16} + \sqrt{25}$
44. $7\sqrt{20y^3} - 3y\sqrt{45y^2}$
45. $9\sqrt{128a^3} - 10a\sqrt{72a}$
46. $\dfrac{\sqrt[3]{4} - \sqrt[3]{2}}{3} - \dfrac{\sqrt[3]{2} + \sqrt[3]{4}}{5}$
47. $\dfrac{-2 + \sqrt{2^2 - 4(-1)}}{2}$
48. $\dfrac{-3 - \sqrt{3^2 - 4(2)}}{2}$
49. $\dfrac{\sqrt{7} + 2}{3} - \dfrac{\sqrt{7} - 2}{4}$

Simplify each of the following quantities. Assume all variables represent positive real numbers.

50. $\dfrac{1}{\sqrt{2}}$
51. $\dfrac{5}{\sqrt[3]{5}}$
52. $\dfrac{\sqrt{3} + 2}{\sqrt{2}}$
53. $\dfrac{1}{\sqrt{2} - 1}$
54. $\dfrac{3}{\sqrt{2} - 3}$
55. $\dfrac{\sqrt{3}}{\sqrt{2} - 5}$
56. $\dfrac{\sqrt{2}}{\sqrt{3} + \sqrt{5}}$
57. $\dfrac{1}{\sqrt{x}}$
58. $\dfrac{1}{\sqrt[3]{y}}$
59. $\dfrac{x - y}{\sqrt{x} - y}$
60. $\sqrt{x}\sqrt{y}$
61. $\sqrt{x}\sqrt[3]{x^2}$
62. $\sqrt[3]{2x^4y^3}$
63. $x(\sqrt{x} - 9)$
64. $2\sqrt{x}(3\sqrt{x} - \sqrt{x^2})$
65. $(-5\sqrt{x^3})(\sqrt{x})$
66. $(2\sqrt{x} - \sqrt{y})(2\sqrt{x} + \sqrt{y})$
67. $(\sqrt{x} + \sqrt{3})^2$
68. $(\sqrt{3xy} - \sqrt{2x})^2$
69. $\dfrac{\sqrt{6} - \sqrt{3}}{\sqrt{3} + \sqrt{2}}$
70. $\dfrac{2\sqrt{a} - 5\sqrt{b}}{3\sqrt{a} + \sqrt{b}}$

Solve each radical equation. Check all solutions in the original equation.

71. $\sqrt{x} = 5$
72. $\sqrt{3x} = 1$
73. $\sqrt{x + 2} = 0$
74. $\sqrt{t - 3} = 2$
75. $\sqrt{2a + 1} = 3\sqrt{a}$
76. $6 + 5\sqrt{2x} = \sqrt{2x} + 6$
77. $\sqrt{3x + 4} = \sqrt{5x - 1}$
78. $x = \sqrt{4x}$
79. $y - 4 = \sqrt{2y}$
80. $\sqrt{x^2 - 8} = x + 4$
81. $\sqrt{6y + 4} + 5 = 1$
82. $\sqrt{t}\sqrt{t - 3} = 2$
83. $4\sqrt{m + 1} = \sqrt{8m + 7}$
84. $\sqrt{x^2 + 5x + 5} = 1$
85. $3 + 5x = 4x - \sqrt{x^2 + 2}$

Find the distance between the given points.

86. $(0, 4)$ and $(1, -5)$
87. $(0, 0)$ and $(5, -3)$
88. $(-2, 3)$ and $(4, -5)$
89. $\left(\dfrac{1}{2}, -3\right)$ and $\left(2, \dfrac{1}{2}\right)$

Simplify each of the following expressions. Assume all variables represent positive real numbers.

90. $(x^{1/4})^{1/2}$
91. $(x^{-5})^{1/3}$
92. $\left(\dfrac{x^{3/4}}{x^{5/3}}\right)^{-1}$
93. $\left(\dfrac{x^{1/2}x^{1/3}}{x^{3/4}x^{2/3}}\right)^{-1/2}$

Solve each of the following applied problems.

94. A rectangular garden has sides of 22 ft and 34 ft. What is the approximate length of the diagonal of the rectangle? (Round your answer to three decimal places.)

95. Two bicycle riders leave Edmond at the same time. One travels north at an average speed of 16 mph, and the other travels west at an average speed of 12 mph. How far apart are they after they have traveled 2 hr?

96. A florist wants to make bouquets of flowers. Each bouquet is to be made up of daisies at $10/bunch and daffodils at $7/bunch. How many bunches of each should she use to make 15 bouquets that she can sell for $8 a bouquet?

97. A nurse needs 20 g of a 52% solution of a certain medicine. He has only 40% and 70% solutions available. How much of each solution should he use to obtain the 20 g of 52% solution?

Chapter Test

Simplify each expression. Assume all variables represent positive real numbers.

1. $\sqrt{36}$
2. $-\sqrt{121}$
3. $\sqrt{28}$
4. $-\sqrt{x^6}$
5. $\sqrt{y^{14}}$
6. $\sqrt[3]{m^5}$
7. $8^{2/3}$
8. $9^{3/2}$

9. $\left(\dfrac{3}{4}\right)^{-1/2}$

10. $-\left(\dfrac{16}{81}\right)^{-3/4}$

11. $\sqrt{\dfrac{121a^{26}}{99a^{13}}}$

12. $-\sqrt[3]{\dfrac{78x^5y^7}{6x^2y^3}}$

13. $-\dfrac{3}{\sqrt{11}}$

14. $\dfrac{-5}{\sqrt{6}-1}$

15. $\dfrac{2}{\sqrt[3]{3}}$

16. $\dfrac{5x}{3\sqrt{x}}$

17. $3\sqrt{8} - 2\sqrt{50} + 4\sqrt{72}$

18. $(\sqrt{3}+\sqrt{2})^2$

19. $\dfrac{2}{\sqrt{3}} + \sqrt{\dfrac{5}{3}} - 2\sqrt{27}$

20. $(3\sqrt{5}+2\sqrt{3})(3\sqrt{5}-2\sqrt{3})$

21. $\dfrac{2\sqrt{3}-4}{\sqrt{3}+2}$

22. $\dfrac{3\sqrt{5}+\sqrt{2}}{5\sqrt{2}+\sqrt{7}}$

Solve each radical equation in Exercises 23–30, and check all solutions in the original equation.

23. $\sqrt{y} = 17$
24. $\sqrt{2y-5} = 7$
25. $\sqrt{3x+1} = \sqrt{x+2}$
26. $\sqrt{x^2-9} = x-3$
27. $x = \sqrt{2x-1} + 2$
28. $x - 6 = \sqrt{x}$
29. $\sqrt{x^2-5x+4} = \sqrt{2x^2-4x+2}$
30. $\sqrt{x-1} = -2$

31. Find the distance between the points $(3, -1)$ and $(4, 6)$. Express your answer in simplest radical form.
32. Find the hypotenuse c of the right triangle shown in the figure. Leave your answer in simplest radical form.

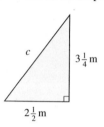

Cumulative Review for Chapters 7–9

Factor each of the following numbers into prime factored form.

1. 36
2. 35
3. 858
4. 2310

Factor each polynomial completely. If not factorable, write prime.

5. $3x - 6$
6. $x^2 + 2x$
7. $-8m - 12$
8. $21x^4y^2 - 14x^3y^3 - 7xy$
9. $-15x^3 + 12x^2 - 3x$
10. $x^2 - 8x + 15$
11. $x^3 + x^2y - 30xy^2$
12. $a(b+3) - 3(b+3)$
13. $y(y-5) - 5(y-5)$
14. $m^2 - 0.7m + 0.12$
15. $(a+b)(a-b) - (a-b)^2$
16. $30 + x - 20x^2$
17. $30ab^3 + 16a^2b^2 - 32a^3b$
18. $x^2 - 16$
19. $y^2 + 6y + 9$
20. $x^3 - 8$
21. $ax - ay + bx - by$
22. $2y(3x-4) - (3x-4)$
23. $16x^4 - 8x^2 + 1$
24. $x^2 + 9$
25. $\dfrac{1}{9}x^2 - \dfrac{1}{4}y^2$
26. $0.36 - 0.25b^2$
27. $27a^3 + 64b^3$
28. $16x^3 - 24x^2y + 9xy^2$

Solve each of the following polynomial equations.

29. $x^2 + 2x - 3 = 0$
30. $a^2 + 4a - 77 = 0$
31. $10y^2 - 3y - 18 = 0$
32. $(x-2)^2 = 9$
33. $a^3 + 8a = 6a^2$
34. $p^2 + 8p - 4 = -11$
35. $x(x^2 - 5x + 4)(x^2 - 4x + 4) = 0$
36. $49a^2 - 81 = 0$

Factor and simplify.

37. $\dfrac{x+3}{2x+6}$
38. $\dfrac{x^2-9}{x^2-6x+9}$
39. $\dfrac{a^2+3a+2}{a^2-4}$
40. $\dfrac{4m^2+4m+1}{4m+2}$
41. $\dfrac{a^3-b^3}{a^2-b^2}$
42. $\dfrac{x^3y - 8x^2y + 15xy}{x^2y - 3xy}$
43. $\dfrac{x^2-y^2}{y-x}$
44. $\dfrac{-6x-12}{8x^2+32x+32}$
45. $\dfrac{a^2+4ab+3b^2}{-a^2-7ab-12b^2}$

Multiply or divide as indicated, and simplify.

46. $\dfrac{3a}{x^2 y} \cdot \dfrac{xy^2}{12a}$

47. $\dfrac{-16s}{18t} \div \dfrac{8s^2}{6t^2}$

48. $\dfrac{x^3 + x^2}{x^2 - 1} \cdot \dfrac{x^2 - 2x + 1}{x^2}$

49. $\dfrac{y - x}{x + y} \cdot \dfrac{x - y}{x + y} \div \dfrac{-1}{x^2 + 2xy + y^2}$

50. $\dfrac{35a^2 b}{18c^2 d} \div \dfrac{8ab}{9cd} \div \dfrac{15a^2}{3cd}$

Simplify.

51. $\dfrac{8}{5x} - \dfrac{3}{5x} + \dfrac{10}{5x}$

52. $\dfrac{1}{x + 2} + \dfrac{2x + 1}{x^2 + 4x + 4}$

53. $\dfrac{x}{x + 4} + \dfrac{4}{3x + 12}$

54. $\dfrac{5y - 1}{y^2 + 2y + 1} + \dfrac{y + 1}{y^2 + 3y + 2}$

Simplify each complex fraction.

55. $\dfrac{\dfrac{1}{x}}{\dfrac{2}{x^2}}$

56. $\dfrac{1 - \dfrac{3}{7}}{1 + \dfrac{5}{14}}$

57. $\dfrac{\dfrac{1}{a} + \dfrac{1}{b}}{\dfrac{1}{a + b}}$

58. $\dfrac{x + \dfrac{1}{x}}{x - \dfrac{1}{x}}$

59. $\dfrac{\dfrac{5}{s} + \dfrac{4}{s^2}}{\dfrac{3}{s} + 1}$

60. $\dfrac{x - 1 - \dfrac{1}{x - 1}}{x + 1 - \dfrac{1}{x - 1}}$

Solve each rational equation and check your solutions. Note any restrictions on the variables.

61. $\dfrac{3}{x} = \dfrac{1}{10}$

62. $\dfrac{p}{5} = \dfrac{-2}{7}$

63. $\dfrac{5}{5x - 4} = \dfrac{1}{x}$

64. $\dfrac{y + 6}{y - 3} + 8 = 0$

65. $\dfrac{2a}{4a + 4} + \dfrac{a}{2a + 2} = \dfrac{a}{a + 1}$

Solve each proportion.

66. $\dfrac{y + 3}{y} = \dfrac{5}{7}$

67. $\dfrac{4}{x + 3} = \dfrac{8}{2x - 3}$

68. $\dfrac{y + 2}{y - 2} = \dfrac{y + 2}{y}$

69. $\dfrac{a + 1}{5a + 5} = \dfrac{1}{a - 1}$

70. If 12 lb of seed is enough to plant 2400 ft² of lawn, how much seed is needed to plant 6000 ft²?

71. If you can drive 198 mi on 11 gal of gas, how far can you drive on 22 gal?

72. Managers at a factory find that 7 out of every 91 items produced is defective. How many defective items can they expect in a production run of 455 items?

73. A survey shows that 3 out of 7 freshmen have already decided on a career. If 896 freshmen are surveyed, how many have decided on a career?

Write an equation for each variation problem; find k and solve for y.

74. If y varies directly with the square of x, and if $y = 12$ when $x = 3$, find y when $x = 7$.

75. If y varies jointly with x and z and inversely with w, and if $y = 10$ when $x = 2$, $z = 3$, and $w = 5$, find y when $x = 3$, $z = 4$, and $w = 2$.

Simplify each of the following radicals. Assume that all variables represent positive real numbers.

76. $\sqrt{36}$

77. $-\sqrt{18}$

78. $\sqrt{x^3}$

79. $\sqrt[3]{\dfrac{1}{27}}$

80. $\sqrt[3]{-64}$

81. $\sqrt{25x^4 y^2}$

82. $\dfrac{\sqrt{18}}{\sqrt{3}}$

83. $\dfrac{\sqrt{30x^2 y}}{\sqrt{2xy}}$

84. $\sqrt[3]{\dfrac{1}{2}} \sqrt[3]{\dfrac{1}{4}}$

85. $\sqrt{0.49}$

86. $\sqrt[3]{0.008}$

87. $3\sqrt{5} - \sqrt{2} + \sqrt{5}$

88. $\sqrt{54} - 5\sqrt{32} - \sqrt{96} + \sqrt{8}$

89. $-3\sqrt{49x^2} + 8\sqrt{28x^3}$

Multiply or divide as indicated, and simplify

90. $\sqrt{3} \sqrt{3^3}$

91. $(\sqrt{5} + \sqrt{2})(\sqrt{5} - \sqrt{2})$

92. $\dfrac{5x}{\sqrt{5x}}, \quad x > 0$

93. $\dfrac{2\sqrt{3} - 5}{\sqrt{3} - \sqrt{2}}$

Solve each equation; check your answers.

94. $\sqrt{x - 3} = 1$

95. $\sqrt{2x + 3} = 5$

96. $\sqrt{4x - 1} = -2$

97. $a + 3 + \sqrt{a^2 + 2} = 0$

98. $2 - \sqrt{x + 8} = -\sqrt{x}$

99. $-7 = -3\sqrt{2a}$

Simplify each of the following quantities. Assume that all variables represent positive real numbers.

100. $3^{1/4} \cdot 3^{1/2}$

101. $\dfrac{8^{2/3}}{4^{1/2}}$

102. $\left(\dfrac{3}{7}\right)^{1/2}\left(\dfrac{3}{7}\right)^{1/2}$

103. $\dfrac{64^{1/2}}{4^{1/2}}$

104. $\left(\dfrac{-1}{8}\right)^{-2/3}$

105. $(x^2)^{1/2}$

106. $\left(\dfrac{x^{1/3}}{x^{1/4}}\right)^{12}$

107. $\left(\dfrac{x^{1/2}y}{x^{1/3}y^{-2}}\right)^{-1}$

Solve each of the following applied problems.

108. Mrs. Orosco invests $10,000. She invests part of it at 5% and the rest at 9%. How much does she invest at each rate if her yearly interest from the two investments is $660?

109. A father and his son are both bricklayers. The son can lay bricks for a fireplace and chimney in 5 days. Together they can lay bricks for an identical fireplace and chimney in 2 days. How long does it take the father to do it alone?

110. One librarian can catalog a group of library books in 8 hr, while a second librarian can catalog the same number of books in 4 hr. A third person catalogs the same number of books in 6 hr. How long will it take if they all work together?

111. Nancy bought $30.64 worth of 19¢, 29¢, and 50¢ stamps. She bought 20 more of the 19¢ stamps than the 50¢ stamps, and she bought twice as many 29¢ stamps as 19¢ stamps. How many of each kind of stamp did she buy?

CHAPTER 10
Quadratic Equations

10.1 Solving Equations by Factoring and the Square Root Method
10.2 Solving Quadratic Equations by Completing the Square
10.3 Solving Quadratic Equations by the Quadratic Formula
10.4 Solving Quadratic Equations with Complex-Number Solutions
10.5 Graphing Quadratic Equations; The Parabola
10.6 More Applications

Application

Maria's Hobby Shop makes and sells wallet-making kits. The amount of profit per day is given by the equation $P(n) = -(n - 5)^2 + 16$, where P = profits in units of $10 and n is the number of kits made and sold. Sketch the profit equation. Let the horizontal axis be the n-axis and the vertical axis the P-axis. Determine the points at which Maria's profit is zero (the break-even points).

What production gives maximum profit? What is the maximum profit? What production causes her to lose money? What is the number of kits she must produce and sell to show a profit of $70? (*Reminder:* P is in units of $10, so a profit of $70 is written as 7.)

If this were your business, how many kits would you produce and sell?

Exercise Set 10.5, Exercise 72

When Frederick the Great wrote to Louis Lagrange that "the greatest king in Europe" wanted the "greatest mathematician in Europe" as part of his court, he may not have been exaggerating in either instance. Lagrange was born of wealthy parents, but by the time he reached his teens the family fortune was gone. Lagrange said, "If I had married rich, I probably would not have cast my lot with mathematics." Some of Lagrange's greatest work occurred between the ages of sixteen and twenty-two. During his lifetime he held numerous posts in prestigious schools, and at one time he served as the president of the French commission that worked out the metric system.

JOSEPH-LOUIS LAGRANGE (1736–1813)

10.1 Solving Equations by Factoring and the Square Root Method

OBJECTIVES

In this section we will learn to
1. solve equations of the form $x^2 = a$; and
2. solve equations of the form $(ax + b)^2 = c$.

In Section 7.6 we solved equations by factoring using the *zero-factor property*, and we have solved numerous problems in the sections that followed by using the same method. Recall that the **zero-factor property** states that if a and b represent any real numbers and if $a \cdot b = 0$, then either $a = 0$ or $b = 0$. Because this method is so important, we will review it in the next two examples.

EXAMPLE 1 Solve: $6x^2 + x - 12 = 0$.

SOLUTION To solve the equation, we find its factors and then set each factor equal to zero:

$$6x^2 + x - 12 = 0$$
$$(2x + 3)(3x - 4) = 0$$
$$2x + 3 = 0 \quad \text{or} \quad 3x - 4 = 0 \quad \text{Zero-factor property}$$
$$2x = -3 \quad \text{or} \quad 3x = 4$$
$$x = \frac{-3}{2} \quad \text{or} \quad x = \frac{4}{3}$$

The solution set is $\left\{\frac{-3}{2}, \frac{4}{3}\right\}$.

EXAMPLE 2 Solve: $x^4 - 29x^2 + 100 = 0$.

SOLUTION We begin by factoring:

$$x^4 - 29x^2 + 100 = 0$$
$$(x^2 - 4)(x^2 - 25) = 0$$
$$(x - 2)(x + 2)(x - 5)(x + 5) = 0$$

Now we use the zero-factor property to find the solutions.

$x - 2 = 0$ or $x + 2 = 0$ or $x - 5 = 0$ or $x + 5 = 0$
$\quad x = 2$ or $\quad x = -2$ or $\quad x = 5$ or $\quad x = -5$

The solution set is $\{-5, -2, 2, 5\}$.

Not all equations can be solved by factoring. In fact, for some quadratic equations, factoring is not always the easiest way to solve the equation, even when it is factorable. Consider the equation $x^2 = 4$. If we were to solve this equation by factoring, we would write

$$x^2 = 4$$
$$x^2 - 4 = 0$$
$$(x - 2)(x + 2) = 0$$
$$x - 2 = 0 \quad \text{or} \quad x + 2 = 0$$
$$x = 2 \quad \text{or} \quad x = -2$$

We can solve equations of this type much more quickly by using the *square root property:*

The Square Root Property

If a is a positive number and

$\quad x^2 = a \quad$ then $\quad x = \pm\sqrt{a}$

The solution to $x^2 = 4$ illustrates that in order to solve a quadratic equation by taking the square root of each side, you must include both a positive and a negative square root in the solution set.

Returning to the equation $x^2 = 4$, which we solved by factoring, let's solve it again using the square root property.

$$x^2 = 4$$
$$x = \pm\sqrt{4}$$
$$x = \pm 2$$

The solution set is $\{-2, 2\}$. Since we know there are two solutions, we can use the sign "$\pm$" to represent both of them. The expression ± 2 is read "positive or negative 2," since it is referring to a number, not an operation. An expression such as "$a \pm b$" is read "a plus or minus b," since $\pm$ now indicates either addition or subtraction. The method we have just used to solve this equation is often called the **square root method**.

EXAMPLE 3   Solve $x^2 = 8$ by the square root method.

SOLUTION $x^2 = 8$
$\qquad\qquad x = \pm\sqrt{8} \qquad$ The square root property

$$x = \pm\sqrt{4 \cdot 2}$$
$$x = \pm 2\sqrt{2}$$

The solution set is $\{-2\sqrt{2}, 2\sqrt{2}\}$.

EXAMPLE 4 Solve $(x - 1)^2 = 4$ by the square root method; check the solutions.

SOLUTION This problem is essentially the same as Example 3; it just looks more complex.

$$(x - 1)^2 = 4$$
$$x - 1 = \pm\sqrt{4} \quad \text{The square root property}$$
$$x - 1 = \pm 2$$
$$x = 1 \pm 2 \quad \text{Add 1 to each side.}$$

There are two solutions.

$$x = 1 + 2 \quad \text{or} \quad x = 1 - 2$$
$$x = 3 \quad \text{or} \quad x = -1$$

CHECK
$(x - 1)^2 = 4 \qquad (x - 1)^2 = 4$
$(-1 - 1)^2 \stackrel{?}{=} 4 \quad \text{or} \quad (3 - 1)^2 \stackrel{?}{=} 4 \qquad x = -1, 3$
$(-2)^2 \stackrel{?}{=} 4 \qquad (2)^2 \stackrel{?}{=} 4$
$4 \stackrel{\checkmark}{=} 4 \qquad 4 \stackrel{\checkmark}{=} 4$

The solution set is $\{-1, 3\}$.

EXAMPLE 5 Solve $(y - 3)^2 = 27$ and check.

SOLUTION
$$(y - 3)^2 = 27$$
$$y - 3 = \pm\sqrt{27} \quad \text{The square root property}$$
$$y - 3 = \pm\sqrt{9 \cdot 3}$$
$$y - 3 = \pm 3\sqrt{3}$$
$$y = 3 \pm 3\sqrt{3}$$

CHECK
$(y - 3)^2 = 27 \qquad (y - 3)^2 = 27$
$(3 - 3\sqrt{3} - 3)^2 \stackrel{?}{=} 27 \quad \text{or} \quad (3 + 3\sqrt{3} - 3)^2 \stackrel{?}{=} 27 \qquad y = 3 \pm \sqrt{3}$
$(-3\sqrt{3})^2 \stackrel{?}{=} 27 \qquad (3\sqrt{3})^2 \stackrel{?}{=} 27$
$27 \stackrel{\checkmark}{=} 27 \qquad 27 \stackrel{\checkmark}{=} 27$

The solution set is $\{3 - 3\sqrt{3}, 3 + 3\sqrt{3}\}$.

EXAMPLE 6 Solve: $(2y + 1)^2 = 17$.

SOLUTION
$$(2y + 1)^2 = 17$$
$$2y + 1 = \pm\sqrt{17} \quad \text{The square root property}$$

$$2y = -1 \pm \sqrt{17} \quad \text{Subtract 1 from each side.}$$
$$y = \frac{-1 \pm \sqrt{17}}{2} \quad \text{Divide each side by 2.}$$

Both solutions check, so the solution set is $\left\{\frac{-1-\sqrt{17}}{2}, \frac{-1+\sqrt{17}}{2}\right\}$. We can also write the solution set $\left\{\frac{-1}{2}-\frac{\sqrt{17}}{2}, \frac{-1}{2}+\frac{\sqrt{17}}{2}\right\}$ by dividing both terms in the numerator by 2.

EXAMPLE 7 Solve: $\left(x - \frac{1}{2}\right)^2 = \frac{4}{9}$.

SOLUTION
$$\left(x - \frac{1}{2}\right)^2 = \frac{4}{9}$$
$$x - \frac{1}{2} = \pm\sqrt{\frac{4}{9}} \quad \text{The square root property}$$
$$x - \frac{1}{2} = \pm\frac{2}{3}$$
$$x = \frac{1}{2} \pm \frac{2}{3}$$
$$x = \frac{3 \pm 4}{6} \quad \text{The LCD is 6.}$$
$$x = \frac{3-4}{6} \quad \text{or} \quad x = \frac{3+4}{6}$$
$$x = \frac{-1}{6} \quad \text{or} \quad x = \frac{7}{6}$$

Both solutions check. The solution set is $\left\{\frac{-1}{6}, \frac{7}{6}\right\}$.

CAUTION
Do not confuse $\sqrt{a}$ with $\pm\sqrt{a}$. The second form arises from solving equations by the square root method. The first form is the **principal square root** of the number a.

EXAMPLE 8 Solve $x^2 = -4$.

SOLUTION Since the square of any real-number is positive, this problem has no real-number solutions. If we were to try to apply the square root property, we would obtain $x = \pm\sqrt{-4}$, which is not a real number.

EXAMPLE 9 The square of a number is 6 more than 5 times the number. Use the zero-factor property to find the number.

SOLUTION   We let $n =$ the number. Then

The square of a number is 6 more than 5 times the number.

$$n^2 = 5n + 6$$
$$n^2 - 5n - 6 = 0 \quad \text{Write in standard form.}$$
$$(n - 6)(n + 1) = 0 \quad \text{Factor the left side of the equation.}$$
$$n - 6 = 0 \quad \text{or} \quad n + 1 = 0 \quad \text{Zero-factor property}$$
$$n = 6 \quad \text{or} \quad n = -1$$

We check the solutions in the words of the problem. The square of 6 is 36, which is 6 more than 5 times 6. So 6 checks. The square of -1 is $(-1)^2 = 1$, which is 6 more than 5 times -1, so -1 checks. There are two numbers, -1 and 6.

Exercise Set 10.1

Each of the following equations is already factored. Solve and check all solutions.

1. $(x + 3)(x - 7) = 0$
2. $(y - 1)(y + 5) = 0$
3. $(3x + 2)(4x - 5) = 0$
4. $(5y - 1)(2y - 3) = 0$
5. $x(7x + 2)(7x - 2) = 0$
6. $y(5y - 4)(5y + 4) = 0$
7. $x(3x - 5)(4x + 7)(2x - 9) = 0$
8. $a(2a - 3)(3a - 1)(2a + 3) = 0$

Solve each equation by factoring. Check all solutions.

9. $p^2 + 10p + 25 = 0$
10. $81x^2 - 49 = 0$
11. $169y^2 - 225 = 0$
12. $t^2 + t + \dfrac{1}{4} = 0$
13. $144a^2 + 8a + \dfrac{1}{9} = 0$
14. $x^2 - 9x + 18 = 0$
15. $40 + y^2 - 22y = 0$
16. $t^3 - 20t - 8t^2 = 0$
17. $3x^3 - 120x - 54x^2 = 0$
18. $2x^3 - 22x^2 - 120x = 0$
19. $5y^2 + 75y - 80 = 0$
20. $2a^4 - 10a^3 - 28a^2 = 0$
21. $8a^3 + 47a^2 - 6a = 0$
22. $25t^3 - 65t^2 + 36t = 0$
23. $10y^3 - 19y^2 + 6y = 0$
24. $2x^4 - 46x^3 - 48x^2 = 0$
25. $3y^3 - 24y^2 - 252y = 0$
26. $4p^5 - 40p^4 - 96p^3 = 0$
27. $5t^4 - 85t^3 - 420t^2 = 0$
28. $3x^5 - 72x^4 + 432x^3 = 0$
29. $225c^2 + 30c + 1 = 0$
30. $72c^3 + 120c^2 + 50c = 0$

Solve each equation by the square root method, if possible, and check your solutions. If no real-number solutions exist, write Not a real number.

31. $x^2 = 25$
32. $x^2 = 49$
33. $y^2 = 81$
34. $y^2 = 64$
35. $p^2 = 1$
36. $p^2 = 2$
37. $a^2 = 12$
38. $a^2 = 48$
39. $x^2 - 6 = 0$
40. $y^2 - 11 = 0$
41. $p^2 + 4 = 0$
42. $p^2 + 9 = 0$
43. $9a^2 - 1 = 0$
44. $25m^2 - 4 = 0$
45. $x^2 = \dfrac{64}{169}$
46. $n^2 = \dfrac{5}{8}$
47. $32t^2 - 15 = 0$
48. $7y^2 - 18 = 0$
49. $6x^2 - 42 = 0$
50. $50x^2 + 28 = 100$
51. $242y^2 - 350 = -62$
52. $400 - 576p^2 = 275$
53. $(x + 3)^2 = 16$
54. $(y - 2)^2 = 4$
55. $(n + 4)^2 = 1$
56. $(p - 5)^2 = 25$
57. $\left(x - \dfrac{1}{3}\right)^2 = \dfrac{4}{9}$
58. $\left(a - \dfrac{7}{2}\right)^2 = \dfrac{25}{4}$

59. $\left(m - \frac{3}{2}\right)^2 = \frac{5}{2}$

60. $\left(p - \frac{3}{7}\right)^2 = \frac{5}{7}$

61. $(2x + 1)^2 = 4$

62. $(7x - 1)^2 = 0$

63. $(2x + 7)^2 = 0$

64. $(3x + 4)^2 = 9$

65. $(2x - 1)^2 = 9$

66. $(5t - 3)^2 = 35$

67. $(4m - 1)^2 = 48$

68. $(11x - 1)^2 = -4$

69. $(3x - 1)^2 = -9$

70. $(2y - 5)^2 = 98$

71. $x^2 = \frac{2x^2}{3} + 2$

72. $9 - p^2 = 4 + \frac{2p^2}{3}$

73. $m^2 - \frac{3}{4} = \frac{1}{4}m^2 + \frac{5}{4}$

74. $5a^2 + 5 = 3a^2 + 55$

75. $20y^2 - 180 = 16 - 5y^2$

76. $m^2 + 1 = 4 + \frac{m^2}{4}$

77. $144 - (7a - 10)^2 = 0$

78. $18 - (3t - 1)^2 = 0$

Solve each of the following applied problems. Check all solutions.

79. A number squared is 12 less than 48. Find the number.

80. If 6 is subtracted from twice the square of a number, the result is 122. Find the number.

81. The square of 1 less than a number is 4. Find the number.

82. Find two numbers with a sum of 15 and a product of 26.

83. Find two consecutive even integers with a product of 224.

84. Find a positive integer such that the integer added to its square is 72.

85. A negative integer added to its square is 56. Find the integer.

86. Find a positive integer such that the integer added to its square is 2.

87. Find a positive number such that 7 less than 8 times the number is equal to the square of the number.

88. Find a positive number such that 3 more than 4 times the number is equal to the negative of the square of the number.

89. Find a negative integer such that 15 added to 8 times the integer added to the square of the integer equals zero.

90. Sherry's garden has an area of 450 ft², and the length of the garden is twice its width. Find the dimensions of the garden.

91. Lisa's room has an area of 154 ft². The length of the room is 3 ft more than the width. Find its dimensions.

92. Dean's Pure Water Company manufactures water purifiers for the home. The company's profit, P, in dollars is given by the formula $P(n) = n^2 + 5n - 50$, where n is the number of purifiers produced. How many purifiers must the company produce and sell in order to break even?

93. Dennis's Tool Shop produces grass edgers. The cost C in dollars of producing n edgers is given by the formula $C(n) = n^2 + 11n + 24$. How many edgers can the company produce at a cost of $204?

Use a calculator, as needed, for Exercises 94–97.

94. Determine if $x = -3$ is a solution to the equation $4x^2 + 7x + 21 = 3x^2 + 9$.

95. Use the Pythagorean theorem to find the distance d shown in the figure. Write your answer in decimal form rounded to the nearest hundredth of a foot.

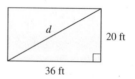

96. A carpenter wants to use a 12-ft ladder to reach the top of a 10-ft wall. How far (to the nearest inch) must the base of the ladder be from the base of the wall? (See the figure.)

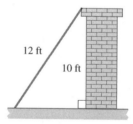

97. Use the given figure of the gable end of a house to find the length r of a rafter.

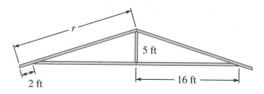

Review Exercises

Simplify each of the following quantities.

98. $\dfrac{2 + \sqrt{2^2 - 4}}{2}$

99. $\dfrac{6 + \sqrt{6^2 - 4(-1)}}{2}$

100. $\dfrac{-2 + \sqrt{18 - 10}}{4}$

101. $\dfrac{-8 - \sqrt{26 - 14}}{2}$

102. $\dfrac{-3 + \sqrt{28 - 10}}{3}$

103. $\dfrac{-5 + \sqrt{24 + 26}}{5}$

Solve each of the following equations.

104. $\sqrt{3x + 6} = x + 2$

105. $\sqrt{4x - 3} = x - 2$

106. $\dfrac{x}{x - 3} = \dfrac{1}{x + 5}$

107. $\dfrac{3}{x + 11} = \dfrac{x}{x - 5}$

10.2 Solving Quadratic Equations by Completing the Square

OBJECTIVES

In this section we will learn to
1. complete the square to create perfect-square trinomials; and
2. solve quadratic equations by completing the square.

In Section 10.1 we solved factorable equations by using the zero-factor property on some and the square root property on others. In this section we will learn how to solve quadratic equations by writing them in a format that allows us to use the square root property. For example, we can't factor

$x^2 + 2x - 2 = 0$

by any of the methods we have learned so far. It is still solvable, however, by *creating* a perfect square and then using the square root property. To do this, we begin by adding 2 to each side of the equation to isolate all terms involving the variable on one side of the equation:

$x^2 + 2x = 2$

Now if we add 1 to each side of the equation, the left side will become a perfect-square trinomial:

$x^2 + 2x + 1 = 2 + 1$
$(x + 1)^2 = 3$

We can now use the square root property to solve for *x*.

$x + 1 = \pm\sqrt{3}$ Square root property
$x = -1 \pm \sqrt{3}$ Subtract 1 from each side.

The process that we used to change $x^2 + 2x$ to $x^2 + 2x + 1$ is called **completing the square.**

You are probably wondering how we knew what number to add to $x^2 + 2x$ to complete the square. The answer to that question can be found by considering what happens when we square the binomial $x + a$:

$$(x + a)^2 = x^2 + 2ax + a^2$$

If the coefficient of the second-degree term is 1, the quantity necessary to complete the square is the square of one-half of the coefficient of the middle term:

$$x^2 + 2ax + a^2 \quad \text{where} \quad a^2 = \left(\frac{1}{2} \cdot 2a\right)^2$$

Further, the a in $(x + a)^2$ is one-half of $2a$, the coefficient of x.

$$(x + a)^2 = x^2 + 2ax + a^2$$

$$\frac{1}{2} \cdot 2a \qquad \left(\frac{1}{2} \cdot 2a\right)^2$$

We can illustrate completing the square in terms of areas of geometric figures. Notice that the area of the figure shown in Figure 10.1(a) is $x^2 + 3x + 3x$, or $x^2 + 6x$, and that the figure is not a square. By filling in the corner with a square 3 units on a side, which has an area of $3^2 = 9$ square units, as shown in Figure 10.1(b), the figure becomes a square. We have completed the square, literally!

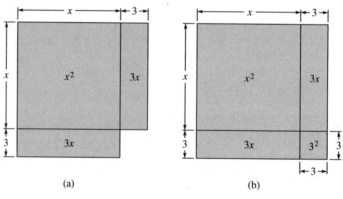

(a)   (b)

FIGURE 10.1

EXAMPLE 1 Complete the square on each expression, and write the answer as the square of a binomial: **(a)** $x^2 + 6x$ **(b)** $x^2 - 8x$ **(c)** $x^2 - 3x$

SOLUTION **(a)** This is the completion illustrated in Figure 10.1.

$$x^2 + 6x \qquad x^2 + 6x + 9 = (x + 3)^2$$
$$\frac{1}{2} \cdot 6 = 3 \qquad 3^2 = 9$$

(b) $x^2 - 8x \qquad x^2 - 8x + 16 = (x - 4)^2$

$$\frac{1}{2} \cdot -8 = -4 \qquad (-4)^2 = 16$$

(c) $x^2 - 3x$ $\quad x^2 - 3x + \dfrac{9}{4} = \left(x - \dfrac{3}{2}\right)^2$

$\dfrac{1}{2} \cdot -3 = -\dfrac{3}{2} \quad \left(-\dfrac{3}{2}\right)^2 = \dfrac{9}{4}$

EXAMPLE 2 Complete the square of each expression, and write the answer as the square of a binomial: **(a)** $x^2 + 4x$ **(b)** $x^2 - 6x$ **(c)** $x^2 - 7x$

SOLUTION **(a)** $x^2 + 4x \quad x^2 + 4x + 4 = (x + 2)^2$

$\dfrac{1}{2} \cdot 4 = 2$

(b) $x^2 - 6x \quad x^2 - 6x + 9 = (x - 3)^2$

$\dfrac{1}{2} \cdot -6 = -3$

(c) $x^2 - 7x \quad x^2 - 7x + \dfrac{49}{4} = \left(x - \dfrac{7}{2}\right)^2$

$\dfrac{1}{2} \cdot -7 = -\dfrac{7}{2}$

EXAMPLE 3 Solve the equation $x^2 + 8x - 9 = 0$ by factoring and by completing the square.

SOLUTION First we solve by factoring.

$$x^2 + 8x - 9 = 0$$
$$(x + 9)(x - 1) = 0$$
$$x + 9 = 0 \quad \text{or} \quad x - 1 = 0 \quad \text{Zero-factor property}$$
$$x = -9 \quad \text{or} \quad x = 1$$

The solutions check. The solution set is $\{-9, 1\}$.

Now we solve by completing the square; we begin by adding 9 to each side of $x^2 + 8x - 9 = 0$.

$$x^2 + 8x = 9 \qquad \text{Isolate the variable terms.}$$
$$x^2 + 8x + 16 = 9 + 16 \qquad \text{Complete the square by adding 16 to each side.}$$
$$(x + 4)^2 = 25 \qquad \text{Write as the square of a binomial.}$$
$$x + 4 = \pm 5 \qquad \text{Square root property}$$
$$x = -4 \pm 5 \qquad \text{Subtract 4 from each side.}$$
$$x = -4 + 5 \quad \text{or} \quad x = -4 - 5$$
$$x = 1 \quad \text{or} \quad x = -9$$

The solution set in Example 3 is the same, whether obtained by factoring or by completing the square. Factoring, however, is not always possible, as the following examples illustrate.

EXAMPLE 4 Solve the equation $x^2 + 6x + 4 = 0$.

SOLUTION The equation is not factorable, so we begin by subtracting 4 from each side.

$$x^2 + 6x = -4$$
$$x^2 + 6x + 9 = -4 + 9 \qquad \text{Complete the square by adding 9 to each side.}$$
$$(x + 3)^2 = 5 \qquad \text{Write as the square of a binomial.}$$
$$x + 3 = \pm\sqrt{5} \qquad \text{Square root property}$$
$$x = -3 \pm \sqrt{5} \qquad \text{Subtract 3 from each side.}$$

We will check one solution, $-3 - \sqrt{5}$; it must be checked in the original equation.

$$x^2 + 6x + 4 = 0$$
$$(-3 - \sqrt{5})^2 + 6(-3 - \sqrt{5}) + 4 \stackrel{?}{=} 0$$
$$9 + 6\sqrt{5} + 5 - 18 - 6\sqrt{5} + 4 \stackrel{?}{=} 0$$
$$0 \stackrel{\checkmark}{=} 0$$

The solution checks. The check of $-3 + \sqrt{5}$ is left to the reader. The solution set is $\{-3 - \sqrt{5}, -3 + \sqrt{5}\}$.

EXAMPLE 5 Solve the equation $2x^2 + 4x - 3 = 0$.

SOLUTION We first write the equation as $2x^2 + 4x = 3$. **In order to complete the square, the lead coefficient** (that of the second-degree term) **must be 1.** We divide each side by 2.

$$2x^2 + 4x = 3$$
$$x^2 + 2x = \frac{3}{2} \qquad \text{Divide each side by 2.}$$
$$x^2 + 2x + 1 = \frac{3}{2} + 1 \qquad \text{Add 1 to each side to complete the square.}$$
$$(x + 1)^2 = \frac{5}{2}$$
$$x + 1 = \pm\sqrt{\frac{5}{2}} \qquad \text{Square root property}$$
$$x + 1 = \pm\sqrt{\frac{5}{2} \cdot \frac{2}{2}} \qquad \text{Simplify the radical.}$$
$$x + 1 = \frac{\pm\sqrt{10}}{2}$$
$$x = -1 \pm \frac{\sqrt{10}}{2} \qquad \text{Subtract 1 from each side.}$$
$$x = \frac{-2 \pm \sqrt{10}}{2} \qquad \text{The LCD is 2.}$$

We can list the solutions separately, as before, or we can leave them with the $\pm$ sign. The check is left to the reader. The solution set is $\left\{\dfrac{-2 \pm \sqrt{10}}{2}\right\}$.

Solving quadratic equations by completing the square is a five-step process.

To Solve a Quadratic Equation by Completing the Square

Step 1 Isolate all terms involving the variable on one side of the equation, using the addition or subtraction property of equality.

Step 2 If the coefficient of the second-degree term is not 1, divide both sides by that coefficient.

Step 3 Add the square of one-half the coefficient of the first-degree term to each side of the equation.

Step 4 Write the perfect-square trinomial obtained in Step 3 as the square of a binomial.

Step 5 Use the square root method on the equation found in Step 4 and solve for the variable. Check the solution in the original equation.

Do You Remember?

Can you match these?

_____ 1. To complete the square of $x^2 + 4x$, add ___.

_____ 2. To complete the square of $x^2 + 3x$, add ___.

_____ 3. To complete the square of a trinomial, the leading coefficient must be ___.

_____ 4. To solve $x^2 + 7x = -2$, add ___ to each side.

_____ 5. To solve $2x^2 - 16x = 13$, add ___ to each side.

a) $\dfrac{9}{4}$

b) $\dfrac{49}{4}$

c) 16

d) 1

e) 4

Answers: 1. e 2. a 3. d 4. b 5. c

Exercise Set 10.2

Complete the square of each expression and factor it as the square of a binomial.

1. $x^2 + 4x$
2. $x^2 + 8x$
3. $y^2 - 10y$
4. $y^2 - 12y$
5. $t^2 - 5t$
6. $a^2 - 7a$
7. $x^2 + 11x$
8. $b^2 - 17b$
9. $r^2 + \frac{1}{2}r$
10. $y^2 + \frac{1}{4}y$

Solve each quadratic equation by completing the square. Check all solutions.

11. $x^2 + 8x - 9 = 0$
12. $y^2 - 10y + 21 = 0$
13. $x^2 + 2x - 5 = 0$
14. $y^2 + 4y + 3 = 0$
15. $x^2 - 6x - 16 = 0$
16. $a^2 + 22a + 21 = 0$
17. $a^2 + 6a = -8$
18. $x^2 - 8x = -16$
19. $-y^2 + 6y = 4$
20. $-p^2 + 4 = 2p$
21. $p^2 - 4 = -2p$
22. $2x^2 + 4x - 7 = 0$
23. $3y^2 - 6y - 5 = 0$
24. $2a^2 = -2 - 5a$
25. $2b^2 + 3 = 6b$
26. $x = 1 - 3x^2$
27. $12y^2 - 4y = 5$
28. $4y^2 - 4y = 1$
29. $-3p^2 + 5p = -2$
30. $-8x^2 = 3x - 6$
31. $2a^2 + 3a = 17$
32. $3a^2 + 4a = 1$
33. $9y^2 = 6y + 9$
34. $2p^2 = 9p + 5$
35. $x^2 + \frac{3}{2}x = \frac{1}{2}$
36. $x^2 + \frac{3}{2}x = \frac{7}{4}$
37. $y^2 - \frac{1}{3}y = \frac{5}{12}$
38. $y^2 - \frac{1}{4}y - \frac{3}{8} = 0$
39. $x^2 - \frac{2}{3}x + \frac{1}{9} = 0$
40. $x^2 + \frac{2}{5}x + \frac{1}{25} = 0$
41. $a^2 + \frac{1}{3}a + \frac{1}{36} = 0$
42. $a^2 + a + \frac{1}{4} = 0$
43. $y^2 - \frac{1}{4}y + \frac{1}{64} = 0$
44. $y^2 + 2y = 6$
45. $y^2 + 4y = 4$
46. $4a^2 + 2a = 1$
47. $a^2 + a = 1$
48. $2y^2 - 4 = y$
49. $y^2 = 3 + 6y$
50. $2x^2 = 3x + 2$
51. $3x^2 - 1 = -2x$
52. $4a^2 = 6 - 5a$
53. $\frac{1}{6}y^2 + y + \frac{1}{6} = 0$
54. $\frac{2}{3}y^2 - y = \frac{1}{3}$
55. $0.4x^2 - x = 0.6$
56. $0.2x^2 - x + 0.1 = 0$
57. $0.05y^2 + 0.1y - 0.05 = 0$
58. $0.2y^2 - y = 2.4$
59. $0.25a^2 = a - 1$
60. $0.2a^2 - a = -1.2$

Review Exercises

Verify if the given quantity is a solution of the accompanying equation.

61. $\dfrac{5 + \sqrt{17}}{4}$; $2x^2 - 5x + 1 = 0$
62. $\dfrac{-3 - \sqrt{17}}{4}$; $2x^2 + 3x - 1 = 0$
63. $\dfrac{7 + \sqrt{13}}{2}$; $3y^2 - 6y - 3 = 0$
64. $\dfrac{-1 - \sqrt{7}}{3}$; $3x^2 = 2 - 3x$
65. $\dfrac{5 - \sqrt{23}}{2}$; $2y^2 = 10y - 1$
66. $\dfrac{-2 + \sqrt{5}}{2}$; $4a^2 + 8a = 1$

Write In Your Own Words

67. Outline the steps in solving $2x^2 - 5x + 5 = 0$ by completing the square.
68. Outline the steps in solving $\sqrt{x} + 3 = 3 - \dfrac{1}{\sqrt{x} + 3}$.

10.3 Solving Quadratic Equations by the Quadratic Formula

OBJECTIVES

In this section we will learn to
1. derive a formula for solving quadratic equations;
2. identify the parts of the formula in a specific equation; and
3. use the quadratic formula to solve equations with both irrational and rational solutions.

In Section 10.2 we solved quadratic equations by completing the square. We can use the method of completing the square to derive a general formula, the **quadratic formula,** that we can then use to solve *any* quadratic equation.

To derive the formula, we start with the general quadratic equation in standard form:

$$ax^2 + bx + c = 0, \quad a > 0$$

To show how we derive the formula, we will simultaneously solve

I. $ax^2 + bx + c = 0$ and **II.** $2x^2 + 3x - 1 = 0$

using the five-step process outlined in Section 10.2. By comparing them, you can see how we use that process in the derivation.

Step 1 Isolate all terms involving the variable on the left side of the equation.

I. $ax^2 + bx + c = 0$
$ax^2 + bx = -c$ Subtract c from each side.

II. $2x^2 + 3x - 1 = 0$
$2x^2 + 3x = 1$ Add 1 to each side.

Step 2 Make the coefficient of the second-degree term 1.

I. $x^2 + \dfrac{b}{a}x = \dfrac{-c}{a}$ Divide each side by a.

II. $x^2 + \dfrac{3}{2}x = \dfrac{1}{2}$ Divide each side by 2.

Step 3 Add the square of one-half the coefficient of the first-degree term to each side of the equation

I. $x^2 + \dfrac{b}{a}x + \dfrac{b^2}{4a^2} = \dfrac{-c}{a} + \dfrac{b^2}{4a^2}$ $\left(\dfrac{1}{2} \cdot \dfrac{b}{a}\right)^2 = \left(\dfrac{b}{2a}\right)^2 = \dfrac{b^2}{4a^2}$

II. $x^2 + \dfrac{3}{2}x + \dfrac{9}{16} = \dfrac{1}{2} + \dfrac{9}{16}$ $\left(\dfrac{1}{2} \cdot \dfrac{3}{2}\right)^2 = \left(\dfrac{3}{4}\right)^2 = \dfrac{9}{16}$

Now we simplify the right side of each equation by combining terms over the LCD.

I. $x^2 + \dfrac{b}{a}x + \dfrac{b^2}{4a^2} = \dfrac{b^2 - 4ac}{4a^2}$ The LCD is $4a^2$.

II. $x^2 + \dfrac{3}{2}x + \dfrac{9}{16} = \dfrac{17}{16}$ The LCD is 16.

Step 4 Express the perfect-square trinomial obtained in Step 3 as the square of a binomial.

I. $\left(x + \dfrac{b}{2a}\right)^2 = \dfrac{b^2 - 4ac}{4a^2}$ **II.** $\left(x + \dfrac{3}{4}\right)^2 = \dfrac{17}{16}$

Step 5 Use the square root method and solve for the variable.

I. $x + \dfrac{b}{2a} = \pm\sqrt{\dfrac{b^2 - 4ac}{4a^2}}$

$x + \dfrac{b}{2a} = \pm\dfrac{\sqrt{b^2 - 4ac}}{2a}$

$x = -\dfrac{b}{2a} \pm \dfrac{\sqrt{b^2 - 4ac}}{2a}$ Subtract $\dfrac{b}{2a}$ from each side.

$= \dfrac{-b \pm \sqrt{b^2 - 4ac}}{2a}$

II. $x + \dfrac{3}{4} = \pm\sqrt{\dfrac{17}{16}}$

$x + \dfrac{3}{4} = \pm\dfrac{\sqrt{17}}{4}$

$x = -\dfrac{3}{4} \pm \dfrac{\sqrt{17}}{4}$ Subtract $\dfrac{3}{4}$ from each side.

$= \dfrac{-3 \pm \sqrt{17}}{4}$

The solution set is

I. $\left\{\dfrac{-b + \sqrt{b^2 - 4ac}}{2a}, \dfrac{-b - \sqrt{b^2 - 4ac}}{2a}\right\}$

II. $\left\{\dfrac{-3 + \sqrt{17}}{4}, \dfrac{-3 - \sqrt{17}}{4}\right\}$

The derivation is now complete. *You must memorize the following formula.*

The Quadratic Formula

The solutions to the general quadratic equation $ax^2 + bx + c = 0$, in standard form, are

$x = \dfrac{-b \pm \sqrt{b^2 - 4ac}}{2a}, \qquad a > 0$

EXAMPLE 1 Identify the values of *a*, *b*, and *c* in the general quadratic equation for each of the following equations.

(a) $3x^2 + 6x + 4 = 0 \quad a = 3, b = 6, c = 4$

(b) $-5x^2 + 4x - 3 = 0$

First we rewrite the equation in standard form with leading coefficient positive by multiplying each side by -1.

$$5x^2 - 4x + 3 = 0 \quad a = 5, b = -4, c = 3$$

(c) $\frac{1}{2}x^2 + \frac{3}{4} = 0$

$$\frac{1}{2}x^2 + \frac{3}{4} = \frac{1}{2}x^2 + 0x + \frac{3}{4} = 0 \quad a = \frac{1}{2}, b = 0, c = \frac{3}{4}$$

EXAMPLE 2 Evaluate the quadratic formula for $a = 2$, $b = 3$, and $c = -2$.

SOLUTION
$$\frac{-b \pm \sqrt{b^2 - 4ac}}{2a} = \frac{-3 \pm \sqrt{3^2 - 4(2)(-2)}}{2(2)}$$
$$= \frac{-3 \pm \sqrt{9 + 16}}{4}$$
$$= \frac{-3 \pm \sqrt{25}}{4}$$
$$= \frac{-3 \pm 5}{4}$$
$$= \frac{-3 + 5}{4} \quad \text{or} \quad \frac{-3 - 5}{4}$$
$$= \frac{1}{2} \quad \text{or} \quad -2$$

EXAMPLE 3 Use the quadratic formula to solve $2x^2 + 4x + 1 = 0$.

SOLUTION To use the quadratic formula, we must first determine the values of *a*, *b*, and *c*.

$$\begin{array}{ccc} ax^2 + & bx + & c = 0 \\ \downarrow & \downarrow & \downarrow \\ 2x^2 + & 4x + & 1 = 0 \end{array}$$

Here $a = 2$, $b = 4$, and $c = 1$. Now we substitute these values into the formula and simplify the results.

$$x = \frac{-b \pm \sqrt{b^2 - 4ac}}{2a}$$
$$= \frac{-4 \pm \sqrt{4^2 - 4(2)(1)}}{2(2)}$$
$$= \frac{-4 \pm \sqrt{16 - 8}}{4}$$

$$= \frac{-4 \pm \sqrt{8}}{4}$$

$$= \frac{-4 \pm \sqrt{4 \cdot 2}}{4}$$

$$= \frac{-4 \pm 2\sqrt{2}}{4}$$

This can be simplified by factoring 2 out of the numerator and then dividing out the common factor 2:

$$x = \frac{2(-2 \pm \sqrt{2})}{2(2)}$$

$$= \frac{-2 \pm \sqrt{2}}{2}$$

We will check $\frac{-2 + \sqrt{2}}{2}$. The check of $\frac{-2 - \sqrt{2}}{2}$ is left to the reader.

CHECK $\qquad 2x^2 + 4x + 1 = 0$

$$2\left(\frac{-2 + \sqrt{2}}{2}\right)^2 + 4\left(\frac{-2 + \sqrt{2}}{2}\right) + 1 \stackrel{?}{=} 0 \qquad \text{Substitute } \frac{-2 + \sqrt{2}}{2} \text{ for } x.$$

$$2\left(\frac{4 - 4\sqrt{2} + 2}{4}\right) + 4\left(\frac{-2 + \sqrt{2}}{2}\right) + 1 \stackrel{?}{=} 0$$

$$\frac{4 - 4\sqrt{2} + 2}{2} + 2(-2 + \sqrt{2}) + 1 \stackrel{?}{=} 0$$

$$2 - 2\sqrt{2} + 1 - 4 + 2\sqrt{2} + 1 \stackrel{?}{=} 0$$

$$0 \stackrel{\checkmark}{=} 0$$

The solution set is $\left\{\frac{-2 - \sqrt{2}}{2}, \frac{-2 + \sqrt{2}}{2}\right\}$.

When the solutions to quadratic equations are irrational numbers, such as those in Example 3, they always appear in **conjugate pairs.** Conjugate pairs differ only in the sign between the two terms, such as $a - b$ and $a + b$. If one checks, the other must check also. If the solutions are rational numbers, both results should be checked, since errors could have been made in their simplification.

EXAMPLE 4 Solve: $3x^2 + 5x = 1$.

SOLUTION First we write the equation in standard form so we can identify the values of a, b, and c.

$$3x^2 + 5x - 1 = 0$$

Here $a = 3$, $b = 5$, and $c = -1$. We substitute these values into the quadratic formula.

$$x = \frac{-b \pm \sqrt{b^2 - 4ac}}{2a}$$

$$= \frac{-5 \pm \sqrt{5^2 - 4(3)(-1)}}{2(3)}$$

$$= \frac{-5 \pm \sqrt{25 + 12}}{6}$$

$$= \frac{-5 \pm \sqrt{37}}{6}$$

There are no common factors in the numerator and the denominator, so the result cannot be reduced. This time we will check $\frac{-5 - \sqrt{37}}{6}$.

CHECK
$$3x^2 + 5x = 1$$

$$3\left(\frac{-5 - \sqrt{37}}{6}\right)^2 + 5\left(\frac{-5 - \sqrt{37}}{6}\right) \stackrel{?}{=} 1$$

$$3\left(\frac{25 + 10\sqrt{37} + 37}{36}\right) + 5\left(\frac{-5 - \sqrt{37}}{6}\right) \stackrel{?}{=} 1$$

$$\frac{25 + 10\sqrt{37} + 37}{12} + \frac{-25 - 5\sqrt{37}}{6} \stackrel{?}{=} 1$$

$$\frac{25 + 10\sqrt{37} + 37 - 50 - 10\sqrt{37}}{12} \stackrel{?}{=} 1 \qquad \text{Combine terms on the left side. The LCD is 12.}$$

$$\frac{12}{12} \stackrel{?}{=} 1$$

$$1 \stackrel{\checkmark}{=} 1$$

The solution checks. The solution set is $\left\{\frac{-5 - \sqrt{37}}{6}, \frac{-5 + \sqrt{37}}{6}\right\}$.

EXAMPLE 5 Solve: $\frac{1}{2}x^2 - \frac{2}{3}x - 1 = 0$.

SOLUTION We could use the quadratic formula directly, with $a = \frac{1}{2}$, $b = -\frac{2}{3}$, and $c = -1$. It is more convenient, however, to clear the equation of fractions first. We do this by multiplying each side by 6, the LCD.

$$6\left(\frac{1}{2}x^2 - \frac{2}{3}x - 1\right) = 6 \cdot 0$$

$$3x^2 - 4x - 6 = 0$$

Now when we solve the equation using the quadratic formula with $a = 3$, $b = -4$, and $c = -6$, we find that the solution set is $\left\{\frac{2 - \sqrt{22}}{3}, \frac{2 + \sqrt{22}}{3}\right\}$.

The checks should be made in the original equation, since errors can occur in eliminating the fractions.

EXAMPLE 6   Solve: $\dfrac{1}{y} + y = 3$, $y \neq 0$.

SOLUTION   We begin by clearing the equation of fractions.

$$\dfrac{1}{y} + y = 3 \qquad \text{Original equation}$$

$$1 + y^2 = 3y \qquad \text{Multiply each side by } y, \text{ the LCD.}$$

$$y^2 - 3y + 1 = 0$$

Now we use the quadratic formula.

$$y = \dfrac{-(-3) \pm \sqrt{(-3)^2 - 4(1)(1)}}{2(1)} \qquad a = 1, b = -3, \text{ and } c = 1$$

$$= \dfrac{3 \pm \sqrt{9 - 4}}{2}$$

$$= \dfrac{3 \pm \sqrt{5}}{2}$$

We will check $\dfrac{3 + \sqrt{5}}{2}$.

$$\dfrac{1}{\dfrac{3 + \sqrt{5}}{2}} + \dfrac{3 + \sqrt{5}}{2} \stackrel{?}{=} 3$$

$$\dfrac{2}{3 + \sqrt{5}} + \dfrac{3 + \sqrt{5}}{2} \stackrel{?}{=} 3$$

$$\dfrac{2}{3 + \sqrt{5}} \cdot \dfrac{3 - \sqrt{5}}{3 - \sqrt{5}} + \dfrac{3 + \sqrt{5}}{2} \stackrel{?}{=} 3 \qquad \text{To rationalize, multiply by } \dfrac{3 - \sqrt{5}}{3 - \sqrt{5}} = 1.$$

$$\dfrac{2(3 - \sqrt{5})}{4} + \dfrac{3 + \sqrt{5}}{2} \stackrel{?}{=} 3$$

$$\dfrac{3 - \sqrt{5}}{2} + \dfrac{3 + \sqrt{5}}{2} \stackrel{?}{=} 3 \qquad \text{Combine terms on the left side. The LCD is 4.}$$

$$\dfrac{6}{2} \stackrel{?}{=} 3$$

$$3 \stackrel{\checkmark}{=} 3$$

EXAMPLE 7   Solve $\dfrac{x^2}{x - 3} - \dfrac{11}{x - 3} = 0$ by using the quadratic formula.

SOLUTION   We begin by observing that x cannot equal 3, since that would make the denominator zero.

$$\dfrac{x^2}{x - 3} - \dfrac{11}{x - 3} = 0$$

$$x^2 - 11 = 0 \qquad \text{Multiply each side by } x - 3, \text{ the LCD.}$$

Here $a = 1$, $b = 0$, and $c = -11$, so

$$x = \frac{-0 \pm \sqrt{0^2 - 4(1)(-11)}}{2(1)} \quad \text{Substitute for } a, b, \text{ and } c.$$

$$= \frac{\pm\sqrt{44}}{2}$$

$$= \frac{\pm 2\sqrt{11}}{2} = \pm\sqrt{11}$$

The solutions check. The solution set is $\{-\sqrt{11}, \sqrt{11}\}$.

Notice that we could have written the equation $x^2 - 11 = 0$ in Example 7 as $x^2 = 11$ and solved it by the square root method. We used the quadratic formula in this case to demonstrate its versatility.

EXAMPLE 8 Rewrite each equation in standard form so that the quadratic formula can be used; identify a, b, and c, but do not solve: **(a)** $3x^2 - 2 = (x - 2)(2x - 1)$ **(b)** $2x + 3 = \sqrt{x + 4}$ **(c)** $\dfrac{6}{y + 2} - \dfrac{4}{y - 2} = 4$, $y \neq 2, -2$

SOLUTION **(a)** $3x^2 - 2 = (x - 2)(2x - 1)$

$3x^2 - 2 = 2x^2 - 5x + 2$

$x^2 + 5x - 4 = 0 \qquad a = 1, b = 5, \text{ and } c = -4$

(b) $\qquad 2x + 3 = \sqrt{x + 4}$

$(2x + 3)^2 = (\sqrt{x + 4})^2 \qquad$ The power property of equality

$4x^2 + 12x + 9 = x + 4$

$4x^2 + 11x + 5 = 0 \qquad a = 4, b = 11, \text{ and } c = 5$

(c) $\qquad \dfrac{6}{y + 2} - \dfrac{4}{y - 2} = 4 \qquad y \neq 2, -2$

$6(y - 2) - 4(y + 2) = 4(y - 2)(y + 2) \qquad$ Clear of fractions by multiplying each side by $(y - 2)(y + 2)$, the LCD.

$6y - 12 - 4y - 8 = 4y^2 - 16 \qquad$ Distributive property

$2y - 20 = 4y^2 - 16 \qquad$ Combine like terms.

$0 = 4y^2 - 2y + 4 \qquad$ Standard form

We divide each side by 2 to simplify the coefficients.

$0 = 2y^2 - y + 2 \qquad a = 2, b = -1, \text{ and } c = 2$

Do You Remember?

Can you match this?

The solution to $ax^2 + bx + c = 0$, $a > 0$

a) $\dfrac{b \pm \sqrt{b^2 - 4ac}}{2a}$ b) $-b \pm \dfrac{\sqrt{b^2 - 4ac}}{2a}$ c) $\dfrac{-b \pm \sqrt{b^2 - 4ac}}{2a}$

d) $\dfrac{-b \pm \sqrt{b^2 + 4ac}}{2a}$ e) $b \pm \dfrac{\sqrt{b^2 - 4ac}}{2a}$

Answer: c

Exercise Set 10.3

Rewrite each equation in the form $ax^2 + bx + c = 0$. Identify a, b, and c. Do not solve.

1. $x^2 - 5x + 4 = 0$
2. $2x^2 + 6x + 5 = 0$
3. $5x + 2 = 3x^2$
4. $5x^2 = x + 4$
5. $x^2 = 10 - 3x$
6. $x^2 - 9 = 0$
7. $11x^2 = 0$
8. $4x^2 - 2 = 7x$

Replace a, b, and c in the quadratic formula $x = \dfrac{-b \pm \sqrt{b^2 - 4ac}}{2a}$ with the given values, and simplify.

9. $a = 1, b = -5, c = 6$
10. $a = 1, b = 4, c = -45$
11. $a = 5, b = -3, c = -14$
12. $a = 8, b = -2, c = -3$
13. $a = 12, b = -4, c = -5$
14. $a = 3, b = -7, c = -6$
15. $a = 1, b = 0, c = -5$
16. $a = 1, b = 0, c = -2$

Solve each equation by the quadratic formula. Simplify and check all solutions.

17. $2x^2 - 3x - 2 = 0$
18. $y^2 - 4y - 21 = 0$
19. $p^2 + 6p - 16 = 0$
20. $p^2 + 5p - 8 = 0$
21. $2t^2 + 13t + 1 = 0$
22. $4t^2 - 9t = 0$
23. $8a^2 - 7a = 0$
24. $3y^2 + 2y - 4 = 0$
25. $t^2 = 20t - 19$
26. $x^2 = 8x + 9$
27. $7x^2 - 36 = 0$
28. $25y^2 - 38 = 0$
29. $2a^2 = 3a + 2$
30. $2y + 4 = 4 - 2y - 2y^2$
31. $-4t + 3 = 8t - 1 - 3t^2$
32. $s + \dfrac{s^2}{2} = 6$
33. $2x = 3 + \dfrac{3}{x}$
34. $p^2 = 3p + \dfrac{1}{2}$
35. $a^2 = \dfrac{8a - 1}{5}$
36. $t + \dfrac{4}{3} = \dfrac{3}{2}t^2$
37. $\dfrac{2}{3}b^2 - \dfrac{4}{9}b - \dfrac{1}{3} = 0$
38. $\dfrac{3}{5}y - \dfrac{2}{5}y^2 + 1 = 0$
39. $9(y + 1)(y + 2) = 8$
40. $(3a - 1)(2a + 5) = a(a - 1)$
41. $\dfrac{x^2}{2x - 3} - \dfrac{5}{2x - 3} = 0$
42. $\dfrac{x^2}{x + 1} - \dfrac{11}{x + 1} = 0$
43. $y + 2 = \dfrac{3}{y + 2}$
44. $y - 1 = \dfrac{5}{y - 1}$
45. $\dfrac{1}{5} - \dfrac{1}{a} = \dfrac{1}{a + 6}$
46. $\dfrac{1}{3} - \dfrac{1}{t} = \dfrac{1}{t + 1}$
47. $0.8y^2 + 0.16y - 0.09 = 0$
48. $0.5x^2 - 0.5x - 0.25 = 0$
49. $0.1x^2 - 0.4x - 0.5 = 0$
50. $0.02y^2 + 0.03y - 0.05 = 0$

Solve each equation by factoring, by the square root property, by completing the square, or by the quadratic formula. Check all solutions.

51. $a^2 - a - 6 = 0$
52. $m^2 + 2m - 8 = 0$
53. $x^2 = 16$
54. $81y^2 - 169 = 0$
55. $6p^2 + 24p = 0$
56. $2m^2 = 4m$
57. $(a - 2)^2 = 3$
58. $(3x - 5)^2 = 36$
59. $\dfrac{24}{x} = 12x - 28$
60. $3a = \dfrac{84 - 9a}{a}$
61. $p^2 = 3p - \dfrac{3}{2}$
62. $t^2 = \dfrac{3}{2}(t + 1)$
63. $(x - 8)(x + 7) = 5$
64. $(x - 2)(x + 1) = 3$
65. $(y - 3)(4y - 5) = 15$
66. $\dfrac{(p + 2)(2p - 3)}{2} = \dfrac{1}{2}$
67. $3a(a + 1) = 2$
68. $6y^2 - 5 = 5y(y + 2)$
69. $(t - 3)^2 = 2(t + 4)$
70. $(4y - 3)^2 = 5(y - 1)$
71. $\dfrac{24}{p + 10} = -1 - \dfrac{-24}{p - 10}$
72. $\dfrac{12}{10a - 1} - 1 = \dfrac{12}{a}$
73. $2x - 1 = \sqrt{5 + 2x}$
74. $\sqrt{x + 2} = x - 2$
75. $\sqrt{x^2 + 7x + 9} = 5$
76. $\sqrt{x^2 - 5x - 1} = 13$

Rewrite each applied problem as a quadratic equation and solve it by any of the four methods discussed.

77. A number added to its reciprocal equals $\tfrac{23}{5}$. Find the number.
78. Find a number such that the difference between it and its reciprocal equals $\tfrac{3}{5}$.
79. The length of a rectangle is 1 less than twice the width. Find the dimensions if the area is 20 m². See the figure.

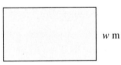
(2w–1) m, w m

80. Find the length of one side of the given square.

81. Find the length of the side h of the given right triangle.

82. A ladder is leaned against a building (see the figure). Find the length of the ladder.

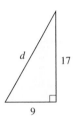

83. Greg and Kevin race their bicycles over a 12-mi route. Greg finishes 8 min ahead of Kevin, and he pedals 3 mph faster than Kevin. What was Greg's speed?
84. Sparkle, Inc., and Cinderella Services clean houses and apartments. Working alone, Sparkle can clean an apartment 1 hr faster than Cinderella. Together they can clean an apartment in $\tfrac{2}{3}$ hour. How long does it take Cinderella to clean an apartment, working alone?

10.4 Solving Quadratic Equations with Complex-Number Solutions

OBJECTIVES In this section we will
1. learn about imaginary and complex numbers;
2. simplify expressions such as $\sqrt{-3}$;
3. find powers of complex numbers;
4. add and subtract complex numbers;
5. multiply and divide complex numbers; and
6. solve quadratic equations with complex-number solutions.

In Section 10.3 we used the quadratic formula to find real-number solutions to quadratic equations. Some equations do not have real-number solutions. If we solve $x^2 + 5x + 8 = 0$ with the quadratic formula, we get

$$x = \frac{-5 \pm \sqrt{5^2 - 4(1)(8)}}{2} \qquad a = 1, b = 5, \text{ and } c = 8$$

$$= \frac{-5 \pm \sqrt{-7}}{2}$$

Since $\sqrt{-7}$ is not a real number, the equation has no real-number solutions. Numbers such as $\sqrt{-7}$ belong to a set called the **complex numbers.** Before defining this set, we first need to understand the concept of an **imaginary number.**

DEFINITION
Imaginary Number

The **imaginary number,** denoted i, is that number such that

$$i^2 = -1 \qquad \text{or} \qquad i = \sqrt{-1}$$

The number i is said to be "imaginary" only because there is no real number that represents $\sqrt{-1}$.

HISTORICAL COMMENT Numbers that we call "imaginary" were called "fictitious and impossible" when mathematicians first tried to cope with them. One attempted explanation was that they represented the side of a square with negative area. Leibniz characterized an imaginary number as "that wonderful creature of an ideal world, almost an amphibian between things that are and things that are not." It was René Descartes in 1637 who originated the name *imaginaire,* and Euler, in 1748, used the letter i to represent $\sqrt{-1}$.

From our previous work, we know that $\sqrt{25} = 5$, but what is meant by $\sqrt{-25}$? In order to answer this question, we need to go back to the product rule for square roots and modify it for products where one of the two radicands is negative.

10.4 · SOLVING QUADRATIC EQUATIONS WITH COMPLEX-NUMBER SOLUTIONS

> **The Product Rule for Square Roots**
> If a and b are real numbers and at least one is not negative, then
> $$\sqrt{a \cdot b} = \sqrt{a}\sqrt{b}$$

We can now determine what is meant by $\sqrt{-25}$.
$$\sqrt{-25} = \sqrt{25(-1)} = \sqrt{25}\sqrt{-1} = 5i$$

EXAMPLE 1 Write each of the following radicals in terms of i.
(a) $\sqrt{-9} = \sqrt{9}\sqrt{-1} = 3i$
(b) $\sqrt{-64} = \sqrt{64}\sqrt{-1} = 8i$
(c) $\sqrt{-7} = \sqrt{7}\sqrt{-1} = \sqrt{7}i$
(d) $\sqrt{-54} = \sqrt{54}\sqrt{-1}$
$= \sqrt{9}\sqrt{6}\sqrt{-1}$
$= 3\sqrt{6}i$

Note: The expression $3\sqrt{6}i$ is sometimes written $3i\sqrt{6}$ to avoid carelessly extending the radical sign over the i.

Example 1 shows that to find the square root of a negative number, all that we need to do is to treat the radicand as though it were a positive number, simplify it, and multiply the result by i. We are now ready to consider the definition of a complex number.

DEFINITION
Complex Number

A **complex number** is any number of the form
$$a + bi$$
where a and b are real numbers and $i = \sqrt{-1}$. In a complex number, a is called the **real part** and b is called the **imaginary part.**

EXAMPLE 2 Each of the following numbers is a complex number. Identify a and b.
(a) $2 + 3i$ $a = 2$ and $b = 3$
(b) $\dfrac{1}{2} - \dfrac{\sqrt{3}}{4}i$ $a = \dfrac{1}{2}$ and $b = -\dfrac{\sqrt{3}}{4}$
(c) $7 = 7 + 0i$ $a = 7$ and $b = 0$
(d) $2i = 0 + 2i$ $a = 0$ and $b = 2$

Example 2(c) illustrates that every real number is a complex number with imaginary part zero. Part (d) illustrates that every imaginary number is a complex number.

> *The set of real numbers and the set of imaginary numbers are subsets of the set of complex numbers.*

We now turn our attention to the methods used to simplify expressions involving complex numbers. We add and subtract complex numbers in the same manner as polynomials: by combining like terms. We use the commutative, associative, and distributive properties to write sums and differences in simplest terms.

> Complex numbers are considered to be in **simplest terms** when written in the form $a + bi$, which is called **standard form.**

EXAMPLE 3   Add: $(2 + 3i) + (5 + 6i)$.

SOLUTION

$(2 + 3i) + (5 + 6i) = (2 + 5) + (3i + 6i)$ Commutative and associative properties
$\qquad\qquad\qquad\quad\ = 7 + (3 + 6)i$ Distributive property
$\qquad\qquad\qquad\quad\ = 7 + 9i$ Standard form

EXAMPLE 4   Subtract: $(5 + 2i) - (3 - 2i)$.

SOLUTION

$(5 + 2i) - (3 - 2i) = 5 + 2i - 3 + 2i$
$\qquad\qquad\qquad\quad\ = (5 - 3) + (2i + 2i)$
$\qquad\qquad\qquad\quad\ = 2 + 4i$ Standard form

Multiplication of complex numbers is also similar to multiplying polynomials. We use the distributive property along with the FOIL method for binomials. Whenever a power of i greater than 1 occurs in the product, it must be simplified.

EXAMPLE 5   Multiply: $2i(3 + 4i)$.

SOLUTION

$2i(3 + 4i) = (2i)(3) + (2i)(4i)$ Distributive property
$\qquad\qquad = 6i + 8i^2$
$\qquad\qquad = 6i + 8(-1)$ $i^2 = -1$
$\qquad\qquad = 6i - 8$
$\qquad\qquad = -8 + 6i$ Standard form

EXAMPLE 6   Multiply: $(2 + 3i)(3 - 2i)$.

SOLUTION

$(2 + 3i)(3 - 2i) = 6 - 4i + 9i - 6i^2$ FOIL method
$\qquad\qquad\qquad\ = 6 - 4i + 9i - (6)(-1)$ $i^2 = -1$

$$= 6 - 4i + 9i + 6$$
$$= 12 + 5i \qquad \text{Standard form}$$

EXAMPLE 7 Multiply: $(3 + 2i)(3 - 2i)$.

SOLUTION These two binomials are conjugates of each other, so we can find the product by following the pattern for the difference of squares: $(a + b)(a - b) = a^2 - b^2$.

$$(3 + 2i)(3 - 2i) = 3^2 - (2i)^2$$
$$= 9 - 4i^2$$
$$= 9 - 4(-1) \qquad i^2 = -1$$
$$= 9 + 4 = 13$$

Notice that the product of the two complex conjugates in Example 7 is a real number. This is always the case.

> The product of two complex conjugates is equal to the sum of the squares of their real and imaginary parts:
> $$(a + bi)(a - bi) = a^2 - b^2i^2 = a^2 + b^2$$

Division by complex numbers is similar to division by radicals and is carried out by *rationalizing* the denominator.

EXAMPLE 8 Divide: $\dfrac{3}{2i}$.

SOLUTION Remember that $i = \sqrt{-1}$ is a radical, so we multiply the numerator and the denominator by i.

$$\frac{3}{2i} \cdot \frac{i}{i} = \frac{3i}{2i^2} = \frac{3i}{-2} \qquad i^2 = -1$$
$$= -\frac{3}{2}i \qquad \text{Standard form}$$

EXAMPLE 9 Divide: $\dfrac{2 + i}{3 - i}$.

SOLUTION To carry out this division, we multiply the numerator and the denominator by the conjugate of the denominator. Since the denominator is $3 - i$, its conjugate is $3 + i$.

$$\frac{2 + i}{3 - i} \cdot \frac{3 + i}{3 + i} = \frac{6 + 2i + 3i + i^2}{10} \qquad \text{FOIL method}$$

$$= \frac{6 + 2i + 3i - 1}{10}$$ The product of complex conjugates is equal to the sum of the squares of the real and imaginary parts.

$$= \frac{5 + 5i}{10} = \frac{5(1 + i)}{10}$$

$$= \frac{1 + i}{2} = \frac{1}{2} + \frac{1}{2}i \quad \text{Standard form}$$

We now return to the quadratic equation that we discussed at the beginning of this section. We solved $x^2 + 5x + 8 = 0$ with the quadratic formula and found the solution to be $x = \frac{-5 \pm \sqrt{-7}}{2}$. We can now write the final result in the form of a complex number:

$$x = \frac{-5 \pm \sqrt{-7}}{2}$$

$$= \frac{-5 \pm \sqrt{7}i}{2}$$

The solution set in standard form is $\left\{ \frac{-5}{2} - \frac{\sqrt{7}}{2}i, \frac{-5}{2} + \frac{\sqrt{7}}{2}i \right\}$.

EXAMPLE 10 Solve: $5y^2 + 4 = 3y$.

SOLUTION We first write the equation in standard form:

$$5y^2 - 3y + 4 = 0$$

Now we use the quadratic formula:

$$y = \frac{-(-3) \pm \sqrt{(-3)^2 - 4(5)(4)}}{2(5)}$$

$$= \frac{3 \pm \sqrt{-71}}{10}$$

$$= \frac{3 \pm \sqrt{71}i}{10}$$

The solution set is $\left\{ \frac{3}{10} - \frac{\sqrt{71}}{10}i, \frac{3}{10} + \frac{\sqrt{71}}{10}i \right\}$.

The number i has interesting properties when raised to powers, assuming that the rules for exponents hold true.

POWERS OF i

$i^1 = i$ $i^5 = i^4 \cdot i = 1 \cdot i = i$

$i^2 = -1$ $i^6 = (i^2)^3 = (-1)^3 = -1$

$i^3 = i^2 \cdot i = -1 \cdot i = -i$ $i^7 = i^6 \cdot i = -1 \cdot i = -i$

$i^4 = (i^2)^2 = (-1)^2 = 1$ $i^8 = (i^4)^2 = (1)^2 = 1$

The pattern i, -1, $-i$, 1 continues without end as the exponent on i increases.

Exercise Set 10.4

Simplify each of the following radicals.

1. $\sqrt{-4}$
2. $\sqrt{-1}$
3. $\sqrt{-9}$
4. $\sqrt{-16}$
5. $\sqrt{-49}$
6. $\sqrt{-81}$
7. $\sqrt{-8}$
8. $\sqrt{-12}$
9. $\sqrt{-32}$
10. $\sqrt{-50}$
11. $\dfrac{\sqrt{-16}}{\sqrt{25}}$
12. $\dfrac{\sqrt{-98}}{\sqrt{49}}$

Add or subtract as indicated, and simplify.

13. $(1 + i) + (1 - i)$
14. $(1 + 3i) + (2 - i)$
15. $(3 + 4i) - (2 + i)$
16. $(5 - 2i) - (3 + 4i)$
17. $(2 + i) - (3 - 4i)$
18. $(8 - 3i) - (2 - 5i)$
19. $(3 - 5i) - (1 + i) - (-2 - 3i)$
20. $(4 + 7i) - (5 - 2i) - (-3 - 4i)$

Find each product and simplify.
Reminder: $\sqrt{-a} = \sqrt{a}\,i,\ a \geq 0$.

21. $\sqrt{-4}\,\sqrt{-2}$
22. $\sqrt{-1}\,\sqrt{-16}$
23. $\sqrt{-49}(3i)$
24. $\sqrt{-64}(5i)$
25. $(1 + i)(1 - i)$
26. $(2 - i)(2 + i)$
27. $(2 + 3i)(2 - 3i)$
28. $(4 - 5i)(4 + 5i)$
29. $(3 + 4i)(2 - 5i)$
30. $(7 + 4i)(5 - 3i)$

Simplify each of the following quotients. Reminder:
$\sqrt{-a} = \sqrt{a}\,i,\ a \geq 0$.

31. $\dfrac{2}{3i}$
32. $\dfrac{-1}{i}$
33. $\dfrac{2}{1 + i}$
34. $\dfrac{3}{1 - i}$
35. $\dfrac{3 + i}{2 - i}$
36. $\dfrac{2 - 5i}{3 + 2i}$
37. $\dfrac{3 - \sqrt{-25}}{2 + \sqrt{-4}}$
38. $\dfrac{-\sqrt{-1}}{2 + \sqrt{-1}}$
39. $-\dfrac{\sqrt{-25}}{1 - \sqrt{-1}}$
40. $-\dfrac{\sqrt{-16}}{5 + \sqrt{-9}}$

Solve each quadratic equation by the square root method.

41. $x^2 = -4$
42. $y^2 = -9$
43. $y^2 = -16$
44. $x^2 = -25$
45. $(a + 1)^2 = -1$
46. $(a - 2)^2 = -8$
47. $(4x - 2)^2 = -32$
48. $(6x + 3)^2 = -27$
49. $(3y + 2)^2 = -18$
50. $(4y - 1)^2 = -20$

Solve each quadratic equation by any of the four methods discussed.

51. $x^2 + 128 = 0$
52. $x^2 + 242 = 0$
53. $y^2 - 4y + 5 = 0$
54. $y^2 - 2y + 3 = 0$
55. $a^2 - 4a + 11 = 0$
56. $a^2 + 6a + 11 = 0$
57. $t^2 - 3t + 5 = 0$
58. $3t^2 + 4t + 4 = 0$
59. $x^2 - 10x + 27 = 0$
60. $x^2 - 4x + 7 = 0$
61. $y^2 - 3y + 6 = 0$
62. $2a^2 - 2a + 1 = 0$
63. $9a^2 = 12a - 5$
64. $4t^2 = 8t - 9$
65. $12r^2 - 12r - 1 = 0$
66. $32r^2 = 16r - 11$
67. $3x^2 + 7 = 3x$
68. $4y^2 + 1 = 2y$

Find the complex-number solutions for each of the following applied problems. There are two solutions (conjugate pairs) to each of Exercises 69–72.

69. A number added to its square equals -3. Find the number.
70. A number subtracted from its square equals -5. Find the number.
71. If you multiply the difference of a number and 1 by the number, the result is -1. Find the number.
72. Twice the square of a number less the number equals -2. Find the number.
73. Show that the pattern exhibited for $i = i$, $i^2 = -1$, $i^3 = -i$, and $i^4 = 1$ continues for i^9 through i^{12}.
74. Based upon the pattern established for powers of i, what do you think i^{16} is equal to?

REVIEW EXERCISES

For each of the following quadratic polynomials:
(a) *complete the ordered pairs* $(x, P(x))$; **(b)** *plot these ordered pairs on a Cartesian coordinate system; and* **(c)** *draw a smooth curve through these points.*

75. $P(x) = x^2 - 4x + 4$; $(0,\ )$, $(1,\ )$, $(2,\ )$, $(3,\ )$, $(4,\ )$
76. $P(x) = x^2$; $(-2,\ ), (-1,\ ), (0,\ ), (1,\ ), (2,\ )$
77. $P(x) = x^2 - 4$; $(-2,\ ), (-1,\ ), (0,\ ), (1,\ ), (2,\ )$
78. $P(x) = x^2 + 3$; $(-2,\ ), (-1,\ ), (0,\ ), (1,\ ), (2,\ )$
79. $P(x) = (x + 1)^2 + 2$; $(-3,\ ), (-2,\ ), (-1,\ ), (0,\ ), (1,\ )$
80. $P(x) = (x - 1)^2 - 3$; $(-1,\ ), (0,\ ), (1,\ ), (2,\ ), (3,\ )$

10.5 Graphing Quadratic Equations; The Parabola

OBJECTIVES

In this section we will learn to
1. find the vertex of a parabola;
2. identify the axis of symmetry of a parabola;
3. graph parabolas using the shifting concept; and
4. graph parabolas using the vertex and x-intercepts.

In Exercise Set 10.4 you were asked to graph quadratic functions of the form $P(x) = y = ax^2 + bx + c$ by finding several ordered pairs that satisfied the equation. In this section, we will take a more systematic approach to graphing quadratic functions and see how slight modifications can change the position and shape of the graph on the coordinate system.

Graphs of equations of the form

$$y = ax^2 + bx + c, \quad a \neq 0$$

are called **parabolas.** Parabolas have many applications in daily living. The reflector of a flashlight is parabolic in shape, as are the mirrors on many of the larger telescopes. The satellite dishes that are so commonly used for TV reception are parabolas. Radar antennae are parabolic and so are the giant antennae that are searching space for evidence of radio transmissions from intelligent beings.

The simplest form of the equation of a parabola is

$$y = x^2$$

which occurs when $b = c = 0$ and $a = 1$. Several ordered pairs satisfying $y = x^2$, as well as the graph of $y = x^2$, are shown in Figure 10.2. The graph is said to *open upward*. The *lowest* point on the parabola, $(0, 0)$, is called the **vertex.** The portion

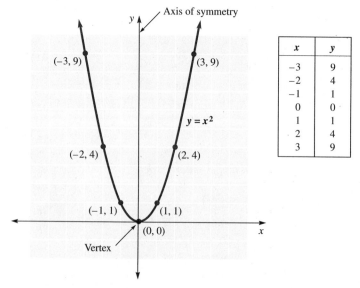

FIGURE 10.2

of the graph to the right of the y-axis is a mirror image of the portion to the left of the y-axis; for this reason, the y-axis is called the *axis of symmetry*.

DEFINITION	When a parabola opens upward or downward, the **axis of symmetry** is that line that passes through the vertex and is perpendicular to the x-axis.
Axis of Symmetry of a Parabola	

EXAMPLE 1 Graph: $y = x^2 + 1$.

SOLUTION We begin by constructing a table of values. The graph is shown in Figure 10.3.

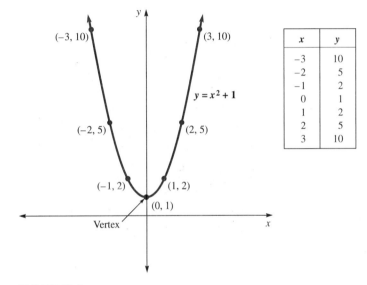

FIGURE 10.3

Notice that the only difference between the graph of $y = x^2$ (Figure 10.2) and that of $y = x^2 + 1$ (Figure 10.3) is that the graph of $y = x^2 + 1$ has been **shifted** 1 unit up. In fact, the graphs of $y = x^2$ and $y = x^2 + k$ will always have the same shape. If k is positive, the graph is shifted up k units, and if k is negative, the graph is shifted down k units.

EXAMPLE 2 Use the graph of $y = x^2$ to determine the amount and direction of shift of each of the following parabolas; then find the vertex:

(a) $y = x^2 - 2$ (b) $y = x^2 + 3$ (c) $y = x^2 - \dfrac{1}{2}$

SOLUTION (a) The graph of $y = x^2 - 2$ is the graph of $y = x^2$ shifted *down* 2 units. The vertex is $(0, -2)$
(b) The graph of $y = x^2 + 3$ is the graph of $y = x^2$ shifted *up* 3 units. The vertex is $(0, 3)$.

(c) The graph of $y = x^2 - \frac{1}{2}$ is the graph of $y = x^2$ shifted down $\frac{1}{2}$ unit. The vertex is $\left(0, -\frac{1}{2}\right)$.

The graphs of the parabolas in Example 2 together with $y = x^2$ are shown in Figure 10.4. They were drawn by using a program called TEMATH and a Macintosh computer. Notice that in each case, the graph has been shifted up or down but the axis of symmetry (the y-axis) remains unchanged.

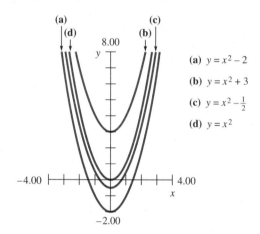

(a) $y = x^2 - 2$
(b) $y = x^2 + 3$
(c) $y = x^2 - \frac{1}{2}$
(d) $y = x^2$

FIGURE 10.4

A parabola can also be shifted horizontally without changing its basic shape.

EXAMPLE 3 Graph: $y = (x - 1)^2$.

SOLUTION We begin by constructing a table of values. Once again, the shape of the parabola is the same as that of $y = x^2$. The axis of symmetry is the line $x = 1$. The graph has been shifted one unit to the *right,* as shown in Figure 10.5. If we had graphed $y = (x + 1)^2$, the graph would have been shifted one unit to the *left* and the axis of symmetry would have been the line $x = -1$.

EXAMPLE 4 Give the vertex and axis of symmetry of each of the following parabolas and graph:
(a) $y = (x - 2)^2$ (b) $y = (x + 2)^2$

SOLUTION (a) The vertex is $(2, 0)$ and the axis of symmetry is the line $x = 2$. The graph is shifted two units to the right.
(b) The vertex is $(-2, 0)$ and the axis of symmetry is the line $x = -2$. The graph is shifted 2 units to the left.

The computer-drawn graphs are shown in Figure 10.6 together with the graph of $y = x^2$.

10.5 · GRAPHING QUADRATIC EQUATIONS; THE PARABOLA

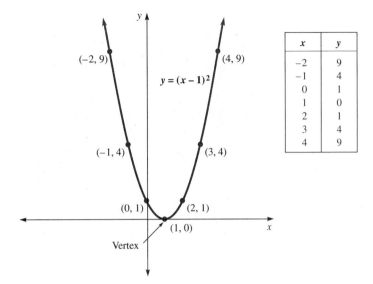

x	y
-2	9
-1	4
0	1
1	0
2	1
3	4
4	9

FIGURE 10.5

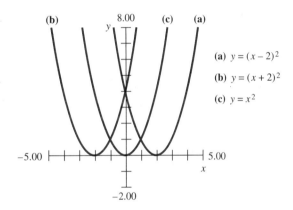

(a) $y = (x - 2)^2$
(b) $y = (x + 2)^2$
(c) $y = x^2$

FIGURE 10.6

EXAMPLE 5   Graph: $y = (x - 1)^2 + 1$.

SOLUTION   The parabola has the same shape as $y = x^2$, but is shifted one unit to the right and one unit up, so the vertex is (1, 1). A few ordered pairs that satisfy the equation are listed in the table of values. The graph is shown in Figure 10.7.

What happens to the graph of a parabola when the leading coefficient is negative? The next example answers that question by comparing their graphs with that of $y = x^2$.

EXAMPLE 6   Graph $y = -x^2$ and $y = x^2$ on the same coordinate system.

SOLUTION   We begin by constructing a table of values for $y = -x^2$. Figure 10.8 shows the two graphs.

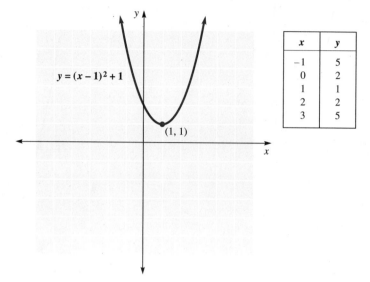

FIGURE 10.7

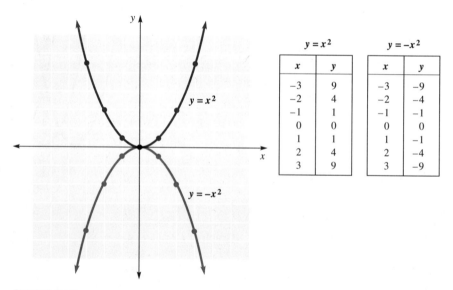

FIGURE 10.8

Example 6 illustrates that *if the lead coefficient of a parabola is **positive**, the parabola opens **upward**, and if the lead coefficient is **negative**, it opens **downward**.* When a parabola opens downward, its vertex is the highest point.

EXAMPLE 7   Graph $y = 2x^2$ and $y = x^2$ on the same coordinate system.

SOLUTION   We construct a table of values for $y = 2x^2$. Figure 10.9 shows the graph. Notice that the vertices of both graphs are located at the origin, but $y = 2x^2$ is narrower than $y = x^2$.

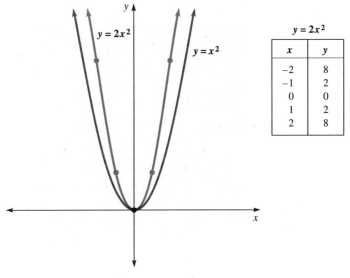

FIGURE 10.9

There are ways to find the vertex of a parabola when its equation is given in the form

$$y = ax^2 + bx + c$$

The most general way is to write the equation in the form

$$y = a(x - h)^2 + k$$

where the vertex is shifted to the point (h, k).

For example, the parabola

$$y = (x - 2)^2 + 4$$

has had its vertex shifted 2 units to the right of the origin and 4 units up, to the point $(2, 4)$. It opens upward because the **leading coefficient** (the coefficient of the x^2-term) is positive.

In working with parabolas, we can use previously learned techniques to find the coordinates of the vertex. For example, the equation $y = x^2 + 6x + 7$ can be rewritten as

$$y = (x + 3)^2 - 2$$

by completing the square, as follows. We begin by writing

$$y = (x^2 + 6x) + 7$$

To complete the square on $x^2 + 6x$, we add 9 to the right side. To keep from throwing the equation out of balance, however, we will immediately subtract 9 from the same side:

$$y = (x^2 + 6x + 9) + 7 - 9$$
$$= (x + 3)^2 - 2$$

This shows that the vertex is shifted 3 units to the left of the origin and 2 units down, to the point $(-3, -2)$.

EXAMPLE 8 Find the vertex and the axis of symmetry of the parabola $y = x^2 + 4x + 8$.

SOLUTION To find the vertex, we complete the square on $x^2 + 4x$ by adding and subtracting 4 from the right side.

$$y = x^2 + 4x + 8$$
$$= (x^2 + 4x + 4) + 8 - 4$$
$$= (x + 2)^2 + 4$$

The vertex is $(-2, 4)$. The axis of symmetry is the line that passes through the vertex and is perpendicular to the x-axis, so it is $x = -2$.

The x-intercepts of a parabola, which is where the graph crosses the x-axis, and the vertex of the parabola are key points in sketching the graph. The next example illustrates how to find them.

EXAMPLE 9 Use the x-intercepts and the vertex to graph the parabola $y = x^2 + x - 6$. Find the equation of the axis of symmetry.

SOLUTION To find where the graph crosses the x-axis, we set $y = 0$.

$$0 = x^2 + x - 6$$
$$0 = (x - 2)(x + 3)$$
$$x - 2 = 0 \quad \text{or} \quad x + 3 = 0$$
$$x = 2 \quad \text{or} \quad x = -3$$

The graph crosses the x-axis at $(2, 0)$ and $(-3, 0)$. To find the coordinates of the vertex, we write the equation of the parabola in the form $y = (x - h)^2 + k$ by completing the square.

$$y = (x^2 + x) - 6$$
$$= \left(x^2 + x + \frac{1}{4}\right) - 6 - \frac{1}{4} \quad \text{Add and subtract } \frac{1}{4} \text{ from the right side.}$$
$$= \left(x + \frac{1}{2}\right)^2 - \frac{25}{4}$$

The vertex is $\left(-\frac{1}{2}, -\frac{25}{4}\right)$. The graph is shown in Figure 10.10 The axis of symmetry passes through the vertex and is perpendicular to the y-axis, so the line is $x = -\frac{1}{2}$.

We can obtain much information about some parabolas by viewing their graphs. The domain and range, the intercepts, the vertex, and the axis of symmetry are a few of the things we can learn from the graph.

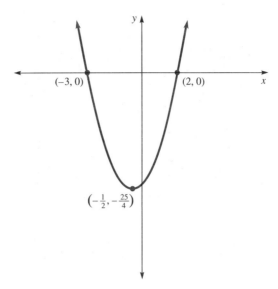

FIGURE 10.10

EXAMPLE 10   Find the domain and range, the intercepts, the vertex, and the axis of symmetry from the graph in Figure 10.11.

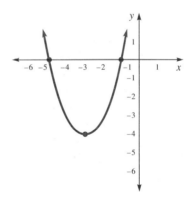

FIGURE 10.11

SOLUTION   The domain is $\{x|x \in R\}$. The lowest point on the graph is $y = -4$ and there is no highest point, so the range is $\{y|y \geq -4\}$. The x-intercepts are $(-5, 0)$ and $(-1, 0)$. The vertex is $(-3, -4)$.

Do You Remember?

Can you match these?

_____ 1. The parabola $y = 2x^2$ opens _____.
_____ 2. The vertex of $y = x^2 - 1$
_____ 3. The vertex of $y = (x - 2)^2 - 3$
_____ 4. The vertex of $y = (x + 2)^2 - 3$
_____ 5. The vertex of $y = (x + 2)^2 + 3$
_____ 6. The x-intercepts of $y = x^2 - 3x - 10$

a) $(-2, -3)$
b) $(2, 3)$
c) $(-2, 3)$
d) $(2, -3)$
e) upward
f) $(1, 0)$
g) $(0, -1)$
h) $(5, 0)$ and $(-2, 0)$
j) downward
k) $(-5, 0)$ and $(2, 0)$

Answers: 1. e 2. g 3. d 4. a 5. c 6. h

Exercise Set 10.5

Sketch the graph of each equation by making a table of ordered pairs using the given values of x. Find each vertex.

1. $y = 3x^2$; $x = -2, -1, 0, 1, 2$
2. $y = -3x^2$; $x = -2, -1, 0, 1, 2$
3. $y = x^2 + 3$; $x = -3, -2, -1, 0, 1, 2, 3$
4. $y = x^2 - 2$; $x = -3, -2, -1, 0, 1, 2, 3$
5. $y = -(x - 3)^2$; $x = 0, 1, 2, 3, 4, 5, 6$
6. $y = -(x + 4)^2$; $x = -7, -6, -5, -4, -3, -2, -1$
7. $y = (x + 1)^2 + 2$; $x = -3, -2, -1, 0, 1$
8. $y = (x - 3)^2 + 2$; $x = 1, 2, 3, 4, 5$
9. $y = -\frac{1}{3}x^2$; $x = -2, -1, 0, 1, 2$
10. $y = \frac{1}{4}x^2$; $x = -2, -1, 0, 1, 2$

Sketch the graph of $y = x^2$ by making a table of ordered pairs. Use the concept of "shifting" to sketch each of the following equations using $y = x^2$ as a model. Determine the coordinates of the vertex and the equation of the axis of symmetry for each parabola.

11. $y = x^2 + 2$
12. $y = x^2 + 1$
13. $y = x^2 - 3$
14. $y = x^2 - 4$
15. $y = (x - 1)^2$
16. $y = (x - 3)^2$
17. $y = (x + 2)^2$
18. $y = (x + 1)^2$
19. $y = (x + 1)^2 - 3$
20. $y = (x + 2)^2 - 4$
21. $y = (x - 2)^2 + 1$
22. $y = (x - 1)^2 + 2$
23. $y = (x + 3)^2 + 2$
24. $y = (x + 4)^2 + 5$

Sketch the graph of each equation by making a table of ordered pairs. Find the coordinates of each vertex.

25. $y = 4x^2$
26. $y = 6x^2$
27. $y = -5x^2$
28. $y = -2x^2$
29. $y = \frac{x^2}{4}$
30. $y = \frac{x^2}{5}$
31. $y = (x - 4)^2$
32. $y = (x + 5)^2$
33. $y = -(x + 1)^2$
34. $y = -(x - 3)^2$
35. $y = (x + 4)^2 - 1$
36. $y = (x + 3)^2 + 3$
37. $y = -x^2 - 3$
38. $y = -x^2 + 5$
39. $3y = -4x^2$
40. $2y = -3x^2$
41. $y = -(x - 1)^2 + 4$
42. $y = -(x - 3)^2 - 2$
43. $y = 3(x + 1)^2 - 2$
44. $y = 2(x - 1)^2 + 3$
45. $y = -\frac{1}{3}(x - 1)^2 - 2$
46. $y = -\frac{1}{2}(x - 1)^2 + 2$

Find the x-intercepts and the coordinates of the vertex. Use that information and other points to sketch each of the following parabolas.

47. $y = x^2 + 4x + 3$

48. $y = x^2 - 2x - 3$

49. $y = x^2 - 6x + 8$

50. $y = x^2 - 4x + 4$

51. $y = x^2 + 6x + 9$

52. $y = x^2 - 4x + 3$

53. $y = x^2 - 2x - 3$

54. $y = x^2 + 3x + 2$

55. $y = -x^2 - 2x - 1$

56. $y = -x^2 - 4x + 5$

Use the given graph to find (a) the x-intercepts (if they exist); (b) the coordinates of the vertex; (c) the axis of symmetry; and (d) the domain and the range.

57.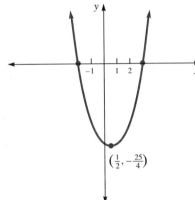

58.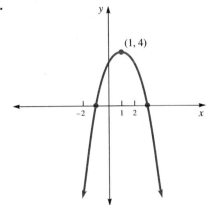

59.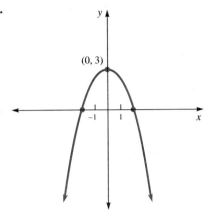

60.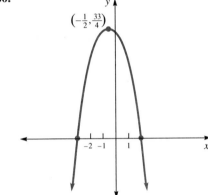

61.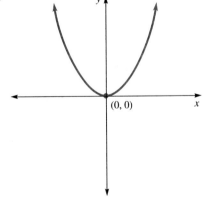

62.

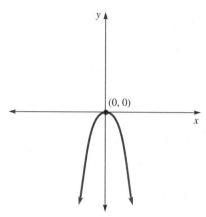

63.

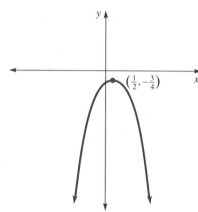

64.

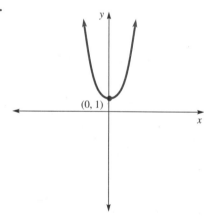

Given the quadratic function $f(x) = x^2 - 4x - 5$, find:

65. $f(0)$ **66.** $f(2)$ **67.** $f(-1)$
68. $f(-2)$ **69.** $f(3)$ **70.** $f(5)$

Solve each of the following applied problems.

71. Mona has invented a new ceramic decorator item that she has decided to make and sell. Her profits per day are given by the equation $P(n) = -2(n - 3)^2 + 8$, where P = profits in units of $10 and n = the number of items produced and sold.

  (a) Sketch the profit equation. Let the horizontal axis be the n-axis and the vertical axis, the P-axis.
  (b) Determine the points at which Mona's profit is zero (the break-even points).
  (c) What production gives maximum profit?
  (d) What is the maximum profit?
  (e) What production level causes her to lose money?
  (f) What is the number of items she must produce and sell to show a profit of $60? (*Reminder:* P is in units of $10, so a profit of $60 is written as 6.)
  (g) If this were your business, how many units would you produce and sell?

72. Maria's Hobby Shop makes and sells wallet-making kits. The amount of profit per day is given by the equation $P(n) = -(n - 5)^2 + 16$, where P = profits in units of $10 and n is the number of kits made and sold.

  (a) Sketch the profit equation. Let the horizontal axis be the n-axis and the vertical axis the P-axis.
  (b) Determine the points at which Maria's profit is zero (the break-even points).
  (c) What production gives maximum profit?
  (d) What is the maximum profit?
  (e) What production causes her to lose money?
  (f) What is the number of kits she must produce and sell to show a profit of $70? (*Reminder:* P is in units of $10, so a profit of $70 is written as 7.)
  (g) If this were your business, how many kits would you produce and sell?

10.6 More Applications

OBJECTIVES In this section we will learn to solve a variety of applied problems by using quadratic equations.

Many of the applications in this section will result in quadratic equations that are factorable. If you have difficulty factoring a particular equation, however, or if it is not factorable, do not hesitate to use one of the other methods. As you prepare to solve the following equations, review the five-step approach to finding solutions to application problems.

EXAMPLE 1 The length of a rectangle is 5 centimeters greater than its width. The area is 50 square centimeters. (See Figure 10.12.) *Find the dimensions of the rectangle.*

SOLUTION The area of the rectangle (50 cm²) is equal to the product of its width and its length, or $A = lw$. We let $x =$ the width and $x + 5 =$ the length. Then

$$(x + 5)x = 50$$
$$x^2 + 5x = 50$$
$$x^2 + 5x - 50 = 0$$
$$(x + 10)(x - 5) = 0$$
$$x + 10 = 0 \quad \text{or} \quad x - 5 = 0$$
$$x = -10 \quad \text{or} \quad x = 5$$

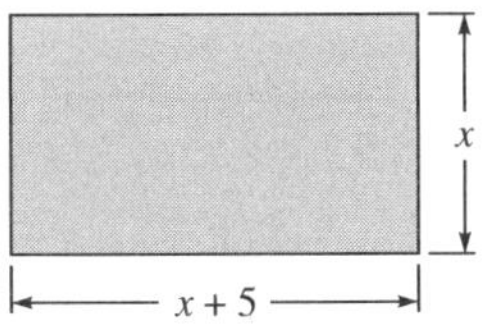

FIGURE 10.12

We reject -10 as an answer because a rectangle cannot have negative width. So the width is $x = 5$ cm and the length is $x + 5 = 10$ cm.

CHECK The length is 5 cm greater than the width; the area is $5 \cdot 10 = 50$ cm². The solution checks.

EXAMPLE 2 A vacationer drove 448 miles. If she had increased her average speed by 8 mph, the trip would have taken 1 hour less. Find her average speed.

SOLUTION This is a distance-rate-time problem. Organizing the given information in a chart is usually the best approach. We let

$s =$ the average speed at the slower rate

$s + 8 =$ the average speed at the faster rate

To complete the entries in the chart, remember that

1. the distance is the same at either speed;
2. the faster speed is 8 mph greater than the slower speed; and
3. time is computed by dividing distance by rate: $t = \dfrac{d}{r}$

	Distance	Rate	Time
Slower Trip	448	s	$\dfrac{448}{s}$
Faster Trip	448	$s + 8$	$\dfrac{448}{s+8}$

To establish an equation, we use the fact that the difference between the two times is 1 hour. It takes longer to travel the distance at the slower average speed, so

Longer time − shorter time = 1 hour

$$\frac{448}{s} - \frac{448}{s+8} = 1$$

$$448(s + 8) - 448s = s(s + 8) \quad \text{Multiply each side by } s(s+8), \text{ the LCD.}$$

$$448s + 3584 - 448s = s^2 + 8s \quad \text{Distributive property}$$

$$0 = s^2 + 8s - 3584 \quad \text{Simplify.}$$

$$0 = (s - 56)(s + 64)$$

$$s - 56 = 0 \quad \text{or} \quad s + 64 = 0$$

$$s = 56 \quad \text{or} \quad s = -64$$

The average speed is 56 mph. We reject −64 because speed cannot be negative.

EXAMPLE 3 What real number added to its square has a sum of 35?

SOLUTION We let x = the real number
x^2 = the square of the real number

Since the sum of the real number and its square is 35, our equation is

$$x + x^2 = 35$$
$$x^2 + x - 35 = 0$$

This equation is not factorable, so we use the quadratic formula to find the solutions.

$$x = \frac{-b \pm \sqrt{b^2 - 4ac}}{2a}$$

$$= \frac{-1 \pm \sqrt{(1)^2 - 4(1)(-35)}}{2(1)} \quad a = 1, b = 1, \text{ and } c = -35$$

$$= \frac{-1 \pm \sqrt{141}}{2}$$

There are two solutions, $\dfrac{-1 - \sqrt{141}}{2}$ or $\dfrac{-1 + \sqrt{141}}{2}$. We will check the first number. The square of the number is

$$\left(\frac{-1 - \sqrt{141}}{2}\right)^2 = \frac{1 + 2\sqrt{141} + 141}{4}$$

$$= \frac{142 + 2\sqrt{141}}{4}$$

$$= \frac{2(71 + \sqrt{141})}{4}$$

$$= \frac{71 + \sqrt{141}}{2}$$

$$\underbrace{\text{A number}}_{\frac{-1 - \sqrt{141}}{2}} \underbrace{\text{plus}}_{+} \underbrace{\text{its square}}_{\frac{71 + \sqrt{141}}{2}} \overset{?}{=} \underbrace{35.}_{\frac{-1 - \sqrt{141} + 71 + \sqrt{141}}{2}}$$

$$\overset{?}{=} \frac{70}{2}$$

$$\overset{\checkmark}{=} 35$$

The check of the other solution is left to the reader.

EXAMPLE 4 The sum S of the first n consecutive natural numbers is given by the formula

$$S = \frac{n(n+1)}{2}$$

For example, the sum of the first 10 natural numbers is

$$1 + 2 + 3 + 4 + \cdots + 10 = \frac{10(10+1)}{2} = 55$$

Starting with 1, how many consecutive natural numbers must be added to produce a sum of 820?

SOLUTION We let $n =$ the number of natural numbers and $820 = S$. Then

$$820 = \frac{n(n+1)}{2}$$

$1640 = n(n+1)$ Multiply each side by 2, the LCD.

$1640 = n^2 + n$

$0 = n^2 + n - 1640$

$0 = (n - 40)(n + 41)$

$n - 40 = 0$ or $n + 41 = 0$

$n = 40$ or $n = -41$

We reject -41 because it is not a natural number. We check 40 in the original equation.

$$S = \frac{n(n+1)}{2}$$

$$820 \overset{?}{=} \frac{40(40+1)}{2}$$

$$820 \stackrel{?}{=} 20(41)$$
$$820 \stackrel{\checkmark}{=} 820$$

The solution checks. The sum of the first 40 natural numbers is 820.

EXAMPLE 5 A rectangle has a perimeter P of 60 feet and an area A of 200 square feet. Find its dimensions.

SOLUTION As a review, we will use two variables to set up the equation. We let $l =$ the length and $w =$ the width. Then two equations are

$$P = 2l + 2w \implies 2l + 2w = 60 \quad \textbf{I}$$
$$A = lw \implies lw = 200 \quad \textbf{II}$$

We will solve equation **I** for l and substitute the result into equation **II**.

$2l + 2w = 60$	Equation **I**
$2l = 60 - 2w$	Subtract $2w$ from each side.
$l = 30 - w$	Divide each side by 2.
$lw = 200$	Equation **II**
$(30 - w)w = 200$	Substitute $30 - w$ for l.
$30w - w^2 = 200$	Distributive property
$-w^2 + 30w = 200$	
$-w^2 + 30w - 200 = 0$	Subtract 200 from each side.
$w^2 - 30w + 200 = 0$	Multiply each side by -1.
$(w - 10)(w - 20) = 0$	
$w = 10$ or $w = 20$	Zero-factor property

If $w = 10$ ft, $l = 30 - 10 = 20$ ft, and if $w = 20$ ft, $l = 30 - 20 = 10$ ft. Since length is generally thought of as being greater than width, we will use the first set of dimensions. The width is 10 ft and the length is 20 ft.

EXAMPLE 6 In 1989 a certain prescription drug cost $24.00 a bottle. In 1993 the patent expired and it became available in a generic variety. Today it costs $1.00 less per ounce and a bottle of the drug, while still $24, holds two more ounces. What was the cost per ounce of the drug in 1989?

SOLUTION We let $c =$ the cost per ounce in 1989

$c - 1 =$ the cost per ounce today

To determine the number of ounces, we divide the total cost by the cost per ounce.

	Cost per Ounce	Total Cost	Number of Ounces
1989	c	$24	$\dfrac{24}{c}$
Today	$c - 1$	$24	$\dfrac{24}{c-1}$

The number of ounces that can be purchased today is 2 greater than in 1989:

$$\frac{24}{c-1} = \frac{24}{c} + 2$$

$24c = 24(c-1) + 2c(c-1)$ Multiply each side by $c(c-1)$, the LCD.
$24c = 24c - 24 + 2c^2 - 2c$ Distributive property
$0 = 2c^2 - 2c - 24$ Subtract $24c$ from each side and put in standard form.
$0 = 2(c-4)(c+3)$
$c - 4 = 0$ or $c + 3 = 0$ Zero-factor property
$c = 4$ or $c = -3$

We reject -3 because cost cannot be negative. In 1989, the drug cost $4/ounce.

Here is a second approach to Example 6:

We let $x =$ the number of ounces purchased in 1989
          $x + 2 =$ the amount purchased today

To determine the cost per ounce then and now, we divide the total cost by the number of ounces purchased.

	NUMBER OF OUNCES	COST PER OUNCE	TOTAL COST
1989	x	$\dfrac{24}{x}$	$24
TODAY	$x + 2$	$\dfrac{24}{x+2}$	$24

Since the drug costs $1 less per ounce today, we can get an equation by subtracting the two costs and setting this difference equal to $1. The greater cost per ounce occurred in 1989.

$$\frac{24}{x} - \frac{24}{x+2} = 1$$

$24(x+2) - 24x = x(x+2)$ Multiply each side by $x(x+2)$, the LCD.
$24x + 48 - 24x = x^2 + 2x$ Distributive property
$0 = x^2 + 2x - 48$ Standard form
$0 = (x-6)(x+8)$
$x - 6 = 0$ or $x + 8 = 0$ Zero-factor property
$x = 6$ or $x = -8$

We reject -8 because the number of ounces cannot be negative. Since 6 ounces could be purchased for $24 in 1989, the cost per ounce then was $\dfrac{\$24}{6} = \$4$. The check is left to the reader.

Exercise Set 10.6

Solve each of the following applied problems.

1. Find two consecutive even numbers with a product of 168.
2. What number added to its square equals 42?
3. One negative number is 7 less than another and their product is 30. Find the two numbers.
4. The sum of two natural numbers is 25. Find the numbers if the sum of their squares is 313.
5. A family is planting a rectangular garden. What is the length of the garden if the perimeter is 94 m and the area is 546 m²?
6. The length and width of a 4 cm-by-2 cm rectangle are each increased by the same amount. The area of the new rectangle is triple the area of the original rectangle. Find the width of the new rectangle.
7. A student jogs 60 mi/week. If he increased his rate by 2 mph, he would spend 1 hr less per week jogging. What is his present speed?
8. A boater leaves a dock and travels 45 km upstream; she returns to the same dock 8 hr later. The engine speed of the boat remains constant and the speed of the current is 3 km/hr. What is the speed of the boat in still water?
9. The longest side (hypotenuse) of a right triangle is 5 ft. The shortest side is 1 ft less than the middle-sized side. Find the lengths of the shortest and middle-sized sides.
10. The longest side (hypotenuse) of a right triangle is 6 m and the second side is double the third side. What are the lengths of the second and third sides?
11. Some students equally share the $600 rent for a vacation house. Two more students join the group, and the rental cost is $25 less per person. How many were in the original group?
12. All members of a class contribute equally to buy a $40 gift for the teacher. If 2 members had been absent, each person's contribution would have had to increase by two-thirds of a dollar to cover the $40. How many are in the class?
13. A trucker makes a 360-mi trip every day at the same speed. One day he has to reduce his average speed by 15 mph because of bad weather; as a result, he is 4 hr late. Find his usual speed and time.
14. Don and Mindy make a 60-mi trip by bicycle. Mindy averages 5 mph faster than Don and, as a result, she arrives 1 hr before he does. How fast does each travel?
15. Two women start on an 18-mi hike. One walks $\frac{3}{4}$ mph faster than the other and, as a result, she travels the distance in 2 hr less time. Find the speed of each.
16. It takes $1\frac{2}{3}$ hr less for Ron's plane to fly 800 km with a tailwind than against it (a headwind). If the plane travels 200 km/hr in still air, what is the speed of the wind?
17. The students in an algebra class decide to go out to dinner together. They have to choose between a club steak dinner or a sirloin tip dinner, which costs $4 less than the steak dinner. They all decide to have the club steak dinner, at a total cost of $240. Had they decided on the sirloin tip dinner, they could have had 3 more dinners for the same total cost. How many dinners did they order?
18. Last week Blake paid $5.76 for an order of halibut. This week he spends the same amount but receives 2 oz more because halibut is on sale for 4¢ less per ounce than last week. How many ounces of halibut did Blake buy last week?
19. To make a box from a rectangular piece of cardboard that is 4 in. longer than it is wide, a 3 in. square is cut out of each corner and the sides are folded up. If the volume of the box is 288 in.³, how long is the box?
20. Strawberry plants are set in rows of uniform length. The number of plants in each row is 18 more than the number of rows. If there is a total of 175 plants, find the number of rows and the number of plants per row.
21. The sum of the first n consecutive even natural numbers (2, 4, 6, 8, . . .) is given by the formula $S = n^2 + n$. How many even natural numbers have to be added together to produce a sum of 272?
22. The sum of the first n consecutive odd natural numbers (1, 3, 5, 7, . . .) is given by the formula $S = n^2$. How many consecutive odd natural numbers have to be added together to produce a sum of 289?
23. A rocket is launched from a pad 6 ft above the ground. The height h in feet above the ground after t seconds is given by the formula $h = -4.9t^2 + 55t + 6$. What is the height (above ground) of the rocket after 1 sec? after 3 sec?
24. The braking distance for a certain car is approximated by $d = 0.3s^2 + 0.9s$, where d is in feet and s is the speed of the car in mph. If the car is traveling at 55 mph, can it stop in 960 ft?
25. The sum of a number and its reciprocal is $\frac{26}{5}$. Find the number.

26. The sum of a number and twice its reciprocal is $\frac{51}{7}$. Find the number.

27. The lengths of the three sides of a right triangle are three consecutive even natural numbers. Find the numbers.

28. The lengths of the three sides of a right triangle are three consecutive natural numbers. Find the numbers.

29. Find the length of a diagonal of a square with a side of $\sqrt{6}$ ft.

30. Find the length of the side of a square with a diagonal of $\sqrt{32}$ ft.

31. The sum of two numbers is 20 and their product is 91. Find the numbers.

32. The sum of two integers is -4 and their product is -77. Find the integers.

33. If a rational number is added to the numerator of the fraction $\frac{5}{6}$ and the denominator is multiplied by the same rational number, the result is $\frac{4}{3}$. Find the rational number.

SUMMARY

10.1 Solving Equations by Factoring and the Square Root Method

One of the easiest ways of solving a quadratic equation is to factor it and use the zero-factor property to find its solutions. When an equation is of the form $x^2 = a$, it can also be solved by using the **square root property,** which says that if a is a positive number and

$$x^2 = a \quad \text{then} \quad x = \pm\sqrt{a}$$

10.2 Solving Quadratic Equations By Completing the Square

When a quadratic equation is not in a form suitable for using the square root property, it can be put in such a form by **completing the square.** If the coefficient of the second-degree term is 1, the quantity necessary to complete the square is the square of one-half of the coefficient of the middle term:

$$x^2 + 2ax + a^2 \quad \text{where} \quad a^2 = \left(\frac{1}{2} \cdot 2a\right)^2$$

To Solve a Quadratic Equation by Completing the Square

Step 1 Isolate all terms involving the variable on one side of the equation, using the addition or subtraction property of equality.

Step 2 If the coefficient of the second-degree term is not 1, divide both sides by that coefficient.

Step 3 Add the square of one-half the coefficient of the first-degree term to each side of the equation.

Step 4 Write the perfect-square trinomial obtained in Step 3 as the square of a binomial.

Step 5 Use the square root method on the equation found in Step 4 and solve for the variable. Check the solution in the original problem.

10.3 Solving Quadratic Equations by the Quadratic Formula

The **quadratic formula** is used to solve the general quadratic equation, $a > 0$, $ax^2 + bx + c = 0$, in standard form. The solutions are

$$x = \frac{-b \pm \sqrt{b^2 - 4ac}}{2a}$$

Solutions obtained by using the quadratic formula always occur in conjugate pairs. If the solutions are irrational numbers, checking only one solution will give reasonable assurance that both solutions are correct.

10.4 Solving Quadratic Equations with Complex-Number Solutions

The **imaginary number** $i = \sqrt{-1}$ is that number such that $i^2 = -1$. It is said to be imaginary only because there is no real number that represents $\sqrt{-1}$.

When i is raised to positive integer powers, only four results occur: i, -1, $-i$, and 1.

Square roots can be multiplied or divided by following the **product rule for square roots,** which says that if a and b are real numbers and at least one is not negative, then

$$\sqrt{a \cdot b} = \sqrt{a}\sqrt{b}$$

If b is a positive number, then $\sqrt{-b} = \sqrt{b}i$.

A **complex number** is any number of the form $a + bi$, where a and b are real numbers. Given this definition, all real numbers are also complex numbers. In the complex number $a + bi$, the number a is called the *real part* and b is referred to as the **imaginary part.**

The set of real numbers and the set of imaginary numbers are both subsets of the set of complex numbers.

Complex numbers are added, subtracted, and multiplied in much the same way as polynomials. They are divided by rationalizing the denominator.

10.5 Graphing Quadratic Equations; The Parabola

The general equation of a parabola is $y = ax^2 + bx + c$. If a is a **positive** number, the graph of the parabola opens **upward,** and if a is **negative,** it opens **downward.** The lowest point on the graph of a parabola that opens upward is called its **vertex,** as is the highest point on one that opens downward. The **axis of symmetry** is the line that passes through the vertex and is perpendicular to the x-axis.

The graph of a parabola can often be determined from that of a known parabola by noting the amount of **shift.** With respect to $y = x^2$,

$y = x^2 + 1$ is shifted up 1 unit

$y = x^2 - 1$ is shifted down 1 unit

$y = (x - 1)^2$ is shifted to the right 1 unit

$y = (x + 1)^2$ is shifted to the left 1 unit

$y = (x - 1)^2 - 1$ is shifted 1 unit to the right and 1 unit down

Cooperative Exercise

The Bowmans are a young couple planning for the future. First, they want to buy a home and their banker says they will need a minimum of $8000 for a down payment. Second, they want to start saving for a retirement annuity (a monthly check after retirement).

I. Presently they are thinking of depositing $100/month in an account at a 6% annual rate of interest compounded monthly. What will the value of this account be in 5 years? Use formula **(a)**.

(a) $A_n = P\left(\dfrac{(1 + i)^n - 1}{i}\right)$

where A_n = the amount at the time immediately after the nth payment;

P = the amount of the monthly payment;

i = the interest rate per month; and

n = the number of months

II. The Bowmans need a minimum of $8000 for a down payment on a house in 5 years. How much should they deposit monthly into an account with a 6% annual rate of interest compounded monthly to provide this $8000? *Hint:* Solve equation **(a)** for *P*.

III. In planning ahead for retirement, they would like to have annuity payments of $1000 per month for 20 years after they retire. How much money will be required at the time of retirement in order to accomplish this?

Assume the money is invested at an annual interest rate of 6% compounded monthly. Use formula **(b)**.

$$\text{(b)} \quad A_n = P\left(\frac{1-(1+i)^{-n}}{i}\right)$$

IV. If they plan to retire at age 60 and they are both presently 20 years old, how much should they deposit each month to have the money available? (See the answer for III.) Assume their account earns 6% annual rate of interest compounded monthly. See the hint in II.

Review Exercises

Solve each of the following equations by factoring.

1. $(2x-3)(x+4)=0$
2. $(5y-2)(3y+4)=0$
3. $x^2+12x+36=0$
4. $x^2-14x+49=0$
5. $8y^2-32=0$
6. $x^2-32=0$
7. $450x^3-242x=0$
8. $75x^4-12x^2=0$
9. $\frac{1}{9}x^2-\frac{49}{16}=0$
10. $\frac{4}{25}y^2-\frac{81}{121}=0$
11. $2y^2-22y-120=0$
12. $36t^3+60t^2+25t=0$
13. $5x^2-10x-15=0$
14. $25y^3-16y=0$
15. $10t^2-17t+3=0$
16. $24x^2-14x-20=0$

Solve each of the following equations by the square root method.

17. $x^2=4$
18. $x^2=144$
19. $x^2=-27$
20. $x^2=-1$
21. $y^2=8$
22. $y^2=50$
23. $x^2-6=18$
24. $x^2-16=24$
25. $t^2+9=0$
26. $t^2+5=-20$

Solve each of the following equations by completing the square. Factor out any common factors first.

27. $x^2-3x-2=0$
28. $x^2+x-3=0$
29. $y^2-2y-5=0$
30. $y^2=8y-16$
31. $2t^2-6t=-3$
32. $t^2+4=6t$
33. $24x^3-8x^2-10x=0$
34. $6x^3-27x^2=15x$
35. $0.1y^2-y=0.3$
36. $0.49y^2-y-1=0$
37. $\frac{1}{3}a^2+2a+\frac{1}{3}=0$
38. $\frac{8}{3}a^2-4a=-\frac{4}{3}$

Solve each of the following equations by the quadratic formula. Factor out any common factors first.

39. $x^2-x-5=0$
40. $x^2+x-3=0$
41. $7t^2-t=0$
42. $8x^3-3x^2=0$
43. $y^2-24=0$
44. $x^2-21=0$
45. $x-\frac{3}{5}=\frac{1}{x}$
46. $x-\frac{1}{3}=\frac{2}{x}$
47. $\frac{y^2}{y-3}-\frac{8}{y-3}=0$
48. $\frac{y^2}{3y-5}=-\frac{2}{5-3y}$
49. $0.4t^2-0.4t-0.16=0$
50. $0.7t^2+0.8t-0.1=0$

Solve each of the following equations by any method.

51. $x(2x-1)(3x+5)=0$
52. $3y(y-1)(2y+7)=0$
53. $(2x-1)^2=4$
54. $(3y+8)^2=1$
55. $\sqrt{x-1}=x-3$
56. $2y+1=\sqrt{3y+5}$
57. $\sqrt{x^2-5x+4}=1$
58. $\sqrt{2t^2-3t-1}=2$
59. $(x+3)^2=2(x-5)$
60. $(y-1)^2=5(y-4)$

Simplify each of the following quantities.

61. $\sqrt{-9}$
62. $\sqrt{-36}$
63. $\sqrt{-18}$
64. $\sqrt{-8}$
65. $(1+i)+(3-i)$
66. $(4-3i)+(2+4i)$
67. $(8-i)-(3-2i)$
68. $(4+3i)-(7+5i)$
69. $(3i)(5i)$
70. $(-2i)(5i)$

71. $(3 + 2i)(3 - 2i)$

72. $(4 + 5i)(4 - 5i)$

73. $\dfrac{i}{4i^2}$

74. $\dfrac{\sqrt{-1}}{1 - \sqrt{-1}}$

75. $\dfrac{3i}{2 + i}$

76. $\dfrac{-\sqrt{-25}}{2 - \sqrt{-9}}$

Solve each of the following equations.

77. $t^2 = -8$
78. $m^2 = -75$
79. $(y - 3)^2 = -20$
80. $3t^2 - 3t = 7$
81. $x^2 - x = -3$
82. $y^2 + 4y = -9$
83. $s^2 + 162 = 0$
84. $3y^3 + 3y = 0$
85. $1 = 2x - 4x^2$
86. $x^2 - x = -5$
87. $2t^2 - 6t = -24$
88. $a^2 + 3a + 7 = 0$

Sketch the graph of each parabola. Find the coordinates of the vertex, the x-intercepts (if they exist), and the equation of the axis of symmetry.

89. $y = x^2 - 2$
90. $y = x^2 + 3$
91. $y = -x^2 + 2$
92. $y = (x - 2)^2$
93. $y = (x + 1)^2$
94. $y = (x + 3)^2 + 2$
95. $y = (x - 2)^2 + 2$
96. $y = x^2 - 3x + 2$
97. $y = x^2 - 4x + 4$

Solve each of the following applied problems.

98. Use the given figure of a rectangle to find x.

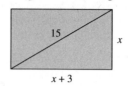

99. Use the given figure of a right triangle to find x.

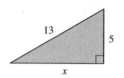

100. Find two consecutive odd numbers that have a product of 143.

101. Twice a positive number added to its square is 7. Find the positive number.

102. Kristyn jogs 40 mi/week. If she were to increase her rate by 1 mph, she would spend 2 hr less per week jogging. What is her average speed?

103. It takes 2 hr less for Enrique to fly 320 mi with a tailwind than it does against it (a headwind). If the plane travels 120 mph in still air, what is the speed of the wind?

Chapter Test

Solve each equation by factoring or by the square root method. Check all solutions.

1. $(7x - 5)(2x + 3) = 0$
2. $36y^2 - 121 = 0$
3. $p^2 - 11p = 0$
4. $10m^3 + 150m^2 - 160m = 0$
5. $t^2 - 7t - 8 = 0$
6. $15a^2 = 19$
7. $(3x - 4)^2 = 7$
8. $20t^2 - 16 = 180 - 5t^2$
9. $p^2 + 12p + 36 = 0$
10. $25r^2 - 60r + 36 = 0$
11. $x^2 = -100$
12. $y^2 + 81 = 0$

Solve each equation by completing the square.

13. $x^2 - 2x - 5 = 0$
14. $3x^2 = 1 - x$
15. $6a^2 - a - 11 = 0$
16. $6y = 3y^2 - 5$
17. $r^2 + \dfrac{5}{4}r - \dfrac{3}{2} = 0$

Solve each equation by the quadratic formula. Factor out any common factors first.

18. $a^2 + a - 1 = 0$
19. $3m^2 - 14 = 0$
20. $6 - 3x^2 = x$
21. $\dfrac{7}{4}t^2 - \dfrac{3}{2}t - 10 = 0$
22. $5p^2 - 7p = 0$
23. $x^2 + x + 3 = 0$
24. $3y^3 + 12y^2 + 39y = 0$

Sketch the graph of each parabola; determine the coordinates of the vertex and the equation of the axis of symmetry.

25. $2y = x^2$
26. $y = (x - 7)^2$

27. $2y = x^2 + 6$
28. $y = (x + 1)^2 + 3$
29. $y = -(x + 4)^2 + 1$
30. $y = -3(x - 1)^2 - 5$

Write a quadratic equation for each applied problem and solve it.

31. The sum of a number and its square is 110. Find the number.
32. In a given right triangle, the hypotenuse is 2 units more than the shortest leg. The other leg is 1 unit more than the shortest leg. Find the dimensions of the triangle.
33. The length of a rectangle is three and one-half times the width. The area is $\frac{27}{2}$ ft². Find its dimensions.
34. David rows a boat 20 mi downstream and returns in 11 hr and 20 min. If he can row at $4\frac{1}{4}$ mph in still water, and he rows at a constant rate, what is the rate of the current?

Cumulative Review for Chapters 1–10

In Exercises 1–5, simplify the given quantities.

1. $\frac{5}{2} + \frac{1}{3} - \frac{5}{6}$
2. $\left(\frac{6}{5} \cdot 4\right) \div \frac{3}{2}$
3. $(7^2 - 4) \cdot 5 \div (3 - 1)(7 - 2)$
4. $[5^2 - 3^2 + 2(1 + 1)] \div 4$
5. $|3 + (-5)| + (-6)$
6. Subtract 4 from -9.
7. Find the difference of 10 and -2.
8. On one play a football team lost 15 yd, and on the next play it gained 7 yd. What is the team's total gain or loss on the two plays?
9. Simplify: $\frac{3}{5} \div (-12)$
10. Simplify: $\left(\frac{2}{3} \cdot 6 + \frac{5}{6} \cdot 8\right) \div \frac{2}{3}$
11. A package of four frozen pie crusts costs $2.80. How much does each pie crust cost?
12. Simplify: $\frac{2}{9} \div 0$
13. Evaluate $\frac{x + y}{3}$ for $x = -11$ and $y = 10$.
14. Solve: The product of a number and 8 is 7 more than the number.
15. State the property illustrated in each of the following equations.
    (a) $3 + 5 = 5 + 3$
    (b) $\left(-\frac{1}{2}\right)(-2) = 1$
    (c) $7(4 - 9) = 7 \cdot 4 - 7 \cdot 9$
    (d) $(x + 3)[a + (-a)] = (x + 3)(0)$
    (e) $1 \cdot (x + y) = x + y$
16. State the property illustrated: $1.7 + (-1.7) = 0$.
17. Simplify: $10 - (3 + y) - 5(-2)$.
18. Simplify: $7(x - 2) - 2x + 14$.

Write Exercises 19 and 20 as algebraic expressions and simplify.

19. A number subtracted from 7 times the sum of itself and 3.
20. Five minus 4 times the sum of b and 15.
21. Simplify: $3.1(10 - pqw^3) - 4.9(3pqw^3 - 5)$.
22. Simplify: $\frac{4}{3}\left(\frac{15}{2} - \frac{21}{4}p^4q\right) - \frac{6}{5}\left(\frac{25}{12}p^4q + 10\right)$.

Solve each of the following equations.

23. $2x + 10 = 3x - 6$
24. $25 - 8x = 20 - 7x$
25. $9.4 - 10.9 = 12.2y - 2.2y$
26. $\frac{x}{2x + 5} = 7$
27. $\frac{2}{3}(2x - 6) = 2 - \frac{1}{3}$
28. $2x + \frac{4}{3} = 6$
29. A 5-ft board is cut into two pieces, with the larger piece $1\frac{1}{2}$ times as the large as the smaller piece. Find the length of each piece.
30. A lady has $2.35 in nickels and dimes in her purse. If she has 7 more dimes than nickels, how many of each coin does she have?
31. Solve: $x - 12.5 \leq -9.5$
32. Solve: $-\frac{9}{8} < x + \frac{7}{8} < \frac{7}{8}$
33. Solve: $-5 < 7 - 3t < 10$
34. Two boys leave school at the same time and bicycle in opposite directions. If the first boy rides at 4 mph and the second at 5 mph, how long does it take them to be 3 mi apart?

35. Find the area of the pictured trapezoid.

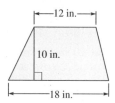

36. Twenty thousand dollars is invested in two securities, one paying 8% and the other 12%. How much is invested at each rate if the yearly interest is $1800?

37. Solve $h = vt - 16t^2$ for v.

38. A company that makes clocks has a weekly cost expressed by $C(x) = 5376 + 150x$, where x is the number of clocks made and sold in a week. The weekly revenue is $R(x) = 598x$. Find the break-even point for the company.

Perform the indicated operations and simplify.

39. $-(-a)^4$

40. $\left(\dfrac{7}{z}\right)^{-2}$

41. $(m^2n^{-2})^{-3}(mn^2)$

42. $\dfrac{(40)(0.015)}{(0.03)(200)}$

43. $\dfrac{1.6(x^2y^3z^2)^3}{0.8(xy^3z^2)^2(x^2y)}$

44. Multiply $3.7x^2y^3$ by $10xy^2$.

45. $7 - 3x^3 + x^3 - 2x^2 + 12$

46. Divide $5xy - 3xy$ by 4.

47. Subtract $a^2 - 9ab - 4b^2$ from $a^2 + 6ab + 4b^2$.

48. $4x^3 - [2 - (3x^3 - 3)]$

49. $(x + 4)(x - 1)$

50. $(2r - 3)(3r - 1) + 3r(1 - 2r)$

51. $(x - 2)^2 - (x + 2)^2$

52. $(w + 3)(w - 3)(w^2 - 9)$

53. $\left(y - \dfrac{3}{4}\right)\left(y + \dfrac{3}{4}\right)$

54. The sum of the square of a number and -3 is equal to the square of the sum of the number and 1. Find the number.

55. Divide $x^2 + x - 6$ by $x - 2$.

56. Divide $y^2 - 6y - 5$ by $y + 1$.

57. Divide $9y^3 - 4y - 1$ by $3y + 1$.

Factor each of the following polynomials. Some may be prime.

58. $5z^6 - 3z^5 + 2z^3$

59. $20a^3x^3 - 15a^2x^2$

60. $s^2 - 13s + 30$

61. $m^2 + 7m + 7$

62. $b^5 - 8b^4 + 12b^3$

63. $12r^2 - 11r + 2$

64. $6x^9 - 8x^8 + 2x^7$

65. $25r^2 - 49s^2$

66. $x^2 + 2xa + a^2$

67. $8p^3 + 27$

68. $u(3 - c) - v^2(3 - c)$

69. $z^2x + z^2w - 16x - 16w$

Solve each of the following quadratic equations. Check your answers.

70. $3v^2 + 6v - 9 = 0$

71. The sum of two numbers is -4 and their product is -12. Find the two numbers.

72. The lengths of the three sides of a right triangle are three consecutive even integers. Find the lengths of the three sides.

Simplify each of the following rational expressions.

73. $\dfrac{3s^2 + 11s + 6}{3s^2 - s - 2}$

74. $\dfrac{ax + ay - 2x - 2y}{a^2 - 4}$

75. $\dfrac{2y^2 + y - 3}{3y - 2} \cdot \dfrac{3y^2 + y - 2}{y^2 - 1}$

76. $\dfrac{1 - x^2}{2x^4 + 2x^3} \div \dfrac{x^2 - 2x + 1}{6x^4}$

77. $\dfrac{1}{y - 1} + \dfrac{1}{3y} - \dfrac{1}{y^2(y - 1)}$

78. $\dfrac{\dfrac{2}{3} - \dfrac{1}{x^2}}{\dfrac{5}{x} + \dfrac{1}{3x^2}}$

In Exercises 79–81, solve each of the equations. Check your answers.

79. $\dfrac{z + 2}{z - 2} = \dfrac{1}{z} + \dfrac{8}{z(z - 2)}$

80. $3x + \dfrac{25}{x - 10} = \dfrac{5(5 - x)}{10 - x}$

81. $\dfrac{a}{a + 2} = \dfrac{2 - a}{a + 7}$

82. It took Mac the same time to drive 450 mi as it took Milt to drive 400 mi. Mac drove 7 mph faster than Milt. How fast did each person drive?

83. Suppose y varies directly as x, and $y = 24$ when $x = 8$. Find y when $x = 5$.

84. If A varies inversely as x and $A = 30$ when $x = \tfrac{1}{2}$, find A when $x = 5$.

85. Graph: $4x - 3y = 12$.

86. Graph: $-2x + y = 4$.

87. Graph: $y = 0$.

88. Find the equation of the line through $(-3, 4)$ and $(2, 1)$.

89. Find the equation of the line through $(6, -2)$ and parallel to the line $4x + 3y = 0$.

90. Graph: $3x - 5y \leq 15$.

91. State the domain and range, and determine if the set of ordered pairs represents a function: $\{(2, 3), (-3, 2), (-1, -2), (1, -1)\}$.

92. Use the vertical line test to determine if the relation in the accompanying figure is a function.

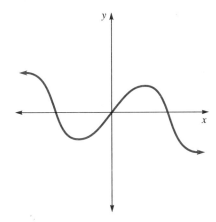

93. State the domain for $f(x) = \dfrac{2}{x + 3}$.

94. Given $g(x) = 4x^2 - 2x + 5$, find $g(2)$.

95. Solve the system: $-3x + 2y = -7$
$6x + 5y = -4$

96. Solve the system: $0.1x + 10y = 30$
$-4.1x - 10y = 10$

97. Use substitution to solve the system:
$-5x + 3y = 6$
$3y = 5x + 6$
Note: If inconsistent or dependent, say so.

98. A company plane flies 225 mi against the wind in 3 hr. The return trip (with the wind) takes 1 hr. Find the speed of the wind and the speed of the plane in still air.

99. Graph the system of inequalities: $3x + 2y \geq -6$
$6x + 7y \leq 42$

100. Use the accompanying figure to find the height h of the triangle.

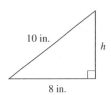

101. Simplify: $\sqrt{81c^3d^8}$.

102. Simplify: $\dfrac{x}{\sqrt{5x}}$.

103. Simplify: $\sqrt{r^{11}} - r^4\sqrt{r^3} + 2\sqrt{r^2}$.

104. Simplify: $\dfrac{2\sqrt{3}}{\sqrt{8}} + \dfrac{2\sqrt{2}}{\sqrt{12}}$.

105. Simplify: $\dfrac{4\sqrt{3} + 3}{\sqrt{3} + 4}$.

106. Solve: $3x - 1 = \sqrt{4x}$.

107. Solve: $\sqrt{16 - 5x} = 4 - \sqrt{x}$.

108. Simplify: $\left(\dfrac{m^{1/3}}{n^{-2/3}}\right)^6$.

109. Solve: $\dfrac{(6x + 12)^2}{3} = 12$.

110. Solve: $y^2 - 4y = 11$.

111. Solve: $4x^2 - x - 1 = 0$.

112. Solve: $x^2 - 2x + 3 = 0$.

113. The sum of a number and its reciprocal is $\frac{13}{6}$. Find the number.

114. Graph: $y = -(x - 2)^2 + 3$.

115. Graph: $y = 2x^2 + 4x - 1$.

ANSWERS

ANSWERS

This answer section provides answers to all odd-numbered problems within the exercise sets, and all of the answers for the chapter reviews, cumulative reviews, and chapter tests. Complete solutions to these problems are available in the Student's Solutions Manual.

Chapter 1

Exercise Set 1.1
1. $6 + 8 = 14$ 3. $16 - 9 = 7$ 5. $26 = 12 + 14$
7. $6 \cdot 9 = 54$ 9. $5 = \frac{30}{6}$ 11. $\frac{56}{7} = 8$ 13. $67 - 51 > 11$
15. $5^2 = 25$ 17. $2^2 - 1 = 3$ 19. $7 < 5 + 4$
21. $5^2 - 4^2 = 3^2$ 23. $\sqrt{9}\sqrt{25} = 15$ 25. $67 < 9^2$
27. $18 > \frac{72}{6}$ 29. $5 + 13 \neq 19$ 31. 19 33. 25 35. 1
37. 16 39. 1 41. 31 43. 48 45. 65 47. 88
49. 25 51. 2 53. 3 55. 2 57. 7 59. 3 61. 3
63. 2, 3, 5 65. 2 67. 2, 3 69. $2^2 \cdot 3$ 71. $2 \cdot 31$
73. $2 \cdot 3 \cdot 5 \cdot 7$ 75. $3^2 \cdot 5^2$ 77. $2^2 \cdot 3^3 \cdot 5$ 79. 1
81. 2 83. $\frac{46}{5}$ 85. $3 < 5$ 87. $6 > 4$ 89. $34
91. 24 hr 93. 21 hr
95. 160 grade points; GPA = 2.963 (to three decimal places)
97. 24 in. 99. no, 299 miles
101. Three times two plus five equals eleven.
103. Twelve equals two squared times three.
105. If the sum of the digits is a multiple of three
107. Three squared equals nine.
109. Seven is less than fifteen.
111. Twelve divided by three and multiplied by 2 equals eight.
113. Six added to the quotient of 18 and 3 equals 12.
115. Quotient 117. Sum
119. Exponent, base: The base 4 is to be used as a factor 5 times.
121. Eight is less than or equal to 10.
123. Five is greater than 1.
125. A number is in prime factored form if the factors are all prime numbers.

Exercise Set 1.2
1. $\frac{16}{25}$ 3. $\frac{4}{5}$ 5. $\frac{2}{7}$ 7. 6 9. 2 11. 15 13. 36
15. 48 17. $\frac{1}{3}$ 19. $\frac{5}{14}$ 21. $\frac{6}{5}$ 23. $\frac{49}{16}$ 25. 1
27. 2 29. $\frac{17}{12}$ 31. $\frac{5}{12}$ 33. $\frac{169}{24}$ 35. $\frac{59}{30}$ 37. 2
39. 4 41. $\frac{19}{16}$ 43. $\frac{121}{400}$ 45. $\frac{7}{24}$ 47. $\frac{169}{56}$ 49. $\frac{295}{36}$
51. $\frac{46}{45}$ 53. $\frac{169}{16}$ 55. $\frac{102}{5}$ 57. $\frac{17}{8}$ 59. $>$ 61. $>$
63. $>$ 65. $<$
67. number line with point at $\frac{1}{2}$ between 0 and 1
69. number line with point at $1\frac{5}{4}$ between 1 and 2
71. number line with point at $\frac{7}{8}$ between 0 and 1
73. number line with point at $2\frac{1}{8}$ between 2 and 3
75. $\frac{5}{7}$ 77. $\frac{10}{3}$ yd 79. $\frac{41}{14}$ yd 81. $\frac{47}{6}$ hr 83. $5.85
85. $\frac{5}{2}$ ft 87. $\frac{3}{8}$ 89. 46 ft 91. $\frac{28}{33}$ 93. $\frac{399}{4}$ gal
95. $\frac{31}{14}$ in. 97. $\frac{651}{344}$ 99. $\frac{650}{1329}$
101. Factor into prime factored form and reduce.
103. A mixed number has a whole number part plus a fractional part.
105. Invert the divisor and multiply.

Exercise Set 1.3
1. 0.25; 25% 3. 0.6; 60% 5. 2.25; 225%
7. 4.14; 414% 9. 2.0; 200% 11. 0.625; 62.5% 13. 0.3
15. 1.0 17. 2.25 19. 1.18 21. 0.008 23. 0.055
25. 4.96 27. 3.02 29. 25.44 31. 0.25 33. 9
35. 3.775 37. 1.175 39. 3.719 41. 2.083 43. 12
45. 30 47. 25% 49. 4.48 51. 1.736 53. 2000
55. 0.0346 57. 1.15 59. 1.9 61. 0.4 63. 8.25
65. 5.05 67. 2 69. 14.94 71. 0.214 73. 0.125
75. 2 77. $25.80 79. $2.50 81. $13.40
83. $10,560; $2,240 85. $308.53 87. $34,285.71
89. 54% 91. $2.68 93. $224
95. Move the decimal point two places to the right and add a percent symbol.
97. Divide the numerator by the denominator.
99. First change to an improper fraction, then divide.

Exercise Set 1.4
1. (a) 18 in. (b) 20.25 in² 3. (a) 60 in. (b) 119.6 in²
5. (a) $(71 + 12\pi)$ ft (b) $(564 + 72\pi)$ ft²
7. (a) $\frac{5}{2}\pi$ yd (b) $\frac{25}{16}\pi$ yd²
9. (a) $\frac{61}{4}$ cm (b) $\frac{387}{32}$ cm²
11. (a) $\frac{365}{24}$ cm (b) $\frac{515}{48}$ cm²
13. (a) 30 m (b) $\frac{225}{4}$ cm²
15. (a) $\frac{2497}{120}$ ft (b) $\frac{213}{80}$ ft²
17. 21.633 in. 19. 9.391 ft 21. 1.549 ft 23. 4.570 in.
25. 6.350 cm 27. 226.980 in² 29. 86.25 ft²
31. 380.133 in² 33. 16.385 in² 35. 62° 37. 55°
39. 41° 41. 125 ft³ 43. 14,400π ft³ 45. 970.3125π ft³
47. 192.9375π in³ 49. 1183.01875 in³ 51. 360 ft²; 40 yd²
53. 208 ft² 55. 2 gal 57. 58.01 acres 59. 1008 boxes
61. 20
63. The area of the triangle is one-half that of the rectangle.
65. A square is a rectangle that has four equal sides.
67. A four-sided figure that has two parallel sides.
69. A four-sided figure 71. 90°
73. An angle with a measure less than 90°
75. An angle with a measure greater than 90° and less than 180°

Review Exercises
1. 23 2. 13 3. 91 4. 7 5. 33 6. 11 7. 9
8. 1 9. 26 10. 1 11. 30 12. 3 13. 13 14. 1
15. $\frac{1}{2}$ 16. $\frac{1}{4}$ 17. $\frac{91}{6}$ 18. 3 19. $\frac{83}{24}$ 20. 1
21. $\frac{5}{16}$ 22. $\frac{65}{144}$ 23. $\frac{151}{12}$ 24. $\frac{139}{40}$ 25. $\frac{113}{24}$

26. $\frac{479}{28}$ 27. $\frac{2}{3}$ 28. $\frac{135}{32}$ 29. $\frac{397}{80}$ 30. 91 31. 9.95
32. 8.51 33. 19.52 34. $\frac{33}{7}$ 35. 3.23 36. 0.307692
37. 7.49 38. 3.926829 39. 17.6 40. $\frac{1}{32}$ 41. 2
42. 4.3 43. 24 44. $14\frac{2}{3}$ 45. $\frac{13}{2}$ 46. 7.525
47. 0.105 48. 2.94 49. 0.3 50. 3 51. 12 52. 0.4
53. $2^2 \cdot 3$ 54. $2^2 \cdot 7$ 55. $2^2 \cdot 5 \cdot 19$ 56. $2^4 \cdot 3^2 \cdot 5$
57. 2, 3, 5 58. 2, 3, 5 59. 3, 5 60. 3 61. 4
62. 15 63. 14 64. 21 65. 45 66. 30 67. 12
68. 30 69. 60% 70. 3.2 71. $\frac{841}{64}$ ft²; $\frac{29}{2}$ ft
72. 1.69π in², 2.6π in. 73. 10.05 m², 16.4 m
74. 7.5 yd², 15 yd 75. $\frac{339}{40}$ ft², $\frac{749}{60}$ ft 76. 243 ft²
77. 5.29 in² 78. 9 km² 79. 6 yd² 80. 9π ft²
81. 0.25π cm² 82. 5 83. 13 84. 12 85. 10 86. 6
87. 5 88. $\frac{1}{3}$ 89. $\frac{2}{3}$ 90. $\frac{4}{5}$ = 80% 91. $\frac{7}{8}$ = 87.5%
92. $1\frac{1}{2}$ cups sugar, 1 cup milk 93. $\frac{1}{2}$ 94. 13.5 gal
95. 20 96. $68.74 97. $65.49 98. $152.52
99. $18.06 100. $346.65 101. $142.31 102. 42.6 mpg
103. $14,485.71 104. $500 105. 40% 106. $430.10
107. $2000 108. $1.33

Chapter Test
1. 11 2. 6 3. 2 4. $\frac{11}{2}$ 5. $\frac{49}{100}$ 6. $\frac{5}{2}$ 7. $\frac{98}{225}$
8. 36.72 9. 0.08 10. $\frac{5}{3}$ 11. 0.7 12. 16.368
13. $4.68 14. 2.81 15. $96.00
16. (a) 16π m² (b) 8π m
17. (a) 5 ft (b) 6 ft² (c) 12 ft
18. (a) $\frac{77}{10}$ yd² (b) $\frac{57}{5}$ yd
19. $\frac{8}{5}$ 20. 6.375 21. 26.5 mpg 22. 15.385% 23. $\frac{7}{6}$ cups
24. $119.60 25. $14.25, $15.25 26. 72% 27. $101.20
28. $153.21

CHAPTER 2

Exercise Set 2.1
1. 4, 5, 6, 7 3. 7, 9, 11, 13, 15 5. 16, 20, 24, 28, 32, 36
7. 20, 25, 30 9. {January, February, . . . , December}
11. 11, 22, 33, 44 13. 1, 2, 3 15. 6, 7, 8, 9
17. Answers will vary. 19. ∅ 21. ∈ 23. ∈ 25. ∉

27. ∈ 29. ⊆ 31. ⊄ 33. ⊆ 35. ⊆
37. {x | x < 6, x is an odd counting number}
39. {x | x ≤ 45, x is a positive multiple of 5}
41. {x | x is an even counting number}
43. {x | x is a day of the week starting with "S"}
45. {x | x is a counting number}
47. {x | x is a counting number less than 801}
49. {x | x is a month of the year with three letters}
51. {1, 2}, {1}, {2}, ∅   53. {shoe, sock}, {shoe}, {sock}, ∅
55. {1, 2, 3}, {1, 2}, {1, 3}, {2, 3}, {1}, {2}, {3}, ∅
57. True 59. False 61. True 63. False 65. False
67. True 69. infinite 71. infinite 73. infinite
75. finite 77. infinite 79. infinite 81. rational
83. integer, rational 85. whole, integer, rational
87. rational 89. rational 91. irrational
93.

95. $0.3\overline{3}$ 97. $0.1\overline{1}$ 99. $0.\overline{428571}$ 101. $0.\overline{27}$
103. $0.\overline{076923}$ 105. 0.5 107. 1 109. π
111. not possible 113. $\frac{5}{0}$ 115. $\frac{1}{2}$
117. A collection of items
119. The set of natural numbers together with zero
121. An element of
123. The set of all x such that x is a real number
125. Something that can change

Exercise Set 2.2

1. $y + 6$ 3. $m + 4$ 5. $R - 6$ 7. $2y$ 9. $\frac{x}{24}$
11. $0.05Q$ 13. $2m + 2$ 15. $\frac{1}{2}D - 12$ 17. 5   19. 16
21. 0   23. 1.24   25. undefined   27. 9   29. 24
31. $\frac{3}{8}$ 33. 7   35. $\frac{173}{36}$ 37. True   39. False
41. True   43. True   45. True   47. True   49. True
51. True   53. False   55. True   57. -2 59. 0
61. 5   63. 21   65. 8   67. -9 69. $2w$ 71. $2h$
73. $2L + 3$ 75. $p - 600$ 77. $100(0.33)$ 79. $4s - 7$
81. $3(x + 9)$ 83. $53(6)$ 85. $1000 - x$ 87. $5s$
89. $2j - 2$ 91. $24.75x + 2(12)$ 93. $x + 0.02x$
95. $c + 0.40c$ 97. $15x + 600$ 99. $D - 300(0.20)$
101. $x - 2(3593) - 3(5000)$
103. $|-9| = 9$, which is greater than 5
105. If $x \geq 0$ then x is positive or zero, so $|x| = x$. If $x < 0$ then $-x$ is positive, so $|x| = -x$.
107. While x can represent different numbers, 5 does not change.

Exercise Set 2.3

1. 25   3. 17   5. 3   7. -26 9. -32 11. -15
13. -40 15. 38   17. 89   19. 54   21. 0.2
23. -2.62 25. $\frac{7}{5}$ 27. $\frac{6}{7}$ 29. 0   31. -1.4 33. $\frac{9}{11}$
35. 38   37. 5   39. -70 41. 64   43. -34 45. 1
47. 3   49. 13   51. 15   53. 94   55. 5   57. -65
59. -29 61. -38 63. 6   65. -49 67. 21   69. 6
71. 9   73. 16   75. 34   77. -89 79. -10.2 81. $\frac{2}{7}$
83. 1   85. 3   87. 3   89. -3 91. -53 93. -3
95. 160   97. 241.95   99. 667.62   101. 0.039   103. $\frac{3}{10}$
105. -10.6 107. 0.84   109. -1
111. their own 41 yd line   113. 5958 ft   115. $-\$52.59$
117. $-19°F$ 119. $13,170 profit   121. 10   123. -15
125. -10 127. -4 129. positive   131. negative
133. positive

Exercise Set 2.4

1. 54   3. -24 5. 0   7. -90 9. 112   11. 0
13. -24 15. -30 17. 48   19. -12 21. 36   23. 0
25. 24   27. $-\frac{1}{6}$ 29. $\frac{2}{3}$ 31. 2   33. -2 35. -4
37. 8   39. -9 41. 27   43. -36 45. 12   47. $-\frac{8}{3}$
49. $\frac{3}{2}$ 51. 1   53. -9 55. -149 57. 0   59. -255
61. 44   63. 19   65. 3   67. undefined   69. -4
71. 0   73. 39   75. 19   77. -68 79. -6 81. 24
83. 6   85. 49   87. -46 89. 1   91. 9
93. undefined   95. $-\frac{1}{8}$ 97. 6   99. 91   101. 18
103. $-\frac{41}{3}$ 105. $\frac{5}{3}$ 107. -2 109. $-\frac{18}{11}$
111. positive   113. negative   115. zero   117. undefined
119. positive   121. positive   123. negative
125. negative
127. Find the sum of their absolute values and prefix the common sign.
129. Change the sign of the number being subtracted and add the resulting numbers.
131. Find the product of their absolute values and attach a negative sign.

Exercise Set 2.5

1. commutative property   3. commutative property
5. additive inverse property   7. associative property
9. multiplicative identity property   11. associative property
13. additive inverse property   15. associative property
17. additive identity property
19. multiplicative identity property
21. additive inverse property   23. associative property
25. associative property   27. additive identity property
29. commutative property   31. associative property
33. $3 \cdot 4 + 3 \cdot 7$ 35. $2x + 2 \cdot 5$ 37. $8p - 8 \cdot 5$

39. $12r + 4 \cdot 12$ **41.** $5(7 + 2)$ **43.** $7(x - 3)$
45. $-3x - 3 \cdot 9$ **47.** $-6(x + 3)$ **49.** $13(x + 2)$
51. $-m + 18$ **53.** $-2(p - 3)$ **55.** $-x - 3y + 4$
57. $(-5)13$ **59.** $3x + 2$ **61.** $-7 + (x + 3)$ **63.** 1
65. $3a - 12$ **67.** $6x$ **69.** x **71.** $21x - 7$ **73.** $3x$
75. $-10y$ **77.** $x - 3y + 7$ **79.** $6(x - 4)$
81. $-3x + 7$ **83.** $3p - 8$ **85.** $2xy$ **87.** $-2x + 2y - 1$
89. $\frac{4}{3}$ **91.** 0 **93.** no: $8 - 1 \neq 1 - 8$ **95.** yes, 0
97. no: $0 \cdot 1 = 0$ **99.** no: $8 - (6 - 2) \neq (8 - 6) - 2$
101. yes **103.** $-y$ **105.** $\frac{1}{x}$ **107.** yes
109. no: $|3| + |-8| \neq |3 - 8|$ **111.** 162 **113.** 530
115. 3 **117.** $\frac{15}{4}$ **119.** 45 **121.** 12

Exercise Set 2.6
1. $2608 **3.** -$168.17 **5.** -$2055 \approx 12\%$ **7.** $376
9. -$45.70 **11.** $805 **13.** $91°$ **15.** $369°$ **17.** -46 ft
19. $20{,}602$ ft **21.** -6 **23.** -27 **25.** -96 **27.** 0
29. 685 mg **31.** $81\%, 1\%$ **33.** -8 **35.** -48.5
37. $354 **39.** 134 cal **41.** 2 sets **43.** -12 lb
45. $\frac{50}{77}$ in² **47.** $S = 96$ in²; $V = 64$ in³ **49.** 125 ft
51. $\frac{88}{441}$ ft³ **53.** 226 ft/sec **55.** 5 sec
57. $585{,}200$ sq. miles ≈ 484 times as big
59. 2536 ft³/sec; $152{,}160$ ft³/min; $9{,}129{,}600$ ft³/hr; 68%
61. 5 strokes **63.** 16 amps **65.** 73 lb/in²
67. Yes, if they are additive inverses; then $|x| = |-x| = x$, where $x > 0$.
69. When n is an even number; when n is an odd number. It will be zero if one or more factors is zero.
71. adding the answer and the number being subtracted.
73. change the sign of the number being subtracted and add.
75. Because there is no unique answer when dividing by zero.

Review Exercises
1. $\{1, 2\}$ **2.** $\{0, 1, 2, 3\}$ **3.** $\{6, 7\}$ **4.** $\{7, 8\}$
5. $\{2, 4, 6, 8\}$ **6.** $\{0, 2, 4, 6\}$ **7.** $\{-2, -1, 0, 1\}$
8. $\{4, 8, 12\}$ **9.** $\{-5, -4, -3\}$ **10.** $\{3, 6\}$
11. $\{0, 1, 2, 3, 4\}$ **12.** $\{-3, -2, -1, 0, 1\}$
13. $\{-3, -2, -1\}$ **14.** $\{5, 10, 15, 20, \ldots\}$
15. $\{\ldots, -12, -8, -4\}$ **16.** $\{6, 12, 18, 24\}$
17. $\{\ldots, -6, -4, -2\}$ **18.** $\emptyset$ **19.** $\{0, 1, 2, 3, \ldots\}$
20. $\{1, 2, 3, \ldots\}$ **21.** $\{x \mid x \in N\}$
22. $\left\{\frac{a}{b} \mid a, b \in J, b \neq 0\right\}$
23. $\{x \mid x \text{ is an even natural number}\}$
24. $\{x \mid x \in N \text{ and a multiple of } 3\}$ **25.** True **26.** True
27. False **28.** True **29.** True

30.

31. $6, 3, -4, 0; x$ if $x \geq 0$ or $-x$ if $x < 0$
32. -20 **33.** 24 **34.** 46 **35.** 29 **36.** $-\frac{1}{6}$ **37.** $-\frac{7}{6}$
38. $\frac{155}{24}$ **39.** -0.82 **40.** -1 **41.** -1.4 **42.** -5
43. 13 **44.** -11 **45.** -3 **46.** 2 **47.** 15 **48.** $\frac{5}{4}$
49. 136 **50.** 180 **51.** 0 **52.** 0 **53.** undefined
54. 0 **55.** 8 **56.** 13 **57.** -2 **58.** -9 **59.** -2
60. $\frac{5}{2}$ **61.** 28 **62.** 30 **63.** $-\frac{16}{13}$ **64.** -24 **65.** 1
66. $\frac{1}{4}$ **67.** -10 **68.** -12 **69.** 0 **70.** $\frac{6}{5}$ **71.** -12
72. $6 + (-3) = 3$ **73.** $-8 - (-5) = -3$ **74.** $\frac{2.5}{0.5} = 5$
75. $\frac{3}{4} \cdot \frac{1}{4} = \frac{3}{16}$ **76.** $32 - (-16) = 48$
77. $|-5| + (-11) = -6$ **78.** $\sqrt{2\frac{3}{4} + 3\frac{1}{2}} = \sqrt{\frac{25}{4}} = \frac{5}{2}$
79. $(7 + (-7))^2 = 0^2 = 0$ **80.** $5 - \frac{0}{7} = 5 - 0 = 5$
81. $\left(6 + \frac{2}{3}\right) - \frac{1}{2}$ **82.** $8[-0.75]$ **83.** $-\frac{1}{13}$
84. 8.12 **85.** $6(x + 3)$ **86.** 0 **87.** 1
88. $[(-6)(-7)](2)$ **89.** $(-5) + 6$ **90.** $-5x + 15$
91. $3217 **92.** $600 **93.** 48 **94.** 1355 ft
95. $80\%, 4\%$ **96.** $V = \frac{3}{4}$ **97.** 7.524 ft³
98. $800; 10\% **99.** $322.40 **100.** $104.26

Chapter Test
1. -14 **2.** -45 **3.** 0 **4.** -5 **5.** 1 **6.** $\frac{5}{18}$ **7.** -6
8. -8 **9.** -4 **10.** undefined **11.** 0 **12.** 0
13. distributive property **14.** commutative property
15. multiplicative inverse property
16. additive inverse property **17.** identity property
18. associative property **19.** commutative property
20. distributive property **21.** identity property
22. associative property **23.** $\{0, 1, 2, 3, \ldots\}$
24. $\{1, 3, 5, 7, \ldots\}$
25. $\{\ldots, -3, -2, -1, 0, 1, 2, 3, \ldots\}$
26. $\{0, 1, 2, 3, 4, 5, 6, 7\}$ **27.** $\emptyset$ **28.** $\{0, 1, 2, 3\}$
29. 6 rafters **30.** 315 yd **31.** $10\frac{1}{3}$ min **32.** 180π ft³

CHAPTER 3

Exercise Set 3.1
1. $11x$ **3.** $8y$ **5.** $9z$ **7.** r **9.** $-5x - 6y$
11. $5x^2 - x$ **13.** $-2y^3 + 6y^2$ **15.** $13p + 5$

17. $4a - b$ **19.** $5y$ **21.** $\frac{5}{6}x - \frac{1}{12}y$ **23.** $-\frac{3}{4}a + \frac{33}{28}b$
25. $-0.5x + 0.8y$ **27.** $3.2a - \frac{4}{3}b - \frac{9}{4}c$ **29.** $\frac{2}{3}c + \frac{1}{4}ab$
31. 11 **33.** -18 **35.** -13 **37.** 7 **39.** 35
41. -161 **43.** 2 **45.** $\frac{1}{2}$ **47.** $\frac{-3}{4}$ **49.** $-x + 9$
51. $7x - 1$ **53.** $3m + 2$ **55.** $5y^2 + y + 3$
57. $-q^2 + 12$ **59.** $-14y - 37$ **61.** $-3x - 27$
63. $-5p - 9$ **65.** $3x + 31$ **67.** $59y - 16$
69. $\frac{1}{3}a + \frac{1}{6}b + \frac{5}{12}c$ **71.** $12.63x - 6.24y$
73. $0.448x + 2.128$ **75.** $-2.2x - y$ **77.** $\frac{1}{2}x + 3x = 3.5x$
79. $\frac{3}{4}x + 4$ **81.** $2(x - 5) = 2x - 10$
83. $9(a - 9) = 9a - 81$ **85.** $5(3 + x) - x = 15 + 4x$
87. $(13 - x) + (7 + 3x) = 20 + 2x$
89. $2w + 2(w + 3) = 4w + 6$
91. $x + (2x - 3) + (2x - 7) = 5x - 10$
93. $5n + 10(n + 3) = (15n + 30)$ cents
95. $10d + 5(d + 3) + 1(d + 9) = (16d + 24)$ cents
97. Terms with the same variables and same exponents on the variables
99. The number in front of the variable(s) in a term
101. Distributive property; Commutative property; Distributive property; Combine numerical coefficients.

Exercise Set 3.2

1. yes **3.** yes **5.** no **7.** yes **9.** yes **11.** {4}
13. {9} **15.** {2} **17.** {−11} **19.** {1} **21.** {10}
23. {−10} **25.** {−7} **27.** {−6} **29.** {7} **31.** {7}
33. {7} **35.** {3} **37.** {0} **39.** {4} **41.** {3} **43.** {4}
45. {3} **47.** {1} **49.** $\left\{\frac{4}{5}\right\}$ **51.** $\left\{\frac{4}{5}\right\}$ **53.** {−20}
55. {26} **57.** {4} **59.** {4} **61.** $\left\{\frac{5}{6}\right\}$ **63.** 6 **65.** 2
67. 26 in. **69.** 5 cm, 8 cm, 13 cm **71.** 4π in.
73. 25π ft^3 **75.** 0 **77.** 6.4 **79.** 48 **81.** $\frac{27}{2}$ **83.** 9
85. $\frac{109}{18}$ **87.** 7 **89.** -9 **91.** -64 **93.** $-\frac{7}{3}$ **95.** -5
97. $\frac{12}{7}$ **99.** $\frac{13}{7}$

Exercise Set 3.3

1. {4} **3.** {−3} **5.** {−5} **7.** {24} **9.** {−40}
11. {28} **13.** {−6} **15.** {12} **17.** {10} **19.** {−25}
21. $\left\{\frac{26}{5}\right\}$ **23.** {0} **25.** {3} **27.** {−1} **29.** $\left\{-\frac{13}{4}\right\}$
31. $\left\{\frac{3}{4}\right\}$ **33.** $\left\{\frac{4}{7}\right\}$ **35.** {0} **37.** $\left\{-\frac{1}{2}\right\}$ **39.** $\left\{\frac{13}{2}\right\}$
41. $\left\{\frac{13}{10}\right\}$ **43.** $\left\{\frac{7}{2}\right\}$ **45.** $\left\{\frac{39}{8}\right\}$ **47.** $\left\{\frac{3}{5}\right\}$ **49.** $\left\{\frac{6}{11}\right\}$
51. {4} **53.** $\left\{\frac{15}{58}\right\}$ **55.** {3} **57.** {2} **59.** $\left\{-\frac{27}{19}\right\}$
61. $\left\{\frac{1}{16}\right\}$ **63.** $\left\{\frac{20}{3}\right\}$ **65.** $\left\{-\frac{35}{23}\right\}$ **67.** $\left\{\frac{81}{17}\right\}$
69. {−10} **71.** $\left\{\frac{17}{8}\right\}$ **73.** $\left\{\frac{35}{12}\right\}$ **75.** {3} **77.** {15}
79. -12 **81.** 17 miles **83.** $4360.00 **85.** 8 cars
87. 70 questions **89.** 2.5 hr **91.** 67 min
93. 850 students **95.** 175 yd **97.** 7.2 m/sec
99. 37.5 ft
101. If the same number is added/subtracted to/from both sides of an equation, the result is an equivalent equation.
103. A sum or difference of a number and products or quotients of numbers and variables raised to powers
105.
$3x - 4 = 8$
$3x - 4 + 4 = 8 + 4$ Addition property
$3x = 12$ Combine like terms.
$\frac{1}{3} \cdot 3x = \frac{1}{3} \cdot 12$ Multiplication property
$x = 4$ Multiply

107.
$2(3 - 2x) = x - 5$
$6 - 4x = x - 5$ Distributive property
$6 - 4x + 4x = x + 4x - 5$ Addition property
$6 = 5x - 5$ Combine like terms.
$6 + 5 = 5x - 5 + 5$ Addition property
$11 = 5x$ Combine like terms.
$\frac{11}{5} = x$ Division property

Exercise Set 3.4

1. $r = \frac{d}{t}$ **3.** $s = \frac{P}{4}$ **5.** $m = \frac{F}{a}$ **7.** $V = IR$ **9.** $p = \frac{Py}{144}$
11. $b = P - a - c$ **13.** $n = m - T$ **15.** $s = \frac{R}{2}$
17. $l = \frac{Fd}{t}$ **19.** $M = \frac{IC}{100}$ **21.** $w = \frac{CL}{100}$
23. 1 mic = 0.001 mg **25.** $I = PO$ **27.** $w = \frac{V}{lh}$
29. $l = \frac{P - 2w}{2}$ **31.** $y_2 = m(x_2 - x_1) + y_1$
33. $h = \frac{2A}{B + b}$ **35.** $F = \frac{W}{d}$ **37.** $\pi = \frac{A}{r^2}$ **39.** $m = \frac{y - y_1}{x - x_1}$
41. $I = 270$
43. $P = 50$ **45.** $A = 80$ **47.** $A = 119$ **49.** $s = 20$

51. $m = 21.25$ mpg **53.** $r = 0.08$ **55.** $A = 84.949$
57. $L = 51.056$ **59.** $m = 31.369$ **61.** $y = 3 - x$
63. $x = \dfrac{c - by}{a}$ **65.** $x = \dfrac{c - b}{a}$ **67.** $x = -5$
69. $y = 7(g - c)$ **71.** $x = \dfrac{5y - 7}{5}$ **73.** $z = \dfrac{-5 - 6x}{18}$
75. $y = 3p - a$ **77.** $t = r - 2B$ **79.** $y = \dfrac{d - 2x}{2}$
81. 13 mpg **83.** 7 hr **85.** 3 hr **87.** 16π in.
89. $120 **91.** $800 **93.** 2 miles **95.** the 10-in. piece
97. 14,142.857 ft³ **99.** {3} **101.** {3} **103.** $\left\{-\dfrac{20}{3}\right\}$
105. ∅ **107.** ∅ **109.** $\left\{-\dfrac{1}{8}\right\}$ **111.** $\left\{-\dfrac{9}{2}\right\}$
113. {29}

Exercise Set 3.5

1. [number line with points at 1, 2]
3. [number line with points at 0, 1, 2, 3, 4, 5]
5. [number line with points at −3, −2, −1, 0, 1]
7. [number line with points at −4, −3, −2, −1]
9. [number line with bracket at 3, arrow right through 5]
11. [number line with parentheses −2 to 1]
13. [number line with bracket 1 to 7]
15. [number line with parenthesis at 8, arrow left]
17. $x > -2$ **19.** $y < 11$ **21.** $p \leq 11$ **23.** $x > 3$
25. $y < 2$ **27.** $x \leq 3$ **29.** $y \geq 2$ **31.** $1 < x < 3.1$
33. $-2 \leq x \leq 3$ **35.** $x < 2$ **37.** $y \geq -1$
39. $-3 \leq x \leq 3$ **41.** $\dfrac{6}{5} \geq m \geq -1$ **43.** $y > -26$
45. $p \leq 1$ **47.** $x \geq -3$ [number line bracket at −3, arrow right]
49. $x < -7$ [number line parenthesis at −7, arrow left]
51. $0 \leq x < 7$ [number line bracket 0, parenthesis 7]
53. $-6 \leq x \leq 2$ [number line bracket −6 to 2]
55. $x > -7$ [number line parenthesis at −7, arrow right]
57. $x \geq 7$ [number line bracket at 7, arrow right]

59. $-8 \leq x < -5$ **61.** $-5 \leq y \leq -2$
63. $-12 < x < 3$ **65.** $-12 \leq x \leq 0$
67. $13 < y < 14$ **69.** $\dfrac{19}{3} \leq x < \dfrac{15}{2}$ **71.** $x \geq 12.1$
73. $x \leq 1.6$ **75.** $x > 14.87$ **77.** $x \leq 42$
79. 19, 20, 21, . . . , 32 **81.** 85 **83.** 96
85. $9.75 \leq P + $3.25 \leq $17.00 **87.** $p \geq 39$ cm
89. $9 < L < 11$; $26 < P < 34$ **91.** $52 < x < 58$

Exercise Set 3.6

1. $x < 4$ [number line, parenthesis at 4]
3. $y \leq -3$ [number line, bracket at −3]
5. $x > -4$ [number line, parenthesis at −4]
7. $p \leq -7$ [number line, bracket at −7]
9. $y < -2$ [number line, parenthesis at −2]
11. $x > 36$ [number line, parenthesis at 36]
13. $x \geq -3$ [number line, bracket at −3]
15. $x \leq 3$ [number line, bracket at 3]
17. $y < 2$ [number line, parenthesis at 2]
19. $x \geq 4$ **21.** $x > -4$ **23.** $-1 < x \leq 2$ **25.** $x < -6$
27. $-6 < x \leq 6$ **29.** $x > 7$ **31.** $x \geq 2$ **33.** $y \leq -8$
35. $x > -1$ **37.** $p \leq 12$ **39.** $y < \dfrac{5}{4}$ **41.** $-2 \leq y \leq 3$
43. $-1 < x \leq 3$ **45.** $x \leq 42$ **47.** $x > \dfrac{9}{2}$ **49.** $x > -\dfrac{20}{21}$
51. $p \leq -5.4$ **53.** $x < -2{,}818$ **55.** $p < 34$
57. $x \geq -1.675$ **59.** $x < 6$ **61.** $x < \dfrac{14}{5}$ **63.** $y < 3$
65. $p < -2$ **67.** $y > 9$ **69.** $x < -\dfrac{9}{4}$ **71.** $x < 0$
73. $y \geq 0$ **75.** $x \geq 7$ **77.** $y \leq 10$ **79.** $-8 < x < 4$
81. $x > 8$ **83.** $w \leq \dfrac{25}{2}$ **85.** $x \geq 10.5\%$ **87.** 158
89. 14 ft **91.** $x \geq $275 **93.** 17 m **95.** $B = \dfrac{2A}{h} - b$
97. $h = \dfrac{2A}{b}$ **99.** $r = \dfrac{C}{2\pi}$ **101.** $x = \dfrac{y}{4}$

Exercise Set 3.7

1. $\{\pm 4\}$ **3.** $\left\{\pm\dfrac{1}{2}\right\}$ **5.** {0} **7.** ∅ **9.** $\{\pm 2\}$

11. $\{\pm 5\}$ **13.** $\{\pm 8\}$ **15.** $\varnothing$ **17.** $\left\{\pm\dfrac{2}{3}\right\}$ **19.** $\{\pm 1\}$
21. $\{-6, 2\}$ **23.** $\{1, 3\}$ **25.** $\{-2, -1\}$ **27.** $\left\{-\dfrac{5}{4}\right\}$
29. $\left\{\dfrac{9}{4}, \dfrac{11}{4}\right\}$ **31.** $\left\{\dfrac{10}{3}, \dfrac{14}{3}\right\}$ **33.** $\{0, 4\}$ **35.** $\left\{-\dfrac{1}{4}, \dfrac{5}{4}\right\}$
37. $\{-6, -4\}$ **39.** $\{-4\}$ **41.** $\{x \mid -3 < x < 3\}$
43. $\{x \mid x > 2 \text{ or } x < -2\}$ **45.** $\{x \mid x \in R\}$
47. $\{x \mid x = 0\}$ **49.** $\{x \mid x \in R\}$ **51.** $\varnothing$
53. $\{x \mid -3 \le x \le -1\}$ **55.** $\{x \mid x \in R\}$
57. $\left\{y \,\Big|\, -\dfrac{13}{4} < y < \dfrac{3}{4}\right\}$ **59.** $\{x \mid x \in R\}$ **61.** $\varnothing$
63. $\left\{x \,\Big|\, x \ge \dfrac{105}{16} \text{ or } x \le \dfrac{87}{16}\right\}$ **65.** $\left\{x \,\Big|\, -\dfrac{7}{8} \le x \le \dfrac{23}{8}\right\}$
67. $\left\{y \,\Big|\, y \ge \dfrac{130}{99} \text{ or } y \le -\dfrac{20}{99}\right\}$ **69.** $\varnothing$
71. $\{x \mid -22 \le x \le 23\}$ **73.** $\{y \mid y \in R\}$
75. $\{y \mid -11 \le y \le 3\}$ **77.** $\varnothing$ **79.** $\{x \mid x \in R\}$
81. $\left\{\dfrac{15}{2}\right\}$ **83.** $\left\{\dfrac{35}{9}\right\}$ **85.** no solution **87.** $\left\{\dfrac{21}{40}\right\}$
89. $\left\{y \,\Big|\, y < \dfrac{1}{2}\right\}$ **91.** $\{0\}$ **93.** $\{0\}$ **95.** $\left\{\dfrac{5}{2}\right\}$
97. $\{9\}$ **99.** $\left\{-\dfrac{2}{5}\right\}$ **101.** $\{1\}$ **103.** $\left\{\dfrac{5}{9}\right\}$ **105.** 9 oz
107. 21 ft **109.** 5.45 miles **111.** $10.13

Exercise Set 3.8
1. 34, 35, 36 **3.** 31, 33, 35 **5.** 6, 7, 8, 9
7. 325 miles, 285 miles, 245 miles, 205 miles
9. 30 in., 27 in., 24 in., 21 in., 18 in.
11. 33 nickels, 3 quarters **13.** 240 at $1.50; 80 at $2.50
15. 2.5 hr **17.** $5908.50 **19.** $118.65
21. 210 at $1.25; 195 at $1.75 **23.** $211.25
25. $w = 6$ ft, $l = 12$ ft **27.** 603 boxes **29.** 6549 lb
31. $19.50 **33.** 43,743 ft^3 **35.** 62,897 miles **37.** no
39. 400 ml **41.** 2 tablets **43.** $w < 4$ ft **45.** 1.2 hr
47. 7200 students **49.** $1217.65 **51.** 160 liters

Review Exercises
1. $x + x + 1$ **2.** $x + x + 2$ **3.** $x(x + 2)$ **4.** $x(x + 3)$
5. $x - \dfrac{1}{x}$ **6.** $x + (-x)$ **7.** $x + (x - 5)$ **8.** $x + \dfrac{1}{2}x$
9. $3x - 4$ **10.** $\dfrac{1}{4}x + 5$ **11.** $2[x + (x + 2)]$
12. $8[x + (x + 2)]$ **13.** $x + (x + 2) + (x + 4)$
14. $x + (x + 2) + (x + 4)$ **15.** $2y + 3$ **16.** $\dfrac{1}{2}(x - 7)$
17. $2(4x + 11)$ **18.** $\$(80 - y)$ **19.** $58\left(5\dfrac{1}{2}\right)$ miles
20. $\$(800 - x)$ **21.** $\{16\}$ **22.** $\{-1\}$ **23.** $\{-4\}$
24. $\{-1\}$ **25.** $\{3\}$ **26.** $\left\{\dfrac{1}{2}\right\}$ **27.** $\{x \mid x < -3\}$
28. $\left\{y \,\Big|\, y < -\dfrac{7}{3}\right\}$ **29.** $\{-3\}$ **30.** $\{t \mid t \ge 5\}$
31. $\left\{m \,\Big|\, m \le -\dfrac{5}{3}\right\}$ **32.** $\left\{s \,\Big|\, s \le \dfrac{5}{4}\right\}$ **33.** $\left\{y \,\Big|\, y \ge \dfrac{26}{3}\right\}$
34. $\{x \mid x > 7\}$ **35.** $\left\{\dfrac{3}{2}\right\}$ **36.** $\left\{\dfrac{21}{11}\right\}$ **37.** $\left\{\dfrac{2}{5}\right\}$
38. $\{-9\}$ **39.** $\left\{\dfrac{63}{20}\right\}$ **40.** $\left\{-\dfrac{7}{34}\right\}$ **41.** $\left\{y \,\Big|\, y < \dfrac{5}{4}\right\}$
42. $\left\{p \,\Big|\, p \le -\dfrac{1}{2}\right\}$ **43.** $\{y \mid y \ge 4\}$ **44.** $\{x \mid x < 5\}$
45. $\pi = \dfrac{C}{2r}$ **46.** $h = \dfrac{2A}{b}$ **47.** $r^2 = \dfrac{V}{\pi h}$
48. $w = \dfrac{V}{lh}$ **49.** $h = \dfrac{2A}{(B + b)}$ **50.** $r = \dfrac{I}{pt}$
51. $V = \dfrac{200}{3}$ **52.** $y = 11$ **53.** $x = \dfrac{1}{3}$ **54.** -41
55. $x = \dfrac{3y + 5}{2}$ **56.** $x = \dfrac{7y + 1}{6}$ **57.** $x = -\dfrac{7}{4}y$
58. $x = \dfrac{y - b}{m}$ **59.** $x = \dfrac{c - by}{a}$ **60.** $x = \dfrac{ay - 4d}{2b}$
61. $x = \dfrac{15y + 36}{20}$ **62.** $x = 11$ **63.** $x = \dfrac{43}{7}$
64. $x > -\dfrac{84}{25}$ **65.** $x = -\dfrac{9}{10}$ **66.** $x \ge \dfrac{5m + 3}{2}$
67. $\{\pm 5\}$ **68.** $\{y \mid -3 < y < 3\}$ **69.** $\{y \mid y \in R\}$
70. $\varnothing$ **71.** $\varnothing$ **72.** $\{\pm 5\}$ **73.** $\left\{x \,\Big|\, x \le -\dfrac{7}{2} \text{ or } x \ge \dfrac{1}{2}\right\}$
74. $\left\{\dfrac{1}{4}\right\}$ **75.** $\{-13, 9\}$ **76.** $\left\{\dfrac{5}{2}\right\}$ **77.** $\dfrac{11}{9}$ **78.** 4
79. 23 **80.** 5, 7, 9 **81.** $220 **82.** $15,652.17 **83.** 50
84. $l < 17$

Chapter Test
1. $4x - 2$ **2.** $2y - 39$ **3.** $p^2 - 2p + 13$
4. $3r^2 + 7r + 29$ **5.** 9 **6.** 8 **7.** $-\dfrac{4}{5}$ **8.** $-\dfrac{269}{12}$
9. $\{11\}$ **10.** $\{3\}$ **11.** $\{-1\}$ **12.** $\left\{-\dfrac{23}{3}\right\}$
13. $\left\{\dfrac{5}{7}\right\}$ **14.** $\left\{\dfrac{5}{3}\right\}$
15. $x = 1$
16. $x \le 4$
17. $x \le 1$

18. $x < -5$

19. $2 \leq x < \dfrac{19}{6}$

20. $\dfrac{9}{2} < x < 7$

21. $\{-2, 12\}$ 22. $\varnothing$ 23. $\{x \mid x \in R\}$
24. $\left\{ x \mid -\dfrac{16}{5} \leq x \leq \dfrac{2}{5} \right\}$ 25. $p = \dfrac{I}{rt}$ 26. $h = \dfrac{A - 2\pi r^2}{2\pi r}$
27. $h = \dfrac{3V}{\pi r^2}$ 28. $h = \dfrac{2A}{B+b}$ 29. 32, 16
30. 10 pennies, 16 nickels, 4 dimes
31. 2nd student: \$7/hr; 1st student: \$14/hr 32. 58 mph

Cumulative Review Chapters 1–3

1. 20 2. 6 3. 12 4. 5 5. 2 6. 12 7. $\dfrac{1}{3}$ 8. 0
9. 2 10. $\dfrac{17}{8}$ 11. 0.375, 37.5% 12. 0.67, $\dfrac{67}{100}$
13. $1.1\overline{1}$ 14. $2 \cdot 3 \cdot 5 \cdot 7$ 15. 5.6 16. 8.4
17. 6.4 in., 2.56 in^2 18. 44 ft, 105 ft^2 19. 12 yd, 6 yd^2
20. 51 m, 126 m^2 21. 18.84 in., 28.26 in^2 22. 64 ft^3
23. 36π mm^3 24. $\dfrac{13}{16}\pi$ ft^3 25. 1.44π in^3
26. 89.91 ft^3 27. True 28. True 29. True 30. False
31. True 32. True 33. True 34. False 35. True
36. True 37. True 38. False 39. -2 40. 4
41. 8 42. -8.1 43. 4.3 44. -14 45. -8 46. $\dfrac{1}{24}$
47. 10 48. 4 49. 1.75 50. undefined
51. commutative property of addition
52. distributive property
53. identity property of addition
54. associative property of multiplication
55. multiplicative inverse property
56. associative property of addition
57. distributive property 58. additive inverse property
59. $17x$ 60. a 61. $6y^2 + y$ 62. $-0.9a - 0.75b + 3$
63. $-18x + 17y$ 64. $a - 6$ 65. $-7y$
66. $-\dfrac{95}{8}y + \dfrac{101}{4}x$ 67. $-0.15x - 0.8y$ 68. $2xy$
69. $\{2\}$ 70. $\{5\}$ 71. $\{3\}$ 72. $\{-1\}$ 73. $\{5\}$
74. $\left\{ -\dfrac{4}{15} \right\}$ 75. $\{4\}$ 76. $\left\{ \dfrac{5}{14} \right\}$ 77. $\left\{ \dfrac{7}{2} \right\}$
78. $\left\{ \dfrac{3}{2} \right\}$ 79. $\{4\}$ 80. $\{0\}$ 81. $r = \dfrac{C}{2\pi}$
82. $x = \dfrac{y - b}{m}$ 83. $F = \dfrac{9}{5}C + 32$ 84. $t = \dfrac{I}{pr}$

85. $h = \dfrac{2A}{B + b}$ 86. $l = \dfrac{P - 2w}{2}$ 87. $x = \dfrac{3y + p}{12}$
88. $y < 9$ 89. $x > 1$ 90. $p \leq 1$ 91. $x \geq -\dfrac{5}{2}$
92. $x < -15$ 93. $-8 < b < -1$ 94. $a \geq \dfrac{17}{5}$
95. $p \leq \dfrac{39}{4}$ 96. $y < -\dfrac{16}{5}$ 97. $\{\pm 4\}$ 98. $\{\pm 2\}$
99. $\{\pm 2\}$ 100. $\{0\}$ 101. $\varnothing$ 102. $x > 3$ or $x < -3$
103. $\{0\}$ 104. $\{x \mid x \in R\}$ 105. $\varnothing$ 106. $y \geq \dfrac{4}{3}$ or $y \leq 0$
107. $\{-1, 9\}$ 108. $y < 0$ or $y > 3$ 109. $\{y \mid y \in R\}$
110. $\left\{ x \mid x \in R \text{ and } x \neq -\dfrac{3}{2} \right\}$ 111. $-18, -17$
112. 12, 14 113. 5 nickels, 8 dimes
114. $l = 30$, $w = 12$ 115. 2.5 hr 116. \$111
117. 10 gal 118. $l = 14$ ft, $w = 11$ ft 119. 12 ft
120. $2\dfrac{1}{2}$ lb

CHAPTER 4

Exercise Set 4.1

1. 1, 4 3. 2, 3 5. 2 7. 4 9. (2, 3) 11. $(-3, -2)$
13. (0, 0) 15. $(-6, 0)$ 17. $(5, -2)$ 19. $(-6, -2)$
21. (7, 1) 23. $(-4, -1)$
25–35.

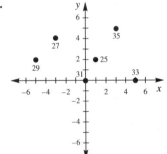

37–49.

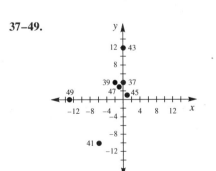

51. $A(2, -4)$, $B(0, 0)$, $C(0, 4)$, $D(-3, -2)$, $E(3, 3)$, $F(5, 1)$, $G(-4, 1)$, $H(-4, -4)$, $I(4, -3)$, $J(-4, 5)$

53. $y = \dfrac{1}{2}x$

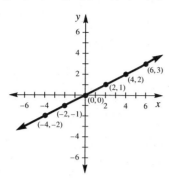

55. $x = 2$

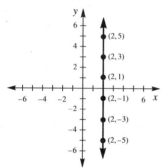

57. infinitely many

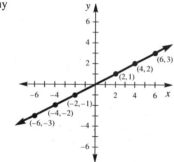

59. $y = -x$

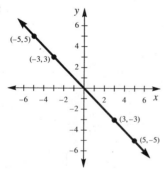

61. $y = x + 1$

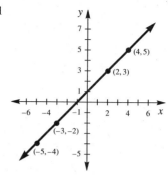

63. $y = 3x - 2$

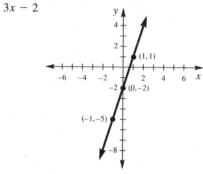

65. $y = x$

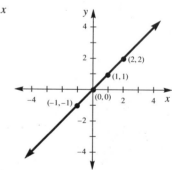

67. $y = -4$

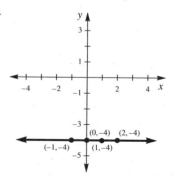

69. $y = 0$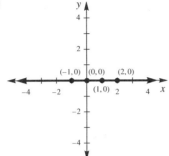

71. $(-2, -2), (0, 0), (3, 3), (-1, -1)$

73. $(-1, 3), (0, 0), \left(-\frac{1}{3}, 1\right) (1, -3)$

75. $(0, -3), (1, -2), (3, 0), (2, -1)$

77. $(0, 5), (1, 8), (-2, -1), \left(-\frac{5}{3}, 0\right)$

79. $(0, -3), (4, 3), (2, 0), (4, 3)$

81. $(1, 0.06), (-19.6, 3), (1.4, 0), (0, 0.2)$

83. $(0, 1.69), (1.32, 0)$ **85.** $\left(0, \frac{1}{2}\right), (-2, 0), \left(4, \frac{3}{2}\right), \left(8, \frac{5}{2}\right)$

87. $\left(0, \frac{4}{3}\right), \left(-\frac{3}{2}, 0\right), \left(\frac{3}{2}, \frac{8}{3}\right), (3, 4)$

89. $(0, 5)$ **91.** $(-4, 1)$ **93.** $(3, -4)$ **95.** $\{-2, 3\}$

97. $\{y \mid y \leq 1 \text{ or } y \geq 7\}$ **99.** $\left\{-\frac{7}{4}\right\}$ **101.** $\left\{-\frac{5}{3}, -1\right\}$

Exercise Set 4.2

1.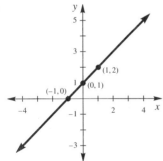

3.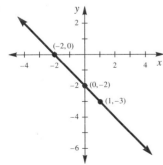

5.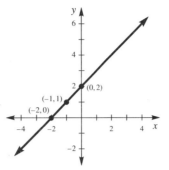

7.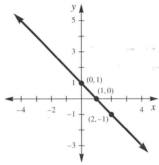

9.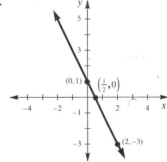

11.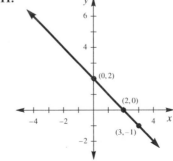

A-12 ANSWERS

13.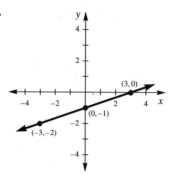

21.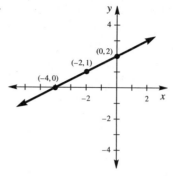

15.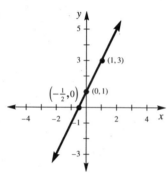

23.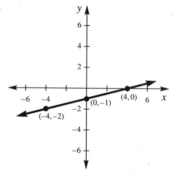

17.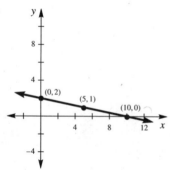

25.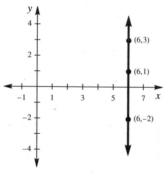

19.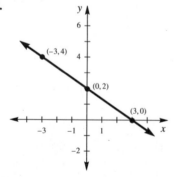

27.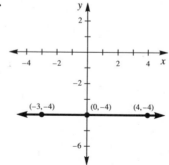

29.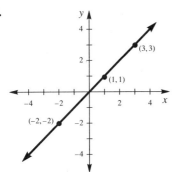

31.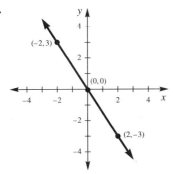

33.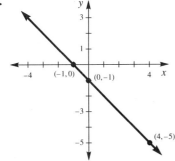

35.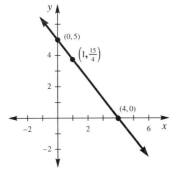

37.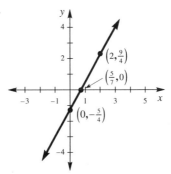

39.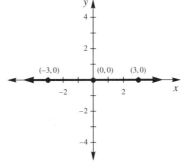

41.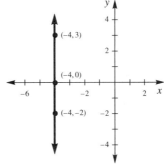

43.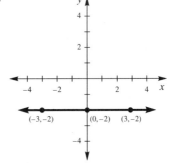

45.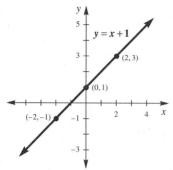

47.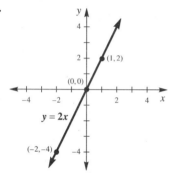

49.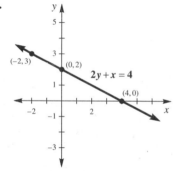

51.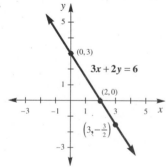

53.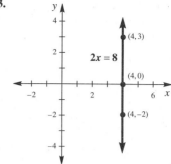

55.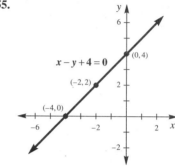

57.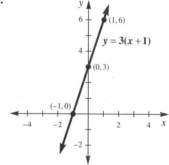

59.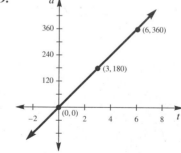

61.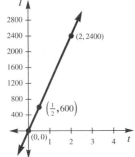

63.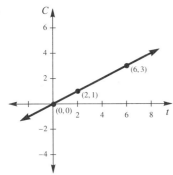

65.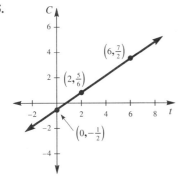

67.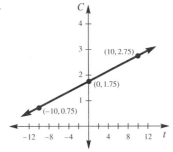

69. (0, 15), (22, 25), (55, 40), (88, 55)

71. (10, 25), (50, 245), (100, 520)

73. (0, 100), (10, 125), (100, 350)

75. $m = -3$, (5, −12), (10, −27), (−2, 9)

77. (1, 2.50), (2, 5.00)

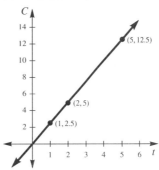

79. where line crosses x-axis, let $y = 0$ and solve for x.

81. The point (0, 0)   **83.** Horizontal lines at $y = k$.

85. Vertical line at $x = k$.

Exercise Set 4.3

1. $-\dfrac{3}{4}$ **3.** $\dfrac{2}{3}$ **5.** Undefined   **7.** 1   **9.** 0   **11.** $-\dfrac{5}{6}$

13. 1   **15.** Undefined

17.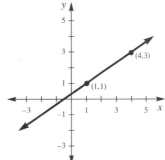

19.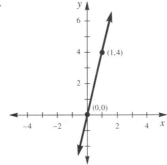

21.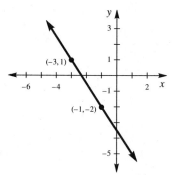

23.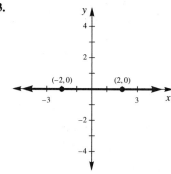

25.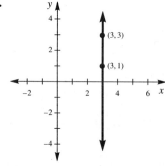

27.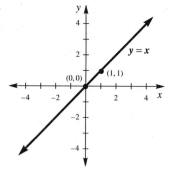

29.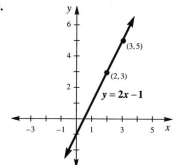

31.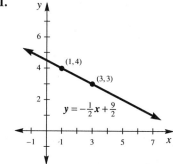

33.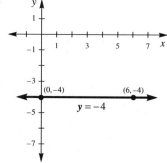

35.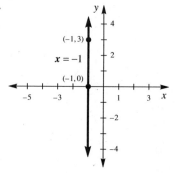

37.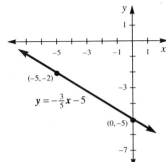

39.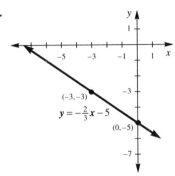

41. $y = x$ 43. $y = -x - 1$ 45. $y = 3$ 47. $y = 2x$

49. $y = x$ 51. $y = \frac{3}{2}x$ 53. $y = x + 2$

55. $y = -3x - 3$ 57. $y = -\frac{3}{2}x + 1$

59. $y = -\frac{1}{5}x$ 61. $y = \frac{4}{3}x$ 63. $y = 0$

65. $y = \frac{3}{4}x - \frac{9}{2}$ 67. $x = 0$ 69. $y = -3$

71. $y = x + 1$ 73. $y = -\frac{1}{2}x + 6$ 75. $y = -x$

77. $y = -2x$ 79. $y = 6$ 81. $x = 1$

83.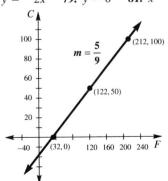

85. Rise over run. Change in y divided by the change in x.
87. m is slope, b is the y-intercept
89. Because division by zero is undefined

91. $x < 1$
93. $1 \leq x < \frac{11}{2}$
95. $x \geq -\frac{2}{9}$
97. $x \leq -\frac{1}{3}$

Exercise Set 4.4

1.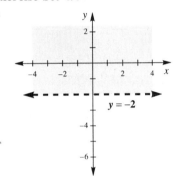

3.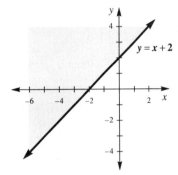

5.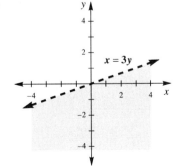

7.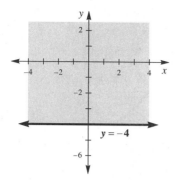

9.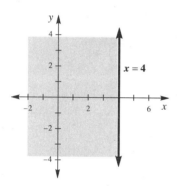

11.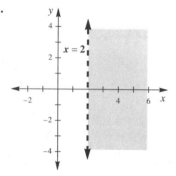

13.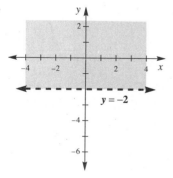

15.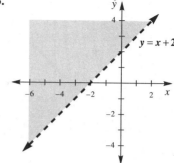

17.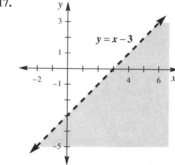

19.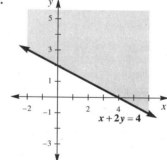

21.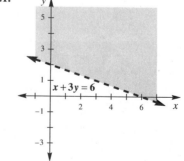

23.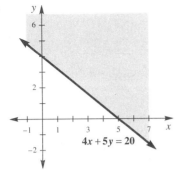

67.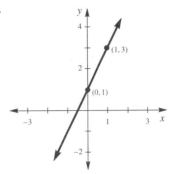

25. $y > 0$ **27.** $x > 3$ **29.** $-4 < y < 3$
31. $x > 0$ or $x < -4$ **33.** $x + y \leq 2$ **35.** $-x + y \leq 1$
37. $2x - 5y \geq 10$ **39.** $3x + 2y < -6$ **41.** $y \leq 4x$
43. $2x + y < 4$
45. The half-plane to the left of and including $x = 2$
47. The half-plane below the x-axis
49. That part of the plane to the right of $x = -3$ and to the left of $x = 4$
51. That part of the plane above $y = -4$ and below $y = 0$
53. $x - y \leq 5$ **55.** $2x + 3y \leq 6$ **57.** $\dfrac{x+y}{2} > 84$
59. $x - 2y < 4$ **61.** $a - b < 4$ **63.** $1852.50 loss
65. 832 racquets **67.** $3x$ hr **69.** $3x + 5y \leq 40$
71. $6x + 9y \leq 54$ **73.** $y = \dfrac{1}{2}x + 4$ **75.** $x = 3$
77. $y = -\dfrac{5}{4}x + \dfrac{5}{2}$ **79.** $y = 2x$ **81.** $y = x + 5$
83. $y = 4x + 8$ **85.** $y = -3x - 18$

69.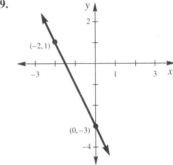

Exercise Set 4.5
1. yes **3.** no **5.** yes **7.** yes **9.** yes **11.** yes
13. no **15.** yes **17.** no **19.** yes **21.** no
23. $D\{x \mid x \in R\}$; $R\{f \mid f(x) \in R\}$
25. $D\{x \mid x \in R\}$; $R\{f \mid f(x) \in R\}$
27. $D\{x \mid x \in R\}$; $R\{f \mid f(x) \in R\}$
29. $D\{x \mid x \in R\}$; $R\{f \mid f(x) \in R\}$
31. $D\{x \mid x \in R\}$; $R\{f \mid f(x) \geq 0\}$
33. $D\{x \mid x \in R\}$; $R\{f \mid f(x) \in R\}$
35. $x \in R$ **37.** $x \in R$ **39.** $x \neq 3$
41. $(2, -1)$, $(0, -5)$, $(-3, -11)$
43. $(4, 5)$, $(-2, -1)$, $(3, \text{undefined})$
45. $(0, -5)$, $(3, 1)$, $(-1, -3)$
47. $(0, 0)$, $(-1, -2)$, $(2, 10)$
49. $(0, 0)$, $(4, 2)$, $(9, 3)$ **51.** yes **53.** yes **55.** no
57. yes **59.** yes **61.** yes
63. (a) $D\{x \mid x \in R\}$; $R\{y \mid y \in R\}$
    (b) $D\{-1 \leq x \leq 1\}$; $R\{0 \leq y \leq 1\}$
65. (a) $D\{-1 \leq x \leq 5\}$; $R\{-5 \leq y \leq 1\}$
    (b) $D\{-2 \leq x \leq 2\}$; $R\{-1 \leq y \leq 1\}$

71.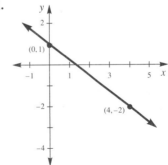

73.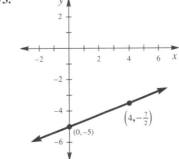

75.

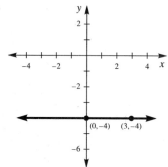

77. A set of ordered pairs where each first element pairs up with exactly one second element

79. No. Each first element pairs with more than one second element

81. If the vertical line only crosses once for each x-value it implies y is a function of x.

83. A function is a special case of a relation. See Exercise 77.

Review Exercises

1. $(0, 1), \left(-\frac{1}{5}, 0\right), (1, 6), (-1, -4)$ **2.** $(-3, 3), (0, 3), (3, 3)$

3. $(-4, -2), (-4, 0), (-4, 3)$

4.

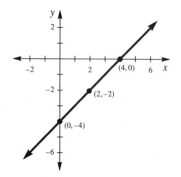

5.

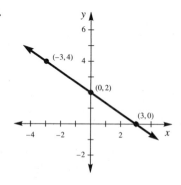

6.

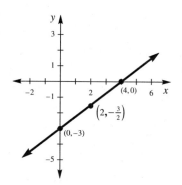

7.

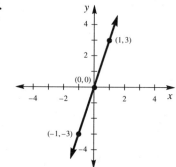

8.

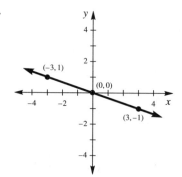

9. $y = 2x - 6$ **10.** $y = -6x + 5$ **11.** $y = 2x - 11$

12. $y = -\frac{2}{3}x$ **13.** $y = \frac{3}{5}x - \frac{11}{5}$ **14.** $y = -\frac{1}{2}x - 4$

15. $x = -3$ **16.** $y = -1$ **17.** $y = -\frac{3}{4}x - \frac{1}{4}$

18. $x = 0$ **19.** $x = 2$ **20.** $y = x + 1$ **21.** $2y - x = 4$

22. $y - 3x = 0$

23.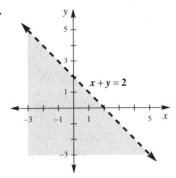

24.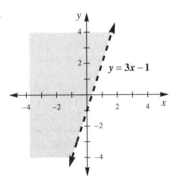

25.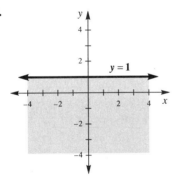

26.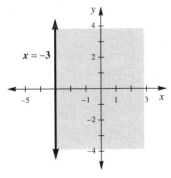

27.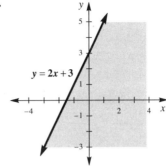

28.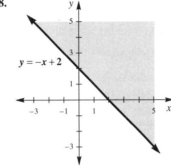

29.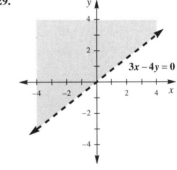

30.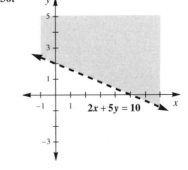

31. yes **32.** no **33.** yes **34.** yes **35.** no **36.** no
37. no **38.** yes **39.** yes **40.** yes
41. D$\{-4, -2, 0, 2\}$; R$\{-2, 0, 2, 4\}$
42. D$\{-2, -1, 0, 2\}$; R$\{-1, 0, 1, 3\}$
43. D$\{x \mid x \in R\}$; R$\{y \mid y = 2\}$
44. D$\{x \mid x \in R\}$; R$\{y \mid y \in R\}$
45. D$\{x \mid x \in R\}$; R$\{y \mid y \in R\}$
46. D$\{x \mid x \in R\}$; R$\{y \mid y \geq 2\}$
47. D$\{x \mid x \in R\}$; R$\{y \mid y \in R\}$
48. D$\{x \mid x \in R\}$; R$\{y \mid y \in R\}$ **49.** function
50. function **51.** function **52.** not a function
53. $f(0) = 3$ **54.** $g(0) = -8$ **55.** $f\left(-\dfrac{3}{2}\right) = 0$
56. $f(-2) = -1$ **57.** $g(-2) = 0$ **58.** $g(2) = 0$
59. D$\{x \mid x \in R, x \neq 5\}$ **60.** D$\left\{x \mid x \in R, x \neq -\dfrac{2}{3}\right\}$
61. $m = \dfrac{1}{6}$ **62.** $m = \dfrac{3}{40}$ **63.** $m = \dfrac{5}{12}$

64.

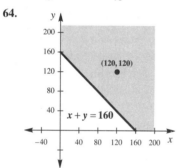

65.

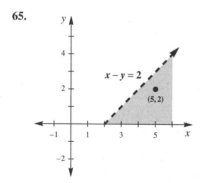

66. $P = $ income $-$ cost $= 500x - 4000$
67. A profit of $1,000 **68.** A loss of $4,000 **69.** $x = 8$
70. A profit of $52,000
71. no for 7; no for 8; yes for 9

Chapter Test

1.

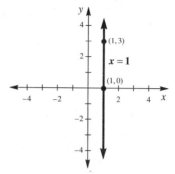

2.

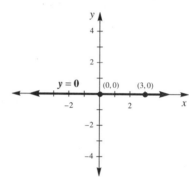

3.

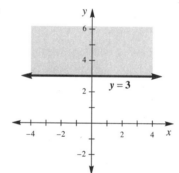

4.

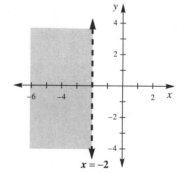

5.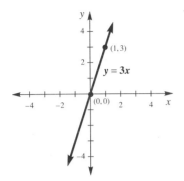

6.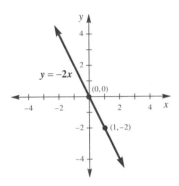

7.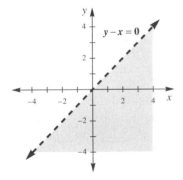

8.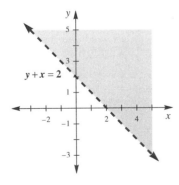

9.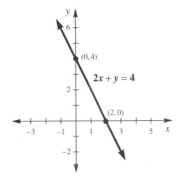

10.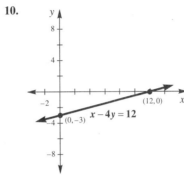

11.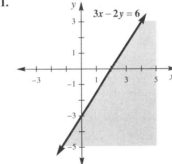

12.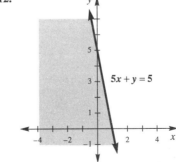

13. $-\dfrac{5}{2}$ **14.** $y = 5$

15.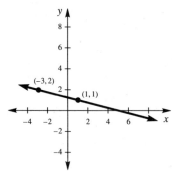

16. undefined **17.** $y = 3x - 17$

18.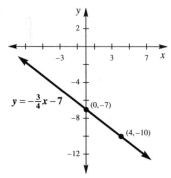

19. the lower side **20.** $y = 0$ **21.** $x = \dfrac{7}{3}$; $y = -\dfrac{7}{5}$

22. $y = -\dfrac{9}{2}x + 4$ **23.** the Cartesian coordinate system

24. the set of all ordered pairs (x, y) such that $y = 2x$

25. $y = 6x$ **26.** 0 **27.** $x = 7$ **28.** $y = -2x$

29. $D\{x \mid x \in R, x \neq 1\}$ **30.** $f(0) = 4$; $f(-2) = 0$

31. (a) $D\{x \mid x \in R\}$; $R\{y \mid y \in R\}$
(b) $D\{-3, -1, 0, 2\}$; $R\{0, 2, 3, 4\}$

32. (a) no (b) yes (c) yes

25.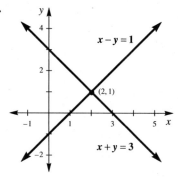

27.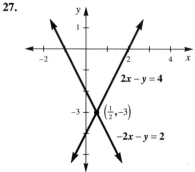

29. dependent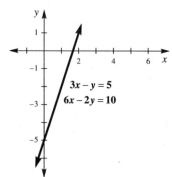

31.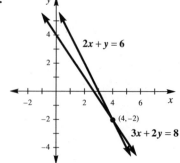

CHAPTER 5

Exercise Set 5.1

1. $(0, 0)$: yes; $(1, 1)$: no **3.** $(0, 4)$: yes; $(-1, 6)$: no
5. $(4, -3)$: yes; $(4, 3)$: no **7.** neither **9.** neither
11. inconsistent **13.** independent **15.** dependent
17. dependent **19.** dependent **21.** independent
23. independent

33. inconsistent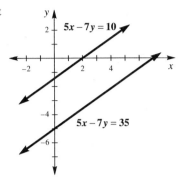

35.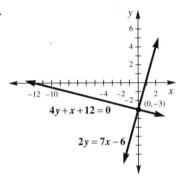

37.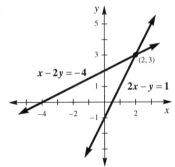

39. dependent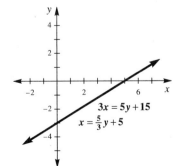

41.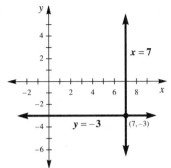

43.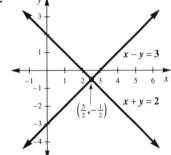

45.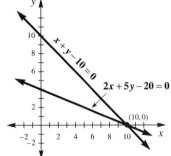

47. inconsistent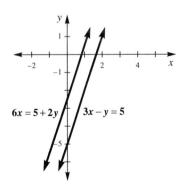

49.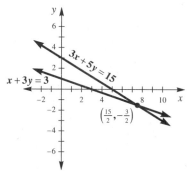

51.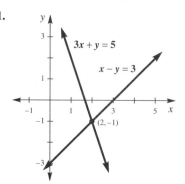

53.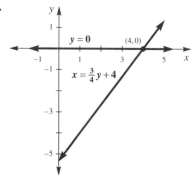

55.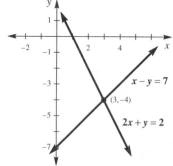

57.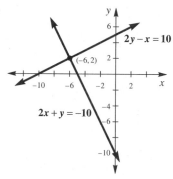

59.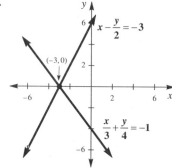

61.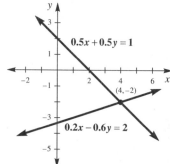

63. The numbers are 20 and 40

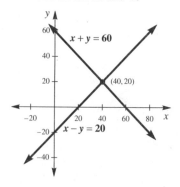

65. The husband is 30 and his wife is 25.

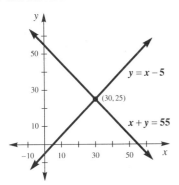

67. Jeff has 15 dimes and Michelle has 25 dimes.

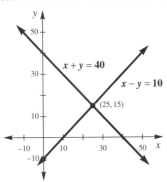

69. The length is 380 and the width is 270.

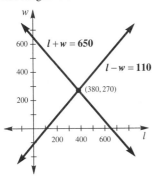

71. One worked 2 hr and the other worked 4 hr.

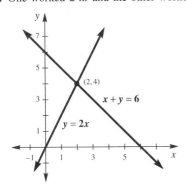

73. a system of equations where the lines coincide
75. one that has no solution (parallel lines)

Exercise Set 5.2

1. $\{(2, 1)\}$ **3.** $\{(1, 2)\}$ **5.** $\{(2, 2)\}$ **7.** $\{(-1, 0)\}$
9. inconsistent   **11.** $\{(-1, -1)\}$ **13.** dependent
15. $\left\{\left(\frac{1}{2}, \frac{1}{3}\right)\right\}$ **17.** $\{(-2, 8)\}$ **19.** $\{(1, 1)\}$
21. $\left\{\left(\frac{18}{5}, -\frac{8}{5}\right)\right\}$ **23.** $\{(2, 2)\}$ **25.** $\left\{\left(\frac{7}{4}, \frac{29}{16}\right)\right\}$
27. $\{(-1, 5)]$ **29.** $\{(-1, -5)\}$ **31.** $\{(4, 6)\}$
33. $\{(2, -1)\}$ **35.** $\{(2, 4)\}$ **37.** $\{(3, 1)\}$ **39.** $\left\{\left(5, \frac{1}{2}\right)\right\}$
41. inconsistent   **43.** $\{(2, -1)\}$ **45.** $\{(0, 2)\}$
47. $\{(-2, 5)\}$ **49.** $\left\{\left(\frac{24}{19}, -\frac{41}{38}\right)\right\}$ **51.** 28, 43
53. 17 cm, 23 cm   **55.** 10 dimes, 110 nickels
57. Larry: 1 hr; Bill: 2 hr   **59.** 10 m, 18 m   **61.** 50°, 130°
63. 64°, 116°   **65.** 30°, 60°   **67.** $\{(0, -5)\}$ **69.** $\{(4, -2)\}$
71. $\{2\}$ **73.** $\left\{-\frac{35}{26}\right\}$ **75.** $\left\{\frac{1}{8}\right\}$ **77.** $\left\{\frac{9}{14}\right\}$ **79.** $\left\{\frac{11}{4}\right\}$
81. $\{2\}$

Exercise Set 5.3

1. $\{(1, 2)\}$ **3.** $\{(4, 2)\}$ **5.** $\{(8, -8)\}$ **7.** $\left\{\left(1, -\frac{1}{4}\right)\right\}$
9. $\left\{\left(\frac{1}{2}, 2\right)\right\}$ **11.** $\{(1, 1)\}$ **13.** $\{(3, 2)\}$ **15.** $\{(3, 4)\}$
17. $\{(1, 1)\}$ **19.** inconsistent   **21.** $\{(2, 2)\}$
23. $\left\{\left(\frac{1}{2}, \frac{1}{2}\right)\right\}$ **25.** dependent   **27.** $\{(-1, 5)\}$
29. $\left\{\left(0, \frac{1}{5}\right)\right\}$ **31.** $\{(-4, -34)\}$ **33.** dependent
35. $\{(-95, -190)\}$ **37.** $\{(4, -3)\}$ **39.** dependent
41. inconsistent   **43.** $\{(5, 10)\}$ **45.** $\{(-4, 3)\}$
47. $\{(5, -1)\}$ **49.** $\{(12, 6)\}$ **51.** $\{(3, 1)\}$
53. $\left\{\left(-2, -\frac{1}{4}\right)\right\}$ **55.** $\{(10, -2)\}$ **57.** $-6, -4$
59. 30 girls, 20 boys
61. 100 cushioned seats, 40 wooden seats   **63.** 15, 23
65. 15°, 75°   **67.** 3 in., 5 in.

69.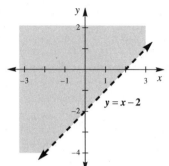

71.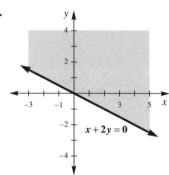

73.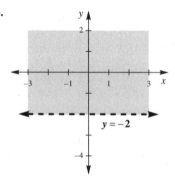

75.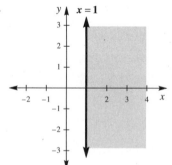

77.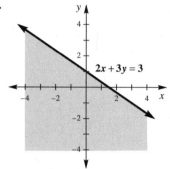

79.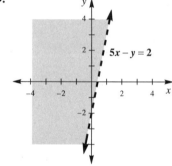

81.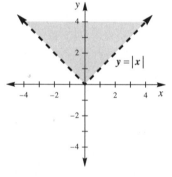

83.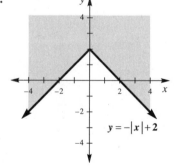

Exercise Set 5.4

1.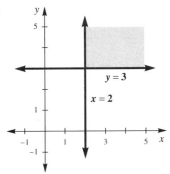

3.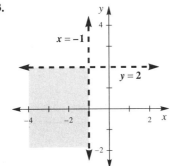

5.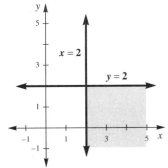

7.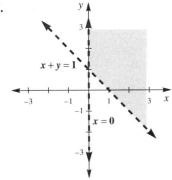

9.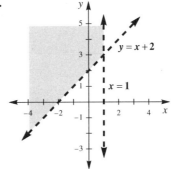

11.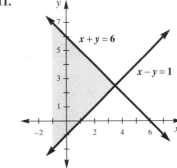

13.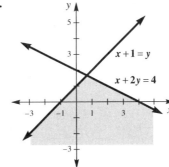

15. The solution set is the empty set.

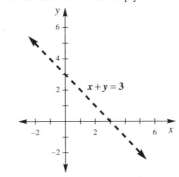

17.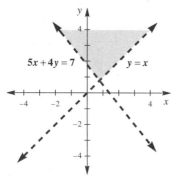

19.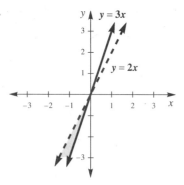

21.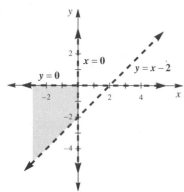

23.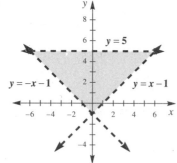

25. $y \geq 3x$
$y \geq -\dfrac{3}{4}x + 3$

27. $y \geq -3x + 6$
$y < \dfrac{1}{2}x - 1$

29. $y > x$
$y < -x + 2$

31.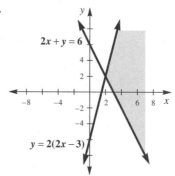

33.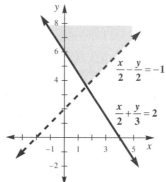

35.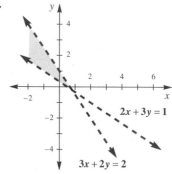

37.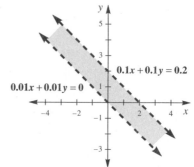

39.

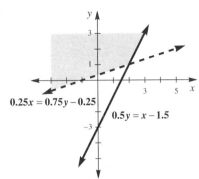

41.

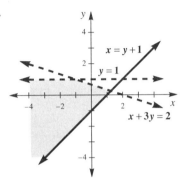

43.

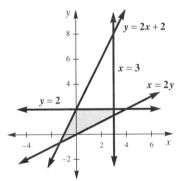

45. 10 yd wide and 30 yd long

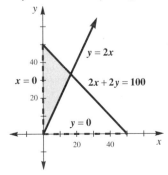

47. 200 pairs of women's shoes and 50 of mens shoes

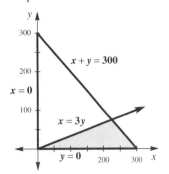

49.

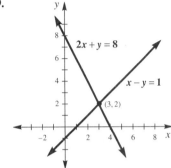

51. $\left\{\left(\frac{9}{2}, -\frac{1}{2}\right)\right\}$ **53.** $\left\{\left(\frac{120}{31}, \frac{14}{31}\right)\right\}$

Exercise Set 5.5

1. 83, 85   3. 40 km/hr, 60 km/hr
5. 30 lb at $1.80, 10 lb at $2.20
7. 40 cl of alcohol, 60 cl of water
9. $105 for skilled, $30 for unskilled
11. $43\frac{1}{5}$ m   13. 60 km/hr, 40 km/hr
15. 15 hr of B work, 9 hr of C work   17. 42 V, 78 V
19. 40 lb at $2.10, 10 lb at $1.60   21. $a = 3000$
23. $19,000 at 13%, $20,000 at 10%
25. 4 oz of A, $6\frac{2}{3}$ oz of B
27. 1690% increase in health expenditures, 202% increase in CPI; $56\frac{1}{3}$% average yearly health expenditure increase, 6.73% average yearly CPI increase
29. middle-sized side: 12 m; shortest side: 8 m   31. $2.99
33. $w = 7$ cm, $l = 20$ cm
35. $600 at 12%, $200 at 16%   37. 5 quarts
39. (a) $x + y \leq 10{,}000$
    $x \geq 3{,}000$

(b)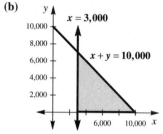

(c) $8,000 and $2,000

41. (a) $x \geq y$
$y \geq 3$
$x + y \leq 10$

(b)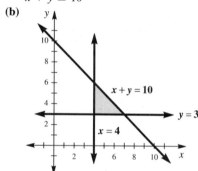

(c) 4 hr in the morning, 6 hr in the afternoon

Review Exercises
1.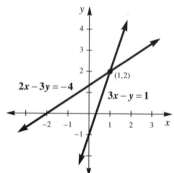

2.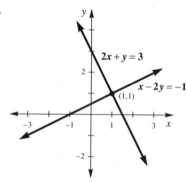

3.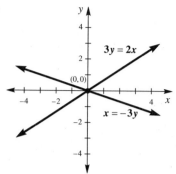

4.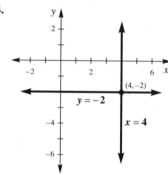

5. $\{(1, -2)\}$ **6.** $\{(-1, 2)\}$ **7.** $\left\{\left(-\dfrac{25}{4}, -3\right)\right\}$

8. dependent **9.** inconsistent **10.** $\left\{\left(\dfrac{4}{9}, -\dfrac{1}{3}\right)\right\}$

11. $\left\{\left(-\dfrac{5}{21}, \dfrac{19}{84}\right)\right\}$ **12.** inconsistent **13.** inconsistent

14. $\{(-4, 7)\}$ **15.** $\{(5, 2)\}$ **16.** $\left\{\left(\dfrac{17}{7}, 1\right)\right\}$

17. $\{(1, 1.1)\}$ **18.** $\left\{\left(2, \dfrac{43}{6}\right)\right\}$ **19.** $\{(-0.2, -0.17)\}$

20. $\{(1, -2)\}$ **21.** $\left\{\left(\dfrac{13}{15}, 1\right)\right\}$ **22.** $\{(-10, 21)\}$

23. $\{(-21, -18)\}$ **24.** $\left\{\left(\dfrac{4}{15}, \dfrac{7}{5}\right)\right\}$ **25.** inconsistent

26. dependent

27.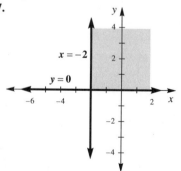

28.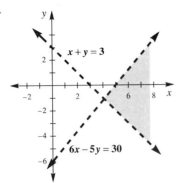

29.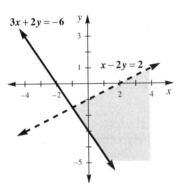

30.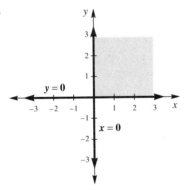

31.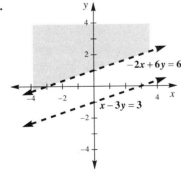

32.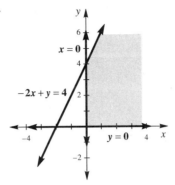

33.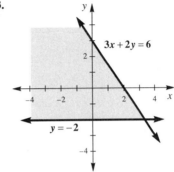

34.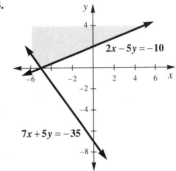

35. $y = -2x + 3$; $y = \frac{1}{3}x + 1$; $\left\{\left(\frac{6}{7}, \frac{9}{7}\right)\right\}$

36. $x = 4$; $y = -3$; $\{(4, -3)\}$

37. $x + 3y = 3$; $2x - 4y = 1$; $\left\{\left(\frac{3}{2}, \frac{1}{2}\right)\right\}$

38. $y = -2x - 1$; $y = -5x - 13$; $\{(-4, 7)\}$

39. $x = -2y$; $y = 3x$; $\{(0, 0)\}$

40. $y = -x + 1$; $y = 5x - 2$; $\left\{\left(\frac{1}{2}, \frac{1}{2}\right)\right\}$

41. $-27, -31$ **42.** 22 m, 33 m

43. 28 nickels, 11 quarters   **44.** 50°, 130°   **45.** 10°, 80°

46. $x + y < 9$; $x - y \leq 3$; $\frac{1}{2}, \frac{1}{2}$

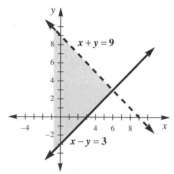

47. 5 ft, 6 ft
48. $x + y \leq 90$; $x - y \leq 30$; 30°, 30°

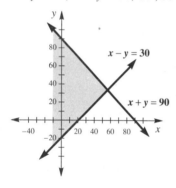

49. $6,000 at 7%, $4,000 at 12%
50. 2.5 lb at $4.50, 2.5 lb at $6.00
51. $2x + y \geq 200$; $x + 3y \geq 300$; 60 g of 1st food, 80 g of 2nd food

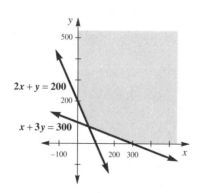

Chapter Test

1.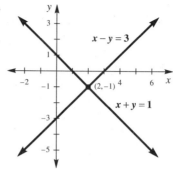

2.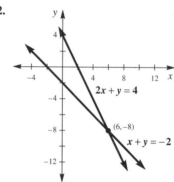

3.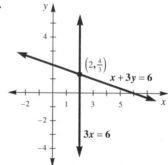

4. inconsistent

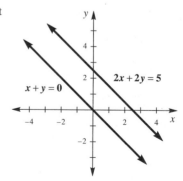

5. dependent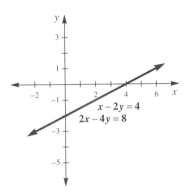

6. $\{(1, -1)\}$ **7.** $\{(1, -2)\}$ **8.** dependent **9.** $\{(1, -2)\}$

10. $\{(2, 1)\}$ **11.** inconsistent **12.** $\{(3, -2)\}$

13. dependent **14.** $\left\{\left(-2, \frac{3}{2}\right)\right\}$ **15.** $\left\{\left(\frac{3}{2}, \frac{1}{2}\right)\right\}$

16. $\{(4, 2)\}$ **17.** $\{(1, 4)\}$ **18.** $\left\{\left(\frac{3}{7}, \frac{5}{7}\right)\right\}$

19. $\left\{\left(-\frac{4}{9}, \frac{7}{9}\right)\right\}$ **20.** $\{(6, 4)\}$

21.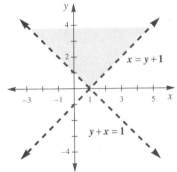

22.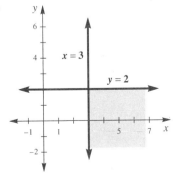

23.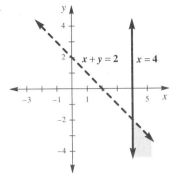

24.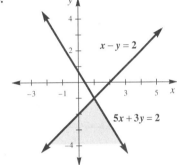

25.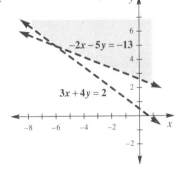

26.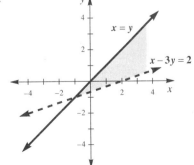

27.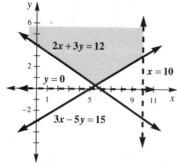

28.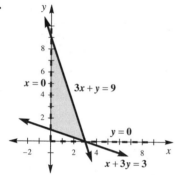

29. (913, 986)   **30.** 26 dimes, 7 quarters
31. 200 km by bus, 380 km by train
32. $7000 at 5%, $5000 at 7%
33. (a) $x + 2y \geq 10$
$5x + 3y \geq 30$

(b)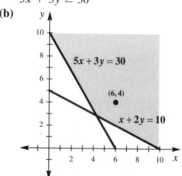

(c) 6 servings of the main dish, 4 servings of the fruit plate

CHAPTER 6

Exercise Set 6.1

1. x is the base, 6 the exponent
3. 2 is the base, 5 the exponent
5. m is the base, 6 the exponent
7. -2 is the base, 7 the exponent

9. 5^2 **11.** 2^4 **13.** $\dfrac{1}{a^3}$ **15.** -2^3 **17.** x^5 **19.** $-x^2$
21. $\dfrac{r^2}{s^3}$ **23.** $-xy + x^2y$ **25.** 125   **27.** $-16,384$ **29.** 1
31. 4   **33.** $\dfrac{1}{64}$ **35.** $\dfrac{1}{8}$ **37.** $-\dfrac{1}{27}$ **39.** 16   **41.** 12
43. $\dfrac{1}{8}$ **45.** 5   **47.** 4   **49.** 54   **51.** $\dfrac{1}{2}$ **53.** -1
55. 2^8 **57.** a^{12} **59.** a^9 **61.** x^6 **63.** $\dfrac{1}{y}$ **65.** 1
67. $\dfrac{1}{5}$ **69.** $\dfrac{100}{9}$ **71.** x^5 **73.** $\dfrac{1}{x^8}$ **75.** $\dfrac{1}{9}$ **77.** 3
79. 2   **81.** 0.5   **83.** $\dfrac{16}{9}$ **85.** 17   **87.** 0   **89.** $6x^6$
91. $10x^6$ **93.** x^5y^9 **95.** $6x^6y^5z^2$ **97.** xy **99.** s^2t
101. $\dfrac{4y^4}{x^2}$ **103.** $x - y$ **105.** 1   **107.** 3   **109.** 72
111. $-\dfrac{7}{9}$

113.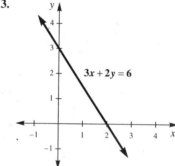

115.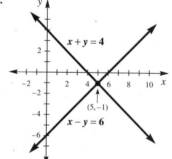

117. $\{(0, 2)\}$ **119.** $\left\{\left(\dfrac{1}{2}, \dfrac{1}{2}\right)\right\}$ **121.** Add exponents: x^7.
123. $7^0 = \dfrac{7^3}{7^3} = 1$

Exercise Set 6.2

1. 1   **3.** -1 **5.** 1   **7.** $\dfrac{1}{8}$ **9.** $\dfrac{1}{9}$ **11.** $\dfrac{1}{9}$ **13.** $\dfrac{3}{2}$

15. 4 **17.** 125 **19.** $\frac{16}{9}$ **21.** $\frac{64}{9}$ **23.** -16 **25.** x^6
27. $\frac{1}{a^{10}}$ **29.** $\frac{x^4}{y^4}$ **31.** $\frac{3^{10}}{b^{10}}$ **33.** b^4 **35.** x^4y^2 **37.** $\frac{y^8}{x^8z^4}$
39. $\frac{y^8}{9x^4}$ **41.** $81x^3$ **43.** $\frac{p^8}{49}$ **45.** 1 **47.** x^8y^2 **49.** x^8y^{12}
51. $\frac{1}{y^2}$ **53.** $25a^4b^4c^4$ **55.** $\frac{x^2}{24}$ **57.** $\frac{-3}{2y^4}$ **59.** $\frac{4y^7}{x^7}$
61. $-4x^5yz^3$ **63.** $16x^{10}$ **65.** $\frac{3a}{2b^9}$ **67.** $\frac{s^5}{t^3}$ **69.** 2304
71. $\frac{125}{64}$ **73.** 16 **75.** 2.6×10^1 **77.** 2×10^{-3}
79. 4.2×10^{-4} **81.** 2.68×10^5 **83.** 2.856×10^6
85. 2.63×10^{-7} **87.** 2600 **89.** 9.23 **91.** 0.00000283
93. 0.0000048 **95.** 2,950,000,000,000 **97.** 7.2×10^7
99. 5.2×10^2 **101.** 7×10^7 **103.** 2×10^{-7}
105. 3×10^2 **107.** 8.65×10^5 miles
109. 6.67×10^{-8} dynes **111.** 6.02×10^{23} molecules
113. 17,600,000 **115.** 0.00000000000000000000003 g
117. 0.000023 in. **119.** 5×10^2 sec **121.** 6 miles

Exercise Set 6.3

1. trinomial; 2nd degree **3.** binomial; 7th degree
5. binomial; 2nd degree **7.** monomial; 2nd degree
9. trinomial; 6th degree **11.** monomial; zero degree
13. yes; 2nd degree
15. no; there is a variable in the denominator
17. no; $x^{1/2}$ is $\sqrt{x}$ **19.** yes, zero degree
21. no; -3 is not a whole number
23. no; there is a variable in the denominator
25. $5x + 10$; 5 **27.** $4x - 1$; 4 **29.** $-y - 1$; -1
31. $x^2 - x + 2$; 1 **33.** $-p^4 + 15p^3 - p^2$; -1
35. $1.25y^2 - 0.35y + 0.6$; 1.25 **37.** $\frac{11}{8}p^2 - \frac{9}{8}p - \frac{7}{2}$; $\frac{11}{8}$
39. $\frac{13}{14}x^2 - \frac{23}{24}x - \frac{1}{4}$; $\frac{13}{14}$ **41.** $5x^2 + 7x$ **43.** $8m^2 + 2$
45. $4x^2 - 5x - 3$ **47.** $12x^3 + x^2 + 12x - 14$
49. $3x^4 + 2x^3 + 2x^2 - 22x + 7$ **51.** $2x^2 + 8$
53. $-p^2 - 12p + 5$ **55.** $-m^2 - m + 1$
57. $-3p^4 - 4p^3 + 15p^2 - 18p$ **59.** $4x^4 - 7x^3 - 5x + 7$
61. $11y + 5$ **63.** $4m + 17$ **65.** $16y^2 + 2y - 4$
67. $4x^2 + 4$ **69.** $9m^2 - 7$ **71.** $-x^2 + 12x - 8$
73. $-7p^2 + 15p + 2$ **75.** $4x^4 - 5x^3 - x^2 + 7x + 2$
77. $m^5 - 4m^4 - 7m^3 + 3m^2 - 15$
79. $2y^4 + 12y^2 - 7y + 7$ **81.** $10m^3 + m^2 + 5m - 4$
83. $4x^3 - 3x^2 + 11$ **85.** $-4y^2 - 12y - 10$
87. $-x^6 + 4x^5 - x^4 + 4x^3 - 21x^2 - 20x + 23$
89. $-m^2p + 7mp + 4mp^2 + 1$ **91.** $\frac{31}{24}x^2 + \frac{3}{4}x - \frac{9}{8}$
93. $-\frac{8}{3}a - \frac{11}{6}$ **95.** $-\frac{17}{24}m^2 - \frac{1}{4}m + \frac{1}{8}$
97. $-1.1x^2 + 2.9x + 13.3$ **99.** $0.001x^2 - 7.12x + 0.35$

101. $1.2x - 21.6$ **103.** $438, $552, $780
105. (a) $P(x) = -0.10x^2 + 101.5x - 250$ (b) $755; $1740
107. The number in front of the variable in the leading term
109. A polynomial where the variable is in descending powers
111. Terms with the same variables and exponents
113. $24x + 12$ **115.** $3x + 17$ **117.** $4x - 13$

Exercise Set 6.4

1. $x^2 + 2x$ **3.** $x^2 - x$ **5.** $x^3 + 6x^2$ **7.** $3x^2 - 2xy$
9. $x^3y^2 + xy^2$ **11.** $x^3y^2 - x^2y^2$ **13.** $y^6 + 6y^4 - 2y^3 + y^2$
15. $-6y^3 + 4y^2 - 8y$ **17.** $-2x^2y^4 + 3x^3y^3 - 2xy^2$
19. $18a^4b^4 + 6a^3b^3 - 18a^3b^2$
21. $-24x^2y + 30xy^3 - 16x^3y + 12x$
23. $-108x^3y + 135x^2y^3 - 72x^4y + 54x^2$
25. $6m^{10} - 30m^9 + 66m^5 - 48m^4 - 42m^3$
27. $-\frac{2}{3}p^{13} + \frac{6}{7}p^{11} - \frac{4}{9}p^9$ **29.** $0.5m^8 - 0.03m^6 - 0.021m^5$
31. $24x^4 - 60x^3 + 84x^2 - 72x + 108$ **33.** $5x$
35. $7p^2 + 10p$ **37.** $-2m^2 + 64m$ **39.** $-12y^2$
41. $4t^3 + 2t - 14$ **43.** $-28s^4 + 30s^2$
45. $-31w^3z^4 - 2z^2$ **47.** $x^2 + 5x + 6$ **49.** $r^2 - 5r + 6$
51. $y^2 + 2y - 24$ **53.** $m^2 + 10m + 25$ **55.** $a^2 - 64$
57. $2x^2 + 13x + 15$ **59.** $9r^2 - 24r + 16$ **61.** $4p^2 - 25$
63. $6m^2 + 11m - 72$ **65.** $10p^2 - p - 24$
67. $-20m^2 + 9m + 18$ **69.** $49x^2 + 126x + 81$
71. $121y^2 - 49$ **73.** $3x^2 - 10xy - 8y^2$
75. $16a^2 - 46ab - 35b^2$ **77.** $81m^2 - 18mn + n^2$
79. $25r^2 - 9s^2$ **81.** $x^4 - 4y^2$ **83.** $x^4 - y^4$
85. $9x^4 + 24x^2y^2 + 16y^4$ **87.** $2x^2 - x - 1$
89. $7p^2 - 6p + 25$ **91.** $m^2 - 10m + 1$
93. $-a^2 - 28a$ **95.** $3y^3 - 4y^2 - 14y$
97. $\frac{1}{4}x^2 - \frac{1}{4}xy + \frac{1}{16}y^2$ **99.** $0.25r^2 + 0.1rs + 0.01s^2$
101. $\frac{9}{4}x^2 - 4xy + \frac{16}{9}y^2$ **103.** $x^3 - 6x^2 + 12x - 8$
105. $8x^3 + 12x^2 + 6x + 1$ **107.** $6x^2 - x - 1$
109. $29x^2 + 35x + 8$
111. (a) $38x^2 - 10x - 6$ (b) $12x^3 + 3x^2 - 9x$

Exercise Set 6.5

1. 8 **3.** $3y$ **5.** $-4a$ **7.** $\frac{11b^9}{a}$ **9.** $\frac{4a^5}{b^3}$ **11.** $-2x^4y$
13. $3x + 1$ **15.** $x + 7$ **17.** $5a - 4$ **19.** $4r^2 - 2r + \frac{3}{2}$
21. $\frac{4}{t} - \frac{8t^2}{3}$ **23.** $\frac{17}{y} + 8 + y$ **25.** $-3q^2 - \frac{21q}{4}$
27. $-15w^2 - 5w + 3$ **29.** $5xy^3 - 2y^3$ **31.** $\frac{-7x^3y^2}{2} + \frac{1}{2}$
33. 96 **35.** $180a^{10} - 150a^8 + 120a^6 - 90a^4 + 120a^2$
37. 56 **39.** $x + 2$ **41.** $p + 1$
43. $6x^3 - 10x^2 + 20x - 60 + \frac{177}{x + 3}$ **45.** $a^2 - a + 1$

47. $y^2 + 2y + 4$ **49.** $p - 3 + \dfrac{17}{2p + 4}$
51. $y^2 - y - 2$ **53.** $4x^3 - x + 2$ **55.** $16s^4 + 4s^2 + 1$
57. $3a + 1$ **59.** $p^3 + 3p^2 - 2p + 1$
61. $-y^2 + 3y - 4 + \dfrac{5}{-5y + 2}$
63. $3m^2 + m - 3 + \dfrac{2m + 6}{m^2 - 5m + 2}$
65. $-3a^2 - 4a + 2 + \dfrac{5}{-a^2 - 3a + 7}$
67. $9y^2 + 6y + 4 + \dfrac{10}{3y - 2}$ **69.** $2x^2 - 4x + 3 + \dfrac{7x - 2}{x^2 - x - 5}$
71. $(x - 4)$ in.   **73.** $(x + 2)$ cm   **75.** yes   **77.** x^2
79. y^2 **81.** $\dfrac{1}{x}$ **83.** $\dfrac{27}{x^3}$ **85.** -7 **87.** $a^8 b^{12}$ **89.** $\dfrac{x^5}{3}$

Exercise Set 6.6
1. $4.18\overline{6}$ ft³   **3.** 314 m²   **5.** 300,000   **7.** 48
9. 196 horsepower   **11.** 46,080,000   **13.** 188.4 in³
15. $10,185   **17.** 1208.9 lb   **19.** 870 gal
21. $(2x - 3)$ units   **23.** $(3x - 1)$ units
25. $V = 8x^3 + 6x^2 + 150x + 125$; $S = 24x^2 + 120x + 150$
27. Let x = the weight of a box of apples, and y = the weight of a box of pears; then $x + y = 70$ and $x - y = 20$. Using the graphing method, the box of apples weighs 45 lb and the box of pears weighs 25 lb.

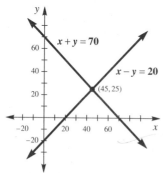

29. Let x = the first number and y = the second number; then $x - y = 15$ and $x = 2y$. By the graphing method, $x = 30$ and $y = 15$.

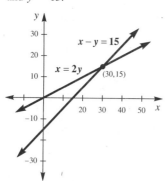

Review Exercises
1. x^{10} **2.** y^{15} **3.** $-4p^2$ **4.** $9x^2$ **5.** $9x^4 y^2$ **6.** 2
7. 1   **8.** $\dfrac{z^4}{9x^4 y^2}$ **9.** $16ab$ **10.** $\dfrac{2}{y}$ **11.** $-\dfrac{n}{2m}$ **12.** $\dfrac{x^3}{8}$
13. $\dfrac{x}{y}$ **14.** $\dfrac{1}{x^3}$ **15.** $\dfrac{7}{12}$ **16.** $-\dfrac{T^9}{8R^6}$ **17.** $-\dfrac{9}{2x^4 y^7}$
18. $-\dfrac{8}{9x}$ **19.** (a) 4.89×10^{-1} (b) 9.1×10^{-4}
20. (a) 0.000184   (b) 720,700
21. (b) $-2x - 3$;   (c) 1st degree;   (d) binomial;   (e) -2
22. (b) $-8y^2 + 8y - 9$;   (c) 2nd degree;   (d) trinomial;   (e) -8
23. (b) $m^3 + m^2 - 7m$;   (c) 3rd degree;   (d) trinomial;   (e) 1
24. (b) $-x + 9$;   (c) 1st degree;   (d) binomial;   (e) -1
25. $5x^2 - 9x + 8$ **26.** $3m^3 - 4m^2 + 4$ **27.** $12t^3 + 3t^2$
28. $-2t^2 - 9$ **29.** $6p^4 - 6p^2 + 1$ **30.** $4x - 7$
31. $-21x + 26$ **32.** $x^3 + x^2 - 5x + 8$
33. $4y^3 - 4y^2 + 16y + 3$ **34.** $-x^2 + 7x + 1$
35. $x^2 - 6$ **36.** $-2y + 4$ **37.** $-s^2 + s + 32$
38. $\dfrac{3}{4}x - \dfrac{3}{4}y$ **39.** $15p^3 - 18p^2 + 21p$
40. $-2x^3 y^3 + 6x^3 y^2$ **41.** $x^2 - 16$ **42.** $x^2 + 6x + 9$
43. $2x^2 - 7x - 15$ **44.** $6rs^2 + 12r^2 s^3 - 6rs^4 - 3rs^3$
45. $25m^2 - 70m + 49$ **46.** $9x^2 - 4y^2$
47. $x^3 + 3x^2 + 3x + 1$ **48.** $-4p^2 - 28pq - 49q^2$
49. $0.25m^2 - 0.04n^2$ **50.** $2.2a^2 - 1.3ab - 1.2b^2$
51. $\dfrac{1}{4}x^2 + \dfrac{1}{3}xy + \dfrac{1}{9}y^2$ **52.** $\dfrac{9}{16}x^2 - \dfrac{49}{64}y^2$
53. $8z^2 + 2z - 3$ **54.** $4q^4 - 12q^2 + 9$
55. $\dfrac{5}{24}x^2 - \dfrac{199}{288}xy + \dfrac{1}{2}y^2$ **56.** $\dfrac{3}{20}x^2 + \dfrac{2}{5}xy - \dfrac{1}{3}y^2$
57. $x + 3$ **58.** $4a^3 + 2a^2 - 3a + 1$ **59.** $\dfrac{7}{2}y - 4$
60. $\dfrac{2}{3} - \dfrac{5}{3}m^3$ **61.** $a - b$ **62.** $-\dfrac{1}{5}x^3 + \dfrac{1}{3}x - \dfrac{1}{15}$
63. $x - 6$ **64.** $y + 4$ **65.** $2a - 1 + \dfrac{3}{a + 4}$
66. $p^2 + p + 1$ **67.** $a^2 + 3a + 9$
68. $x^4 + x^3 + x^2 + x + 1$ **69.** $-y^3 + 2y - 7 + \dfrac{-2}{-3y + 4}$
70. $m^2 - m + 2$ **71.** $\dfrac{1}{4} + 2y + 4y^2$
72. $0.64x^2 - 0.64x + 0.16$
73. $A = 4x^2 + 12x + 9$; $P = 8x + 12$
74. $A = 20x^2 + 11x - 4$; $P = 18x + 6$
75. sum = $2n$; product = $n^2 - 1$
76. sum = $3n + 6$; product = $n^3 + 6n^2 + 8n$
77. $A = \pi(x^2 - 8x + 16)$; $C = \pi(2x - 8)$
78. $S = 6x^2 + 36x + 54$; $V = x^3 + 9x^2 + 27x + 27$
79. $x = 25$ **80.** n

Chapter Test

1. 1   2. 1   3. x^{12} 4. x^2 5. $-36x^6$ 6. $-6x^5y^5$
7. $\dfrac{1}{x^{12}y^{18}}$ 8. $\dfrac{4x^4}{3}$ 9. $49x^7y^8z$ 10. $-8x^{11}y^9z$ 11. $\dfrac{y^{18}}{4x^{18}}$
12. $\dfrac{1}{3x-2y}$ 13. 6.3×10^3 14. 8.2×10^{-3} 15. 6,210
16. 0.000293   17. $9x^2 - 4x - 5$
18. $23x^3 - 2x^2 - 2x + 10$ 19. $-x^2 - 3x - 6$
20. $4y^3 + 17y^2 - 13y + 6$
21. $4x^2yz - xy^2z - 9xyz^2 - 8xyz$ 22. $-m^3 + 5m - 7$
23. $x^2 - 25$ 24. $4y^2 + 12y + 9$ 25. $12p^2 + 11p - 15$
26. $-2x^2 + 4x - 1$ 27. $8x - 7y - \dfrac{9}{7}$ 28. $x - 7$
29. $x^2 + 2x + 1$ 30. $3x + 22 + \dfrac{99}{x-5}$ 31. 82.64 lb
32. $S = 6x^2 - 48x + 96$; $V = x^3 - 12x^2 + 48x - 64$

Cumulative Review Chapters 4–6

1. $(0, 3), (4, 0), (-4, 6)$ 2. $(0, -6), (5, 0), (10, 6)$
3. $(0, 3), (-3, 0), (-6, -3)$
4. $(0, 5), \left(-\dfrac{1}{2}, 3\right), \left(-\dfrac{3}{2}, -1\right)$

5.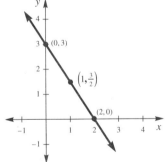

6.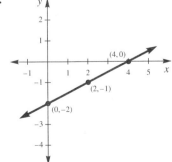

7.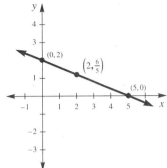

8.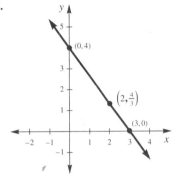

9. -1 10. $\dfrac{3}{2}$ 11. undefined   12. 0   13. $y = 2x - 4$
14. $y = -\dfrac{2}{3}x + \dfrac{4}{3}$ 15. $y = x + 1$ 16. $x = -6$
17. $y = -4$
18.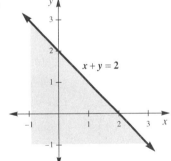

19.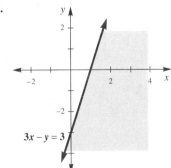

20.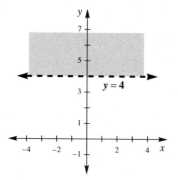

21.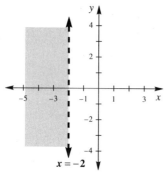

22. $-1, -3, 2$ **23.** $-2, 7, -1$ **24.** $-6, -1, -11$

25. $-\dfrac{2}{3}, -1,$ undefined

26. function: $D\{x \mid x \in R\}$; $R\{y \mid y \in R\}$
27. function: $D\{x \mid x \in R\}$; $R\{y \mid y = 3\}$
28. not a function: $D\{x \mid x = -7\}$; $R\{y \mid y \in R\}$
29. function: $D\{x \mid x \in R\}$; $R\{y \mid y \in R\}$
30. function: $D\{x \mid x \in R, x \neq 0\}$; $R\{y \mid y \in R, y \neq 0\}$
31. $\{(3, 1)\}$

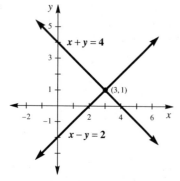

32. $\{(-1, 1)\}$ **33.** $\left\{\left(\dfrac{19}{17}, \dfrac{7}{17}\right)\right\}$

34. $\left\{\left(\dfrac{11}{3}, \dfrac{2}{3}\right)\right\}$ **35.** $\{(1, 1)\}$

36. $\{(1, 2)\}$

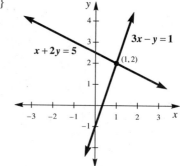

37.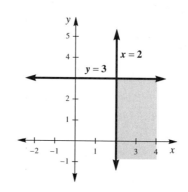

38.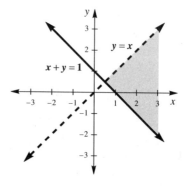

39.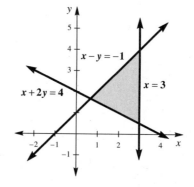

40.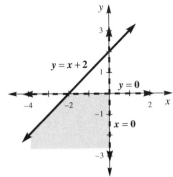

41. 25 **42.** −9 **43.** 54 **44.** 1 **45.** $\frac{1}{5}$ **46.** 32
47. x^7 **48.** $12x^5$ **49.** $\frac{3}{2}$ **50.** $(x-4)^8$ **51.** b^2
52. $24a$ **53.** 9 **54.** x^{12} **55.** $\frac{x^8}{9y^2}$ **56.** 1
57. 8.3×10^3 **58.** 6.4×10^1 **59.** 1.2×10^{-3}
60. 4×10^{-1} **61.** 69,000 **62.** 0.00123 **63.** 0.08
64. binomial, 2nd degree **65.** monomial, 3rd degree
66. trinomial, 2nd degree **67.** $4x - 3$ **68.** -4
69. $9x^2 - 7x - 5$ **70.** $13x^2 - 11x + 17$
71. $4y^3 - 3y^2 + 11$ **72.** $-0.5x^3 - 2.7x^2 - 1.7x$
73. $6y^2 - 12y$ **74.** $x^2 - 64$ **75.** $4x^2 - 12x + 9$
76. $8y^2 - 19y - 56$ **77.** $9a^2 - 16b^2$
78. $6a^2 + 22a + 12$ **79.** $p^3 - 5p^2 + 7p - 3$
80. $4x - 5$ **81.** $-6y^2 + 4y - 1$ **82.** $3x - 7 + \frac{13}{x+2}$
83. $a + 3$ **84.** $a^2 + 2a + 1 + \frac{1}{a+1}$
85. $2y^2 - 4y + 3 + \frac{7y - 2}{y^2 - y - 5}$ **86.** $x^2 + 2x + 4$
87. 2 and 9 **88.** 6 and 9
89. boat: 8 mph; current: 3 mph
90. 8 lb at $1.90, 12 lb at $2.40 **91.** 12 and 15
92. $3,000 at 6%, $7,000 at 8% **93.** $l = 40$ ft, $w = 30$ ft

CHAPTER 7

Exercise Set 7.1
1. $2 \cdot 3$ **3.** $2 \cdot 19$ **5.** prime **7.** $2 \cdot 2 \cdot 2 \cdot 3 \cdot 11$
9. $2 \cdot 3 \cdot 3 \cdot 5$ **11.** $7 \cdot 7$ **13.** prime
15. $2 \cdot 3 \cdot 3 \cdot 5 \cdot 5$ **17.** 12 **19.** 7 **21.** 10 **23.** $3x$
25. $5x$ **27.** 16 **29.** x **31.** $x + 2$ **33.** $y - 4$
35. $x - 5$ **37.** $p + 7$ **39.** $r - 6$ **41.** $7x - 4$
43. $4x + 9$ **45.** $2a^2 - 3$ **47.** $3x^2 - 4x - 5$
49. $2s^2t - 3st - 1$ **51.** $1 - 3y^2z^3$ **53.** $2x - 1$
55. $y - 3$ **57.** $2(x + 3)$ **59.** prime **61.** $-4(2p + 3)$
63. $3x(x - 3)$ **65.** $ax(x - 1)$ **67.** $-y(5y + 12)$
69. $\pi r(r - 2)$ **71.** $2x(2x^2 - 4x + 1)$
73. $6x^2y^5(2x - 3y)$ **75.** $-13y^5(3y^3 + 4y + 6)$
77. $-2x^2y^3(11x^4y - 20x^2 + 4y^2)$
79. $2a(6a^3 - 8a^2 + 12a - 9)$
81. $2ab(9a^4b - 5a^3 + 2ab^2 + 1)$
83. $-6r^3s^2(10r^2 + 3s + 2s^2)$ **85.** $3x^n(2x - 1)$
87. $9a^{3n}(3a^2 - 2)$ **89.** $13x^3(7x^{4n} + 1)$
91. $24s^{n+1}(3s^{n+6} - 1)$ **93.** $7x^{n-8}(7 + 2x^{n+11})$ **95.** $x^2 - 1$
97. $y^2 - 6y + 9$ **99.** $p^2 - 6p - 7$
101. $3a^2 - 15a + 18$ **103.** $-2s^2 - 2s + 12$
105. $-3x^3 + 33x^2 - 90x$ **107.** $x^4 + 10x^3 + 21x^2$
109. $-2x^4 + 30s^3 - 108s^2$

Exercise Set 7.2
1. 2 **3.** $n + 3$ **5.** $(v - 3)(v + 2)$ **7.** $(y - 2)(y - 9)$
9. $x - 8$ **11.** $(x + 10)(x + 1)$ **13.** $(x + 2)(x + 1)$
15. $(p - 2)(p + 1)$ **17.** $(v - 5)(v - 3)$
19. $(x + 9)(x - 8)$ **21.** $(z - 7)^2$ **23.** $3(m^2 + 4m + 5)$
25. $5(b^2 + 11b + 12)$ **27.** $x(x - 10)(x + 3)$
29. $(a - 9)(a + 7)$ **31.** $-2(p + 13)(p + 2)$
33. prime **35.** $3s^3(s - 5)(s - 1)$ **37.** $(a + 14)(a + 3)$
39. $-3(x - 5)(x - 2)$ **41.** prime **43.** $(x - 2y)(x - y)$
45. $(p - 12q)(p - 2q)$ **47.** $(a + 5b)(a - 3b)$
49. $3m^2(m - 6)(m + 5)$ **51.** $2x^2(x - 9)(x + 8)$
53. $10a(a + 6b)(a + 5b)$ **55.** $-3(x + 45)(x + 1)$
57. prime **59.** $9x(x + 6)(x + 2)$ **61.** $7(t - 4s)(t - 2s)$
63. $(x - 2)(x + 3)$ **65.** $(p + 3)(m - n)$
67. $(y - 4)(y + 1)$ **69.** $x^2(x + 1)^2$ **71.** $m(m - 7)^2$
73. $-6y^2(y + 2)^2$ **75.** $(x + y)(x + y - 1)$ **77.** $(x - y)^2$
79. $(m + n)(-2n)$ **81.** $(x + 7)(x + 2)$ **83.** $p^2(p + 6)$
85. $(x + 2)^2(x - 1)$ **87.** $\left(x - \frac{1}{2}\right)\left(\frac{1}{2}x - 1\right)$
89. $\left(r + \frac{1}{2}\right)\left(\frac{3}{5}r + 1\right)$ **91.** $\frac{1}{2}(s + 2)(s - 2)$
93. $(x + 0.2)(x + 1.1)$ **95.** 5 or 7 **97.** 0 or 8
99. 4 or -6 **101.** 9 or -16 **103.** $2x^2 - 3x - 2$
105. $6p^2 + 13p + 6$ **107.** $4y^2 + 6y - 40$
109. $-8x^3 - 18x^2 + 5x$ **111.** $-24y^4 - 28y^3 + 96y^2$

Exercise Set 7.3
1. $(2y + 1)(y + 2)$ **3.** $(5x - 8)(x + 1)$ **5.** $(3x - 2)(x - 2)$
7. $(2s - 3)(x + 2)$ **9.** prime **11.** $(8y - 1)(y + 6)$
13. $(3 - 4x)(2 - x)$ **15.** prime **17.** $(11b + 2)(b - 1)$
19. $(2n + 5)(2n + 1)$ **21.** $-1(2y + 3)(y + 1)$
23. $-2(p - 5)(p - 1)$ **25.** $25m^2(25 - m)$
27. $-2(21r + 17)(r - 2)$ **29.** $25x^2(9 + 16x)$
31. $2(2y + 5)^2$ **33.** $-2xy(5x - 1)(6x + 1)$
35. $4(7 - 4d)(1 + 2d)$ **37.** $(2y - 7)^2$
39. $-4x(x + 2)(x + 1)$ **41.** $-1(30a^2 + a + 20)$ or prime
43. $4(4a + 7)(a - 2)$ **45.** $2y(4y + 1)(y + 3)$
47. $2x(5x - 13y)(3x + 5y)$ **49.** $(8y + 5z)^2$
51. $26m^4n^4(m + 4n)$ **53.** $2st(8s - 9t)(5s + 4t)$

55. prime **57.** $-10x^3y(11x - 3)(x - 1)$
59. $(3x + y)(2x - 3y)$ **61.** $(5y - 2)(3y - 7)$ **63.** prime
65. $(4x + 7y)(2x - 5y)$ **67.** $(10a + b)(a - 4b)$
69. $(6x + 5y)(4x - 3y)$ **71.** $2xy(4x + 3y)(4x - 5y)$
73. $4ab(3a - 2b)^2$ **75.** $-2y^2(5y + 2)(y - 7)$
77. $-4x^3(10x^2 + x - 5)$
79. $(-1, 0); (0, -4); (1, -6); \left(\dfrac{3}{2}, -\dfrac{25}{4}\right); (2, -6);$
$(3, -4); (4, 0)$

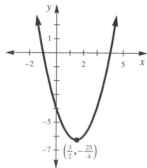

81. $(-3, 2); (-2, 0); (-1, 0); (0, 2); (1, 6); \left(-\dfrac{3}{2}, -\dfrac{1}{4}\right)$

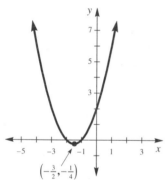

83. $(-5, -9); (-4, -4); (-3, -1); (-2, 0); (-1, -1); (0, -4);$
$(1, -9)$

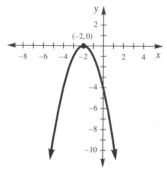

85. $x^2 + 2x + 1$ **87.** $x^2 - 1$ **89.** $y^2 - 10y + 25$
91. $4x^2 - 4xy + y^2$ **93.** $25x^2 - 4y^2$ **95.** $36m^2 - n^2$
97. $-16s^2 - 40st - 25t^2$ **99.** $x^3 + 3x^2 + 3x + 1$
101. $r^3 - 9r^2 + 27r - 27$ **103.** $x^3 + 3x^2y + 3xy^2 + y^3$

Exercise Set 7.4

1. $(x + 1)(x - 1)$ **3.** $(y + 4)(y - 4)$ **5.** $(t + 5)(t - 5)$
7. $(r + 8)(r - 8)$ **9.** $(p + 11)(p - 11)$
11. $3(y + 3)(y - 3)$ **13.** $2(5 + t)(5 - t)$
15. $5y(y + 1)(y - 1)$ **17.** $(a + b)(a - b)$
19. $5(a + 3b)(a - 3b)$ **21.** $(y^2 + 4)(y + 2)(y - 2)$
23. $(x^2 + y^2)(x + y)(x - y)$ **25.** $(4 + s^2)(2 + s)(2 - s)$
27. $2xy(8y^2 - 13x^2)$ **29.** prime **31.** $(y + 2)^2$
33. $(b - 4)^2$ **35.** $(r + 7)^2$ **37.** $(s - 9)^2$ **39.** $(12 + p)^2$
41. $(8 - q)^2$ **43.** prime **45.** $3(y + 4)^2$ **47.** $3(b + 5)^2$
49. $(m + 4n)^2$ **51.** $(3p + 2)^2$ **53.** $(5p - 6q)^2$
55. $(y + 3)^2(y - 3)^2$ **57.** $(3x^2 - 1)^2$ **59.** $(6x^2 + 7y^2)^2$
61. $(c + d)(x - y)$ **63.** $(y + 2)(y - 2)(y - 3)$
65. $(x + 1)(x - 1)(x + 7)$ **67.** $(x + c)(y + 2)$
69. $(r - t)(x - s)$ **71.** $(r + 1)(r - 1)(z + x)$
73. $(3cd - 1)(ab - 1)$ **75.** $(2x - 1)(5y - 1)$
77. $(y + 1)(y^2 - y + 1)$ **79.** $(y - 3)(y^2 + 3y + 9)$
81. $(b + 5)(b^2 - 5b + 25)$
83. $(4p - q)(16p^2 + 4pq + q^2)$
85. $3(a - 2b)(a^2 + 2ab + 4b^2)$
87. $-3(2p + 3q)(4p^2 - 6pq + 9q^2)$ **89.** $-7(2x + 1)$
91. $12a$ **93.** $-(2y + 1)$ **95.** $-20x$ **97.** See text.
99. Multiply the right-hand side.
101. There are no polynomials with a product of $x^2 - 5x + 3$ if they are restricted to integers only.

Exercise Set 7.5

1. $3(x + 1)(x - 1)$ **3.** $xy(y + 3x)(y - 3x)$
5. $(y - 4)(y + 1)$ **7.** $(a + 2b)(a^2 - 2ab + 4b^2)$
9. $(6 + 7k)(6 - 7k)$ **11.** $(b - 2)(b^2 + 2b + 4)$
13. $y^2(y - 1)$ **15.** $(x + 5)(x - 3)$ **17.** prime
19. $2(5 + 3y)(1 - 3y)$ **21.** $(x^2 + 9)(x + 3)(x - 3)$
23. $(5 + 2a)(25 - 10a + 4a^2)$ **25.** $(a - b)(a + b + 5)$
27. $3(3x - 5)(x + 2)$ **29.** $(xy + 1)(x^2y^2 - xy + 1)$
31. $(a + 7)(b + 3)$ **33.** $2(3 - 8x)(1 + 2x)$
35. $2(2q + 3)^2$ **37.** $5(p^2 + p - 1)$
39. $(a^2 - 10)(a + 3)$ **41.** $x(x + 5)(x - 5)$
43. $5(b - c)(b^2 + bc + c^2)$ **45.** $6ab(3a + 2b + 5)$
47. $(8p - 1)(p - 3)$ **49.** $(5 - p)(6 - p)$
51. $-2c(c + 5)(c - 1)$ **53.** $xy(y + 2)(y - 2)$
55. $-2(t + 9)(t - 8)$ **57.** $2(10 + 3a)(3 - a)$
59. $(a - b)(rs + t)$ **61.** $(x + 13)(x - 13)$ **63.** prime
65. $-3(t + 5)(t - 4)$ **67.** $(4p + 7q)^2$
69. $8(1 + m)(1 - m)$ **71.** $(r - 3)(r^2 + 3r + 9)$
73. $y(y - 2)(y^2 + 2y + 4)$ **75.** prime
77. $6x(9 - x)(1 - x)$
79. $(x + 1)(x^2 - x + 1)(a + b)(a^2 - ab + b^2)$
81. $(3y - 2x)^2$ **83.** $\left(x + \dfrac{1}{3}\right)\left(x - \dfrac{1}{3}\right)$
85. $\left(\dfrac{1}{2}a + \dfrac{4}{3}b\right)\left(\dfrac{1}{4}a^2 - \dfrac{2}{3}ab + \dfrac{16}{9}b^2\right)$
87. $(0.4x + 0.5y)(0.4x - 0.5y)$

89. $(0.7 + 0.9t)(0.7 - 0.9t)$ 91. 6   93. 12   95. 2
97. 8   99. 9

Exercise Set 7.6

1. $\{-3, 1\}$ 3. $\{-6, 2\}$ 5. $\{0, 5\}$ 7. $\left\{-\frac{6}{5}, \frac{3}{2}\right\}$
9. $\left\{-7, 0, \frac{1}{2}\right\}$ 11. $\left\{0, \frac{3}{8}, \frac{9}{5}\right\}$ 13. $\left\{-\frac{3}{2}, 0, 5\right\}$
15. $\left\{-4, \frac{1}{7}, \frac{3}{2}\right\}$ 17. $\{-8, 1\}$ 19. $\{-2, 6\}$ 21. $\{-2, -1\}$
23. $\{2, 3\}$ 25. $\{-2, 3\}$ 27. $\{-8, 3\}$ 29. $\{-1, 3\}$
31. $\{-2, 4\}$ 33. $\{-4, -2\}$ 35. $\{1, 11\}$ 37. $\left\{-\frac{1}{2}, \frac{2}{3}\right\}$
39. $\left\{-\frac{1}{2}, \frac{5}{3}\right\}$ 41. $\{-3, 3\}$ 43. $\{-1, 0, 1\}$
45. $\left\{-3, \frac{4}{5}, 1\right\}$ 47. $\{1\}$ 49. $\{-2, 0, 4\}$
51. $\left\{-\frac{7}{2}, 0, \frac{7}{2}\right\}$ 53. $\left\{-12, \frac{11}{2}\right\}$ 55. $\{3, 5\}$ 57. $\{0, 5\}$
59. $\{0\}$ 61. $\{-2, 3\}$ 63. $\{-1, 3\}$ 65. $\left\{-\frac{7}{3}, \frac{7}{3}\right\}$
67. $\left\{-\frac{5}{3}, 0\right\}$ 69. $\{-5, 6\}$ 71. $\left\{-\frac{1}{3}, -\frac{1}{4}, 0\right\}$
73. $\{-6, -3, 2, 5\}$ 75. 8 and 9
77. 0 and 5 or -5 and 0   79. $\frac{1}{2}$ and $\frac{7}{2}$
81. $w = 4$ in., $l = 6$ in.   83. 3 sec

Exercise Set 7.7

1. 7 and 9 or -7 and -9 3. 8 and 10
5. 11 and 13 or -11 and -13
7. $w = 6$ m, $l = 10$ m, $P = 32$ m
9. the side of the first square is 25 yd, the area is 625 yd^2; the side of the second square is 35 yd, the area is 1225 yd^2
11. $p = \$20$ 13. $t = 2$ sec and $t = 5$ sec
15. 21 rows and 25 trees/row   17. 5 ft wide
19. the tank is 10 cm by 10 cm by 4 cm; the piece of sheet metal is 18 cm by 18 cm
21. $h = 4$ in.   23. $a = 3$ m, $b = 4$ m, $c = 5$ m
25. 5 mph   27. 3, 4, and 5   29. $\frac{4}{3}$, 1, and $\frac{5}{3}$
31. 1 and 12 or -1 and -12 33. $w = 3$ ft, $l = 3$ ft
35. 8 m   37. 24 ft   39. 2 cm and 3 cm

Review Exercises

1. $2 \cdot 2 \cdot 17$ 2. prime   3. $3 \cdot 3 \cdot 5 \cdot 47$ 4. $7 \cdot 13$
5. 7   6. $5x$ 7. 2   8. $3xy^2$ 9. $6(3x^2 - 6x + 10)$
10. $-9ab(3a - 7b + 9)$ 11. $15s^3(3s^2 - s + 2)$
12. prime   13. $(x + 9)(x - 9)$ 14. prime

15. $4(2m + 3n)(2m - 3n)$ 16. $(x - 3)^2$ 17. $(y + 7)^2$
18. $(t^2 + 1)(t + 1)(t - 1)$ 19. $-(4 - s)^2$
20. $(3y + 2)(3y - 2)^2$ 21. $(x + y)(a - b)$
22. $-(x + 1)(x - 1)^2$ 23. $(m^2 + 9)(m + 3)(m - 3)$
24. $(x - 8)^2$ 25. prime   26. prime   27. $(x + 6)(x - 1)$
28. $(y - 6)(y - 4)$ 29. $3(t + 7)(t - 3)$
30. $2(z^2 - 6z + 4)$ 31. $2(x - 2y)(x - 3y)$
32. $(y + 2)(y - 2)^2$ 33. $6p^2(p - 5)(p + 2)$
34. $2x(3x - 5y)(x - 2y)$ 35. $(z - 2)(z^2 + 2z + 4)$
36. $(m + 3)(m^2 - 3m + 9)$ 37. $(x + 2y)(x + b)$
38. $(a + d)(b + c)(b - c)$ 39. $\left(8x + \frac{1}{5}\right)\left(8x - \frac{1}{5}\right)$
40. $\left(6y + \frac{z}{7}\right)\left(6y - \frac{z}{7}\right)$ 41. $(0.5x - 0.2)(0.5x - 0.2)$
42. $(0.9m + 0.4n)(0.9m - 0.4n)$
43. $x^5(x + 10)(x - 10)$ 44. prime
45. $3(3a - 5b)(a - 2b)$ 46. $(2p + 3q)(4p^2 - 6pq + 9q^2)$
47. $-(9 + 12x + 16x^2)$ 48. prime   49. $(a + b)^3(a - b)^3$
50. $(11x + 12y)(11x - 12y)$ 51. $\{0, 2\}$ 52. $\{-3, 4\}$
53. $\left\{\frac{5}{2}, \frac{2}{3}\right\}$ 54. $\{-6, 0\}$ 55. $\left\{-\frac{7}{2}, 0, 5\right\}$
56. $\left\{-\frac{1}{4}, \frac{3}{2}\right\}$ 57. $\{0, 7\}$ 58. $\{\pm 4\}$ 59. $\{-4\}$
60. $\{-2, 5\}$ 61. $\{-2, 0, 4\}$ 62. $\left\{-4, -\frac{3}{2}\right\}$
63. $\{-3, 7\}$ 64. $\left\{\frac{1}{2}, -\frac{2}{5}\right\}$ 65. $\left\{-2, \frac{1}{2}, \frac{3}{4}\right\}$
66. $\left\{\pm\frac{1}{10}\right\}$ 67. $\{-7, 0, 7\}$ 68. $\left\{-\frac{5}{3}, \frac{4}{7}\right\}$
69. $\left\{-\frac{1}{3}, 4\right\}$ 70. $\left\{\pm\frac{5}{3}\right\}$ 71. 12 and 13
72. 14 and 16   73. -11 and -13 74. 6 and -6
75. $l = 16$ m, $w = 11$ m, $P = 54$ m
76. the hypotenuse is 17 ft, $P = 40$ ft, $A = 60$ ft^2

Chapter Test

1. $x(x - 3)$ 2. $5x^3(x + 6)$ 3. $-3y^2(3y^2 - 4y + 7)$
4. $-7a^3b^4(3b^2 - 9a + 11a^2 - 12ab)$ 5. $(3x - 5)^2$
6. $(p + 2)(p - 2)$ 7. $(7y + 8z)^2$
8. $(x - 4)(x^2 + 4x + 16)$ 9. $3(y + 2z)(y^2 - 2yz + 4z^2)$
10. $(5x - 3)(x + 1)(x - 1)$ 11. $2(4a^2 + 6a - 5)$
12. $3(x + 3)(x - 3)$ 13. $2(m + 6)^2$ 14. prime
15. $x(6x + 7y)^2$ 16. $5(y - 12z)(y - 2z)$
17. $y(8y^2 - 5xy + 13x^2)$ 18. $(x^2 + 4)(x + 2)(x - 2)$
19. $-3(a + 45)(a + 1)$ 20. $16(b + 5c)(b - 5c)$
21. prime   22. $-15(x + 2)(x - 2)$
23. $2s(5t + 3s)(13t - 5s)$ 24. $-(5x + 2y)(2x + 13y)$
25. $\left\{-\frac{3}{2}, 0, 1\right\}$ 26. $\left\{-\frac{7}{4}, 2\right\}$ 27. $\{0, 2, 4\}$
28. $\{\pm 2, \pm 3\}$ 29. $\{1\}$ 30. -9 and -8 31. 23

32. $w = 15$ m, $l = 30$ m
33. the sides have length, 3, 4, and 5, $P = 12$, $A = 6$; or the sides have length 8, 15, and 17, $P = 40$, $A = 60$

Chapter 8

Exercise Set 8.1

1. $\frac{1}{2}$ **3.** $\frac{13}{21}$ **5.** $-\frac{3}{4}$ **7.** in lowest terms **9.** $\frac{7}{8}$
11. $\frac{3}{2}, x \neq 0$ **13.** $-\frac{4s}{t}, s \neq 0, t \neq 0$
15. $\frac{3n^2}{10m}, m \neq 0, n \neq 0$ **17.** $-\frac{1}{4y^3}, x \neq 0, y \neq 0$
19. $-a, a \neq 0$ **21.** $\frac{1}{9}$ **23.** $2, x \neq -1$
25. $\frac{x+1}{x-1}, x \neq -1, 1$ **27.** $-\frac{5}{4}, x \neq 0, 2$
29. in lowest terms, $a \neq -b$ **31.** $\frac{x}{2}, x \neq 1$
33. $-(x^2 + xy + y^2)$ **35.** $\frac{a-1}{a+5}$ **37.** $\frac{2y-1}{2}$
39. $\frac{-3}{x+4}$ **41.** $\frac{4(p+2)}{3}$ **43.** $\frac{-1(a+b)}{a^2+ab+b^2}$ **45.** -1
47. in lowest terms **49.** $\frac{a+1}{a-1}$ **51.** in lowest terms
53. $\frac{x-4y}{x-y}$ **55.** -2 **57.** $x+1$ **59.** $\frac{x+4}{x+3}$ **61.** $\frac{1}{5-a}$
63. $\frac{1}{3s^2}$ **65.** in lowest terms **67.** $\frac{1}{2}$ **69.** $\frac{8}{5}$ **71.** $\frac{2}{3}$
73. $\frac{4}{3}$ **75.** $\frac{1}{10}$ **77.** $\frac{3}{4}$ **79.** $\frac{23}{35}$ **81.** -2 **83.** $\frac{2}{5}$
85. undefined **87.** 0
89. The quotient of two polynomials
91. Factor the numerator and denominator, then divide out all common factors.

Exercise Set 8.2

1. $\frac{8}{25}$ **3.** 42 **5.** $\frac{x}{4}$ **7.** $\frac{40y}{147}$ **9.** 1 **11.** $-\frac{5}{3}$ **13.** $\frac{3}{2}$
15. $-\frac{3t}{2s^2}$ **17.** $\frac{2ac}{5b}$ **19.** $\frac{15ay}{8x}$ **21.** $-\frac{2y^4}{3a}$ **23.** $\frac{ab}{2cx^2}$
25. -1 **27.** $\frac{81x^2b}{100a^2}$ **29.** $\frac{135}{512ab}$ **31.** $\frac{2(x-3)}{(x+2)^2}$
33. $\frac{1}{a-b}$ **35.** 2 **37.** $\frac{3ary}{7s^2}$ **39.** $\frac{1}{3x^2(a+1)}$
41. $\frac{4x(x+y)}{3(2x+3y)}$ **43.** $\frac{b^4c^4n^2p^2}{m^4a^2}$ **45.** 1
47. $\frac{x-3}{(x+3)(x-4)}$ **49.** $\frac{5(3x+4)(x+4)}{9(5x^2+4)}$ **51.** 1

53. $\frac{2t^2(t-9)}{t-3}$ **55.** $\frac{1}{2}$ **57.** undefined **59.** 0
61. $\frac{5}{24}$ **63.** $\frac{19}{5}$ **65.** $\frac{13}{7}$ **67.** $\frac{5}{18}$ **69.** $\frac{1}{20}$ **71.** $\frac{7}{9}$
73. $\frac{1}{20}$ **75.** $-\frac{1}{8}$ **77.** $\frac{21}{5}$ **79.** $\frac{ad+bc}{bd}$
81. $x + 2 + \frac{-1}{x+2}$ **83.** $x^4 + x^3 + x^2 + x + 1$
85. $y^2 + 4y + 7 + \frac{9}{y-1}$
87. Factor both the numerators and denominators and divide out common factors; then multiply.
89. Answers may vary.

Exercise Set 8.3

1. 8 **3.** 15 **5.** $90x$ **7.** $36x^2y^2$ **9.** $504xy$
11. $2(x+1)$ **13.** $(y+2)(y-2)(y+1)$
15. $a(a-3)(a-2)(a+2)$ **17.** $\frac{9}{4}$ **19.** $\frac{5x}{4}$ **21.** $\frac{1}{n}$
23. $s-1$ **25.** $\frac{m+3}{7st}$ **27.** $-\frac{1}{ab}$ **29.** 1 **31.** -2
33. $\frac{4}{x-2}$ **35.** $\frac{-3}{y-3}$ **37.** $\frac{4-3x}{3(x+4)}$ **39.** $\frac{a^2+3a+1}{a(a+1)}$
41. $\frac{3y-5}{2(y+2)(y-2)}$ **43.** $\frac{-2}{(x-1)^2(x+1)}$
45. $\frac{8p+88}{(p+16)(p+8)(p+1)}$ **47.** $\frac{2(2x^2+2x-1)}{(x+1)^2(x+2)}$
49. $\frac{4n^2-15n-7}{(n+2)(n-4)}$ **51.** $\frac{-y-5}{y(y+3)}$ **53.** $\frac{a^2-3a+14}{2(a-3)}$
55. $\frac{2}{n-1}$ **57.** $\frac{2n+7}{(n+2)(2n-3)}$ **59.** $\frac{3}{p+3}$
61. $\frac{6}{5(t-1)}$ **63.** $\frac{9x^3+22x^2-4x-30}{x^2(x-5)(x+3)}$
65. $\frac{-4a^3+20a^2+15}{a(a-5)}$ **67.** $\frac{n+11}{(n+3)(n-2)}$ **69.** $\frac{6}{5}$
71. $\frac{5}{9}$ **73.** 5 **75.** $\frac{1}{2}$ **77.** $\frac{x-2}{x+1}$ **79.** 1
81. Add the two numerators and place the sum over the same denominator.
83. Factor the numbers into primes, then multiply each distinct prime factor the greatest number of times it occurs in any of the expressions.

Exercise Set 8.4

1. $\frac{2}{3}$ **3.** $\frac{2}{3}$ **5.** -4 **7.** $-\frac{35}{32}$ **9.** $\frac{y}{2x}$ **11.** $\frac{5x}{3}$
13. $\frac{7x^2}{4}$ **15.** $-\frac{1}{2a}$ **17.** $\frac{13}{2s}$ **19.** $\frac{8x}{7y}$ **21.** $\frac{11}{4b}$ **23.** $\frac{1}{2}$
25. $-\frac{18}{65}$ **27.** $\frac{5}{79}$ **29.** $\frac{1-5x}{1+3x}$ **31.** n **33.** $\frac{x}{x+1}$
35. $\frac{7(a^2+3)}{a(a^2+2)}$ **37.** $\frac{3p+1}{p}$ **39.** $-\frac{3}{xy}$ **41.** $\frac{n(5m+2n)}{m(7m^2-2n^3)}$

43. $\dfrac{3s+4}{s(4-s)}$ 45. $\dfrac{a-2}{a-3}$ 47. $\dfrac{(2t+1)(t-5)}{(4t-3)(t-7)}$ 49. $\dfrac{x+y}{x-y}$
51. $\dfrac{a^2-a-14}{a^2-3a}$ 53. 1 55. $\dfrac{rR}{r+R}$ 57. $\left\{\dfrac{5}{3}\right\}$
59. $\left\{\dfrac{42}{19}\right\}$ 61. $\{-3, 0, 2\}$ 63. $\{\pm 7\}$ 65. $\{-3, 0, 4\}$
67. $\{-2, 2, 3\}$

Exercise Set 8.5

1. $\left\{\dfrac{3}{2}\right\}$ 3. $\{-5\}$ 5. $\{-1\}$ 7. $\left\{\dfrac{8}{5}\right\}$ 9. $\{1\}$
11. $\left\{\dfrac{35}{3}\right\}$ 13. $\left\{-\dfrac{44}{5}\right\}$ 15. $\{5\}$ 17. $\left\{\dfrac{28}{11}\right\}$
19. $\varnothing$; $x \neq 0$ 21. $\{-4\}$ 23. $\varnothing$; $p \neq 2$ 25. $\varnothing$; $k \neq \dfrac{1}{2}$
27. $\left\{\dfrac{19}{7}\right\}$ 29. $\{2\}$ 31. $\{-2\}$ 33. $\{1\}$
35. True for all x except $x = -1$ 37. $\{8\}$ 39. $\dfrac{9}{2}$
41. 36 43. $w = 7, l = 42$ 45. $9\dfrac{1}{3}$ hr
47. 24 km/hr, 64 km/hr 49. 3 51. $53\dfrac{1}{3}$ hr
53. at 7 P.M. 55. $\left\{3, -\dfrac{1}{3}\right\}$ 57. $\{-2, -5\}$ 59. $\{4, 5\}$
61. $P_2 = \dfrac{P_1 V_2}{V_1}$ 63. $M_1 = \dfrac{d^2 F}{GM_2}$ 65. $a = \dfrac{2b-3}{3b-1}$
67. $y = \dfrac{b(x-h)}{a} + k$ 69. $x = \dfrac{2y+1}{3-y}$
71. (a) $p = \dfrac{fq}{q-f}$ (b) $\dfrac{28}{13}$ cm

Exercise Set 8.6

1. $\dfrac{2}{1}$ 3. $\dfrac{3}{14}$ 5. $\dfrac{2}{3}$ 7. $\dfrac{2}{13}$ 9. $\dfrac{19}{6}$ 11. $\dfrac{3}{5}$ 13. $\dfrac{5}{4}$
15. $\dfrac{5}{24}$ 17. $\dfrac{7}{4}$ 19. $\dfrac{1}{5}$ 21. $\dfrac{5}{4}$ 23. $\{2\}$ 25. $\{72\}$
27. $\left\{\dfrac{49}{5}\right\}$ 29. $\left\{\dfrac{27}{2}\right\}$ 31. $\left\{\dfrac{68}{3}\right\}$ 33. $\left\{-\dfrac{9}{20}\right\}$
35. $\left\{-4, \dfrac{1}{2}\right\}$ 37. $\{-1, 9\}$ 39. $\{-2, 1\}$ 41. $\left\{-\dfrac{2}{3}\right\}$
43. 500 miles 45. 900 ml 47. $472
49. Dennis: $4,827; Dean: $8,045 51. 60 lb
53. 680 shirts
55. $\dfrac{7}{6}$ cups of flour, $\dfrac{4}{3}$ teaspoons of baking powder, $\dfrac{2}{3}$ of a tablespoon of sugar, 4 tablespoons of butter, 1 cup of milk
57. 432 cm^3 59. 6 in. 61. $\dfrac{24}{5}$ qt 63. $x = 8, y = 10$
65. $x = \dfrac{44}{3}, y = \dfrac{64}{3}$ 67. $x = 8, y = 10$
69. When written in the form $a : b = c : d$, the quantities a and d are the extreme (outside) numbers, and b and c are the mean (middle) numbers.

Exercise Set 8.7

1. $C = kd$ 3. $d = krt$ 5. $y = \dfrac{k}{x}$ 7. $V = ks^2$
9. $T = \dfrac{k}{\sqrt{n}}$ 11. $y = \dfrac{kx}{z^2}$ 13. $y = kx$; $k = 4$
15. $M = kab$; $k = 4$ 17. $S = \dfrac{k}{t^2}$; $k = 50$
19. $y = \dfrac{ku}{v}$; $k = 14$ 21. 28 23. 3 25. 288 27. 28
29. 14 31. 114 33. 512 ft 35. 422.4 lb
37. $\dfrac{144}{5}$ ft 39. $\dfrac{13}{2}$ in. 41. 36π ft^3 43. 4.5%
45. P varies jointly with m and n.
47. V varies directly with x^2 and inversely with y^3 and z.
49. V varies jointly with r^2 and h.

Review Exercises

1. -5 2. $\dfrac{x^2}{6}$ 3. $-\dfrac{63}{10}$ 4. $-\dfrac{48x^2}{7}$ 5. $\dfrac{2}{7a^2}$ 6. $\dfrac{4a^2 x}{3y}$
7. $\dfrac{y(y+3)}{y-4}$ 8. $\dfrac{48}{a+b}$ 9. $\dfrac{x+2}{x-2}$ 10. $200(y^2 - 16)$
11. -1 12. $\dfrac{(x+2)(5x-1)}{3x-4}$ 13. $\dfrac{1}{x+2}$ 14. $\dfrac{135}{512xy}$
15. $\dfrac{(x-4)(x-1)}{x+4}$ 16. $\dfrac{x-y}{2x-y}$ 17. $\dfrac{1}{3x}$ 18. $x - y$ 19. $\dfrac{7}{5}$
20. $\dfrac{2x+1}{5x}$ 21. $\dfrac{4x-9}{3y-4}$ 22. $\dfrac{-2x-1}{x^2-9}$ 23. $\dfrac{3x-11}{x(x-4)}$
24. $\dfrac{8x+8}{(x-3)(x+2)}$ 25. $\dfrac{2x}{(x-4)^2(x+4)}$ 26. $\dfrac{22-x}{x(x-7)}$
27. $\dfrac{-x^3 + 20x^2 - 30x - 5}{5x}$ 28. $\dfrac{11}{x^2-9}$ 29. 2 30. $\dfrac{3}{5}$
31. $\dfrac{1}{3y}$ 32. $\dfrac{22}{27}$ 33. $\dfrac{1-6y}{2+2y}$ 34. $-y$ 35. $\dfrac{x^2}{x+7}$
36. $\dfrac{5m+1}{m}$ 37. $\dfrac{11}{xy}$ 38. $\dfrac{a+b}{a}$ 39. $\dfrac{t^2-2}{t^2-2t}$ 40. 1
41. $\left\{\dfrac{5}{2}\right\}$ 42. $\left\{-\dfrac{9}{20}\right\}$ 43. $\left\{\dfrac{7}{2}\right\}$ 44. $\varnothing$ 45. $\varnothing$
46. $\left\{-\dfrac{9}{2}\right\}$ 47. $\{-11\}$ 48. $\left\{\dfrac{10}{7}\right\}$ 49. $\varnothing$ 50. $\{3\}$
51. $\varnothing$ 52. $\dfrac{38}{49}$ 53. $\dfrac{7}{11}$ 54. $\dfrac{143}{4}$ mph 55. $\left\{\dfrac{35}{9}\right\}$
56. $\left\{\dfrac{8}{5}\right\}$ 57. $\left\{-\dfrac{22}{21}\right\}$ 58. $\left\{-\dfrac{51}{83}\right\}$ 59. $\{-2, 1\}$
60. $\{-1, 9\}$ 61. $y = \dfrac{15}{4}$ 62. $A = 169\pi$ ft^2

63. $h = \frac{48}{11}$ m   **64.** $\frac{1}{3}x + 6 = 5; \{-3\}$
65. $\frac{3+x}{4x} = \frac{2}{3}; \left\{\frac{9}{5}\right\}$ **66.** $\frac{1}{5} + \frac{1}{x} = \frac{2}{5}$; 5 hrs
67. $45(x + 2) = 60x$; 3:00 P.M.

Chapter Test

1. $\frac{3}{5}$ **2.** $-\frac{9x}{24}$ **3.** $\frac{2}{3a-4}$ **4.** $\frac{3(x-2)}{2(x+2)}$ **5.** $\frac{-1}{m+7}$
6. $\frac{3x-2y}{3x+4y}$ **7.** $3p - 1$ **8.** $5a - 8b - 1$ **9.** $y + 1$
10. $\frac{4x^2+2x-11}{2x-3}$ **11.** $\frac{1}{2}$ **12.** $\frac{2x}{x+1}$ **13.** 1   **14.** 1
15. $\frac{32-3x}{12y}$ **16.** 3   **17.** 4   **18.** $\frac{9x}{(x-6)(x-1)(x+2)}$
19. $\frac{2x+x^3}{3+5x^2}$ **20.** $x + 1$ **21.** $\{22\}$ **22.** $\varnothing$ **23.** 60 min
24. 300 km/hr   **25.** $\frac{5}{3}$ **26.** $\frac{11}{5}$ **27.** $58\frac{1}{3}$ mph   **28.** $\frac{2}{5}$
29. 210 grade points   **30.** $90
31. $V = khr^2$, $k = \pi$, $V = 27\pi$ in³
32. $b = \frac{kA}{h}$, $k = 2$, $b = 24$ m

CHAPTER 9

Exercise Set 9.1

1. 2   **3.** 4   **5.** 1   **7.** 9   **9.** 7   **11.** -3 **13.** -10
15. 2   **17.** 4   **19.** -1 **21.** -3 **23.** -1 **25.** 4
27. not a real number   **29.** not a real number
31. not a real number   **33.** $\frac{2}{3}$ **35.** $-\frac{4}{5}$ **37.** $\frac{1}{2}$ **39.** $\frac{1}{2}$
41. x **43.** x^2 **45.** t^6 **47.** x **49.** $-x^3$ **51.** x^6
53. $-m^8$ **55.** x^2 **57.** $-m^6$ **59.** y^2 **61.** $9p^3$
63. $3a^{17}$ **65.** $7y$ **67.** $-11t^2$ **69.** $2xy$ **71.** $9m^2n^2p$
73. $\frac{3x}{4y}$ **75.** $\frac{2x}{y}$ **77.** $\frac{2t^2}{s}$ **79.** $x + 1$ **81.** $x + 2$
83. $x + 5$ **85.** 1.732   **87.** 3.536   **89.** -1.414
91. 2.933   **93.** 2.179   **95.** 9.203   **97.** 5   **99.** 12 ft
101. The cube root of 8   **103.** The positive square root
105. an irrational number

Exercise Set 9.2

1. $2\sqrt{2}$ **3.** $2\sqrt[3]{2}$ **5.** $-2\sqrt{7}$ **7.** $-3\sqrt[3]{2}$ **9.** $3\sqrt{5}$
11. $2\sqrt[4]{2}$ **13.** $\frac{\sqrt{6}}{5}$ **15.** $-\frac{4}{7}$ **17.** $12\sqrt{2}$ **19.** $4\sqrt[3]{5}$
21. $50\sqrt{2}$ **23.** 192   **25.** $\frac{11}{8}$ **27.** $-\frac{3}{4}$ **29.** $2\sqrt[5]{2}$
31. $\frac{12\sqrt{2}}{5}$ **33.** $-96\sqrt{2}$ **35.** $7\sqrt{3}$ **37.** 45   **39.** $\frac{2}{5}$
41. $\frac{9}{7}$ **43.** $\frac{\sqrt{11}}{4}$ **45.** $5\sqrt{15}$ **47.** $2\sqrt{2}$ **49.** x
51. $y\sqrt{y}$ **53.** y^2 **55.** $x^2\sqrt[3]{x}$ **57.** p **59.** m
61. $2c\sqrt{2c}$ **63.** $-a\sqrt[3]{b^2}$ **65.** $3xy^2\sqrt{y}$ **67.** $3xy\sqrt[3]{2}$
69. $6\sqrt{2}$ **71.** $\frac{2\sqrt{3}}{3}$ **73.** $2\sqrt[3]{9}$ **75.** 0.4   **77.** 0.2
79. 6   **81.** 2   **83.** $\frac{\sqrt[3]{2}}{5}$ **85.** $\frac{-\sqrt[3]{5}}{4}$ **87.** $\frac{2\sqrt[4]{2}}{3}$
89. $\frac{\sqrt{3}}{2x}$ **91.** $\frac{2\sqrt[4]{a}}{3b^2c}$ **93.** $\frac{-1}{2a^2}$ **95.** 0.577   **97.** 0.655
99. 2.449   **101.** 2.646   **103.** $2x^3 + x^2 + 2x + 3$
105. $x^2 - 2x + 1$ **107.** $-16y^2 + 16y + 6$ **109.** 8.602 ft
111. 5 cm   **113.** 6.403 yd   **115.** 4.507 m   **117.** 3 hr

Exercise Set 9.3

1. $15\sqrt{2}$ **3.** $2\sqrt{7}$ **5.** $-2\sqrt{11}$ **7.** $16\sqrt{y}$ **9.** $-\sqrt{m}$
11. $18 + 4\sqrt{7}$ **13.** $-2\sqrt{5}$ **15.** $6\sqrt{y}$
17. $3\sqrt{5} - 2\sqrt{7}$ **19.** $2\sqrt{a} + 2\sqrt{b}$
21. $\sqrt{3} - 4\sqrt{2} + \sqrt{7}$ **23.** $2\sqrt{x}$ **25.** $-3\sqrt{7}$
27. $17\sqrt{2} - 14\sqrt{3}$ **29.** $\sqrt{5}$ **31.** $\sqrt{3}$ **33.** $\frac{37\sqrt{11}}{30}$
35. $-2\sqrt[3]{6}$ **37.** $-\sqrt[4]{3}$ **39.** $2\sqrt[3]{3} - 5\sqrt{3}$
41. $2\sqrt{7} - \sqrt[4]{7}$ **43.** $2y\sqrt{2y}$ **45.** $2\sqrt{p} - \sqrt{pt}$
47. 0.9   **49.** -0.09 **51.** $-5\sqrt{11}$
53. $-6\sqrt{3} - 6\sqrt{5} + 10\sqrt{10}$ **55.** $8y\sqrt{3y}$ **57.** $2x\sqrt{2x}$
59. $\frac{28\sqrt{2y}}{45}$ **61.** $53ab\sqrt{6a}$ **63.** $23\sqrt{t} - 12\sqrt{3t}$
65. $\frac{-2\sqrt{6}+8}{15}$ **67.** $\frac{5\sqrt[3]{3}-\sqrt[3]{5}}{6}$ **69.** $11\sqrt[3]{7y}$
71. $16y\sqrt[3]{5} - 2\sqrt[3]{4y^2}$ **73.** $1 + \sqrt{3}$ **75.** $2 - \sqrt{2}$
77. $5 - 4\sqrt{3}$ **79.** 1   **81.** $-2 - \sqrt{2}$
83. $-28x^3y^2 + 32x^2y^3 - 36xy^4$ **85.** $4x^2 + 28x + 49$
87. $16y^2 - 9$ **89.** $-20y^3 + 245y$
91. $16x^4y^2 + 40x^3y^3 + 25x^2y^4$ **93.** $2x + 3$ **95.** 1.866
97. -1.235 **99.** 0.191   **101.** 1.335   **103.** 4 quarters
105. 4 hr   **107.** 500 mph

Exercise Set 9.4

1. $\sqrt{6}$ **3.** $2\sqrt{3}$ **5.** $32\sqrt{5}$ **7.** -18 **9.** -200
11. $2\sqrt{3} - 10$ **13.** $\sqrt{15} + 2\sqrt{3}$ **15.** $\sqrt{70} - 3\sqrt{7}$
17. $\sqrt{30} - \sqrt{42}$ **19.** $6\sqrt{6} + 15\sqrt{2}$ **21.** $-42 + 15\sqrt{21}$
23. 0   **25.** -22 **27.** $27 - 8\sqrt{11}$ **29.** 42
31. $56 - 12\sqrt{3}$ **33.** $\sqrt{30} - \sqrt{15} + 2\sqrt{3} - \sqrt{6}$ **35.** $\frac{\sqrt{3}}{3}$
37. $\sqrt{3}$ **39.** $\frac{\sqrt{3}}{3}$ **41.** $\frac{\sqrt{2}}{2}$ **43.** $\frac{\sqrt{15}}{3}$
45. $\frac{\sqrt{15}+2\sqrt{3}}{3}$ **47.** $\frac{\sqrt{5}-5}{15}$
49. $\frac{6\sqrt{2}+3\sqrt{6}+2\sqrt{3}+3}{3}$ **51.** $\frac{-(\sqrt{6}+\sqrt{10})}{2}$

53. $\dfrac{29 + 12\sqrt{5}}{11}$ **55.** $\dfrac{-109 + 28\sqrt{15}}{11}$ **57.** x^2 **59.** $y^2\sqrt{y}$
61. $3x\sqrt{x}$ **63.** $-6y$ **65.** $12a^3\sqrt{a}$ **67.** $3\sqrt{x}+6$
69. $x\sqrt{x}-7x$ **71.** $6x-2x\sqrt{x}$ **73.** $x-y$
75. $9m-12\sqrt{mn}+4n$ **77.** $121s-25t$ **79.** $\dfrac{\sqrt{x}}{x}$
81. $\sqrt{y}$ **83.** $\sqrt{3x}$ **85.** $\sqrt{x+y}$ **87.** $\dfrac{x-\sqrt{x}}{x}$
89. $\dfrac{\sqrt{2x}+2x}{2x}$ **91.** $\dfrac{2\sqrt{p}+p\sqrt{2}}{p}$ **93.** $\dfrac{4y+12\sqrt{y}+9}{4y-9}$
95. $\dfrac{6t-23\sqrt{tz}+20z}{9t-16z}$ **97.** yes **99.** no **101.** yes
103. $\{\pm 2\}$ **105.** $\{-1, 5\}$ **107.** $\{2, 6\}$ **109.** $\left\{\dfrac{2}{3}\right\}$
111. $\left\{-\dfrac{3}{4}, 0, \dfrac{3}{4}\right\}$ **113.** -4 **115.** $-\dfrac{5}{3}$
117. $733\dfrac{1}{3}$ gal of the 3% milk; $166\dfrac{2}{3}$ gal of the 30% milk

Exercise Set 9.5

1. $\{81\}$ **3.** $\{32\}$ **5.** $\varnothing$ **7.** $\varnothing$ **9.** $\left\{\dfrac{25}{12}\right\}$ **11.** $\left\{\dfrac{49}{192}\right\}$
13. $\{11\}$ **15.** $\{1\}$ **17.** $\varnothing$ **19.** $\varnothing$ **21.** $\{17\}$ **23.** $\{12\}$
25. $\{8\}$ **27.** $\left\{\dfrac{1}{7}\right\}$ **29.** $\varnothing$ **31.** $\left\{\dfrac{2}{9}\right\}$ **33.** $\{0, 18\}$
35. $\{4\}$ **37.** $\varnothing$ **39.** $\left\{\dfrac{3}{2}\right\}$ **41.** $\varnothing$ **43.** $\{0\}$ **45.** $\{6\}$
47. $\left\{\dfrac{3}{2}\right\}$ **49.** $\left\{-\dfrac{9}{7}\right\}$ **51.** $\{9\}$ **53.** $\left\{-1, -\dfrac{4}{5}\right\}$
55. $\{0, 2\}$ **57.** $\{2\}$ **59.** $\{8\}$ **61.** $\{9\}$ **63.** $\left\{\dfrac{1}{4}\right\}$
65. $\{9\}$ **67.** $\{-2, -5\}$ **69.** $\varnothing$ **71.** $\{\pm 1\}$ **73.** $\{\pm 2\}$
75. $\{-4, 0\}$ **77.** 5 **79.** 17 **81.** $\sqrt{5}$ **83.** $\dfrac{\sqrt{73}}{4}$
85. $\sqrt{30}$ **87.** $a^2 = C^2 - b^2$ **89.** $s = \dfrac{T^2 g}{2}$
91. $n = \dfrac{s^2}{E^2}$ **93.** $E = C^2 m$ **95.** 45 **97.** 0 or 1
99. -1 or 3 **101.** 4 and 5 **103.** $T = 17.763$ sec
105. $A = 3.678$ in². **107.** No, since $\sqrt{x} \geq 0$.
109. Yes, since $\sqrt[3]{x^3} = x$ for all real values of x.
111. An extraneous solution is one that does not satisfy the original equation.

Exercise Set 9.6

1. $\sqrt{4}$ **3.** $\sqrt{16}$ **5.** $\sqrt[4]{3}$ **7.** $-\sqrt[3]{8^2}$ **9.** $\sqrt{x}$
11. $4\sqrt{x}$ **13.** $x^{1/2}$ **15.** $7^{1/3}$ **17.** $-8^{2/3}$ **19.** $27^{3/4}$
21. $-x^{1/2}$ **23.** $x^{2/3}$ **25.** $4y^{3/5}$ **27.** $-8x^{1/2}y$ **29.** $x^{3/4}y^{7/4}$
31. 2 **33.** 2 **35.** -4 **37.** 8 **39.** -8 **41.** 4

Review
1. 7 **2.** -6
6. not a real number
11. 0.2 **12.** 0.2 **13.** x
17. $-x$ **18.** x **19.** $-y$
22. $2\sqrt{14}$ **23.** $2x\sqrt[3]{2}$ **24.**
26. $2xy\sqrt[4]{2x}$ **27.** $-3\sqrt[3]{4x}$ **28.** 0
30. $\dfrac{\sqrt{11}}{3a}$ **31.** $\dfrac{\sqrt[3]{2ab^2}}{5b}$ **32.** $\dfrac{2}{3xy}$ **33.**
35. $4\sqrt{x}$ **36.** $4\sqrt{y}$ **37.** $2\sqrt{3}+4\sqrt{5}$
38. $5\sqrt{x}-3\sqrt{y}$ **39.** $8\sqrt[3]{4}$ **40.** $-7\sqrt[3]{6}$
41. $-2x\sqrt{x}$ **42.** $-5\sqrt{11}$ **43.** $-2\sqrt{3}+24\sqrt{2}+1$
44. $14y\sqrt{5y}-9y^2\sqrt{5}$ **45.** $12a\sqrt{2a}$ **46.** $\dfrac{2\sqrt[3]{4}-8\sqrt[3]{2}}{15}$
47. $-1+\sqrt{2}$ **48.** -2 **49.** $\dfrac{\sqrt{7}+14}{12}$ **50.** $\dfrac{\sqrt{2}}{2}$
51. $\sqrt[3]{25}$ **52.** $\dfrac{\sqrt{6}+2\sqrt{2}}{2}$ **53.** $\sqrt{2}+1$
54. $-\dfrac{(3\sqrt{2}+9)}{7}$ **55.** $-\dfrac{(\sqrt{6}+5\sqrt{3})}{23}$ **56.** $\dfrac{\sqrt{10}-\sqrt{6}}{2}$
57. $\dfrac{\sqrt{x}}{x}$ **58.** $\dfrac{\sqrt[3]{y^2}}{y}$ **59.** $\sqrt{x-y}$ **60.** $\sqrt{xy}$ **61.** $x\sqrt[6]{x}$
62. $xy\sqrt[3]{2x}$ **63.** $x\sqrt{x}-9x$ **64.** $6x-2x\sqrt{x}$ **65.** $-5x^2$
66. $4x-y$ **67.** $x+2\sqrt{3x}+3$ **68.** $3xy-2x\sqrt{6y}+2x$
69. $3\sqrt{2}-2\sqrt{3}-3+\sqrt{6}$ **70.** $\dfrac{6a-17\sqrt{ab}+5b}{9a-b}$
71. $\{25\}$ **72.** $\left\{\dfrac{1}{3}\right\}$ **73.** $\{-2\}$ **74.** $\{7\}$ **75.** $\left\{\dfrac{1}{7}\right\}$
76. $\{0\}$ **77.** $\left\{\dfrac{5}{2}\right\}$ **78.** $\{0, 4\}$ **79.** $\{8\}$ **80.** $\{-3\}$
81. $\varnothing$ **82.** $\{4\}$ **83.** $\left\{\dfrac{1}{4}\right\}$ **84.** $\{-1, -4\}$ **85.** $\varnothing$
86. $\sqrt{82}$ **87.** $\sqrt{34}$ **88.** 10 **89.** $\dfrac{\sqrt{58}}{2}$ **90.** $x^{1/8}$
91. $\dfrac{1}{x^{5/3}}$ **92.** $x^{11/12}$ **93.** $x^{7/24}$ **94.** ≈ 40.497 ft

5. y^7 6. $m\sqrt[3]{m^2}$
11. $\dfrac{a^6\sqrt{11a}}{3}$
14. $-1 - \sqrt{6}$
17. $20\sqrt{2}$ 18. $5 + 2\sqrt{6}$
20. 33 21. $8\sqrt{3} - 14$
$\dfrac{\sqrt{35} + 10 - \sqrt{14}}{43}$ 23. {289}
25. $\left\{\dfrac{1}{2}\right\}$ 26. {3} 27. {5} 28. {9}
{−2, 1} 30. ∅ 31. $5\sqrt{2}$ 32. $\dfrac{\sqrt{269}}{4}$

Cumulative Review Chapters 7–9
1. $2^2 \cdot 3^2$ 2. $5 \cdot 7$ 3. $2 \cdot 3 \cdot 11 \cdot 13$ 4. $2 \cdot 3 \cdot 7 \cdot 11$
5. $3(x - 2)$ 6. $x(x + 2)$ 7. $-4(2m + 3)$
8. $7xy(3x^3y - 2x^2y^2 - 1)$ 9. $-3x(5x^2 - 4x + 1)$
10. $(x - 3)(x - 5)$ 11. $x(x + 6y)(x - 5y)$
12. $(a - 3)(b + 3)$ 13. $(y - 5)^2$
14. $(m - 0.3)(m - 0.4)$ 15. $(a - b)(2b)$
16. $(6 + 5x)(5 - 4x)$ 17. $2ab(5b - 4a)(3b + 4a)$
18. $(x + 4)(x - 4)$ 19. $(y + 3)^2$
20. $(x - 2)(x^2 + 2x + 4)$ 21. $(a + b)(x - y)$
22. $(2y - 1)(3x - 4)$ 23. $(2x + 1)^2(2x - 1)^2$
24. prime 25. $\left(\dfrac{1}{3}x + \dfrac{1}{2}y\right)\left(\dfrac{1}{3}x - \dfrac{1}{2}y\right)$
26. $(0.6 + 0.5b)(0.6 - 0.5b)$
27. $(3a + 4b)(9a^2 - 12ab + 16b^2)$ 28. $x(4x - 3y)^2$
29. {−3, 1} 30. {−11, 7} 31. $\left\{-\dfrac{6}{5}, \dfrac{3}{2}\right\}$ 32. {−1, 5}
33. {0, 2, 4} 34. {−7, −1} 35. {0, 1, 2, 4} 36. $\left\{\pm\dfrac{9}{7}\right\}$
37. $\dfrac{1}{2}$ 38. $\dfrac{x + 3}{x - 3}$ 39. $\dfrac{a + 1}{a - 2}$ 40. $\dfrac{2m + 1}{2}$
41. $\dfrac{a^2 + ab + b^2}{a + b}$ 42. $x - 5$ 43. $-(x + y)$
44. $\dfrac{-3}{4(x + 2)}$ 45. $-\dfrac{(a + b)}{a + 4b}$ 46. $\dfrac{y}{4x}$ 47. $\dfrac{-2t}{3s}$
48. $x - 1$ 49. $(x - y)^2$ 50. $\dfrac{7d}{16a}$ 51. $\dfrac{3}{x}$
52. $\dfrac{3x + 3}{(x + 2)^2}$ 53. $\dfrac{3x + 4}{3(x + 4)}$ 54. $\dfrac{6y^2 + 11y - 1}{(y + 1)^2(y + 2)}$
55. $\dfrac{x}{2}$ 56. $\dfrac{8}{19}$ 57. $\dfrac{(a + b)^2}{ab}$ 58. $\dfrac{x^2 + 1}{x^2 - 1)}$

59. $\dfrac{5s + 4}{s^2 + 3s}$ 60. $\dfrac{x^2 - 2x}{x^2 - 2}$ 61. {30} 62. $\left\{-\dfrac{10}{7}\right\}$
63. ∅ 64. {2} 65. $\{a \mid a \in R, a \neq -1\}$
66. $\left\{-\dfrac{21}{2}\right\}$ 67. ∅ 68. {−2} 69. {6} 70. 30 lb
71. 396 miles 72. 35 items 73. 384 freshmen
74. $y = kx^2, k = \dfrac{4}{3}, y = \dfrac{196}{3}$ 75. $y = \dfrac{kxz}{w}, k = \dfrac{25}{3}, y = 50$
76. 6 77. $-3\sqrt{2}$ 78. $x\sqrt{x}$ 79. $\dfrac{1}{3}$ 80. −4
81. $5x^2y$ 82. $\sqrt{6}$ 83. $\sqrt{15x}$ 84. $\dfrac{1}{2}$ 85. 0.7
86. 0.2 87. $4\sqrt{5} - \sqrt{2}$ 88. $-\sqrt{6} - 18\sqrt{2}$
89. $-21x + 16x\sqrt{7x}$ 90. 9 91. 3 92. $\sqrt{5x}$
93. $6 - 5\sqrt{3} - 5\sqrt{2} + 2\sqrt{6}$ 94. {4} 95. {11}
96. ∅ 97. ∅ 98. {1} 99. $\left\{\dfrac{49}{18}\right\}$ 100. $3^{3/4}$
101. 2 102. $\dfrac{3}{7}$ 103. 4 104. 4 105. x 106. x
107. $\dfrac{1}{x^{1/6}y^3}$ 108. $6,000 at 5%, $4,000 at 9%
109. $3\dfrac{1}{3}$ days 110. $1\dfrac{11}{13}$ hr
111. 12 of the 50-cent stamps, 32 of the 19-cent stamps, and 64 of the 29-cent stamps

CHAPTER 10

Exercise Set 10.1
1. {−3, 7} 3. $\left\{-\dfrac{2}{3}, \dfrac{5}{4}\right\}$ 5. $\left\{0, -\dfrac{2}{7}, \dfrac{2}{7}\right\}$
7. $\left\{0, \dfrac{5}{3}, -\dfrac{7}{4}, \dfrac{9}{2}\right\}$ 9. {−5} 11. $\left\{\dfrac{15}{13}, -\dfrac{15}{13}\right\}$
13. $\left\{-\dfrac{1}{36}\right\}$ 15. {2, 20} 17. {0, −2, 20}
19. {1, −16} 21. $\left\{0, \dfrac{1}{8}, -6\right\}$ 23. $\left\{0, \dfrac{2}{5}, \dfrac{3}{2}\right\}$
25. {0, −6, 14} 27. {0, −4, 21} 29. $\left\{-\dfrac{1}{15}\right\}$
31. {5, −5} 33. {9, −9} 35. {1, −1}
37. $\{2\sqrt{3}, -2\sqrt{3}\}$ 39. $\{\sqrt{6}, -\sqrt{6}\}$
41. Not a real number 43. $\left\{\dfrac{1}{3}, -\dfrac{1}{3}\right\}$ 45. $\left\{\dfrac{8}{13}, -\dfrac{8}{13}\right\}$
47. $\left\{\dfrac{\sqrt{30}}{8}, -\dfrac{\sqrt{30}}{8}\right\}$ 49. $\{\sqrt{7}, -\sqrt{7}\}$ 51. $\left\{\dfrac{12}{11}, -\dfrac{12}{11}\right\}$
53. {1, −7} 55. {−3, −5} 57. $\left\{1, -\dfrac{1}{3}\right\}$

59. $\left\{\dfrac{3+\sqrt{10}}{2}, \dfrac{3-\sqrt{10}}{2}\right\}$ **61.** $\left\{-\dfrac{3}{2}, \dfrac{1}{2}\right\}$ **63.** $\left\{-\dfrac{7}{2}\right\}$
65. $\{2, -1\}$ **67.** $\left\{\dfrac{1+4\sqrt{3}}{4}, \dfrac{1-4\sqrt{3}}{4}\right\}$
69. Not a real number **71.** $\{\sqrt{6}, -\sqrt{6}\}$
73. $\left\{\dfrac{2\sqrt{6}}{3}, -\dfrac{2\sqrt{6}}{3}\right\}$ **75.** $\left\{\dfrac{14}{5}, -\dfrac{14}{5}\right\}$ **77.** $\left\{-\dfrac{2}{7}, \dfrac{22}{7}\right\}$
79. 6 or -6 **81.** -1 or, 3
83. 14 and 16 or -16 and -14 **85.** -8 **87.** 1 or 7
89. -5 or -3 **91.** $w = 11$, $l = 14$ **93.** 9 edgers
95. 41.18 ft **97.** 18.763 ft **99** $3 + \sqrt{10}$
101. $-4 - \sqrt{3}$ **103.** $-1 + \sqrt{2}$ **105.** $\{7\}$
107. $\{-5, -3\}$

Exercise Set 10.2
1. $x^2 + 4x + 4 = (x+2)^2$ **3.** $y^2 - 10y + 25 = (y-5)^2$
5. $t^2 - 5t + \dfrac{25}{4} = \left(t - \dfrac{5}{2}\right)^2$ **7.** $x^2 + 11x + \dfrac{121}{4} = \left(x + \dfrac{11}{2}\right)^2$
9. $r^2 + \dfrac{1}{2}r + \dfrac{1}{16} = \left(r + \dfrac{1}{4}\right)^2$ **11.** $\{1, -9\}$
13. $\{-1 + \sqrt{6}, -1 - \sqrt{6}\}$ **15.** $\{8, -2\}$ **17.** $\{-2, -4\}$
19. $\{3 + \sqrt{5}, 3 - \sqrt{5}\}$ **21.** $\{-1 + \sqrt{5}, -1 - \sqrt{5}\}$
23. $\left\{\dfrac{3 + 2\sqrt{6}}{3}, \dfrac{3 - 2\sqrt{6}}{3}\right\}$ **25.** $\left\{\dfrac{3 + \sqrt{3}}{2}, \dfrac{3 - \sqrt{3}}{2}\right\}$
27. $\left\{\dfrac{5}{6}, -\dfrac{1}{2}\right\}$ **29.** $\left\{2, -\dfrac{1}{3}\right\}$
31. $\left\{\dfrac{-3 + \sqrt{145}}{4}, \dfrac{-3 - \sqrt{145}}{4}\right\}$ **33.** $\left\{\dfrac{1 + \sqrt{10}}{3}, \dfrac{1 - \sqrt{10}}{3}\right\}$
35. $\left\{\dfrac{-3 + \sqrt{17}}{4}, \dfrac{-3 - \sqrt{17}}{4}\right\}$ **37.** $\left\{\dfrac{5}{6}, -\dfrac{1}{2}\right\}$ **39.** $\left\{\dfrac{1}{3}\right\}$
41. $\left\{-\dfrac{1}{6}\right\}$ **43.** $\left\{\dfrac{1}{8}\right\}$ **45.** $\{-2 \pm 2\sqrt{2}\}$
47. $\left\{\dfrac{-1 \pm \sqrt{5}}{2}\right\}$ **49.** $\{3 \pm 2\sqrt{3}\}$ **51.** $\left\{-1, \dfrac{1}{3}\right\}$
53. $\{-3 \pm 2\sqrt{2}\}$ **55.** $\left\{-\dfrac{1}{2}, 3\right\}$ **57.** $\{-1 \pm \sqrt{2}\}$
59. $\{2\}$ **61.** yes **63.** no **65.** yes
67.
$2x^2 - 5x + 5 = 0$ Original equation
$2x^2 - 5x = -5$ Isolate terms involving the variable on the left side.
$x^2 - \dfrac{5}{2}x = -\dfrac{5}{2}$ Make the leading coefficient 1.
$x^2 - \dfrac{5}{2}x + \dfrac{25}{16} = -\dfrac{5}{2} + \dfrac{25}{16}$ Add the space of $\dfrac{1}{2}$ the coefficient of x to each side.
$\left(x - \dfrac{5}{4}\right)^2 = -\dfrac{15}{16}$ Write the left side as the square of a binomial.
$x - \dfrac{5}{4} = \dfrac{\pm \sqrt{-15}}{4}$ Take the square root of both sides.
$x = \dfrac{5 \pm \sqrt{-15}}{4}$ Add $\dfrac{5}{4}$ to both sides.

Exercise Set 10.3
1. $a = 1$, $b = -5$, $c = 4$ **3.** $a = 3$, $b = -5$, $c = -2$
5. $a = 1$, $b = 3$, $c = -10$ **7.** $a = 11$, $b = 0$, $c = 0$
9. 2, 3 **11.** $2, \dfrac{-7}{5}$ **13.** $-\dfrac{1}{2}, \dfrac{5}{6}$ **15.** $\sqrt{5}, -\sqrt{5}$
17. $\left\{2, -\dfrac{1}{2}\right\}$ **19.** $\{2, -8\}$
21. $\left\{\dfrac{-13 + \sqrt{161}}{4}, \dfrac{-13 - \sqrt{161}}{4}\right\}$ **23.** $\left\{0, \dfrac{7}{8}\right\}$
25. $\{19, 1\}$ **27.** $\left\{\dfrac{6\sqrt{7}}{7}, -\dfrac{6\sqrt{7}}{7}\right\}$ **29.** $\left\{2, -\dfrac{1}{2}\right\}$
31. $\left\{\dfrac{6 + 2\sqrt{6}}{3}, \dfrac{6 - 2\sqrt{6}}{3}\right\}$ **33.** $\left\{\dfrac{3 + \sqrt{33}}{4}, \dfrac{3 - \sqrt{33}}{4}\right\}$
35. $\left\{\dfrac{4 + \sqrt{11}}{5}, \dfrac{4 - \sqrt{11}}{5}\right\}$ **37.** $\left\{\dfrac{2 + \sqrt{22}}{6}, \dfrac{2 - \sqrt{22}}{6}\right\}$
39. $\left\{\dfrac{-9 + \sqrt{41}}{6}\right\}$ **41.** $\{\pm \sqrt{5}\}$ **43.** $\{-2 \pm \sqrt{3}\}$
45. $\{2 \pm \sqrt{34}\}$ **47.** $\left\{-\dfrac{9}{20}, \dfrac{1}{4}\right\}$ **49.** $\{-1, 5\}$
51. $\{3, -2\}$ **53.** $\{-4, 4\}$ **55.** $\{0, -4\}$
57. $\{2 + \sqrt{3}, 2 - \sqrt{3}\}$ **59.** $\left\{3, -\dfrac{2}{3}\right\}$
61. $\left\{\dfrac{3 + \sqrt{3}}{2}, \dfrac{3 - \sqrt{3}}{2}\right\}$ **63.** $\left\{\dfrac{1 + 7\sqrt{5}}{2}, \dfrac{1 - 7\sqrt{5}}{2}\right\}$
65. $\left\{0, \dfrac{17}{4}\right\}$ **67.** $\left\{\dfrac{-3 + \sqrt{33}}{6}, \dfrac{-3 - \sqrt{33}}{6}\right\}$
69. $\{4 + \sqrt{15}, 4 - \sqrt{15}\}$ **71.** $\{2\sqrt{145}, -2\sqrt{145}\}$
73. $\{2\}$ **75.** $\left\{\dfrac{-7 + \sqrt{113}}{2}, \dfrac{-7 - \sqrt{113}}{2}\right\}$
77. $\dfrac{23 \pm \sqrt{429}}{10}$ **79.** $w = \dfrac{1 + \sqrt{161}}{4}$, $l = \dfrac{-1 + \sqrt{161}}{2}$
81. $h = 4\sqrt{5}$ **83.** 18 mph

Exercise Set 10.4
1. $2i$ **3.** $3i$ **5.** $7i$ **7.** $2\sqrt{2}i$ **9.** $4\sqrt{2}i$ **11.** $\dfrac{4i}{5}$
13. 2 **15.** $1 + 3i$ **17.** $-1 + 5i$ **19.** $4 - 3i$
21. $-2\sqrt{2}$ **23.** -21 **25.** 2 **27.** 13 **29.** $26 - 7i$
31. $-\dfrac{2i}{3}$ **33.** $1 - i$ **35.** $1 + i$ **37.** $\dfrac{-1 - 4i}{2}$
39. $\dfrac{5 - 5i}{2}$ **41.** $\{\pm 2i\}$ **43.** $\{\pm 4i\}$ **45.** $\{-1 \pm i\}$

47. $\left\{\dfrac{1 \pm 2\sqrt{2}i}{2}\right\}$ **49.** $\left\{\dfrac{-2 \pm 3\sqrt{2}i}{3}\right\}$ **51.** $\{\pm 8\sqrt{2}i\}$

53. $\{2 \pm i\}$ **55.** $\{2 \pm \sqrt{7}i\}$ **57.** $\left\{\dfrac{3 \pm \sqrt{11}i}{2}\right\}$

59. $\{5 \pm \sqrt{2}i\}$ **61.** $\left\{\dfrac{3 \pm \sqrt{15}i}{2}\right\}$ **63.** $\left\{\dfrac{2 \pm i}{3}\right\}$

65. $\left\{\dfrac{3 \pm 2\sqrt{3}}{6}\right\}$ **67.** $\left\{\dfrac{3 \pm 5\sqrt{3}i}{6}\right\}$

69. $\left\{\dfrac{-1 \pm \sqrt{11}i}{2}\right\}$ **71.** $\left\{\dfrac{1 \pm \sqrt{3}i}{2}\right\}$

73. $i^9 = i$; $i^{10} = -1$; $i^{11} = -i$; $i^{12} = 1$

75. $(0, 4)$; $(1, 1)$; $(2, 0)$; $(3, 1)$; $(4, 4)$

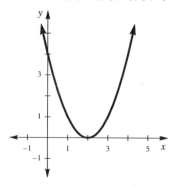

77. $(-2, 0)$; $(-1, -3)$; $(0, -4)$; $(1, -3)$; $(2, 0)$

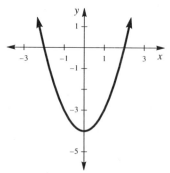

79. $(-3, 6)$; $(-2, 3)$; $(-1, 2)$; $(0, 3)$; $(1, 6)$

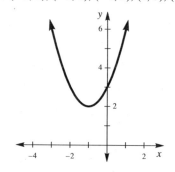

Exercise Set 10.5

1.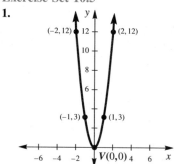

3.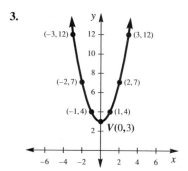

5.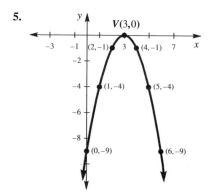

7.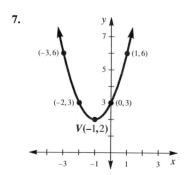

9.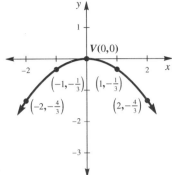

11.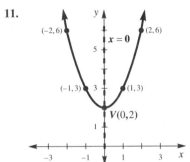

13.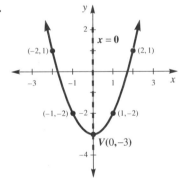

15.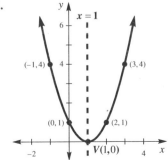

17.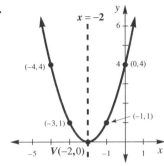

19.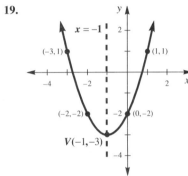

21.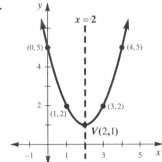

23.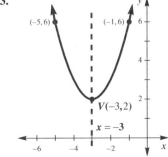

A-52 ANSWERS

25.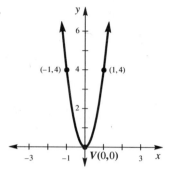

27.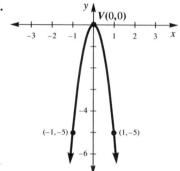

29.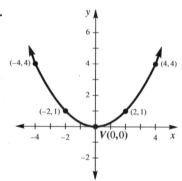

31.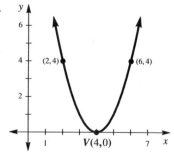

33.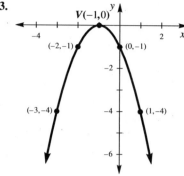

35.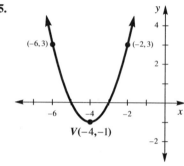

37.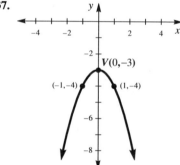

39.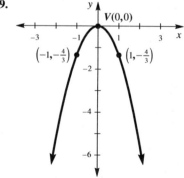

41.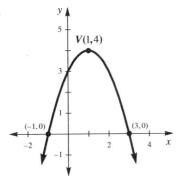

43.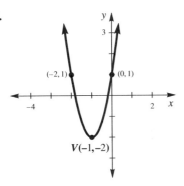

45.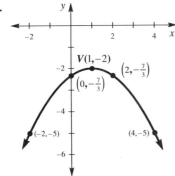

47.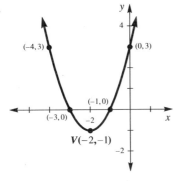

49.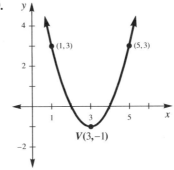

51.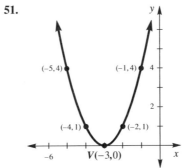

53.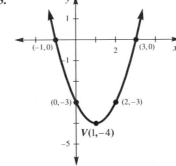

55.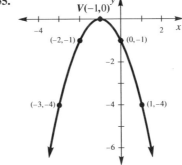

57. (a) $(3, 0)$ and $(-2, 0)$ (b) $\left(\frac{1}{2}, -\frac{25}{4}\right)$ (c) $x = \frac{1}{2}$
  (d) $D\{x \mid x \in R\}$; $R\left\{y \mid y \geq -\frac{25}{4}\right\}$

59. (a) $(-2, 0)$ and $(2, 0)$ (b) $(0, 3)$ (c) $x = 0$
  (d) $D\{x \mid x \in R\}$; $R\{y \mid y \leq 3\}$

61. (a) $(0, 0)$ (b) $(0, 0)$ (c) $x = 0$
  (d) $D\{x \mid x \in R\}$; $R\{y \mid y \geq 0\}$

63. (a) Does not cross x-axis (b) $\left(\frac{1}{2}, -\frac{3}{4}\right)$ (c) $x = \frac{1}{2}$
  (d) $D\{x \mid x \in R\}$; $R\left\{y \mid y \leq -\frac{3}{4}\right\}$

65. $f(0) = -5$ **67.** $f(-1) = 0$ **69.** $f(3) = -8$

71. (a)

(Graph showing parabola with vertex $V(3, 8)$, opening downward, crossing n-axis at $(1, 0)$ and $(5, 0)$, with P on vertical axis)

  (b) $n = 1$ item or $n = 5$ items
  (c) $n = 3$ items
  (d) $\$80$
  (e) $n = 0$ items or $n > 5$ items
  (f) $n = 2$ items or $n = 4$ items
  (g) 3 units

Exercise Set 10.6

1. 12 and 14 or -14 and -12 **3.** -3 and -10 **5.** 26 m
7. 10 mph   **9.** 3 ft, 4 ft   **11.** 6 students
13. 45 mph, 8 hr   **15.** 2.25 mph, 3 mph   **17.** 12 dinners
19. 12 in.   **21.** 16 numbers   **23.** 56.1 ft; 126.9 ft
25. $\frac{1}{5}$ or 5   **27.** 6, 8, and 10   **29.** $2\sqrt{3}$ **31.** 7 and 13
33. $\frac{5}{7}$

Review Exercises

1. $\left\{-4, \frac{3}{2}\right\}$ **2.** $\left\{-\frac{4}{3}, \frac{2}{5}\right\}$ **3.** $\{-6\}$ **4.** $\{7\}$ **5.** $\{\pm 2\}$
6. $\{\pm 4\sqrt{2}\}$ **7.** $\left\{0, \pm\frac{11}{15}\right\}$ **8.** $\left\{0, \pm\frac{2}{5}\right\}$ **9.** $\left\{\pm\frac{21}{4}\right\}$
10. $\left\{\pm\frac{45}{22}\right\}$ **11.** $\{-4, 15\}$ **12.** $\left\{0, -\frac{5}{6}\right\}$ **13.** $\{-1, 3\}$
14. $\left\{0, \pm\frac{4}{5}\right\}$ **15.** $\left\{\frac{3}{2}, \frac{1}{5}\right\}$ **16.** $\left\{-\frac{2}{3}, \frac{5}{4}\right\}$ **17.** $\{\pm 2\}$

18. $\{\pm 12\}$ **19.** $\{\pm 3\sqrt{3}i\}$ **20.** $\{\pm i\}$ **21.** $\{\pm 2\sqrt{2}\}$
22. $\{\pm 5\sqrt{2}\}$ **23.** $\{\pm 2\sqrt{6}\}$ **24.** $\{\pm 2\sqrt{10}\}$ **25.** $\{\pm 3i\}$
26. $\{\pm 5i\}$ **27.** $\left\{\frac{3 \pm \sqrt{17}}{2}\right\}$ **28.** $\left\{\frac{-1 \pm \sqrt{13}}{2}\right\}$
29. $\{1 \pm \sqrt{6}\}$ **30.** $\{4\}$ **31.** $\left\{\frac{3 \pm \sqrt{3}}{2}\right\}$ **32.** $\{3 \pm \sqrt{5}\}$
33. $\left\{0, -\frac{1}{2}, \frac{5}{6}\right\}$ **34.** $\left\{-\frac{1}{2}, 0, 5\right\}$ **35.** $\{5 \pm 2\sqrt{7}\}$
36. $\left\{\frac{50 \pm 10\sqrt{74}}{49}\right\}$ **37.** $\{-3 \pm 2\sqrt{2}\}$ **38.** $\left\{\frac{1}{2}, 1\right\}$
39. $\left\{\frac{1 \pm \sqrt{21}}{2}\right\}$ **40.** $\left\{\frac{-1 \pm \sqrt{13}}{2}\right\}$ **41.** $\left\{0, \frac{1}{7}\right\}$
42. $\left\{0, \frac{3}{8}\right\}$ **43.** $\{\pm 2\sqrt{6}\}$ **44.** $\{\pm \sqrt{21}\}$
45. $\left\{\frac{3 \pm \sqrt{109}}{10}\right\}$ **46.** $\left\{\frac{1 \pm \sqrt{73}}{6}\right\}$ **47.** $\{\pm 2\sqrt{2}\}$
48. $\{\pm \sqrt{2}\}$ **49.** $\left\{\frac{5 \pm \sqrt{65}}{10}\right\}$ **50.** $\left\{\frac{-4 \pm \sqrt{23}}{7}\right\}$
51. $\left\{-\frac{5}{3}, 0, \frac{1}{2}\right\}$ **52.** $\left\{-\frac{7}{2}, 0, 1\right\}$ **53.** $\left\{-\frac{1}{2}, \frac{3}{2}\right\}$
54. $\left\{-3, -\frac{7}{3}\right\}$ **55.** $\{5\}$ **56.** $\left\{\frac{-1 \pm \sqrt{65}}{8}\right\}$
57. $\left\{\frac{5 \pm \sqrt{13}}{2}\right\}$ **58.** $\left\{-1, \frac{5}{2}\right\}$ **59.** $\{-2 \pm \sqrt{15}i\}$
60. $\left\{\frac{7 \pm \sqrt{35}i}{2}\right\}$ **61.** $3i$ **62.** $6i$ **63.** $3\sqrt{2}i$
64. $2\sqrt{2}i$ **65.** 4   **66.** $6 + i$ **67.** $5 + i$
68. $-3 - 2i$ **69.** -15 **70.** 10   **71.** 13   **72.** 41
73. $-\frac{i}{4}$ **74.** $\frac{-1 + i}{2}$ **75.** $\frac{3 + 6i}{5}$ **76.** $\frac{15 - 10i}{13}$
77. $\{\pm 2\sqrt{2}i\}$ **78.** $\{\pm 5\sqrt{3}i\}$ **79.** $\{3 \pm 2\sqrt{5}i\}$
80. $\left\{\frac{3 \pm \sqrt{93}}{6}\right\}$ **81.** $\left\{\frac{1 \pm \sqrt{11}i}{2}\right\}$ **82.** $\{-2 \pm \sqrt{5}i\}$
83. $\{\pm 9\sqrt{2}i\}$ **84.** $\{0, \pm i\}$ **85.** $\left\{\frac{1 \pm \sqrt{3}i}{4}\right\}$
86. $\left\{\frac{1 \pm \sqrt{19}i}{2}\right\}$ **87.** $\left\{\frac{3 \pm \sqrt{39}i}{2}\right\}$ **88.** $\left\{\frac{-3 \pm \sqrt{19}i}{2}\right\}$

89.

(Graph showing upward parabola with vertex $V(0, -2)$, passing through $(-\sqrt{2}, 0)$, $(\sqrt{2}, 0)$, $(-2, 2)$, $(2, 2)$; axis of symmetry $x = 0$)

90.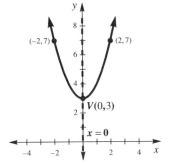

91.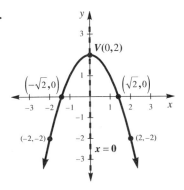

92.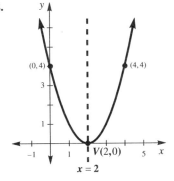

93.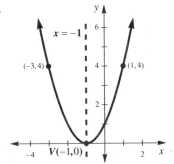

94.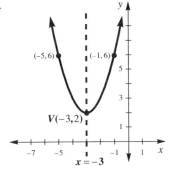

95.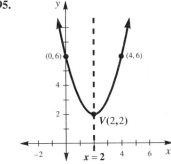

96.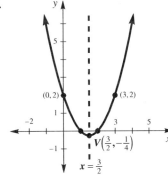

97.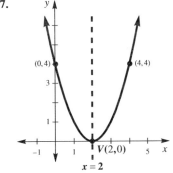

98. $x = 9$ **99.** $x = 12$ **100.** 11 and 13 or -13 and -11
101. $-1 + 2\sqrt{2}$ **102.** 4 mph **103.** 40 mph

Chapter Test

1. $\left\{\dfrac{5}{7}, -\dfrac{3}{2}\right\}$ **2.** $\left\{\dfrac{11}{6}, -\dfrac{11}{6}\right\}$ **3.** $\{0, 11\}$ **4.** $\{0, 1, -16\}$

5. $\{8, -1\}$ **6.** $\left\{\dfrac{\sqrt{285}}{15}, -\dfrac{\sqrt{285}}{15}\right\}$ **7.** $\left\{\dfrac{4+\sqrt{7}}{3}, \dfrac{4-\sqrt{7}}{3}\right\}$

8. $\left\{\dfrac{14}{5}, -\dfrac{14}{5}\right\}$ **9.** $\{-6\}$ **10.** $\left\{\dfrac{6}{5}\right\}$ **11.** $\{\pm 10i\}$

12. $\{\pm 9i\}$ **13.** $\{1 \pm \sqrt{6}\}$ **14.** $\left\{\dfrac{-1 \pm \sqrt{13}}{6}\right\}$

15. $\left\{\dfrac{1 \pm \sqrt{265}}{12}\right\}$ **16.** $\left\{\dfrac{3 \pm 2\sqrt{6}}{3}\right\}$ **17.** $\left\{\dfrac{3}{4}, -2\right\}$

18. $\left\{\dfrac{-1 \pm \sqrt{5}}{2}\right\}$ **19.** $\left\{\dfrac{\pm \sqrt{42}}{3}\right\}$ **20.** $\left\{\dfrac{-1 \pm \sqrt{73}}{6}\right\}$

21. $\left\{\dfrac{20}{7}, -2\right\}$ **22.** $\left\{0, \dfrac{7}{5}\right\}$ **23.** $\left\{\dfrac{-1 \pm \sqrt{11}i}{2}\right\}$

24. $\{0, -2 \pm 3i\}$

25.

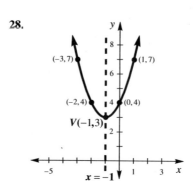

26.

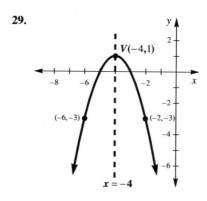

27.

28.

29.

30.

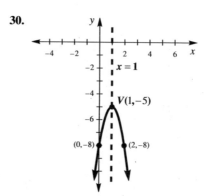

31. 10 or -11 **32.** 3, 4, and 5 **33.** $w = \dfrac{3\sqrt{21}}{7}, l = \dfrac{3\sqrt{21}}{2}$

34. $\dfrac{7}{4}$ mph

Cumulative Review Chapters 1–10

1. 2 **2.** $\dfrac{16}{5}$ **3.** $\dfrac{1125}{2}$ **4.** 5 **5.** -4 **6.** -13 **7.** 12

8. 8-yd loss **9.** $-\dfrac{1}{20}$ **10.** 16 **11.** 70 cents

12. undefined **13.** $-\dfrac{1}{3}$ **14.** 1

15. (a) commutative property of addition
 (b) multiplicative inverse property
 (c) distributive property
 (d) additive inverse property
 (e) multiplicative identity property
16. additive inverse property **17.** $17 - y$ **18.** $5x$
19. $6n + 21$ **20.** $-55 - 4b$ **21.** $55.5 - 17.8pqw^3$
22. $-2 - \dfrac{19}{2}p^4q$ **23.** $\{16\}$ **24.** $\{5\}$ **25.** $\{-0.15\}$
26. $\left\{-\dfrac{35}{13}\right\}$ **27.** $\left\{\dfrac{17}{4}\right\}$ **28.** $\left\{\dfrac{7}{3}\right\}$ **29.** 2 ft, 3 ft
30. 11 nickels, 18 dimes **31.** $x \leq 3$ **32.** $-2 < x < 0$
33. $-1 < t < 4$ **34.** $\dfrac{1}{3}$ hr **35.** 150 in^2
36. \$15,000 at 8%, \$5,000 at 12% **37.** $v = \dfrac{h + 16t^2}{t}$
38. 12 clocks/week **39.** $-a^4$ **40.** $\dfrac{z^2}{49}$ **41.** $\dfrac{n^8}{m^5}$
42. 0.1 **43.** $2x^2y^2z^2$ **44.** $37x^3y^5$ **45.** $-2x^3 - 2x^2 + 19$
46. $\dfrac{1}{2}xy$ **47.** $15ab + 8b^2$ **48.** $7x^3 - 5$
49. $x^2 + 3x - 4$ **50.** $-8r + 3$ **51.** $-8x$
52. $w^4 - 18w^2 + 81$ **53.** $y^2 - \dfrac{9}{16}$ **54.** -2 **55.** $x + 3$
56. $y - 7 + \dfrac{2}{y+1}$ **57.** $3y^2 - y - 1$
58. $z^3(5z^3 - 3z^2 + 2)$ **59.** $5a^2x^2(4ax - 3)$
60. $(s - 3)(s - 10)$ **61.** prime **62.** $b^3(b - 6)(b - 2)$
63. $(4r - 1)(3r - 2)$ **64.** $2x^7(x - 1)(3x - 1)$
65. $(5r + 7s)(5r - 7s)$ **66.** $(x + a)^2$
67. $(2p + 3)(4p^2 - 6p + 9)$ **68.** $(3 - c)(u - v^2)$
69. $(x + w)(z + 4)(z - 4)$ **70.** $\{-3, 1\}$ **71.** $-6, 2$
72. 6, 8, 10 **73.** $\dfrac{s+3}{s-1}$ **74.** $\dfrac{x+y}{a+2}$ **75.** $2y + 3$
76. $\dfrac{3x}{1-x}$ **77.** $\dfrac{4y+3}{3y^2}$ **78.** $\dfrac{2x^2-3}{15x+1}$ **79.** $\{-3\}$
80. $\left\{\dfrac{5}{3}\right\}$ **81.** $\left\{-4, \dfrac{1}{2}\right\}$ **82.** Mac: 63 mph; Milt: 56 mph
83. $y = 15$ **84.** $A = 3$

85.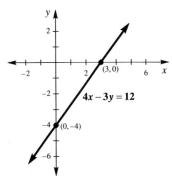

86.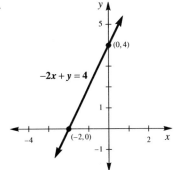

87.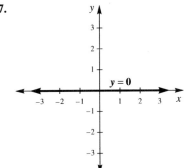

88. $3x + 5y = 11$ **89.** $4x + 3y = 18$

90.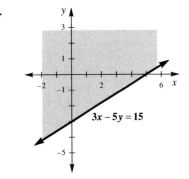

91. function: D{−3, −1, 1, 2}; R{−2, −1, 2, 3}   **92.** yes
93. D{x | x ∈ R, x ≠ −3}   **94.** g(2) = 17   **95.** {(1, −2)}
96. {(−10, 3.1)}   **97.** dependent
98. wind: 75 mph; plane: 150 mph
99.

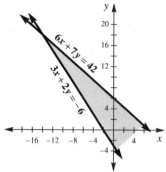

114.

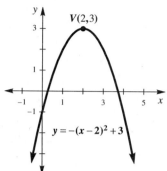

100. h = 6 in.   **101.** $9cd^4\sqrt{c}$ **102.** $\dfrac{\sqrt{5x}}{5}$ **103.** $2r$
104. $\dfrac{5\sqrt{6}}{6}$ **105.** $\sqrt{3}$ **106.** {1}   **107.** $\left\{0, \dfrac{16}{9}\right\}$
108. m^2n^4 **109.** {−3, −1}   **110.** $\{2 \pm \sqrt{15}\}$
111. $\left\{\dfrac{1 \pm \sqrt{17}}{8}\right\}$ **112.** $\{1 \pm \sqrt{2}i\}$ **113.** $\dfrac{2}{3}$ or $\dfrac{3}{2}$

115. $y = 2(x+1)^2 - 3$

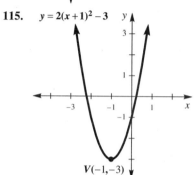

Index

Absolute value, 65–66, 100, 153
Absolute value equations, 153–155, 168
Absolute value inequalities
  explanation of, 155
  interpreting, 157–158
  solutions to, 156–157, 168
Acute angles, 39
Addition
  of decimals, 22
  of fractions, 15–17
  of polynomials, 296–297, 323
  of radicals, 455–458
  of rational expressions, 394–400, 434–435
  of signed numbers, 69–72, 74
Addition property
  of equality, 115–119, 124, 167, 240–241, 274
  of inequality, 139–141, 167
Additive identity, 88
Additive inverse, 64, 89, 101
Agnesi, Maria, 107
Algebra
  basic concepts of, 3–9, 43–45
  use of letters in, 61
Algebraic expressions
  explanation of, 107
  simplifying, 107–111, 167
Analytical engine, 3
Angles, 39
Applied problems
  examples of, 94–97
  factoring used to solve, 368–371, 376
  historical comment about, 266
  involving polynomials, 318–320
  properties of equality and inequality used to solve, 160–164
  proportions used to solve, 422–425
  quadratic equations used to solve, 527–531
  ratios used to solve, 422–425
  rules of exponents used to solve, 318–319
  steps to solve, 94, 102, 160, 368
  systems of equations used to solve, 231, 245–246, 255, 266–270
  systems of linear inequalities used to solve, 270–271
Area
  explanation of, 34, 45
  measurement of, 37, 45
Arithmetic, basic concepts of, 3–9, 43–45
Associative property, 87–88, 101
Axes, 175, 224
Axis of abscissas, 176
Axis of symmetry, 517, 534

Babbage, Charles, 3
Base, 4, 34
Binomials. *See also* Polynomials
  explanation of, 296
  FOIL method to multiply, 305–307, 323, 461
  squares of, 306–307, 323
Boundary line, 204, 209, 225
Byron, Ada, 3

Calculators. *See also* Graphics calculators; Scientific calculators
  adding and subtracting signed numbers using, 74
  decimal point placement when using, 26
  early development of, 3
  evaluating expressions involving exponents using, 286
  graphing straight lines and other equations on, 200
  scientific notation on, 292–293
  square roots on, 38, 443–445

Cartesian coordinate system, 175–179. *See also* Rectangular coordinate system
Check points, 183, 224
Chuquet, 281
Circles, 35
Circumference, 35
Commutative property, 86–87, 101
Completing the square
  explanation of, 495
  used to solve quadratic equations, 496–499, 533
Complex fractions
  explanation of, 402, 435
  simplifying, 402–406
Complex numbers
  addition and subtraction of, 512
  division of, 513–514
  explanation of, 510, 511, 534
  multiplication of, 512–513
  solutions to quadratic equations using, 514, 534
  standard form for, 512
Composite numbers, 332
Compound fractions. *See* Complex fractions
Compound inequalities
  explanation of, 141, 168
  solutions to, 141–143
Computers
  for graphing straight lines and other equations, 200
  used to solve systems of equations, 240, 246–247
Conditional equations, 118, 167
Conjugate pairs, 504
Conjugates, 462, 465, 513
Consistent systems, 234, 236
Constant of variation, 428, 429, 436
Constant term, 296
Constants, 52, 100

I-1

Coordinate
  of number line, 55
  of point, 175
Cube roots, 444, 451, 481
Cubes, factoring the sum and difference of two, 351–353

Decimal points, 21, 26
Decimals
  adding and subtracting, 22–23
  changed to percent, 28–29
  converting percent to, 27–28
  division of, 23–24, 44
  elements of, 21–22
  historical comment about, 21
  multiplication of, 22–23, 44
  rounding off, 39
  square roots involving, 267
Degree, 295
Denominators, rationalizing, 463–465, 481
Dependent systems
  elimination method and, 244–245
  explanation of, 234, 236, 273
  substitution method and, 254
Descartes, René, 175, 231, 281, 510
Descending powers, 296
Diagrams, Venn, 102
Diameter, 35
Difference, 62
Difference of two squares, 351
Direct variation, 428, 436
Discount formula, 376
Distance formula, 471–473, 481
Distributive property
  explanation of, 89–91, 102
  simplifying algebraic expressions by using, 107–109
Divisibility, 7, 8
Division
  algebraic method of writing, 62
  of decimals, 23–24
  of exponents, 282–286
  of fractions, 15, 17
  of polynomials, 311–316, 324
  of radicals, 450, 462–466
  of rational expressions, 390–392, 434
  of real numbers, 80–83
  simplifying complex fractions by, 402–403
  by zero, 55, 82

Division property
  of equality, 123–129, 167, 450
  of inequality, 146–150, 168
  of radicals, 450
Domain, 215, 217, 225
Double negative property, 64, 101

Einstein, Albert, 51
Elements, of sets, 51–52, 100
Elimination method
  explanation of, 240, 241
  historical comment about, 240
  to solve systems of linear equations, 241–247, 274
  substitution method vs., 254
Empty sets, 52, 100
Equality
  addition property of, 115–119, 124, 167
  division property of, 123–129, 167
  modified addition property of, 240–241, 274
  multiplication property of, 123–129, 167
  power property of, 468, 481
  squaring property of, 468–471
  substitution property of, 249–250, 274
  subtraction property of, 115–119, 167
  symbols of, 122
Equations
  absolute value, 153–155, 168
  conditional, 118
  explanation of, 113–114, 167
  fractions in, 126
  linear. See Linear equations
  of lines, 195–199
  literal, 131–135
  rational, 408–416, 435
  root of, 113
Euclid, 359
Euler, Leonhard, 381, 510
Expanded notation, 21
Exponents
  definition of negative, 290
  definition of zero, 290
  division with, 282–286
  explanation of, 4, 281
  historical comment about notation for, 281
  multiplication with, 281–282, 322

  nonpositive, 284–285
  power rules for, 288–290, 323
  product rule for, 282, 290, 322
  quotient rule for, 283, 290
  rational, 475–478
  simplest terms for, 285–286
  used to solve applied problems, 318–319
  used with signed numbers, 79
Extraneous solutions, 412, 468, 481
Extremes, in proportions, 420, 421, 435

Factors/factoring. See also Greatest common factor
  explanation of, 331
  greatest common, 331–336, 341–342, 374
  by grouping, 342, 353–354, 374–375
  quadratic equations solved by, 362–366, 375, 533
  set of, 331
  to solve applied problems, 368–371, 376
  suggestions for successful, 356–358, 375
  of trinomials of form $ax^2 + bx + c$, 344–348, 374
  of trinomials of form $x^2 + bx + c$, 338–342, 374
Fermat, Pierre de, 231
Fibonacci, 68
Finite sets, 52, 100
FOIL method
  to check results of factoring trinomials, 338–339
  for multiplication of complex numbers, 512
  to multiply two binomials, 305–307, 323
  used with radicals, 461
Formulas, recognizing and using common, 131–135
Fourier, Jean Baptiste, 231
Fourth roots, 444, 481
Fractions
  adding and subtracting, 15–17
  changing signs of, 385
  comparison of, 17–18
  complex, 402–406, 435
  converted to decimals, 24–25, 44

Fractions (*cont.*)
  division of, 15, 17
  explanation of, 12, 44, 381
  fundamental principle of, 12–14, 44
  historical comment about, 11
  improper, 12
  multiplication of, 14
  proper, 12, 44
Functions
  equation as relation or, 217
  explanation of, 216–217, 225
  notation for, 218–219, 225, 299
Fundamental property of rational expressions, 382–383, 434

General form, of linear equations, 195, 225
Geometry
  historical comment about, 33
  principles of, 34–39, 45–46
Germain, Sophie, 441
Graphics calculators. *See also* Calculators
  adding and subtracting signed numbers using, 74
  finding square roots on, 443–445
  solutions to linear equations in one variable using, 128
  used to solve systems of equations, 240, 246–247
Greater than, 138, 167
Greatest common factor (GCF)
  explanation of, 331–332, 374
  factoring trinomials after factoring out, 341–342
  removal of, 334–336
Grouping, 342, 353–354, 374–375

Half-plane
  explanation of, 204, 225
  graphs of inequalities in two variables as, 217
Harriot, Thomas, 137, 359
Height, 34
Hindu-Arabic numerals, 51
Horizontal axis, 176
Horizontal line, slope of, 192
Hypotenuse, 35

Identities, 118–119, 167
Identity property, 88, 101
Imaginary numbers
  explanation of, 510–512, 534
  historical comment about, 510

Improper fractions, 12
Inconsistent systems
  elimination method and, 244, 245
  explanation of, 234–236, 273
  solved by substitution method, 254
Independent systems, 234–236, 273
Index, 444, 480
Inequalities. *See also* Linear inequalities
  absolute value, 155–158, 168
  addition and subtraction properties of, 139–143, 167
  compound, 141–143, 168
  explanation of, 138, 167
  multiplication and division properties of, 145–150, 168
  symbols of, 137, 138
Infinite sets, 52, 100
Integers, set of, 54–56, 100
Intervals, on number line, 138
Inverse property, 88–89, 101
Inverse variation, 428, 436
Irrational numbers
  conjugate pairs of, 504
  explanation of, 57, 100
  square roots and, 442, 443, 480

Joint variation, 428, 436

Kovalevskaya, Sofya, 441

Lagrange, Joseph-Louis, 489
Leading coefficient, 296
Least common denominator (LCD)
  explanation of, 16
  of rational expressions, 395–399, 434–435
  simplifying complex fractions by finding, 403–406
Leibniz, Gottfried Wilhelm, 175, 510
Less than, 138, 167
Linear equations
  in one variable
    explanation of, 114–115, 167
    steps to solve, 126
    using addition and subtraction properties of equality to solve, 115–119
    using multiplication and division properties of equality to solve, 123–129

systems of
  elimination method to solve, 240–247, 274
  explanation of, 231–232, 273
  graphing method to solve, 232–238, 240, 273–274
  historical comment about, 240
  solutions to applied problems by using, 231, 245–246, 255, 266–270
  substitution method to solve, 249–255
  used to solve applied problems, 266–270
in two variables
  explanation of, 182, 224
  graphs of, 183, 224
  identifying intercepts in, 183–187
  using check points in, 183–184
Linear inequalities
  systems of
    applied problems used to solve, 270–271, 275
    using graphs to solve, 257–263, 275
  in two variables
    boundary line of, 204, 206, 207, 209
    graphs of, 204–210, 217, 225
Lines
  equations of, 195–199, 225, 231
  slope of, 189–195, 224–225
  in system, 234
Listing method, 52, 100
Literal equations
  explanation of, 167
  solutions to, 131–135

Mathematics
  history of women specializing in, 441
  as language, 3
Means, in proportions, 420, 421, 435
Members, of sets. *See* Elements
Mixed numbers, 12
Monomials
  division of polynomials by, 311–312, 324
  explanation of, 296
  greatest common factor of, 332–334, 374

Monomials (*cont.*)
   multiplication of polynomials by, 302–303
Multiplication
   algebraic method of writing, 62
   of decimals, 22–23
   of exponents, 281–282
   of fractions, 14
   of polynomials, 302–308, 323–324
   of radicals, 449, 460–462
   of rational expressions, 388–390, 434
   of real numbers, 77–80
Multiplication property
   of equality, 123–129, 167
   of inequality, 146–150, 168
   of radicals, 449
Multiplicative identity, 88
Multiplicative inverse, 89

Natural numbers, set of, 54, 100
Negative numbers
   explanation of, 54
   historical comment about, 68
   odd roots of, 445
Negative y-coordinates, 175
Newton, Sir Isaac, 107
Nonpositive exponents, 284–285
Notation
   expanded, 21
   for functions, 218–219, 225
   historical comments about, 281, 302, 448–449
   set-builder, 52, 100
nth roots, 444, 481
Number lines
   adding signed numbers by using, 69, 70
   additive inverses on, 64
   explanation of, 54–55
   solution to inequalities as interval on, 138
   subtracting signed numbers by using, 72–74
Numbers
   complex, 510, 511, 534
   composite, 332
   imaginary, 510–512, 534
   irrational, 57, 100, 442, 443, 480, 504
   mixed, 12
   negative, 54
   positive, 54

   prime, 8, 332
   rational, 55–57
   real, 53–58
Numerical coefficients, 107, 167

Obtuse angles, 39
Order of operations
   examples using, 5–6
   explanation of, 5, 43–44
Ordered pairs
   coordinates of, 176–177
   explanation of, 175, 204, 224
   steps to graph, 175–176
   in system of equations, 231–232
Ordinates
   axis of, 176
   explanation of, 224
Origin
   of number line, 55
   in rectangular coordinate system, 175, 224
Oughtred, William, 137

Parabolas
   axis of symmetry of, 517
   explanation of, 516, 534
   shift method to graph, 518–522, 534
   vertex and x-intercepts to graph, 522, 523, 534
Parallel lines
   equation of, 199
   explanation of, 198
   system of, 234
Percent
   changing decimals to, 28–29
   concept of, 26–27
   converted to decimals, 27–28
   explanation of, 27, 45
Percent equation
   explanation of, 29, 45
   use of, 29–30
Perfect squares, 442, 480
Perfect-square trinomials, 350–351
Perimeter, 34, 45
Perpendicular, 35
Perpendicular lines, 198, 199
π (pi), 33
Poincaré, Henri, 281
Point-slope form, 195–197, 199, 225
Polynomials
   addition of, 296–297, 323
   degree of, in one variable, 296
   division of, 311–316, 324

   explanation of, 295–296
   function notation used for, 299
   multiplication of, 302–308, 323–324
   prime, 340
   subtraction of, 297–299, 323
Positive numbers, 54
Positive y-coordinates, 175
Power property of equality, 468, 481
Power rules for exponents, 288–290, 323
Prime factored form, 8–9, 332
Prime numbers, 8, 332
Prime polynomials, 340
Product, 62
Product rule
   for exponents, 282, 290, 322
   for square roots, 510–511, 534
Product rule for square roots, 510–511, 534
Proper fractions, 12, 44
Proportions
   applied problems using, 422–425
   explanation of, 420–423, 435
Pythagoras, 38
Pythagorean Theorem, 38, 46, 447, 471–472, 481

Quadrants, 175, 224
Quadratic equations
   completing the square used to solve, 495–499, 533
   complex-number solutions to solve, 510–514, 534
   explanation of, 360, 375
   factoring and square root method used to solve, 489–493
   graphs of, 516–523, 534
   historical comment about, 359
   quadratic formula used to solve, 501–507, 533
   used to solve applied problems, 527–531
   using factoring to solve, 362–366, 375
   using zero-factor property to solve, 360–361, 363, 489–490, 492–493
Quadratic formula
   derivation of, 501–502
   explanation of, 502
   used to solve quadratic equations, 503–507, 533

Quotient, 62
Quotient rule for exponents, 283, 290

Radical sign, 442, 449, 480
Radicals. *See also* Square roots
   addition and subtraction of, 455–458, 481
   division of, 450, 462–466, 481
   explanation of, 442
   multiplication of, 449, 460–462, 481
   notation for, 442, 449
   rational exponents and, 475–478, 482
   simplifying, 450–453, 457–458, 481
   solutions to equations using, 468–473, 481
Radicand, 442, 446, 480
Radius, 35
Range, 215, 225
Rational equations
   finding solutions to, 408–416
   historical problem involving, 408, 416
   steps to solve, 409–410, 435
Rational exponents, 475–478, 482
Rational expressions
   addition and subtraction of, 394–400, 434–435
   division of, 390–392, 434
   explanation of, 381, 433, 480
   finding sum or difference of, 397–398
   fundamental property of, 382–383, 434
   least common denominator of, 395–399, 434–435
   multiplication of, 388–390, 434
   simplifying, 382–385, 433–434
   solutions to equations involving, 408–416
   undefined, 381
Rational numbers
   explanation of, 55–57, 100
   squares of, 442, 480
Rationalizing the denominators, 463–465, 481
Ratios
   applied problems using, 422–425
   explanation of, 419–420, 435
Real numbers
   addition of, 69–72, 74, 101

   division of, 80–83, 101
   explanation of, 53–58
   multiplication of, 77–80, 101
   properties of, 86–91, 101–102
   set of, 53, 57–58
   subtraction of, 72–74, 101
Reciprocal, 89
Recorde, Robert, 122
Rectangular coordinate system
   explanation of, 175, 224
   graphing polynomials by plotting on, 348
   locating points on, 177–179
   parts of, 175–177
Relation
   equation as function or, 217
   explanation of, 215, 225
Riese, Adam, 61
Right angles, 39
Right triangles, 35
Roman numerals, 51
Roots. *See also* Radicals; Square roots
   cube, 444, 451, 481
   explanation of, 441–442, 480–481
   fourth, 444, 481
   nth, 444
Roster, 52
Rounding digit, 39, 46
Rule method, 52, 100
Russell, Bertrand, 441

Scientific calculators. *See also* Calculators
   adding and subtracting signed numbers using, 74
   finding square roots on, 443–445
   solutions to linear equations in one variable using, 127
Scientific notation
   on calculators, 292–293
   explanation of, 291–292, 323
Set of integers
   explanation of, 54, 55
   as subset of set of rational numbers, 56
Set of natural numbers, 54
Set of real numbers, 53, 57–58
Set of whole numbers, 54
Set-builder notation, 52, 100
Sets
   explanation of, 51, 100
   methods of writing, 52
   types of, 52

Similar triangles, 423, 435
Simple interest, 376
Slope
   of line, 189–195, 224–225
   negative, 192
   positive, 191
Slope formula, 194
Slope-intercept form
   explanation of, 197–199, 225
   for two equations in linear system, 274
Solution set
   explanation of, 116
   of linear inequalities in two variables, 204
   of systems of linear inequalities, 257–263
Solutions
   to equalities, 113
   extraneous, 412, 468, 481
   to inequalities, 138
Somerville, Mary, 441
Square root property, 490–492, 533
Square roots. *See also* Radicals; Roots
   involving decimals, 26
   irrational, 442
   principal, 441, 480
   product rule for, 510–511, 534
   solutions to equations using, 468–471
   using calculators to find, 38, 443
Squares
   of binomials, 306–307, 323
   difference of two, 351
Squaring property of equality, 468–471
Standard form, for complex numbers, 512
Stevin, Simon, 21
Subsets, 53, 100
Substitution method
   elimination method *vs.*, 254
   explanation of, 249
   steps to solve system of equations by, 250, 251, 274
   used to solve systems of linear equations, 251–255
Substitution property of equality
   explanation of, 249–250
   used to solve system of linear equations, 274
Subtraction
   of decimals, 22

Subtraction (*cont.*)
   of fractions, 15–17
   of polynomials, 297–299, 323
   of radicals, 455–458
   of rational expressions, 394–400, 434–435
   of signed numbers, 72–74
Subtraction property
   of equality, 115–119, 167
   of inequality, 139–141, 167
Sum, 62
Supscripts, 190
Symbols
   algebraic, 62–64
   description of common, 3–4
   development of mathematical, 61
   translation of written statements into, 4–7

Table of values, construction of, 183
Target zones, 225
TEMATH, 518
Terms
   constant, 296
   explanation of, 107, 167, 295
   like, 108
   in proportions, 420, 435

Test point, 206, 225
Trapezoid, 34
Triangles
   right, 35
   similar, 423, 435
Trinomials. *See also* Polynomials
   explanation of, 296
   factoring, 338–342
   factoring of form $ax^2 + bx + c$, 344–348, 374
   factoring of form $x^2 + bx + c$, 338–342, 374
   factoring perfect-square, 350–351

Variables, 52, 100
Variation
   applications of different types of, 429–431
   constant of, 428, 429, 436
   direct, 428, 436
   inverse, 428, 436
   joint, 428, 436
Venn, John, 102
Venn diagrams, 102
Vertex
   explanation of, 34–35
   of parabola, 516, 521–523

Vertical axis, 176
Vertical line, slope of, 194
Vertical line test, 217–218
Viète, François, 113
Volume
   explanation of, 36–37
   measurement of, 37–38, 45

Whole numbers, 54, 100

x-axis, 175, 224
x-coordinates, 175, 224
x-intercepts
   explanation of, 183, 224
   of parabola, 522, 523

y-axis, 175, 224
y-intercepts, 183, 224

Zero
   division by, 55, 82
   invention of, 51
Zero-factor property
   explanation of, 360
   to solve quadratic equations, 360–361, 363, 489–490, 492–493

Complex Numbers

$$i = \sqrt{-1}; \quad i^2 = -1$$

If a and b are real numbers, then:

$$(a + bi) + (c + di) = (a + c) + (b + d)i$$
$$(a + bi)(c + di) = (ac - bd) + (ad + bc)i$$

$a + bi$ and $a - bi$ are **complex conjugates.**

$$(a + bi)(a - bi) = a^2 + b^2$$

The Quadratic Formula

The quadratic formula $\quad x = \dfrac{-b \pm \sqrt{b^2 - 4ac}}{2a} \quad$ for $ax^2 + bx + c = 0;\ a \neq 0$

The Slope of a Line

$$m = \frac{\text{change in } y \text{ values}}{\text{change in } x \text{ values}} = \frac{y_2 - y_1}{x_2 - x_1}, \quad x_2 \neq x_1$$

The Midpoint of a Line Segment

$$(x_m, y_m) = \left(\frac{x_1 + x_2}{2}, \frac{y_1 + y_2}{2} \right)$$

Linear Equations

General form	$ax + by = c$
Point–slope form	$y - y_1 = m(x - x_1)$
Slope–intercept form	$y = mx + b$
Horizontal line	$y = $ constant
Vertical line	$x = $ constant

Variation

Direct	$y = kx$	y varies directly as x
Inverse	$y = \dfrac{k}{x}$	y varies inversely as x
Joint	$y = kxz$	y varies jointly with x and z

The Distance Formula

$$d = \sqrt{(x_2 - x_1)^2 + (y_2 - y_1)^2}$$

Abbreviations

second: sec	minute: min	hour: hr
inch: in.	foot: ft	yard: yd
square inch: in.^{2}	square foot: ft^2	square yard: yd^2
cubic inch: in.^{3}	cubic foot: ft^3	cubic yard: yd^3
meter: m	centimeter: cm	liter: l
square meter: m^2	square centimeter: cm^2	centiliter: cl
cubic meter: m^3	cubic centimeter: cm^3	gallon: gal
kilometer: km	mile: mi	pounds: lb
kilometer per hour: km/hr	miles per hour: mph	greatest common divisor: GCD
pint: pt	quart: qt	
amperes: amps	volts: V	
least common denominator: LCD	greatest common factor: GCF	